ALGEBRA

Factors and Zeros of Polynomials:

Given the polynomial $p(x) = a_n x^n + a_{n-1} x^{n-1} + \cdots + a_1 x + a_0$. If $p(b) = 0$, then b is a *zero* of the polynomial and a *root* of the equation $p(x) = 0$. Furthermore, $(x - b)$ is a *factor* of the polynomial.

Fundamental Theorem of Algebra:

An nth degree polynomial has n (not necessarily distinct) zeros.

Quadratic Formula:

If $p(x) = ax^2 + bx + c$, $a \neq 0$ and $b^2 - 4ac \geq 0$, then the real zeros of p are $x = \left(-b \pm \sqrt{b^2 - 4ac}\right)/2a$.

Example

If $p(x) = x^2 + 3x - 1$, then $p(x) = 0$ if
$$x = \frac{-3 \pm \sqrt{13}}{2}$$

Special Factors:

$x^2 - a^2 = (x - a)(x + a)$
$x^3 - a^3 = (x - a)(x^2 + ax + a^2)$
$x^3 + a^3 = (x + a)(x^2 - ax + a^2)$
$x^4 - a^4 = (x - a)(x + a)(x^2 + a^2)$
$x^4 + a^4 = (x^2 + \sqrt{2}ax + a^2)(x^2 - \sqrt{2}ax + a^2)$
$x^n - a^n = (x - a)(x^{n-1} + ax^{n-2} + \cdots + a^{n-1})$, for n odd
$x^n + a^n = (x + a)(x^{n-1} - ax^{n-2} + \cdots + a^{n-1})$, for n odd
$x^{2n} - a^{2n} = (x^n - a^n)(x^n + a^n)$

Examples

$x^2 - 9 = (x - 3)(x + 3)$
$x^3 - 8 = (x - 2)(x^2 + 2x + 4)$
$x^3 + 4 = \left(x + \sqrt[3]{4}\right)\left(x^2 - \sqrt[3]{4}x + \sqrt[3]{16}\right)$
$x^4 - 4 = \left(x - \sqrt{2}\right)\left(x + \sqrt{2}\right)(x^2 + 2)$
$x^4 + 4 = (x^2 + 2x + 2)(x^2 - 2x + 2)$
$x^5 - 1 = (x - 1)(x^4 + x^3 + x^2 + x + 1)$
$x^7 + 1 = (x + 1)(x^6 - x^5 + x^4 - x^3 + x^2 - x + 1)$
$x^6 - 1 = (x^3 - 1)(x^3 + 1)$

Binomial Theorem:

$(x + a)^2 = x^2 + 2ax + a^2$
$(x - a)^2 = x^2 - 2ax + a^2$
$(x + a)^3 = x^3 + 3ax^2 + 3a^2 x + a^3$
$(x - a)^3 = x^3 - 3ax^2 + 3a^2 x - a^3$
$(x + a)^4 = x^4 + 4ax^3 + 6a^2 x^2 + 4a^3 x + a^4$
$(x - a)^4 = x^4 - 4ax^3 + 6a^2 x^2 - 4a^3 x + a^4$
$(x + a)^n = x^n + nax^{n-1} + \dfrac{n(n-1)}{2!} a^2 x^{n-2} + \cdots + na^{n-1}x + a^n$
$(x - a)^n = x^n - nax^{n-1} + \dfrac{n(n-1)}{2!} a^2 x^{n-2} - \cdots \pm na^{n-1}x \mp a^n$

Examples

$(x + 3)^2 = x^2 + 6x + 9$
$(x^2 - 5)^2 = x^4 - 10x^2 + 25$
$(x + 2)^3 = x^3 + 6x^2 + 12x + 8$
$(x - 1)^3 = x^3 - 3x^2 + 3x - 1$
$\left(x + \sqrt{2}\right)^4 = x^4 + 4\sqrt{2}x^3 + 12x^2 + 8\sqrt{2}x + 4$
$(x - 4)^4 = x^4 - 16x^3 + 96x^2 - 256x + 256$
$(x + 1)^5 = x^5 + 5x^4 + 10x^3 + 10x^2 + 5x + 1$
$(x - 1)^6 = x^6 - 6x^5 + 15x^4 - 20x^3 + 15x^2 - 6x + 1$

Rational Zero Test:

If $p(x) = a_n x^n + a_{n-1} x^{n-1} + \cdots + a_1 x + a_0$ has integer coefficients, then every *rational* root of $p(x) = 0$ is of the form $x = r/s$, where r is a factor of a_0 and s is a factor of a_n.

Example

If $p(x) = 2x^4 - 7x^3 + 5x^2 - 7x + 3$, then the only possible *rational* roots are $x = \pm 1$, $\pm \frac{1}{2}$, ± 3, and $\pm \frac{3}{2}$. By testing, we find the two rational roots to be $\frac{1}{2}$ and 3.

Factoring by Grouping:

$acx^3 + adx^2 + bcx + bd = ax^2(cx + d) + b(cx + d)$
$ = (ax^2 + b)(cx + d)$

Example

$3x^3 - 2x^2 - 6x + 4 = x^2(3x - 2) - 2(3x - 2)$
$ = (x^2 - 2)(3x - 2)$

Arithmetic Operations:

$$ab + ac = a(b + c)$$

$$\frac{a}{b} + \frac{c}{d} = \frac{ad + bc}{bd}$$

$$\frac{a + b}{c} = \frac{a}{c} + \frac{b}{c}$$

$$\frac{\left(\frac{a}{b}\right)}{\left(\frac{c}{d}\right)} = \frac{ad}{bc}$$

$$a\left(\frac{b}{c}\right) = \frac{ab}{c}$$

$$\frac{a - b}{c - d} = \frac{b - a}{d - c}$$

$$\frac{ab + ac}{a} = b + c, \; a \neq 0$$

$$\frac{\left(\frac{a}{b}\right)}{c} = \frac{a}{bc}$$

$$\frac{a}{\left(\frac{b}{c}\right)} = \frac{ac}{b}$$

Exponents and Radicals:

$$a^0 = 1, \; a \neq 0$$

$$\frac{a^x}{a^y} = a^{x-y}$$

$$\left(\frac{a}{b}\right)^x = \frac{a^x}{b^x}$$

$$\sqrt[n]{a^m} = a^{m/n} = \left(\sqrt[n]{a}\right)^m$$

$$a^{-x} = \frac{1}{a^x}$$

$$(a^x)^y = a^{xy}$$

$$\sqrt{a} = a^{1/2}$$

$$\sqrt[n]{ab} = \sqrt[n]{a} \sqrt[n]{b}$$

$$a^x a^y = a^{x+y}$$

$$(ab)^x = a^x b^x$$

$$\sqrt[n]{a} = a^{1/n}$$

$$\sqrt[n]{\left(\frac{a}{b}\right)} = \frac{\sqrt[n]{a}}{\sqrt[n]{b}}$$

Algebraic Errors to Avoid:

$\dfrac{a}{x + b} \neq \dfrac{a}{x} + \dfrac{a}{b}$ (To see this error, let $a = b = x = 1$.)

$\sqrt{x^2 + a^2} \neq x + a$ (To see this error, let $x = 3$ and $a = 4$.)

$a - b(x - 1) \neq a - bx - b$ [Remember to distribute negative signs. The equation should be $a - b(x - 1) = a - bx + b$.]

$\dfrac{\left(\frac{x}{a}\right)}{b} \neq \dfrac{bx}{a}$ [To divide fractions, invert and multiply. The equation should be
$$\dfrac{\left(\frac{x}{a}\right)}{b} = \dfrac{\left(\frac{x}{a}\right)}{\left(\frac{b}{1}\right)} = \left(\frac{x}{a}\right)\left(\frac{1}{b}\right) = \frac{x}{ab}.]$$

$\sqrt{-x^2 + a^2} \neq -\sqrt{x^2 - a^2}$ (The negative sign cannot be factored out of the square root.)

$\dfrac{a + bx}{a} \neq 1 + bx$ (This is one of many examples of incorrect cancellation. The equation should be $\dfrac{a + bx}{a} = \dfrac{a}{a} + \dfrac{bx}{a} = 1 + \dfrac{bx}{a}$.)

$\dfrac{1}{x^{1/2} - x^{1/3}} \neq x^{-1/2} - x^{-1/3}$ (This error is a more complex version of the first error.)

$(x^2)^3 \neq x^5$ [This equation should be $(x^2)^3 = x^2 x^2 x^2 = x^6$.]

Conversion Table:

1 centimeter ≈ 0.394 inch	1 joule ≈ 0.738 foot-pound	1 mile ≈ 1.609 kilometers
1 meter ≈ 39.370 inches	1 gram ≈ 0.035 ounce	1 gallon ≈ 3.785 liters
≈ 3.281 feet	1 kilogram ≈ 2.205 pounds	1 pound ≈ 4.448 newtons
1 kilometer ≈ 0.621 mile	1 inch ≈ 2.540 centimeters	1 foot-lb ≈ 1.356 joules
1 liter ≈ 0.264 gallon	1 foot ≈ 30.480 centimeters	1 ounce ≈ 28.350 grams
1 newton ≈ 0.225 pound	≈ 0.305 meter	1 pound ≈ 0.454 kilogram

FIFTH EDITION

TRIGONOMETRY

Ron Larson

Robert P. Hostetler
The Pennsylvania State University
The Behrend College

With the assistance of
David C. Falvo
The Pennsylvania State University
The Behrend College

HOUGHTON MIFFLIN COMPANY
BOSTON   NEW YORK

Editor-in-Chief: Jack Shira
Managing Editor: Cathy Cantin
Senior Associate Editor: Maureen Ross
Associate Editor: Laura Wheel
Assistant Editor: Carolyn Johnson
Supervising Editor: Karen Carter
Project Editor: Patty Bergin
Editorial Assistant: Kate Hartke
Art Supervisor: Gary Crespo
Marketing Manager: Michael Busnach
Senior Manufacturing Coordinator: Sally Culler
Composition and Art: Meridian Creative Group

Custom Publishing Editor: Teresa Hoens
Custom Publishing Production Manager: Kathleen McCourt
Project Coordinator: Andrea Wagner

Cover Design: Kelly Melin

This book contains select works from existing Houghton Mifflin Company resources and was produced by Houghton Mifflin Custom Publishing for collegiate use. As such, those adopting and/or contributing to this work are responsible for editorial content, accuracy, continuity and completeness.

Copyright © 2001 by Houghton Mifflin Company. 2003 Impression. All rights reserved.

No part of this work may be reproduced or transmitted in any form or by any means, electronic or mechanical, including photocopying and recording, or by any information storage or retrieval system without the prior written permission of Houghton Mifflin Company unless such copying is expressly permitted by federal copyright law. Address inquiries to College Permissions, Houghton Mifflin Company, 222 Berkeley Street, Boston, MA 02116-3764.

Printed in the United States of America.

ISBN: 0-618-44966-3
N02943

1 2 3 4 5 6 7 8 9 – PP – 05 04 03

Houghton Mifflin
Custom Publishing

222 Berkeley Street • Boston, MA 02116

Address all correspondence and order information to the above address.

Contents

A Word from the Authors (Preface) vi
Features Highlights viii
An Introduction to Graphing Utilities xv

Chapter P ▶ Prerequisites 1

- P.1 Real Numbers 2
- P.2 Solving Equations 12
- P.3 The Cartesian Plane and Graphs of Equations 25
- P.4 Linear Equations in Two Variables 38
- P.5 Functions 53
- P.6 Analyzing Graphs of Functions 67
- P.7 Shifting, Reflecting, and Stretching Graphs 80
- P.8 Combinations of Functions 91
- P.9 Inverse Functions 99

Chapter Summary 109 **Review Exercises** 111
Chapter Project: A Graphical Approach to Maximization 116
Chapter Test 117

Chapter 1 ▶ Trigonometry 118

- 1.1 Radian and Degree Measure 120
- 1.2 Trigonometric Functions: The Unit Circle 131
- 1.3 Right Triangle Trigonometry 138
- 1.4 Trigonometric Functions of Any Angle 149
- 1.5 Graphs of Sine and Cosine Functions 158
- 1.6 Graphs of Other Trigonometric Functions 169
- 1.7 Inverse Trigonometric Functions 180
- 1.8 Applications and Models 190

Chapter Summary 201 **Review Exercises** 202
Chapter Project: Analyzing a Graph 206
Chapter Test 207

Chapter 2 ▶ Analytic Trigonometry 208

- 2.1 Using Fundamental Identities 210
- 2.2 Verifying Trigonometric Identities 218
- 2.3 Solving Trigonometric Equations 225
- 2.4 Sum and Difference Formulas 236
- 2.5 Multiple-Angle and Product-to-Sum Formulas 243

Chapter Summary 254 **Review Exercises** 255
Chapter Project: Projectile Motion 258
Chapter Test 259

Chapter 3 ▶ Additional Topics in Trigonometry 260

- 3.1 Law of Sines 262
- 3.2 Law of Cosines 271
- 3.3 Vectors in the Plane 279
- 3.4 Vectors and Dot Products 292

Chapter Summary 300 **Review Exercises** 301
Chapter Project: Adding Vectors Graphically 304
Chapter Test 305 **Cumulative Test:** Chapters 1–3 306

Chapter 4 ▶ Complex Numbers 308

- 4.1 Complex Numbers 310
- 4.2 Complex Solutions of Equations 317
- 4.3 Trigonometric Form of a Complex Number 324
- 4.4 DeMoivre's Theorem 331

Chapter Summary 337 **Review Exercises** 338
Chapter Project: Using a Graphing Utility to Create Fractals 340
Chapter Test 341

Chapter 5 ▶ Exponential and Logarithmic Functions 342

- 5.1 Exponential Functions and Their Graphs 344
- 5.2 Logarithmic Functions and Their Graphs 356
- 5.3 Properties of Logarithms 367
- 5.4 Exponential and Logarithmic Equations 373
- 5.5 Exponential and Logarithmic Models 384

Chapter Summary 397 **Review Exercises** 398
Chapter Project: Graphical Approach to Compound Interest 402
Chapter Test 403

Chapter 6 ▶ Topics in Analytic Geometry 404

    6.1 Lines 406
    6.2 Introduction to Conics: Parabolas 413
    6.3 Ellipses 422
    6.4 Hyperbolas 431
    6.5 Rotation of Conics 440
    6.6 Parametric Equations 448
    6.7 Polar Coordinates 456
    6.8 Graphs of Polar Equations 462
    6.9 Polar Equations of Conics 470
    Chapter Summary 477   **Review Exercises** 478
    Chapter Project: Polar, Rectangular, and Parametric Forms 482
    Chapter Test 483   **Cumulative Test:** Chapters 4–6 484

Appendices

Appendix A Proofs of Selected Theorems A1

Appendix B Concepts in Statistics A12

    B.1 Representing Data A12
    B.2 Measures of Central Tendency and Dispersion A16
    B.3 Least Squares Regression A25

Answers to Odd-Numbered Exercises and Tests A29

Index of Applications A111

Index A113

A Word from the Authors

Welcome to *Trigonometry*, Fifth Edition. In this revision we focus on student success, accessibility, and flexibility.

Student Success: During the past 30 years of teaching and writing, we have learned many things about the teaching and learning of mathematics. We have found that students are most successful when they know what they are expected to learn and why it is important to learn. With that in mind, we have restructured the Fifth Edition to include a thematic study thread in every chapter.

Each chapter begins with a study guide called *How to Study This Chapter*, which includes a comprehensive overview of the chapter concepts (*The Big Picture*), a list of *Important Vocabulary* that is integral to learning *The Big Picture* concepts, a list of study resources, and a general study tip. The study guide allows students to get organized and prepare for the chapter.

An old pedagogical recipe goes something like this: "First I'm going to tell you what I'm going to teach you, then I will teach it to you, and finally I will go over what I taught you." Following this recipe, we have also included a set of learning objectives in every section that outlines what students are expected to learn, followed by an interesting real-life application that illustrates why it is important to learn the concepts in that section. Finally, the chapter summary (*What did you learn?*), which reinforces the section objectives, and the chapter *Review Exercises*, which are correlated to the chapter summary, provide additional study support at the conclusion of each chapter.

Our new *Student Success Organizer* supplement takes this study thread one step further, providing a content-based study aid.

Accessibility: Over the years we have taken care to write our texts for the student. We have paid careful attention to the presentation, using precise mathematical language and clear writing, to create an effective learning tool. We believe that every student can learn mathematics and we are committed to providing a text that makes the mathematics within it accessible to all students. In the Fifth Edition, we have revised and improved many text features designed for this purpose. The *Technology*, *Exploration*, and *Study Tip* features have been expanded. *Chapter Tests*, which give students an opportunity for self-assessment, now follow every chapter in the Fifth Edition. The exercise sets now include both *Synthesis* exercises, which check students' conceptual understanding, and *Review* exercises, which reinforce skills learned in previous sections and chapters. Also, students have access to several media resources that accompany this text—videotapes, *Interactive Trigonometry* CD-ROM, and a *Trigonometry* website—that provide additional text-specific support.

Flexibility: From the time we first began writing in the early 1970s, we have always viewed part of our authoring role as that of providing instructors with flexible teaching programs. The optional features within the text allow instructors with different pedagogical approaches to design their courses to meet both their instructional needs and the needs of their students. Instructors who stress applications and problem solving, or exploration and technology, or more traditional methods, will be able to use this text successfully. In addition, we provide several print and media resources to support instructors, including a new *Instructor Success Organizer*.

We hope you enjoy the Fifth Edition.

Ron Larson

Robert P. Hostetler

Features Highlights

Student Success Tools

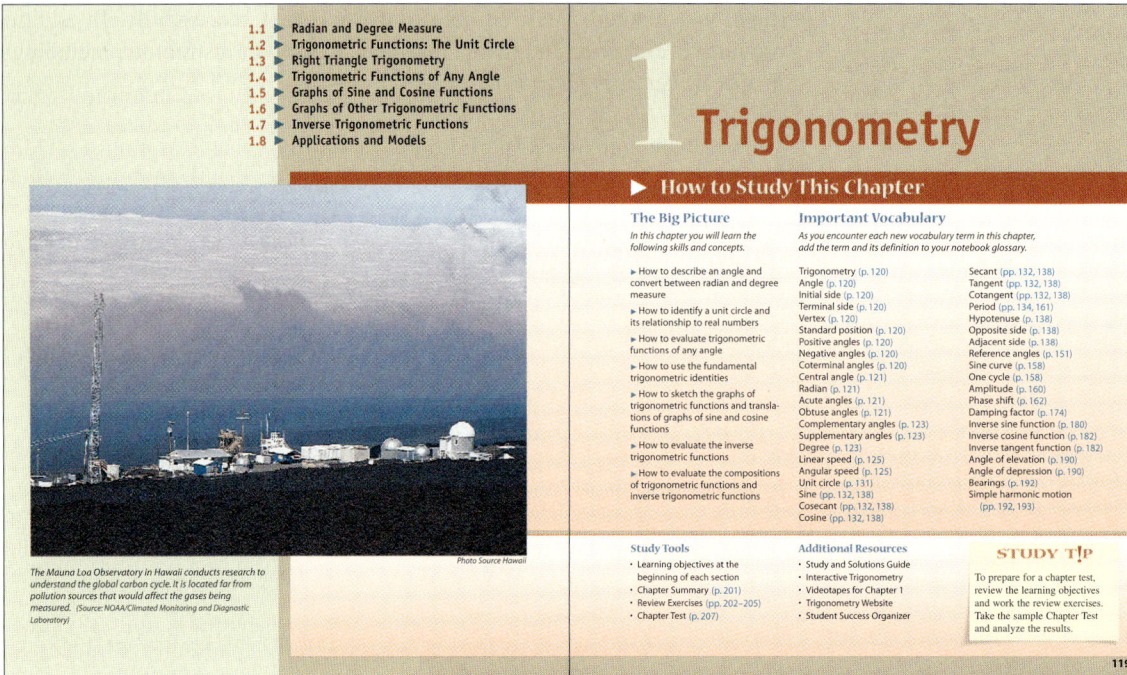

▶ **"How to Study This Chapter"**

The new chapter-opening study guide includes:

- *The Big Picture*—an objective-based overview of the main concepts of the chapter
- *Important Vocabulary*—mathematical terms integral to learning *The Big Picture* concepts
- *Study Tools*
- *Additional Resources*
- *Study Tip*

New Section Openers include:

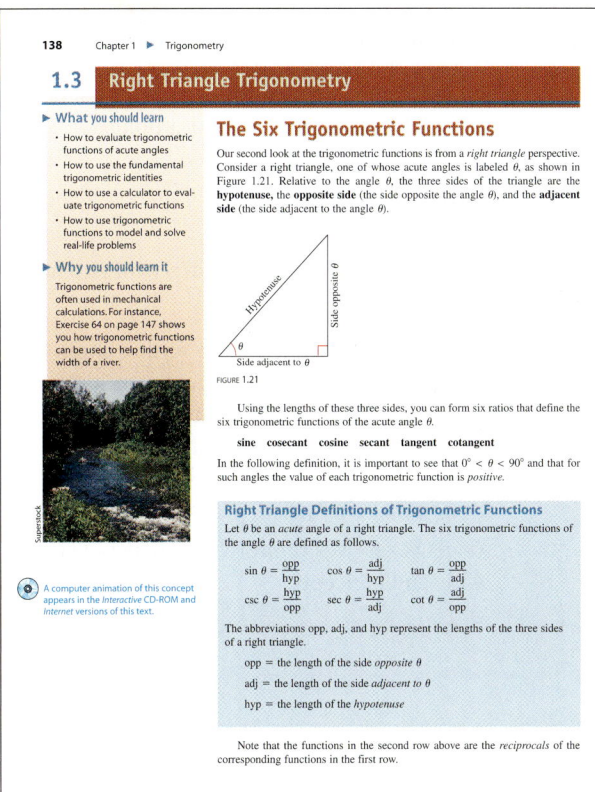

▶ **"What you should learn"**

Objectives outline the main concepts and help keep students focused on *The Big Picture*.

▶ **"Why you should learn it"**

A real-life application or a reference to other branches of mathematics illustrates the relevance of the section's content.

▶ **"What did you learn?" Summary**

The chapter summary provides a concise, section-by-section review of the section objectives. These objectives are correlated to the chapter Review Exercises.

Revised Exercises and Applications

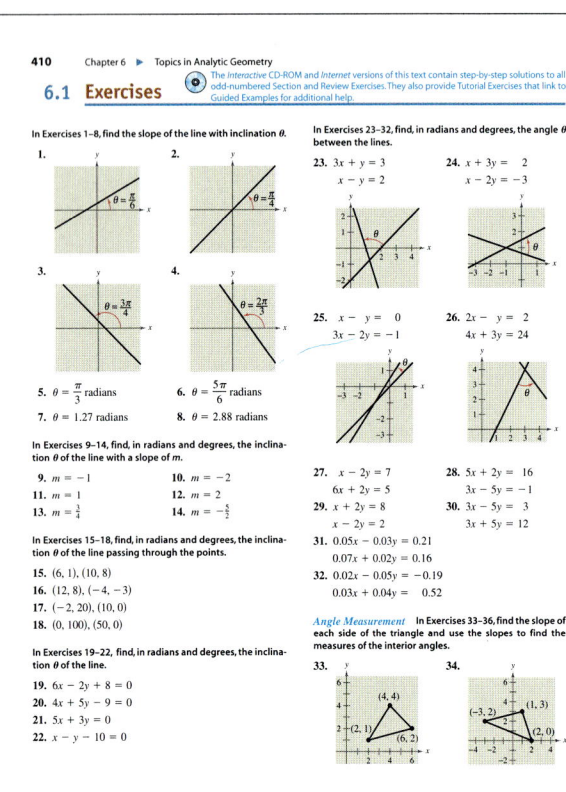

▶ **Exercises**

- Each exercise set contains a variety of computational, conceptual, and applied problems.
- Each exercise set is carefully graded in difficulty to allow students to gain confidence as they progress.
- Each exercise set now concludes with two new types of exercises:
  - *Synthesis* exercises promote further exploration of mathematical concepts, critical thinking skills, and writing about mathematics. These exercises require students to synthesize the main concepts presented in the section and chapter.
  - *Review* exercises reinforce previously learned skills and concepts.

Features Highlights

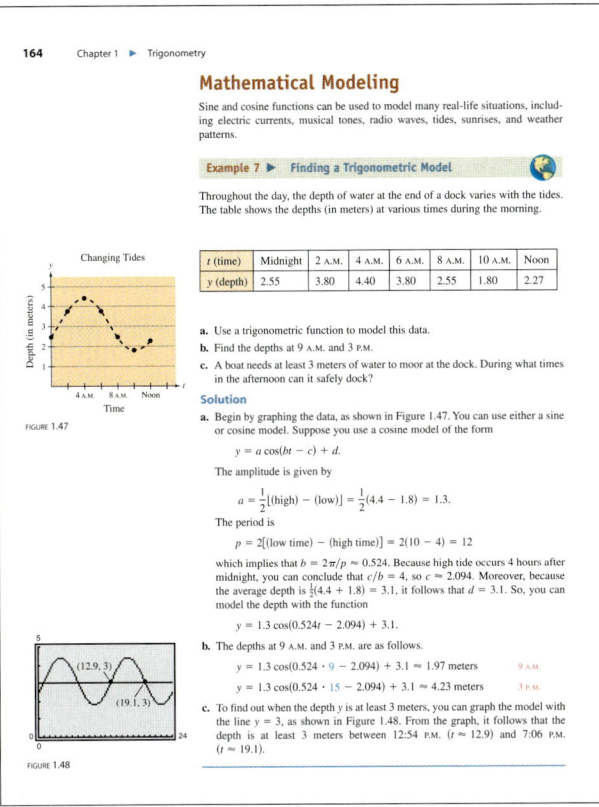

▶ Real-Life Applications

- A wide variety of real-life applications, many using current, real data, are integrated throughout examples and exercises.

- The icon 🌎 indicates an example that involves a real-life application.

▶ Algebra of Calculus

- Special emphasis is given to the algebraic techniques used in calculus.
- Algebra of Calculus examples and exercises are integrated throughout the text.
- The symbol 🔴 indicates an example or exercise in which the Algebra of Calculus is featured.

Flexibility and Accessibility

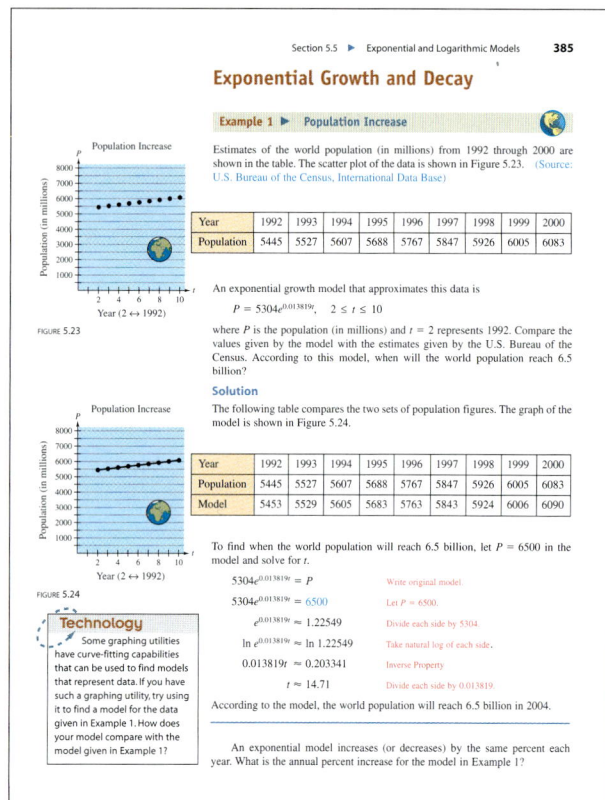

Technology

- Point-of-use instructions for using graphing utilities appear in the margins, encouraging the use of graphing technology as a tool for visualization of mathematical concepts, for verification of other solution methods, and for facilitation of computations.

- The use of technology is optional in this text. This feature and related exercises can easily be omitted without loss of continuity in coverage. Exercises that require the use of a graphing utility are identified by the symbol.

► Examples

- Each example was carefully chosen to illustrate a particular mathematical concept or problem-solving skill.

- All examples contain step-by-step solutions, most with side-by-side explanations that lead students through the solution process.

◄ Exploration ►

- Before introduction of selected topics, *Exploration* engages students in active discovery of mathematical concepts and relationships, often through the power of technology.

- *Exploration* strengthens students' critical thinking skills and helps them develop an intuitive understanding of theoretical concepts.

- *Exploration* is an optional feature and can be omitted without loss of continuity in coverage.

► Additional Features

Carefully crafted learning tools designed to create a rich learning environment can be found throughout the text. These learning tools include Study Tips, Historical Notes, Writing About Mathematics, Chapter Projects, Chapter Review Exercises, Chapter Tests, Cumulative Tests, and an extensive art program.

Supplements

Resources

Website (*college.hmco.com*)
Many additional text-specific study and interactive features for students and instructors can be found at the Houghton Mifflin website.

For the Student

Student Success Organizer

Study and Solutions Guide by Dianna L. Zook (Indiana University/Purdue University–Fort Wayne)

Graphing Technology Guide by Benjamin N. Levy and Laurel Technical Services

Instructional Videotapes by Dana Mosely

Instructional Videotapes for Graphing Calculators by Dana Mosely

For the Instructor

Instructor's Annotated Edition

Instructor Success Organizer

Complete Solutions Guide by Dianna L. Zook (Indiana University/Purdue University–Fort Wayne), Laurel Technical Services, and Mike Jones

Test Item File

Problem Solving, Modeling, and Data Analysis Labs by Wendy Metzger (Palomar College)

Computerized Testing (Windows, Macintosh)

HMClassPrep Instructor's CD-ROM

Acknowledgments

We would like to thank the many people who have helped us at various stages of this project to prepare the text and supplements package. Their encouragement, criticisms, and suggestions have been invaluable to us.

Fifth Edition Reviewers

James Alsobrook, Southern Union State Community College; Sherry Biggers, Clemson University; Charles Biles, Humboldt State University; Randall Boan, Aims Community College; Jeremy Carr, Pensacola Junior College; D. J. Clark, Portland Community College; Donald Clayton, Madisonville Community College; Linda Crabtree, Metropolitan Community College; David DeLatte, University of North Texas; Gregory Dlabach, Northeastern Oklahoma A & M College; Joseph Lloyd Harris, Gulf Coast Community College; Jeff Heiking, St. Petersburg Junior College; Celeste Hernandez, Richland College; Heidi Howard, Florida Community College at Jacksonville; Wanda Long, St. Charles County Community College; Wayne F. Mackey, University of Arkansas; Rhonda MacLeod, Florida State University; M. Maheswaran, University of Wisconsin–Marathon County; Valerie Miller, Georgia State University; Katharine Muller, Cisco Junior College; Bonnie Oppenheimer, Mississippi University for Women; James Pohl, Florida Atlantic University; Hari Pulapaka, Valdosta State University; Michael Russo, Suffolk County Community College; Cynthia Floyd Sikes, Georgia Southern University; Susan Schindler, Baruch College–CUNY; Stanley Smith, Black Hills State University. In addition, we would like to thank all the college algebra instructors who took the time to respond to our survey.

We would like to extend a special thanks to Hari Pulapaka for his contributions to this revision.

We would like to thank the staff of Larson Texts, Inc. and the staff of Meridian Creative Group, who assisted in proofreading the manuscript, preparing and proofreading the art package, and typesetting the supplements.

On a personal level, we are grateful to our wives, Deanna Gilbert Larson and Eloise Hostetler, for their love, patience, and support. Also, a special thanks goes to R. Scott O'Neil.

If you have suggestions for improving this text, please feel free to write to us. Over the past two decades we have received many useful comments from both instructors and students, and we value these comments very much.

Ron Larson
Robert P. Hostetler

An Introduction to Graphing Utilities

Graphing utilities such as graphing calculators and computers with graphing software are very valuable tools for visualizing mathematical principles, verifying solutions to equations, exploring mathematical ideas, and developing mathematical models. Although graphing utilities are extremely helpful in learning mathematics, their use does not mean that learning algebra is any less important. In fact, the combination of knowledge of mathematics and the use of graphing utilities allows you to explore mathematics more easily and to a greater depth. If you are using a graphing utility in this course, it is up to you to learn its capabilities and to practice using this tool to enhance your mathematical learning.

In this text there are many opportunities to use a graphing utility, some of which are described below.

Some Uses of a Graphing Utility
A graphing utility can be used to

- check or validate answers to problems obtained using algebraic methods.
- discover and explore algebraic properties, rules, and concepts.
- graph functions, and approximate solutions to equations involving functions.
- efficiently perform complicated mathematical procedures such as those found in many real-life applications.
- find mathematical models for sets of data.

In this introduction, the features of graphing utilities are discussed from a generic perspective. To learn how to use the features of a specific graphing utility, consult your user's manual or the website for this text found at *college.hmco.com*. Additionally, keystroke guides are available for most graphing utilities, and your college library may have a videotape on how to use your graphing utility.

The Equation Editor

Many graphing utilities are designed to act as "function graphers." In this course, you will study functions and their graphs in detail. You may recall from previous courses that a function can be thought of as a rule that describes the relationship between two variables. These rules are frequently written in terms of x and y. For example, the equation $y = 3x + 5$ represents y as a function of x.

Many graphing utilities have an equation editor that requires an equation to be written in "$y =$" form in order to be entered, as shown in Figure 1. (You should note that your equation editor screen may not look like the screen shown in Figure 1.) To determine exactly how to enter an equation into your graphing utility, consult your user's manual.

FIGURE 1

FIGURE 2

The Table Feature

Most graphing utilities are capable of displaying a table of values with x-values and one or more corresponding y-values. These tables can be used to check solutions of an equation and to generate ordered pairs to assist in graphing an equation.

To use the *table* feature, enter an equation into the equation editor in "$y =$" form. The table may have a setup screen, which allows you to select the starting x-value and the table step or x-increment. You may then have the option of automatically generating values for x and y or building your own table using the *ask* mode. In the *ask* mode, you enter a value for x and the graphing utility displays the y-value.

For example, enter the equation

$$y = \frac{3x}{x + 2}$$

into the equation editor, as shown in Figure 2. In the table setup screen, set the table to start at $x = -4$ and set the table step to 1. When you view the table, notice that the first x-value is -4 and each value after it increases by 1. Also notice that the Y_1 column gives the resulting y-value for each x-value, as shown in Figure 3. The table shows that the y-value when $x = -2$ is ERROR. This means that the variable x may not take on the value -2 in this equation.

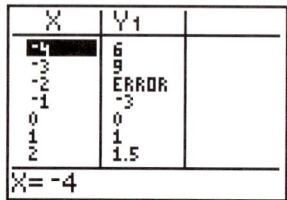

FIGURE 3

With the same equation in the equation editor, set the table to *ask* mode. In this mode you do not need to set the starting x-value or the table step, because you are entering any value you choose for x. You may enter any real value for x—an integer, fraction, decimal, irrational number, and so forth. If you enter $x = 1 + \sqrt{3}$, the graphing utility may rewrite the number as a decimal approximation, as shown in Figure 4. You can continue to build your own table by entering additional x-values in order to generate y-values.

If you have several equations in the equation editor, the table may generate y-values for each equation.

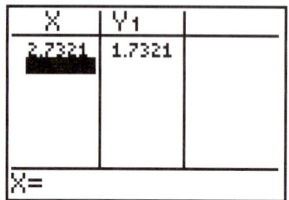

FIGURE 4

Creating a Viewing Window

A **viewing window** for a graph is a rectangular portion of the coordinate plane. A viewing window is determined by the following six values.

Xmin = the smallest value of x
Xmax = the largest value of x
Xscl = the number of units per tick mark on the x-axis
Ymin = the smallest value of y
Ymax = the largest value of y
Yscl = the number of units per tick mark on the y-axis

When you enter these six values into a graphing utility, you are setting the viewing window. Some graphing utilities have a standard viewing window, as shown in Figure 5.

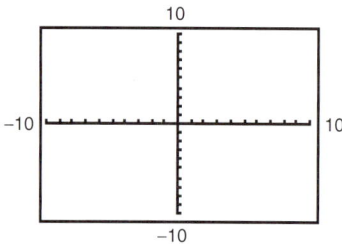

FIGURE 5

By choosing different viewing windows for a graph, it is possible to obtain very different impressions of the graph's shape. For instance, Figure 6 shows four different viewing windows for the graph of

$$y = 0.1x^4 - x^3 + 2x^2.$$

Of these, the view shown in part (a) is the most complete.

(a)

(b)

(c)

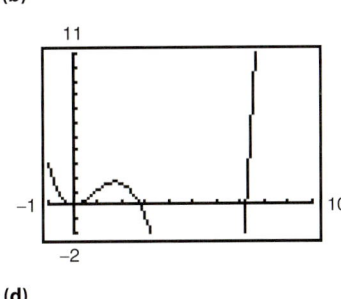
(d)

FIGURE 6

On most graphing utilities, the display screen is two-thirds as high as it is wide. On such screens, you can obtain a graph with a true geometric perspective by using a **square setting**—one in which

$$\frac{\text{Ymax} - \text{Ymin}}{\text{Xmax} - \text{Xmin}} = \frac{2}{3}.$$

One such setting is shown in Figure 7. Notice that the x and y tick marks are equally spaced on a square setting, but not on a standard setting.

To see how the viewing window affects the geometric perspective, graph the semicircles $y_1 = \sqrt{9 - x^2}$ and $y_2 = -\sqrt{9 - x^2}$ in a standard viewing window. Then graph y_1 and y_2 in a square window. Note the difference in the shapes of the circles.

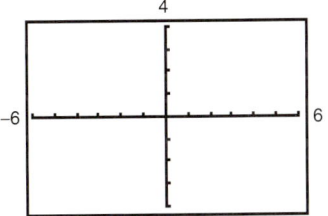

FIGURE 7

Zoom and Trace Features

When you graph an equation, you can move from point to point along its graph using the *trace* feature. As you trace the graph, the coordinates of each point are displayed, as shown in Figure 8. The *trace* feature combined with the *zoom* feature allows you to obtain better and better approximations of desired points on a graph. For instance, you can use the *zoom* feature of a graphing utility to approximate the x-intercept(s) of a graph [the point(s) where the graph crosses the x-axis]. Suppose you want to approximate the x-intercept(s) of the graph of

$$y = 2x^3 - 3x + 2.$$

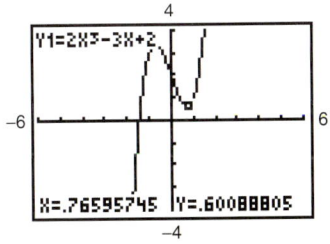

FIGURE 8

Begin by graphing the equation, as shown in Figure 9(a). From the viewing window shown, the graph appears to have only one x-intercept. This intercept lies between -2 and -1. By zooming in on the intercept, you can improve the approximation, as shown in Figure 9(b). To three decimal places, the solution is $x \approx -1.476$.

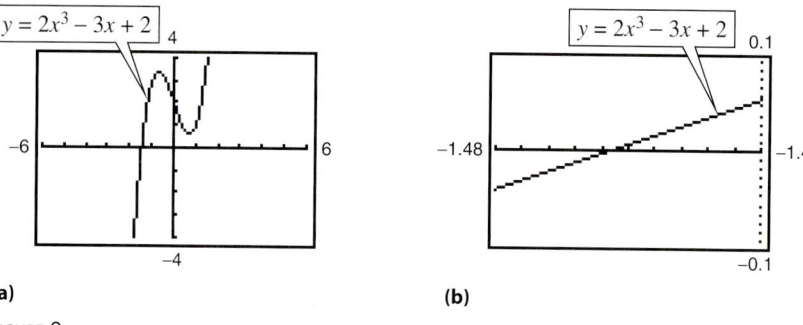

(a) **(b)**

FIGURE 9

Here are some suggestions for using the *zoom* feature.

1. With each successive zoom-in, adjust the x-scale so that the viewing window shows at least one tick mark on each side of the x-intercept.
2. The error in your approximation will be less than the distance between two scale marks.
3. The *trace* feature can usually be used to add one more decimal place of accuracy without changing the viewing window.

Figure 10(a) shows the graph of $y = x^2 - 5x + 3$. Figures 10(b) and 10(c) show "zoom-in views" of the two x-intercepts. From these views, you can approximate the x-intercepts to be $x \approx 0.697$ and $x \approx 4.303$.

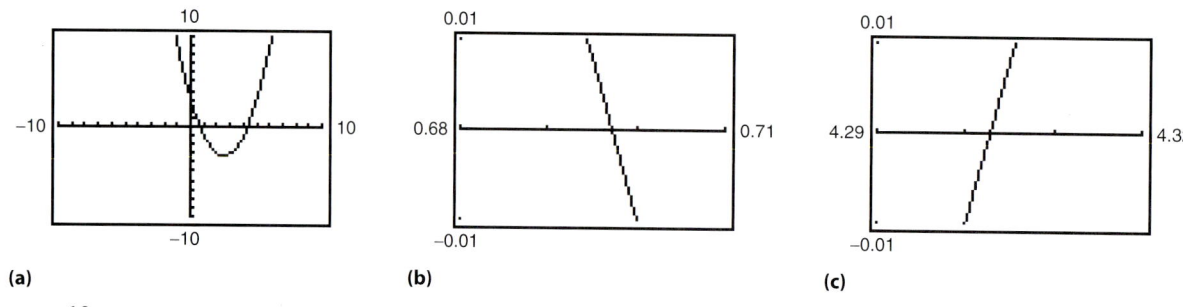

(a) **(b)** **(c)**

FIGURE 10

Zero or Root Feature

Using the *zero* or *root* feature, you can find the real zeros of functions of the various types studied in this text—polynomial, exponential, logarithmic, and trigonometric functions. To find the zeros of a function such as $f(x) = \frac{3}{4}x - 2$, first enter the function as $y_1 = \frac{3}{4}x - 2$. Then use the *zero* or *root* feature, which may require entering lower and upper bound estimates of the root, as shown in Figure 11.

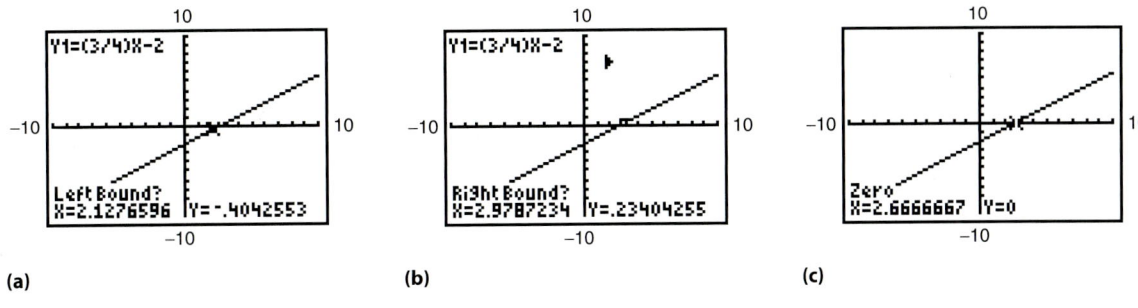

(a)　(b)　(c)

FIGURE 11

In Figure 11(c), you can see that the zero is $x = 2.6666667 \approx 2\frac{2}{3}$.

Intersect Feature

To find the points of intersection of two graphs, you can use the *intersect* feature. For instance, to find the points of intersection of the graphs of $y_1 = -x + 2$ and $y_2 = x + 4$, enter these two functions and use the *intersect* feature, as shown in Figure 12.

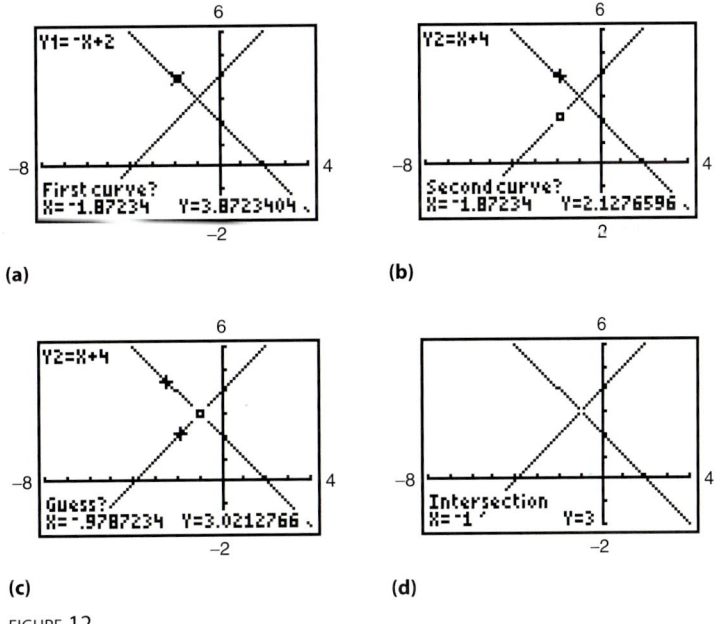

FIGURE 12

From Figure 12(d), you can see that the point of intersection is $(-1, 3)$.

Regression Capabilities

Throughout the text, you will be asked to use the regression capabilities of a graphing utility to find models for sets of data. Most graphing utilities have built-in regression programs for the following.

Regression	Form of Model
Linear	$y = ax + b$
Quadratic	$y = ax^2 + bx + c$
Cubic	$y = ax^3 + bx^2 + cx + d$
Quartic	$y = ax^4 + bx^3 + cx^2 + dx + e$
Logarithmic	$y = a + b \ln(x)$
Exponential	$y = ab^x$
Power	$y = ax^b$
Logistic	$y = \dfrac{c}{1 + ae^{-bx}}$
Sine	$y = a \sin(bx + c) + d$

For instance, you can find the linear regression model for the average hourly wages y (in dollars per hour) of production workers in manufacturing industries from 1987 through 1997 shown in the table. (Source: U.S. Bureau of Labor Statistics)

Year	1987	1988	1989	1990	1991	1992
y	9.91	10.19	10.48	10.83	11.18	11.46

Year	1993	1994	1995	1996	1997
y	11.74	12.06	12.37	12.78	13.17

FIGURE 13

First, let $x = 0$ correspond to 1990 and enter the data into the list editor, as shown in Figure 13. Note that the list in the first column contains the years and the list in the second column contains the hourly wages that correspond to the years. Run your graphing utility's built-in linear regression program to obtain the coefficients a and b for the model $y = ax + b$, as shown in Figure 14. So, a linear model for the data is

$$y \approx 0.321x + 10.83.$$

When you run some regression programs, you may obtain an "r-value," which gives a measure of how well the model fits the data. The closer the value of $|r|$ is to 1, the better the fit. For the data in the table above, $r \approx 0.999$, which implies that the model is a very good fit.

FIGURE 14

Trigonometry

Fifth Edition

- **P.1** ▶ Real Numbers
- **P.2** ▶ Solving Equations
- **P.3** ▶ The Cartesian Plane and Graphs of Equations
- **P.4** ▶ Linear Equations in Two Variables
- **P.5** ▶ Functions
- **P.6** ▶ Analyzing Graphs of Functions
- **P.7** ▶ Shifting, Reflecting, and Stretching Graphs
- **P.8** ▶ Combinations of Functions
- **P.9** ▶ Inverse Functions

Mark Joseph/Tony Stone Images

The average cost of a new domestic car increased from $18,064 in 1996 to $18,580 in 1997 even though sales (demand) and production (supply) of new domestic cars declined. (Source: U.S. Bureau of Economic Analysis)

P Prerequisites

▶ How to Study This Chapter

The Big Picture
In this chapter you will learn the following skills and concepts.

- ▶ How to represent, classify, and order real numbers
- ▶ How to evaluate algebraic expressions and solve linear and nonlinear equations
- ▶ How to plot points and use the Distance and Midpoint Formulas
- ▶ How to find and use the slopes of lines to write and graph linear equations in two variables
- ▶ How to evaluate functions and find their domains
- ▶ How to analyze graphs of functions, identify transformations of functions, and find inverses of functions
- ▶ How to find arithmetic combinations and compositions of functions

Important Vocabulary
As you encounter each new vocabulary term in this chapter, add the term and its definition to your notebook glossary.

Real numbers (p. 2)
Real number line (p. 2)
Inequality (p. 3)
Absolute value (p. 5)
Algebraic expression (p. 6)
Equation (p. 12)
Extraneous solution (p. 14)
Quadratic equation (p. 15)
Rectangular coordinate system (p. 25)
Ordered pair (p. 25)
Distance Formula (p. 27)
Midpoint Formula (p. 28)
Graph of an equation (p. 29)
Intercepts (p. 30)
Symmetry (p. 31)
Slope (pp. 38, 41)
Slope-intercept form (p. 38)
Point-slope form (p. 43)
Parallel (p. 45)
Perpendicular (p. 45)

Relation (p. 53)
Function (p. 53)
Domain (p. 53)
Range (p. 53)
Independent variable (p. 54)
Dependent variable (p. 54)
Implied domain (p. 57)
Vertical Line Test (p. 68)
Zeros of a function (p. 69)
Relative minimum (p. 71)
Relative maximum (p. 71)
Even function (p. 74)
Odd function (p. 74)
Rigid transformation (p. 85)
Nonrigid transformation (p. 85)
Arithmetic combinations of functions (p. 91)
Composition of functions (p. 93)
Inverse function (p. 99)
Horizontal Line Test (p. 102)

Study Tools
- Learning objectives at the beginning of each section
- Chapter Summary (pp. 109–110)
- Review Exercises (pp. 111–115)
- Chapter Test (p. 117)

Additional Resources
- Study and Solutions Guide
- Interactive Trigonometry
- Videotapes for Chapter P
- Trigonometry Website
- Student Success Organizer

STUDY T!P
During class, take notes on definitions, examples, concepts, and rules—whatever you identify as important to the instructor. Then, as soon after class as possible, review your notes.

P.1 Real Numbers

▶ **What you should learn**
- How to represent and classify real numbers
- How to order real numbers and use inequalities
- How to find the absolute values of real numbers and find the distance between two real numbers
- How to evaluate algebraic expressions
- How to use the basic rules and properties of algebra

▶ **Why you should learn it**
Real numbers are used to represent many real-life quantities, such as the variance of a budget (see Exercises 83–86 on page 10).

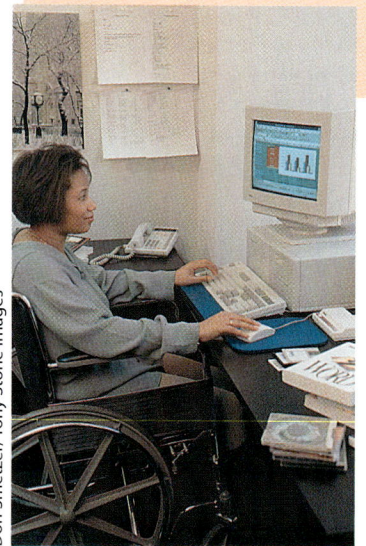

Real Numbers

Real numbers are used in everyday life to describe quantities such as age, miles per gallon, container size, and population. Real numbers are represented by symbols such as

$$-5, 9, 0, \frac{4}{3}, 0.666\ldots, 28.21, \sqrt{2}, \pi, \text{ and } \sqrt[3]{-32}.$$

Here are some important subsets of the real numbers.

$\{1, 2, 3, 4, \ldots\}$ Set of natural numbers

$\{0, 1, 2, 3, 4, \ldots\}$ Set of whole numbers

$\{\ldots, -3, -2, -1, 0, 1, 2, 3, \ldots\}$ Set of integers

A real number is **rational** if it can be written as the ratio p/q of two integers, where $q \neq 0$. For instance, the numbers

$$\frac{1}{3} = 0.3333\ldots = 0.\overline{3}, \quad \frac{1}{8} = 0.125, \text{ and } \frac{125}{111} = 1.126126\ldots = 1.\overline{126}$$

are rational. The decimal representation of a rational number either repeats (as in $\frac{173}{55} = 3.1\overline{45}$) or terminates (as in $\frac{1}{2} = 0.5$). A real number that cannot be written as the ratio of two integers is called **irrational.** Irrational numbers have infinite nonrepeating decimal representations. For instance, the numbers

$$\sqrt{2} \approx 1.4142136 \quad \text{and} \quad \pi \approx 3.1415927$$

are irrational. (The symbol $\approx$ means "is approximately equal to.")

Real numbers are represented graphically by a **real number line.** The point 0 on the real number line is the **origin.** Numbers to the right of 0 are positive, and numbers to the left of 0 are negative, as shown in Figure P.1. The term **nonnegative** describes a number that is either positive or zero.

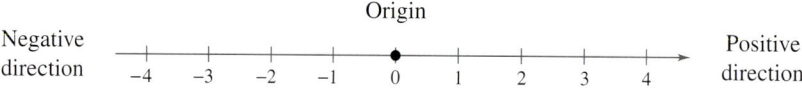

FIGURE P.1 *The Real Number Line*

As illustrated in Figure P.2, there is a *one-to-one correspondence* between real numbers and points on the real number line.

Every real number corresponds to exactly one point on the real number line.
FIGURE P.2 *One-to-One Correspondence*

Every point on the real number line corresponds to exactly one real number.

Ordering Real Numbers

One important property of real numbers is that they are *ordered*.

> ### Definition of Order on the Real Number Line
> If a and b are real numbers, a is *less than* b if $b - a$ is positive. The **order** of a and b is denoted by the **inequality**
> $$a < b.$$
> This relationship can also be described by saying that b is *greater than* a and writing $b > a$. The inequality $a \leq b$ means that a is *less than or equal to* b, and the inequality $b \geq a$ means that b is *greater than or equal to* a. The symbols $<, >, \leq$, and $\geq$ are *inequality symbols*.

FIGURE P.3 $a < b$ if and only if a lies to the left of b.

Geometrically, this definition implies that $a < b$ if and only if a lies to the *left* of b on the real number line, as shown in Figure P.3.

Example 1 ▶ Interpreting Inequalities

Describe the subset of real numbers represented by each inequality.

a. $x \leq 2$ **b.** $-2 \leq x < 3$

Solution

a. The inequality $x \leq 2$ denotes all real numbers less than or equal to 2, as shown in Figure P.4(a).

b. The inequality $-2 \leq x < 3$ means that $x \geq -2$ and $x < 3$. This "double inequality" denotes all real numbers between -2 and 3, including -2 but not including 3, as shown in Figure P.4(b).

(a)

(b)

FIGURE P.4

Inequalities can be used to describe subsets of real numbers called **intervals**. In the bounded intervals below, the real numbers a and b are the **endpoints** of each interval.

A computer animation of this example appears in the *Interactive* CD-ROM and *Internet* versions of this text.

Bounded Intervals on the Real Number Line

Notation	Interval Type	Inequality	Graph
$[a, b]$	Closed	$a \leq x \leq b$	[───] a b
(a, b)	Open	$a < x < b$	(───) a b
$[a, b)$		$a \leq x < b$	[───) a b
$(a, b]$		$a < x \leq b$	(───] a b

The symbols ∞, **positive infinity,** and $-\infty$, **negative infinity,** do not represent real numbers. They are simply convenient symbols used to describe the unboundedness of an interval such as $(1, \infty)$ or $(-\infty, 3]$.

Unbounded Intervals on the Real Number Line

Notation	Interval Type	Inequality	Graph
$[a, \infty)$		$x \geq a$	
(a, ∞)	Open	$x > a$	
$(-\infty, b]$		$x \leq b$	
$(-\infty, b)$	Open	$x < b$	
$(-\infty, \infty)$	Entire real line		

Example 2 ▶ Using Inequalities to Represent Intervals

Use inequality notation to describe each of the following.

a. c is at most 2.
b. m is at least -3.
c. All x in the interval $(-3, 5]$

Solution

a. The statement "c is at most 2" can be represented by $c \leq 2$.
b. The statement "m is at least -3" can be represented by $m \geq -3$.
c. "All x in the interval $(-3, 5]$" can be represented by $-3 < x \leq 5$.

Example 3 ▶ Interpreting Intervals

Give a verbal description of each interval.

a. $(-1, 0)$ **b.** $[2, \infty)$ **c.** $(-\infty, 0)$

Solution

a. This interval consists of all real numbers that are greater than -1 and less than 0.
b. This interval consists of all real numbers that are greater than or equal to 2.
c. This interval consists of all real numbers that are less than zero (the negative real numbers).

The **Law of Trichotomy** states that for any two real numbers a and b, *precisely* one of three relationships is possible:

$$a = b, \quad a < b, \quad \text{or} \quad a > b. \qquad \text{Law of Trichotomy}$$

Exploration

Absolute value expressions can be evaluated on a graphing utility. When an expression such as $|3 - 8|$ is evaluated, parentheses should surround the expression, as in abs$(3 - 8)$. Evaluate each expression below. What can you conclude?

a. $|6|$ **b.** $|-1|$
c. $|5 - 2|$ **d.** $|2 - 5|$

Absolute Value and Distance

The **absolute value** of a real number is its *magnitude*, or the distance between the origin and the point representing the real number on the real number line.

Definition of Absolute Value

If a is a real number, then the absolute value of a is

$$|a| = \begin{cases} a, & \text{if } a \geq 0 \\ -a, & \text{if } a < 0. \end{cases}$$

Notice in this definition that the absolute value of a real number is never negative. For instance, if $a = -5$, then $|-5| = -(-5) = 5$. The absolute value of a real number is either positive or zero. Moreover, 0 is the only real number whose absolute value is 0. So, $|0| = 0$.

Example 4 Evaluating the Absolute Value of a Number

Evaluate $\dfrac{|x|}{x}$ for (a) $x > 0$ and (b) $x < 0$.

Solution

a. If $x > 0$, then $|x| = x$ and $\dfrac{|x|}{x} = \dfrac{x}{x} = 1$.

b. If $x < 0$, then $|x| = -x$ and $\dfrac{|x|}{x} = \dfrac{-x}{x} = -1$.

Properties of Absolute Values

1. $|a| \geq 0$ 2. $|-a| = |a|$
3. $|ab| = |a||b|$ 4. $\left|\dfrac{a}{b}\right| = \dfrac{|a|}{|b|}$, $b \neq 0$

Distance Between Two Points on the Real Line

Let a and b be real numbers. The **distance between a and b** is

$$d(a, b) = |b - a| = |a - b|.$$

Absolute value can be used to define the distance between two points on the real number line. For instance, the distance between -3 and 4 is

$$|-3 - 4| = |-7| = 7,$$

as shown in Figure P.5.

FIGURE P.5 *The distance between -3 and 4 is 7.*

Algebraic Expressions

One characteristic of algebra is the use of letters to represent numbers. The letters are **variables,** and combinations of letters and numbers are **algebraic expressions.** Here are a few examples of algebraic expressions.

$$5x, \quad 2x - 3, \quad \frac{4}{x^2 + 2}, \quad 7x + y$$

> **Definition of an Algebraic Expression**
>
> An **algebraic expression** is a collection of letters (**variables**) and real numbers (**constants**) combined using the operations of addition, subtraction, multiplication, division, and exponentiation.

The **terms** of an algebraic expression are those parts that are separated by *addition.* For example,

$$x^2 - 5x + 8 = x^2 + (-5x) + 8$$

has three terms: x^2 and $-5x$ are the **variable terms** and 8 is the **constant term.** The numerical factor of a variable term is the **coefficient** of the variable term. For instance, the coefficient of $-5x$ is -5, and the coefficient of x^2 is 1.

To **evaluate** an algebraic expression, substitute numerical values for each of the variables in the expression. Here are two examples.

Expression	Value of Variable	Substitute	Value of Expression
$-3x + 5$	$x = 3$	$-3(3) + 5$	$-9 + 5 = -4$
$3x^2 + 2x - 1$	$x = -1$	$3(-1)^2 + 2(-1) - 1$	$3 - 2 - 1 = 0$

When an algebraic expression is evaluated, the **Substitution Principle** is used. It states that "If $a = b$, then a can be replaced by b in any expression involving a." In the first evaluation shown above, for instance, 3 is *substituted* for x in the expression $-3x + 5$.

A computer animation of this concept appears in the *Interactive* CD-ROM and *Internet* versions of this text.

Basic Rules of Algebra

There are four arithmetic operations with real numbers: *addition, multiplication, subtraction,* and *division,* denoted by the symbols $+, \times$ or $\cdot, -,$ and $\div$. Of these, addition and multiplication are the two primary operations. Subtraction and division are the inverse operations of addition and multiplication, respectively.

Subtraction: Add the opposite. *Division:* Multiply by the reciprocal.

$$a - b = a + (-b) \qquad \text{If } b \neq 0, \text{ then } a/b = a\left(\frac{1}{b}\right) = \frac{a}{b}.$$

> **STUDY TIP**
>
> Try to formulate a verbal description of each property. For instance, the first property states that *the order in which two real numbers are added does not affect their sum.*

In these definitions, $-b$ is the **additive inverse** (or opposite) of b, and $1/b$ is the **multiplicative inverse** (or reciprocal) of b. In the fractional form a/b, a is the **numerator** of the fraction and b is the **denominator.**

Because the properties of real numbers on page 7 are true for variables and algebraic expressions as well as for real numbers, they are often called the **Basic Rules of Algebra.**

Basic Rules of Algebra

Let a, b, and c be real numbers, variables, or algebraic expressions.

Property		Example
Commutative Property of Addition:	$a + b = b + a$	$4x + x^2 = x^2 + 4x$
Commutative Property of Multiplication:	$ab = ba$	$(4 - x)x^2 = x^2(4 - x)$
Associative Property of Addition:	$(a + b) + c = a + (b + c)$	$(x + 5) + x^2 = x + (5 + x^2)$
Associative Property of Multiplication:	$(ab)c = a(bc)$	$(2x \cdot 3y)(8) = (2x)(3y \cdot 8)$
Distributive Properties:	$a(b + c) = ab + ac$	$3x(5 + 2x) = 3x \cdot 5 + 3x \cdot 2x$
	$(a + b)c = ac + bc$	$(y + 8)y = y \cdot y + 8 \cdot y$
Additive Identity Property:	$a + 0 = a$	$5y^2 + 0 = 5y^2$
Multiplicative Identity Property:	$a \cdot 1 = a$	$(4x^2)(1) = 4x^2$
Additive Inverse Property:	$a + (-a) = 0$	$5x^3 + (-5x^3) = 0$
Multiplicative Inverse Property:	$a \cdot \dfrac{1}{a} = 1, \quad a \neq 0$	$(x^2 + 4)\left(\dfrac{1}{x^2 + 4}\right) = 1$

Because subtraction is defined as "adding the opposite," the Distributive Properties are also true for subtraction. For instance, the "subtraction form" of $a(b + c) = ab + ac$ is

$$a(b - c) = ab - ac.$$

STUDY TIP

Be sure you see the difference between the *opposite of a number* and a *negative number*. If a is already negative, then its opposite, $-a$, is positive. For instance, if $a = -5$, then

$$-a = -(-5) = 5.$$

Properties of Negation

Let a and b be real numbers, variables, or algebraic expressions.

Property	Example
1. $(-1)a = -a$	$(-1)7 = -7$
2. $-(-a) = a$	$-(-6) = 6$
3. $(-a)b = -(ab) = a(-b)$	$(-5)3 = -(5 \cdot 3) = 5(-3)$
4. $(-a)(-b) = ab$	$(-2)(-x) = 2x$
5. $-(a + b) = (-a) + (-b)$	$-(x + 8) = (-x) + (-8)$
	$\quad = -x - 8$

Properties of Equality

Let a, b, and c be real numbers, variables, or algebraic expressions.

1. If $a = b$, then $a + c = b + c$.	Add c to each side.
2. If $a = b$, then $ac = bc$.	Multiply each side by c.
3. If $a + c = b + c$, then $a = b$.	Subtract c from each side.
4. If $ac = bc$ and $c \neq 0$, then $a = b$.	Divide each side by c.

> **STUDY TIP**
>
> The "or" in the Zero-Factor Property includes the possibility that either or both factors may be zero. This is an **inclusive or,** and it is the way the word "or" is generally used in mathematics.

Properties of Zero

Let a and b be real numbers, variables, or algebraic expressions.

1. $a + 0 = a$ and $a - 0 = a$
2. $a \cdot 0 = 0$
3. $\dfrac{0}{a} = 0, \quad a \neq 0$
4. $\dfrac{a}{0}$ is undefined.
5. **Zero-Factor Property:** If $ab = 0$, then $a = 0$ or $b = 0$.

Properties and Operations of Fractions

Let a, b, c, and d be real numbers, variables, or algebraic expressions such that $b \neq 0$ and $d \neq 0$.

1. **Equivalent Fractions:** $\dfrac{a}{b} = \dfrac{c}{d}$ if and only if $ad = bc$.
2. **Rules of Signs:** $-\dfrac{a}{b} = \dfrac{-a}{b} = \dfrac{a}{-b}$ and $\dfrac{-a}{-b} = \dfrac{a}{b}$
3. **Generate Equivalent Fractions:** $\dfrac{a}{b} = \dfrac{ac}{bc}, \quad c \neq 0$
4. **Add or Subtract with Like Denominators:** $\dfrac{a}{b} \pm \dfrac{c}{b} = \dfrac{a \pm c}{b}$
5. **Add or Subtract with Unlike Denominators:** $\dfrac{a}{b} \pm \dfrac{c}{d} = \dfrac{ad \pm bc}{bd}$
6. **Multiply Fractions:** $\dfrac{a}{b} \cdot \dfrac{c}{d} = \dfrac{ac}{bd}$
7. **Divide Fractions:** $\dfrac{a}{b} \div \dfrac{c}{d} = \dfrac{a}{b} \cdot \dfrac{d}{c} = \dfrac{ad}{bc}, \quad c \neq 0$

> **STUDY TIP**
>
> In Property 1, the phrase "if and only if" implies two statements. One statement is: If $a/b = c/d$, then $ad = bc$. The other statement is: If $ad = bc$, where $b \neq 0$ and $d \neq 0$, then $a/b = c/d$.

Example 5 ▶ Properties and Operations of Fractions

a. Equivalent fractions: $\dfrac{x}{5} = \dfrac{3 \cdot x}{3 \cdot 5} = \dfrac{3x}{15}$

b. Divide fractions: $\dfrac{7}{x} \div \dfrac{3}{2} = \dfrac{7}{x} \cdot \dfrac{2}{3} = \dfrac{14}{3x}$

c. Add fractions with unlike denominators: $\dfrac{x}{3} + \dfrac{2x}{5} = \dfrac{5 \cdot x + 3 \cdot 2x}{15} = \dfrac{11x}{15}$

If a, b, and c are integers such that $ab = c$, then a and b are **factors** or **divisors** of c. A **prime number** is an integer that has exactly two positive factors: itself and 1, such as 2, 3, 5, 7, and 11. The numbers 4, 6, 8, 9, and 10 are **composite** because they can be written as the product of two or more prime numbers. The number 1 is neither prime nor composite. The **Fundamental Theorem of Arithmetic** states that every positive integer greater than 1 can be written as the product of prime numbers in precisely one way (disregarding order). For instance, the *prime factorization* of 24 is $24 = 2 \cdot 2 \cdot 2 \cdot 3$.

P.1 Exercises

The *Interactive* CD-ROM and *Internet* versions of this text contain step-by-step solutions to all odd-numbered Section and Review Exercises. They also provide Tutorial Exercises that link to Guided Examples for additional help.

In Exercises 1–6, determine which numbers are (a) natural numbers, (b) integers, (c) rational numbers, and (d) irrational numbers.

1. $-9, -\frac{7}{2}, 5, \frac{2}{3}, \sqrt{2}, 0, 1, -4, 2, -11$
2. $\sqrt{5}, -7, -\frac{7}{3}, 0, 3.12, \frac{5}{4}, -3, 12, 5$
3. $2.01, 0.666\ldots, -13, 0.010110111\ldots, 1, -6$
4. $2.3030030003\ldots, 0.7575, -4.63, \sqrt{10}, -75, 15, 31$
5. $-\pi, -\frac{1}{3}, \frac{6}{3}, \frac{1}{2}\sqrt{2}, -7.5, -1, 8, -22$
6. $25, -17, -\frac{12}{5}, \sqrt{9}, 3.12, \frac{1}{2}\pi, 7, -11.1, 13$

In Exercises 7–10, use a calculator to find the decimal form of the rational number. If it is a nonterminating decimal, write the repeating pattern.

7. $\frac{5}{8}$
8. $\frac{1}{3}$
9. $\frac{41}{333}$
10. $\frac{6}{11}$

In Exercises 11–16, use a calculator with a fraction feature to write the rational number as the ratio of two integers.

11. 4.1
12. 8.5
13. $10.\overline{2}$
14. $5.\overline{45}$
15. $-2.01\overline{2}$
16. $-1.6\overline{5}$

In Exercises 17 and 18, approximate the numbers and place the correct symbol (< or >) between them.

17.
18.

In Exercises 19–24, plot the two real numbers on the real number line. Then place the appropriate inequality symbol (< or >) between them.

19. $-4, -8$
20. $-3.5, 1$
21. $\frac{3}{2}, 7$
22. $1, \frac{16}{3}$
23. $\frac{5}{6}, \frac{2}{3}$
24. $-\frac{8}{7}, -\frac{3}{7}$

In Exercises 25–34, verbally describe the subset of real numbers represented by the inequality. Then sketch the subset on the real number line. State whether the interval is bounded or unbounded.

25. $x \leq 5$
26. $x \geq -2$
27. $x < 0$
28. $x > 3$
29. $x \geq 4$
30. $x < 2$
31. $-2 < x < 2$
32. $0 \leq x \leq 5$
33. $-1 \leq x < 0$
34. $0 < x \leq 6$

In Exercises 35 and 36, use a calculator to order the numbers from smallest to largest.

35. $\frac{7071}{5000}, \frac{584}{413}, \sqrt{2}, \frac{47}{33}, \frac{127}{90}$
36. $\frac{26}{15}, \sqrt{3}, 1.7320, \frac{381}{220}, \frac{2103}{1214}$

In Exercises 37–44, use inequality notation to describe the set.

37. All x in the interval $(-2, 4]$
38. All y in the interval $[-6, 0)$
39. y is nonnegative.
40. y is no more than 25.
41. t is at least 10 and at most 22.
42. k is less than 5 but no less than -3.
43. The dog's weight W is more than 65 pounds.
44. The annual rate of inflation r is expected to be at least 2.5% but no more than 5%.

In Exercises 45–48, give a verbal description of the interval.

45. $[0, 8)$
46. $[-5, 7]$
47. $(-6, \infty)$
48. $(-\infty, 4]$

In Exercises 49–58, evaluate the expression.

49. $|-10|$
50. $|0|$
51. $|3 - x|$
52. $|4 - x|$
53. $|-1| - |-2|$
54. $-3 - |-3|$
55. $\frac{-5}{|-5|}$
56. $-3|-3|$
57. $\frac{|x+2|}{x+2}, \quad x < -2$
58. $\frac{|x-1|}{x-1}, \quad x > 1$

In Exercises 59–64, place the correct symbol (<, >, or =) between the pair of real numbers.

59. $|-3| \quad\square\quad -|-3|$
60. $|-4| \quad\square\quad |4|$
61. $-5 \quad\square\quad -|5|$
62. $-|-6| \quad\square\quad |-6|$
63. $-|-2| \quad\square\quad -|2|$
64. $-(-2) \quad\square\quad -2$

In Exercises 65–72, find the distance between a and b.

65.
    $a = -1$, $b = 3$

66.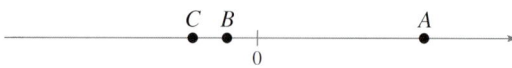
    $a = -4$, $b = -\frac{3}{2}$

67. $a = 126$, $b = 75$
68. $a = -126$, $b = -75$
69. $a = -\frac{5}{2}$, $b = 0$
70. $a = \frac{1}{4}$, $b = \frac{11}{4}$
71. $a = \frac{16}{5}$, $b = \frac{112}{75}$
72. $a = 9.34$, $b = -5.65$

In Exercises 73 and 74, use the real numbers A, B, and C shown on the number line. Determine the sign of each expression.

```
       C  B                 A
───────●──●────────────────●──────
          0
```

73. (a) $-A$
    (b) $B - A$

74. (a) $-C$
    (b) $A - C$

In Exercises 75–82, use absolute value notation to describe the situation.

75. While traveling, you pass milepost 7, then milepost 18. How far do you travel during that time period?
76. While traveling, you pass milepost 103, then milepost 86. How far do you travel during that time period?
77. The temperature was 60° at noon, then 23° at midnight. What was the change in temperature over the 12-hour period?
78. The temperature was 48° last night at midnight, then 82° at noon today. What was the change in temperature over the 12-hour period?
79. The distance between x and 5 is no more than 3.
80. The distance between x and -10 is at least 6.
81. y is at least 6 units from 0.
82. y is at most 2 units from a.

Budget Variance In Exercises 83–86, the accounting department of a company is checking to see whether the actual expenses of a department differ from the budgeted expenses by more than $500 or by more than 5%. Fill in the missing parts of the table, and determine whether the actual expense passes the "budget variance test."

		Budgeted Expense, b	Actual Expense, a	$|a - b|$	$0.05b$
83.	Wages	$112,700	$113,356		
84.	Utilities	$9,400	$9,772		
85.	Taxes	$37,640	$37,335		
86.	Insurance	$2,575	$2,613		

Federal Deficit In Exercises 87–90, use the bar graph, which shows the receipts of the federal government (in billions of dollars) for selected years from 1960 through 1998. In each exercise you are given the outlay of the federal government. Find the magnitude of the surplus or deficit for the year. *(Source: U.S. Treasury Department)*

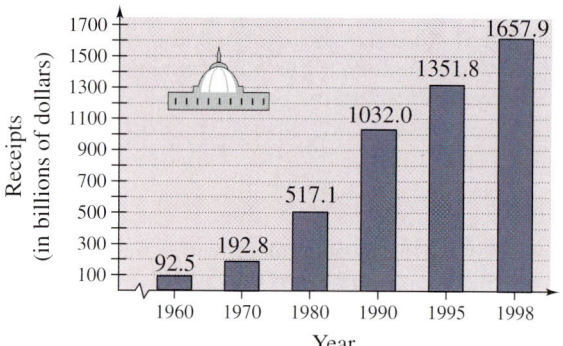

		Receipts	Outlay	$|Receipts - Outlay|$
87.	1960		$92.2 billion	
88.	1980		$590.9 billion	
89.	1990		$1253.2 billion	
90.	1998		$1667.8 billion	

In Exercises 91–96, identify the terms. Then identify the coefficients of the variable terms of the expression.

91. $7x + 4$
92. $6x^3 - 5x$
93. $\sqrt{3}x^2 - 8x - 11$
94. $3\sqrt{3}x^2 + 1$
95. $4x^3 + \dfrac{x}{2} - 5$
96. $3x^4 - \dfrac{x^2}{4}$

In Exercises 97–102, evaluate the expression for each value of x. (If not possible, state the reason.)

	Expression	Values	
97.	$4x - 6$	(a) $x = -1$	(b) $x = 0$
98.	$9 - 7x$	(a) $x = -3$	(b) $x = 3$
99.	$x^2 - 3x + 4$	(a) $x = -2$	(b) $x = 2$

Expression	Values	
100. $-x^2 + 5x - 4$	(a) $x = -1$	(b) $x = 1$
101. $\dfrac{x+1}{x-1}$	(a) $x = 1$	(b) $x = -1$
102. $\dfrac{x}{x+2}$	(a) $x = 2$	(b) $x = -2$

In Exercises 103–112, identify the rule(s) of algebra illustrated by the equation.

103. $x + 9 = 9 + x$

104. $2\left(\frac{1}{2}\right) = 1$

105. $\dfrac{1}{h+6}(h+6) = 1, \quad h \neq -6$

106. $(x + 3) - (x + 3) = 0$

107. $2(x + 3) = 2x + 6$

108. $(z - 2) + 0 = z - 2$

109. $1 \cdot (1 + x) = 1 + x$

110. $x + (y + 10) = (x + y) + 10$

111. $x(3y) = (x \cdot 3)y = (3x)y$

112. $\frac{1}{7}(7 \cdot 12) = \left(\frac{1}{7} \cdot 7\right)12 = 1 \cdot 12 = 12$

In Exercises 113–120, perform the operation(s). (Write fractional answers in simplest form.)

113. $\frac{3}{16} + \frac{5}{16}$

114. $\frac{6}{7} - \frac{4}{7}$

115. $\frac{5}{8} - \frac{5}{12} + \frac{1}{6}$

116. $\frac{10}{11} + \frac{6}{33} - \frac{13}{66}$

117. $12 \div \frac{1}{4}$

118. $-\left(6 \cdot \frac{4}{8}\right)$

119. $\frac{2x}{3} - \frac{x}{4}$

120. $\frac{5x}{6} \cdot \frac{2}{9}$

In Exercises 121–124, use a calculator to evaluate the expression. (Round your answer to two decimal places.)

121. $-3 + \frac{3}{7}$

122. $3\left(-\frac{5}{12} + \frac{3}{8}\right)$

123. $\dfrac{11.46 - 5.37}{3.91}$

124. $\dfrac{\frac{1}{5}(-8 - 9)}{-\frac{1}{3}}$

125. (a) Use a calculator to complete the table.

n	1	0.5	0.01	0.0001	0.000001
$5/n$					

(b) Use the result from part (a) to make a conjecture about the value of $5/n$ as n approaches 0.

126. (a) Use a calculator to complete the table.

n	1	10	100	10,000	100,000
$5/n$					

(b) Use the result from part (a) to make a conjecture about the value of $5/n$ as n increases without bound.

Synthesis

127. *Exploration* Consider $|u + v|$ and $|u| + |v|$.
   (a) Are the values of the expressions always equal? If not, under what conditions are they unequal?
   (b) If the two expressions are not equal for certain values of u and v, is one of the expressions always greater than the other? Explain.

128. *Think About It* Is there a difference between saying that a real number is positive and saying that a real number is nonnegative? Explain.

129. *Think About It* Because every even number is divisible by 2, is it possible that there exist any even prime numbers? Explain.

True or False? **In Exercises 130 and 131, determine whether the statement is true or false. Justify your answer.**

130. If $a < b$, then $\dfrac{1}{a} < \dfrac{1}{b}$, where $a \neq b \neq 0$.

131. Because $\dfrac{a+b}{c} = \dfrac{a}{c} + \dfrac{b}{c}$, then $\dfrac{c}{a+b} = \dfrac{c}{a} + \dfrac{c}{b}$.

132. *Writing* Describe the differences among the sets of natural numbers, integers, rational numbers, and irrational numbers.

133. *Writing* You may hear it said that to take the absolute value of a real number you simply remove any negative sign and make the number positive. Can it ever be true that $|a| = -a$ for a real number a? Explain.

P.2 Solving Equations

▶ **What you should learn**
- How to identify different types of equations
- How to solve linear equations in one variable
- How to solve quadratic equations by factoring, extracting square roots, and using the Quadratic Formula
- How to solve polynomial equations of degree three or greater
- How to solve equations involving radicals
- How to solve equations involving absolute values

▶ **Why you should learn it**
Linear equations are used in many real-life applications. For example, Exercises 185 and 186 on page 23 show how linear equations can model the relationship between the length of a thigh bone and the height of a person, helping researchers learn about ancient cultures.

M. Greenlar/The Image Works

Equations and Solutions of Equations

An **equation** in x is a statement that two algebraic expressions are equal. For example,

$$3x - 5 = 7, \quad x^2 - x - 6 = 0, \text{ and } \sqrt{2x} = 4$$

are equations. To **solve** an equation in x means to find all values of x for which the equation is true. Such values are **solutions.** For instance, $x = 4$ is a solution of the equation

$$3x - 5 = 7,$$

because $3(4) - 5 = 7$ is a true statement.

The solutions of an equation depend on the kinds of numbers being considered. For instance, in the set of rational numbers, $x^2 = 10$ has no solution because there is no rational number whose square is 10. However, in the set of real numbers, the equation has the two solutions $\sqrt{10}$ and $-\sqrt{10}$.

An equation that is true for *every* real number in the domain of the variable is called an **identity.** For example,

$$x^2 - 9 = (x + 3)(x - 3) \qquad \text{Identity}$$

is an identity because it is a true statement for any real value of x, and

$$\frac{x}{3x^2} = \frac{1}{3x} \qquad \text{Identity}$$

where $x \neq 0$, is an identity because it is true for any nonzero real value of x.

An equation that is true for just *some* (or even none) of the real numbers in the domain of the variable is called a **conditional equation.** For example, the equation

$$x^2 - 9 = 0 \qquad \text{Conditional equation}$$

is conditional because $x = 3$ and $x = -3$ are the only values in the domain that satisfy the equation. The equation $2x - 4 = 2x + 1$ is conditional because there are no real values of x for which the equation is true. Learning to solve conditional equations is the primary focus of this section.

Linear Equations

Definition of Linear Equation

A **linear equation** in one variable x is an equation that can be written in the standard form

$$ax + b = 0$$

where a and b are real numbers with $a \neq 0$.

A linear equation has exactly one solution. To see this, consider the following steps. (Remember that $a \neq 0$.)

$ax + b = 0$ Write original equation.

$ax = -b$ Subtract b from each side.

$x = -\dfrac{b}{a}$ Divide each side by a.

To solve a conditional equation in x, isolate x on one side of the equation by a sequence of **equivalent** (and usually simpler) **equations,** each having the same solution(s) as the original equation. The operations that yield equivalent equations come from the Substitution Principle and simplification techniques.

Generating Equivalent Equations

An equation can be transformed into an *equivalent equation* by one or more of the following steps.

	Given Equation	Equivalent Equation
1. Remove symbols of grouping, combine like terms, or simplify fractions on one or both sides of the equation.	$2x - x = 4$	$x = 4$
2. Add (or subtract) the same quantity to (from) *each* side of the equation.	$x + 1 = 6$	$x = 5$
3. Multiply (or divide) *each* side of the equation by the same *nonzero* quantity.	$2x = 6$	$x = 3$
4. Interchange the two sides of the equation.	$2 = x$	$x = 2$

Example 1 ▶ Solving a Linear Equation

$3x - 6 = 0$ Write original equation.

$3x = 6$ Add 6 to each side.

$x = 2$ Divide each side by 3.

Check

After solving an equation, you should check each solution in the original equation.

$3x - 6 = 0$ Write original equation.

$3(2) - 6 \stackrel{?}{=} 0$ Substitute 2 for x.

$0 = 0$ Solution checks. ✓

So, the solution is 2.

To solve an equation involving fractional expressions, find the least common denominator of all terms and multiply every term by this LCD.

Example 2 ▶ An Equation Involving Fractional Expressions

Solve $\dfrac{x}{3} + \dfrac{3x}{4} = 2$.

Solution

$$\dfrac{x}{3} + \dfrac{3x}{4} = 2 \qquad \text{Write original equation.}$$

$$(12)\dfrac{x}{3} + (12)\dfrac{3x}{4} = (12)2 \qquad \text{Multiply each term by the LCD of 12.}$$

$$4x + 9x = 24 \qquad \text{Divide out and multiply.}$$

$$13x = 24 \qquad \text{Combine like terms.}$$

$$x = \dfrac{24}{13} \qquad \text{Divide each side by 13.}$$

The solution is $\dfrac{24}{13}$. Check this in the original equation.

When multiplying or dividing an equation by a *variable* quantity, it is possible to introduce an extraneous solution. An **extraneous solution** is one that does not satisfy the original equation.

Example 3 ▶ An Equation with an Extraneous Solution

Solve $\dfrac{1}{x-2} = \dfrac{3}{x+2} - \dfrac{6x}{x^2 - 4}$.

Solution

The LCD is $x^2 - 4$, or $(x+2)(x-2)$. Multiply each term by this LCD.

$$\dfrac{1}{x-2}(x+2)(x-2) = \dfrac{3}{x+2}(x+2)(x-2) - \dfrac{6x}{x^2-4}(x+2)(x-2)$$

$$x + 2 = 3(x-2) - 6x, \qquad x \ne \pm 2$$

$$x + 2 = 3x - 6 - 6x$$

$$x + 2 = -3x - 6$$

$$4x = -8$$

$$x = -2$$

In the original equation, $x = -2$ yields a denominator of zero. Therefore, $x = -2$ is an extraneous solution, and the original equation has *no solution*.

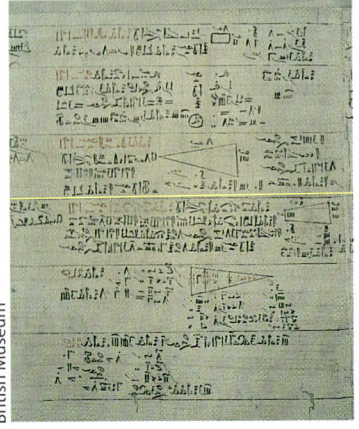

Historical Note

This ancient Egyptian papyrus, discovered in 1858, contains one of the earliest examples of mathematical writing in existence. The papyrus itself dates back to around 1650 B.C., but it is actually a copy of writings from two centuries earlier. The algebraic equations on the papyrus were written in words. Diophantus, a Greek who lived around A.D. 250, is often called the Father of Algebra. He was the first to use abbreviated word forms in equations.

Quadratic Equations

A **quadratic equation** in x is an equation that can be written in the general form

$$ax^2 + bx + c = 0$$

where a, b, and c are real numbers, with $a \neq 0$. A quadratic equation in x is also known as a **second-degree polynomial equation in x.**

You should be familiar with the following four methods for solving quadratic equations.

Solving a Quadratic Equation

Factoring: If $ab = 0$, then $a = 0$ or $b = 0$.

Example:
$$x^2 - x - 6 = 0$$
$$(x - 3)(x + 2) = 0$$
$$x - 3 = 0 \implies x = 3$$
$$x + 2 = 0 \implies x = -2$$

Square Root Principle: If $u^2 = c$, where $c > 0$, then $u = \pm\sqrt{c}$.

Example:
$$(x + 3)^2 = 16$$
$$x + 3 = \pm 4$$
$$x = -3 \pm 4$$
$$x = 1 \quad \text{or} \quad x = -7$$

Completing the Square: If $x^2 + bx = c$, then

$$x^2 + bx + \left(\frac{b}{2}\right)^2 = c + \left(\frac{b}{2}\right)^2$$

$$\left(x + \frac{b}{2}\right)^2 = c + \frac{b^2}{4}.$$

Example:
$$x^2 + 6x = 5$$
$$x^2 + 6x + 3^2 = 5 + 3^2$$
$$(x + 3)^2 = 14$$
$$x + 3 = \pm\sqrt{14}$$
$$x = -3 \pm \sqrt{14}$$

Quadratic Formula: If $ax^2 + bx + c = 0$, then $x = \dfrac{-b \pm \sqrt{b^2 - 4ac}}{2a}$.

Example:
$$2x^2 + 3x - 1 = 0$$
$$x = \frac{-3 \pm \sqrt{3^2 - 4(2)(-1)}}{2(2)}$$
$$= \frac{-3 \pm \sqrt{17}}{4}$$

Example 4 ▶ Solving a Quadratic Equation by Factoring

a. $2x^2 + 9x + 7 = 3$ Write original equation.

$2x^2 + 9x + 4 = 0$ Write in general form.

$(2x + 1)(x + 4) = 0$ Factor.

$2x + 1 = 0 \implies x = -\dfrac{1}{2}$ Set 1st factor equal to 0.

$x + 4 = 0 \implies x = -4$ Set 2nd factor equal to 0.

The solutions are $-\tfrac{1}{2}$ and -4. Check these in the original equation.

b. $6x^2 - 3x = 0$ Write original equation.

$3x(2x - 1) = 0$ Factor.

$3x = 0 \implies x = 0$ Set 1st factor equal to 0.

$2x - 1 = 0 \implies x = \dfrac{1}{2}$ Set 2nd factor equal to 0.

The solutions are 0 and $\tfrac{1}{2}$. Check these in the original equation.

Note that the method of solution in Example 4 is based on the Zero-Factor Property given in Section P.1. Be sure you see that this property works *only* for equations written in general form (in which the right side of the equation is zero). Therefore, all terms must be collected on one side *before* factoring. For instance, in the equation

$$(x - 5)(x + 2) = 8$$

it is *incorrect* to set each factor equal to 8. Can you solve this equation correctly?

Example 5 ▶ Extracting Square Roots

Solve each equation by extracting square roots.

a. $4x^2 = 12$ **b.** $(x - 3)^2 = 7$

Solution

a. $4x^2 = 12$ Write original equation.

$x^2 = 3$ Divide each side by 4.

$x = \pm\sqrt{3}$ Extract square roots.

The solutions are $\sqrt{3}$ and $-\sqrt{3}$. Check these in the original equation.

b. $(x - 3)^2 = 7$ Write original equation.

$x - 3 = \pm\sqrt{7}$ Extract square roots.

$x = 3 \pm \sqrt{7}$ Add 3 to each side.

The solutions are $3 \pm \sqrt{7}$. Check these in the original equation.

A computer animation of this example appears in the *Interactive* CD-ROM and *Internet* versions of this text.

Example 6 ▶ The Quadratic Formula: Two Distinct Solutions

Use the Quadratic Formula to solve

$$x^2 + 3x = 9.$$

Solution

$$x^2 + 3x = 9 \qquad \text{Write original equation.}$$

$$x^2 + 3x - 9 = 0 \qquad \text{Write in general form.}$$

$$x = \frac{-b \pm \sqrt{b^2 - 4ac}}{2a} \qquad \text{Quadratic Formula}$$

$$x = \frac{-3 \pm \sqrt{(3)^2 - 4(1)(-9)}}{2(1)} \qquad \text{Substitute 1 for } a, \text{ 3 for } b, \text{ and } -9 \text{ for } c.$$

$$x = \frac{-3 \pm \sqrt{45}}{2} \qquad \text{Simplify.}$$

$$x = \frac{-3 \pm 3\sqrt{5}}{2} \qquad \text{Simplify.}$$

The equation has two solutions:

$$x = \frac{-3 + 3\sqrt{5}}{2} \quad \text{and} \quad x = \frac{-3 - 3\sqrt{5}}{2}.$$

Check these in the original equation.

Example 7 ▶ The Quadratic Formula: One Solution

Use the Quadratic Formula to solve

$$8x^2 - 24x + 18 = 0.$$

Solution

$$8x^2 - 24x + 18 = 0 \qquad \text{Write original equation.}$$

$$4x^2 - 12x + 9 = 0 \qquad \text{Divide out common factor of 2.}$$

$$x = \frac{-b \pm \sqrt{b^2 - 4ac}}{2a} \qquad \text{Quadratic Formula}$$

$$x = \frac{-(-12) \pm \sqrt{(-12)^2 - 4(4)(9)}}{2(4)} \qquad \text{Substitute.}$$

$$x = \frac{12 \pm \sqrt{0}}{8} \qquad \text{Simplify.}$$

$$x = \frac{3}{2} \qquad \text{Simplify.}$$

This quadratic equation has only one solution: $\frac{3}{2}$. Check this in the original equation.

Polynomial Equations of Higher Degree

The methods used to solve quadratic equations can sometimes be extended to polynomials of higher degree.

> **STUDY TIP**
>
> A common mistake that is made in solving an equation such as that in Example 8 is to divide both sides of the equation by the variable factor x^2. This loses the solution $x = 0$. When solving an equation, be sure to write the equation in general form, then factor the equation and set each factor equal to zero. Don't divide both sides of an equation by a variable factor in an attempt to simplify the equation.

Example 8 ▶ Solving a Polynomial Equation by Factoring

Solve $3x^4 = 48x^2$.

Solution

First write the polynomial equation in general form with zero on one side, factor the other side, and then set each factor equal to zero.

$3x^4 = 48x^2$	Write original equation.
$3x^4 - 48x^2 = 0$	Write in general form.
$3x^2(x^2 - 16) = 0$	Factor.
$3x^2(x + 4)(x - 4) = 0$	Write in factored form.
$3x^2 = 0 \implies x = 0$	Set 1st factor equal to 0.
$x + 4 = 0 \implies x = -4$	Set 2nd factor equal to 0.
$x - 4 = 0 \implies x = 4$	Set 3rd factor equal to 0.

You can check these solutions by substituting in the original equation, as follows.

Check

$3x^4 = 48x^2$	Write original equation.
$3(0)^4 = 48(0)^2$	0 checks. ✓
$3(-4)^4 = 48(-4)^2$	-4 checks. ✓
$3(4)^4 = 48(4)^2$	4 checks. ✓

After checking, you can conclude that the solutions are 0, -4, and 4.

Example 9 ▶ Solving a Polynomial Equation by Factoring

Solve $x^3 - 3x^2 - 3x + 9 = 0$.

Solution

$x^3 - 3x^2 - 3x + 9 = 0$	Write original equation.
$x^2(x - 3) - 3(x - 3) = 0$	Factor by grouping.
$(x - 3)(x^2 - 3) = 0$	Distributive Property
$x - 3 = 0 \implies x = 3$	Set 1st factor equal to 0.
$x^2 - 3 = 0 \implies x = \pm\sqrt{3}$	Set 2nd factor equal to 0.

The solutions are 3, $\sqrt{3}$, and $-\sqrt{3}$.

Radical Equations

The steps involved in solving the remaining equations in this section will often introduce *extraneous solutions*. Extraneous solutions occur during operations such as squaring each side of an equation, raising each side of an equation to a rational power, and multiplying each side by a variable quantity. So, when you use any of these operations, checking is crucial.

Example 10 ▶ **Solving an Equation Involving a Rational Exponent**

$$4x^{3/2} - 8 = 0 \qquad \text{Write original equation.}$$
$$4x^{3/2} = 8 \qquad \text{Add 8 to each side.}$$
$$x^{3/2} = 2 \qquad \text{Divide each side by 4.}$$
$$x = 2^{2/3} \qquad \text{Raise each side to the } \tfrac{2}{3} \text{ power.}$$
$$x \approx 1.587 \qquad \text{Round to three decimal places.}$$

The solution is $2^{2/3}$. Check this in the original equation.

> **STUDY TIP**
>
> The essential operations in Example 10 are isolating the factor with the rational exponent and raising each side to the reciprocal power. In Example 11, this is equivalent to isolating the square root and squaring each side.

Example 11 ▶ **Solving Equations Involving Radicals**

a.
$$\sqrt{2x + 7} - x = 2 \qquad \text{Write original equation.}$$
$$\sqrt{2x + 7} = x + 2 \qquad \text{Add } x \text{ to each side.}$$
$$2x + 7 = x^2 + 4x + 4 \qquad \text{Square each side.}$$
$$0 = x^2 + 2x - 3 \qquad \text{Write in general form.}$$
$$0 = (x + 3)(x - 1) \qquad \text{Factor.}$$
$$x + 3 = 0 \implies x = -3 \qquad \text{Set 1st factor equal to 0.}$$
$$x - 1 = 0 \implies x = 1 \qquad \text{Set 2nd factor equal to 0.}$$

By checking these values, you can determine that the only solution is 1.

b.
$$\sqrt{2x - 5} - \sqrt{x - 3} = 1 \qquad \text{Write original equation.}$$
$$\sqrt{2x - 5} = \sqrt{x - 3} + 1 \qquad \text{Add } \sqrt{x - 3} \text{ to each side.}$$
$$2x - 5 = x - 3 + 2\sqrt{x - 3} + 1 \qquad \text{Square each side.}$$
$$2x - 5 = x - 2 + 2\sqrt{x - 3} \qquad \text{Combine like terms.}$$
$$x - 3 = 2\sqrt{x - 3} \qquad \text{Add } -x + 2 \text{ to each side.}$$
$$x^2 - 6x + 9 = 4(x - 3) \qquad \text{Square each side.}$$
$$x^2 - 10x + 21 = 0 \qquad \text{Write in general form.}$$
$$(x - 3)(x - 7) = 0 \qquad \text{Factor.}$$
$$x - 3 = 0 \implies x = 3 \qquad \text{Set 1st factor equal to 0.}$$
$$x - 7 = 0 \implies x = 7 \qquad \text{Set 2nd factor equal to 0.}$$

The solutions are 3 and 7. Check these in the original equation.

Absolute Value Equations

To solve an equation involving an absolute value, remember that the expression inside the absolute value signs can be positive or negative. This results in *two* separate equations, each of which must be solved. For instance, the equation

$$|x - 2| = 3$$

results in the two equations $x - 2 = 3$ and $-(x - 2) = 3$ which implies that the equation has two solutions: 5 and -1.

Example 12 ▶ Solving an Equation Involving Absolute Value

Solve $|x^2 - 3x| = -4x + 6$.

Solution

Because the variable expression inside the absolute value signs can be positive or negative, you must solve the following two equations.

First Equation

$x^2 - 3x = -4x + 6$	Use positive expression.
$x^2 + x - 6 = 0$	Write in general form.
$(x + 3)(x - 2) = 0$	Factor.
$x + 3 = 0 \implies x = -3$	Set 1st factor equal to 0.
$x - 2 = 0 \implies x = 2$	Set 2nd factor equal to 0.

Second Equation

$-(x^2 - 3x) = -4x + 6$	Use negative expression.
$x^2 - 7x + 6 = 0$	Write in general form.
$(x - 1)(x - 6) = 0$	Factor.
$x - 1 = 0 \implies x = 1$	Set 1st factor equal to 0.
$x - 6 = 0 \implies x = 6$	Set 2nd factor equal to 0.

Check

$\|(-3)^2 - 3(-3)\| \stackrel{?}{=} -4(-3) + 6$	Substitute -3 for x.
$18 = 18$	-3 checks. ✓
$\|(2)^2 - 3(2)\| \stackrel{?}{=} -4(2) + 6$	Substitute 2 for x.
$2 \neq -2$	2 does not check.
$\|(1)^2 - 3(1)\| \stackrel{?}{=} -4(1) + 6$	Substitute 1 for x.
$2 = 2$	1 checks. ✓
$\|(6)^2 - 3(6)\| \stackrel{?}{=} -4(6) + 6$	Substitute 6 for x.
$18 \neq -18$	6 does not check.

The solutions are -3 and 1.

P.2 Exercises

In Exercises 1–10, determine whether the equation is an identity or a conditional equation.

1. $2(x - 1) = 2x - 2$
2. $3(x + 2) = 5x + 4$
3. $-6(x - 3) + 5 = -2x + 10$
4. $3(x + 2) - 5 = 3x + 1$
5. $4(x + 1) - 2x = 2(x + 2)$
6. $-7(x - 3) + 4x = 3(7 - x)$
7. $x^2 - 8x - 5 = (x - 4)^2 - 11$
8. $x^2 + 2(3x - 2) = x^2 + 6x - 4$
9. $3 + \dfrac{1}{x + 1} = \dfrac{4x}{x + 1}$
10. $\dfrac{5}{x} + \dfrac{3}{x} = 24$

In Exercises 11–26, solve the equation and check your solution.

11. $x + 11 = 15$
12. $7 - x = 19$
13. $7 - 2x = 25$
14. $7x + 2 = 23$
15. $8x - 5 = 3x + 20$
16. $7x + 3 = 3x - 17$
17. $2(x + 5) - 7 = 3(x - 2)$
18. $3(x + 3) = 5(1 - x) - 1$
19. $x - 3(2x + 3) = 8 - 5x$
20. $9x - 10 = 5x + 2(2x - 5)$
21. $\dfrac{5x}{4} + \dfrac{1}{2} = x - \dfrac{1}{2}$
22. $\dfrac{x}{5} - \dfrac{x}{2} = 3 + \dfrac{3x}{10}$
23. $\tfrac{3}{2}(z + 5) - \tfrac{1}{4}(z + 24) = 0$
24. $\dfrac{3x}{2} + \tfrac{1}{4}(x - 2) = 10$
25. $0.25x + 0.75(10 - x) = 3$
26. $0.60x + 0.40(100 - x) = 50$

In Exercises 27–48, solve the equation and check your solution. (If not possible, explain why.)

27. $x + 8 = 2(x - 2) - x$
28. $8(x + 2) - 3(2x + 1) = 2(x + 5)$
29. $\dfrac{100 - 4x}{3} = \dfrac{5x + 6}{4} + 6$
30. $\dfrac{17 + y}{y} + \dfrac{32 + y}{y} = 100$
31. $\dfrac{5x - 4}{5x + 4} = \dfrac{2}{3}$
32. $\dfrac{10x + 3}{5x + 6} = \dfrac{1}{2}$
33. $10 - \dfrac{13}{x} = 4 + \dfrac{5}{x}$
34. $\dfrac{15}{x} - 4 = \dfrac{6}{x} + 3$
35. $\dfrac{x}{x + 4} + \dfrac{4}{x + 4} + 2 = 0$
36. $3 = 2 + \dfrac{2}{z + 2}$
37. $\dfrac{1}{x} + \dfrac{2}{x - 5} = 0$
38. $\dfrac{7}{2x + 1} - \dfrac{8x}{2x - 1} = -4$
39. $\dfrac{2}{(x - 4)(x - 2)} = \dfrac{1}{x - 4} + \dfrac{2}{x - 2}$
40. $\dfrac{4}{x - 1} + \dfrac{6}{3x + 1} = \dfrac{15}{3x + 1}$
41. $\dfrac{1}{x - 3} + \dfrac{1}{x + 3} = \dfrac{10}{x^2 - 9}$
42. $\dfrac{1}{x - 2} + \dfrac{3}{x + 3} = \dfrac{4}{x^2 + x - 6}$
43. $\dfrac{3}{x^2 - 3x} + \dfrac{4}{x} = \dfrac{1}{x - 3}$
44. $\dfrac{6}{x} - \dfrac{2}{x + 3} = \dfrac{3(x + 5)}{x^2 + 3x}$
45. $(x + 2)^2 + 5 = (x + 3)^2$
46. $(x + 1)^2 + 2(x - 2) = (x + 1)(x - 2)$
47. $(x + 2)^2 - x^2 = 4(x + 1)$
48. $(2x + 1)^2 = 4(x^2 + x + 1)$

In Exercises 49–54, write the quadratic equation in general form.

49. $2x^2 = 3 - 8x$
50. $x^2 = 16x$
51. $(x - 3)^2 = 3$
52. $13 - 3(x + 7)^2 = 0$
53. $\tfrac{1}{5}(3x^2 - 10) = 18x$
54. $x(x + 2) = 5x^2 + 1$

In Exercises 55–68, solve the quadratic equation for x by factoring.

55. $6x^2 + 3x = 0$
56. $9x^2 - 1 = 0$
57. $x^2 - 2x - 8 = 0$
58. $x^2 - 10x + 9 = 0$
59. $x^2 + 10x + 25 = 0$
60. $4x^2 + 12x + 9 = 0$
61. $3 + 5x - 2x^2 = 0$
62. $2x^2 = 19x + 33$
63. $x^2 + 4x = 12$
64. $-x^2 + 8x = 12$

65. $\frac{3}{4}x^2 + 8x + 20 = 0$ **66.** $\frac{1}{8}x^2 - x - 16 = 0$
67. $x^2 + 2ax + a^2 = 0$ **68.** $(x + a)^2 - b^2 = 0$

In Exercises 69–82, solve the equation by extracting square roots. List both the exact solution and the decimal solution rounded to two decimal places.

69. $x^2 = 49$ **70.** $x^2 = 169$
71. $x^2 = 11$ **72.** $x^2 = 32$
73. $3x^2 = 81$ **74.** $9x^2 = 36$
75. $(x - 12)^2 = 16$ **76.** $(x + 13)^2 = 25$
77. $(x + 2)^2 = 14$ **78.** $(x - 5)^2 = 30$
79. $(2x - 1)^2 = 18$ **80.** $(4x + 7)^2 = 44$
81. $(x - 7)^2 = (x + 3)^2$ **82.** $(x + 5)^2 = (x + 4)^2$

In Exercises 83–92, solve the quadratic equation by completing the square.

83. $x^2 - 2x = 0$ **84.** $x^2 + 4x = 0$
85. $x^2 + 4x - 32 = 0$ **86.** $x^2 - 2x - 3 = 0$
87. $x^2 + 6x + 2 = 0$ **88.** $x^2 + 8x + 14 = 0$
89. $9x^2 - 18x = -3$ **90.** $9x^2 - 12x = 14$
91. $8 + 4x - x^2 = 0$ **92.** $4x^2 - 4x - 99 = 0$

In Exercises 93–116, use the Quadratic Formula to solve the equation.

93. $2x^2 + x - 1 = 0$ **94.** $2x^2 - x - 1 = 0$
95. $16x^2 + 8x - 3 = 0$ **96.** $25x^2 - 20x + 3 = 0$
97. $2 + 2x - x^2 = 0$ **98.** $x^2 - 10x + 22 = 0$
99. $x^2 + 14x + 44 = 0$ **100.** $6x = 4 - x^2$
101. $x^2 + 8x - 4 = 0$ **102.** $4x^2 - 4x - 4 = 0$
103. $12x - 9x^2 = -3$ **104.** $16x^2 + 22 = 40x$
105. $9x^2 + 24x + 16 = 0$
106. $36x^2 + 24x - 7 = 0$
107. $4x^2 + 4x = 7$ **108.** $16x^2 - 40x + 5 = 0$
109. $28x - 49x^2 = 4$ **110.** $3x + x^2 - 1 = 0$
111. $8t = 5 + 2t^2$ **112.** $25h^2 + 80h + 61 = 0$
113. $(y - 5)^2 = 2y$ **114.** $(z + 6)^2 = -2z$
115. $\frac{1}{2}x^2 + \frac{3}{8}x = 2$ **116.** $\left(\frac{5}{7}x - 14\right)^2 = 8x$

In Exercises 117–124, use the Quadratic Formula to solve the equation. (Round your answers to three decimal places.)

117. $5.1x^2 - 1.7x - 3.2 = 0$
118. $2x^2 - 2.50x - 0.42 = 0$
119. $-0.067x^2 - 0.852x + 1.277 = 0$
120. $-0.005x^2 + 0.101x - 0.193 = 0$
121. $422x^2 - 506x - 347 = 0$
122. $1100x^2 + 326x - 715 = 0$
123. $12.67x^2 + 31.55x + 8.09 = 0$
124. $-3.22x^2 - 0.08x + 28.651 = 0$

In Exercises 125–134, solve the equation for x by any convenient method.

125. $x^2 - 2x - 1 = 0$ **126.** $11x^2 + 33x = 0$
127. $(x + 3)^2 = 81$ **128.** $x^2 - 14x + 49 = 0$
129. $x^2 - x - \frac{11}{4} = 0$ **130.** $x^2 + 3x - \frac{3}{4} = 0$
131. $(x + 1)^2 = x^2$ **132.** $a^2x^2 - b^2 = 0$
133. $3x + 4 = 2x^2 - 7$
134. $4x^2 + 2x + 4 = 2x + 8$

In Exercises 135–152, find all real solutions of the equation. Check your solutions in the original equation.

135. $4x^4 - 18x^2 = 0$ **136.** $20x^3 - 125x = 0$
137. $x^4 - 81 = 0$ **138.** $x^6 - 64 = 0$
139. $x^3 + 216 = 0$ **140.** $27x^3 - 512 = 0$
141. $5x^3 + 30x^2 + 45x = 0$
142. $9x^4 - 24x^3 + 16x^2 = 0$
143. $x^3 - 3x^2 - x + 3 = 0$
144. $x^3 + 2x^2 + 3x + 6 = 0$
145. $x^4 - x^3 + x - 1 = 0$
146. $x^4 + 2x^3 - 8x - 16 = 0$
147. $x^4 - 4x^2 + 3 = 0$
148. $x^4 + 5x^2 - 36 = 0$
149. $4x^4 - 65x^2 + 16 = 0$
150. $36t^4 + 29t^2 - 7 = 0$
151. $x^6 + 7x^3 - 8 = 0$
152. $x^6 + 3x^3 + 2 = 0$

In Exercises 153–170, find all solutions of the equation. Check your solutions in the original equation.

153. $\sqrt{2x} - 10 = 0$ **154.** $4\sqrt{x} - 3 = 0$
155. $\sqrt{x - 10} - 4 = 0$ **156.** $\sqrt{5 - x} - 3 = 0$
157. $\sqrt[3]{2x + 5} + 3 = 0$ **158.** $\sqrt[3]{3x + 1} - 5 = 0$
159. $-\sqrt{26 - 11x} + 4 = x$
160. $x + \sqrt{31 - 9x} = 5$
161. $\sqrt{x + 1} = \sqrt{3x + 1}$
162. $\sqrt{x + 5} = \sqrt{x - 5}$

163. $(x - 5)^{3/2} = 8$ 164. $(x + 3)^{3/2} = 8$
165. $(x + 3)^{2/3} = 8$ 166. $(x + 2)^{2/3} = 9$
167. $(x^2 - 5)^{3/2} = 27$
168. $(x^2 - x - 22)^{3/2} = 27$
169. $3x(x - 1)^{1/2} + 2(x - 1)^{3/2} = 0$
170. $4x^2(x - 1)^{1/3} + 6x(x - 1)^{4/3} = 0$

In Exercises 171–184, find all solutions of the equation. Check your solutions in the original equation.

171. $\dfrac{20 - x}{x} = x$ 172. $\dfrac{4}{x} - \dfrac{5}{3} = \dfrac{x}{6}$

173. $\dfrac{1}{x} - \dfrac{1}{x + 1} = 3$ 174. $\dfrac{x}{x^2 - 4} + \dfrac{1}{x + 2} = 3$

175. $x = \dfrac{3}{x} + \dfrac{1}{2}$ 176. $4x + 1 = \dfrac{3}{x}$

177. $\dfrac{4}{x + 1} - \dfrac{3}{x + 2} = 1$ 178. $\dfrac{x + 1}{3} - \dfrac{x + 1}{x + 2} = 0$

179. $|2x - 1| = 5$ 180. $|3x + 2| = 7$
181. $|x| = x^2 + x - 3$ 182. $|x^2 + 6x| = 3x + 18$
183. $|x + 1| = x^2 - 5$ 184. $|x - 10| = x^2 - 10x$

Anthropology In Exercises 185 and 186, use the following information. The relationship between the length of an adult's thigh bone and the height of the adult can be approximated by the linear equations

$y = 0.432x - 10.44$ Female

$y = 0.449x - 12.15$ Male

where y is the length of the femur (thigh bone) in inches and x is the height in inches (see figure).

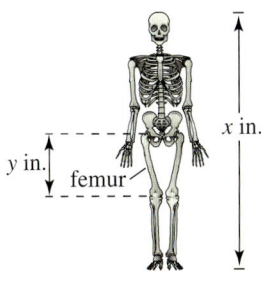

185. An anthropologist discovers a thigh bone belonging to an adult human female. The bone is 16 inches long. Estimate the height of the female.

186. From the foot bones of an adult human male, an anthropologist estimates that the person's height was 69 inches. A few feet away from the site where the foot bones were discovered, the anthropologist discovered a male adult thigh bone that was 19 inches long. Is it likely that both the foot bones and the thigh bone came from the same person?

187. *Operating Cost* A delivery company has a fleet of vans. The annual operating cost per van is

$C = 0.32m + 2500$

where m is the number of miles traveled by a van in a year. What number of miles will yield an annual operating cost that is equal to $10,000?

188. *Flood Control* Suppose a river has risen 8 feet above its flood stage. The water begins to recede at a rate of 3 inches per hour. Write a mathematical model that shows the number of feet above flood stage after t hours. If the water continually recedes at this rate, when will the river be 1 foot above its flood stage?

189. *Floor Space* The floor of a one-story building is 14 feet longer than it is wide. The building has 1632 square feet of floor space.

 (a) Draw a diagram that gives a visual representation of the floor space. Represent the width as w and show the length in terms of w.
 (b) Write a quadratic equation in terms of w.
 (c) Find the length and width of the building floor.

190. *Packaging* An open box with a square base (see figure) is to be constructed from 84 square inches of material. What should be the dimensions of the base if the height of the box is to be 2 inches? (*Hint:* The surface area is $S = x^2 + 4xh$.)

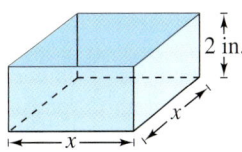

191. *Geometry* The hypotenuse of an isosceles right triangle is 5 centimeters long. How long are its sides?

192. *Geometry* An equilateral triangle has a height of 10 inches. How long are its sides? (*Hint:* Use the height of the triangle to partition the triangle into two congruent right triangles.)

193. *Flying Speed* Two planes leave simultaneously from the same airport, one flying due north and the other due east (see figure). The northbound plane is flying 50 miles per hour faster than the eastbound plane. After 3 hours the planes are 2440 miles apart. Find the speed of each plane.

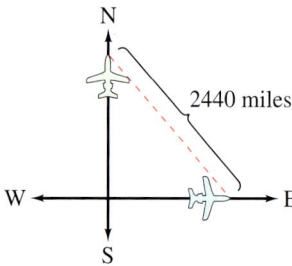

194. *Airline Passengers* An airline offers daily flights between Chicago and Denver. The total monthly cost of these flights is

$$C = \sqrt{0.2x + 1}$$

where C is the cost (in millions of dollars) and x is the number of passengers (in thousands). The total cost of the flights for a certain month is 2.5 million dollars. How many passengers flew that month?

195. *Economics* The demand equation for a certain product is $p = 20 - 0.0002x$, where p is the price per unit and x is the number of units sold. The total revenue for selling x units is

$$\text{Revenue} = xp = x(20 - 0.0002x).$$

How many units must be sold to produce a revenue of $500,000?

196. *Economics* The demand equation for a certain product is modeled by

$$p = 40 - \sqrt{0.01x + 1}$$

where x is the number of units demanded per day and p is the price per unit. Approximate the demand if the price is $37.55.

197. *Saturated Steam* The temperature T (in degrees Fahrenheit) of saturated steam increases as pressure increases. This relationship is approximated by

$$T = 75.82 - 2.11x + 43.51\sqrt{x}, \quad 5 \le x \le 40$$

where x is the absolute pressure (in pounds per square inch). Approximate the pressure if the temperature of the steam is 240°F.

Synthesis

True or False? In Exercises 198–200, determine whether the statement is true or false. Justify your answer.

198. The equation $x(3 - x) = 10$ is a linear equation.

199. If $(2x - 3)(x + 5) = 8$, then $2x - 3 = 8$ or $x + 5 = 8$.

200. When solving an absolute value equation, you will always have to check more than one solution.

201. To solve the equation

$$2x^2 + 3x = 15x$$

a student divides both sides by x and solves the equation $2x + 3 = 15$. The resulting solution ($x = 6$) satisfies the given equation. Is there an error? Explain.

202. *Think About It* What is meant by "equivalent equations"? Give an example of two equivalent equations.

203. *Writing* In your own words, describe the steps used to transform an equation into an equivalent equation.

204. *Exploration* Solve $3(x + 4)^2 + (x + 4) - 2 = 0$ in two ways.

(a) Let $u = x + 4$, and solve the resulting equation for u. Then solve the u-solution for x.

(b) Expand and collect like terms in the equation, and solve the resulting equation for x.

(c) Which method is easier? Explain.

205. *Exploration* Solve the equations, given that a and b are not zero.

(a) $ax^2 + bx = 0$

(b) $ax^2 - ax = 0$

In Exercises 206 and 207, consider an equation of the form $x + |x - a| = b$, where a and b are constants.

206. *Exploration* Find a and b if the solution to the equation is $x = 9$. (There are many correct answers.)

207. *Writing* Write a short paragraph listing the steps required to solve this equation involving absolute value.

P.3 The Cartesian Plane and Graphs of Equations

▶ **What you should learn**

- How to plot points in the Cartesian plane
- How to use the Distance Formula to find the distance between two points
- How to use the Midpoint Formula to find the midpoint of a line segment
- How to sketch graphs of equations
- How to find *x*- and *y*-intercepts of graphs
- How to use symmetry to sketch graphs of equations
- How to find equations and sketch graphs of circles

▶ **Why you should learn it**

The graph of an equation can help you see relationships between real-life quantities. For example, Exercise 106 on page 37 shows how a graph can be used to estimate the life expectancies of children who are born in the years 2002 and 2004.

The Cartesian Plane

Just as you can represent real numbers by points on a real number line, you can represent ordered pairs of real numbers by points in a plane called the **rectangular coordinate system,** or the **Cartesian plane,** named after the French mathematician René Descartes (1596–1650).

The Cartesian plane is formed by using two real number lines intersecting at right angles, as shown in Figure P.6. The horizontal real number line is usually called the **x-axis,** and the vertical real number line is usually called the **y-axis.** The point of intersection of these two axes is the **origin,** and the two axes divide the plane into four parts called **quadrants.**

FIGURE P.6　　　FIGURE P.7

Each point in the plane corresponds to an **ordered pair** (x, y) of real numbers x and y, called **coordinates** of the point. The **x-coordinate** represents the directed distance from the y-axis to the point, and the **y-coordinate** represents the directed distance from the x-axis to the point, as shown in Figure P.7.

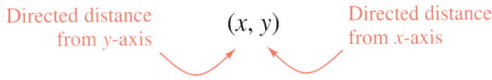

The notation (x, y) denotes both a point in the plane and an open interval on the real number line. The context will tell you which meaning is intended.

Example 1 ▶ Plotting Points in the Cartesian Plane

Plot the points $(-1, 2)$, $(3, 4)$, $(0, 0)$, $(3, 0)$, and $(-2, -3)$.

Solution

To plot the point $(-1, 2)$, imagine a vertical line through -1 on the x-axis and a horizontal line through 2 on the y-axis. The intersection of these two lines is the point $(-1, 2)$. The other four points can be plotted in a similar way, as shown in Figure P.8.

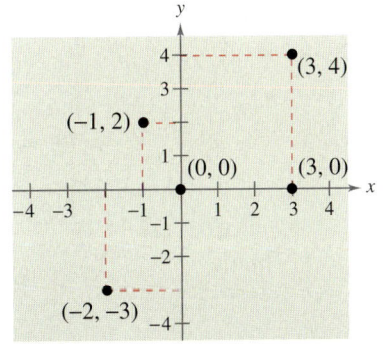

FIGURE P.8

The beauty of a rectangular coordinate system is that it allows you to see relationships between two variables. It would be difficult to overestimate the importance of Descartes's introduction of coordinates to the plane. Today, his ideas are in common use in virtually every scientific and business-related field.

Example 2 ▶ Sketching a Scatter Plot

A computer animation of this example appears in the Interactive CD-ROM and Internet versions of this text.

The amount A (in millions of dollars) spent on archery equipment in the United States from 1988 through 1997 is given in the table, where t represents the year. Sketch a scatter plot of the data. *(Source: National Sporting Goods Association)*

t	1988	1989	1990	1991	1992	1993	1994	1995	1996	1997
A	235	261	265	270	334	285	306	287	272	273

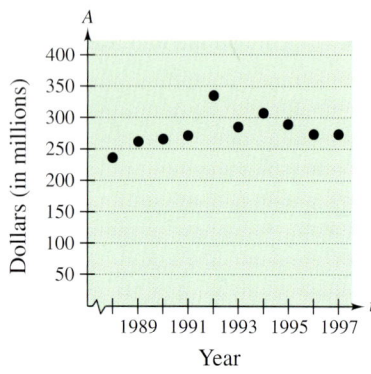

FIGURE P.9

Solution

To sketch a *scatter plot* of the data given in the table, you simply represent each pair of values by an ordered pair (t, A) and plot the resulting points, as shown in Figure P.9. For instance, the first pair of values is represented by the ordered pair (1988, 235). Note that the break in the t-axis indicates that the numbers between 0 and 1988 have been omitted.

In Example 2, you could have let $t = 1$ represent the year 1988. In that case, the horizontal axis would not have been broken, and the tick marks would have been labeled 1 through 10 (instead of 1988 through 1997).

Technology

The scatter plot in Example 2 is only one way to represent the data graphically. Two other techniques are shown at the right. The first is a bar graph and the second is a line graph. All three graphical representations were created with a computer. If you have access to a graphing utility, try using it to represent graphically the data given in Example 2.

Bar graph

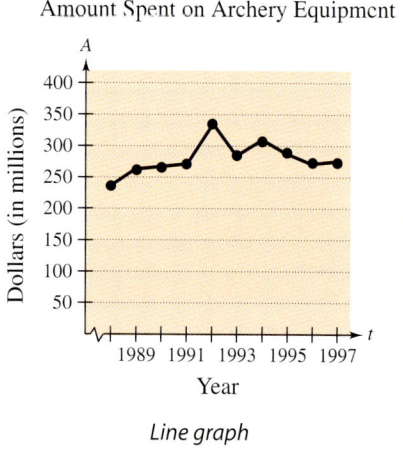

Line graph

Section P.3 ▶ The Cartesian Plane and Graphs of Equations   27

The Distance Formula

Recall from the Pythagorean Theorem that, for a right triangle with hypotenuse of length c and sides of lengths a and b, you have

$$a^2 + b^2 = c^2 \qquad \text{Pythagorean Theorem}$$

as shown in Figure P.10. (The converse is also true. That is, if $a^2 + b^2 = c^2$, then the triangle is a right triangle.)

Suppose you want to determine the distance d between two points (x_1, y_1) and (x_2, y_2) in the plane. With these two points, a right triangle can be formed, as shown in Figure P.11. The length of the vertical side of the triangle is $|y_2 - y_1|$, and the length of the horizontal side is $|x_2 - x_1|$. By the Pythagorean Theorem, you can write

$$d^2 = |x_2 - x_1|^2 + |y_2 - y_1|^2$$
$$d = \sqrt{|x_2 - x_1|^2 + |y_2 - y_1|^2}$$
$$d = \sqrt{(x_2 - x_1)^2 + (y_2 - y_1)^2}.$$

This result is the **Distance Formula.**

FIGURE P.10

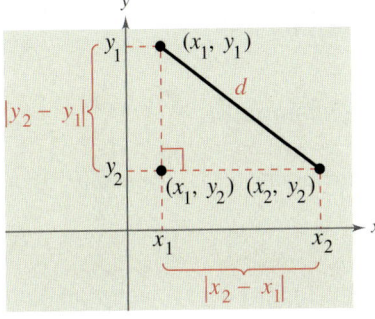

FIGURE P.11

The Distance Formula

The distance d between the points (x_1, y_1) and (x_2, y_2) in the plane is

$$d = \sqrt{(x_2 - x_1)^2 + (y_2 - y_1)^2}.$$

Example 3 ▶ Finding a Distance

Find the distance between the points $(-2, 1)$ and $(3, 4)$.

Solution

Let $(x_1, y_1) = (-2, 1)$ and $(x_2, y_2) = (3, 4)$. Then apply the Distance Formula.

$$d = \sqrt{(x_2 - x_1)^2 + (y_2 - y_1)^2} \qquad \text{Distance Formula}$$
$$= \sqrt{[3 - (-2)]^2 + (4 - 1)^2} \qquad \text{Substitute for } x_1, y_1, x_2, \text{ and } y_2.$$
$$= \sqrt{(5)^2 + (3)^2} \qquad \text{Simplify.}$$
$$= \sqrt{34}$$
$$\approx 5.83 \qquad \text{Use a calculator.}$$

Note in Figure P.12 that a distance of 5.83 looks about right.
You can use the Pythagorean Theorem to check that the distance is correct.

$$d^2 \stackrel{?}{=} 3^2 + 5^2 \qquad \text{Pythagorean Theorem}$$
$$\left(\sqrt{34}\right)^2 \stackrel{?}{=} 3^2 + 5^2 \qquad \text{Substitute for } d.$$
$$34 = 34 \qquad \text{Distance checks.} \;\checkmark$$

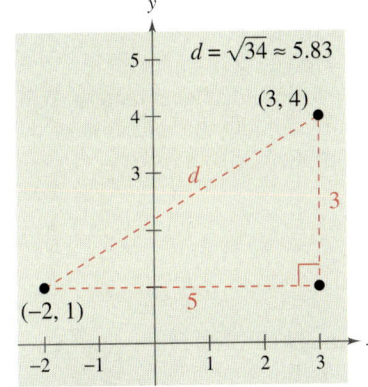

FIGURE P.12

The Midpoint Formula

To find the **midpoint** of the line segment that joins two points in a coordinate plane, you can simply find the average values of the respective coordinates of the two endpoints using the **Midpoint Formula**. (See Appendix A for a proof of the Midpoint Formula.)

The Midpoint Formula

The midpoint of the segment joining the points (x_1, y_1) and (x_2, y_2) is given by the Midpoint Formula

$$\text{Midpoint} = \left(\frac{x_1 + x_2}{2}, \frac{y_1 + y_2}{2}\right).$$

Example 4 ▶ Finding a Line Segment's Midpoint

Find the midpoint of the line segment joining the points $(-5, -3)$ and $(9, 3)$, as shown in Figure P.13.

Solution

Let $(x_1, y_1) = (-5, -3)$ and $(x_2, y_2) = (9, 3)$.

$$\begin{aligned}\text{Midpoint} &= \left(\frac{x_1 + x_2}{2}, \frac{y_1 + y_2}{2}\right) && \text{Midpoint Formula} \\ &= \left(\frac{-5 + 9}{2}, \frac{-3 + 3}{2}\right) && \text{Substitute for } x_1, y_1, x_2, \text{ and } y_2. \\ &= (2, 0) && \text{Simplify.}\end{aligned}$$

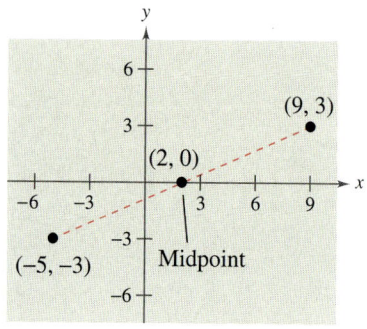

FIGURE P.13

Example 5 ▶ Estimating Annual Sales

Winn-Dixie Stores had annual sales of $1.30 billion in 1996 and $1.36 billion in 1998. Without knowing any additional information, what would you estimate the 1997 sales to have been? (Source: Winn-Dixie Stores, Inc.)

Solution

One solution to the problem is to assume that sales followed a linear pattern. With this assumption, you can estimate the 1997 sales by finding the midpoint of the segment connecting the points (1996, 1.30) and (1998, 1.36).

$$\begin{aligned}\text{Midpoint} &= \left(\frac{1996 + 1998}{2}, \frac{1.30 + 1.36}{2}\right) \\ &= (1997, 1.33)\end{aligned}$$

So, you would estimate the 1997 sales to have been about $1.33 billion, as shown in Figure P.14. (The actual 1997 sales were $1.32 billion.)

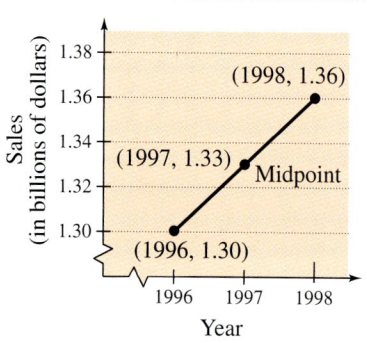

FIGURE P.14

The Graph of an Equation

Earlier in this section, you used a coordinate system to represent graphically the relationship between two quantities. There, the graphical picture consisted of a collection of points in a coordinate plane (see Example 2).

Frequently, a relationship between two quantities is expressed as an **equation in two variables**. For instance, $y = 7 - 3x$ is an equation in x and y. An ordered pair (a, b) is a **solution** or **solution point** of an equation in x and y if the equation is true when a is substituted for x and b is substituted for y. For instance, $(1, 4)$ is a solution of $y = 7 - 3x$ because $4 = 7 - 3(1)$ is a true statement.

In the remainder of this section, you will review some basic procedures for sketching the graph of an equation in two variables. The **graph of an equation** is the set of all points that are solutions of the equation.

> The *Interactive* CD-ROM and *Internet* versions of this text offer a built-in graphing calculator, which can be used in the Examples, Explorations, Technology notes, and Exercises.

Example 6 ▶ Sketching the Graph of an Equation

Sketch the graph of

$$y = x^2 - 2.$$

Solution

Begin by constructing a table of values.

x	-2	-1	0	1	2	3
$y = x^2 - 2$	2	-1	-2	-1	2	7

Next, plot the points given in the table, as shown in Figure P.15(a). Finally, connect the points with a smooth curve, as shown in Figure P.15(b).

> **STUDY TIP**
>
> One of your goals in this course is to learn to classify the basic shape of a graph from its equation. For instance, you will learn that a *linear equation* has the form
>
> $$y = mx + b$$
>
> and its graph is a straight line. Similarly, the *quadratic equation* in Example 6 has the form
>
> $$y = ax^2 + bx + c$$
>
> and its graph is a parabola.

(a)

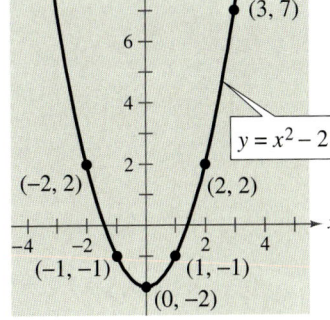
(b)

FIGURE P.15

> A computer animation of this concept appears in the *Interactive* CD-ROM and *Internet* versions of this text.

Intercepts of a Graph

It is often easy to determine the solution points that have zero as either the x-coordinate or the y-coordinate. These points are called **intercepts** because they are the points at which the graph intersects the x- or y-axis. It is possible for a graph to have no intercepts or several intercepts, as shown in Figure P.16.

No x-intercept
One y-intercept

Three x-intercepts
One y-intercept

One x-intercept
Two y-intercepts

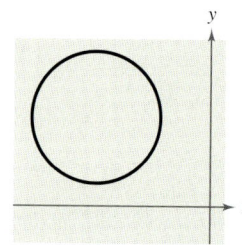
No intercepts

FIGURE P.16

Note that an x-intercept is written as the ordered pair $(x, 0)$ and a y-intercept is written as the ordered pair $(0, y)$.

Finding Intercepts

1. To find x-intercepts, let y be zero and solve the equation for x.
2. To find y-intercepts, let x be zero and solve the equation for y.

Example 7 ▶ Finding x- and y-Intercepts

Find the x- and y-intercepts of the graph of each equation.

a. $y = x^3 - 4x$ **b.** $y^2 = x + 4$

Solution

a. Let $y = 0$. Then

$$0 = x^3 - 4x = x(x^2 - 4)$$

has solutions $x = 0$ and $x = \pm 2$.

  x-intercepts: $(0, 0), (2, 0), (-2, 0)$

Let $x = 0$. Then $y = (0)^3 - 4(0) = 0$.

  y-intercept: $(0, 0)$ (See Figure P.17.)

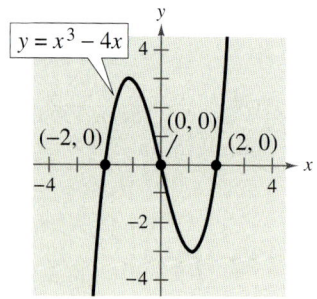
FIGURE P.17

b. Let $y = 0$. Then $(0)^2 = x + 4$, and $-4 = x$.

  x-intercept: $(-4, 0)$

Let $x = 0$. Then $y^2 = 0 + 4 = 4$ has solutions $y = \pm 2$.

  y-intercepts: $(0, 2), (0, -2)$ (See Figure P.18.)

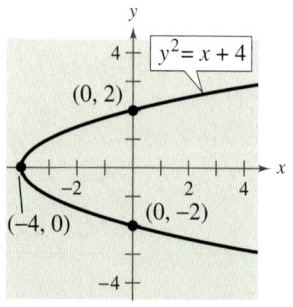
FIGURE P.18

Symmetry

The graphs shown in Figures P.15(b), P.17, and P.18 each have **symmetry** with respect to one of the coordinate axes or with respect to the origin.

Figure P.15(b)	$y = x^2 - 2$	y-Axis symmetry
Figure P.17	$y = x^3 - 4x$	Origin symmetry
Figure P.18	$y^2 = x + 4$	x-Axis symmetry

Symmetry with respect to the x-axis means that if the Cartesian plane were folded along the x-axis, the portion of the graph above the x-axis would coincide with the portion below the x-axis. Symmetry with respect to the y-axis or the origin can be described in a similar manner, as shown in Figure P.19.

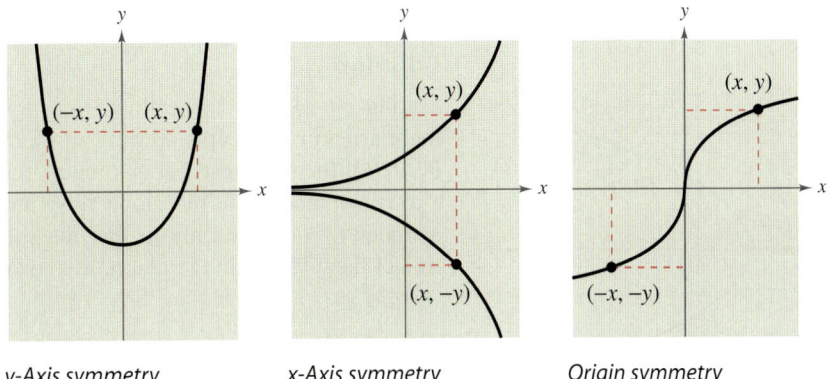

y-Axis symmetry   x-Axis symmetry   Origin symmetry

FIGURE P.19

A computer animation of this concept appears in the *Interactive* CD-ROM and *Internet* versions of this text.

Knowing the symmetry of a graph *before* attempting to sketch it is helpful, because then you need only half as many solution points to sketch the graph. There are three basic types of symmetry. A graph is *symmetric with respect to the y-axis* if, whenever (x, y) is on the graph, $(-x, y)$ is also on the graph. A graph is *symmetric with respect to the x-axis* if, whenever (x, y) is on the graph, $(x, -y)$ is also on the graph. A graph is *symmetric with respect to the origin* if, whenever (x, y) is on the graph, $(-x, -y)$ is also on the graph.

Example 8 ▶ Testing for Symmetry

The graph of $y = x^2 - 2$ is symmetric with respect to the y-axis because the point $(-x, y)$ satisfies the equation.

$y = x^2 - 2$	Write original equation.
$y = (-x)^2 - 2$	Substitute $(-x, y)$ for (x, y).
$y = x^2 - 2$	Replacement yields equivalent equation.

See Figure P.20.

y-Axis symmetry

FIGURE P.20

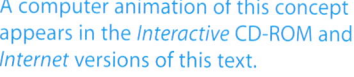

A computer animation of this concept appears in the *Interactive* CD-ROM and *Internet* versions of this text.

Tests for Symmetry

1. The graph of an equation is symmetric with respect to the *y-axis* if replacing x with $-x$ yields an equivalent equation.
2. The graph of an equation is symmetric with respect to the *x-axis* if replacing y with $-y$ yields an equivalent equation.
3. The graph of an equation is symmetric with respect to the *origin* if replacing x with $-x$ *and* y with $-y$ yields an equivalent equation.

Example 9 ▶ Using Intercepts and Symmetry as Sketching Aids

Use intercepts and symmetry to sketch the graph of

$$x - y^2 = 1.$$

Solution

Letting $x = 0$, you can see that $-y^2 = 1$ or $y^2 = -1$ has no real solutions. So, there are no y-intercepts. Letting $y = 0$, you obtain $x = 1$. So, the x-intercept is $(1, 0)$. Of the three tests for symmetry, the only one that is satisfied is the test for x-axis symmetry. So, the graph is symmetric with respect to the x-axis. Using symmetry, you need only to find the solution points above the x-axis and then reflect them to obtain the graph, as shown in Figure P.21.

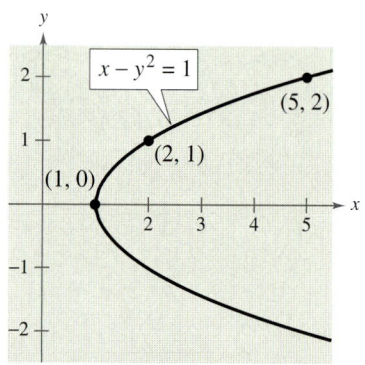

FIGURE P.21

y	0	1	2
$x = y^2 + 1$	1	2	5

Example 10 ▶ Sketching the Graph of an Equation

Sketch the graph of

$$y = |x - 1|.$$

Solution

Letting $x = 0$ yields $y = 1$, which means that $(0, 1)$ is the y-intercept. Letting $y = 0$ yields $x = 1$, which means that $(1, 0)$ is the x-intercept. This equation fails all three tests for symmetry and consequently its graph is not symmetric with respect to either axis or to the origin. The absolute value sign indicates that y is always nonnegative.

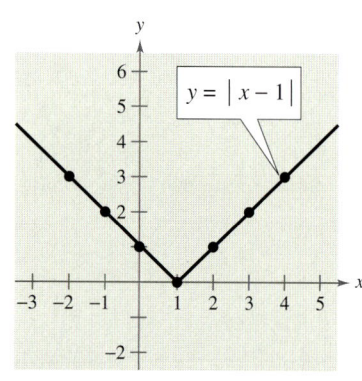

FIGURE P.22

x	-2	-1	0	1	2	3	4
$y = |x - 1|$	3	2	1	0	1	2	3

The graph is shown in Figure P.22.

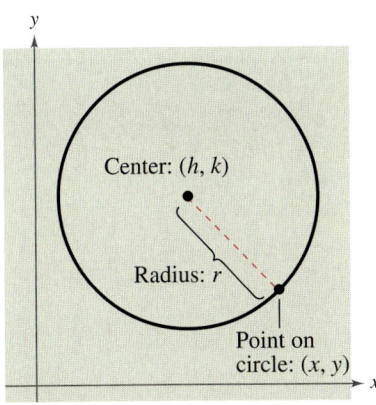

FIGURE P.23

Throughout this course, you will learn to recognize several types of graphs from their equations. For instance, you will learn to recognize that the graph of a second-degree equation of the form

$$y = ax^2 + bx + c$$

is a parabola (see Example 6). Another easily recognized graph is that of a **circle**.

Circles

Consider the circle shown in Figure P.23. A point (x, y) is on the circle if and only if its distance from the center (h, k) is r. By the Distance Formula,

$$\sqrt{(x - h)^2 + (y - k)^2} = r.$$

By squaring both sides of this equation, you obtain the **standard form of the equation of a circle**.

Standard Form of the Equation of a Circle

The point (x, y) lies on the circle of radius r and center (h, k) if and only if

$$(x - h)^2 + (y - k)^2 = r^2.$$

From this result, you can see that the standard form of the equation of a circle *with its center at the origin*, $(h, k) = (0, 0)$, is simply

$$x^2 + y^2 = r^2. \qquad \text{Circle with center at origin}$$

STUDY TIP

To find the correct values for h and k, it may be helpful to rewrite the quantities $(x + 1)^2$ and $(y - 2)^2$.

$(x + 1)^2 = [x - (-1)]^2$,
$h = -1$

$(y - 2)^2 = [y - (2)]^2$,
$k = 2$

Example 11 ▶ Finding the Equation of a Circle

The point $(3, 4)$ lies on a circle whose center is at $(-1, 2)$, as shown in Figure P.24. Find the standard form of the equation of this circle.

Solution

The radius of the circle is the distance between $(-1, 2)$ and $(3, 4)$.

$r = \sqrt{(x - h)^2 + (y - k)^2}$ Distance Formula

$r = \sqrt{[3 - (-1)]^2 + (4 - 2)^2}$ Substitute for x, y, h, and k.

$= \sqrt{(4)^2 + (2)^2}$ Simplify.

$= \sqrt{16 + 4}$ Simplify.

$= \sqrt{20}$ Radius

Using $(h, k) = (-1, 2)$ and $r = \sqrt{20}$, the equation of the circle is

$(x - h)^2 + (y - k)^2 = r^2$

$[x - (-1)]^2 + (y - 2)^2 = \left(\sqrt{20}\right)^2$ Substitute for h, k, and r.

$(x + 1)^2 + (y - 2)^2 = 20.$ Standard form

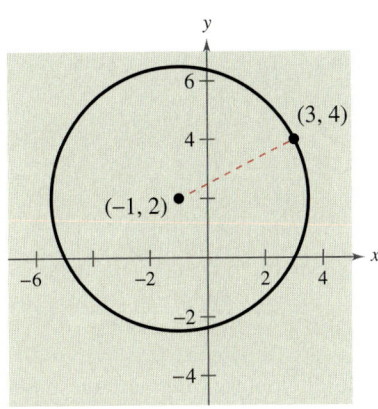

FIGURE P.24

P.3 Exercises

In Exercises 1–4, sketch the polygon with the indicated vertices.

1. Triangle: $(-1, 1), (2, -1), (3, 4)$
2. Triangle: $(0, 3), (-1, -2), (4, 8)$
3. Square: $(2, 4), (5, 1), (2, -2), (-1, 1)$
4. Parallelogram: $(5, 2), (7, 0), (1, -2), (-1, 0)$

In Exercises 5 and 6, approximate the coordinates of the points.

5.

6.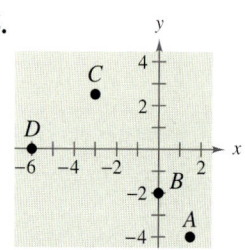

In Exercises 7–10, find the coordinates of the point.

7. The point is located 3 units to the left of the y-axis and 4 units above the x-axis.
8. The point is located 8 units below the x-axis and 4 units to the right of the y-axis.
9. The point is located 5 units below the x-axis and the coordinates of the point are equal.
10. The point is on the x-axis and 12 units to the left of the y-axis.

In Exercises 11–20, determine the quadrant(s) in which (x, y) is located so that the condition(s) is (are) satisfied.

11. $x > 0$ and $y < 0$
12. $x < 0$ and $y < 0$
13. $x = -4$ and $y > 0$
14. $x > 2$ and $y = 3$
15. $y < -5$
16. $x > 4$
17. $(x, -y)$ is in the second quadrant.
18. $(-x, y)$ is in the fourth quadrant.
19. $xy > 0$
20. $xy < 0$

In Exercises 21 and 22, sketch a scatter plot of the data given in the table.

21. *Meteorology* The table shows the lowest temperature of record y (in degrees Fahrenheit) in Duluth, Minnesota, for each month x, where $x = 1$ represents January. (Source: NOAA)

x	1	2	3	4	5	6
y	-39	-33	-29	-5	17	27

x	7	8	9	10	11	12
y	35	32	22	8	-23	-34

TABLE FOR 21

22. *Business* The table shows the number y of Wal-Mart stores for each year x from 1992 through 1999. (Source: Wal-Mart Stores, Inc.)

x	1992	1993	1994	1995
y	2136	2440	2759	2943

x	1996	1997	1998	1999
y	3054	3406	3630	3815

In Exercises 23–26, (a) find the length of each side of the right triangle, and (b) show that these lengths satisfy the Pythagorean Theorem.

23.

24.

25.

26.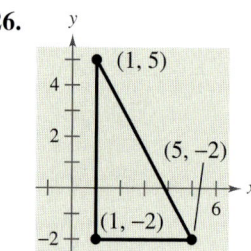

In Exercises 27–38, (a) plot the points, (b) find the distance between the points, and (c) find the midpoint of the line segment joining the points.

27. $(1, 1), (9, 7)$
28. $(1, 12), (6, 0)$
29. $(-4, 10), (4, -5)$
30. $(-7, -4), (2, 8)$

31. $(-1, 2), (5, 4)$
32. $(2, 10), (10, 2)$
33. $\left(\frac{1}{2}, 1\right), \left(-\frac{5}{2}, \frac{4}{3}\right)$
34. $\left(-\frac{1}{3}, -\frac{1}{3}\right), \left(-\frac{1}{6}, -\frac{1}{2}\right)$
35. $(6.2, 5.4), (-3.7, 1.8)$
36. $(-16.8, 12.3), (5.6, 4.9)$
37. $(-36, -18), (48, -72)$
38. $(1.451, 3.051), (5.906, 11.360)$

39. *Rock and Roll Hall of Fame* The graph below shows the numbers of recording artists who were elected to the Rock and Roll Hall of Fame in the years from 1986 through 1999.

(a) Describe any trends in the data. From these trends, predict the number of artists elected in 2001.

(b) Why do you think the numbers elected in 1986 and 1987 were greater than in other years?

40. *Make a Conjecture* Plot the points $(2, 1), (-3, 5)$, and $(7, -3)$ on a rectangular coordinate system. Then change the sign of the x-coordinate of each point and plot the three new points on the same rectangular coordinate system. Make a conjecture about the location of a point when each of the following occurs.

(a) The sign of the x-coordinate is changed.

(b) The sign of the y-coordinate is changed.

(c) The signs of both the x- and y-coordinates are changed.

41. *Sports* In a football game, a quarterback throws a pass from the 15-yard line, 10 yards from the sideline, as shown in the figure. The pass is caught on the 40-yard line, 45 yards from the same sideline. How long is the pass?

FIGURE FOR 41

42. *Flying Distance* A plane flies in a straight line to a city that is 100 kilometers east and 150 kilometers north of the point of departure. How far does it fly?

43. *Business* Starbucks Corp. had annual sales of $696.5 million in 1996 and $1308.7 million in 1998. Use the Midpoint Formula to estimate the sales in 1997. (Source: Starbucks Corp.)

44. *Business* Lands' End, Inc. had annual sales of $1118.7 million in 1996 and $1371.4 million in 1998. Use the Midpoint Formula to estimate the sales in 1997. (Source: Lands' End, Inc.)

In Exercises 45–48, determine whether each point lies on the graph of the equation.

Equation	Points	
45. $y = \sqrt{x + 4}$	(a) $(0, 2)$	(b) $(5, 3)$
46. $y = x^2 - 3x + 2$	(a) $(2, 0)$	(b) $(-2, 8)$
47. $y = 4 - |x - 2|$	(a) $(1, 5)$	(b) $(6, 0)$
48. $y = \frac{1}{3}x^3 - 2x^2$	(a) $\left(2, -\frac{16}{3}\right)$	(b) $(-3, 9)$

In Exercises 49 and 50, complete the table. Use the resulting solution points to sketch the graph of the equation.

49. $y = \frac{3}{4}x - 1$

x	−2	0	1	$\frac{4}{3}$	2
y					

50. $y = x^2 - 3x$

x	−1	0	1	2	3
y					

In Exercises 51–54, find the x- and y-intercepts of the graph of the equation.

51. $y = 16 - 4x^2$
52. $y = (x + 2)^2$
53. $y = 2x^3 - 5x^2$
54. $y^2 = x + 1$

In Exercises 55–58, use a graphing utility to graph the equation. Use a standard setting. Approximate any intercepts.

55. $y = 3 - \frac{1}{2}x$

56. $y = x^2 + x - 2$

57. $y = \dfrac{2x}{x-1}$

58. $y = (6-x)\sqrt{x}$

In Exercises 59–66, check for symmetry with respect to both axes and the origin.

59. $x^2 - y = 0$

60. $x - y^2 = 0$

61. $y = x^3$

62. $y = x^4 - x^2 + 3$

63. $y = \dfrac{x}{x^2+1}$

64. $y = \sqrt{9 - x^2}$

65. $xy^2 + 10 = 0$

66. $xy = 4$

In Exercises 67–70, assume that the graph has the indicated type of symmetry. Sketch the complete graph of the equation.

67.

y-Axis symmetry

68.

x-Axis symmetry

69.

Origin symmetry

70.
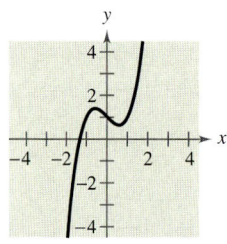
y-Axis symmetry

In Exercises 71–74, match the equation with its graph. [The graphs are labeled (a), (b), (c), and (d).]

(a)

(b)

(c)

(d)
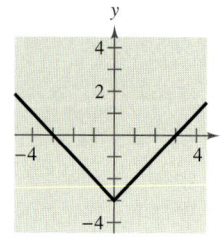

71. $y = 1 - x$

72. $y = x^2 - 2x$

73. $y = x^3 - x + 1$

74. $y = |x| - 3$

In Exercises 75–86, use intercepts and symmetry to sketch the graph of the equation.

75. $y = -3x + 1$

76. $y = 2x - 3$

77. $y = x^2 - 2x$

78. $y = -x^2 - 2x$

79. $y = x^3 + 3$

80. $y = x^3 - 1$

81. $y = \sqrt{x-3}$

82. $y = \sqrt{1-x}$

83. $y = |x - 6|$

84. $y = 1 - |x|$

85. $x = y^2 - 1$

86. $x = y^2 - 5$

In Exercises 87–94, find the standard form of the equation of the specified circle.

87. Center: $(0, 0)$; Radius: 4

88. Center: $(0, 0)$; Radius: 5

89. Center: $(2, -1)$; Radius: 4

90. Center: $(-7, -4)$; Radius: 7

91. Center: $(-1, 2)$; Solution point: $(0, 0)$

92. Center: $(3, -2)$; Solution point: $(-1, 1)$

93. Endpoints of a diameter: $(0, 0)$, $(6, 8)$

94. Endpoints of a diameter: $(-4, -1)$, $(4, 1)$

In Exercises 95–100, find the center and radius, and sketch the graph of the equation.

95. $x^2 + y^2 = 25$

96. $x^2 + y^2 = 16$

97. $(x-1)^2 + (y+3)^2 = 9$

98. $x^2 + (y-1)^2 = 1$

99. $\left(x - \frac{1}{2}\right)^2 + \left(y - \frac{1}{2}\right)^2 = \frac{9}{4}$

100. $(x-2)^2 + (y+1)^2 = 3$

101. *Business* A manufacturing plant purchases a new molding machine for $225,000. The depreciated value y after t years is

$$y = 225{,}000 - 20{,}000t, \quad 0 \le t \le 8.$$

Sketch the graph of the equation.

The symbol indicates an exercise or part of an exercise in which you are instructed to use a graphing utility.

102. *Consumerism* You purchase a jet ski for $8100. The depreciated value y after t years is

$$y = 8100 - 929t, \quad 0 \leq t \leq 6.$$

Sketch the graph of the equation.

103. *Geometry* A rectangle of length x and width w has a perimeter of 12 meters.

(a) Draw a rectangle that gives a visual representation of the problem. Use the specified variables to label the sides of the rectangle.

(b) Show that the width of the rectangle is $w = 6 - x$ and its area is $A = x(6 - x)$.

(c) Use a graphing utility to graph the area equation.

(d) From the graph in part (c), estimate the dimensions of the rectangle that yield a maximum area.

104. *Geometry* A rectangle of length x and width w has a perimeter of 22 yards.

(a) Draw a rectangle that gives a visual representation of the problem. Use the specified variables to label the sides of the rectangle.

(b) Show that the width of the rectangle is $w = 11 - x$ and its area is $A = x(11 - x)$.

(c) Use a graphing utility to graph the area equation. Be sure to adjust your window settings.

(d) From the graph in part (c), estimate the dimensions of the rectangle that yield a maximum area.

Data Analysis In Exercises 105 and 106, (a) sketch a scatter plot of the data, (b) graph the model for the data and compare the scatter plot and graph, and (c) use the model to estimate the values of y for the years 2002 and 2004.

105. *Federal Debt* The table shows the per capita federal debt of the United States for several years. (Source: U.S. Treasury Department, U.S. Bureau of the Census)

Year	1950	1960	1970	1980
Per capita debt	$1688	$1572	$1807	$3981

Year	1990	1994	1997	1998
Per capita debt	$12,848	$15,750	$20,063	$20,513

A model for the per capita debt during this period is

$$y = 0.223t^3 - 0.733t^2 - 78.255t + 1837.433$$

where y represents the per capita debt and t is the time in years, with $t = 0$ corresponding to 1950.

106. *Population Statistics* The table shows the life expectancy of a child (at birth) in the United States for selected years from 1920 to 2000. (Source: U.S. National Center for Health Statistics, U.S. Bureau of the Census)

Year	1920	1930	1940	1950
Life expectancy	54.1	59.7	62.9	68.2

Year	1960	1970	1980	1990	2000
Life expectancy	69.7	70.8	73.7	75.4	76.4

A model for the life expectancy during this period is

$$y = \frac{66.93 + t}{1 + 0.01t}$$

where y represents the life expectancy and t is the time in years, with $t = 0$ corresponding to 1950.

Synthesis

True or False? In Exercises 107–109, determine whether the statement is true or false. Justify your answer.

107. The points $(-8, 4)$, $(2, 11)$, and $(-5, 1)$ represent the vertices of an isosceles triangle.

108. In order to find the y-intercepts of the graph of an equation, let $y = 0$ and solve the equation for x.

109. The graph of a linear equation has, at most, one x-intercept.

110. *Think About It* When plotting points on the rectangular coordinate system, is it true that the scales on the x- and y-axes must be the same? Explain.

111. Prove that the diagonals of the parallelogram in the figure intersect at their midpoints.

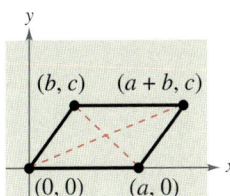

112. *Think About It* Find a and b if the graph of $y = ax^2 + bx^3$ is symmetric with respect to (a) the y-axis and (b) the origin. (There are many correct answers.)

P.4 Linear Equations in Two Variables

▶ **What you should learn**
- How to use slope to graph linear equations in two variables
- How to find slopes of lines
- How to write linear equations in two variables
- How to use slope to identify parallel and perpendicular lines
- How to use linear equations in two variables to model and solve real-life problems

▶ **Why you should learn it**

Linear equations in two variables can be used to model and solve real-life problems. For instance, Exercise 112 on page 51 shows how to use a linear equation to model the average annual salaries of major league baseball players from 1988 to 1998.

 A computer simulation of this concept appears in the *Interactive* CD-ROM and *Internet* versions of this text.

Using Slope

The simplest mathematical model for relating two variables is the **linear equation in two variables** $y = mx + b$. The equation is called *linear* because its graph is a line. (In mathematics, the term *line* means *straight line*.) By letting $x = 0$, you can see that the line crosses the y-axis at $y = b$, as shown in Figure P.25. In other words, the y-intercept is $(0, b)$. The steepness or slope of the line is m.

$$y = mx + b$$

Slope ↑   ↑ y-Intercept

The **slope** of a nonvertical line is the number of units the line rises (or falls) vertically for each unit of horizontal change from left to right, as shown in Figure P.25.

 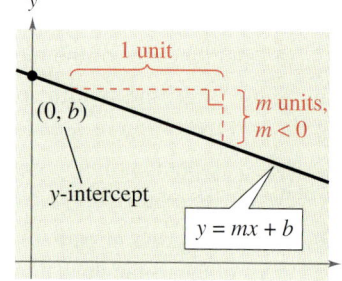

Positive slope, line rises.   Negative slope, line falls.
FIGURE P.25

A linear equation that is written in the form $y = mx + b$ is said to be written in **slope-intercept form.**

The Slope-Intercept Form of the Equation of a Line

The graph of the equation

$$y = mx + b$$

is a line whose slope is m and whose y-intercept is $(0, b)$.

◀ **Exploration** ▶

Use a graphing utility to compare the slopes of the lines $y = mx$ where $m = 0.5, 1, 2,$ and 4. Which line rises most quickly? Now, let $m = -0.5, -1, -2,$ and -4. Which line falls most quickly? Use a square setting to obtain a true geometric perspective. What can you conclude about the slope and the "rate" at which the line rises or falls?

Section P.4 ▶ Linear Equations in Two Variables 39

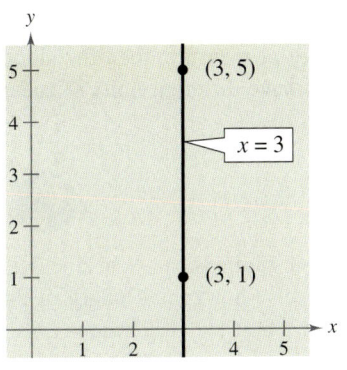

FIGURE P.26 *Slope is undefined.*

Once you have determined the slope and the y-intercept of a line, it is a relatively simple matter to sketch its graph. In the following example, note that none of the lines is vertical. A vertical line has an equation of the form

$x = a.$ Vertical line

The equation of a vertical line cannot be written in the form $y = mx + b$ because the slope of a vertical line is undefined, as indicated in Figure P.26.

Example 1 ▶ Graphing a Linear Equation

Sketch the graph of each linear equation.

a. $y = 2x + 1$
b. $y = 2$
c. $x + y = 2$

Solution

a. Because $b = 1$, the y-intercept is $(0, 1)$. Moreover, because the slope is $m = 2$, the line *rises* 2 units for each unit the line moves to the right, as shown in Figure P.27(a).

b. By writing this equation in the form $y = (0)x + 2$, you can see that the y-intercept is $(0, 2)$ and the slope is zero. A zero slope implies that the line is horizontal—that is, it doesn't rise *or* fall, as shown in Figure P.27(b).

c. By writing this equation in slope-intercept form

$x + y = 2$ Write original equation.
$y = -x + 2$ Subtract x from each side.
$y = (-1)x + 2$ Write in slope-intercept form.

you can see that the y-intercept is $(0, 2)$. Moreover, because the slope is $m = -1$, the line *falls* 1 unit for each unit the line moves to the right, as shown in Figure P.27(c).

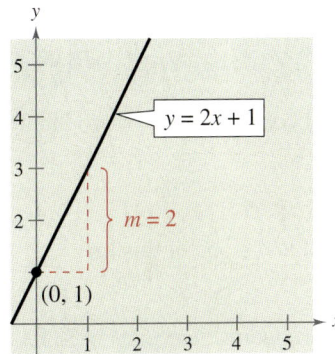

(a) When *m* is positive, the line rises.

FIGURE P.27

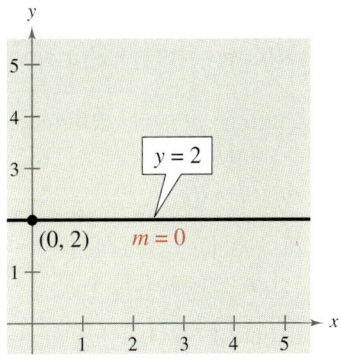

(b) When *m* is 0, the line is horizontal.

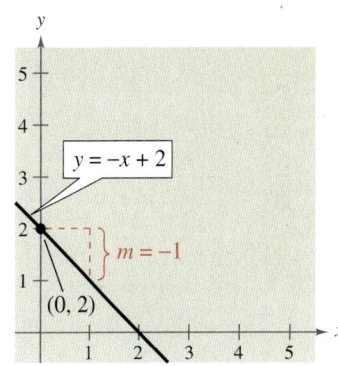

(c) When *m* is negative, the line falls.

In real-life problems, the slope of a line can be interpreted as either a *ratio* or a *rate*. If the *x*-axis and *y*-axis have the same unit of measure, then the slope has no units and is a **ratio**. If the *x*-axis and *y*-axis have different units of measure, then the slope is a **rate** or **rate of change.**

Example 2 Using Slope as a Ratio

The maximum recommended slope of a wheelchair ramp is $\frac{1}{12}$. A business is installing a wheelchair ramp that rises 22 inches over a horizontal length of 24 feet. Is the ramp steeper than recommended? (Source: Americans with Disabilities Act Handbook)

Solution

The horizontal length of the ramp is 24 feet or $12(24) = 288$ inches, as shown in Figure P.28. So, the slope of the ramp is

$$\text{Slope} = \frac{\text{vertical change}}{\text{horizontal change}}$$

$$= \frac{22 \text{ in.}}{288 \text{ in.}}$$

$$\approx 0.076.$$

Because $\frac{1}{12} \approx 0.083$, the slope of the ramp is not steeper than recommended.

FIGURE P.28

Example 3 Using Slope as a Rate of Change

A manufacturing company determines that the total cost in dollars of producing *x* units of a product is

$$C = 25x + 3500. \qquad \text{Cost equation}$$

Describe the practical significance of the *y*-intercept and slope of this line.

Solution

The *y*-intercept (0, 3500) tells you that the cost of producing zero units is $3500. This is the *fixed cost* of production—it includes costs that must be paid regardless of the number of units produced. The slope of $m = 25$ tells you that the cost of producing each unit is $25, as shown in Figure P.29. Economists call the cost per unit the *marginal cost*. If the production increases by 1 unit, then the "margin," or extra amount of cost, is $25. So, the cost increases at a rate of $25 per unit.

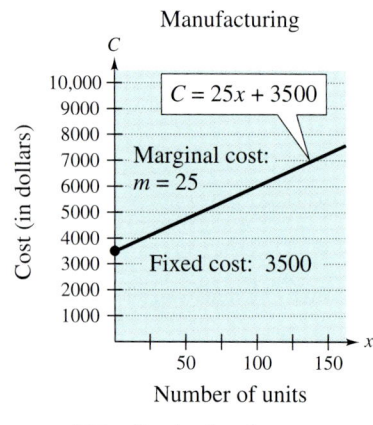

FIGURE P.29 *Production Cost*

Finding the Slope of a Line

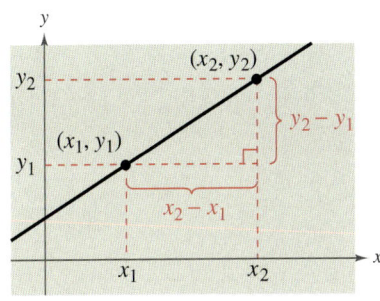

FIGURE P.30

Given an equation of a line, you can find its slope by writing the equation in slope-intercept form. If you are not given an equation, you can still find the slope of a line. For instance, suppose you want to find the slope of the line passing through the points (x_1, y_1) and (x_2, y_2), as shown in Figure P.30. As you move from left to right along this line, a change of $(y_2 - y_1)$ units in the vertical direction corresponds to a change of $(x_2 - x_1)$ units in the horizontal direction.

$$y_2 - y_1 = \text{the change in } y = \text{rise}$$

and

$$x_2 - x_1 = \text{the change in } x = \text{run}$$

The ratio of $(y_2 - y_1)$ to $(x_2 - x_1)$ represents the slope of the line that passes through the points (x_1, y_1) and (x_2, y_2).

$$\text{Slope} = \frac{\text{change in } y}{\text{change in } x}$$

$$= \frac{\text{rise}}{\text{run}}$$

$$= \frac{y_2 - y_1}{x_2 - x_1}$$

The Slope of a Line Passing Through Two Points

The **slope** m of the nonvertical line through (x_1, y_1) and (x_2, y_2) is

$$m = \frac{y_2 - y_1}{x_2 - x_1}$$

where $x_1 \neq x_2$.

Historical Note
René Descartes (1596–1650) was largely responsible for the development of analytic geometry. His use of the symbol m for slope came from the French verb *monter*, meaning to mount, to climb, or to rise.

When this formula is used for slope, the *order of subtraction* is important. Given two points on a line, you are free to label either one of them as (x_1, y_1) and the other as (x_2, y_2). However, once you have done this, you must form the numerator and denominator using the same order of subtraction.

$$m = \frac{y_2 - y_1}{x_2 - x_1} \qquad m = \frac{y_1 - y_2}{x_1 - x_2} \qquad m = \frac{y_2 - y_1}{x_1 - x_2}$$

Correct Correct Incorrect

For instance, the slope of the line passing through the points $(3, 4)$ and $(5, 7)$ can be calculated as

$$m = \frac{7 - 4}{5 - 3} = \frac{3}{2}$$

or

$$m = \frac{4 - 7}{3 - 5} = \frac{-3}{-2} = \frac{3}{2}.$$

The *Interactive* CD-ROM and *Internet* versions of this text show every example with its solution; clicking on the *Try It!* button brings up similar problems. Guided Examples and Integrated Examples show step-by-step solutions to additional examples. Integrated Examples are related to several concepts in the section.

Example 4 ▶ Finding the Slope of a Line Through Two Points

Find the slope of the line passing through each pair of points. (See Figure P.31.)

a. $(-2, 0)$ and $(3, 1)$ **b.** $(-1, 2)$ and $(2, 2)$
c. $(0, 4)$ and $(1, -1)$ **d.** $(3, 4)$ and $(3, 1)$

Solution

a. Letting $(x_1, y_1) = (-2, 0)$ and $(x_2, y_2) = (3, 1)$, you obtain a slope of

$$m = \frac{y_2 - y_1}{x_2 - x_1} = \frac{1 - 0}{3 - (-2)} = \frac{1}{5}.$$

b. The slope of the line passing through $(-1, 2)$ and $(2, 2)$ is

$$m = \frac{2 - 2}{2 - (-1)} = \frac{0}{3} = 0.$$

c. The slope of the line passing through $(0, 4)$ and $(1, -1)$ is

$$m = \frac{-1 - 4}{1 - 0} = \frac{-5}{1} = -5.$$

d. The slope of the vertical line passing through $(3, 4)$ and $(3, 1)$ is

$$m = \frac{1 - 4}{3 - 3} = \frac{-3}{0}.$$

Because division by 0 is undefined, the slope is undefined.

STUDY TIP

In Figure P.31, note the relationships between slope and the description of the line.

a. Positive slope; line rises from left to right
b. Zero slope; line is horizontal
c. Negative slope; line falls from left to right
d. Undefined slope; line is vertical

(a)

(b)

(c)

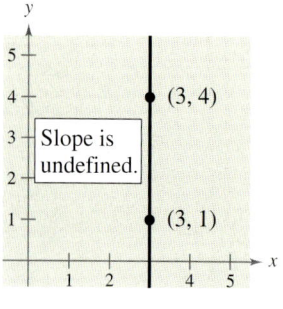
(d)

FIGURE P.31

Writing Linear Equations in Two Variables

If (x_1, y_1) is a point on a line of slope m and (x, y) is *any other* point on the line, then

$$\frac{y - y_1}{x - x_1} = m.$$

This equation, involving the variables x and y, can be rewritten in the form

$$y - y_1 = m(x - x_1),$$

which is the **point-slope form** of the equation of a line.

> ### Point-Slope Form of the Equation of a Line
> The equation of the line with slope m passing through the point (x_1, y_1) is
> $$y - y_1 = m(x - x_1).$$

The point-slope form is most useful for *finding* the equation of a line. You should remember this formula.

Example 5 ▶ Using the Point-Slope Form

Find the slope-intercept form of the equation of the line that has a slope of 3 and passes through the point $(1, -2)$, as shown in Figure P.32.

Solution

Use the point-slope form with $m = 3$ and $(x_1, y_1) = (1, -2)$.

$y - y_1 = m(x - x_1)$	Point-slope form
$y - (-2) = 3(x - 1)$	Substitute for m, x_1, and y_1.
$y + 2 = 3x - 3$	Simplify.
$y = 3x - 5$	Write in slope-intercept form.

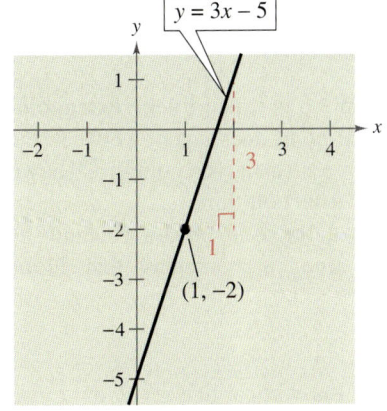

FIGURE P.32

The point-slope form can be used to find an equation of the line passing through points (x_1, y_1) and (x_2, y_2). To do this, first find the slope of the line

$$m = \frac{y_2 - y_1}{x_2 - x_1}, \quad x_1 \neq x_2$$

and then use the point-slope form to obtain the equation

$$y - y_1 = \frac{y_2 - y_1}{x_2 - x_1}(x - x_1). \qquad \text{Two-point form}$$

This is sometimes called the **two-point form** of the equation of a line.

Example 6 ▶ Predicting Cash Flow Per Share

The cash flow per share for Yahoo! Inc. was $0.03 in 1997 and $0.30 in 1998. Using only this information, write a linear equation that gives the cash flow per share in terms of the year. Then predict the cash flow for 1999. *(Source: Yahoo! Inc.)*

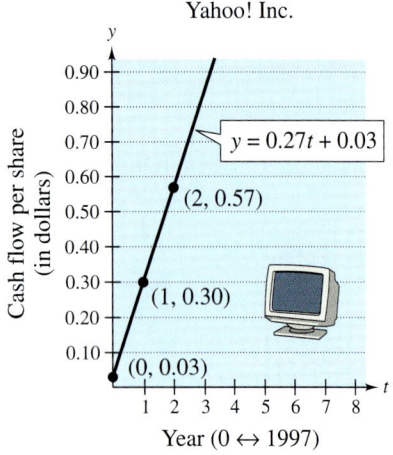

FIGURE P.33

Solution

Let $t = 0$ represent 1997. Then the two given values are represented by the data points $(0, 0.03)$ and $(1, 0.30)$. The slope of the line through these points is

$$m = \frac{0.30 - 0.03}{1 - 0}$$

$$= 0.27.$$

Using the point-slope form, you can find the equation that relates the cash flow y and the year t to be

$$y = 0.27t + 0.03.$$

According to this equation, the cash flow in 1999 was $0.57, as shown in Figure P.33. (In this case, the prediction is quite good—the actual cash flow in 1999 was $0.55.)

The prediction method illustrated in Example 6 is called **linear extrapolation**. Note in Figure P.34(a) that an extrapolated point does not lie between the given points. When the estimated point lies between two given points, as shown in Figure P.34(b), the procedure is called **linear interpolation**.

Because the slope of a vertical line is not defined, its equation cannot be written in slope-intercept form. However, every line has an equation that can be written in the **general form**

$$Ax + By + C = 0 \qquad \text{General form}$$

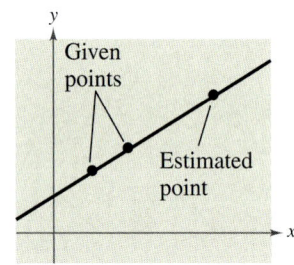

(a) Linear extrapolation

where A and B are not both zero. For instance, the vertical line given by $x = a$ can be represented by the general form

$$x - a = 0.$$

Equations of Lines

1. General form: $\qquad Ax + By + C = 0$
2. Vertical line: $\qquad x = a$
3. Horizontal line: $\qquad y = b$
4. Slope-intercept form: $\qquad y = mx + b$
5. Point-slope form: $\qquad y - y_1 = m(x - x_1)$
6. Two-point form: $\qquad y - y_1 = \dfrac{y_2 - y_1}{x_2 - x_1}(x - x_1)$

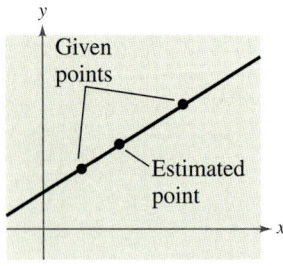

(b) Linear interpolation

FIGURE P.34

Technology

On a graphing utility, lines will not appear to have the correct slope unless you use a viewing window that has a square setting. For instance, try graphing the lines in Example 7 using the standard setting $-10 \leq x \leq 10$ and $-10 \leq y \leq 10$. Then reset the viewing window with the square setting $-9 \leq x \leq 9$ and $-6 \leq y \leq 6$. On which setting do the lines $y = \frac{2}{3}x - \frac{5}{3}$ and $y = -\frac{3}{2}x + 2$ appear perpendicular?

Parallel and Perpendicular Lines

Slope can be used to decide whether two nonvertical lines in a plane are parallel, perpendicular, or neither.

Parallel and Perpendicular Lines

1. Two distinct nonvertical lines are **parallel** if and only if their slopes are equal. That is, $m_1 = m_2$.

2. Two nonvertical lines are **perpendicular** if and only if their slopes are negative reciprocals of each other. That is, $m_1 = -1/m_2$.

Example 7 ▶ Finding Parallel and Perpendicular Lines

Find the slope-intercept form of the equation of the line that passes through the point $(2, -1)$ and is (a) parallel to and (b) perpendicular to the line $2x - 3y = 5$.

Solution

By writing the equation of the given line in slope-intercept form

$2x - 3y = 5$ Write original equation.

$-3y = -2x + 5$ Subtract $2x$ from each side.

$y = \frac{2}{3}x - \frac{5}{3}$ Write in slope-intercept form.

you can see that it has a slope of $m = \frac{2}{3}$, as shown in Figure P.35.

a. Any line parallel to the given line must also have a slope of $\frac{2}{3}$. So, the line through $(2, -1)$ that is parallel to the given line has the following equation.

$y - (-1) = \frac{2}{3}(x - 2)$ Write in point-slope form.

$3(y + 1) = 2(x - 2)$ Multiply each side by 3.

$3y + 3 = 2x - 4$ Distributive Property

$2x - 3y - 7 = 0$ Write in general form.

$y = \frac{2}{3}x - \frac{7}{3}$ Write in slope-intercept form.

b. Any line perpendicular to the given line must have a slope of $-1/(2/3)$ or $-3/2$. So, the line through $(2, -1)$ that is perpendicular to the given line has the following equation.

$y - (-1) = -\frac{3}{2}(x - 2)$ Write in point-slope form.

$2(y + 1) = -3(x - 2)$ Multiply each side by 2.

$2y + 2 = -3x + 6$ Distributive Property

$3x + 2y - 4 = 0$ Write in general form.

$y = -\frac{3}{2}x + 2$ Write in slope-intercept form.

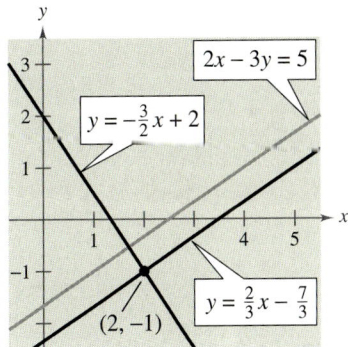

FIGURE P.35

Application

Most business expenses can be deducted in the same year they occur. One exception is the cost of property that has a useful life of more than 1 year. Such costs must be *depreciated* over the useful life of the property. If the *same amount* is depreciated each year, the procedure is called *linear* or *straight-line depreciation*. The *book value* is the difference between the original value and the total amount of depreciation accumulated to date.

Example 8 ▶ Straight-Line Depreciation

Your company has purchased a $12,000 machine that has a useful life of 8 years. The salvage value at the end of 8 years is $2000. Write a linear equation that describes the book value of the machine each year.

Solution

Let V represent the value of the machine at the end of year t. You can represent the initial value of the machine by the data point $(0, 12,000)$ and the salvage value of the machine by the data point $(8, 2000)$. The slope of the line is

$$m = \frac{2000 - 12,000}{8 - 0} = -\$1250$$

which represents the annual depreciation in *dollars per year*. Using the point-slope form, you can write the equation of the line as follows.

$V - 12,000 = -1250(t - 0)$ Write in point-slope form.

$V = -1250t + 12,000$ Write in slope-intercept form.

The table shows the book value at the end of each year, and the graph of the equation is shown in Figure P.36.

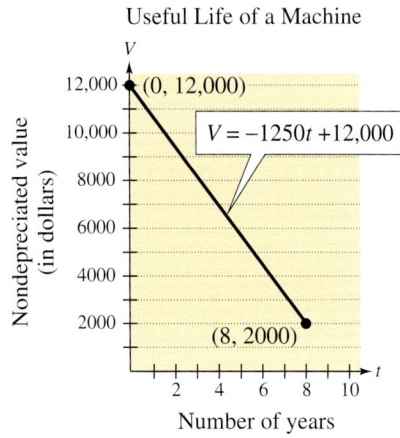

FIGURE P.36 *Straight-Line Depreciation*

t	0	1	2	3	4	5	6	7	8
V	12,000	10,750	9500	8250	7000	5750	4500	3250	2000

Writing ABOUT MATHEMATICS

x	y
0	80.90
1	72.74
2	68.68
3	61.48

x	y
4	56.21
5	51.00
6	47.70
7	42.78

Modeling Linear Data The table at the left shows the average monthly cellular phone bill y (in dollars) for subscribers in the United States from 1990 through 1997, where x is the time in years, with $x = 0$ corresponding to 1990. (Source: Cellular Telecommunications Industry Association)

Sketch a scatter plot of the data, and use a straightedge to sketch the best-fitting line through the points. Find the equation of the line. Explain the procedure you used. Write a short paragraph explaining the meaning of the slope and the y-intercept of the line in terms of the data. Compare the values obtained using your model with the actual values in the table. Use your model to estimate the average monthly bill in the year 2000.

P.4 Exercises

In Exercises 1 and 2, identify the line that has each slope.

1. (a) $m = \frac{2}{3}$
   (b) m is undefined.
   (c) $m = -2$

2. (a) $m = 0$
   (b) $m = -\frac{3}{4}$
   (c) $m = 1$

In Exercises 11–18, plot the points and find the slope of the line passing through the pair of points.

11. $(-3, -2), (1, 6)$
12. $(2, 4), (4, -4)$
13. $(-6, -1), (-6, 4)$
14. $(0, -10), (-4, 0)$
15. $\left(\frac{11}{2}, -\frac{4}{3}\right), \left(-\frac{3}{2}, -\frac{1}{3}\right)$
16. $\left(\frac{7}{8}, \frac{3}{4}\right), \left(\frac{5}{4}, -\frac{1}{4}\right)$
17. $(4.8, 3.1), (-5.2, 1.6)$
18. $(-1.75, -8.3), (2.25, -2.6)$

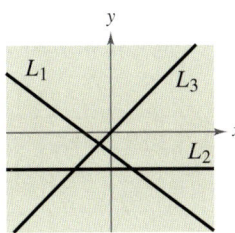

In Exercises 3 and 4, sketch the graph of the line through the point with each indicated slope on the same set of coordinate axes.

Point	Slopes			
3. $(2, 3)$	(a) 0	(b) 1	(c) 2	(d) -3
4. $(-4, 1)$	(a) 3	(b) -3	(c) $\frac{1}{2}$	(d) Undefined

In Exercises 19–28, use the point on the line and the slope of the line to find three additional points through which the line passes. (There are many correct answers.)

Point	Slope
19. $(2, 1)$	$m = 0$
20. $(-4, 1)$	m is undefined.
21. $(5, -6)$	$m = 1$
22. $(10, -6)$	$m = -1$
23. $(-8, 1)$	m is undefined.
24. $(-3, -1)$	$m = 0$
25. $(-5, 4)$	$m = 2$
26. $(0, -9)$	$m = -2$
27. $(7, -2)$	$m = \frac{1}{2}$
28. $(-1, -6)$	$m = -\frac{1}{2}$

In Exercises 5–10, estimate the slope of the line.

5.

6.

7.

8.

9.

10.
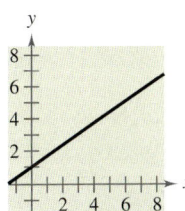

In Exercises 29–32, determine whether the lines L_1 and L_2 passing through the pairs of points are parallel, perpendicular, or neither.

29. L_1: $(0, -1), (5, 9)$
    L_2: $(0, 3), (4, 1)$

30. L_1: $(-2, -1), (1, 5)$
    L_2: $(1, 3), (5, -5)$

31. L_1: $(3, 6), (-6, 0)$
    L_2: $(0, -1), \left(5, \frac{7}{3}\right)$

32. L_1: $(4, 8), (-4, 2)$
    L_2: $(3, -5), \left(-1, \frac{1}{3}\right)$

33. **Business** The following are the slopes of lines representing annual sales y in terms of time x in years. Use the slopes to interpret any change in annual sales for a 1-year increase in time.
    (a) The line has a slope of $m = 135$.
    (b) The line has a slope of $m = 0$.
    (c) The line has a slope of $m = -40$.

34. *Business* The following are the slopes of lines representing daily revenues y in terms of time x in days. Use the slopes to interpret any change in daily revenues for a 1-day increase in time.

(a) The line has a slope of $m = 400$.

(b) The line has a slope of $m = 100$.

(c) The line has a slope of $m = 0$.

35. *Business* The graph shows the earnings per share of stock for the Kellogg Company for the years 1988 through 1998. (Source: Kellogg Company)

(a) Use the slopes to determine the years when the earnings per share showed the greatest increase and decrease.

(b) Find the slope of the line segment connecting the years 1988 and 1998.

(c) Interpret the meaning of the slope in part (b) in the context of the problem.

36. *Business* The graph shows the dividends declared per share of stock for the Colgate-Palmolive Company for the years 1988 through 1998. (Source: Colgate-Palmolive Company)

(a) Use the slopes to determine the years when the dividends declared per share showed the greatest increase and the smallest increase.

(b) Find the slope of the line segment connecting the years 1988 and 1998.

(c) Interpret the meaning of the slope in part (b) in the context of the problem.

37. *Road Grade* From the top of a mountain road, a surveyor takes several horizontal measurements x and several vertical measurements y, as shown in the table.

x	300	600	900	1200	1500	1800	2100
y	-25	-50	-75	-100	-125	-150	-175

(a) Sketch a scatter plot of the data.

(b) Use a straightedge to sketch the best-fitting line through the points.

(c) Find an equation for the line you sketched in part (b).

(d) Interpret the meaning of the slope of the line in part (c) in the context of the problem.

(e) The surveyor needs to put up a road sign that indicates the steepness of the road. For instance, a surveyor would put a sign that states "8% grade" on a road with a downhill grade that has a slope of $-\frac{8}{100}$. What should the sign state for the road in this problem?

38. *Road Grade* You are driving on a road that has a 6% uphill grade. This means that the slope of the road is $\frac{6}{100}$. Approximate the amount of vertical change in your position if you drive 200 feet.

39. *Height of an Attic* The "rise to run" in determining the steepness of the roof on a house is 3 to 4 (see figure). Determine the maximum height in the attic if the house is 32 feet wide.

In Exercises 40–45, find the slope and *y*-intercept (if possible) of the equation of the line. Sketch a graph of the line.

40. $y = 5x + 3$
41. $y = x - 10$
42. $5x - 2 = 0$
43. $3y + 5 = 0$
44. $7x + 6y = 30$
45. $2x + 3y = 9$

In Exercises 46–57, find the slope-intercept form of the equation of the line that passes through the given point and has the indicated slope. Sketch a graph of the line.

Point	Slope
46. $(0, -2)$	$m = 3$
47. $(0, 10)$	$m = -1$
48. $(-3, 6)$	$m = -2$
49. $(0, 0)$	$m = 4$
50. $(4, 0)$	$m = -\frac{1}{3}$
51. $(-2, -5)$	$m = \frac{3}{4}$
52. $(6, -1)$	m is undefined.
53. $(-10, 4)$	$m = 0$
54. $(4, \frac{5}{2})$	$m = \frac{4}{3}$
55. $(-\frac{1}{2}, \frac{3}{2})$	$m = -3$
56. $(-5.1, 1.8)$	$m = 5$
57. $(2.3, -8.5)$	$m = -\frac{5}{2}$

In Exercises 58–67, find the slope-intercept form of the equation of the line passing through the points. Sketch a graph of the line.

58. $(5, -1), (-5, 5)$
59. $(4, 3), (-4, -4)$
60. $(-8, 1), (-8, 7)$
61. $(-1, 4), (6, 4)$
62. $(2, \frac{1}{2}), (\frac{1}{2}, \frac{5}{4})$
63. $(1, 1), (6, -\frac{2}{3})$
64. $(-\frac{1}{10}, -\frac{3}{5}), (\frac{9}{10}, -\frac{9}{5})$
65. $(\frac{3}{4}, \frac{3}{2}), (-\frac{4}{3}, \frac{7}{4})$
66. $(1, 0.6), (-2, -0.6)$
67. $(-8, 0.6), (2, -2.4)$

In Exercises 68–73, use the *intercept form* to find the equation of the line with the given intercepts. The intercept form of the equation of a line with intercepts $(a, 0)$ and $(0, b)$ is

$$\frac{x}{a} + \frac{y}{b} = 1, \quad a \neq 0, \ b \neq 0.$$

68. *x*-intercept: $(2, 0)$
   y-intercept: $(0, 3)$
69. *x*-intercept: $(-3, 0)$
   y-intercept: $(0, 4)$
70. *x*-intercept: $\left(-\frac{1}{6}, 0\right)$
   y-intercept: $\left(0, -\frac{2}{3}\right)$
71. *x*-intercept: $\left(\frac{2}{3}, 0\right)$
   y-intercept: $(0, -2)$
72. Point on line: $(1, 2)$
   x-intercept: $(c, 0)$
   y-intercept: $(0, c), \quad c \neq 0$
73. Point on line: $(-3, 4)$
   x-intercept: $(d, 0)$
   y-intercept: $(0, d), \quad d \neq 0$

In Exercises 74–81, write the slope-intercept forms of the equations of the lines through the given point (a) parallel to the given line and (b) perpendicular to the given line.

Point	Line
74. $(2, 1)$	$4x - 2y = 3$
75. $(-3, 2)$	$x + y = 7$
76. $\left(-\frac{2}{3}, \frac{7}{8}\right)$	$3x + 4y = 7$
77. $\left(\frac{7}{8}, \frac{3}{4}\right)$	$5x + 3y = 0$
78. $(-1, 0)$	$y = -3$
79. $(2, 5)$	$x = 4$
80. $(2.5, 6.8)$	$x - y = 4$
81. $(-3.9, -1.4)$	$6x + 2y = 9$

Graphical Interpretation In Exercises 82–85, identify any relationships that exist among the lines, and then use a graphing utility to graph the three equations in the same viewing window. Adjust the viewing window so that the slope appears visually correct.

82. (a) $y = 2x$ (b) $y = -2x$ (c) $y = \frac{1}{2}x$
83. (a) $y = \frac{2}{3}x$ (b) $y = -\frac{3}{2}x$ (c) $y = \frac{2}{3}x + 2$
84. (a) $y = -\frac{1}{2}x$ (b) $y = -\frac{1}{2}x + 3$ (c) $y = 2x - 4$
85. (a) $y = x - 8$ (b) $y = x + 1$ (c) $y = -x + 3$

Rate of Change In Exercises 86 and 87, you are given the dollar value of a product in 2001 and the rate at which the value of the product is expected to change during the next 5 years. Use this information to write a linear equation that gives the dollar value V of the product in terms of the year t. (Let $t = 1$ represent 2001.)

	2001 Value	Rate
86.	$2540	$125 increase per year
87.	$156	$4.50 increase per year

Graphical Interpretation In Exercises 88–91, match the description of the situation with its graph. Also determine the slope of each graph and interpret the slope in the context of the situation. [The graphs are labeled (a), (b), (c), and (d).]

(a)

(b)

(c)

(d)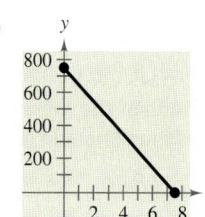

88. A person is paying $20 per week to a friend to repay a $200 loan.
89. An employee is paid $8.50 per hour plus $2 for each unit produced per hour.
90. A sales representative receives $30 per day for food plus $0.32 for each mile traveled.
91. A word processor that was purchased for $750 depreciates $100 per year.

In Exercises 92–95, find a relationship between x and y such that (x, y) is equidistant from the two points.

92. $(4, -1), (-2, 3)$
93. $(6, 5), (1, -8)$
94. $\left(3, \frac{5}{2}\right), (-7, 1)$
95. $\left(-\frac{1}{2}, -4\right), \left(\frac{7}{2}, \frac{5}{4}\right)$

96. *Cash Flow per Share* The cash flow per share for America Online, Inc. was $0.26 in 1998 and $0.70 in 1999. Write a linear equation that gives the cash flow per share in terms of the year. Let $t = 0$ represent 1998. Then predict the cash flows for the years 2000 and 2001. (Source: America Online, Inc.)

97. *Number of Stores* In 1996 there were 3927 J.C. Penney stores and in 1997 there were 3981 stores. Write a linear equation that gives the number of stores in terms of the year. Let $t = 0$ represent 1996. Then predict the numbers of stores for the years 1999 and 2000. (Source: J.C. Penney Co.)

98. *Temperature* Find the equation of the line that shows the relationship between the temperature in degrees Celsius C and degrees Fahrenheit F. Remember that water freezes at 0° Celsius (32° Fahrenheit) and boils at 100° Celsius (212° Fahrenheit).

99. *Temperature* Use the result of Exercise 98 to complete the table.

C		-10°	10°			177°
F	0°			68°	90°	

100. *Annual Salary* Your salary was $28,500 in 1998 and $32,900 in 2000. If your salary follows a linear growth pattern, what will your salary be in 2003?

101. *College Enrollment* A small college had 2546 students in 1998 and 2702 students in 2000. If the enrollment follows a linear growth pattern, how many students will the college have in 2004?

102. *Business* A business purchases a piece of equipment for $875. After 5 years the equipment will be outdated and have no value. Write a linear equation giving the value V of the equipment during the 5 years it will be used.

103. *Business* A business purchases a piece of equipment for $25,000. After 10 years the equipment will have to be replaced. Its value at that time is expected to be $2000. Write a linear equation giving the value V of the equipment during the 10 years it will be used.

104. *Sales* A store is offering a 15% discount on all items. Write a linear equation giving the sale price S for an item with a list price L.

105. *Hourly Wage* A manufacturer pays its assembly line workers $11.50 per hour. In addition, workers receive a piecework rate of $0.75 per unit produced. Write a linear equation for the hourly wage W in terms of the number of units x produced per hour.

106. *Business* A contractor purchases a piece of equipment for $36,500. The equipment requires an average expenditure of $5.25 per hour for fuel and maintenance, and the operator is paid $11.50 per hour.

(a) Write a linear equation giving the total cost C of operating this equipment for t hours. (Include the purchase cost of the equipment.)

(b) Assuming that customers are charged $27 per hour of machine use, write an equation for the revenue R derived from t hours of use.

(c) Use the formula for profit $(P = R - C)$ to write an equation for the profit derived from t hours of use.

(d) Use the result of part (c) to find the break-even point—that is, the number of hours this equipment must be used to yield a profit of 0 dollars.

107. *Rental Demand* A real estate office handles an apartment complex with 50 units. When the rent per unit is $580 per month, all 50 units are occupied. However, when the rent is $625 per month, the average number of occupied units drops to 47. Assume that the relationship between the monthly rent p and the demand x is linear.

(a) Write the equation of the line giving the demand x in terms of the rent p.

(b) Use this equation to predict the number of units occupied if the rent is $655.

(c) Predict the number of units occupied if the rent is $595.

108. *Geometry* The length and width of a rectangular garden are 15 meters and 10 meters, respectively. A walkway of width x surrounds the garden.

(a) Draw a diagram that gives a visual representation of the problem.

(b) Write the equation for the perimeter y of the walkway in terms of x.

(c) Use a graphing utility to graph the equation for the perimeter.

(d) Determine the slope of the graph in part (c). For each additional 1-meter increase in the width of the walkway, determine the increase in its perimeter.

109. *Monthly Salary* A salesperson receives a monthly salary of $2500 plus a commission of 7% of sales. Write a linear equation for the salesperson's monthly wage W in terms of monthly sales S.

110. *Business Costs* A sales representative of a company using a personal car receives $120 per day for lodging and meals plus $0.31 per mile driven. Write a linear equation giving the daily cost C to the company in terms of x, the number of miles driven.

111. *Investment* An inheritance of $12,000 is invested in two different mutual funds. A less risky fund pays $2\frac{1}{2}\%$ simple interest and a more risky fund pays 4% simple interest.

(a) If x dollars is invested in the fund paying $2\frac{1}{2}\%$, how much money is invested in the fund paying 4%?

(b) Write the total annual interest y in terms of x.

(c) Use a graphing utility to graph the function in part (b) over the interval $0 \leq x \leq 12,000$.

(d) Explain why the slope of the line in part (c) is negative.

112. *Sports* The average annual salaries of major league baseball players (in thousands of dollars) from 1988 to 1998 are shown in the scatter plot. Find the equation of the line that you think best fits these data. (Let y represent the average salary and let t represent the year, with $t = 0$ corresponding to 1988.) (Source: Major League Baseball Player Relations Committee)

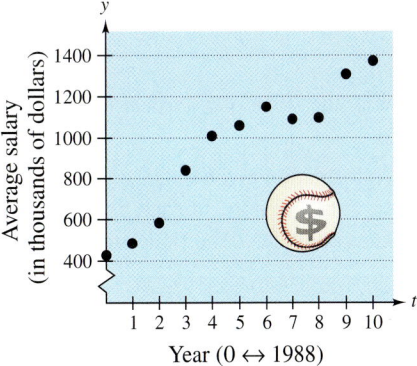

113. *Data Analysis* An instructor gives regular 20-point quizzes and 100-point exams in a mathematics course. Average scores for six students, given as data points (x, y) where x is the average quiz score and y is the average test score, are $(18, 87)$, $(10, 55)$, $(19, 96)$, $(16, 79)$, $(13, 76)$, and $(15, 82)$. [*Note*: There are many correct answers for parts (b)–(d).]

(a) Sketch a scatter plot of the data.

(b) Use a straightedge to sketch the best-fitting line through the points.

(c) Find an equation for the line sketched in part (b).

(d) Use the equation in part (c) to estimate the average test score for a person with an average quiz score of 17.

(e) If the instructor adds 4 points to the average test score of everyone in the class, describe the changes in the positions of the plotted points and the change in the equation of the line.

Synthesis

True or False? In Exercises 114 and 115, determine whether the statement is true or false. Justify your answer.

114. A line with a slope of $-\frac{5}{7}$ is steeper than a line with a slope of $-\frac{6}{7}$.

115. The line through $(-8, 2)$ and $(-1, 4)$ and the line through $(0, -4)$ and $(-7, 7)$ are parallel.

116. Explain how you could show that the points $A(2, 3)$, $B(2, 9)$, and $C(7, 2)$ are the vertices of a right triangle.

117. Explain why the slope of a vertical line is said to be undefined.

118. With the information given in the graphs, is it possible to determine the slope of each line? Is it possible that the lines could have the same slope? Explain.

(a)

(b)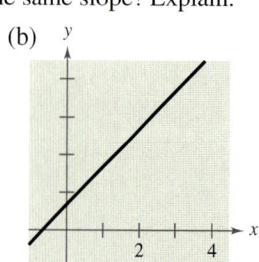

119. The slopes of two lines are -4 and $\frac{5}{2}$. Which is steeper? Explain.

120. The value V of a machine t years after it is purchased is

$$V = -4000t + 58,500, \quad 0 \le t \le 5.$$

Explain what the V-intercept and slope measure.

121. *Writing* Write a brief paragraph explaining whether or not any pair of points on a line can be used to calculate the slope of the line.

122. *Think About It* Is it possible for two lines with positive slopes to be perpendicular? Explain.

P.5 Functions

What you should learn
- How to decide whether relations between two variables are functions
- How to use function notation and evaluate functions
- How to find the domains of functions
- How to use functions to model and solve real-life problems

Why you should learn it
Functions can be used to model and solve real-life problems. For instance, Exercise 98 on page 65 shows how to use a function to find the force of water against the face of a dam.

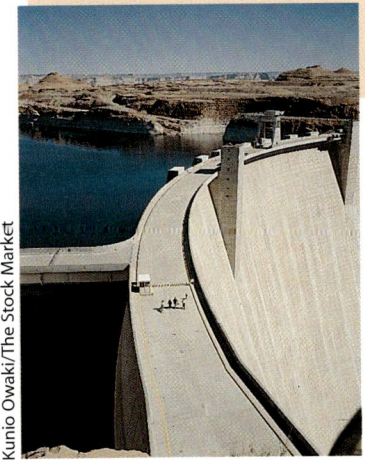

Introduction to Functions

Many everyday phenomena involve two quantities that are related to each other by some rule of correspondence. The mathematical term for such a rule of correspondence is a **relation**. Here are two examples.

1. The simple interest I earned on $1000 for 1 year is related to the annual interest rate r by the formula $I = 1000r$.
2. The area A of a circle is related to its radius r by the formula $A = \pi r^2$.

Not all relations have simple mathematical formulas. For instance, people commonly match up NFL starting quarterbacks with touchdown passes and hours of the day with temperature. In cases 1 and 2 above, however, there is some relation that matches each item from one set with exactly one item from a different set. Such a relation is called a **function**.

Definition of a Function

A **function** f from a set A to a set B is a relation that assigns to each element x in the set A exactly one element y in the set B. The set A is the **domain** (or set of inputs) of the function f, and the set B contains the **range** (or set of outputs).

To help understand this definition, look at the function in Figure P.37.

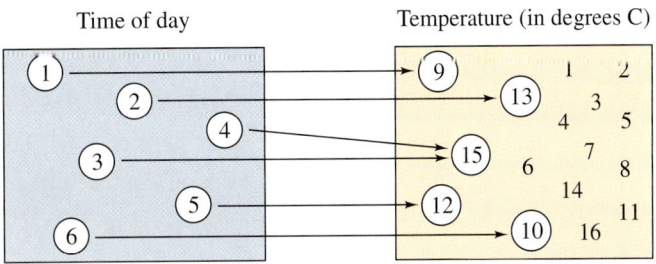

Set A is the domain.
Inputs: 1, 2, 3, 4, 5, 6

Set B contains the range.
Outputs: 9, 10, 12, 13, 15

FIGURE P.37

This function can be represented by the following ordered pairs.

$$\{(1, 9°), (2, 13°), (3, 15°), (4, 15°), (5, 12°), (6, 10°)\}$$

In each ordered pair, the first coordinate is the input and the second coordinate is the output. In this example, note the following characteristics of a function.

1. Each element in A must be matched with an element of B.
2. Some elements in B may not be matched with any element in A.
3. Two or more elements of A may be matched with the same element of B.

The converse of the third statement is not true. That is, an element of A (the domain) cannot be matched with two different elements of B.

Functions are commonly represented in four ways.

1. *Verbally* by a sentence that describes how the input variable is related to the output variable
2. *Numerically* by a table or a list of ordered pairs that matches input values with output values
3. *Graphically* by points on a graph in a coordinate plane in which the input values are represented by the horizontal axis and the output values are represented by the vertical axis
4. *Algebraically* by an equation in two variables

In the following example, you are asked to decide whether or not the given relation is a function. To do this, you must decide whether each input value is matched with exactly one output value. If any input value is matched with two or more output values, the relation is not a function.

Example 1 ▶ Testing for Functions

Decide whether the description represents y as a function of x.

a. The input value x is the number of representatives from a state, and the output value y is the number of senators.

b.

Input x	2	2	3	4	5
Output y	11	10	8	5	1

c.

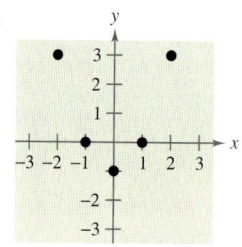

FIGURE P.38

Solution

a. This verbal description *does* describe y as a function of x. Regardless of the value of x, the value of y is always 2. Such functions are called *constant functions*.

b. This table *does not* describe y as a function of x. The input value 2 is matched with two different y-values.

c. The graph in Figure P.38 *does* describe y as a function of x. No input value is matched with two output values.

Representing functions by sets of ordered pairs is common in *discrete mathematics*. In algebra, however, it is more common to represent functions by equations or formulas involving two variables. For instance, the equation

$$y = x^2 \qquad \text{\textit{y} is a function of \textit{x}.}$$

represents the variable y as a function of the variable x. In this equation, x is the **independent variable** and y is the **dependent variable**. The domain of the function is the set of all values taken on by the independent variable x, and the range of the function is the set of all values taken on by the dependent variable y.

Example 2 ▶ Testing for Functions Represented Algebraically

Which of the equations represent(s) y as a function of x?

a. $x^2 + y = 1$
b. $-x + y^2 = 1$

Solution

To determine whether y is a function of x, try to solve for y in terms of x.

a. Solving for y yields

$$x^2 + y = 1 \qquad \text{Write original equation.}$$
$$y = 1 - x^2. \qquad \text{Solve for } y.$$

To each value of x there corresponds exactly one value of y. So, y is a function of x.

b. Solving for y yields

$$-x + y^2 = 1 \qquad \text{Write original equation.}$$
$$y^2 = 1 + x \qquad \text{Add } x \text{ to each side.}$$
$$y = \pm\sqrt{1 + x}. \qquad \text{Solve for } y.$$

The $\pm$ indicates that to a given value of x there correspond two values of y. So, y is not a function of x.

Historical Note

Leonhard Euler (1707–1783), a Swiss mathematician, is considered to have been the most prolific and productive mathematician in history. One of his greatest influences on mathematics was his use of symbols, or notation. The function notation $y = f(x)$ was introduced by Euler.

Function Notation

When an equation is used to represent a function, it is convenient to name the function so that it can be referenced easily. For example, you know that the equation $y = 1 - x^2$ describes y as a function of x. Suppose you give this function the name "f." Then you can use the following **function notation**.

Input	Output	Equation
x	$f(x)$	$f(x) = 1 - x^2$

The symbol $f(x)$ is read as *the value of f at x* or simply *f of x*. The symbol $f(x)$ corresponds to the y-value for a given x. So, you can write $y = f(x)$. Keep in mind that f is the *name* of the function, whereas $f(x)$ is the *value* of the function at x. For instance, the function

$$f(x) = 3 - 2x$$

has *function values* denoted by $f(-1), f(0), f(2)$, and so on. To find these values, substitute the specified input values into the given equation.

For $x = -1$, $\quad f(-1) = 3 - 2(-1) = 3 + 2 = 5.$
For $x = 0$, $\quad f(0) = 3 - 2(0) = 3 - 0 = 3.$
For $x = 2$, $\quad f(2) = 3 - 2(2) = 3 - 4 = -1.$

Although f is often used as a convenient function name and x is often used as the independent variable, you can use other letters. For instance,

$$f(x) = x^2 - 4x + 7, \quad f(t) = t^2 - 4t + 7, \quad \text{and} \quad g(s) = s^2 - 4s + 7$$

all define the same function. In fact, the role of the independent variable is that of a "placeholder." Consequently, the function could be described by

$$f(\;\;\;) = (\;\;\;)^2 - 4(\;\;\;) + 7.$$

Example 3 ▶ Evaluating a Function

Let $g(x) = -x^2 + 4x + 1$ and find

a. $g(2)$ **b.** $g(t)$ **c.** $g(x + 2)$.

Solution

a. Replacing x with 2 in $g(x) = -x^2 + 4x + 1$ yields the following.

$$g(2) = -(2)^2 + 4(2) + 1 = -4 + 8 + 1 = 5$$

b. Replacing x with t yields the following.

$$g(t) = -(t)^2 + 4(t) + 1 = -t^2 + 4t + 1$$

c. Replacing x with $x + 2$ yields the following.

$$\begin{aligned} g(x + 2) &= -(x + 2)^2 + 4(x + 2) + 1 \\ &= -(x^2 + 4x + 4) + 4x + 8 + 1 \\ &= -x^2 - 4x - 4 + 4x + 8 + 1 \\ &= -x^2 + 5 \end{aligned}$$

> **STUDY TIP**
>
> In Example 3, note that $g(x + 2)$ is not equal to $g(x) + g(2)$. In general, $g(u + v) \neq g(u) + g(v)$.

A function defined by two or more equations over a specified domain is called a **piecewise-defined function.**

Example 4 ▶ A Piecewise-Defined Function

Evaluate the function when $x = -1, 0,$ and 1.

$$f(x) = \begin{cases} x^2 + 1, & x < 0 \\ x - 1, & x \geq 0 \end{cases}$$

Solution

Because $x = -1$ is less than 0, use $f(x) = x^2 + 1$ to obtain

$$f(-1) = (-1)^2 + 1 = 2.$$

For $x = 0$, use $f(x) = x - 1$ to obtain

$$f(0) = (0) - 1 = -1.$$

For $x = 1$, use $f(x) = x - 1$ to obtain

$$f(1) = (1) - 1 = 0.$$

The Domain of a Function

> **Technology**
>
> Use a graphing utility to graph $y = \sqrt{4 - x^2}$. What is the domain of this function? Then graph $y = \sqrt{x^2 - 4}$. What is the domain of this function? Do the domains of these two functions overlap? If so, for what values?

The domain of a function can be described explicitly or it can be *implied* by the expression used to define the function. The **implied domain** is the set of all real numbers for which the expression is defined. For instance, the function

$$f(x) = \frac{1}{x^2 - 4} \quad \text{Domain excludes } x\text{-values that result in division by zero.}$$

has an implied domain that consists of all real x other than $x = \pm 2$. These two values are excluded from the domain because division by zero is undefined. Another common type of implied domain is that used to avoid even roots of negative numbers. For example, the function

$$f(x) = \sqrt{x} \quad \text{Domain excludes } x\text{-values that result in even roots of negative numbers.}$$

is defined only for $x \geq 0$. So, its implied domain is the interval $[0, \infty)$. In general, the domain of a function *excludes* values that would cause division by zero *or* that would result in the even root of a negative number.

Example 5 ▶ Finding the Domain of a Function

Find the domain of each function.

a. f: $\{(-3, 0), (-1, 4), (0, 2), (2, 2), (4, -1)\}$ **b.** $g(x) = \dfrac{1}{x + 5}$

c. Volume of a sphere: $V = \frac{4}{3}\pi r^3$ **d.** $h(x) = \sqrt{4 - x^2}$

Solution

a. The domain of f consists of all first coordinates in the set of ordered pairs.

$$\text{Domain} = \{-3, -1, 0, 2, 4\}$$

b. Excluding x-values that yield zero in the denominator, the domain of g is the set of all real numbers $x \neq -5$.

c. Because this function represents the volume of a sphere, the values of the radius r must be positive. So, the domain is the set of all real numbers r such that $r > 0$.

d. This function is defined only for x-values for which

$$4 - x^2 \geq 0.$$

By solving this inequality, you can conclude that $-2 \leq x \leq 2$. So, the domain is the interval $[-2, 2]$.

In Example 5(c), note that the domain of a function may be implied by the physical context. For instance, from the equation $V = \frac{4}{3}\pi r^3$, you would have no reason to restrict r to positive values, but the physical context implies that a sphere cannot have a negative radius.

FIGURE P.39

Applications

Example 6 ▶ The Dimensions of a Container

You work in the marketing department of a soft-drink company and are experimenting with a new soft-drink can that is slightly narrower and taller than a standard can. For your experimental can, the ratio of the height to the radius is 4, as shown in Figure P.39.

a. Express the volume of the can as a function of the radius r.

b. Express the volume of the can as a function of the height h.

Solution

a. $V(r) = \pi r^2 h = \pi r^2 (4r) = 4\pi r^3$ Write V as a function of r.

b. $V(h) = \pi \left(\dfrac{h}{4}\right)^2 h = \dfrac{\pi h^3}{16}$ Write V as a function of h.

Example 7 ▶ The Path of a Baseball

A baseball is hit at a point 3 feet above ground at a velocity of 100 feet per second and an angle of 45°. The path of the baseball is given by the function

$$f(x) = -0.0032x^2 + x + 3$$

where y and x are measured in feet, as shown in Figure P.40. Will the baseball clear a 10-foot fence located 300 feet from home plate?

Solution

When $x = 300$, the height of the baseball is

$$f(300) = -0.0032(300)^2 + 300 + 3$$
$$= 15 \text{ feet.}$$

So, the ball will clear the fence.

FIGURE P.40

In the equation in Example 7, the height of the baseball is a function of the distance from home plate.

Example 8 ▶ Direct Mail Advertising

The money C (in billions of dollars) spent for direct mail advertising in the United States increased in a linear pattern from 1990 to 1992, as shown in Figure P.41. Then, in 1993, the money spent took a jump and, until 1996, increased in a *different* linear pattern. These two patterns can be approximated by the function

$$C(t) = \begin{cases} 23.40 + 1.01t, & 0 \leq t \leq 2 \\ 19.85 + 2.50t, & 3 \leq t \leq 6 \end{cases}$$

where $t = 0$ represents 1990. Use this function to approximate the total amount spent for direct mail advertising between 1990 and 1996. *(Source: McCann-Erickson)*

FIGURE P.41

Solution

From 1990 to 1992, use the formula $C(t) = 23.40 + 1.01t$.

$$\underbrace{\$23.40,}_{1990} \underbrace{\$24.41,}_{1991} \underbrace{\$25.42}_{1992}$$

From 1993 to 1996, use the formula $C(t) = 19.85 + 2.50t$.

$$\underbrace{\$27.35,}_{1993} \underbrace{\$29.85,}_{1994} \underbrace{\$32.35,}_{1995} \underbrace{\$34.85}_{1996}$$

The total of these seven amounts is $197.63, which implies that the total amount spent was approximately $197,630,000,000.

One of the basic definitions in calculus employs the ratio

$$\frac{f(x+h) - f(x)}{h}, \quad h \neq 0.$$

This ratio is called a **difference quotient**, as illustrated in Example 9.

Example 9 ▶ Evaluating a Difference Quotient

For $f(x) = x^2 - 4x + 7$, find $\dfrac{f(x+h) - f(x)}{h}$.

Solution

$$\frac{f(x+h) - f(x)}{h} = \frac{[(x+h)^2 - 4(x+h) + 7] - (x^2 - 4x + 7)}{h}$$

$$= \frac{x^2 + 2xh + h^2 - 4x - 4h + 7 - x^2 + 4x - 7}{h}$$

$$= \frac{2xh + h^2 - 4h}{h}$$

$$= \frac{h(2x + h - 4)}{h}$$

$$= 2x + h - 4, \quad h \neq 0$$

The symbol indicates an example or exercise that highlights algebraic techniques specifically used in calculus.

Summary of Function Terminology

Function: A **function** is a relationship between two variables such that to each value of the independent variable there corresponds exactly one value of the dependent variable.

Function Notation: $y = f(x)$

 f is the *name* of the function.

 y is the **dependent variable.**

 x is the **independent variable.**

 $f(x)$ is the *value of the function at x.*

Domain: The **domain** of a function is the set of all values (inputs) of the independent variable for which the function is defined. If x is in the domain of f, f is said to be *defined* at x. If x is not in the domain of f, f is said to be *undefined* at x.

Range: The **range** of a function is the set of all values (outputs) assumed by the dependent variable (that is, the set of all function values).

Implied Domain: If f is defined by an algebraic expression and the domain is not specified, the **implied domain** consists of all real numbers for which the expression is defined.

Writing ABOUT MATHEMATICS

Modeling with Piecewise-Defined Functions The table below shows the monthly revenue y (in thousands of dollars) for one year of a landscaping business, with $x = 1$ representing January.

x	1	2	3	4	5	6
y	5.2	5.6	6.6	8.3	11.5	15.8

x	7	8	9	10	11	12
y	12.8	10.1	8.6	6.9	4.5	2.7

A mathematical model that represents these data is

$$f(x) = \begin{cases} -1.97x + 26.33 \\ 0.5x^2 - 1.47x + 6.3 \end{cases}.$$

What is the domain of each part of the piecewise-defined function? How can you tell? Explain your reasoning.

Find $f(5)$ and $f(11)$, and interpret your results in the context of the problem. How do these model values compare with the actual data values?

P.5 Exercises

In Exercises 1–4, is the relationship a function?

1. Domain   Range

2. Domain   Range

3. Domain   Range

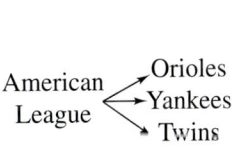

4. Domain   Range
   (Year)   (North Atlantic tropical storms and hurricanes)

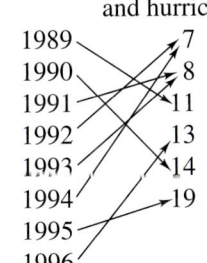

In Exercises 5–8, does the table describe a function? Explain your reasoning.

5.
Input value	−2	−1	0	1	2
Output value	−8	−1	0	1	8

6.
Input value	0	1	2	1	0
Output value	−4	−2	0	2	4

7.
Input value	10	7	4	7	10
Output value	3	6	9	12	15

8.
Input value	0	3	9	12	15
Output value	3	3	3	3	3

In Exercises 9 and 10, which sets of ordered pairs represent functions from *A* to *B*? Explain.

9. $A = \{0, 1, 2, 3\}$ and $B = \{-2, -1, 0, 1, 2\}$
   (a) $\{(0, 1), (1, -2), (2, 0), (3, 2)\}$
   (b) $\{(0, -1), (2, 2), (1, -2), (3, 0), (1, 1)\}$
   (c) $\{(0, 0), (1, 0), (2, 0), (3, 0)\}$
   (d) $\{(0, 2), (3, 0), (1, 1)\}$

10. $A = \{a, b, c\}$ and $B = \{0, 1, 2, 3\}$
    (a) $\{(a, 1), (c, 2), (c, 3), (b, 3)\}$
    (b) $\{(a, 1), (b, 2), (c, 3)\}$
    (c) $\{(1, a), (0, a), (2, c), (3, b)\}$
    (d) $\{(c, 0), (b, 0), (a, 3)\}$

Circulation of Newspapers   In Exercises 11 and 12, use the graph, which shows the circulation (in millions) of daily newspapers in the United States. (Source: Editor & Publisher Company)

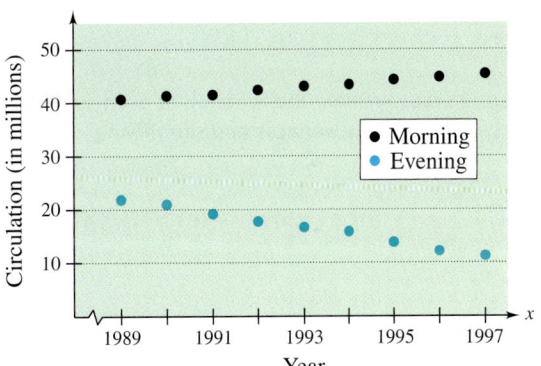

11. Is the circulation of morning newspapers a function of the year? Is the circulation of evening newspapers a function of the year? Explain.

12. Let $f(x)$ represent the circulation of evening newspapers in year x. Find $f(1994)$.

In Exercises 13–22, determine whether the equation represents *y* as a function of *x*.

13. $x^2 + y^2 = 4$
14. $x = y^2$
15. $x^2 + y = 4$
16. $x + y^2 = 4$

17. $2x + 3y = 4$
18. $(x - 2)^2 + y^2 = 4$
19. $y^2 = x^2 - 1$
20. $y = \sqrt{x + 5}$
21. $y = |4 - x|$
22. $|y| = 4 - x$

In Exercises 23 and 24, fill in the blanks using the specified function and the given values of the independent variable.

23. $f(s) = \dfrac{1}{s + 1}$

(a) $f(4) = \dfrac{1}{() + 1}$

(b) $f(0) = \dfrac{1}{() + 1}$

(c) $f(4x) = \dfrac{1}{() + 1}$

(d) $f(x + c) = \dfrac{1}{() + 1}$

24. $g(x) = x^2 - 2x$

(a) $g(2) = ()^2 - 2()$

(b) $g(-3) = ()^2 - 2()$

(c) $g(t + 1) = ()^2 - 2()$

(d) $g(x + c) = ()^2 - 2()$

In Exercises 25–36, evaluate the function at each specified value of the independent variable and simplify.

25. $f(x) = 2x - 3$
  (a) $f(1)$ (b) $f(-3)$ (c) $f(x - 1)$

26. $g(y) = 7 - 3y$
  (a) $g(0)$ (b) $g\left(\tfrac{7}{3}\right)$ (c) $g(s + 2)$

27. $V(r) = \tfrac{4}{3}\pi r^3$
  (a) $V(3)$ (b) $V\left(\tfrac{3}{2}\right)$ (c) $V(2r)$

28. $h(t) = t^2 - 2t$
  (a) $h(2)$ (b) $h(1.5)$ (c) $h(x + 2)$

29. $f(y) = 3 - \sqrt{y}$
  (a) $f(4)$ (b) $f(0.25)$ (c) $f(4x^2)$

30. $f(x) = \sqrt{x + 8} + 2$
  (a) $f(-8)$ (b) $f(1)$ (c) $f(x - 8)$

31. $q(x) = \dfrac{1}{x^2 - 9}$
  (a) $q(0)$ (b) $q(3)$ (c) $q(y + 3)$

32. $q(t) = \dfrac{2t^2 + 3}{t^2}$
  (a) $q(2)$ (b) $q(0)$ (c) $q(-x)$

33. $f(x) = \dfrac{|x|}{x}$
  (a) $f(2)$ (b) $f(-2)$ (c) $f(x - 1)$

34. $f(x) = |x| + 4$
  (a) $f(2)$ (b) $f(-2)$ (c) $f(x^2)$

35. $f(x) = \begin{cases} 2x + 1, & x < 0 \\ 2x + 2, & x \geq 0 \end{cases}$
  (a) $f(-1)$ (b) $f(0)$ (c) $f(2)$

36. $f(x) = \begin{cases} x^2 + 2, & x \leq 1 \\ 2x^2 + 2, & x > 1 \end{cases}$
  (a) $f(-2)$ (b) $f(1)$ (c) $f(2)$

In Exercises 37–42, complete the table.

37. $f(x) = x^2 - 3$

x	-2	-1	0	1	2
$f(x)$					

38. $g(x) = \sqrt{x - 3}$

x	3	4	5	6	7
$g(x)$					

39. $h(t) = \tfrac{1}{2}|t + 3|$

t	-5	-4	-3	-2	-1
$h(t)$					

40. $f(s) = \dfrac{|s - 2|}{s - 2}$

s	0	1	$\tfrac{3}{2}$	$\tfrac{5}{2}$	4
$f(s)$					

41. $f(x) = \begin{cases} -\tfrac{1}{2}x + 4, & x \leq 0 \\ (x - 2)^2, & x > 0 \end{cases}$

x	-2	-1	0	1	2
$f(x)$					

42. $h(x) = \begin{cases} 9 - x^2, & x < 3 \\ x - 3, & x \geq 3 \end{cases}$

x	1	2	3	4	5
$h(x)$					

In Exercises 43–50, find all real values of x such that $f(x) = 0$.

43. $f(x) = 15 - 3x$
44. $f(x) = 5x + 1$
45. $f(x) = \dfrac{3x - 4}{5}$
46. $f(x) = \dfrac{12 - x^2}{5}$
47. $f(x) = x^2 - 9$
48. $f(x) = x^2 - 8x + 15$
49. $f(x) = x^3 - x$
50. $f(x) = x^3 - x^2 - 4x + 4$

In Exercises 51–54, find the value(s) of x for which $f(x) = g(x)$.

51. $f(x) = x^2$, $g(x) = x + 2$
52. $f(x) = x^2 + 2x + 1$, $g(x) = 3x + 3$
53. $f(x) = \sqrt{3x} + 1$, $g(x) = x + 1$
54. $f(x) = x^4 - 2x^2$, $g(x) = 2x^2$

In Exercises 55–68, find the domain of the function.

55. $f(x) = 5x^2 + 2x - 1$
56. $g(x) = 1 - 2x^2$
57. $h(t) = \dfrac{4}{t}$
58. $s(y) = \dfrac{3y}{y + 5}$
59. $g(y) = \sqrt{y - 10}$
60. $f(t) = \sqrt[3]{t + 4}$
61. $f(x) = \sqrt[4]{1 - x^2}$
62. $f(x) = \sqrt[4]{x^2 + 3x}$
63. $g(x) = \dfrac{1}{x} - \dfrac{3}{x + 2}$
64. $h(x) = \dfrac{10}{x^2 - 2x}$
65. $f(s) = \dfrac{\sqrt{s - 1}}{s - 4}$
66. $f(x) = \dfrac{\sqrt{x + 6}}{6 + x}$
67. $f(x) = \dfrac{\sqrt[3]{x - 4}}{x}$
68. $f(x) = \dfrac{x - 5}{x^2 - 9}$

In Exercises 69–72, assume that the domain of f is the set $A = \{-2, -1, 0, 1, 2\}$. Determine the set of ordered pairs that represents the function f.

69. $f(x) = x^2$
70. $f(x) = \dfrac{2x}{x^2 + 1}$
71. $f(x) = \sqrt{x + 2}$
72. $f(x) = |x + 1|$

Exploration In Exercises 73–76, determine the function from

$f(x) = cx$, $g(x) = cx^2$, $h(x) = c\sqrt{|x|}$, and $r(x) = \dfrac{c}{x}$

and the value of the constant c that will make the function fit the data in the table.

73.

x	-4	-1	0	1	4
y	-32	-2	0	-2	-32

74.

x	-4	-1	0	1	4
y	-1	$-\frac{1}{4}$	0	$\frac{1}{4}$	1

75.

x	-4	-1	0	1	4
y	-8	-32	Undef.	32	8

76.

x	-4	-1	0	1	4
y	6	3	0	3	6

In Exercises 77–84, find the difference quotient and simplify your answer.

77. $f(x) = x^2 - x + 1$, $\dfrac{f(2 + h) - f(2)}{h}$, $h \neq 0$

78. $f(x) = 5x - x^2$, $\dfrac{f(5 + h) - f(5)}{h}$, $h \neq 0$

79. $f(x) = x^3$, $\dfrac{f(x + c) - f(x)}{c}$, $c \neq 0$

80. $f(x) = 2x$, $\dfrac{f(x + c) - f(x)}{c}$, $c \neq 0$

81. $g(x) = 3x - 1$, $\dfrac{g(x) - g(3)}{x - 3}$, $x \neq 3$

82. $f(t) = \dfrac{1}{t}$, $\dfrac{f(t) - f(1)}{t - 1}$, $t \neq 1$

83. $f(x) = \sqrt{5x}$, $\dfrac{f(x) - f(5)}{x - 5}$, $x \neq 5$

84. $f(x) = x^{2/3} + 1$, $\dfrac{f(x) - f(8)}{x - 8}$, $x \neq 8$

85. *Geometry* Express the area A of a square as a function of its perimeter P.

86. *Geometry* Express the area A of a circle as a function of its circumference C.

The symbol indicates an example or exercise that highlights algebraic techniques specifically used in calculus.

87. *Geometry* Express the area A of an isosceles triangle with a height of 8 inches and a base of b inches as a function of the length s of one of its two equal sides.

88. *Geometry* Express the area A of an equilateral triangle as a function of the length s of its sides.

89. *Maximum Volume* An open box of maximum volume is to be made from a square piece of material 24 centimeters on a side by cutting equal squares from the corners and turning up the sides (see figure).

(a) Complete six rows of a table. (The first two rows are shown.) Use the result to estimate the maximum volume.

Height x	Width	Volume V
1	$24 - 2(1)$	$1[24 - 2(1)]^2 = 484$
2	$24 - 2(2)$	$2[24 - 2(2)]^2 = 800$

(b) Plot the points (x, V). Is V a function of x?

(c) If V is a function of x, write the function and determine its domain.

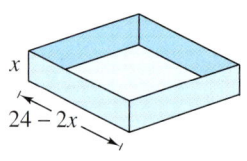

90. *Maximum Profit* The cost per unit in the production of a certain radio model is $60. The manufacturer charges $90 per unit for orders of 100 or less. To encourage large orders, the manufacturer reduces the charge by $0.15 per radio for each unit ordered in excess of 100 (for example, there would be a charge of $87 per radio for an order size of 120).

(a) Complete six rows of the table. Use the result to estimate the maximum profit.

Units x	Price p	Profit P
110	$90 - 10(0.15)$	$xp - 110(60)$
120	$90 - 20(0.15)$	$xp - 120(60)$

(b) Plot the points (x, P). Is P a function of x?

(c) If P is a function of x, write the function and determine its domain.

91. *Geometry* A right triangle is formed in the first quadrant by the x- and y-axes and a line through the point $(2, 1)$ (see figure). Write the area A of the triangle as a function of x, and determine the domain of the function.

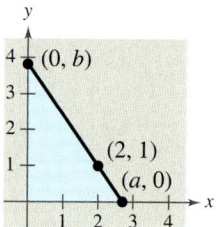

92. *Geometry* A rectangle is bounded by the x-axis and the semicircle $y = \sqrt{36 - x^2}$ (see figure). Write the area A of the rectangle as a function of x, and determine the domain of the function.

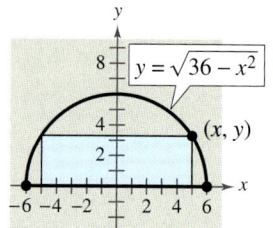

93. *Average Price* The average price p (in thousands of dollars) of a new mobile home in the United States from 1974 to 1997 (see figure) can be approximated by the model

$$p(t) = \begin{cases} 17.27 + 1.036t, & -6 \leq t \leq 11 \\ -4.807 + 2.882t - 0.011t^2, & 12 \leq t \leq 17 \end{cases}$$

where $t = 0$ represents 1980. Use this model to find the average prices of a mobile home in 1978, 1988, 1993, and 1997. (Source: U.S. Bureau of Census, Construction Reports)

94. *Postal Regulations* A rectangular package to be sent by the U.S. Postal Service can have a maximum combined length and girth (perimeter of a cross section) of 108 inches (see figure).

(a) Write the volume V of the package as a function of x.

(b) What is the domain of the function?

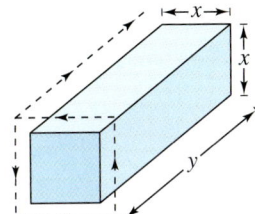

(c) Use a graphing utility to graph your function. Be sure to use the appropriate window setting.

(d) What dimensions will maximize the volume of the package? Explain your answer.

95. *Business* A company produces a product for which the variable cost is $12.30 per unit and the fixed costs are $98,000. The product sells for $17.98. Let x be the number of units produced and sold.

(a) The total cost for a business is the sum of the variable cost and the fixed costs. Write the total cost C as a function of the number of units produced.

(b) Write the revenue R as a function of the number of units sold.

(c) Write the profit P as a function of the number of units sold. (*Note*: $P = R - C$.)

96. *Total Average Cost* The inventor of a new game believes that the variable cost for producing the game is $0.95 per unit and the fixed costs are $6000. The inventor sells each game for $1.69. Let x be the number of games sold.

(a) The total cost for a business is the sum of the variable cost and the fixed costs. Write the total cost C as a function of the number of games sold.

(b) Write the average cost per unit $\overline{C} = C/x$ as a function of x.

97. *Transportation* For groups of 80 or more people, a charter bus company determines the rate per person according to the formula

$$\text{Rate} = 8 - 0.05(n - 80), \quad n \geq 80$$

where the rate is given in dollars and n is the number of people.

(a) Express the revenue R for the bus company as a function of n.

(b) Use the function in part (a) to complete the table. What can you conclude?

n	90	100	110	120	130	140	150
$R(n)$							

98. *Physics* The force F (in tons) of water against the face of a dam is estimated by the function

$$F(y) = 149.76\sqrt{10}\,y^{5/2}$$

where y is the depth of the water in feet.

(a) Complete the table. What can you conclude from the table?

y	5	10	20	30	40
$F(y)$					

(b) Use the table to approximate the depth at which the force against the dam is 1,000,000 tons. How could you find a better estimate?

99. *Height of a Balloon* A balloon carrying a transmitter ascends vertically from a point 3000 feet from the receiving station.

(a) Draw a diagram that gives a visual representation of the problem. Let h represent the height of the balloon and let d represent the distance between the balloon and the receiving station.

(b) Express the height of the balloon as a function of d. What is the domain of the function?

100. Lynx Population A study was done on the lynx population of the Yukon Territory in Canada. The graph shows the lynx population from 1988 through 1995 in a 350-square-kilometer region of the Yukon Territory. Let $f(t)$ represent the number of lynx in year t. *(Source: Kluane Boreal Forest Ecosystem Project)*

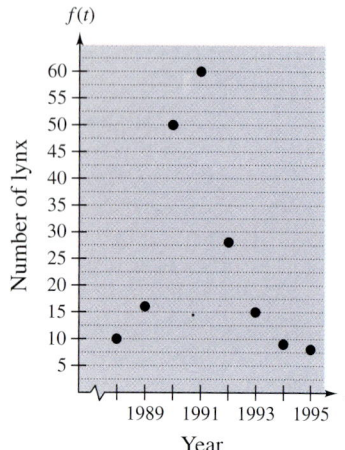

(a) Find
$$\frac{f(1994) - f(1991)}{1994 - 1991}$$
and interpret the result in the context of the problem.

(b) An approximate formula for the function $f(t)$ is
$$N(x) = \frac{434x + 4387}{45x^2 - 55x + 100}$$
where N is the number of lynx and x is time in years, with $x = 0$ corresponding to 1990. Complete the table and compare the result with the data.

x	-2	-1	0	1
N				

x	2	3	4	5
N				

101. Path of a Ball The height y (in feet) of a baseball thrown by a child is
$$y = -\frac{1}{10}x^2 + 3x + 6,$$
where x is the horizontal distance (in feet) from where the ball was thrown. Will the ball fly over the head of another child 30 feet away trying to catch the ball? (Assume that the child who is trying to catch the ball holds a baseball glove at a height of 5 feet.)

Synthesis

True or False? In Exercises 102 and 103, determine whether the statement is true or false. Justify your answer.

102. The domain of the function $f(x) = x^4 - 1$ is $(-\infty, \infty)$, and the range of $f(x)$ is $(0, \infty)$.

103. The set of ordered pairs $\{(-8, -2), (-6, 0), (-4, 0), (-2, 2), (0, 4), (2, -2)\}$ represents a function.

104. Does the relationship shown in the figure represent a function from set A to set B? Explain.

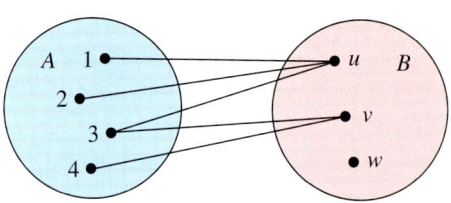

105. *Think About It* In your own words, explain the meanings of *domain* and *range*.

106. *Think About It* Describe an advantage of function notation.

P.6 Analyzing Graphs of Functions

▶ What you should learn
- How to use the Vertical Line Test for functions
- How to find the zeros of functions
- How to determine intervals on which functions are increasing or decreasing
- How to identify and graph linear functions
- How to identify and graph step functions and other piecewise-defined functions
- How to identify even and odd functions

▶ Why you should learn it
Graphs of functions can help you visualize relationships between variables in real life. For instance, Exercise 103 on page 77 shows how the graph of a step function visually represents the cost of a telephone call.

The Graph of a Function

In Section P.5, you studied functions from an algebraic point of view. In this section, you will study functions from a graphical perspective.

The **graph of a function** f is the collection of ordered pairs $(x, f(x))$ such that x is in the domain of f. As you study this section, remember that

$x =$ the directed distance from the y-axis

$f(x) =$ the directed distance from the x-axis

as shown in Figure P.42.

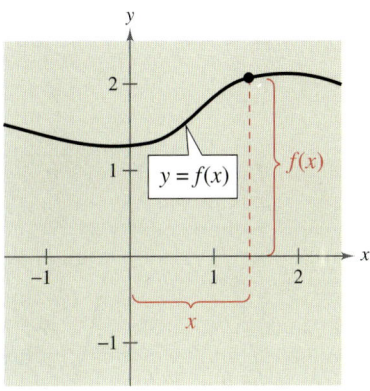

FIGURE P.42

Example 1 ▶ Finding the Domain and Range of a Function

Use the graph of the function f, shown in Figure P.43, to find (a) the domain of f, (b) the function values $f(-1)$ and $f(2)$, and (c) the range of f.

Solution

a. The closed dot at $(-1, -5)$ indicates that $x = -1$ is in the domain of f, whereas the open dot at $(4, 0)$ indicates $x = 4$ is not in the domain. So, the domain of f is all x in the interval $[-1, 4)$.

b. Because $(-1, -5)$ is a point on the graph of f, it follows that $f(-1) = -5$. Similarly, because $(2, 4)$ is a point on the graph of f, it follows that $f(2) = 4$.

c. Because the graph does not extend below $f(-1) = -5$ or above $f(2) = 4$, the range of f is the interval $[-5, 4]$.

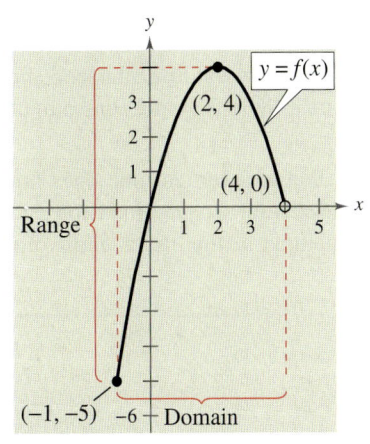

FIGURE P.43

The use of dots (open or closed) at the extreme left and right points of a graph indicates that the graph does not extend beyond these points. If no such dots are shown, assume that the graph extends beyond these points.

By the definition of a function, at most one *y*-value corresponds to a given *x*-value. This means that the graph of a function cannot have two or more different points with the same *x*-coordinate, and no two points on the graph of a function can be vertically above and below each other. It follows, then, that a vertical line can intersect the graph of a function at most once. This observation provides a convenient visual test called the **Vertical Line Test** for functions.

> ### Vertical Line Test for Functions
> A set of points in a coordinate plane is the graph of *y* as a function of *x* if and only if no vertical line intersects the graph at more than one point.

Example 2 ▶ Vertical Line Test for Functions

Use the Vertical Line Test to decide whether the graphs in Figure P.44 represent *y* as a function of *x*.

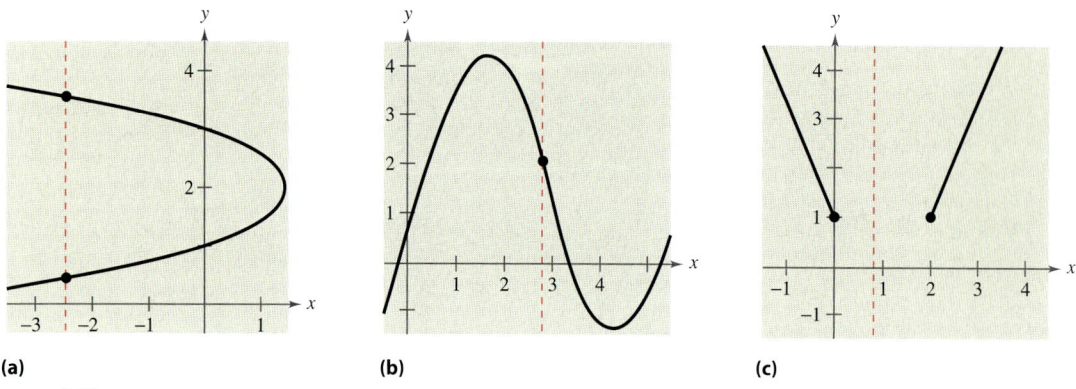

(a)   (b)   (c)

FIGURE P.44

Solution

a. This *is not* a graph of *y* as a function of *x* because you can find a vertical line that intersects the graph twice. That is, for a particular input *x*, there is more than one output *y*.

b. This *is* a graph of *y* as a function of *x* because every vertical line intersects the graph at most once. That is, for a particular input *x*, there is at most one output *y*.

c. This *is* a graph of *y* as a function of *x*. (Note that if a vertical line does not intersect the graph, it simply means that the function is undefined for that particular value of *x*.) That is, for a particular input *x*, there is at most one output *y*.

Zeros of a Function

If the graph of a function of x has an x-intercept at $(a, 0)$, then a is a **zero** of the function.

Zeros of a Function

The **zeros of a function** f of x are the x-values for which $f(x) = 0$.

Example 3 ▶ Finding Zeros of a Function

Find the zeros of each function.

a. $f(x) = 3x^2 + x - 10$ **b.** $g(x) = \sqrt{10 - x^2}$ **c.** $h(t) = \dfrac{2t - 3}{t + 5}$

Solution

To find the zeros of a function, set the function equal to zero and solve for the independent variable.

a.
$3x^2 + x - 10 = 0$	Set $f(x)$ equal to 0.
$(3x - 5)(x + 2) = 0$	Factor.
$3x - 5 = 0 \implies x = \tfrac{5}{3}$	Set 1st factor equal to 0.
$x + 2 = 0 \implies x = -2$	Set 2nd factor equal to 0.

The zeros of f are $\tfrac{5}{3}$ and -2. In Figure P.45(a), note that the graph of f has $\left(\tfrac{5}{3}, 0\right)$ and $(-2, 0)$ as its x-intercepts.

b.
$\sqrt{10 - x^2} = 0$	Set $g(x)$ equal to 0.
$10 - x^2 = 0$	Square each side.
$10 = x^2$	Add x^2 to each side.
$\pm\sqrt{10} = x$	Extract square root.

The zeros of g are $-\sqrt{10}$ and $\sqrt{10}$. In Figure P.45(b), note that the graph of g has $\left(-\sqrt{10}, 0\right)$ and $\left(\sqrt{10}, 0\right)$ as its x-intercepts.

c.
$\dfrac{2t - 3}{t + 5} = 0$	Set $h(t)$ equal to 0.
$2t - 3 = 0$	Set numerator equal to 0.
$2t = 3$	Add 3 to each side.
$t = \dfrac{3}{2}$	Divide each side by 2.

The zero of h is $\tfrac{3}{2}$. In Figure P.45(c), note that the graph of h has $\left(\tfrac{3}{2}, 0\right)$ as its x-intercept.

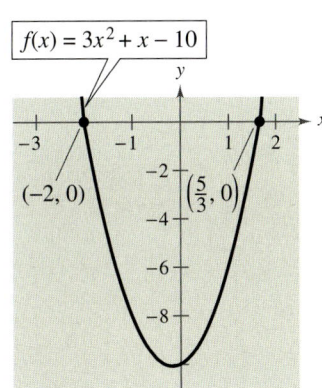

(a) Zeros of f: $x = -2, x = \tfrac{5}{3}$

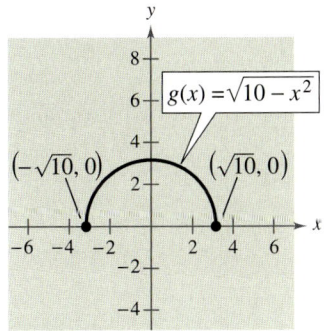

(b) Zeros of g: $x = \pm\sqrt{10}$

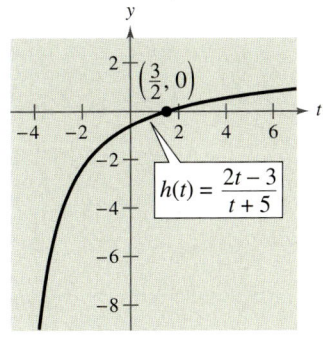

(c) Zero of h: $t = \tfrac{3}{2}$

FIGURE P.45

Increasing and Decreasing Functions

The more you know about the graph of a function, the more you know about the function itself. Consider the graph shown in Figure P.46. As you move from *left to right*, this graph decreases, then is constant, and then increases.

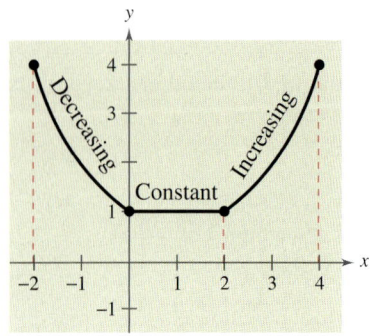

FIGURE P.46

Increasing, Decreasing, and Constant Functions

A function f is **increasing** on an interval if, for any x_1 and x_2 in the interval, $x_1 < x_2$ implies $f(x_1) < f(x_2)$.

A function f is **decreasing** on an interval if, for any x_1 and x_2 in the interval, $x_1 < x_2$ implies $f(x_1) > f(x_2)$.

A function f is **constant** on an interval if, for any x_1 and x_2 in the interval, $f(x_1) = f(x_2)$.

Example 4 ▶ Increasing and Decreasing Functions

In Figure P.47, use the graphs to describe the increasing or decreasing behavior of each function.

Solution

a. This function is increasing over the entire real line.

b. This function is increasing on the interval $(-\infty, -1)$, decreasing on the interval $(-1, 1)$, and increasing on the interval $(1, \infty)$.

c. This function is increasing on the interval $(-\infty, 0)$, constant on the interval $(0, 2)$, and decreasing on the interval $(2, \infty)$.

(a)

(b)

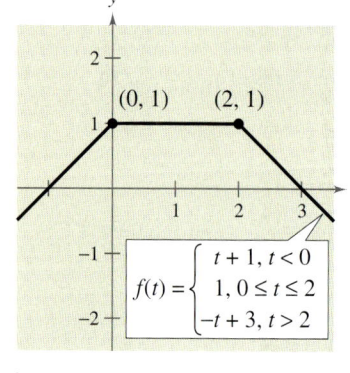

(c)

FIGURE P.47

To decide whether a function is increasing, decreasing, or constant, you can evaluate the function for several values of x. For instance, the table below verifies that the function in Example 4(a) is increasing over the entire real line.

x	-100	-10	-1	0	1	10	100
$f(x) = x^3$	$-1{,}000{,}000$	-1000	-1	0	1	1000	1,000,000

The points at which a function changes its increasing, decreasing, or constant behavior are helpful in determining the **relative maximum** or **relative minimum** values of the function.

Definition of Relative Minimum and Relative Maximum

A function value $f(a)$ is called a **relative minimum** of f if there exists an interval (x_1, x_2) that contains a such that

$$x_1 < x < x_2 \quad \text{implies} \quad f(a) \leq f(x).$$

A function value $f(a)$ is called a **relative maximum** of f if there exists an interval (x_1, x_2) that contains a such that

$$x_1 < x < x_2 \quad \text{implies} \quad f(a) \geq f(x).$$

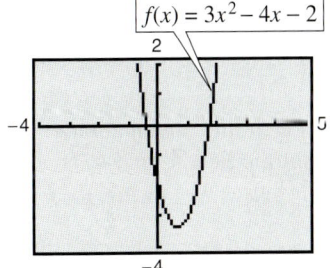

FIGURE P.48

Figure P.48 shows several different examples of relative minimums and relative maximums. By writing a second-degree polynomial in standard form $y = a(x - h)^2 + k$, you can find the exact point (h, k) at which it has a relative minimum or relative maximum. For the time being, however, you can use a graphing utility to find reasonable approximations of these points.

Example 5 ▶ Approximating a Relative Minimum

Use a graphing utility to approximate the relative minimum of the function $f(x) = 3x^2 - 4x - 2$.

Solution

The graph of f is shown in Figure P.49. By using the zoom and trace features of a graphing utility, you can estimate that the function has a relative minimum at the point

$$(0.67, -3.33). \qquad \text{Relative minimum}$$

By writing this second-degree function in standard form, $f(x) = 3\left(x - \frac{2}{3}\right)^2 - \frac{10}{3}$ you can determine that the exact point at which the relative minimum occurs is $\left(\frac{2}{3}, -\frac{10}{3}\right)$.

FIGURE P.49

Note in Example 5 that you can also use the table feature of a graphing utility to approximate numerically the relative minimum of the function. Using a table that begins at 0.6 and increments the value of x by 0.01, you can approximate the minimum of $f(x) = 3x^2 - 4x - 2$ to be $(0.67, -3.33)$.

Technology

If you use a graphing utility to estimate the x- and y-values of a relative minimum or relative maximum, the automatic zoom feature will often produce graphs that are nearly flat. To overcome this problem, you can manually change the vertical setting of the viewing window. The graph will stretch vertically if the values of Ymin and Ymax are closer together.

(a)

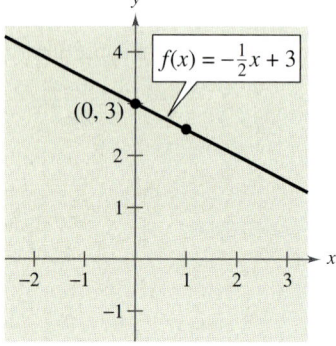

(b)

FIGURE P.50

Linear Functions

A **linear function** of x is a function of the form

$$f(x) = mx + b. \qquad \text{Linear function}$$

In Section P.4, you learned that the graph of such a function is a line that has a slope of m and a y-intercept of $(0, b)$.

Example 6 ▶ Graphing a Linear Function

Sketch the graph of the linear function

$$f(x) = -\frac{1}{2}x + 3.$$

Solution

The graph of this function is a line that has a slope of $m = -\frac{1}{2}$ and a y-intercept of $(0, 3)$. To sketch the line, plot the y-intercept. Then, because the slope is $-\frac{1}{2}$, move 1 unit to the right and $\frac{1}{2}$ unit *down* and plot a second point, as shown in Figure P.50(a). Finally, draw the line that passes through these two points, as shown in Figure P.50(b).

Example 7 ▶ Writing a Linear Function

Write the linear function f for which $f(1) = 3$ and $f(4) = 0$.

Solution

The graph of a linear function is a line. So, you need to find an equation of the line that passes through $(1, 3)$ and $(4, 0)$. The slope of the line is

$$m = \frac{y_2 - y_1}{x_2 - x_1}$$

$$= \frac{0 - 3}{4 - 1}$$

$$= -1.$$

Using the point-slope form of the equation of a line, you can write

$$y - y_1 = m(x - x_1) \qquad \text{Point-slope form}$$

$$y - 3 = -1(x - 1) \qquad \text{Substitute for } y_1, m, \text{ and } x_1.$$

$$y = -x + 4. \qquad \text{Simplify.}$$

So, the linear function is $f(x) = -x + 4$. You can check this result as follows.

$$f(1) = -(1) + 4 = 3 \checkmark$$

$$f(4) = -(4) + 4 = 0 \checkmark$$

Step Functions and Piecewise-Defined Functions

The **greatest integer function** is denoted by $[\![x]\!]$ and is defined as follows.

$[\![x]\!]$ = the greatest integer less than or equal to x

The graph of this function is shown in Figure P.51. Note that the graph of the greatest integer function jumps vertically 1 unit at each integer and is constant (a horizontal line segment) between each pair of consecutive integers. The greatest integer function is an example of a category of functions called **step functions** whose graphs resemble a set of stair steps. Some values of the greatest integer function are as follows.

$[\![-1]\!] = -1$ $[\![-0.5]\!] = -1$ $[\![0]\!] = 0$

$[\![0.5]\!] = 0$ $[\![1]\!] = 1$ $[\![1.5]\!] = 1$

The range of the greatest integer function is the set of all integers. (If you use a graphing utility to sketch a step function, you should set the utility to *dot* mode rather than *connected* mode.)

In Section P.5, you learned that a piecewise-defined function is a function that is defined by two or more equations over a specified domain. To sketch the graph of a piecewise-defined function, you need to sketch the graph of each equation on the appropriate portion of the domain.

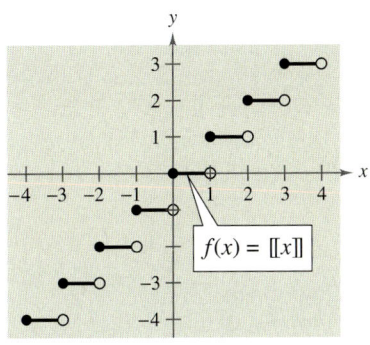

FIGURE P.51

> **Technology**
> When graphing a step function, you should set your graphing utility to dot mode.

Example 8 ▶ Graphing a Piecewise-Defined Function

Sketch the graph of $f(x) = \begin{cases} 2x + 3, & x \leq 1 \\ -x + 4, & x > 1 \end{cases}$.

Solution

This piecewise-defined function is composed of two linear functions. At $x = 1$ and to the left of $x = 1$ the graph is the line $y = 2x + 3$, and to the right of $x = 1$ the graph is the line $y = -x + 4$, as shown in Figure P.52.

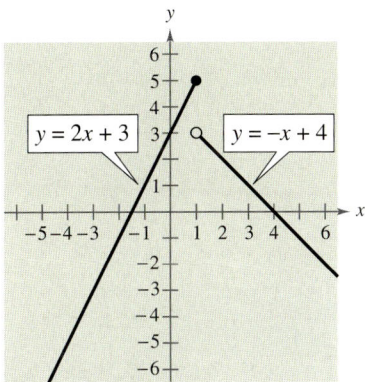

FIGURE P.52

Even and Odd Functions

A computer animation of this concept appears in the Interactive CD-ROM and Internet versions of this text.

In Section P.3, you studied different types of symmetry of a graph. In the terminology of functions, a function is said to be **even** if its graph is symmetric with respect to the y-axis and to be **odd** if its graph is symmetric with respect to the origin. The symmetry tests in Section P.3 yield the following tests for even and odd functions.

Tests for Even and Odd Functions

A function $y = f(x)$ is **even** if, for each x in the domain of f,
$$f(-x) = f(x).$$
A function $y = f(x)$ is **odd** if, for each x in the domain of f,
$$f(-x) = -f(x).$$

Exploration

Graph each of the following functions with a graphing utility. Determine whether the function is *even, odd,* or *neither.*

$f(x) = x^2 - x^4$

$g(x) = 2x^3 + 1$

$h(x) = x^5 - 2x^3 + x$

$j(x) = 2 - x^6 - x^8$

$k(x) = x^5 - 2x^4 + x - 2$

$p(x) = x^9 + 3x^5 - x^3 + x$

What do you notice about the equations of functions that are odd? What do you notice about the equations of functions that are even? Can you describe a way to identify a function as odd or even by inspecting the equation? Can you describe a way to identify a function as neither odd nor even by inspecting the equation?

Example 9 ▶ Even and Odd Functions

a. The function $g(x) = x^3 - x$ is odd because $g(-x) = -g(x)$, as follows.

$g(-x) = (-x)^3 - (-x)$ Substitute $-x$ for x.

$\quad\quad = -x^3 + x$ Simplify.

$\quad\quad = -(x^3 - x)$ Distributive Property

$\quad\quad = -g(x)$ Test for odd function.

b. The function $h(x) = x^2 + 1$ is even because $h(-x) = h(x)$, as follows.

$h(-x) = (-x)^2 + 1$ Substitute $-x$ for x.

$\quad\quad = x^2 + 1$ Simplify.

$\quad\quad = h(x)$ Test for even function.

The graphs of these two functions are shown in Figure P.53.

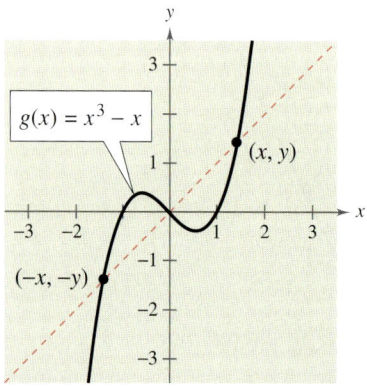

(a) Symmetric to Origin: Odd Function

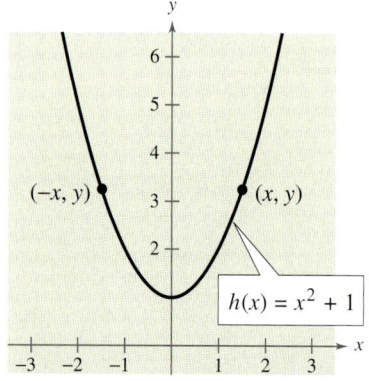

(b) Symmetric to y-Axis: Even Function

FIGURE P.53

P.6 Exercises

In Exercises 1–8, find the domain and range of the function.

1. $f(x) = \frac{2}{3}x - 4$
2. $f(x) = x^3 - 3x + 2$

 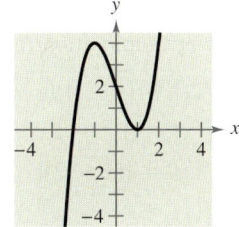

3. $f(x) = 1 - x^2$
4. $f(x) = \sqrt{x - 1}$

 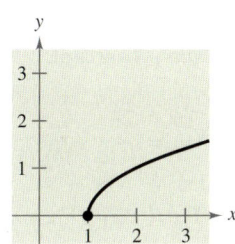

5. $f(x) = \sqrt{x^2 - 1}$
6. $f(x) = \frac{1}{2}|x - 2|$

 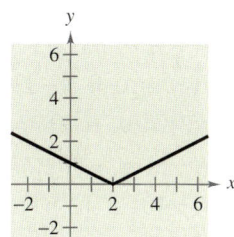

7. $h(x) = \sqrt{16 - x^2}$
8. $g(x) = \dfrac{|x - 1|}{x - 1}$

 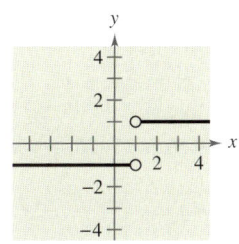

In Exercises 9–14, use the Vertical Line Test to determine whether y is a function of x.

9. $y = \frac{1}{2}x^2$
10. $y = \frac{1}{4}x^3$

11. $x - y^2 = 1$
12. $x^2 + y^2 = 25$

13. $x^2 = 2xy - 1$
14. $x = |y + 2|$

 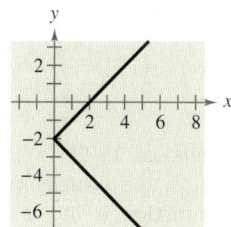

In Exercises 15–22, find the zeros of the function by factoring.

15. $f(x) = 2x^2 - 7x - 30$
16. $f(x) = 3x^2 + 22x - 16$
17. $f(x) = \dfrac{x}{9x^2 - 4}$
18. $f(x) = \dfrac{x^2 - 9x + 14}{4x}$
19. $f(x) = \frac{1}{2}x^3 - x$
20. $f(x) = x^3 - 4x^2 - 9x + 36$
21. $f(x) = 4x^3 - 24x^2 - x + 6$
22. $f(x) = 9x^4 - 25x^2$

In Exercises 23–28, find the zeros of the function algebraically. Verify your results graphically.

23. $f(x) = 3 + \dfrac{5}{x}$

24. $f(x) = x(x - 7)$

25. $f(x) = \sqrt{2x + 11}$

26. $f(x) = \sqrt{3x - 14} - 8$

27. $f(x) = \dfrac{3x - 1}{x - 6}$

28. $f(x) = \dfrac{2x^2 - 9}{3 - x}$

In Exercises 29–32, (a) determine the intervals over which the function is increasing, decreasing, or constant, and (b) determine whether the function is even, odd, or neither.

29. $f(x) = \tfrac{3}{2}x$

30. $f(x) = x^2 - 4x$

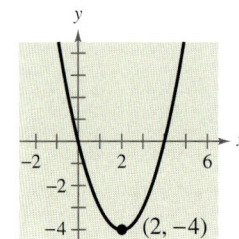

31. $f(x) = x^3 - 3x^2 + 2$ **32.** $f(x) = \sqrt{x^2 - 1}$

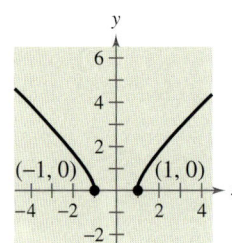

In Exercises 33–50, (a) use a graphing utility to graph the function and visually determine the intervals over which the function is increasing, decreasing, or constant, and (b) make a table of values to verify whether the function is increasing, decreasing, or constant.

33. $f(x) = 3$

34. $g(x) = x$

35. $f(x) = 5 - 3x$

36. $f(x) = -x - 10$

37. $g(s) = \dfrac{s^2}{4}$

38. $h(x) = x^2 - 4$

39. $f(t) = -t^4$

40. $f(x) = 3x^4 - 6x^2$

41. $f(x) = \sqrt{1 - x}$

42. $f(x) = x\sqrt{x + 3}$

43. $f(x) = x^{3/2}$

44. $f(x) = x^{2/3}$

45. $g(t) = \sqrt[3]{t - 1}$

46. $f(x) = \sqrt[4]{x + 5}$

47. $f(x) = |x + 2|$

48. $f(x) = |x + 1| + |x - 1|$

49. $f(x) = \begin{cases} x + 3, & x \le 0 \\ 3, & 0 < x \le 2 \\ 2x - 1, & x > 2 \end{cases}$

50. $f(x) = \begin{cases} 2x + 1, & x \le -1 \\ x^2 - 2, & x > -1 \end{cases}$

In Exercises 51–56, use a graphing utility to approximate the relative minimum/relative maximum of each function.

51. $f(x) = (x - 4)(x + 2)$

52. $f(x) = 3x^2 - 2x - 5$

53. $f(x) = x(x - 2)(x + 3)$

54. $f(x) = x^3 - 3x^2 - x + 1$

55. $f(x) = 2x^3 - 5x^2 - 4x - 1$

56. $f(x) = 8x^4 - 3x - 1$

In Exercises 57–64, sketch the graph of the linear function. Label the y-intercept.

57. $f(x) = 2x - 1$

58. $f(x) = 11 - 3x$

59. $f(x) = -x - \tfrac{3}{4}$

60. $f(x) = 3x - \tfrac{5}{2}$

61. $f(x) = -\tfrac{1}{6}x - \tfrac{5}{2}$

62. $f(x) = \tfrac{5}{6} - \tfrac{2}{3}x$

63. $f(x) = -1.8 + 2.5x$

64. $f(x) = 10.2 + 3.1x$

In Exercises 65–72, write the linear function that has the indicated function values.

65. $f(1) = 4,\ f(0) = 6$

66. $f(-3) = -8,\ f(1) = 2$

67. $f(5) = -4,\ f(-2) = 17$

68. $f(3) = 9,\ f(-1) = -11$

69. $f(-5) = -5,\ f(5) = -1$

70. $f(-10) = 12,\ f(16) = -1$

71. $f(\tfrac{1}{2}) = -6,\ f(4) = -3$

72. $f(\tfrac{2}{3}) = -\tfrac{15}{2},\ f(-4) = -11$

In Exercises 73 and 74, sketch the graph of the function. Describe how it differs from the graph of $f(x) = [\![x]\!]$.

73. $f(x) = [\![x]\!] - 2$

74. $f(x) = [\![x - 1]\!]$

In Exercises 75–78, graph the function.

75. $f(x) = \begin{cases} 2x + 3, & x < 0 \\ 3 - x, & x \ge 0 \end{cases}$

76. $f(x) = \begin{cases} \sqrt{4 + x}, & x < 0 \\ \sqrt{4 - x}, & x \ge 0 \end{cases}$

77. $f(x) = \begin{cases} x^2 + 5, & x \leq 1 \\ -x^2 + 4x + 3, & x > 1 \end{cases}$

78. $f(x) = \begin{cases} 1 - (x - 1)^2, & x \leq 2 \\ \sqrt{x - 2}, & x > 2 \end{cases}$

In Exercises 79–90, graph the function and determine the interval(s) for which $f(x) \geq 0$.

79. $f(x) = 4 - x$
80. $f(x) = 4x + 2$
81. $f(x) = x^2 - 9$
82. $f(x) = x^2 - 4x$
83. $f(x) = 1 - x^4$
84. $f(x) = \sqrt{x + 2}$
85. $f(x) = x^2 + 1$
86. $f(x) = -(1 + |x|)$
87. $f(x) = -5$
88. $f(x) = \frac{1}{2}(2 + |x|)$
89. $f(x) = \begin{cases} 1 - 2x^2, & x \leq -2 \\ -x + 8, & x > -2 \end{cases}$
90. $f(x) = \begin{cases} \sqrt{x - 5}, & x > 5 \\ x^2 + x - 1, & x \leq 5 \end{cases}$

In Exercises 91 and 92, use a graphing utility to graph the function. State the domain and range of the function. Describe the pattern of the graph.

91. $s(x) = 2\left(\frac{1}{4}x - [\![\frac{1}{4}x]\!]\right)$
92. $g(x) = 2\left(\frac{1}{4}x - [\![\frac{1}{4}x]\!]\right)^2$

In Exercises 93–98, determine whether the function is even, odd, or neither.

93. $f(x) = x^6 - 2x^2 + 3$
94. $h(x) = x^3 - 5$
95. $g(x) = x^3 - 5x$
96. $f(x) = x\sqrt{1 - x^2}$
97. $f(t) = t^2 + 2t - 3$
98. $g(s) = 4s^{2/3}$

Think About It In Exercises 99–102, find the coordinates of a second point on the graph of a function f if the given point is on the graph and the function is (a) even and (b) odd.

99. $\left(-\frac{3}{2}, 4\right)$
100. $\left(-\frac{5}{3}, -7\right)$
101. $(4, 9)$
102. $(5, -1)$

103. ***Communications*** The cost of using a telephone calling card is $1.05 for the first minute and $0.38 for each additional minute or portion of a minute.

  (a) A customer needs a model for the cost C of using a calling card for a call lasting t minutes. Which of the following is the appropriate model? Explain.

  $C_1(t) = 1.05 + 0.38[\![t - 1]\!]$

  $C_2(t) = 1.05 - 0.38[\![-(t - 1)]\!]$

  (b) Graph the appropriate model. Determine the cost of a call lasting 18 minutes and 45 seconds.

104. ***Delivery Charges*** Suppose that the cost of sending an overnight package from New York to Atlanta is $9.80 for a package weighing up to but not including 1 pound and $2.50 for each additional pound or portion of a pound. Use the greatest integer function to create a model for the cost C of overnight delivery of a package weighing x pounds, $x > 0$. Sketch the graph of the function.

105. ***Electronics*** The number of lumens (time rate of flow of light) L from a fluorescent lamp can be approximated by the model

$$L = -0.294x^2 + 97.744x - 664.875, \quad 20 \leq x \leq 90$$

where x is the wattage of the lamp. Use a graphing utility to graph the function and estimate the wattage necessary to obtain 2000 lumens.

In Exercises 106–109, write the height h of the rectangle as a function of x.

106.
107.

108.
109.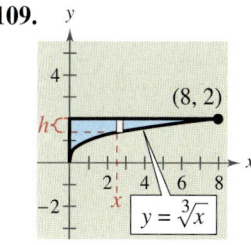

In Exercises 110–113, write the length L of the rectangle as a function of y.

110.
111.

112.

113.

FIGURE FOR 115

114. *Data Analysis* The table shows the amount y (in billions of dollars) of the merchandise trade balance of the United States for the years 1990 through 1997. The merchandise trade balance is the difference between the values of exports and imports. A negative merchandise trade balance indicates that imports exceeded exports. (Source: U.S. International Trade Administration)

Year	1990	1991	1992	1993
y	-101.7	-66.7	-84.5	-115.6

Year	1994	1995	1996	1997
y	-150.6	-158.7	-170.2	-181.5

(a) Use the regression feature of a graphing utility to find a cubic model (a model of the form $y = ax^3 + bx^2 + cx + d$) for the data. Let x be the time (in years), with $x = 0$ corresponding to 1990.

(b) What is the domain of the model?

(c) Use a graphing utility to graph the data and the model in the same viewing window.

(d) For which year does the model most accurately estimate the actual data? During which year is it least accurate?

115. *Geometry* Corners of equal size are cut from a square with sides of length 8 meters (see figure).

(a) Write the area A of the resulting figure as a function of x. Determine the domain of the function.

(b) Use a graphing utility to graph the area function over its domain. Use the graph to find the range of the function.

(c) Identify the resulting figure for the maximum value of x in the domain of the function. What is the length of each side of the figure?

116. *Coordinate Axis Scale* Each function models the specified data for the years 1993 through 2000, with $t = 3$ corresponding to 1993. Estimate a reasonable scale for the vertical axis (e.g., hundreds, thousands, millions, etc.) of the graph and justify your answer. (There are many correct answers.)

(a) $f(t)$ represents the average salary of college professors.

(b) $f(t)$ represents the U.S. population.

(c) $f(t)$ represents the percent of the civilian work force that is unemployed.

117. *Fluid Flow* The intake pipe of a 100-gallon tank has a flow rate of 10 gallons per minute, and two drainpipes have flow rates of 5 gallons per minute each. The figure shows the volume V of fluid in the tank as a function of time t. Determine the combination of the input pipe and drain pipes in which the fluid is flowing in specific subintervals of the 1 hour of time shown on the graph. (There are many correct answers.)

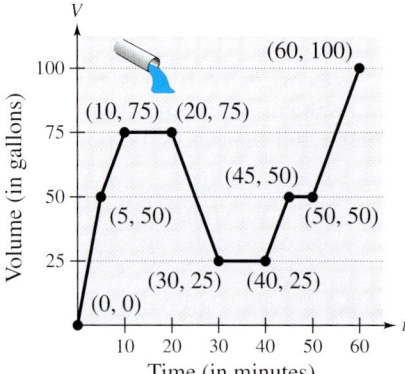

Synthesis

True or False? **In Exercises 118 and 119, determine whether the statement is true or false. Justify your answer.**

118. A function with a square root cannot have a domain that is the set of real numbers.

119. A piecewise-defined function will always have at least one x-intercept or at least one y-intercept.

120. Prove that a function of the following form is odd.
$$y = a_{2n+1}x^{2n+1} + a_{2n-1}x^{2n-1} + \cdots + a_3 x^3 + a_1 x$$

121. Prove that a function of the following form is even.
$$y = a_{2n}x^{2n} + a_{2n-2}x^{2n-2} + \cdots + a_2 x^2 + a_0$$

122. If f is an even function, determine whether g is even, odd, or neither. Explain.
 (a) $g(x) = -f(x)$ (b) $g(x) = f(-x)$
 (c) $g(x) = f(x) - 2$ (d) $g(x) = f(x-2)$

123. *Think About It* Does the graph in Exercise 11 represent x as a function of y? Explain.

124. *Think About It* Does the graph in Exercise 12 represent x as a function of y? Explain.

125. *Writing* Use a graphing utility to graph each function. Write a paragraph describing any similarities and differences you observe among the graphs.
 (a) $y = x$ (b) $y = x^2$
 (c) $y = x^3$ (d) $y = x^4$
 (e) $y = x^5$ (f) $y = x^6$

126. *Conjecture* Use the results of Exercise 125 to make a conjecture about the graphs of $y = x^7$ and $y = x^8$. Use a graphing utility to graph the functions and compare the results with your conjecture.

P.7 Shifting, Reflecting, and Stretching Graphs

▶ **What you should learn**
- How to recognize graphs of common functions
- How to use vertical and horizontal shifts to sketch graphs of functions
- How to use reflections to sketch graphs of functions
- How to use nonrigid transformations to sketch graphs of functions

▶ **Why you should learn it**

Knowing the graphs of common functions and knowing how to shift, reflect, and stretch graphs of functions can help you sketch a wide variety of simple functions by hand. This skill is useful in sketching graphs of functions that model real-life data, such as in Exercise 59 on page 89, where you are asked to sketch a function that models the amount of fuel used by trucks from 1980 through 1996.

Chuck Keeler/The Stock Market

Summary of Graphs of Common Functions

One of the goals of this text is to enable you to recognize the basic shapes of the graphs of different types of functions. For instance, from your study of lines in Section P.4, you can determine the basic shape of the graph of the linear function $f(x) = mx + b$. Specifically, you know that the graph of this function is a line whose slope is m and whose y-intercept is b.

The six graphs shown in Figure P.54 represent the most commonly used functions in algebra. Familiarity with the basic characteristics of these simple graphs will help you analyze the shapes of more complicated graphs.

(a) Constant Function

(b) Identity Function

(c) Absolute Value Function

(d) Square Root Function

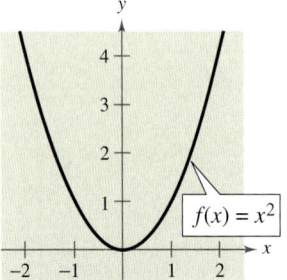

(e) Quadratic Function

(f) Cubic Function

FIGURE P.54

A computer animation of this concept appears in the *Interactive* CD-ROM and *Internet* versions of this text.

Shifting Graphs

Many functions have graphs that are simple transformations of the common graphs summarized on page 80. For example, you can obtain the graph of

$$h(x) = x^2 + 2$$

by shifting the graph of $f(x) = x^2$ *up* 2 units, as shown in Figure P.55. In function notation, h and f are related as follows.

$$h(x) = x^2 + 2$$
$$= f(x) + 2 \qquad \text{Upward shift of 2}$$

Similarly, you can obtain the graph of

$$g(x) = (x - 2)^2$$

by shifting the graph of $f(x) = x^2$ to the *right* 2 units, as shown in Figure P.56. In this case, the functions g and f have the following relationship.

$$g(x) = (x - 2)^2$$
$$= f(x - 2) \qquad \text{Right shift of 2}$$

FIGURE P.55

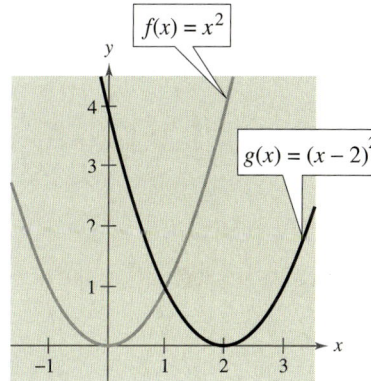

FIGURE P.56

The following list summarizes this discussion about horizontal and vertical shifts.

Vertical and Horizontal Shifts

Let c be a positive real number. **Vertical and horizontal shifts** in the graph of $y = f(x)$ are represented as follows.

1. Vertical shift c units *upward*: $\qquad h(x) = f(x) + c$
2. Vertical shift c units *downward*: $\qquad h(x) = f(x) - c$
3. Horizontal shift c units to the *right*: $\qquad h(x) = f(x - c)$
4. Horizontal shift c units to the *left*: $\qquad h(x) = f(x + c)$

In items 3 and 4, be sure you see that $h(x) = f(x - c)$ corresponds to a *right* shift and $h(x) = f(x + c)$ corresponds to a *left* shift for $c > 0$.

Some graphs can be obtained from a combination of vertical and horizontal shifts, as demonstrated in Example 1(b). Vertical and horizontal shifts generate a *family of functions*, each with the same shape but at different locations in the plane.

Example 1 ▶ Shifts in the Graph of a Function

Use the graph of $f(x) = x^3$ to sketch the graph of each function.

a. $g(x) = x^3 - 1$
b. $h(x) = (x + 2)^3 + 1$

Solution

Relative to the graph of $f(x) = x^3$, the graph of $g(x) = x^3 - 1$ is a downward shift of 1 unit. The graph of $h(x) = (x + 2)^3 + 1$ involves a left shift of 2 units *and* an upward shift of 1 unit. The graphs of both functions are compared with the graph of $f(x) = x^3$ in Figure P.57.

(a)

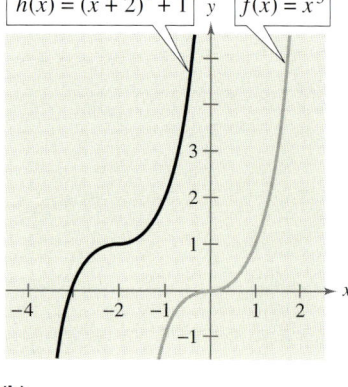
(b)

FIGURE P.57

In part (b) of Figure P.57, notice that the same result is obtained if the vertical shift precedes the horizontal shift *or* if the horizontal shift precedes the vertical shift.

◀ Exploration ▶

Graphing utilities are ideal tools for exploring translations of functions. Graph f, g, and h on the same screen. Before looking at the graphs, try to predict how the graphs of g and h relate to the graph of f.

a. $f(x) = x^2$, $g(x) = (x - 4)^2$, $h(x) = (x - 4)^2 + 3$
b. $f(x) = x^2$, $g(x) = (x + 1)^2$, $h(x) = (x + 1)^2 - 2$
c. $f(x) = x^2$, $g(x) = (x + 4)^2$, $h(x) = (x + 4)^2 + 2$

Section P.7 ▶ Shifting, Reflecting, and Stretching Graphs 83

Reflecting Graphs

The second common type of transformation is a **reflection**. For instance, if you consider the x-axis to be a mirror, the graph of

$$h(x) = -x^2$$

is the mirror image (or reflection) of the graph of $f(x) = x^2$, as shown in Figure P.58.

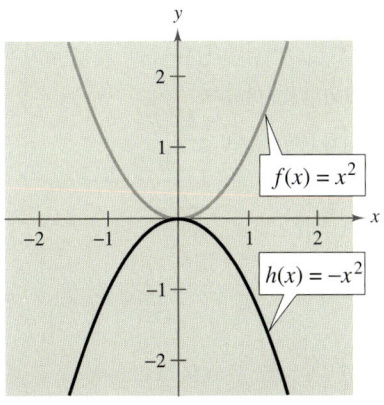

FIGURE P.58

Reflections in the Coordinate Axes

Reflections in the coordinate axes of the graph of $y = f(x)$ are represented as follows.

1. Reflection in the x-axis: $h(x) = -f(x)$
2. Reflection in the y-axis: $h(x) = f(-x)$

A computer animation of this concept appears in the *Interactive* CD-ROM and *Internet* versions of this text.

Example 2 ▶ Finding Equations from Graphs

The graph of the function

$$f(x) = x^4$$

is shown in Figure P.59. Each of the graphs in Figure P.60 is a transformation of the graph of f. Find an equation for each of these functions.

FIGURE P.59

(a)

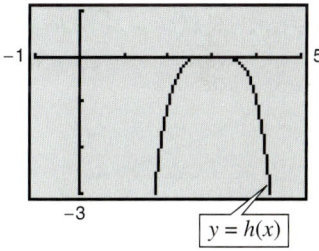

(b)

FIGURE P.60

Solution

a. The graph of g is a reflection in the x-axis *followed by* an upward shift of 2 units of the graph of $f(x) = x^4$. So, the equation for g is

$$g(x) = -x^4 + 2.$$

b. The graph of h is a horizontal shift of 3 units to the right *followed by* a reflection in the x-axis of the graph of $f(x) = x^4$. So, the equation for h is

$$h(x) = -(x - 3)^4.$$

Example 3 ▶ Reflections and Shifts

Compare the graph of each function with the graph of $f(x) = \sqrt{x}$.

a. $g(x) = -\sqrt{x}$ **b.** $h(x) = \sqrt{-x}$ **c.** $k(x) = -\sqrt{x+2}$

Solution

a. The graph of g is a reflection of the graph of f in the x-axis because
$$g(x) = -\sqrt{x}$$
$$= -f(x).$$

b. The graph of h is a reflection of the graph of f in the y-axis because
$$h(x) = \sqrt{-x}$$
$$= f(-x).$$

c. The graph of k is a left shift of 2 units, followed by a reflection in the x-axis because
$$k(x) = -\sqrt{x+2}$$
$$= -f(x+2).$$

The graphs of all three functions are shown in Figure P.61.

(a)

(b)

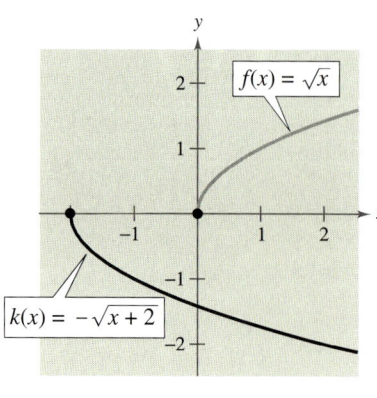
(c)

FIGURE P.61

When sketching the graphs of functions involving square roots, remember that the domain must be restricted to exclude negative numbers inside the radical. For instance, here are the domains of the functions in Example 3.

Domain of $g(x) = -\sqrt{x}$: $x \geq 0$
Domain of $h(x) = \sqrt{-x}$: $x \leq 0$
Domain of $k(x) = -\sqrt{x+2}$: $x \geq -2$

Nonrigid Transformations

 A computer animation of this concept appears in the *Interactive* CD-ROM and *Internet* versions of this text.

Horizontal shifts, vertical shifts, and reflections are **rigid transformations** because the basic shape of the graph is unchanged. These transformations change only the *position* of the graph in the *xy*-plane. **Nonrigid transformations** are those that cause a *distortion*—a change in the shape of the original graph. For instance, a nonrigid transformation of the graph of $y = f(x)$ is represented by $g(x) = cf(x)$, where the transformation is a **vertical stretch** if $c > 1$ and a **vertical shrink** if $0 < c < 1$.

Exploration

Sketch the graphs of $f(x) = 2x^2$ and $h(x) = x^2$ on the same set of axes. Describe the effect of multiplying x^2 by a number greater than 1. Then sketch the graphs of $g(x) = \frac{1}{2}x^2$ and $h(x) = x^2$ on the same set of axes. Describe the effect of multiplying x^2 by a number less than 1. Can you think of an easy way to remember this generalization?

Example 4 ▶ Nonrigid Transformations

Compare the graph of each function with the graph of $f(x) = |x|$.

a. $h(x) = 3|x|$

b. $g(x) = \frac{1}{3}|x|$

Solution

a. Relative to the graph of $f(x) = |x|$, the graph of
$$h(x) = 3|x|$$
$$= 3f(x)$$
is a vertical stretch (each *y*-value is multiplied by 3) of the graph of f.

b. Similarly, the function
$$g(x) = \frac{1}{3}|x|$$
$$= \frac{1}{3}f(x)$$
indicates that the graph of g is a vertical shrink $\left(\text{each } y\text{-value is multiplied by } \frac{1}{3}\right)$ of the graph of f.

The graphs of both functions are shown in Figure P.62.

(a)

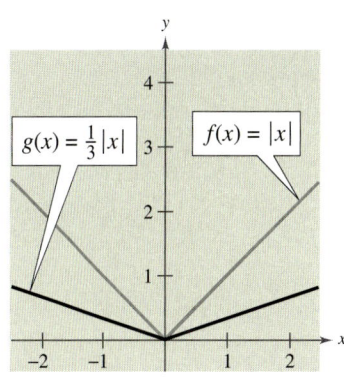

(b)

FIGURE P.62

P.7 Exercises

1. Sketch (on the same set of coordinate axes) a graph of each function for $c = -2, 0,$ and 2.
   (a) $f(x) = x^3 + c$
   (b) $f(x) = (x - c)^3$

2. Sketch (on the same set of coordinate axes) a graph of each function for $c = -1, 1,$ and 3.
   (a) $f(x) = |x| + c$
   (b) $f(x) = |x - c|$
   (c) $f(x) = |x + 4| + c$

3. Sketch (on the same set of coordinate axes) a graph of each function for $c = -3, -1, 1,$ and 3.
   (a) $f(x) = \sqrt{x} + c$
   (b) $f(x) = \sqrt{x - c}$
   (c) $f(x) = \sqrt{x - 3} + c$

4. Sketch (on the same set of coordinate axes) a graph of each function for $c = -3, -1, 1,$ and 3.
   (a) $f(x) = \begin{cases} x^2 + c, & x < 0 \\ -x^2 + c, & x \geq 0 \end{cases}$
   (b) $f(x) = \begin{cases} (x + c)^2, & x < 0 \\ -(x + c)^2, & x \geq 0 \end{cases}$

5. Use the graph of f to sketch each graph.
   (a) $y = f(x) + 2$
   (b) $y = f(x - 2)$
   (c) $y = 2f(x)$
   (d) $y = -f(x)$
   (e) $y = f(x + 3)$
   (f) $y = f(-x)$

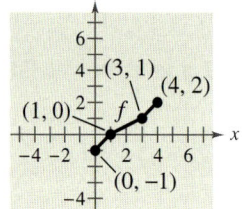

6. Use the graph of f to sketch each graph.
   (a) $y = f(-x)$
   (b) $y = f(x) + 4$
   (c) $y = 2f(x)$
   (d) $y = -f(x - 4)$
   (e) $y = f(x) - 3$
   (f) $y = -f(x) - 1$

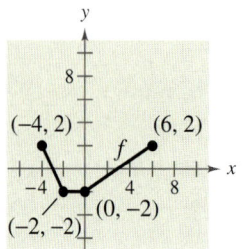

7. Use the graph of f to sketch each graph.
   (a) $y = f(x) - 1$
   (b) $y = f(x - 1)$
   (c) $y = f(-x)$
   (d) $y = f(x + 1)$
   (e) $y = -f(x - 2)$
   (f) $y = \frac{1}{2}f(x)$

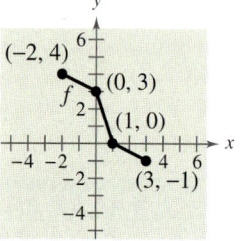

8. Use the graph of f to sketch each graph.
   (a) $y = f(x - 5)$
   (b) $y = -f(x) + 3$
   (c) $y = \frac{1}{3}f(x)$
   (d) $y = -f(x + 1)$
   (e) $y = f(-x)$
   (f) $y = f(x) - 10$

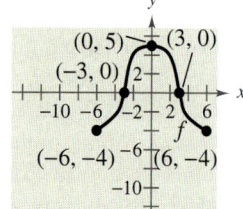

9. Use the graph of $f(x) = x^2$ to write an equation for each function whose graph is shown.
   (a)

   (b)

   (c)

   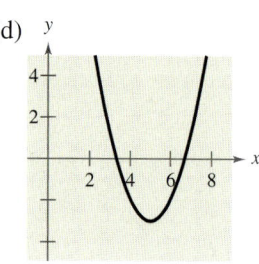

   (d)

10. Use the graph of $f(x) = x^3$ to write an equation for each function whose graph is shown.

(a)
(b)
(c)
(d)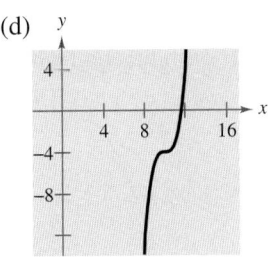

11. Use the graph of $f(x) = |x|$ to write an equation for each function whose graph is shown.

(a)
(b)
(c)
(d)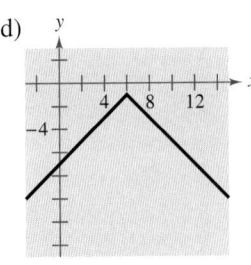

12. Use the graph of $f(x) = \sqrt{x}$ to write an equation for each function whose graph is shown.

(a)
(b)

(c)
(d)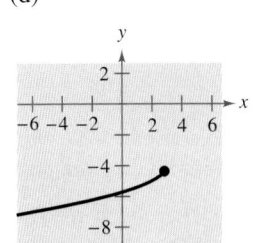

In Exercises 13–18, identify the common function and the transformation shown in the graph. Write an equation for the function shown in the graph.

13.
14.

15.
16.

17.
18.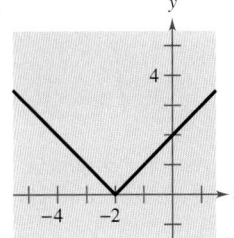

In Exercises 19–34, describe the transformation that occurs in the function. Then sketch its graph.

19. $f(x) = 12 - x^2$ **20.** $f(x) = (x - 8)^2$
21. $f(x) = x^3 + 7$ **22.** $f(x) = -x^3 - 1$
23. $f(x) = 2 - (x + 5)^2$ **24.** $f(x) = -(x + 10)^2 + 5$
25. $f(x) = (x - 1)^3 + 2$ **26.** $f(x) = (x + 3)^3 - 10$
27. $f(x) = -|x| - 2$ **28.** $f(x) = 6 - |x + 5|$

29. $f(x) = -|x + 4| + 8$
30. $f(x) = |-x + 3| + 9$
31. $f(x) = \sqrt{x - 9}$
32. $f(x) = \sqrt{x + 4} + 8$
33. $f(x) = \sqrt{7 - x} - 2$
34. $f(x) = -\sqrt{x + 1} - 6$

In Exercises 35–42, write an equation for the function that is described by the given characteristics.

35. The shape of $f(x) = x^2$, but moved 2 units to the right and 8 units down
36. The shape of $f(x) = x^2$, but moved 3 units to the left, 7 units up, and reflected in the x-axis
37. The shape of $f(x) = x^3$, but moved 13 units to the right
38. The shape of $f(x) = x^3$, but moved 6 units to the left, 6 units down, and reflected in the y-axis
39. The shape of $f(x) = |x|$, but moved 10 units up and reflected in the x-axis
40. The shape of $f(x) = |x|$, but moved 1 unit to the left and 7 units down
41. The shape of $f(x) = \sqrt{x}$, but moved 6 units to the left and reflected in both the x-axis and the y-axis
42. The shape of $f(x) = \sqrt{x}$, but moved 9 units down and reflected in both the x-axis and the y-axis

43. Use the graph of $f(x) = x^2$ to write an equation for each function whose graph is shown.

(a)   (b)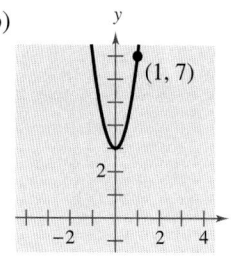

44. Use the graph of $f(x) = x^3$ to write an equation for each function whose graph is shown.

(a)   (b)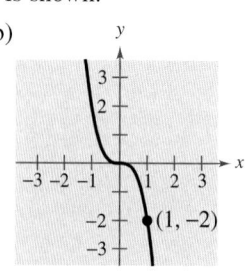

45. Use the graph of $f(x) = |x|$ to write an equation for each function whose graph is shown.

(a)   (b)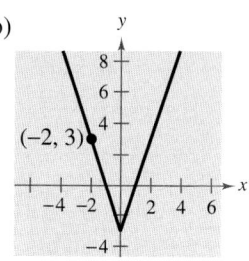

46. Use the graph of $f(x) = \sqrt{x}$ to write an equation for each function whose graph is shown.

(a)   (b)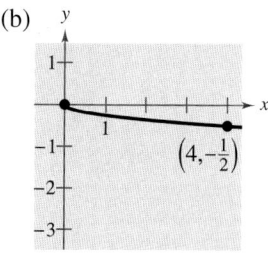

In Exercises 47–52, identify the common function and the transformation shown in the graph. Write an equation for the function shown in the graph. Then use a graphing utility to verify your answer.

47.   48.

49.   50.

51.   **52.**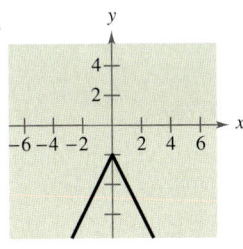

Graphical Analysis   In Exercises 53–56, use the viewing window shown to write a possible equation for the transformation of the common function.

53.

54.

55.

56.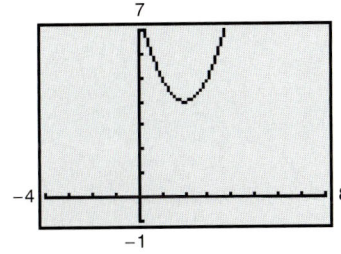

Graphical Reasoning   In Exercises 57 and 58, use the graph of *f* to sketch the graph of *g*.

57.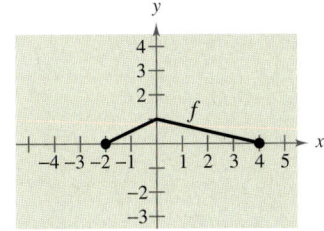

(a) $g(x) = f(x) + 2$ (b) $g(x) = f(x) - 1$
(c) $g(x) = f(-x)$ (d) $g(x) = -2f(x)$

58.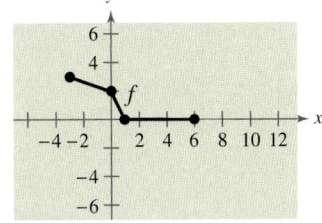

(a) $g(x) = f(x) - 5$ (b) $g(x) = f(x) + \frac{1}{2}$
(c) $g(x) = f(-x)$ (d) $g(x) = -4f(x)$

59. *Fuel Use* The amount of fuel F (in billions of gallons) used by trucks from 1980 through 1996 can be approximated by the function

$$F = f(t) = 20.46 + 0.04t^2$$

where $t = 0$ represents 1980. (Source: U.S. Federal Highway Administration)

(a) Describe the transformation of the common function $f(x) = x^2$. Then sketch the graph over the interval $0 \le t \le 16$.

(b) Rewrite the function so that $t = 0$ represents 1990. Explain how you got your answer.

60. *Finance* The amount (in trillions of dollars) of mortgage debt M outstanding in the United States from 1985 through 1997 can be approximated by the function

$$M = f(t) = 1.5\sqrt{t} - 1.25$$

where $t = 5$ represents 1985. (Source: Board of Governors of the Federal Reserve System)

(a) Describe the transformation of the common function $f(x) = \sqrt{x}$. Then sketch the graph over the interval $5 \le t \le 17$.

(b) Rewrite the function so that $t = 5$ represents 1995. Explain how you got your answer.

Synthesis

In Exercises 61 and 62, determine whether the statement is true or false. Justify your answer.

61. The graphs of $f(x) = |x| + 6$ and $f(x) = |-x| + 6$ are identical.

62. If the graph of the common function $f(x) = x^2$ is moved 6 units to the right, 3 units up, and reflected in the x-axis, then the point $(-2, 19)$ will lie on the graph of the transformation.

63. *Describing Profits* Management originally predicted that the profits from the sales of a new product would be approximated by the graph of the function f shown. The actual profits are shown by the function g along with a verbal description. Use the concepts of transformations of graphs to write g in terms of f.

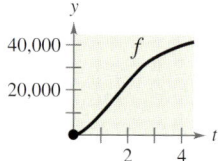

(a) The profits were only three-fourths as large as expected.

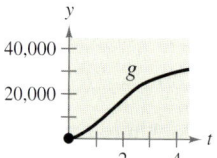

(b) The profits were consistently $10,000 greater than predicted.

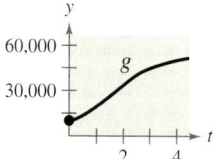

(c) There was a 2-year delay in the introduction of the product. After sales began, profits grew as expected.

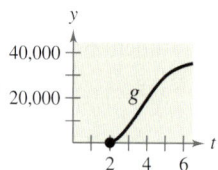

64. Explain why the graph of $y = -f(x)$ is a reflection of the graph of $y = f(x)$ about the x-axis.

65. The graph of $y = f(x)$ passes through the points $(0, 1)$, $(1, 2)$, and $(2, 3)$. Find the corresponding points on the graph of $y = f(x + 2) - 1$.

66. *Think About It* You can use two methods to graph a function: plotting points, or translating a common function as shown in this section. Which method of graphing do you prefer to use? Explain.

P.8 Combinations of Functions

▶ **What you should learn**
- How to add, subtract, multiply, and divide functions
- How to find compositions of one function with another function
- How to use combinations of functions to model and solve real-life problems

▶ **Why you should learn it**

Combinations of functions can be used to model and solve real-life problems. For instance, Exercises 33 and 34 on page 97 show how to use combinations of functions to analyze U.S. health expenditures.

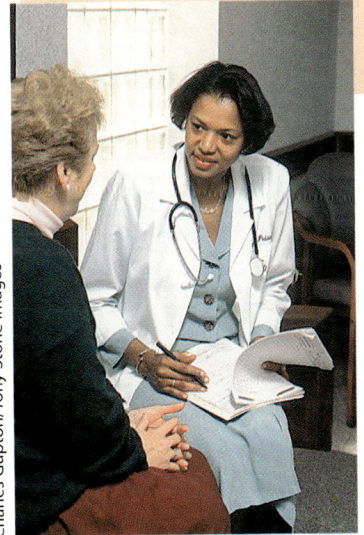

A computer animation of this concept appears in the *Interactive* CD-ROM and *Internet* versions of this text.

Arithmetic Combinations of Functions

Just as two real numbers can be combined by the operations of addition, subtraction, multiplication, and division to form other real numbers, two *functions* can be combined to create new functions. For example, the functions $f(x) = 2x - 3$ and $g(x) = x^2 - 1$ can be combined to form the sum, difference, product, and quotient of f and g.

$$f(x) + g(x) = (2x - 3) + (x^2 - 1)$$
$$= x^2 + 2x - 4 \qquad \text{Sum}$$
$$f(x) - g(x) = (2x - 3) - (x^2 - 1)$$
$$= -x^2 + 2x - 2 \qquad \text{Difference}$$
$$f(x)g(x) = (2x - 3)(x^2 - 1)$$
$$= 2x^3 - 3x^2 - 2x + 3 \qquad \text{Product}$$
$$\frac{f(x)}{g(x)} = \frac{2x - 3}{x^2 - 1}, \quad x \neq \pm 1 \qquad \text{Quotient}$$

The domain of an **arithmetic combination** of functions f and g consists of all real numbers that are common to the domains of f and g. In the case of the quotient $f(x)/g(x)$, there is the further restriction that $g(x) \neq 0$.

Sum, Difference, Product, and Quotient of Functions

Let f and g be two functions with overlapping domains. Then, for all x common to both domains, the *sum, difference, product,* and *quotient* of f and g are defined as follows.

1. *Sum:* $\quad (f + g)(x) = f(x) + g(x)$
2. *Difference:* $\quad (f - g)(x) = f(x) - g(x)$
3. *Product:* $\quad (fg)(x) = f(x) \cdot g(x)$
4. *Quotient:* $\quad \left(\dfrac{f}{g}\right)(x) = \dfrac{f(x)}{g(x)}, \quad g(x) \neq 0$

Example 1 ▶ Finding the Sum of Two Functions

Given $f(x) = 2x + 1$ and $g(x) = x^2 + 2x - 1$, find $(f + g)(x)$. Then evaluate the sum when $x = -1$.

Solution

$$(f + g)(x) = f(x) + g(x)$$
$$= (2x + 1) + (x^2 + 2x - 1)$$
$$= x^2 + 4x$$

When $x = -1$, the value of the sum is $(f + g)(-1) = (-1)^2 + 4(-1) = -3$.

Example 2 ▶ Finding the Difference of Two Functions

Given $f(x) = 2x + 1$ and $g(x) = x^2 + 2x - 1$, find $(f - g)(x)$. Then evaluate the difference when $x = 2$.

Solution

The difference of f and g is

$$(f - g)(x) = f(x) - g(x)$$
$$= (2x + 1) - (x^2 + 2x - 1)$$
$$= -x^2 + 2.$$

When $x = 2$, the value of this difference is

$$(f - g)(2) = -(2)^2 + 2$$
$$= -2.$$

In Examples 1 and 2, both f and g have domains that consist of all real numbers. So, the domains of $(f + g)$ and $(f - g)$ are also the set of all real numbers. Remember that any restrictions on the domains of f and g must be considered when forming the sum, difference, product, or quotient of f and g.

Example 3 ▶ Finding the Domains of Quotients of Functions

Find the domains of $\left(\dfrac{f}{g}\right)(x)$ and $\left(\dfrac{g}{f}\right)(x)$ for the functions

$$f(x) = \sqrt{x} \quad \text{and} \quad g(x) = \sqrt{4 - x^2}.$$

Solution

The quotient of f and g is

$$\left(\dfrac{f}{g}\right)(x) = \dfrac{f(x)}{g(x)} = \dfrac{\sqrt{x}}{\sqrt{4 - x^2}}$$

and the quotient of g and f is

$$\left(\dfrac{g}{f}\right)(x) = \dfrac{g(x)}{f(x)} = \dfrac{\sqrt{4 - x^2}}{\sqrt{x}}.$$

The domain of f is $[0, \infty)$ and the domain of g is $[-2, 2]$. The intersection of these domains is $[0, 2]$. So, the domains of $\left(\dfrac{f}{g}\right)$ and $\left(\dfrac{g}{f}\right)$ are as follows.

$$\text{Domain of } \left(\dfrac{f}{g}\right): [0, 2) \qquad \text{Domain of } \left(\dfrac{g}{f}\right): (0, 2]$$

Can you see why these two domains differ slightly?

Section P.8 ▶ Combinations of Functions 93

A computer animation of this concept appears in the *Interactive* CD-ROM and *Internet* versions of this text.

Composition of Functions

Another way of combining two functions is to form the **composition** of one with the other. For instance, if $f(x) = x^2$ and $g(x) = x + 1$, the composition of f with g is

$$f(g(x)) = f(x + 1)$$
$$= (x + 1)^2.$$

This composition is denoted as $(f \circ g)$.

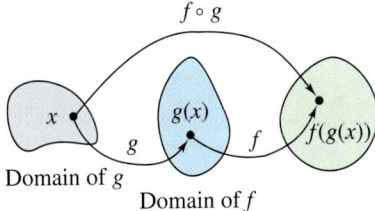

FIGURE P.63

Definition of Composition of Two Functions

The **composition** of the function f with the function g is

$$(f \circ g)(x) = f(g(x)).$$

The domain of $(f \circ g)$ is the set of all x in the domain of g such that $g(x)$ is in the domain of f. (See Figure P.63.)

STUDY TIP

The following tables of values help illustrate the composition of the functions f and g given in Example 4.

x	0	1	2	3
$g(x)$	4	3	0	-5

$g(x)$	4	3	0	-5
$f(g(x))$	6	5	2	-3

x	0	1	2	3
$f(g(x))$	6	5	2	-3

Note that the first two tables can be combined (or "composed") to produce the values given in the third table.

Example 4 ▶ Composition of Functions

Given $f(x) = x + 2$ and $g(x) = 4 - x^2$, find the following.

a. $(f \circ g)(x)$ **b.** $(g \circ f)(x)$ **c.** $(g \circ f)(-2)$

Solution

a. The composition of f with g is as follows.

$(f \circ g)(x) = f(g(x))$ Definition of $f \circ g$
$= f(4 - x^2)$ Definition of $g(x)$
$= (4 - x^2) + 2$ Definition of $f(x)$
$= -x^2 + 6$ Simplify.

b. The composition of g with f is as follows.

$(g \circ f)(x) = g(f(x))$ Definition of $g \circ f$
$= g(x + 2)$ Definition of $f(x)$
$= 4 - (x + 2)^2$ Definition of $g(x)$
$= 4 - (x^2 + 4x + 4)$ Expand.
$= -x^2 - 4x$ Simplify.

Note that, in this case, $(f \circ g)(x) \neq (g \circ f)(x)$.

c. Using the result of part (b), you can write the following.

$(g \circ f)(-2) = -(-2)^2 - 4(-2)$ Substitute.
$= -4 + 8$ Simplify.
$= 4$ Simplify.

Example 5 ▶ Finding the Domain of a Composite Function

Find the composition $(f \circ g)(x)$ for the functions
$$f(x) = x^2 - 9 \quad \text{and} \quad g(x) = \sqrt{9 - x^2}.$$
Then find the domain of $(f \circ g)$.

Solution

$$\begin{aligned}
(f \circ g)(x) &= f(g(x)) \\
&= f\left(\sqrt{9 - x^2}\right) \\
&= \left(\sqrt{9 - x^2}\right)^2 - 9 \\
&= 9 - x^2 - 9 \\
&= -x^2
\end{aligned}$$

From this, it might appear that the domain of the composition is the set of all real numbers. This, however, is not true because the domain of g is $-3 \leq x \leq 3$.

In Examples 4 and 5 you formed the composition of two given functions. In calculus, it is also important to be able to identify two functions that make up a given composite function. For instance, the function h given by
$$h(x) = (3x - 5)^3$$
is the composition of f with g, where $f(x) = x^3$ and $g(x) = 3x - 5$. That is,
$$h(x) = (3x - 5)^3 = [g(x)]^3 = f(g(x)).$$
Basically, to "decompose" a composite function, look for an "inner" and an "outer" function. In the function h above, $g(x) = 3x - 5$ is the inner function and $f(x) = x^3$ is the outer function.

Example 6 ▶ Identifying a Composite Function

Express the function
$$h(x) = \frac{1}{(x - 2)^2}$$
as a composition of two functions.

Solution

One way to write h as a composition of two functions is to take the inner function to be $g(x) = x - 2$ and the outer function to be
$$f(x) = \frac{1}{x^2} = x^{-2}.$$
Then you can write
$$h(x) = \frac{1}{(x - 2)^2} = (x - 2)^{-2} = f(x - 2) = f(g(x)).$$

Exploration

You are buying an automobile whose price is $18,500. Which of the following options would you choose? Explain.

a. You are given a factory rebate of $2000, followed by a dealer discount of 10%.

b. You are given a dealer discount of 10%, followed by a factory rebate of $2000.

Let $f(x) = x - 2000$ and let $g(x) = 0.9x$. Which option is represented by the composite $f(g(x))$? Which is represented by the composite $g(f(x))$?

Application

Example 7 ▶ Bacteria Count

The number N of bacteria in a refrigerated food is

$$N(T) = 20T^2 - 80T + 500, \quad 2 \leq T \leq 14$$

where T is the temperature of the food. When the food is removed from refrigeration, the temperature is

$$T(t) = 4t + 2, \quad 0 \leq t \leq 3$$

where t is the time in hours. (a) Find the composite $N(T(t))$ and interpret its meaning in context. (b) Find the time when the bacterial count reaches 2000.

Solution

a. $N(T(t)) = 20(4t + 2)^2 - 80(4t + 2) + 500$

$= 20(16t^2 + 16t + 4) - 320t - 160 + 500$

$= 320t^2 + 320t + 80 - 320t - 160 + 500$

$= 320t^2 + 420$

The composite function $N(T(t))$ represents the number of bacteria in the food as a function of time.

b. The bacterial count will reach 2000 when $320t^2 + 420 = 2000$. Solve this equation to find that the count will reach 2000 when $t \approx 2.2$ hours. When you solve this equation, note that the negative value is rejected because it is not in the domain of the composite function.

Writing ABOUT MATHEMATICS

Analyzing Arithmetic Combinations of Functions

a. Use the graphs of f and $(f + g)$ in Figure P.64 to make a table showing the values of $g(x)$ when $x = 1, 2, 3, 4, 5,$ and 6. Explain your reasoning.

b. Use the graphs of f and $(f - h)$ in Figure P.64 to make a table showing the values of $h(x)$ when $x = 1, 2, 3, 4, 5,$ and 6. Explain your reasoning.

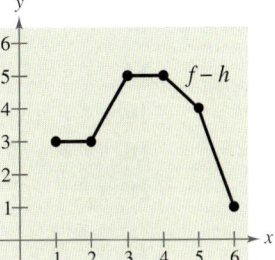

FIGURE P.64

P.8 Exercises

In Exercises 1–4, use the graphs of f and g to graph $h(x) = (f + g)(x)$.

1.

2.

3.

4.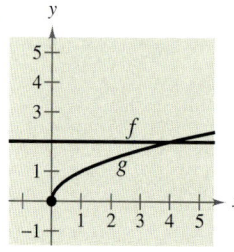

In Exercises 5–12, find (a) $(f + g)(x)$, (b) $(f - g)(x)$, (c) $(fg)(x)$, and (d) $(f/g)(x)$. What is the domain of f/g?

5. $f(x) = x + 2$, $\quad g(x) = x - 2$
6. $f(x) = 2x - 5$, $\quad g(x) = 2 - x$
7. $f(x) = x^2$, $\quad g(x) = 2 - x$
8. $f(x) = 2x - 5$, $\quad g(x) = 4$
9. $f(x) = x^2 + 6$, $\quad g(x) = \sqrt{1 - x}$
10. $f(x) = \sqrt{x^2 - 4}$, $\quad g(x) = \dfrac{x^2}{x^2 + 1}$
11. $f(x) = \dfrac{1}{x}$, $\quad g(x) = \dfrac{1}{x^2}$
12. $f(x) = \dfrac{x}{x + 1}$, $\quad g(x) = x^3$

In Exercises 13–24, evaluate the indicated function for $f(x) = x^2 + 1$ and $g(x) = x - 4$.

13. $(f + g)(2)$
14. $(f - g)(-1)$
15. $(f - g)(0)$
16. $(f + g)(1)$
17. $(f - g)(3t)$
18. $(f + g)(t - 2)$
19. $(fg)(6)$
20. $(fg)(-6)$
21. $\left(\dfrac{f}{g}\right)(5)$
22. $\left(\dfrac{f}{g}\right)(0)$
23. $\left(\dfrac{f}{g}\right)(-1) - g(3)$
24. $(2f)(5)$

In Exercises 25–28, graph the functions f, g, and $f + g$ on the same set of coordinate axes.

25. $f(x) = \tfrac{1}{2}x$, $\quad g(x) = x - 1$
26. $f(x) = \tfrac{1}{3}x$, $\quad g(x) = -x + 4$
27. $f(x) = x^2$, $\quad g(x) = -2x$
28. $f(x) = 4 - x^2$, $\quad g(x) = x$

Graphical Reasoning In Exercises 29 and 30, use a graphing utility to sketch the graphs of f, g, and $f + g$ in the same viewing window. Which function contributes most to the magnitude of the sum when $0 \le x \le 2$? Which function contributes most to the magnitude of the sum when $x > 6$?

29. $f(x) = 3x$, $\quad g(x) = -\dfrac{x^3}{10}$
30. $f(x) = \dfrac{x}{2}$, $\quad g(x) = \sqrt{x}$

31. *Stopping Distance* While traveling in a car at x miles per hour, you are required to stop quickly to avoid an accident. The distance the car travels during your reaction time is given by $R(x) = \tfrac{3}{4}x$. The distance traveled while you are braking is
$$B(x) = \dfrac{1}{15}x^2.$$
Find the function that represents the total stopping distance T. Graph the functions R, B, and T on the same set of coordinate axes for $0 \le x \le 60$.

32. *Business* You own two restaurants. From 1995 to 2000, the sales R_1 (in thousands of dollars) for one restaurant can be modeled by
$$R_1 = 480 - 8t - 0.8t^2, \quad t = 0, 1, 2, 3, 4, 5$$
where $t = 0$ represents 1995. During the same 6-year period, the sales R_2 (in thousands of dollars) for the second restaurant can be modeled by
$$R_2 = 254 + 0.78t, \quad t = 0, 1, 2, 3, 4, 5.$$
Write a function that represents the total sales for the two restaurants. Use a graphing utility to graph the total sales function.

Health Care Costs In Exercises 33 and 34, use the table, which gives the total amount spent (in billions of dollars) on health services and supplies in the United States (including Puerto Rico) for the years 1990 through 1996. The variables y_1, y_2, and y_3 represent out-of-pocket payments, insurance premiums, and other types of payments, respectively. (Source: U.S. Health Care Financing Administration)

Year	1990	1991	1992	1993	1994	1995	1996
y_1	144.4	151.6	159.5	163.6	164.8	166.7	171.2
y_2	238.6	259.4	282.5	303.3	315.6	326.9	337.3
y_3	21.8	24.0	25.1	27.3	29.6	31.7	32.4

33. Use a graphing utility to find a mathematical model for each of the variables. Let $t = 0$ represent 1990. Find a quadratic model for y_1 and linear models for y_2 and y_3.

34. Use a graphing utility to graph y_1, y_2, y_3, and $y_1 + y_2 + y_3$ in the same viewing window. Use the model to estimate the total amount spent on health services and supplies in the year 2000.

35. *Graphical Reasoning* An electronically controlled thermostat in a home is programmed to lower the temperature automatically during the night. The temperature in the house T (in degrees Fahrenheit) is given in terms of t, the time in hours on a 24-hour clock (see figure).

(a) Explain why T is a function of t.

(b) Approximate $T(4)$ and $T(15)$.

(c) Suppose the thermostat were reprogrammed to produce a temperature H where $H(t) = T(t - 1)$. How would this change the temperature?

(d) Suppose the thermostat were reprogrammed to produce a temperature H where $H(t) = T(t) - 1$. How would this change the temperature?

36. *Think About It* Write a piecewise-defined function that represents the graph in Exercise 35.

In Exercises 37–40, find (a) $f \circ g$, (b) $g \circ f$, and (c) $f \circ f$.

37. $f(x) = x^2$, $g(x) = x - 1$

38. $f(x) = \sqrt[3]{x - 1}$, $g(x) = x^3 + 1$

39. $f(x) = 3x + 5$, $g(x) = 5 - x$

40. $f(x) = x^3$, $g(x) = \dfrac{1}{x}$

In Exercises 41–50, find (a) $f \circ g$ and (b) $g \circ f$. Find the domain of each function and each composite function.

41. $f(x) = \sqrt{x + 4}$, $g(x) = x^2$

42. $f(x) = \sqrt[3]{x - 5}$, $g(x) = x^3 + 1$

43. $f(x) = \frac{1}{3}x - 3$, $g(x) = 3x + 1$

44. $f(x) = x^2 + 1$, $g(x) = \sqrt{x}$

45. $f(x) = x^4$, $g(x) = x^4$

46. $f(x) = \sqrt{x}$, $g(x) = 2x - 3$

47. $f(x) = |x|$, $g(x) = x + 6$

48. $f(x) = x^{2/3}$, $g(x) = x^6$

49. $f(x) = \dfrac{1}{x}$, $g(x) = x + 3$

50. $f(x) = \dfrac{3}{x^2 - 1}$, $g(x) = x + 1$

In Exercises 51–54, use the graphs of f and g to evaluate the functions.

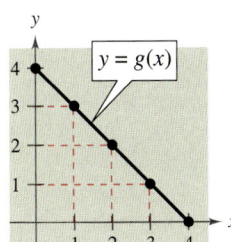

51. (a) $(f + g)(3)$ (b) $\left(\dfrac{f}{g}\right)(2)$

52. (a) $(f - g)(1)$ (b) $(fg)(4)$

53. (a) $(f \circ g)(2)$ (b) $(g \circ f)(2)$

54. (a) $(f \circ g)(1)$ (b) $(g \circ f)(3)$

In Exercises 55–62, find two functions f and g such that $(f \circ g)(x) = h(x)$. (There is more than one correct answer.)

55. $h(x) = (2x + 1)^2$
56. $h(x) = (1 - x)^3$
57. $h(x) = \sqrt[3]{x^2 - 4}$
58. $h(x) = \sqrt{9 - x}$
59. $h(x) = \dfrac{1}{x + 2}$
60. $h(x) = \dfrac{4}{(5x + 2)^2}$
61. $h(x) = \dfrac{-x^2 + 3}{4 - x^2}$
62. $h(x) = \dfrac{27x^3 + 6x}{10 - 27x^3}$

63. *Geometry* A square concrete foundation was prepared as a base for a cylindrical tank (see figure).

    (a) Express the radius r of the tank as a function of the length x of the sides of the square.

    (b) Express the area A of the circular base of the tank as a function of the radius r.

    (c) Find and interpret $(A \circ r)(x)$.

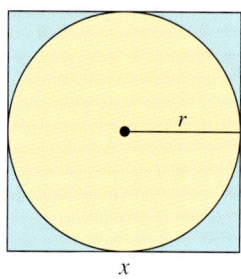

64. *Physics* A pebble is dropped into a calm pond, causing ripples in the form of concentric circles. The radius (in feet) of the outer ripple is $r(t) = 0.6t$, where t is the time in seconds after the pebble strikes the water. The area of the circle is given by the function $A(r) = \pi r^2$. Find and interpret $(A \circ r)(t)$.

65. *Economics* The weekly cost of producing x units in a manufacturing process is given by the function

    $C(x) = 60x + 750.$

    The number of units produced in t hours is $x(t) = 50t$. Find and interpret $(C \circ x)(t)$.

Synthesis

True or False? In Exercises 66 and 67, determine whether the statement is true or false. Justify your answer.

66. If $f(x) = x + 1$ and $g(x) = 6x$, then $(f \circ g)(x) = (g \circ f)(x)$.

67. If you are given two functions $f(x)$ and $g(x)$, you can calculate $(f \circ g)(x)$ if and only if the range of g is a subset of the domain of f.

68. *Think About It* You are a sales representative for an automobile manufacturer. You are paid an annual salary, plus a bonus of 3% of your sales over $500,000. Consider the two functions

    $f(x) = x - 500,000 \quad \text{and} \quad g(x) = 0.03x.$

    If x is greater than $500,000, which of the following represents your bonus? Explain your reasoning.

    (a) $f(g(x))$ (b) $g(f(x))$

69. Prove that the product of two odd functions is an even function, and that the product of two even functions is an even function.

70. *Conjecture* Use examples to hypothesize whether the product of an odd function and an even function is even or odd. Then prove your hypothesis.

P.9 Inverse Functions

▶ **What you should learn**
- How to find inverse functions informally and verify that two functions are inverses of each other
- How to use graphs of functions to decide whether functions have inverses
- How to find inverse functions algebraically

▶ **Why you should learn it**

Inverse functions can be used to model and solve real-life problems. For instance, Exercise 81 on page 107 shows how the inverse of a function can be used to determine when damage to a diesel engine may occur.

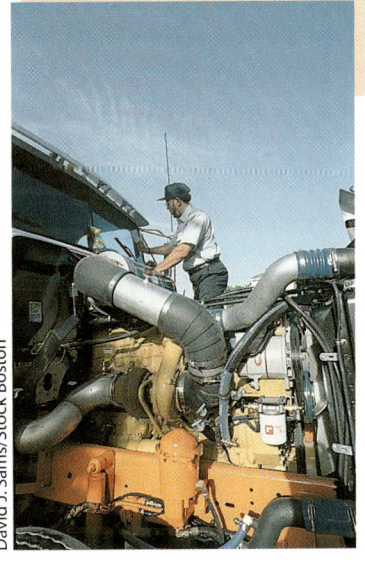

The Inverse of a Function

Recall from Section P.5 that a function can be represented by a set of ordered pairs. For instance, the function $f(x) = x + 4$ from the set $A = \{1, 2, 3, 4\}$ to the set $B = \{5, 6, 7, 8\}$ can be written as follows.

$$f(x) = x + 4: \{(1, 5), (2, 6), (3, 7), (4, 8)\}$$

In this case, by interchanging the first and second coordinates of each of these ordered pairs, you can form the **inverse function** of f, which is denoted by f^{-1}. It is a function from the set B to the set A, and can be written as follows.

$$f^{-1}(x) = x - 4: \{(5, 1), (6, 2), (7, 3), (8, 4)\}$$

Note that the domain of f is equal to the range of f^{-1}, and vice versa, as shown in Figure P.65. Also note that the functions f and f^{-1} have the effect of "undoing" each other. In other words, when you form the composition of f with f^{-1} or the composition of f^{-1} with f, you obtain the identity function.

$$f(f^{-1}(x)) = f(x - 4) = (x - 4) + 4 = x$$

$$f^{-1}(f(x)) = f^{-1}(x + 4) = (x + 4) - 4 = x$$

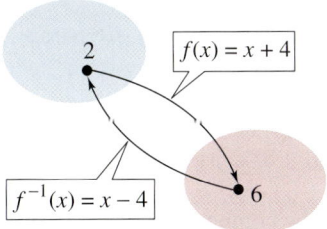

FIGURE P.65

Example 1 ▶ Finding Inverse Functions Informally

Find the inverse of $f(x) = 4x$. Then verify that both $f(f^{-1}(x))$ and $f^{-1}(f(x))$ are equal to the identity function.

Solution

The function f *multiplies* each input by 4. To "undo" this function, you need to *divide* each input by 4. So, the inverse function of $f(x) = 4x$ is

$$f^{-1}(x) = \frac{x}{4}.$$

You can verify that both $f(f^{-1}(x))$ and $f^{-1}(f(x))$ are equal to the identity function as follows.

$$f(f^{-1}(x)) = f\left(\frac{x}{4}\right) = 4\left(\frac{x}{4}\right) = x \qquad f^{-1}(f(x)) = f^{-1}(4x) = \frac{4x}{4} = x$$

Exploration

Consider the functions

$$f(x) = x + 2$$

and

$$f^{-1}(x) = x - 2.$$

Evaluate $f(f^{-1}(x))$ and $f^{-1}(f(x))$ for the indicated values of x. What can you conclude about the functions?

x	-10	0	7	45
$f(f^{-1}(x))$				
$f^{-1}(f(x))$				

Definition of the Inverse of a Function

Let f and g be two functions such that

$$f(g(x)) = x \quad \text{for every } x \text{ in the domain of } g$$

and

$$g(f(x)) = x \quad \text{for every } x \text{ in the domain of } f.$$

Under these conditions, the function g is the **inverse** of the function f. The function g is denoted by f^{-1} (read "f-inverse"). So,

$$f(f^{-1}(x)) = x \quad \text{and} \quad f^{-1}(f(x)) = x.$$

The domain of f must be equal to the range of f^{-1}, and the range of f must be equal to the domain of f^{-1}.

Don't be confused by the use of -1 to denote the inverse function f^{-1}. In this text, whenever f^{-1} is written, it *always* refers to the inverse of the function f and *not* to the reciprocal of $f(x)$.

If the function g is the inverse of the function f, it must also be true that the function f is the inverse of the function g. For this reason, you can say that the functions f and g are *inverses of each other*.

Example 2 Verifying Inverse Functions

Which of the functions is the inverse of $f(x) = \dfrac{5}{x-2}$?

$$g(x) = \dfrac{x-2}{5} \qquad h(x) = \dfrac{5}{x} + 2$$

Solution

By forming the composition of f with g, you have

$$f(g(x)) = f\left(\dfrac{x-2}{5}\right)$$

$$= \dfrac{5}{\left(\dfrac{x-2}{5}\right) - 2} \qquad \text{Substitute } \dfrac{x-2}{5} \text{ for } x.$$

$$= \dfrac{25}{x - 12} \qquad \text{Simplify.}$$

$$\neq x.$$

Because this composition is not equal to the identity function x, it follows that g is *not* the inverse of f. By forming the composition of f with h, you have

$$f(h(x)) = f\left(\dfrac{5}{x} + 2\right) = \dfrac{5}{\left(\dfrac{5}{x} + 2\right) - 2} = \dfrac{5}{\left(\dfrac{5}{x}\right)} = x.$$

So, it appears that h is the inverse of f. You can confirm this by showing that the composition of h with f is also equal to the identity function. (Try doing this.)

The Graph of the Inverse of a Function

A computer animation of this concept appears in the *Interactive* CD-ROM and *Internet* versions of this text.

The graphs of a function f and its inverse f^{-1} are related to each other in the following way. If the point (a, b) lies on the graph of f, then the point (b, a) must lie on the graph of f^{-1}, and vice versa. This means that the graph of f^{-1} is a *reflection* of the graph of f in the line $y = x$, as shown in Figure P.66.

Example 3 ▶ The Graphs of f and f^{-1}

Sketch the graphs of the inverse functions $f(x) = 2x - 3$ and $f^{-1}(x) = \frac{1}{2}(x + 3)$ on the same rectangular coordinate system and show that the graphs are reflections of each other in the line $y = x$.

Solution

The graphs of f and f^{-1} are shown in Figure P.67. It appears that the graphs are reflections of each other in the line $y = x$. You can further verify this reflective property by testing a few points on each graph. Note in the following list that if the point (a, b) is on the graph of f, the point (b, a) is on the graph of f^{-1}.

Graph of $f(x) = 2x - 3$	Graph of $f^{-1}(x) = \frac{1}{2}(x + 3)$
$(-1, -5)$	$(-5, -1)$
$(0, -3)$	$(-3, 0)$
$(1, -1)$	$(-1, 1)$
$(2, 1)$	$(1, 2)$
$(3, 3)$	$(3, 3)$

FIGURE P.66

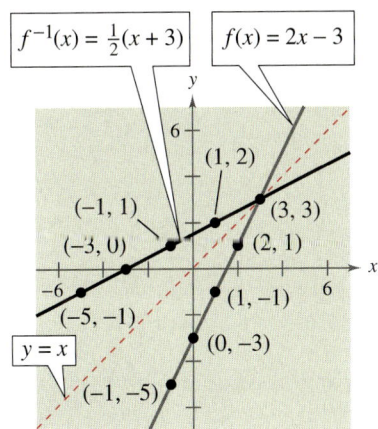
FIGURE P.67

Example 4 ▶ Finding Inverse Functions Graphically

Sketch the graphs of the inverse functions $f(x) = x^2$, $x \geq 0$ and $f^{-1}(x) = \sqrt{x}$ on the same rectangular coordinate system and show that the graphs are reflections of each other in the line $y = x$.

Solution

The graphs of f and f^{-1} are shown in Figure P.68. It appears that the graphs are reflections of each other in the line $y = x$. You can further verify this reflective property by testing a few points on each graph. Note in the following list that if the point (a, b) is on the graph of f, the point (b, a) is on the graph of f^{-1}.

Graph of $f(x) = x^2$, $x \geq 0$	Graph of $f^{-1}(x) = \sqrt{x}$
$(0, 0)$	$(0, 0)$
$(1, 1)$	$(1, 1)$
$(2, 4)$	$(4, 2)$
$(3, 9)$	$(9, 3)$

Try showing that $f(f^{-1}(x)) = x$ and $f^{-1}(f(x)) = x$.

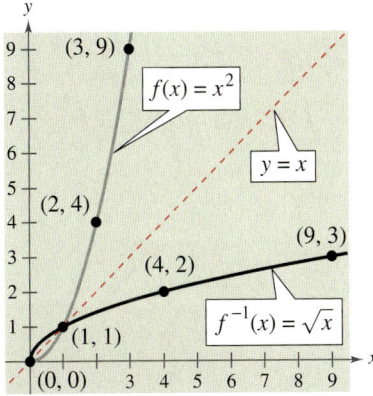
FIGURE P.68

The reflective property of the graphs of inverse functions gives you a nice *geometric* test for determining whether a function has an inverse. This test is called the **Horizontal Line Test** for inverse functions.

Horizontal Line Test for Inverse Functions

A function f has an inverse function if and only if no *horizontal* line intersects the graph of f at more than one point.

STUDY TIP

Not every function has an inverse function. Consider the following table of values for the function $f(x) = x^2$.

x	-2	-1	0	1	2	3
$f(x)$	4	1	0	1	4	9

The table of values made up by interchanging the rows of the table does not represent a function because the input $x = 4$ is matched with two different outputs: $y = -2$ and $y = 2$.

x	4	1	0	1	4	9
y	-2	-1	0	1	2	3

So, $f(x) = x^2$ does not have an inverse function.

Example 5 ▶ Applying the Horizontal Line Test

a. The graph of the function $f(x) = x^3 - 1$ is shown in Figure P.69(a). Because no horizontal line intersects the graph of f at more than one point, you can conclude that f *does* have an inverse function.

b. The graph of the function $f(x) = x^2 - 1$ is shown in Figure P.69(b). Because it is possible to find a horizontal line that intersects the graph of f at more than one point, you can conclude that f *does not* have an inverse function.

(a)

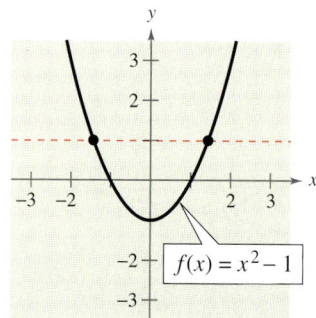
(b)

FIGURE P.69

Finding the Inverse of a Function Algebraically

For simple functions (such as the one in Example 1), you can find inverse functions by inspection. For more complicated functions, however, it is best to use the following guidelines. The key step in these guidelines is Step 3—interchanging the roles of x and y. This step corresponds to the fact that inverse functions have ordered pairs with the coordinates reversed.

Finding the Inverse of a Function

1. Use the Horizontal Line Test to decide whether f has an inverse.
2. In the equation for $f(x)$, replace $f(x)$ by y.
3. Interchange the roles of x and y, and solve for y.
4. Replace y by $f^{-1}(x)$ in the new equation.
5. Verify that f and f^{-1} are inverses of each other by showing that the domain of f is equal to the range of f^{-1}, the range of f is equal to the domain of f^{-1}, and $f(f^{-1}(x)) = x = f^{-1}(f(x))$.

Example 6 ▶ Finding the Inverse of a Function Algebraically

Find the inverse of $f(x) = \dfrac{5 - 3x}{2}$.

Solution

The graph of f is a line, as shown in Figure P.70. This graph passes the Horizontal Line Test. So, you know that f has an inverse.

$$f(x) = \dfrac{5 - 3x}{2} \qquad \text{Write original function.}$$

$$y = \dfrac{5 - 3x}{2} \qquad \text{Replace } f(x) \text{ by } y.$$

$$x = \dfrac{5 - 3y}{2} \qquad \text{Interchange } x \text{ and } y.$$

$$2x = 5 - 3y \qquad \text{Multiply each side by 2.}$$

$$3y = 5 - 2x \qquad \text{Isolate the } y\text{-term.}$$

$$y = \dfrac{5 - 2x}{3} \qquad \text{Solve for } y.$$

$$f^{-1}(x) = \dfrac{5 - 2x}{3} \qquad \text{Replace } y \text{ by } f^{-1}(x).$$

Note that both f and f^{-1} have domains and ranges that consist of the entire set of real numbers. Check that $f(f^{-1}(x)) = x$ and $f^{-1}(f(x)) = x$.

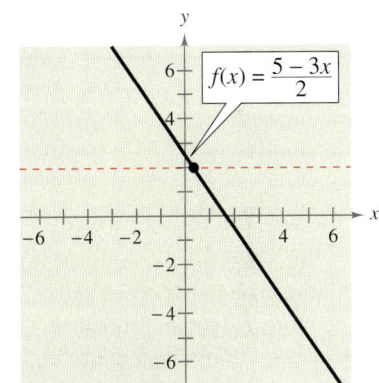

FIGURE P.70

Example 7 Finding the Inverse of a Function

Find the inverse of $f(x) = \sqrt[3]{x+1}$.

Solution

The graph of f is a curve, as shown in Figure P.71. Because this graph passes the Horizontal Line Test, you know that f has an inverse function.

FIGURE P.71

$$f(x) = \sqrt[3]{x+1} \quad \text{Write original function.}$$
$$y = \sqrt[3]{x+1} \quad \text{Replace } f(x) \text{ by } y.$$
$$x = \sqrt[3]{y+1} \quad \text{Interchange } x \text{ and } y.$$
$$x^3 = y + 1 \quad \text{Cube each side.}$$
$$x^3 - 1 = y \quad \text{Solve for } y.$$
$$x^3 - 1 = f^{-1}(x) \quad \text{Replace } y \text{ by } f^{-1}(x).$$

Both f and f^{-1} have domains and ranges that consist of the entire set of real numbers.

Check

$$f(f^{-1}(x)) = f(x^3 - 1)$$
$$= \sqrt[3]{(x^3 - 1) + 1}$$
$$= \sqrt[3]{x^3}$$
$$= x \checkmark$$

$$f^{-1}(f(x)) = f^{-1}\left(\sqrt[3]{x+1}\right)$$
$$= \left(\sqrt[3]{x+1}\right)^3 - 1$$
$$= x + 1 - 1$$
$$= x \checkmark$$

You can verify the result of Example 7 numerically as shown in the tables below.

x	-28	-9	-2	-1	0	7	26
$f(x)$	-3	-2	-1	0	1	2	3

x	-3	-2	-1	0	1	2	3
$f^{-1}(x)$	-28	-9	-2	-1	0	7	26

Writing ABOUT MATHEMATICS

The Existence of an Inverse Function Write a short paragraph describing why the following functions do or do not have inverse functions.

a. Let x represent the retail price of an item (in dollars), and let $f(x)$ represent the sales tax on the item. Assume that the sales tax is 6% of the retail price *and* that the sales tax is rounded to the nearest cent. Does this function have an inverse? (*Hint:* Can you undo this function? For instance, if you know that the sales tax is $0.12, can you determine *exactly* what the retail price is?)

b. Let x represent the temperature in degrees Celsius, and let $f(x)$ represent the temperature in degrees Fahrenheit. Does this function have an inverse? (*Hint:* The formula for converting from degrees Celsius to degrees Fahrenheit is $F = \frac{9}{5}C + 32$.)

P.9 Exercises

In Exercises 1–4, match the graph of the function with the graph of its inverse. [The graphs of the inverse functions are labeled (a), (b), (c), and (d).]

(a)

(b)

(c)

(d)

1.

2.

3.

4.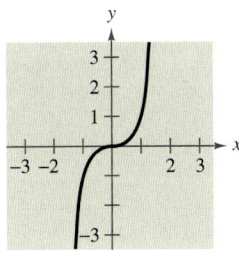

In Exercises 5–12, find the inverse of f informally. Verify that $f(f^{-1}(x)) = x$ and $f^{-1}(f(x)) = x$.

5. $f(x) = 6x$
6. $f(x) = \frac{1}{3}x$
7. $f(x) = x + 9$
8. $f(x) = x - 4$
9. $f(x) = 3x + 1$
10. $f(x) = \dfrac{x - 1}{5}$
11. $f(x) = \sqrt[3]{x}$
12. $f(x) = x^5$

In Exercises 13–24, show that f and g are inverse functions (a) algebraically and (b) graphically.

13. $f(x) = 2x$, $\quad g(x) = \dfrac{x}{2}$
14. $f(x) = x - 5$, $\quad g(x) = x + 5$
15. $f(x) = 5x + 1$, $\quad g(x) = \dfrac{x - 1}{5}$
16. $f(x) = 3 - 4x$, $\quad g(x) = \dfrac{3 - x}{4}$
17. $f(x) = x^3$, $\quad g(x) = \sqrt[3]{x}$
18. $f(x) = \dfrac{1}{x}$, $\quad g(x) = \dfrac{1}{x}$
19. $f(x) = \sqrt{x - 4}$, $\quad g(x) = x^2 + 4, \quad x \geq 0$
20. $f(x) = 1 - x^3$, $\quad g(x) = \sqrt[3]{1 - x}$
21. $f(x) = 9 - x^2, \; x \geq 0$, $\quad g(x) = \sqrt{9 - x}, \; x \leq 9$
22. $f(x) = \dfrac{1}{1 + x}, \quad x \geq 0$

   $g(x) = \dfrac{1 - x}{x}, \quad 0 < x \leq 1$

23. $f(x) = \dfrac{x - 1}{x + 5}$, $\quad g(x) = -\dfrac{5x + 1}{x - 1}$
24. $f(x) = \dfrac{x + 3}{x - 2}$, $\quad g(x) = \dfrac{2x + 3}{x - 1}$

In Exercises 25 and 26, does the function have an inverse?

25.

x	−1	0	1	2	3	4
$f(x)$	−2	1	2	1	−2	−6

26.

x	−3	−2	−1	0	2	3
$f(x)$	10	6	4	1	−3	−10

In Exercises 27 and 28, use the table of values for $y = f(x)$ to complete a table for $y = f^{-1}(x)$.

27.
x	-2	-1	0	1	2	3
$f(x)$	-2	0	2	4	6	8

28.
x	-3	-2	-1	0	1	2
$f(x)$	-10	-7	-4	-1	2	5

In Exercises 29–32, does the function have an inverse?

29.

30.

31.

32.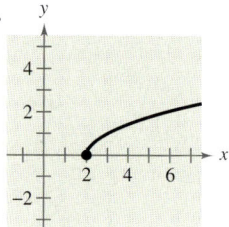

In Exercises 33–38, use a graphing utility to graph the function and use the Horizontal Line Test to determine whether the function has an inverse.

33. $g(x) = \dfrac{4 - x}{6}$

34. $f(x) = 10$

35. $h(x) = |x + 4| - |x - 4|$

36. $g(x) = (x + 5)^3$

37. $f(x) = -2x\sqrt{16 - x^2}$

38. $f(x) = \frac{1}{8}(x + 2)^2 - 1$

In Exercises 39–54, find the inverse of the function f. Then graph both f and f^{-1} on the same set of coordinate axes.

39. $f(x) = 2x - 3$
40. $f(x) = 3x + 1$
41. $f(x) = x^5 - 2$
42. $f(x) = x^3 + 1$
43. $f(x) = \sqrt{x}$
44. $f(x) = x^2$, $x \geq 0$
45. $f(x) = \sqrt{4 - x^2}$, $0 \leq x \leq 2$
46. $f(x) = x^2 - 2$, $x \leq 0$
47. $f(x) = \dfrac{4}{x}$
48. $f(x) = -\dfrac{2}{x}$
49. $f(x) = \dfrac{x + 1}{x - 2}$
50. $f(x) = \dfrac{x - 3}{x + 2}$
51. $f(x) = \sqrt[3]{x - 1}$
52. $f(x) = x^{3/5}$
53. $f(x) = \dfrac{6x + 4}{4x + 5}$
54. $f(x) = \dfrac{8x - 4}{2x + 6}$

In Exercises 55–68, determine whether the function has an inverse. If it does, find the inverse.

55. $f(x) = x^4$
56. $f(x) = \dfrac{1}{x^2}$
57. $g(x) = \dfrac{x}{8}$
58. $f(x) = 3x + 5$
59. $p(x) = -4$
60. $f(x) = \dfrac{3x + 4}{5}$
61. $f(x) = (x + 3)^2$, $x \geq -3$
62. $q(x) = (x - 5)^2$
63. $f(x) = \begin{cases} x + 3, & x < 0 \\ 6 - x, & x \geq 0 \end{cases}$
64. $f(x) = \begin{cases} -x, & x \leq 0 \\ x^2 - 3x, & x > 0 \end{cases}$
65. $h(x) = \dfrac{1}{x}$
66. $f(x) = |x - 2|$, $x \leq 2$

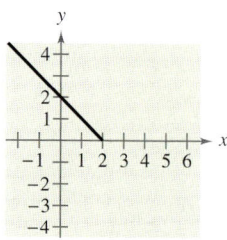

67. $f(x) = \sqrt{2x + 3}$
68. $f(x) = \sqrt{x - 2}$

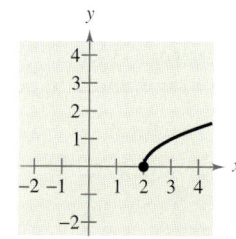

In Exercises 69–74, use the functions $f(x) = \frac{1}{8}x - 3$ and $g(x) = x^3$ to find the indicated value or function.

69. $(f^{-1} \circ g^{-1})(1)$
70. $(g^{-1} \circ f^{-1})(-3)$
71. $(f^{-1} \circ f^{-1})(6)$
72. $(g^{-1} \circ g^{-1})(-4)$
73. $(f \circ g)^{-1}$
74. $g^{-1} \circ f^{-1}$

In Exercises 75–78, use the functions $f(x) = x + 4$ and $g(x) = 2x - 5$ to find the specified function.

75. $g^{-1} \circ f^{-1}$
76. $f^{-1} \circ g^{-1}$
77. $(f \circ g)^{-1}$
78. $(g \circ f)^{-1}$

79. *Hourly Wage* Your wage is $8.00 per hour plus $0.75 for each unit produced per hour. So, your hourly wage y in terms of the number of units produced is

$$y = 8 + 0.75x.$$

(a) Find the inverse of the function.

(b) What does each variable represent in the inverse function?

(c) Determine the number of units produced when your hourly wage is $22.25.

80. *Cost* Suppose you need a total of 50 pounds of two commodities costing $1.25 and $1.60 per pound, respectively.

(a) Verify that the total cost is

$$y = 1.25x + 1.60(50 - x)$$

where x is the number of pounds of the less expensive commodity.

(b) Find the inverse of the cost function. What does each variable represent in the inverse function?

(c) Use the context of the problem to determine the domain of the inverse function.

(d) Determine the number of pounds of the less expensive commodity purchased if the total cost is $73.

81. *Diesel Mechanics* The function

$$y = 0.03x^2 + 245.50, \quad 0 < x < 100$$

approximates the exhaust temperature y in degrees Fahrenheit where x is the percent load for a diesel engine.

(a) Find the inverse of the function. What does each variable represent in the inverse function?

(b) Use a graphing utility to graph the inverse function.

(c) Determine the percent load interval if the exhaust temperature of the engine must not exceed 500 degrees Fahrenheit.

82. *New Car Sales* The total value of new car sales f (in billions of dollars) in the United States from 1992 through 1997 is shown in the table. The time (in years) is given by t, with $t = 2$ corresponding to 1992. (Source: National Automobile Dealers Association)

t	2	3	4	5	6	7
$f(t)$	333.8	377.3	430.6	456.2	490.0	507.5

(a) Does f^{-1} exist?

(b) If f^{-1} exists, what does it mean in the context of the problem?

(c) If f^{-1} exists, find $f^{-1}(456.2)$.

83. If the table in Exercise 82 were extended to 1998 and if the total value of new car sales for that year was $430.6 billion, would f^{-1} exist? Explain.

84. *Cellular Phones* The average local bill (in dollars) for cellular phones in the United States from 1990 to 1997 is shown in the table. The time (in years) is given by t, with $t = 0$ corresponding to 1990. (Source: Cellular Telecommunications Industry Association)

t	0	1	2	3
$f(t)$	80.90	72.74	68.68	61.48

t	4	5	6	7
$f(t)$	56.21	51.00	47.70	42.78

(a) Find $f^{-1}(51)$.

(b) What does f^{-1} mean in the context of the problem?

(c) Use the regression feature of a graphing utility to find a linear model for the data, $y = mx + b$. Round m and b to two decimal places.

(d) Algebraically find the inverse of the linear model in part (c).

(e) Use the inverse of the linear model you found in part (d) to approximate $f^{-1}(11)$.

85. *Soft Drink Consumption* The per capita consumption of regular soft drinks f (in gallons) in the United States from 1991 through 1996 is shown in the table. The time (in years) is given by t, with $t = 1$ corresponding to 1991. (Source: U.S. Department of Agriculture)

t	1	2	3	4	5	6
$f(t)$	36.3	36.9	38.4	39.5	39.8	40.2

(a) Does f^{-1} exist? If so, what does it represent in the context of the problem?

(b) If f^{-1} exists, what is $f^{-1}(39.8)$?

Synthesis

True or False? In Exercises 86–89, determine whether the statement is true or false. Justify your answer.

86. If f is an even function, f^{-1} exists.

87. If the inverse of f exists and the y-intercept of the graph of f exists, the y-intercept of f is an x-intercept of f^{-1}.

88. If $f(x) = x^n$ where n is odd, f^{-1} exists.

89. There exists no function f such that $f = f^{-1}$.

In Exercises 90–93, use the graph of the function f to create a table of values for the given points. Then create a second table that can be used to find f^{-1} and sketch the graph of f^{-1} if possible.

90.

91.

92.

93.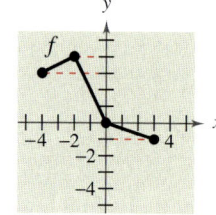

94. *Think About It* The function
$$f(x) = k(2 - x - x^3)$$
has an inverse, and $f^{-1}(3) = -2$. Find k.

Chapter Summary

What did you learn?

Section P.1	Review Exercises
☐ How to represent and classify real numbers	1, 2
☐ How to order real numbers and use inequalities	3–6
☐ How to find the absolute values of real numbers and find the distance between two real numbers	7–10
☐ How to evaluate algebraic expressions	11, 12
☐ How to use the basic rules and properties of algebra	13–22

Section P.2	
☐ How to identify different types of equations	23, 24
☐ How to solve linear equations in one variable	25–29
☐ How to solve quadratic equations by factoring, extracting square roots, and using the Quadratic Formula	30–35
☐ How to solve polynomial equations of degree three or greater	36, 37
☐ How to solve equations involving radicals	38–41, 44
☐ How to solve equations involving absolute values	42, 43

Section P.3	
☐ How to plot points in the Cartesian plane	45–54
☐ How to use the Distance Formula to find the distance between two points	51, 52
☐ How to use the Midpoint Formula to find the midpoint of a line segment	53, 54
☐ How to sketch graphs of equations	55–58
☐ How to find x- and y-intercepts of graphs	59–62
☐ How to use symmetry to sketch graphs of equations	63–66
☐ How to find equations and sketch graphs of circles	67–72

Section P.4	
☐ How to use slope to graph linear equations in two variables	73–82
☐ How to find slopes of lines	83–88
☐ How to write linear equations in two variables	89–96
☐ How to use slope to identify parallel and perpendicular lines	97, 98
☐ How to use linear equations in two variables to model and solve real-life problems	99–102

Section P.5

	Review Exercises
☐ How to decide whether relations between two variables are functions	103–108
☐ How to use function notation and evaluate functions	109, 110
☐ How to find the domains of functions	111–116
☐ How to use functions to model and solve real-life problems	117–120

Section P.6

☐ How to use the Vertical Line Test for functions	121–124
☐ How to find the zeros of functions	125–128
☐ How to determine intervals on which functions are increasing or decreasing	129–132
☐ How to identify and graph linear functions	133–136
☐ How to identify and graph step functions and other piecewise-defined functions	137, 138
☐ How to identify even and odd functions	139–142

Section P.7

☐ How to recognize graphs of common functions	143–146
☐ How to use vertical and horizontal shifts to sketch graphs of functions	147–150
☐ How to use reflections to sketch graphs of functions	151–154
☐ How to use nonrigid transformations to sketch graphs of functions	155–158

Section P.8

☐ How to add, subtract, multiply, and divide functions	159–162
☐ How to find compositions of one function with another function	163–166
☐ How to use combinations of functions to model and solve real-life problems	167, 168

Section P.9

☐ How to find inverse functions informally and verify that two functions are inverses of each other	169–172
☐ How to use graphs of functions to decide whether functions have inverses	173–178
☐ How to find inverse functions algebraically	179–184

Review Exercises

P.1 In Exercises 1 and 2, determine which numbers in the set are (a) natural numbers, (b) integers, (c) rational numbers, and (d) irrational numbers.

1. $\{11, -14, -\frac{8}{9}, \frac{5}{2}, \sqrt{6}, 0.4\}$
2. $\{\sqrt{15}, -22, -\frac{10}{3}, 0, 5.2, \frac{3}{7}\}$

In Exercises 3 and 4, use a calculator to find the decimal form of each rational number. If it is a nonterminating decimal, write the repeating pattern. Then plot the numbers on the real number line and place the appropriate inequality sign (< or >) between them.

3. (a) $\frac{5}{6}$ (b) $\frac{7}{8}$ 4. (a) $\frac{9}{25}$ (b) $\frac{5}{7}$

In Exercises 5 and 6, give a verbal description of the subset of real numbers represented by the inequality, and sketch the subset on the real number line.

5. $x \leq 7$
6. $x > 1$

In Exercises 7 and 8, find the distance between a and b.

7. $a = -92, b = 63$
8. $a = -112, b = -6$

In Exercises 9 and 10, use absolute value notation to describe the statement.

9. The distance between x and 25 is no more than 10.
10. The distance between z and -16 is greater than 8.

In Exercises 11 and 12, evaluate the expression for each value of x. (If not possible, state the reason.)

Expression	Values	
11. $-x^2 + x - 1$	(a) $x = 1$	(b) $x = -1$
12. $\dfrac{x}{x-3}$	(a) $x = -3$	(b) $x = 3$

In Exercises 13–16, identify the rule of algebra illustrated by the equation.

13. $2x + (3x - 10) = (2x + 3x) - 10$
14. $(t + 4)(2t) = (2t)(t + 4)$
15. $\dfrac{2}{y+4} \cdot \dfrac{y+4}{2} = 1, \quad y \neq -4$
16. $0 + (a - 5) = a - 5$

In Exercises 17–22, perform the operation(s) without using a calculator.

17. $|-3| + 4(-2) - 6$
18. $\dfrac{|-10|}{-10}$
19. $\dfrac{5}{18} \div \dfrac{10}{3}$
20. $(16 - 8) \div 4$
21. $6[4 - 2(6 + 8)]$
22. $-4[16 - 3(7 - 10)]$

P.2 In Exercises 23 and 24, determine whether the equation is an identity or a conditional equation.

23. $6 - (x - 2)^2 = 2 + 4x - x^2$
24. $3(x - 2) + 2x = 2(x + 3)$

In Exercises 25–28, solve the equation (if possible) and check your solution.

25. $3x - 2(x + 5) = 10$
26. $4x + 2(7 - x) = 5$
27. $4(x + 3) - 3 = 2(4 - 3x) - 4$
28. $\frac{1}{2}(x - 3) - 2(x + 1) = 5$

29. *Mixture Problem* A car radiator contains 10 liters of a 30% antifreeze solution. How many liters will have to be replaced with pure antifreeze if the resulting solution is to be 50% antifreeze?

In Exercises 30–35, use any method to solve the equation.

30. $15 + x - 2x^2 = 0$
31. $2x^2 - x - 28 = 0$
32. $6 = 3x^2$
33. $(x - 8)^2 = 15$
34. $x^2 - 12x + 30 = 0$
35. $-2x^2 - 5x + 27 = 0$

In Exercises 36–43, find all solutions of the equation. Check your solutions in the original equation.

36. $x^4 - 5x^2 + 6 = 0$
37. $4x^3 - 6x^2 = 0$
38. $\sqrt{x + 4} = 3$
39. $\sqrt{3x - 2} = 4 - x$
40. $(x - 1)^{2/3} - 25 = 0$
41. $(x + 2)^{3/4} = 27$
42. $|x - 5| = 10$
43. $|x^2 - 6| = x$

44. *Economics* The demand equation for a product is
$$p = 42 - \sqrt{0.001x + 2}$$
where x is the number of units demanded per day and p is the price per unit. Find the demand if the price is set at $29.95.

P.3 *Geometry* In Exercises 45 and 46, plot the points and verify that the points form the polygon.

45. Right triangle: $(2, 3), (13, 11), (5, 22)$
46. Parallelogram: $(1, 2), (8, 3), (9, 6), (2, 5)$

In Exercises 47–50, determine the quadrant(s) in which (x, y) is located so that the condition(s) is (are) satisfied.

47. $x > 0$ and $y = -2$
48. $y > 0$
49. $(-x, y)$ is in the third quadrant.
50. $xy = 4$

In Exercises 51 and 52, (a) plot the points and (b) find the distance between the points.

51. $(-3, 8), (1, 5)$
52. $(5.6, 0), (0, 8.2)$

In Exercises 53 and 54, (a) plot the points and (b) find the midpoint of the line segment joining the points.

53. $(-2, 6), (4, -3)$
54. $(0, -1.2), (-3.6, 0)$

In Exercises 55–58, sketch the graph *by hand*.

55. $y - 2x - 3 = 0$
56. $y = |x - 8|$
57. $y + 2x^2 = 0$
58. $y = \sqrt{x + 2}$

In Exercises 59–62, find the *x*- and *y*-intercepts of the graph of the equation.

59. $y = 2x - 9$
60. $y = 4 - (x - 4)^2$
61. $y = x\sqrt{9 - x^2}$
62. $y = |x + 2| + |3 - x|$

In Exercises 63–66, use intercepts and symmetry to sketch the graph of the equation.

63. $y = 5 - x^2$
64. $y = -6 - x^3$
65. $y = \sqrt{x + 5}$
66. $y = |x| + 9$

In Exercises 67–70, find the center and radius of the circle and sketch its graph.

67. $(x + 2)^2 + y^2 = 16$
68. $x^2 + (y - 8)^2 = 81$
69. $\left(x - \frac{1}{2}\right)^2 + (y + 1)^2 = 36$
70. $(x + 4)^2 + \left(y - \frac{3}{2}\right)^2 = 100$

71. Find the standard form of the equation of the circle for which the endpoints of a diameter are $(0, 0)$ and $(4, -6)$.

72. Find the standard form of the equation of the circle for which the endpoints of a diameter are $(-2, -3)$ and $(4, -10)$.

P.4 In Exercises 73–80, sketch the graph of the linear equation.

73. $y = -2x - 7$
74. $y = 4x - 3$
75. $y = 6$
76. $x = -3$
77. $y = 3x + 13$
78. $y = -10x + 9$
79. $y = -\frac{5}{2}x - 1$
80. $y = \frac{5}{6}x + 5$

In Exercises 81 and 82, use the point on the line and the slope of the line to find three additional points through which the line passes. (There are many correct answers.)

	Point	Slope
81.	$(2, -1)$	$m = \frac{1}{4}$
82.	$(-3, 5)$	$m = -\frac{3}{2}$

In Exercises 83–86, plot the points and find the slope of the line passing through the pair of points.

83. $(3, -4), (-7, 1)$
84. $(-1, 8), (6, 5)$
85. $(-4.5, 6), (2.1, 3)$
86. $(-3, 2), (8, 2)$

In Exercises 87 and 88, use the concept of slope to find *t* such that the three points are on the same line.

87. $(-2, 5), (0, t), (1, 1)$
88. $(-6, 1), (1, t), (10, 5)$

In Exercises 89–92, find an equation of the line that passes through the points.

89. $(0, 0), (0, 10)$
90. $(2, 5), (-2, -1)$
91. $(-1, 4), (2, 0)$
92. $(11, -2), (6, -1)$

In Exercises 93–96, find an equation of the line that passes through the given point and has the specified slope. Sketch the graph of the line.

	Point	Slope
93.	$(0, -5)$	$m = \frac{3}{2}$
94.	$(-2, 6)$	$m = 0$
95.	$(10, -3)$	$m = -\frac{1}{2}$
96.	$(-8, 5)$	Undefined

In Exercises 97 and 98, write an equation of the line through the point (a) parallel to the given line and (b) perpendicular to the given line.

	Point	Line
97.	$(3, -2)$	$5x - 4y = 8$
98.	$(-8, 3)$	$2x + 3y = 5$

Rate of Change In Exercises 99 and 100, you are given the dollar value of a product in the year 2000 *and* the rate at which the value of the item is expected to change during the next 5 years. Write a linear equation that gives the dollar value V of the product in terms of the year t. (Let $t = 0$ represent 2000.)

	2000 Value	Rate
99.	$12,500	$850 increase per year
100.	$72.95	$5.15 increase per year

101. *Business* During the second and third quarters of the year, a business had sales of $160,000 and $185,000, respectively. If the growth of sales follows a linear pattern, estimate sales during the fourth quarter.

102. *Inflation* The dollar value of a product in 1999 is $85, and the product is expected to increase in value at a rate of $3.75 per year.
 (a) Write a linear equation that gives the dollar value V of the product in terms of the year t. (Let $t = 9$ represent 1999.)
 (b) Use a graphing utility to graph the sales equation.
 (c) Move the cursor along the graph of the sales model to estimate the dollar value of the product in 2005.

P.5 In Exercises 103 and 104, determine which of the sets of ordered pairs represents a function from A to B. Give reasons for your answers.

103. $A = \{10, 20, 30, 40\}$ and $B = \{0, 2, 4, 6\}$
 (a) $\{(20, 4), (40, 0), (20, 6), (30, 2)\}$
 (b) $\{(10, 4), (20, 4), (30, 4), (40, 4)\}$
 (c) $\{(40, 0), (30, 2), (20, 4), (10, 6)\}$
 (d) $\{(20, 2), (10, 0), (40, 4)\}$

104. $A = \{u, v, w\}$ and $B = \{-2, -1, 0, 1, 2\}$
 (a) $\{(v, -1), (u, 2), (w, 0), (u, -2)\}$
 (b) $\{(u, -2), (v, 2), (w, 1)\}$
 (c) $\{(u, 2), (v, 2), (w, 1), (w, 1)\}$
 (d) $\{(w, -2), (v, 0), (w, 2)\}$

In Exercises 105–108, determine whether the equation represents y as a function of x.

105. $16x - y^4 = 0$
106. $2x - y - 3 = 0$
107. $y = \sqrt{1 - x}$
108. $|y| = x + 2$

In Exercises 109 and 110, evaluate the function as indicated. Simplify your answers.

109. $f(x) = x^2 + 1$
 (a) $f(2)$
 (b) $f(-4)$
 (c) $f(t^2)$
 (d) $-f(x)$

110. $g(x) = x^{4/3}$
 (a) $g(8)$
 (b) $g(t + 1)$
 (c) $\dfrac{g(8) - g(1)}{8 - 1}$
 (d) $g(-x)$

In Exercises 111–116, determine the domain of the function. Verify your result with a graph.

111. $f(x) = \sqrt{25 - x^2}$
112. $f(x) = 3x + 4$
113. $g(s) = \dfrac{5}{3s - 9}$
114. $f(x) = \sqrt{x^2 + 8x}$
115. $h(x) = \dfrac{x}{x^2 - x - 6}$
116. $h(t) = |t + 1|$

117. *Physics* The velocity of a ball thrown vertically upward from ground level is $v(t) = -32t + 48$, where t is the time in seconds and v is the velocity in feet per second.
 (a) Find the velocity when $t = 1$.
 (b) Find the time when the ball reaches its maximum height. [*Hint:* Find the time when $v(t) = 0$.]
 (c) Find the velocity when $t = 2$.

118. *Total Cost* A company produces a product for which the variable cost is $5.35 per unit and the fixed costs are $16,000. The company sells the product for $8.20 and can sell all that it produces.
 (a) Find the total cost as a function of x, the number of units produced.
 (b) Find the profit as a function of x.

119. *Geometry* A wire 24 inches long is to be cut into four pieces to form a rectangle with one side of length x.
 (a) Express the area A of the rectangle as a function of x.
 (b) Determine the domain of the function.

120. *Mixture Problem* From a full 50-liter container of a 40% concentration of acid, x liters is removed and replaced with 100% acid.

(a) Write the amount of acid in the final mixture as a function of x.

(b) Determine the domain and range of the function.

(c) Determine x if the final mixture is 50% acid.

P.6 In Exercises 121–124, use the Vertical Line Test to determine whether y is a function of x.

121. $y = (x - 3)^2$

122. $y = -\frac{3}{5}x^3 - 2x + 1$

 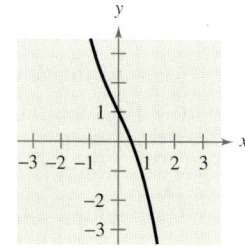

123. $x - 4 = y^2$

124. $x = -|4 - y|$

 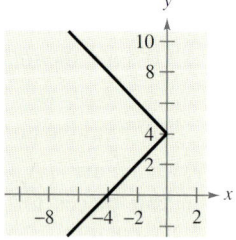

In Exercises 125–128, find the zeros of the function.

125. $f(x) = 3x^2 - 16x + 21$

126. $f(x) = 5x^2 + 4x - 1$

127. $f(x) = \dfrac{8x + 3}{11 - x}$

128. $f(x) = x^3 - x^2 - 25x + 25$

Graphical Analysis In Exercises 129–132, use a graphing utility to graph the function. Approximate the open intervals over which the function is increasing, decreasing, or constant.

129. $g(x) = |x + 2| - |x - 2|$

130. $f(x) = (x^2 - 4)^2$

131. $h(x) = 4x^3 - x^4$

132. $g(x) = \sqrt[3]{x(x + 3)^2}$

In Exercises 133–136, write the linear function f so that the following are true. Then use a graphing utility to graph the function.

133. $f(2) = -6, \ f(-1) = 3$

134. $f(0) = -5, \ f(4) = -8$

135. $f\left(-\frac{4}{5}\right) = 2, \ f\left(\frac{11}{5}\right) = 7$

136. $f(3.3) = 5.6, \ f(-4.7) = -1.4$

In Exercises 137 and 138, graph the function.

137. $f(x) = \begin{cases} 5x - 3, & x \geq -1 \\ -4x + 5, & x < -1 \end{cases}$

138. $f(x) = \begin{cases} x^2 - 2, & x < -2 \\ 5, & -2 \leq x \leq 0 \\ 8x - 5, & x > 0 \end{cases}$

In Exercises 139–142, determine whether the function is even, odd, or neither.

139. $f(x) = x^5 + 4x - 7$ **140.** $f(x) = x^4 - 20x^2$

141. $f(x) = 2x\sqrt{x^2 + 3}$ **142.** $f(x) = \sqrt[5]{6x^2}$

P.7 In Exercises 143–146, identify the common function and describe the transformation shown in the graph.

143.   **144.**

145.   **146.**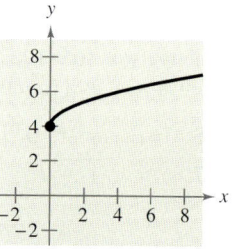

Review Exercises

In Exercises 147–158, identify the transformation of the graph of f and sketch the graph of h.

147. $f(x) = x^2$, $h(x) = x^2 - 9$
148. $f(x) = x^3$, $h(x) = (x - 2)^3 + 2$
149. $f(x) = \sqrt{x}$, $h(x) = \sqrt{x - 7}$
150. $f(x) = |x|$, $h(x) = |x + 3| - 5$
151. $f(x) = x^2$, $h(x) = -(x + 3)^2 + 1$
152. $f(x) = x^3$, $h(x) = -(x - 5)^3 - 5$
153. $f(x) = \sqrt{x}$, $h(x) = -\sqrt{x + 1} + 9$
154. $f(x) = |x|$, $h(x) = -|-x + 4| + 6$
155. $f(x) = x^2$, $h(x) = -2(x + 1)^2 - 3$
156. $f(x) = x^3$, $h(x) = -\frac{1}{3}x^3$
157. $f(x) = \sqrt{x}$, $h(x) = -2\sqrt{x - 4}$
158. $f(x) = |x|$, $h(x) = \frac{1}{2}|x| - 1$

P.8 In Exercises 159–166, let $f(x) = 3 - 2x$, $g(x) = \sqrt{x}$, and $h(x) = 3x^2 + 2$. Find the indicated value.

159. $(f - g)(4)$
160. $(f + g)(4)$
161. $(fh)(1)$
162. $\left(\dfrac{f}{h}\right)(0)$
163. $(h \circ g)(7)$
164. $(f \circ h)(3)$
165. $(g \circ f)(-2)$
166. $(g \circ h)(-1)$

Data Analysis In Exercises 167 and 168, use the table, which shows the total values (in billions of dollars) of U.S. imports from Mexico and Canada for the years 1992 through 1997. The variables y_1 and y_2 represent the total values of imports from Mexico and Canada, respectively. (Source: U.S. Bureau of the Census)

Year	1992	1993	1994	1995	1996	1997
y_1	35.2	39.9	49.5	62.1	74.3	85.9
y_2	98.6	111.2	128.4	144.4	155.9	168.2

167. Use a graphing utility to find quadratic models for each of the variables. Let $t = 2$ represent 1992.
168. Use a graphing utility to graph y_1, y_2, and $y_1 + y_2$ in the same viewing window. Use the model to estimate the total value of U.S. imports from Canada and Mexico in 2002.

P.9 In Exercises 169–172, find the inverse of f informally. Verify that $f(f^{-1}(x)) = x = f^{-1}(f(x))$.

169. $f(x) = 6x$
170. $f(x) = \frac{1}{12}x$
171. $f(x) = x - 7$
172. $f(x) = x + 5$

In Exercises 173–178, use a graphing utility to graph each function and determine whether the function has an inverse.

173. $f(x) = 3x^3 - 5$
174. $f(x) = -\frac{1}{4}x^2 - 3$
175. $f(x) = -\sqrt{4 - x}$
176. $f(x) = -x\sqrt{25 - 3x^2}$
177. $f(x) = -|x + 2| + |7 - x|$
178. $f(x) = \dfrac{x^2 + 5}{x - 3}$

In Exercises 179–182, (a) find f^{-1}, (b) sketch the graphs of f and f^{-1} on the same coordinate system, and (c) verify that $f^{-1}(f(x)) = x = f(f^{-1}(x))$.

179. $f(x) = \frac{1}{2}x - 3$
180. $f(x) = 5x - 7$
181. $f(x) = \sqrt{x + 1}$
182. $f(x) = x^3 + 2$

In Exercises 183 and 184, restrict the domain of the function f to an interval over which the function is increasing and determine f^{-1} over that interval.

183. $f(x) = 2(x - 4)^2$
184. $f(x) = |x - 2|$

Synthesis

True or False? In Exercises 185–188, determine whether the statement is true or false. Justify your answer.

185. $\dfrac{x^3 - 1}{x - 1} = x^2 + x + 1$ for all values of x.

186. The Quadratic Formula can be used to solve any quadratic equation, but not all quadratic equations can be factored.

187. Relative to the graph of $f(x) = \sqrt{x}$, the function $h(x) = -\sqrt{x + 9} - 13$ is shifted 9 units to the left and 13 units down, then is reflected in the x-axis.

188. If f and g are two inverse functions, then the domain of g is equal to the range of f.

189. *Error Analysis* What is wrong with the following solution?

$$|11x - 4| = -26$$
$$11x - 4 = -26 \quad \text{or} \quad 11x - 4 = 26$$
$$11x = -22 \qquad\qquad 11x = 30$$
$$x = -2 \qquad\qquad x = \dfrac{30}{11}$$

190. Explain how to tell whether a relation between two variables is a function.

191. Explain the difference between the Vertical Line Test and the Horizontal Line Test.

Chapter Project ▶ A Graphical Approach to Maximization

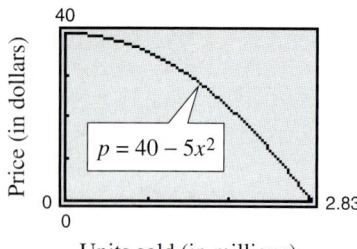

In business, a **demand function** gives the price per unit p in terms of the number of units sold x. The demand function whose graph is shown at the left is

$$p = 40 - 5x^2, \quad 0 \leq x \leq \sqrt{8} \qquad \text{Demand function}$$

where x is measured in millions of units. Note that as the price decreases, the number of units sold increases. The **revenue** R (in millions of dollars) is determined by multiplying the number of units sold by the price per unit. So,

$$R = xp = x(40 - 5x^2), \quad 0 \leq x \leq \sqrt{8}. \qquad \text{Revenue function}$$

Example ▶ Finding the Maximum Revenue

Use a graphing utility to sketch the graph of the revenue function $R = 40x - 5x^3$ for $0 \leq x \leq \sqrt{8}$. How many units should be sold to obtain maximum revenue? What price per unit should be charged to obtain maximum revenue?

Solution

To begin, you need to determine a viewing window that will display the part of the graph that is important to this problem. The domain is given, so you can set the x-boundaries of the graph between 0 and $\sqrt{8}$. To determine the y-boundaries, however, you need to experiment a little. After calculating several values of R, you could decide to use y-boundaries between 0 and 50, as shown in the graph at the left. Next, you can use the trace key to find that the maximum revenue of about \$43.5 million occurs when x is approximately 1.64 million units. To find the price per unit that corresponds to this maximum revenue, you can substitute $x = 1.64$ into the demand function to obtain

$$p = 40 - 5(1.64)^2 \approx \$26.55.$$

Chapter Project Investigations

1. For the demand function $p = 40 - 5x^2$, match each of the points $(0, 40)$ and $(\sqrt{8}, 0)$ with statement (a) or (b). Explain your reasoning.
   (a) No one will buy the product at this price.
   (b) You can't give more than this number away.

2. Use a graphing utility to zoom in on the maximum point of the revenue function in the example. (Use a setting of $1.62 \leq x \leq 1.65$ and $43.5 \leq y \leq 43.6$.) Use the trace feature to improve the accuracy of the approximation obtained in the example. Do you think this improved accuracy is appropriate in the context of this particular problem? Does it change the price?

3. For the revenue function discussed in the example, the cost of producing each unit is \$15, so the total cost of producing x million units is $C = 15x$. Use a graphing utility to graph the profit function $P = R - C$ to determine how many units should be sold to obtain maximum profit. What price per unit should be charged to obtain maximum profit?

Chapter Test

Take this test as you would take a test in class. After you are done, check your work against the answers given in the back of the book.

1. Place < or > between the real numbers $-\frac{10}{3}$ and $-|-4|$.
2. Find the distance between the real numbers -5.4 and $3\frac{3}{4}$.

In Exercises 3–5, use intercepts and symmetry to sketch the complete graph of the equation.

3. $y = 4 - \frac{3}{4}|x|$
4. $y = 4 - (x-2)^2$
5. $(x-3)^2 + y^2 = 9$

In Exercises 6–9, solve the equation (if possible).

6. $\frac{2}{3}(x-1) + \frac{1}{4}x = 10$
7. $(x-3)(x+2) = 14$
8. $2\sqrt{x} - \sqrt{2x+1} = 1$
9. $|3x - 1| = 7$

10. Find an equation of the line that passes through the point $(3, 8)$ and is (a) parallel to and (b) perpendicular to the line $-4x + 7y = -5$.
11. Evaluate the function $f(x) = |x+2| - 15$ at each specified value.
    (a) $f(-8)$
    (b) $f(14)$
    (c) $f(x-6)$

In Exercises 12 and 13, determine the domain of the function.

12. $f(x) = \sqrt{100 - x^2}$
13. $f(x) = |-x+6| + 2$

In Exercises 14–16, (a) use a graphing utility to graph the function, (b) approximate the open intervals over which the function is increasing, decreasing, or constant, and (c) determine whether the function is even, odd, or neither.

14. $f(x) = 2x^6 + 5x^4 - x^2$
15. $f(x) = 4x\sqrt{3-x}$
16. $f(x) = |x+5|$

17. Sketch the graph of $f(x) = \begin{cases} 3x + 7, & x \leq -3 \\ 4x^2 - 1, & x > -3 \end{cases}$.

In Exercises 18–20, identify the common function in the transformation. Then sketch a graph of the function.

18. $h(x) = -x^3 - 7$
19. $h(x) = -\sqrt{x+5} + 8$
20. $h(x) = \frac{1}{4}|x+1| - 3$

In Exercises 21–23, let $f(x) = 3x^2 - 7$ and $g(x) = -x^2 - 4x + 5$. Find the indicated value.

21. $(f+g)(2)$
22. $(fg)(0)$
23. $(g \circ f)(-1)$

In Exercises 24–26, find the inverse function, if possible.

24. $f(x) = x^3 + 8$
25. $f(x) = |x^2 - 3| + 6$
26. $f(x) = \frac{3x\sqrt{x}}{8}$

27. It costs a company $58 to produce 6 units of a product and $78 to produce 10 units. How much does it cost to produce 25 units, assuming that the cost function is linear?

1.1 ▶ **Radian and Degree Measure**
1.2 ▶ **Trigonometric Functions: The Unit Circle**
1.3 ▶ **Right Triangle Trigonometry**
1.4 ▶ **Trigonometric Functions of Any Angle**
1.5 ▶ **Graphs of Sine and Cosine Functions**
1.6 ▶ **Graphs of Other Trigonometric Functions**
1.7 ▶ **Inverse Trigonometric Functions**
1.8 ▶ **Applications and Models**

Photo Source Hawaii

The Mauna Loa Observatory in Hawaii conducts research to understand the global carbon cycle. It is located far from pollution sources that would affect the gases being measured. (Source: NOAA/Climated Monitoring and Diagnostic Laboratory)

1 Trigonometry

▶ How to Study This Chapter

The Big Picture
In this chapter you will learn the following skills and concepts.

- ▶ How to describe an angle and convert between radian and degree measure
- ▶ How to identify a unit circle and its relationship to real numbers
- ▶ How to evaluate trigonometric functions of any angle
- ▶ How to use the fundamental trigonometric identities
- ▶ How to sketch the graphs of trigonometric functions and translations of graphs of sine and cosine functions
- ▶ How to evaluate the inverse trigonometric functions
- ▶ How to evaluate the compositions of trigonometric functions and inverse trigonometric functions

Important Vocabulary
As you encounter each new vocabulary term in this chapter, add the term and its definition to your notebook glossary.

Trigonometry (p. 120)
Angle (p. 120)
Initial side (p. 120)
Terminal side (p. 120)
Vertex (p. 120)
Standard position (p. 120)
Positive angles (p. 120)
Negative angles (p. 120)
Coterminal angles (p. 120)
Central angle (p. 121)
Radian (p. 121)
Acute angles (p. 121)
Obtuse angles (p. 121)
Complementary angles (p. 123)
Supplementary angles (p. 123)
Degree (p. 123)
Linear speed (p. 125)
Angular speed (p. 125)
Unit circle (p. 131)
Sine (pp. 132, 138)
Cosecant (pp. 132, 138)
Cosine (pp. 132, 138)
Secant (pp. 132, 138)
Tangent (pp. 132, 138)
Cotangent (pp. 132, 138)
Period (pp. 134, 161)
Hypotenuse (p. 138)
Opposite side (p. 138)
Adjacent side (p. 138)
Reference angles (p. 151)
Sine curve (p. 158)
One cycle (p. 158)
Amplitude (p. 160)
Phase shift (p. 162)
Damping factor (p. 174)
Inverse sine function (p. 180)
Inverse cosine function (p. 182)
Inverse tangent function (p. 182)
Angle of elevation (p. 190)
Angle of depression (p. 190)
Bearings (p. 192)
Simple harmonic motion (pp. 192, 193)

Study Tools
- Learning objectives at the beginning of each section
- Chapter Summary (p. 201)
- Review Exercises (pp. 202–205)
- Chapter Test (p. 207)

Additional Resources
- Study and Solutions Guide
- Interactive Trigonometry
- Videotapes for Chapter 1
- Trigonometry Website
- Student Success Organizer

STUDY T!P
To prepare for a chapter test, review the learning objectives and work the review exercises. Take the sample Chapter Test and analyze the results.

1.1 Radian and Degree Measure

▶ What you should learn

- How to describe angles
- How to use radian measure
- How to use degree measure
- How to use angles to model and solve real-life problems

▶ Why you should learn it

You can use angles to model and solve real-life problems. For instance, in Exercises 91–94 on page 129, you are asked to use angles to find the distance between two cities on the same longitude.

Ernest Jones

Angles

As derived from the Greek language, the word **trigonometry** means "measurement of triangles." Initially, trigonometry dealt with relationships among the sides and angles of triangles and was used in the development of astronomy, navigation, and surveying. With the development of calculus and the physical sciences in the 17th century, a different perspective arose—one that viewed the classic trigonometric relationships as *functions* with the set of real numbers as their domains. Consequently, the applications of trigonometry expanded to include a vast number of physical phenomena involving rotations and vibrations. These phenomena include sound waves, light rays, planetary orbits, vibrating strings, pendulums, and orbits of atomic particles.

The approach in this text incorporates *both* perspectives, starting with angles and their measure.

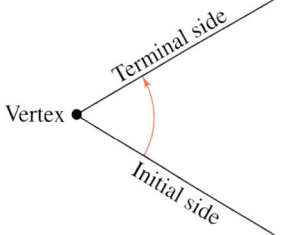

(a) Angle   (b) Angle in Standard Position

FIGURE 1.1

An **angle** is determined by rotating a ray (half-line) about its endpoint. The starting position of the ray is the **initial side** of the angle, and the position after rotation is the **terminal side**, as shown in Figure 1.1(a). The endpoint of the ray is the **vertex** of the angle. This perception of an angle fits a coordinate system in which the origin is the vertex and the initial side coincides with the positive x-axis. Such an angle is in **standard position,** as shown in Figure 1.1(b). **Positive angles** are generated by counterclockwise rotation, and **negative angles** by clockwise rotation, as shown in Figure 1.2. Angles are labeled with Greek letters α (alpha), β (beta), and θ (theta), as well as uppercase letters A, B, and C. In Figure 1.3, note that angles α and β have the same initial and terminal sides. Such angles are **coterminal.**

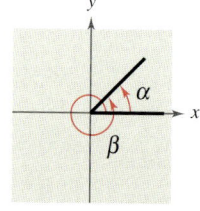

FIGURE 1.2   FIGURE 1.3

Section 1.1 ▶ Radian and Degree Measure   121

Radian Measure

The **measure of an angle** is determined by the amount of rotation from the initial side to the terminal side. One way to measure angles is in *radians*. This type of measure is especially useful in calculus. To define a radian, you can use a **central angle** of a circle, one whose vertex is the center of the circle, as shown in Figure 1.4.

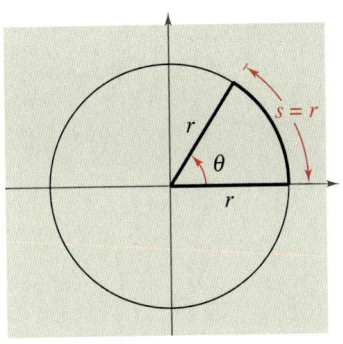

Arc length = radius when θ = 1 radian
FIGURE **1.4**

> ### Definition of Radian
> One **radian** is the measure of a central angle θ that intercepts an arc s equal in length to the radius r of the circle. See Figure 1.4.

Because the circumference of a circle is $2\pi r$ units, it follows that a central angle of one full revolution (counterclockwise) corresponds to an arc length of

$$s = 2\pi r.$$

Moreover, because $2\pi \approx 6.28$, there are just over six radius lengths in a full circle, as shown in Figure 1.5. In general, the radian measure of a central angle θ is obtained by dividing the arc length s by r. That is, $s/r = \theta$, where θ *is measured in radians*. Because the units of measure for s and r are the same, this ratio is unitless—it is simply a real number.

Because the radian measure of an angle of one full revolution is 2π, you can obtain the following.

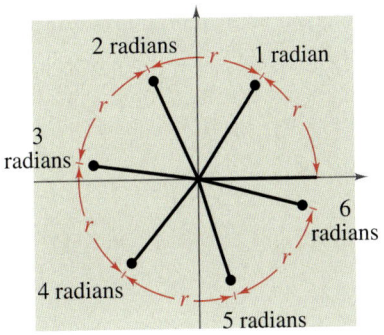

FIGURE **1.5**

$$\frac{1}{2} \text{ revolution} = \frac{2\pi}{2} = \pi \text{ radians}$$

$$\frac{1}{4} \text{ revolution} = \frac{2\pi}{4} = \frac{\pi}{2} \text{ radians}$$

$$\frac{1}{6} \text{ revolution} = \frac{2\pi}{6} = \frac{\pi}{3} \text{ radians}$$

These and other common angles are shown in Figure 1.6.

FIGURE **1.6**

> ### STUDY T!P
> One revolution around a circle of radius r corresponds to an angle of 2π radians because
> $$\frac{s}{r} = \frac{2\pi r}{r} = 2\pi \text{ radians}.$$

Recall that the four quadrants in a coordinate system are numbered I, II, III, and IV. Figure 1.7 shows which angles between 0 and 2π lie in each of the four quadrants. Note that angles between 0 and $\pi/2$ are **acute** and that angles between $\pi/2$ and π are **obtuse**.

STUDY TIP

The phrase "the terminal side of θ lies in a quadrant" is often abbreviated by simply saying that "θ lies in a quadrant." The terminal sides of the "quadrant angles" 0, $\pi/2$, π, and $3\pi/2$ do not lie within quadrants.

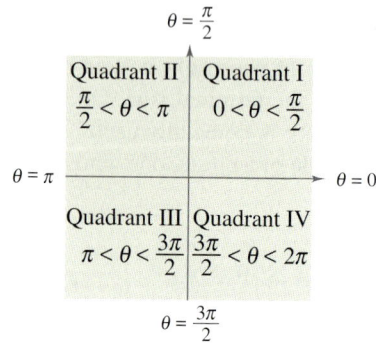

FIGURE 1.7

Two angles are coterminal if they have the same initial and terminal sides. For instance, the angles 0 and 2π are coterminal, as are the angles $\pi/6$ and $13\pi/6$. You can find an angle that is coterminal to a given angle θ by adding or subtracting 2π (one revolution), as demonstrated in Example 1. A given angle θ has infinitely many coterminal angles. For instance, $\theta = \pi/6$ is coterminal with

$$\frac{\pi}{6} + 2n\pi$$

where n is an integer.

Example 1 ▶ Sketching and Finding Coterminal Angles

a. For the positive angle $13\pi/6$, subtract 2π to obtain a coterminal angle

$$\frac{13\pi}{6} - 2\pi = \frac{\pi}{6}. \quad \text{See Figure 1.8(a).}$$

b. For the positive angle $3\pi/4$, subtract 2π to obtain a coterminal angle

$$\frac{3\pi}{4} - 2\pi = -\frac{5\pi}{4}. \quad \text{See Figure 1.8(b).}$$

c. For the negative angle $-2\pi/3$, add 2π to obtain a coterminal angle

$$-\frac{2\pi}{3} + 2\pi = \frac{4\pi}{3}. \quad \text{See Figure 1.8(c).}$$

(a)

(b)

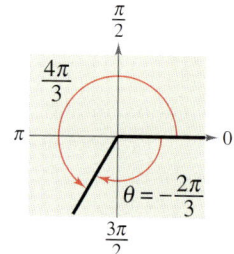
(c)

FIGURE 1.8

Two positive angles α and β are **complementary** (complements of each other) if their sum is $\pi/2$. Two positive angles are **supplementary** (supplements of each other) if their sum is π. See Figure 1.9.

Complementary Angles

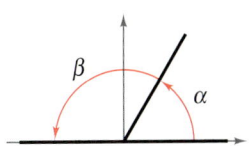
Supplementary Angles

FIGURE 1.9

Example 2 ▶ Complementary and Supplementary Angles

If possible, find the complement and the supplement of (a) $2\pi/5$ and (b) $4\pi/5$.

Solution

a. The complement of $2\pi/5$ is

$$\frac{\pi}{2} - \frac{2\pi}{5} = \frac{5\pi}{10} - \frac{4\pi}{10} = \frac{\pi}{10}.$$

The supplement of $2\pi/5$ is

$$\pi - \frac{2\pi}{5} = \frac{3\pi}{5}.$$

b. Because $4\pi/5$ is greater than $\pi/2$, it has no complement. (Remember to use only positive angles for complements.) The supplement is

$$\pi - \frac{4\pi}{5} = \frac{\pi}{5}.$$

FIGURE 1.10

Degree Measure

A second way to measure angles is in terms of degrees. A measure of **1 degree** (1°) is equivalent to a rotation of $1/360$ of a complete revolution about the vertex. To measure angles, it is convenient to mark degrees on the circumference of a circle, as shown in Figure 1.10. So, a full revolution (counterclockwise) corresponds to 360°, a half revolution to 180°, a quarter revolution to 90°, and so on.

Because 2π radians corresponds to one complete revolution, degrees and radians are related by the equations

$$360° = 2\pi \text{ rad} \quad \text{and} \quad 180° = \pi \text{ rad}.$$

From the latter equation, you obtain

$$1° = \frac{\pi}{180} \text{ rad} \quad \text{and} \quad 1 \text{ rad} = \left(\frac{180°}{\pi}\right)$$

which lead to the conversion rules at the top of the next page.

Conversions Between Degrees and Radians

1. To convert degrees to radians, multiply degrees by $\dfrac{\pi \text{ rad}}{180°}$.

2. To convert radians to degrees, multiply radians by $\dfrac{180°}{\pi \text{ rad}}$.

To apply these two conversion rules, use the basic relationship $\pi \text{ rad} = 180°$. (See Figure 1.11.)

The Interactive CD-ROM and Internet versions of this text show every example with its solution; clicking on the Try It! button brings up similar problems. Guided Examples and Integrated Examples show step-by-step solutions to additional examples. Integrated Examples are related to several concepts in the section.

$\dfrac{\pi}{6}$ $\dfrac{\pi}{4}$ $\dfrac{\pi}{3}$
30°   45°   60°

$\dfrac{\pi}{2}$ π 2π
90°   180°   360°

FIGURE 1.11

When no units of angle measure are specified, *radian measure is implied*. For instance, if you write $\theta = \pi$ or $\theta = 2$, you should mean $\theta = \pi$ radians or $\theta = 2$ radians.

Technology

With calculators it is convenient to use *decimal* degrees to denote fractional parts of degrees. Historically, however, fractional parts of degrees were expressed in *minutes* and *seconds*, using the prime (′) and double prime (″) notations, respectively. That is,

$1' = $ one minute $= \frac{1}{60}(1°)$

$1'' = $ one second $= \frac{1}{3600}(1°)$

Consequently, an angle of 64 degrees, 32 minutes, and 47 seconds is represented by $\theta = 64°\,32'\,47''$.

Many calculators have special keys for converting an angle in degrees, minutes, and seconds (D° M′ S″) into decimal degree form, and vice versa.

Example 3 ▶ Converting from Degrees to Radians

a. $135° = (135 \text{ deg})\left(\dfrac{\pi \text{ rad}}{180 \text{ deg}}\right) = \dfrac{3\pi}{4} \text{ rad}$ Multiply by $\pi/180$.

b. $540° = (540 \text{ deg})\left(\dfrac{\pi \text{ rad}}{180 \text{ deg}}\right) = 3\pi \text{ rad}$ Multiply by $\pi/180$.

c. $-270° = (-270 \text{ deg})\left(\dfrac{\pi \text{ rad}}{180 \text{ deg}}\right) = -\dfrac{3\pi}{2} \text{ rad}$ Multiply by $\pi/180$.

Example 4 ▶ Converting from Radians to Degrees

a. $-\dfrac{\pi}{2} \text{ rad} = \left(-\dfrac{\pi}{2} \text{ rad}\right)\left(\dfrac{180 \text{ deg}}{\pi \text{ rad}}\right) = -90°$ Multiply by $180/\pi$.

b. $\dfrac{9\pi}{2} \text{ rad} = \left(\dfrac{9\pi}{2} \text{ rad}\right)\left(\dfrac{180 \text{ deg}}{\pi \text{ rad}}\right) = 810°$ Multiply by $180/\pi$.

c. $2 \text{ rad} = (2 \text{ rad})\left(\dfrac{180 \text{ deg}}{\pi \text{ rad}}\right) = \dfrac{360}{\pi} \approx 114.59°$ Multiply by $180/\pi$.

If you have a calculator with a "radian-to-degree" conversion key, try using it to verify the result shown in part (c) of Example 4.

Applications

The *radian measure* formula, $\theta = s/r$, can be used to measure arc length along a circle. Specifically, for a circle of radius r, a central angle θ intercepts an arc of length s given by

$$s = r\theta \qquad \text{Length of circular arc}$$

where θ is measured in radians.

Example 5 ▶ Finding Arc Length

A circle has a radius of 4 inches. Find the length of the arc intercepted by a central angle of 240°, as shown in Figure 1.12.

Solution

To use the formula $s = r\theta$, first convert 240° to radian measure.

$$240° = (240 \text{ deg})\left(\frac{\pi \text{ rad}}{180 \text{ deg}}\right) \qquad \text{Convert from degrees to radians.}$$

$$= \frac{4\pi}{3} \text{ rad} \qquad \text{Simplify.}$$

Then, using a radius of $r = 4$ inches, you can find the arc length to be

$$s = r\theta \qquad \text{Length of circular arc}$$

$$= 4\left(\frac{4\pi}{3}\right) \qquad \text{Substitute for } r \text{ and } \theta.$$

$$= \frac{16\pi}{3} \qquad \text{Simplify.}$$

$$\approx 16.76 \text{ inches.} \qquad \text{Use a calculator.}$$

Note that the units for $r\theta$ are determined by the units for r because θ is given in radian measure and therefore has no units.

FIGURE 1.12

The formula for the length of a circular arc can be used to analyze the motion of a particle moving at a *constant speed* along a circular path.

Linear and Angular Speed

Consider a particle moving at a constant speed along a circular arc of radius r. If s is the length of the arc traveled in time t, then the **linear speed** of the particle is

$$\text{Linear speed} = \frac{\text{arc length}}{\text{time}} = \frac{s}{t}.$$

Moreover, if θ is the angle (in radian measure) corresponding to the arc length s, then the **angular speed** of the particle is

$$\text{Angular speed} = \frac{\text{central angle}}{\text{time}} = \frac{\theta}{t}.$$

FIGURE 1.13

Example 6 ▶ Finding Linear Speed

The second hand of a clock is 10.2 centimeters long, as shown in Figure 1.13. Find the linear speed of the tip of this second hand as it passes around the clock face.

Solution

In one revolution, the arc length traveled is

$$s = 2\pi r$$
$$= 2\pi(10.2) \quad \text{Substitute for } r.$$
$$= 20.4\pi \text{ centimeters.}$$

The time required for the second hand to travel this distance is

$$t = 1 \text{ minute} = 60 \text{ seconds.}$$

So, the linear speed of the tip of the second hand is

$$\text{Linear speed} = \frac{s}{t}$$
$$= \frac{20.4\pi \text{ centimeters}}{60 \text{ seconds}}$$
$$\approx 1.068 \text{ cm/sec.}$$

Example 7 ▶ Finding Angular and Linear Speed

A lawn roller with a 10-inch radius (see Figure 1.14) makes 1.2 revolutions per second.

a. Find the angular speed of the roller in radians per second.
b. Find the speed of the tractor that is pulling the roller.

Solution

a. Because each revolution generates 2π radians, it follows that the roller turns $(1.2)(2\pi) = 2.4\pi$ radians per second. In other words, the angular speed is

$$\text{Angular speed} = \frac{\theta}{t}$$
$$= \frac{2.4\pi \text{ radians}}{1 \text{ second}}$$
$$= 2.4\pi \text{ rad/sec.}$$

b. The linear speed is

$$\text{Linear speed} = \frac{s}{t}$$
$$= \frac{r\theta}{t}$$
$$= \frac{10(2.4\pi) \text{ inches}}{1 \text{ second}}$$
$$\approx 75.4 \text{ in./sec.}$$

FIGURE 1.14

1.1 Exercises

In Exercises 1–6, estimate the angle to the nearest one-half radian.

1.   2.

3.   4.

5.   6.

In Exercises 7–12, determine the quadrant in which each angle lies. (The angle measure is given in radians.)

7. (a) $\dfrac{\pi}{5}$ (b) $\dfrac{7\pi}{5}$

8. (a) $\dfrac{11\pi}{8}$ (b) $\dfrac{9\pi}{8}$

9. (a) $-\dfrac{\pi}{12}$ (b) $-\dfrac{11\pi}{9}$

10. (a) -1 (b) -2

11. (a) 3.5 (b) 2.25

12. (a) 6.02 (b) -4.25

In Exercises 13–16, sketch each angle in standard position.

13. (a) $\dfrac{5\pi}{4}$ (b) $\dfrac{2\pi}{3}$

14. (a) $-\dfrac{7\pi}{4}$ (b) $-\dfrac{5\pi}{2}$

15. (a) $\dfrac{11\pi}{6}$ (b) 7π

16. (a) 4 (b) -3

In Exercises 17–20, determine two coterminal angles (one positive and one negative) for each angle. Give your answers in radians.

17. (a)   (b)

18. (a)   (b)

19. (a)   (b)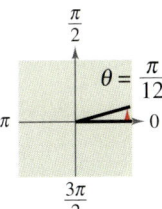

20. (a) $-\dfrac{9\pi}{4}$ (b) $-\dfrac{2\pi}{15}$

In Exercises 21–24, find (if possible) the complement and supplement of each angle.

21. (a) $\dfrac{\pi}{3}$ (b) $\dfrac{3\pi}{4}$

22. (a) $\dfrac{\pi}{12}$ (b) $\dfrac{11\pi}{12}$

23. (a) 1 (b) 2

24. (a) 3 (b) 1.5

In Exercises 25–28, express each angle in radian measure as a multiple of π. (Do not use a calculator.)

25. (a) $30°$ (b) $150°$

26. (a) $315°$ (b) $120°$

27. (a) $-20°$ (b) $-240°$

28. (a) $-270°$ (b) $144°$

In Exercises 29–36, convert the measure from degrees to radians. Round to three decimal places.

29. 115° **30.** 87.4°
31. −216.35° **32.** −48.27°
33. 532° **34.** 345°
35. −0.83° **36.** 0.54°

In Exercises 37–40, express each angle in degree measure. (Do not use a calculator.)

37. (a) $3\pi/2$ (b) $7\pi/6$
38. (a) $-7\pi/12$ (b) $\pi/9$
39. (a) $7\pi/3$ (b) $-11\pi/30$
40. (a) $11\pi/6$ (b) $34\pi/15$

In Exercises 41–48, convert the measure from radians to degrees. Round to three decimal places.

41. $\dfrac{\pi}{7}$ **42.** $\dfrac{5\pi}{11}$
43. $\dfrac{15\pi}{8}$ **44.** $\dfrac{13\pi}{2}$
45. -4.2π **46.** 4.8π
47. -2 **48.** -0.57

In Exercises 49–54, estimate the number of degrees in the angle.

49.   **50.**

51.   **52.**

53.   **54.**

In Exercises 55–58, determine the quadrant in which each angle lies.

55. (a) 130° (b) 285°
56. (a) 8.3° (b) 257° 30′
57. (a) −132° 50′ (b) −336°
58. (a) −260° (b) −3.4°

In Exercises 59–62, sketch each angle in standard position.

59. (a) 30° (b) 150°
60. (a) −270° (b) −120°
61. (a) 405° (b) 480°
62. (a) −750° (b) −600°

In Exercises 63–66, determine two coterminal angles (one positive and one negative) for each angle. Give your answers in degrees.

63. (a) (b)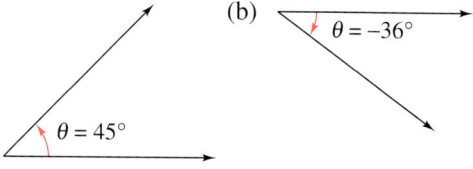

$\theta = 45°$, $\theta = -36°$

64. (a) (b)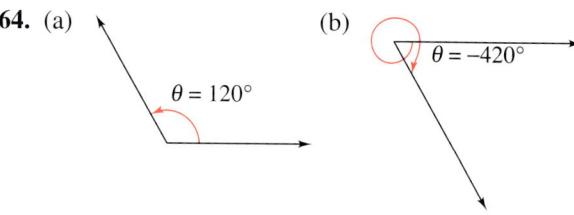

$\theta = 120°$, $\theta = -420°$

65. (a) (b)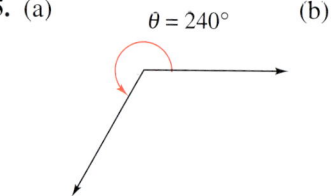

$\theta = 240°$

66. (a) −420° (b) 230°

In Exercises 67–70, find (if possible) the complement and supplement of each angle.

67. (a) 18° (b) 115°
68. (a) 3° (b) 64°
69. (a) 79° (b) 150°
70. (a) 130° (b) 170°

In Exercises 71–74, convert each angle measure to decimal degree form.

71. (a) 54° 45′ (b) −128° 30′
72. (a) 245° 10′ (b) 2° 12′
73. (a) 85° 18′ 30″ (b) 330° 25″
74. (a) −135° 36″ (b) −408° 16′ 20″

In Exercises 75–78, convert each angle measure to D° M′ S″ form.

75. (a) 240.6°   (b) −145.8°
76. (a) −345.12°   (b) 0.45°
77. (a) 2.5°   (b) −3.58°
78. (a) −0.355°   (b) 0.7865°

In Exercises 79–82, find the angle in radians.

79.

80.

81.

82.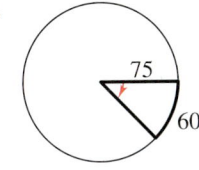

In Exercises 83–86, find the radian measure of the central angle of a circle of radius r that intercepts an arc of length s.

	Radius	Arc Length
83.	27 inches	6 inches
84.	14 feet	8 feet
85.	14.5 centimeters	25 centimeters
86.	80 kilometers	160 kilometers

In Exercises 87–90, find the length of the arc on a circle of radius r intercepted by a central angle θ.

	Radius	Central Angle
87.	15 inches	180°
88.	9 feet	60°
89.	3 meters	1 radian
90.	20 centimeters	$\pi/4$ radian

Distance Between Cities In Exercises 91–94, find the distance between the cities. Assume that the earth is a sphere of radius 4000 miles and that the cities are on the same longitude (one city is due north of the other).

	City	Latitude
91.	Dallas, Texas	32° 47′ 9″ N
	Omaha, Nebraska	41° 15′ 42″ N
92.	San Francisco, California	37° 46′ 39″ N
	Seattle, Washington	47° 36′ 32″ N

	City	Latitude
93.	Miami, Florida	25° 46′ 37″ N
	Erie, Pennsylvania	42° 7′ 15″ N
94.	Johannesburg, South Africa	26° 10′ S
	Jerusalem, Israel	31° 47′ N

95. *Difference in Latitudes* Assuming that the earth is a sphere of radius 6378 kilometers, what is the difference in latitude of two cities, one of which is 400 kilometers due north of the other?

96. *Difference in Latitudes* Assuming that the earth is a sphere of radius 6378 kilometers, what is the difference in latitude of two cities, one of which is 500 kilometers due north of the other?

97. *Instrumentation* The pointer on a voltmeter is 6 centimeters in length (see figure). Find the angle through which the pointer rotates when it moves 2.5 centimeters on the scale.

FIGURE FOR 97   FIGURE FOR 98

98. *Electric Hoist* An electric hoist is being used to lift a beam (see figure). The diameter of the drum on the hoist is 10 inches, and the beam must be raised 2 feet. Find the number of degrees through which the drum must rotate.

99. *Angular Speed* A car is moving at a rate of 65 miles per hour, and the diameter of its wheels is 2.5 feet.

(a) Find the number of revolutions per minute the wheels are rotating.

(b) Find the angular speed of the wheels in radians per minute.

100. *Angular Speed* A 2-inch-diameter pulley on an electric motor that runs at 1700 revolutions per minute is connected by a belt to a 4-inch-diameter pulley on a saw arbor.

(a) Find the angular speed (in radians per minute) of each pulley.

(b) Find the revolutions per minute of the saw.

101. *Floppy Disk* The radius of the magnetic disk in a 3.5-inch diskette is 1.68 inches. Find the linear speed of a point on the circumference of the disk if it is rotating at a speed of 360 revolutions per minute.

102. *Speed of a Bicycle* The radii of the sprocket assemblies and the wheel of the bicycle in the figure are 4 inches, 2 inches, and 14 inches, respectively. If the cyclist is pedaling at a rate of 1 revolution per second, find the speed of the bicycle in (a) feet per second and (b) miles per hour.

Synthesis

True or False? In Exercises 103 and 104, determine whether the statement is true or false. Justify your answer.

103. A measurement of 4 radians corresponds to two complete revolutions from the initial to the terminal side of an angle.

104. The difference of the measures of two coterminal angles is always a multiple of $360°$ if expressed in degrees and is always a multiple of 2π radians if expressed in radians.

105. In your own words, explain the meanings of (a) an angle in standard position, (b) a negative angle, (c) coterminal angles, and (d) an obtuse angle.

106. A fan motor turns at a given angular speed. How does the speed of the tips of the blades change if a fan of greater diameter is installed on the motor? Explain.

107. *Think About It* Is a degree or a radian the larger unit of measure? Explain.

108. *Writing* If the radius of a circle is increasing and the magnitude of a central angle is held constant, how is the length of the intercepted arc changing? Explain your reasoning.

109. *Geometry* Prove that the area of a circular sector of radius r with central angle θ is $A = \frac{1}{2}\theta r^2$, where θ is measured in radians.

Review

In Exercises 110–113, sketch the graphs of $y = x^5$ and the specified transformation.

110. $f(x) = (x - 2)^5$ **111.** $f(x) = x^5 - 4$

112. $f(x) = 2 - x^5$ **113.** $f(x) = -(x + 3)^5$

In Exercises 114–121, simplify the radical expression.

114. $\dfrac{4}{4\sqrt{2}}$ **115.** $\dfrac{2}{\sqrt{3}}$

116. $\dfrac{2\sqrt{3}}{\sqrt{6}}$ **117.** $\dfrac{5\sqrt{5}}{2\sqrt{10}}$

118. $\sqrt{2^2 + 6^2}$ **119.** $\sqrt{18^2 + 12^2}$

120. $\sqrt{18^2 - 6^2}$ **121.** $\sqrt{17^2 - 9^2}$

1.2 Trigonometric Functions: The Unit Circle

▶ **What you should learn**
- How to identify a unit circle and its relationship to real numbers
- How to evaluate trigonometric functions using the unit circle
- How to use the domain and period to evaluate sine and cosine functions
- How to use a calculator to evaluate trigonometric functions

▶ **Why you should learn it**

Trigonometric functions are used to model the movement of an oscillating weight. For instance, in Exercises 59 and 60 on page 137, the displacement from equilibrium of an oscillating weight suspended by a spring is modeled as a function of time.

A computer animation of this concept appears in the *Interactive* CD-ROM and *Internet* versions of this text.

The Unit Circle

The two historical perspectives of trigonometry incorporate different methods for introducing the trigonometric functions. Our first introduction to these functions is based on the unit circle.

Consider the **unit circle** given by

$$x^2 + y^2 = 1 \qquad \text{Unit circle}$$

as shown in Figure 1.15.

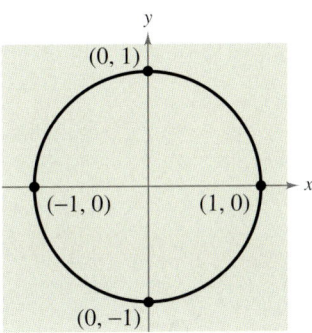

FIGURE 1.15

Imagine that the real number line is wrapped around this circle, with positive numbers corresponding to a counterclockwise wrapping and negative numbers corresponding to a clockwise wrapping, as shown in Figure 1.16.

 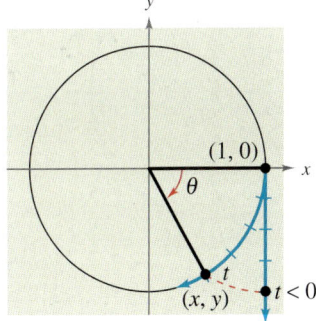

FIGURE 1.16

As the real number line is wrapped around the unit circle, each real number t corresponds to a point (x, y) on the circle. For example, the real number 0 corresponds to the point $(1, 0)$. Moreover, because the unit circle has a circumference of 2π, the real number 2π also corresponds to the point $(1, 0)$.

In general, each real number t also corresponds to a central angle θ (in standard position) whose radian measure is t. With this interpretation of t, the arc length formula $s = r\theta$ (with $r = 1$) indicates that the real number t is the length of the arc intercepted by the angle θ, given in radians.

The Trigonometric Functions

From the preceding discussion, it follows that the coordinates x and y are two functions of the real variable t. You can use these coordinates to define the six trigonometric functions of t.

sine	cosecant
cosine	secant
tangent	cotangent

These six functions are normally abbreviated sin, csc, cos, sec, tan, and cot, respectively.

Definitions of Trigonometric Functions

Let t be a real number and let (x, y) be the point on the unit circle corresponding to t.

$$\sin t = y \qquad \csc t = \frac{1}{y}, \quad y \neq 0$$

$$\cos t = x \qquad \sec t = \frac{1}{x}, \quad x \neq 0$$

$$\tan t = \frac{y}{x}, \quad x \neq 0 \qquad \cot t = \frac{x}{y}, \quad y \neq 0$$

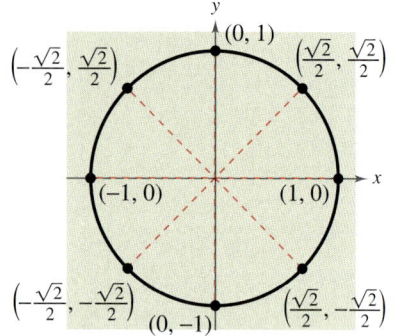

FIGURE 1.17

Note that the functions in the second column are the *reciprocals* of the corresponding functions in the first column.

In the definitions of the trigonometric functions, note that the tangent and secant are not defined when $x = 0$. For instance, because $t = \pi/2$ corresponds to $(x, y) = (0, 1)$, it follows that $\tan(\pi/2)$ and $\sec(\pi/2)$ are *undefined*. Similarly, the cotangent and cosecant are not defined when $y = 0$. For instance, because $t = 0$ corresponds to $(x, y) = (1, 0)$, cot 0 and csc 0 are *undefined*.

In Figure 1.17, the unit circle has been divided into eight equal arcs, corresponding to t-values of

$$0, \frac{\pi}{4}, \frac{\pi}{2}, \frac{3\pi}{4}, \pi, \frac{5\pi}{4}, \frac{3\pi}{2}, \frac{7\pi}{4}, \text{ and } 2\pi.$$

Similarly, in Figure 1.18, the unit circle has been divided into 12 equal arcs, corresponding to t-values of

$$0, \frac{\pi}{6}, \frac{\pi}{3}, \frac{\pi}{2}, \frac{2\pi}{3}, \frac{5\pi}{6}, \pi, \frac{7\pi}{6}, \frac{4\pi}{3}, \frac{3\pi}{2}, \frac{5\pi}{3}, \frac{11\pi}{6}, \text{ and } 2\pi.$$

Using the (x, y) coordinates in Figures 1.17 and 1.18, you can easily evaluate the trigonometric functions for common t-values. This procedure is demonstrated in Examples 1 and 2.

FIGURE 1.18

A computer animation of this concept appears in the *Interactive* CD-ROM and *Internet* versions of this text.

Example 1 ▶ Evaluating Trigonometric Functions

Evaluate the six trigonometric functions at each real number.

a. $t = \dfrac{\pi}{6}$ **b.** $t = \dfrac{5\pi}{4}$ **c.** $t = 0$ **d.** $t = \pi$

Solution

For each t-value, begin by finding the corresponding point (x, y) on the unit circle. Then use the definitions of trigonometric functions listed on page 132.

a. $t = \pi/6$ corresponds to the point $(x, y) = \left(\sqrt{3}/2, \, 1/2\right)$.

$$\sin \dfrac{\pi}{6} = y = \dfrac{1}{2} \qquad\qquad \csc \dfrac{\pi}{6} = \dfrac{1}{y} = 2$$

$$\cos \dfrac{\pi}{6} = x = \dfrac{\sqrt{3}}{2} \qquad\qquad \sec \dfrac{\pi}{6} = \dfrac{1}{x} = \dfrac{2}{\sqrt{3}} = \dfrac{2\sqrt{3}}{3}$$

$$\tan \dfrac{\pi}{6} = \dfrac{y}{x} = \dfrac{1/2}{\sqrt{3}/2} = \dfrac{1}{\sqrt{3}} \qquad\qquad \cot \dfrac{\pi}{6} = \dfrac{x}{y} = \sqrt{3}$$

b. $t = 5\pi/4$ corresponds to the point $(x, y) = \left(-\sqrt{2}/2, \, -\sqrt{2}/2\right)$.

$$\sin \dfrac{5\pi}{4} = y = -\dfrac{\sqrt{2}}{2} \qquad\qquad \csc \dfrac{5\pi}{4} = \dfrac{1}{y} = -\dfrac{2}{\sqrt{2}} = -\sqrt{2}$$

$$\cos \dfrac{5\pi}{4} = x = -\dfrac{\sqrt{2}}{2} \qquad\qquad \sec \dfrac{5\pi}{4} = \dfrac{1}{x} = -\dfrac{2}{\sqrt{2}} = -\sqrt{2}$$

$$\tan \dfrac{5\pi}{4} = \dfrac{y}{x} = \dfrac{-\sqrt{2}/2}{-\sqrt{2}/2} = 1 \qquad\qquad \cot \dfrac{5\pi}{4} = \dfrac{x}{y} = 1$$

c. $t = 0$ corresponds to the point $(x, y) = (1, 0)$.

$$\sin 0 = y = 0 \qquad\qquad \csc 0 = \dfrac{1}{y} \text{ is undefined.}$$

$$\cos 0 = x = 1 \qquad\qquad \sec 0 = \dfrac{1}{x} = 1$$

$$\tan 0 = \dfrac{y}{x} = \dfrac{0}{1} = 0 \qquad\qquad \cot 0 = \dfrac{x}{y} \text{ is undefined.}$$

d. $t = \pi$ corresponds to the point $(x, y) = (-1, 0)$.

$$\sin \pi = y = 0 \qquad\qquad \csc \pi = \dfrac{1}{y} \text{ is undefined.}$$

$$\cos \pi = x = -1 \qquad\qquad \sec \pi = \dfrac{1}{x} = -1$$

$$\tan \pi = \dfrac{y}{x} = \dfrac{0}{-1} = 0 \qquad\qquad \cot \pi = \dfrac{x}{y} \text{ is undefined.}$$

Exploration

With your graphing utility in radian and parametric modes, enter the equations

X1T = cos T   and   Y1T = sin T

and use the following settings.

Tmin = 0, Tmax = 6.3,
Tstep = 0.1
Xmin = -1.5, Xmax = 1.5,
Xscl = 1
Ymin = -1, Ymax = 1,
Yscl = 1

1. Graph the entered equations and describe the graph.
2. Use the trace key to move the cursor around the graph. What do the *t*-values represent? What do the *x*- and *y*-values represent?
3. What are the least and greatest values for *x* and *y*?

Example 2 ▶ Evaluating Trigonometric Functions

Evaluate the six trigonometric functions at $t = -\frac{\pi}{3}$.

Solution

Moving *clockwise* around the unit circle, it follows that $t = -\pi/3$ corresponds to the point $(x, y) = \left(1/2, -\sqrt{3}/2\right)$.

$$\sin\left(-\frac{\pi}{3}\right) = -\frac{\sqrt{3}}{2} \qquad \csc\left(-\frac{\pi}{3}\right) = -\frac{2}{\sqrt{3}}$$

$$\cos\left(-\frac{\pi}{3}\right) = \frac{1}{2} \qquad \sec\left(-\frac{\pi}{3}\right) = 2$$

$$\tan\left(-\frac{\pi}{3}\right) = -\sqrt{3} \qquad \cot\left(-\frac{\pi}{3}\right) = -\frac{1}{\sqrt{3}}$$

Domain and Period of Sine and Cosine

The *domain* of the sine and cosine functions is the set of all real numbers. To determine the *range* of these two functions, consider the unit circle shown in Figure 1.19. Because $r = 1$, it follows that $\sin t = y$ and $\cos t = x$. Moreover, because (x, y) is on the unit circle, you know that $-1 \le y \le 1$ and $-1 \le x \le 1$. So, the values of sine and cosine also range between -1 and 1.

$$-1 \le y \le 1 \qquad \text{and} \qquad -1 \le x \le 1$$
$$-1 \le \sin t \le 1 \qquad\qquad -1 \le \cos t \le 1$$

Adding 2π to each value of t in the interval $[0, 2\pi]$ completes a second revolution around the unit circle, as shown in Figure 1.20. The values of $\sin(t + 2\pi)$ and $\cos(t + 2\pi)$ correspond to those of $\sin t$ and $\cos t$. Similar results can be obtained for repeated revolutions (positive or negative) on the unit circle. This leads to the general result

$$\sin(t + 2\pi n) = \sin t$$

and

$$\cos(t + 2\pi n) = \cos t$$

for any integer n and real number t. Functions that behave in such a repetitive (or cyclic) manner are called **periodic**.

Definition of Periodic Function

A function f is **periodic** if there exists a positive real number c such that

$$f(t + c) = f(t)$$

for all t in the domain of f. The smallest number c for which f is periodic is called the **period** of f.

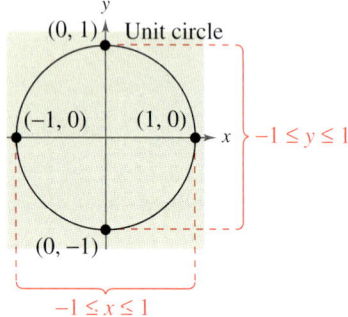

FIGURE 1.19

FIGURE 1.20

STUDY TIP

From the definition of periodic function it follows that the sine and cosine functions are periodic and have a period of 2π. The other four trigonometric functions are also periodic, and more will be said about this in Section 1.6.

Example 3 ▶ **Using the Period to Evaluate the Sine and Cosine**

a. Because $\dfrac{13\pi}{6} = 2\pi + \dfrac{\pi}{6}$, you have

$$\sin\dfrac{13\pi}{6} = \sin\left(2\pi + \dfrac{\pi}{6}\right) = \sin\dfrac{\pi}{6} = \dfrac{1}{2}.$$

b. Because $-\dfrac{7\pi}{2} = -4\pi + \dfrac{\pi}{2}$, you have

$$\cos\left(-\dfrac{7\pi}{2}\right) = \cos\left(-4\pi + \dfrac{\pi}{2}\right) = \cos\dfrac{\pi}{2} = 0.$$

Recall from Section P.6 that a function f is *even* if $f(-t) = f(t)$, and is *odd* if $f(-t) = -f(t)$.

Even and Odd Trigonometric Functions

The cosine and secant functions are *even*.

$$\cos(-t) = \cos t \qquad \sec(-t) = \sec t$$

The sine, cosecant, tangent, and cotangent functions are *odd*.

$$\sin(-t) = -\sin t \qquad \csc(-t) = -\csc t$$
$$\tan(-t) = -\tan t \qquad \cot(-t) = -\cot t$$

Evaluating Trigonometric Functions with a Calculator

When evaluating a trigonometric function with a calculator, you need to set the calculator to the desired *mode* of measurement (degrees or radians).

Most calculators do not have keys for the cosecant, secant, and cotangent functions. To evaluate these functions, you can use the $\boxed{x^{-1}}$ key with their respective reciprocal functions sine, cosine, and tangent. For example, to evaluate $\csc(\pi/8)$, use the fact that

$$\csc\dfrac{\pi}{8} = \dfrac{1}{\sin(\pi/8)}$$

and enter the following keystroke sequence in radian mode.

$\boxed{(}$ $\boxed{\text{SIN}}$ $\boxed{(}$ $\boxed{\pi}$ $\boxed{\div}$ $\boxed{8}$ $\boxed{)}$ $\boxed{)}$ $\boxed{x^{-1}}$ $\boxed{\text{ENTER}}$ Display 2.6131259

Technology

When evaluating trigonometric functions with a calculator, remember to enclose all fractional angle measures in parentheses. For instance, if you want to evaluate $\sin\theta$ for $\theta = \pi/6$, you should enter

$\boxed{\text{SIN}}$ $\boxed{(}$ $\boxed{\pi}$ $\boxed{\div}$ $\boxed{6}$ $\boxed{)}$ $\boxed{\text{ENTER}}$

rather than

$\boxed{\text{SIN}}$ $\boxed{\pi}$ $\boxed{\div}$ $\boxed{6}$ $\boxed{\text{ENTER}}$.

The first set of keystrokes yields the correct value of 0.5. The second set yields the incorrect value of 0.

Example 4 ▶ **Using a Calculator**

Function	Mode	Calculator Keystrokes	Display
a. $\sin 2\pi/3$	Radian	$\boxed{\text{SIN}}$ $\boxed{(}$ $\boxed{2}$ $\boxed{\pi}$ $\boxed{\div}$ $\boxed{3}$ $\boxed{)}$ $\boxed{\text{ENTER}}$	0.8660254
b. $\cot 1.5$	Radian	$\boxed{(}$ $\boxed{\text{TAN}}$ $\boxed{1.5}$ $\boxed{)}$ $\boxed{x^{-1}}$ $\boxed{\text{ENTER}}$	0.0709148

1.2 Exercises

In Exercises 1–4, determine the exact values of the six trigonometric functions of the angle θ.

1. $\left(-\frac{8}{17}, \frac{15}{17}\right)$

2. $\left(\frac{12}{13}, \frac{5}{13}\right)$

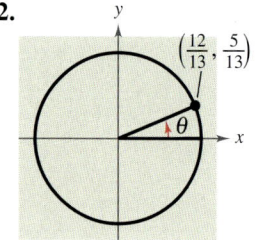

3. $\left(\frac{12}{13}, -\frac{5}{13}\right)$

4. $\left(-\frac{4}{5}, -\frac{3}{5}\right)$

In Exercises 5–12, find the point (x, y) on the unit circle that corresponds to the real number t.

5. $t = \frac{\pi}{4}$
6. $t = \frac{\pi}{3}$
7. $t = \frac{7\pi}{6}$
8. $t = \frac{5\pi}{4}$
9. $t = \frac{4\pi}{3}$
10. $t = \frac{5\pi}{3}$
11. $t = \frac{3\pi}{2}$
12. $t = \pi$

In Exercises 13–22, evaluate (if possible) the sine, cosine, and tangent of the real number.

13. $t = \frac{\pi}{4}$
14. $t = \frac{\pi}{3}$
15. $t = -\frac{\pi}{6}$
16. $t = -\frac{\pi}{4}$
17. $t = -\frac{7\pi}{4}$
18. $t = -\frac{4\pi}{3}$
19. $t = \frac{11\pi}{6}$
20. $t = \frac{5\pi}{3}$
21. $t = -\frac{3\pi}{2}$
22. $t = -2\pi$

In Exercises 23–28, evaluate (if possible) the six trigonometric functions of the real number.

23. $t = \frac{3\pi}{4}$
24. $t = \frac{5\pi}{6}$
25. $t = \frac{\pi}{2}$
26. $t = \frac{3\pi}{2}$
27. $t = -\frac{\pi}{3}$
28. $t = -\frac{3\pi}{2}$

In Exercises 29–36, evaluate the trigonometric function using its period as an aid.

29. $\sin 5\pi$
30. $\cos 5\pi$
31. $\cos \frac{8\pi}{3}$
32. $\sin \frac{9\pi}{4}$
33. $\cos(-3\pi)$
34. $\sin -3\pi$
35. $\sin\left(-\frac{9\pi}{4}\right)$
36. $\cos\left(-\frac{8\pi}{3}\right)$

In Exercises 37–42, use the value of the trigonometric function to evaluate the indicated functions.

37. $\sin t = \frac{1}{3}$
    (a) $\sin(-t)$
    (b) $\csc(-t)$

38. $\sin(-t) = \frac{3}{8}$
    (a) $\sin t$
    (b) $\csc t$

39. $\cos(-t) = -\frac{1}{5}$
    (a) $\cos t$
    (b) $\sec(-t)$

40. $\cos t = -\frac{3}{4}$
    (a) $\cos(-t)$
    (b) $\sec(-t)$

41. $\sin t = \frac{4}{5}$
    (a) $\sin(\pi - t)$
    (b) $\sin(t + \pi)$

42. $\cos t = \frac{4}{5}$
    (a) $\cos(\pi - t)$
    (b) $\cos(t + \pi)$

In Exercises 43–52, use a calculator to evaluate the expression. Round to four decimal places.

43. $\sin \frac{\pi}{4}$
44. $\tan \frac{\pi}{3}$
45. $\csc 1.3$
46. $\cot 1$
47. $\cos(-1.7)$
48. $\cos(-2.5)$
49. $\csc 0.8$
50. $\sec 1.8$
51. $\sec 22.8$
52. $\sin(-0.9)$

Estimation
In Exercises 53 and 54, use the figure and a straightedge to approximate the value of each trigonometric function.

53. (a) $\sin 5$ (b) $\cos 2$
54. (a) $\sin 0.75$ (b) $\cos 2.5$

Estimation
In Exercises 55 and 56, use the figure and a straightedge to approximate the solution of each equation, where $0 \leq t < 2\pi$.

55. (a) $\sin t = 0.25$ (b) $\cos t = -0.25$
56. (a) $\sin t = -0.75$ (b) $\cos t = 0.75$

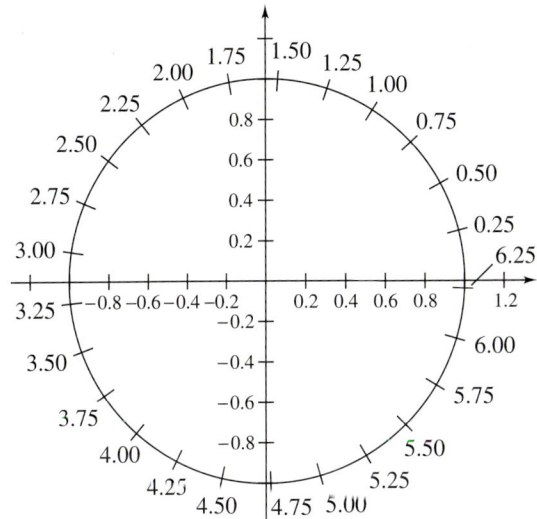

FIGURE FOR 53–56

57. Verify that $\cos 2t \neq 2 \cos t$ by approximating $\cos 1.5$ and $2 \cos 0.75$.

58. Verify that $\sin(t_1 + t_2) \neq \sin t_1 + \sin t_2$ by approximating $\sin 0.25$, $\sin 0.75$, and $\sin 1$.

59. *Harmonic Motion* The displacement from equilibrium of an oscillating weight suspended by a spring is
$$y(t) = \frac{1}{4} \cos 6t$$
where y is the displacement in feet and t is the time in seconds. Find the displacement when (a) $t = 0$, (b) $t = \frac{1}{4}$, and (c) $t = \frac{1}{2}$.

60. *Water Wave* The cross-section of a water wave is
$$y(x) = 3 \sin \frac{\pi}{4} x$$
where y is the vertical height of the water wave (in feet) and x is the distance to the wave (in feet). Find the height when (a) $x = 1$, (b) $x = 2$, and (c) $x = 4$.

Synthesis

True or False? In Exercises 61 and 62, determine whether the statement is true or false. Justify your answer.

61. Because $\sin(-t) = -\sin t$, it can be said that the sine of a negative angle is a negative number.

62. $\tan a = \tan(a - 6\pi)$

63. *Exploration* Let (x_1, y_1) and (x_2, y_2) be points on the unit circle corresponding to $t = t_1$ and $t = \pi - t_1$, respectively.

(a) Identify the symmetry of the points (x_1, y_1) and (x_2, y_2).

(b) Make a conjecture about any relationship between $\sin t_1$ and $\sin(\pi - t_1)$.

(c) Make a conjecture about any relationship between $\cos t_1$ and $\cos(\pi - t_1)$.

64. *Exploration* Let (x_1, y_1) and (x_2, y_2) be points on the unit circle corresponding to $t = t_1$ and $t = t_1 + \pi$, respectively.

(a) Identify the symmetry of the points (x_1, y_1) and (x_2, y_2).

(b) Make a conjecture about any relationship between $\sin t_1$ and $\sin(t_1 + \pi)$.

(c) Make a conjecture about any relationship between $\cos t_1$ and $\cos(t_1 + \pi)$.

65. Use the unit circle to verify that the cosine and secant functions are even and that the sine, cosecant, tangent, and cotangent functions are odd.

66. *Think About It* Because $f(t) = \sin t$ is an odd function and $g(t) = \cos t$ is an even function, what can be said about the function $h(t) = f(t)g(t)$?

67. *Think About It* Because $f(t) = \sin t$ and $g(t) = \tan t$ are odd functions, what can be said about the function $h(t) = f(t)g(t)$?

Review

In Exercises 68–71, find the inverse of the one-to-one function f.

68. $f(x) = \frac{1}{2}(3x - 2)$
69. $f(x) = \frac{1}{4}x^3 + 1$
70. $f(x) = \sqrt{x^2 - 4}, \quad x \geq 2$
71. $f(x) = \frac{2x}{x + 1}, \quad x > -1$

1.3 Right Triangle Trigonometry

▶ **What you should learn**
- How to evaluate trigonometric functions of acute angles
- How to use the fundamental trigonometric identities
- How to use a calculator to evaluate trigonometric functions
- How to use trigonometric functions to model and solve real-life problems

▶ **Why you should learn it**

Trigonometric functions are often used in mechanical calculations. For instance, Exercise 64 on page 147 shows you how trigonometric functions can be used to help find the width of a river.

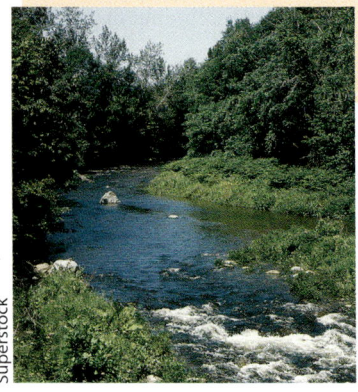

A computer animation of this concept appears in the *Interactive* CD-ROM and *Internet* versions of this text.

The Six Trigonometric Functions

Our second look at the trigonometric functions is from a *right triangle* perspective. Consider a right triangle, one of whose acute angles is labeled θ, as shown in Figure 1.21. Relative to the angle θ, the three sides of the triangle are the **hypotenuse,** the **opposite side** (the side opposite the angle θ), and the **adjacent side** (the side adjacent to the angle θ).

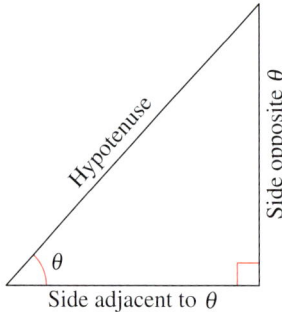

FIGURE 1.21

Using the lengths of these three sides, you can form six ratios that define the six trigonometric functions of the acute angle θ.

sine   cosecant   cosine   secant   tangent   cotangent

In the following definition, it is important to see that $0° < \theta < 90°$ and that for such angles the value of each trigonometric function is *positive*.

Right Triangle Definitions of Trigonometric Functions

Let θ be an *acute* angle of a right triangle. The six trigonometric functions of the angle θ are defined as follows.

$$\sin \theta = \frac{\text{opp}}{\text{hyp}} \qquad \cos \theta = \frac{\text{adj}}{\text{hyp}} \qquad \tan \theta = \frac{\text{opp}}{\text{adj}}$$

$$\csc \theta = \frac{\text{hyp}}{\text{opp}} \qquad \sec \theta = \frac{\text{hyp}}{\text{adj}} \qquad \cot \theta = \frac{\text{adj}}{\text{opp}}$$

The abbreviations opp, adj, and hyp represent the lengths of the three sides of a right triangle.

opp = the length of the side *opposite* θ

adj = the length of the side *adjacent to* θ

hyp = the length of the *hypotenuse*

Note that the functions in the second row above are the *reciprocals* of the corresponding functions in the first row.

Section 1.3 ▶ Right Triangle Trigonometry

Example 1 ▶ Evaluating Trigonometric Functions

Use the triangle in Figure 1.22 to find the values of the six trigonometric functions of θ.

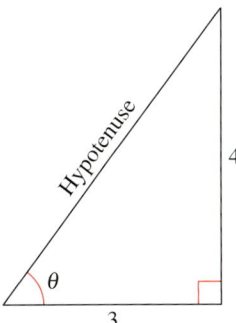

FIGURE 1.22

Solution

By the Pythagorean Theorem, $(\text{hyp})^2 = (\text{opp})^2 + (\text{adj})^2$, it follows that

$$\text{hyp} = \sqrt{4^2 + 3^2} = \sqrt{25} = 5.$$

So, the six trigonometric functions of θ are

$$\sin\theta = \frac{\text{opp}}{\text{hyp}} = \frac{4}{5} \qquad \cos\theta = \frac{\text{adj}}{\text{hyp}} = \frac{3}{5} \qquad \tan\theta = \frac{\text{opp}}{\text{adj}} = \frac{4}{3}$$

$$\csc\theta = \frac{\text{hyp}}{\text{opp}} = \frac{5}{4} \qquad \sec\theta = \frac{\text{hyp}}{\text{adj}} = \frac{5}{3} \qquad \cot\theta = \frac{\text{adj}}{\text{opp}} = \frac{3}{4}.$$

Historical Note
Georg Joachim Rhaeticus (1514–1576) was the leading Teutonic mathematical astronomer of the 16th century. He was the first to define the trigonometric functions as ratios of the sides of a right triangle.

In Example 1, you were given the lengths of two sides of the right triangle, but not the angle θ. It is more common in trigonometry to be asked to find the trigonometric functions of a *given* acute angle θ. To do this, you can construct a right triangle having θ as one of its angles.

Example 2 ▶ Evaluating Trigonometric Functions of 45°

Find the values of sin 45°, cos 45°, and tan 45°.

Solution

Construct a right triangle having 45° as one of its acute angles, as shown in Figure 1.23. Choose the length of the adjacent side to be 1. From geometry, you know that the other acute angle is also 45°. So, the triangle is isosceles and the length of the opposite side is also 1. Using the Pythagorean Theorem, you find the length of the hypotenuse to be $\sqrt{2}$.

$$\sin 45° = \frac{\text{opp}}{\text{hyp}} = \frac{1}{\sqrt{2}} = \frac{\sqrt{2}}{2} \qquad \cos 45° = \frac{\text{adj}}{\text{hyp}} = \frac{1}{\sqrt{2}} = \frac{\sqrt{2}}{2}$$

$$\tan 45° = \frac{\text{opp}}{\text{adj}} = \frac{1}{1} = 1$$

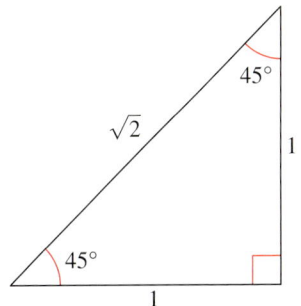

FIGURE 1.23

Example 3 ▶ **Evaluating Trigonometric Functions of 30° and 60°**

Use the equilateral triangle shown in Figure 1.24 to find the values of sin 60°, cos 60°, sin 30°, and cos 30°.

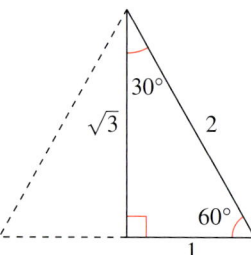

FIGURE 1.24

Solution

Try using the Pythagorean Theorem and the equilateral triangle in Figure 1.24 to verify the lengths of the sides given in Figure 1.24. For $\theta = 60°$, you have adj = 1, opp = $\sqrt{3}$, and hyp = 2. Therefore,

$$\sin 60° = \frac{\text{opp}}{\text{hyp}} = \frac{\sqrt{3}}{2} \quad \text{and} \quad \cos 60° = \frac{\text{adj}}{\text{hyp}} = \frac{1}{2}.$$

For $\theta = 30°$, adj = $\sqrt{3}$, opp = 1, and hyp = 2. So,

$$\sin 30° = \frac{\text{opp}}{\text{hyp}} = \frac{1}{2} \quad \text{and} \quad \cos 30° = \frac{\text{adj}}{\text{hyp}} = \frac{\sqrt{3}}{2}.$$

Because the angles 30°, 45°, and 60° ($\pi/6$, $\pi/4$, and $\pi/3$) occur frequently in trigonometry, you should learn to construct the triangles shown in Figures 1.23 and 1.24.

Sines, Cosines, and Tangents of Special Angles

$$\sin 30° = \sin \frac{\pi}{6} = \frac{1}{2} \qquad \cos 30° = \cos \frac{\pi}{6} = \frac{\sqrt{3}}{2} \qquad \tan 30° = \tan \frac{\pi}{6} = \frac{\sqrt{3}}{3}$$

$$\sin 45° = \sin \frac{\pi}{4} = \frac{\sqrt{2}}{2} \qquad \cos 45° = \cos \frac{\pi}{4} = \frac{\sqrt{2}}{2} \qquad \tan 45° = \tan \frac{\pi}{4} = 1$$

$$\sin 60° = \sin \frac{\pi}{3} = \frac{\sqrt{3}}{2} \qquad \cos 60° = \cos \frac{\pi}{3} = \frac{1}{2} \qquad \tan 60° = \tan \frac{\pi}{3} = \sqrt{3}$$

A computer animation of this concept appears in the *Interactive* CD-ROM and *Internet* versions of this text.

In the box, note that $\sin 30° = \frac{1}{2} = \cos 60°$. This occurs because 30° and 60° are complementary angles. In general, it can be shown from the right triangle definitions that *cofunctions of complementary angles are equal.* That is, if θ is an acute angle, the following relationships are true.

$$\sin(90° - \theta) = \cos \theta \qquad \cos(90° - \theta) = \sin \theta$$
$$\tan(90° - \theta) = \cot \theta \qquad \cot(90° - \theta) = \tan \theta$$
$$\sec(90° - \theta) = \csc \theta \qquad \csc(90° - \theta) = \sec \theta$$

Trigonometric Identities

In trigonometry, a great deal of time is spent studying relationships between trigonometric functions (identities).

Fundamental Trigonometric Identities

Reciprocal Identities

$$\sin\theta = \frac{1}{\csc\theta} \qquad \cos\theta = \frac{1}{\sec\theta} \qquad \tan\theta = \frac{1}{\cot\theta}$$

$$\csc\theta = \frac{1}{\sin\theta} \qquad \sec\theta = \frac{1}{\cos\theta} \qquad \cot\theta = \frac{1}{\tan\theta}$$

Quotient Identities

$$\tan\theta = \frac{\sin\theta}{\cos\theta} \qquad \cot\theta = \frac{\cos\theta}{\sin\theta}$$

Pythagorean Identities

$$\sin^2\theta + \cos^2\theta = 1 \qquad 1 + \tan^2\theta = \sec^2\theta$$

$$1 + \cot^2\theta = \csc^2\theta$$

Note that $\sin^2\theta$ represents $(\sin\theta)^2$, $\cos^2\theta$ represents $(\cos\theta)^2$, and so on.

Example 4 ▶ Applying Trigonometric Identities

Let θ be an acute angle such that $\sin\theta = 0.6$. Find the values of (a) $\cos\theta$ and (b) $\tan\theta$ using trigonometric identities.

Solution

a. To find the value of $\cos\theta$, use the Pythagorean identity

$$\sin^2\theta + \cos^2\theta = 1.$$

So, you have

$$(0.6)^2 + \cos^2\theta = 1 \qquad \text{Substitute 0.6 for } \sin\theta.$$

$$\cos^2\theta = 1 - (0.6)^2 = 0.64 \qquad \text{Subtract } (0.6)^2 \text{ from each side.}$$

$$\cos\theta = \sqrt{0.64} = 0.8. \qquad \text{Extract the positive square root.}$$

b. Now, knowing the sine and cosine of θ, you can find the tangent of θ to be

$$\tan\theta = \frac{\sin\theta}{\cos\theta}$$

$$= \frac{0.6}{0.8}$$

$$= 0.75.$$

Try using the definitions of $\cos\theta$ and $\tan\theta$, and the triangle shown in Figure 1.25, to check these results.

FIGURE 1.25

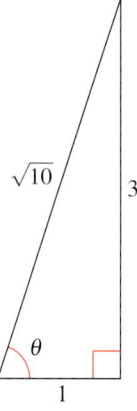

FIGURE 1.26

Example 5 ▶ Applying Trigonometric Identities

Let θ be an acute angle such that $\tan \theta = 3$. Find the values of each trigonometric function using trigonometric identities.

a. $\cot \theta$ **b.** $\sec \theta$

Solution

a. $\cot \theta = \dfrac{1}{\tan \theta} = \dfrac{1}{3}$ Reciprocal identity

b. $\sec^2 \theta = 1 + \tan^2 \theta$ Pythagorean identity

$\sec^2 \theta = 1 + 3^2$

$\sec^2 \theta = 10$

$\sec \theta = \sqrt{10}$

Try using the definitions of $\cot \theta$ and $\sec \theta$, and the triangle shown in Figure 1.26, to check these results.

Evaluating Trigonometric Functions with a Calculator

To use a calculator to evaluate trigonometric functions of angles measured in degrees, first set the calculator to degree mode and then proceed as demonstrated in Section 1.2. For instance, you can find values of $\cos 28°$ and $\sec 28°$ as follows.

Function	*Graphing Calculator Keystrokes*	*Display*
$\cos 28°$	[COS] 28 [ENTER]	0.8829476
$\sec 28°$	[(] [COS] 28 [)] [x^{-1}] [ENTER]	1.1325701

Throughout this text, angles are assumed to be measured in radians unless noted otherwise. For example, $\sin 1$ means the sine of 1 radian and $\sin 1°$ means the sine of 1 degree.

Example 6 ▶ Using a Calculator

Use a calculator to evaluate $\sec(5° \, 40' \, 12'')$.

Solution

Begin by converting to decimal form.

$$5° \, 40' \, 12'' = 5° + \left(\dfrac{40}{60}\right)° + \left(\dfrac{12}{3600}\right)° = 5.67°$$

Then, use a calculator to evaluate $\sec 5.67°$.

$$\sec(5° \, 40' \, 12'') = \sec 5.67° = \dfrac{1}{\cos 5.67°} \approx 1.00492$$

The *Interactive* CD-ROM and *Internet* versions of this text offer a built-in graphing calculator, which can be used with the Examples, Explorations, Technology notes, and Exercises.

Applications Involving Right Triangles

Many applications of trigonometry involve a process called **solving right triangles.** In this type of application, you are usually given one side of a right triangle and one of the acute angles and asked to find one of the other sides, *or* you are given two sides and asked to find one of the acute angles.

Example 7 ▶ Using Trigonometry to Solve a Right Triangle

A surveyor is standing 50 feet from the base of a large tree, as shown in Figure 1.27. The surveyor measures the angle of elevation to the top of the tree as $71.5°$. How tall is the tree?

Solution

From Figure 1.27, you see that

$$\tan 71.5° = \frac{\text{opp}}{\text{adj}} = \frac{y}{x}$$

where $x = 50$ and y is the height of the tree. So, the height of the tree is

$$y = x \tan 71.5°$$
$$\approx 50(2.98868)$$
$$\approx 149.4 \text{ feet.}$$

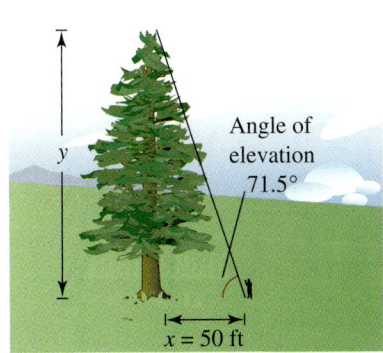

FIGURE 1.27

Example 8 ▶ Using Trigonometry to Solve a Right Triangle

A person is 200 yards from a river. Rather than walking directly to the river, the person walks 400 yards along a straight path to the river's edge. Find the acute angle θ between this path and the river's edge, as illustrated in Figure 1.28.

FIGURE 1.28

Solution

From Figure 1.28, you can see that the sine of the angle θ is

$$\sin \theta = \frac{\text{opp}}{\text{hyp}} = \frac{200}{400} = \frac{1}{2}.$$

Therefore, $\theta = 30°$.

In Example 8, you were able to recognize that the acute angle that satisfies the equation $\sin \theta = \frac{1}{2}$ is $\theta = 30°$. Suppose, however, that you were given the equation $\sin \theta = 0.6$ and asked to find the acute angle θ. Because

$$\sin 30° = \frac{1}{2}$$

$$= 0.5000$$

and

$$\sin 45° = \frac{1}{\sqrt{2}}$$

$$\approx 0.7071$$

you might guess that θ lies somewhere between $30°$ and $45°$. A more precise value of θ can be found using the *inverse sine* key on a calculator. To do this, you can use the following keystroke sequence in degree mode.

[sin⁻¹] .6 [ENTER]   Display 36.8699

So, you can conclude that if $\sin \theta = 0.6$, then $\theta \approx 36.87°$.

Example 9 ▶ Solving a Right Triangle

Specifications for a loading dock ramp require a rise of 1 foot for each 3 feet of horizontal length. In Figure 1.29, find the lengths of sides b and c and find the measure of θ.

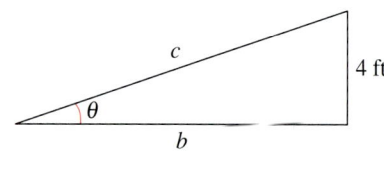

FIGURE **1.29**

Solution

From the given specifications, you can write

$$\frac{\text{rise}}{\text{run}} = \frac{1}{3} = \frac{4 \text{ ft}}{b \text{ ft}}$$

which implies that $b = 12$ feet. Using the Pythagorean Theorem, you can write

$$c^2 = a^2 + b^2 \qquad \text{Pythagorean Theorem}$$

$$c^2 = 4^2 + 12^2 \qquad \text{Substitute for } a \text{ and } b.$$

$$c^2 = 160 \qquad \text{Simplify.}$$

$$c = 4\sqrt{10}. \qquad \text{Extract positive square root.}$$

So, $c = 4\sqrt{10} \approx 12.65$ feet. To solve for θ, you can write

$$\tan \theta = \frac{4}{12}$$

$$= \frac{1}{3}.$$

Then, using the calculator keystrokes in degree mode

[tan⁻¹] [(] 1 [÷] 3 [)] [ENTER]

you obtain $\theta \approx 18.43°$.

1.3 Exercises

In Exercises 1–4, find the exact values of the six trigonometric functions of the angle θ given in the figure. (Use the Pythagorean Theorem to find the third side of the triangle.)

1.

2.

3.

4.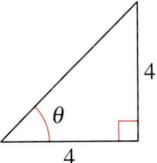

In Exercises 5–8, find the exact values of the six trigonometric functions of the angle θ for each of the two triangles. Explain why the function values are the same.

5.

6.

7.

8.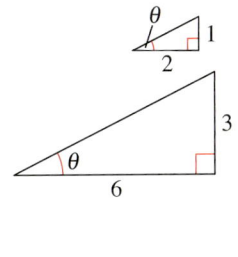

In Exercises 9–16, sketch a right triangle corresponding to the trigonometric function of the acute angle θ. Use the Pythagorean Theorem to determine the third side and then find the other five trigonometric functions of θ.

9. $\sin \theta = \frac{3}{4}$
10. $\cos \theta = \frac{5}{7}$
11. $\sec \theta = 2$
12. $\cot \theta = 5$
13. $\tan \theta = 3$
14. $\sec \theta = 6$
15. $\cot \theta = \frac{3}{2}$
16. $\csc \theta = \frac{17}{4}$

In Exercises 17–22, use the given function value(s), and trigonometric identities (including the cofunction identities), to find the indicated trigonometric functions.

17. $\sin 60° = \frac{\sqrt{3}}{2}$, $\cos 60° = \frac{1}{2}$

   (a) $\tan 60°$ (b) $\sin 30°$
   (c) $\cos 30°$ (d) $\cot 60°$

18. $\sin 30° = \frac{1}{2}$, $\tan 30° = \frac{\sqrt{3}}{3}$

   (a) $\csc 30°$ (b) $\cot 60°$
   (c) $\cos 30°$ (d) $\cot 30°$

19. $\csc \theta = \frac{\sqrt{13}}{2}$, $\sec \theta = \frac{\sqrt{13}}{3}$

   (a) $\sin \theta$ (b) $\cos \theta$
   (c) $\tan \theta$ (d) $\sec(90° - \theta)$

20. $\sec \theta = 5$, $\tan \theta = 2\sqrt{6}$

   (a) $\cos \theta$ (b) $\cot \theta$
   (c) $\cot(90° - \theta)$ (d) $\sin \theta$

21. $\cos \alpha = \frac{1}{3}$

   (a) $\sec \alpha$ (b) $\sin \alpha$
   (c) $\cot \alpha$ (d) $\sin(90° - \alpha)$

22. $\tan \beta = 5$

   (a) $\cot \beta$ (b) $\cos \beta$
   (c) $\tan(90° - \beta)$ (d) $\csc \beta$

In Exercises 23–26, evaluate the trigonometric function by memory or by constructing an appropriate triangle for the given special angle.

23. (a) $\cos 60°$ (b) $\csc 30°$ (c) $\tan 60°$
24. (a) $\cot 45°$ (b) $\cos 45°$ (c) $\csc 45°$
25. (a) $\sin 45°$ (b) $\cos 30°$ (c) $\tan 30°$
26. (a) $\sin 60°$ (b) $\tan 45°$ (c) $\sec 30°$

In Exercises 27–36, use a calculator to evaluate each function. Round your answers to four decimal places. (Be sure the calculator is in the correct angle mode.)

27. (a) $\sin 10°$ (b) $\cos 80°$
28. (a) $\tan 23.5°$ (b) $\cot 66.5°$
29. (a) $\sin 16.35°$ (b) $\csc 16.35°$

30. (a) $\cos 16° 18'$ (b) $\sin 73° 56'$
31. (a) $\sec 42° 12'$ (b) $\csc 48° 7'$
32. (a) $\cos 4° 50' 15''$ (b) $\sec 4° 50' 15''$
33. (a) $\cot 11° 15'$ (b) $\tan 11° 15'$
34. (a) $\sec 56° 8' 10''$ (b) $\cos 56° 8' 10''$
35. (a) $\csc 32° 40' 3''$ (b) $\tan 44° 28' 16''$
36. (a) $\sec\left(\frac{9}{5} \cdot 20 + 32\right)°$ (b) $\cot\left(\frac{9}{5} \cdot 30 + 32\right)°$

In Exercises 37–42, find the values of θ in degrees $(0° < \theta < 90°)$ and radians $(0 < \theta < \pi/2)$ without the aid of a calculator.

37. (a) $\sin \theta = \dfrac{1}{2}$ (b) $\csc \theta = 2$
38. (a) $\cos \theta = \dfrac{\sqrt{2}}{2}$ (b) $\tan \theta = 1$
39. (a) $\sec \theta = 2$ (b) $\cot \theta = 1$
40. (a) $\tan \theta = \sqrt{3}$ (b) $\cos \theta = \dfrac{1}{2}$
41. (a) $\csc \theta = \dfrac{2\sqrt{3}}{3}$ (b) $\sin \theta = \dfrac{\sqrt{2}}{2}$
42. (a) $\cot \theta = \dfrac{\sqrt{3}}{3}$ (b) $\sec \theta = \sqrt{2}$

In Exercises 43–46, find the values of θ in degrees $(0° < \theta < 90°)$ and radians $(0 < \theta < \pi/2)$ by using a calculator.

43. (a) $\sin \theta = 0.0145$ (b) $\sin \theta = 0.4565$
44. (a) $\cos \theta = 0.9848$ (b) $\cos \theta = 0.8746$
45. (a) $\tan \theta = 0.0125$ (b) $\tan \theta = 2.3545$
46. (a) $\sin \theta = 0.3746$ (b) $\cos \theta = 0.3746$

In Exercises 47–50, solve for x, y, or r, as indicated.

47. Solve for x.

48. Solve for y.

49. Solve for x.

50. Solve for r.

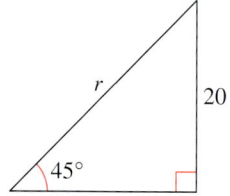

In Exercises 51–60, use trigonometric identities to transform the left side of the equation into the right side.

51. $\tan \theta \cot \theta = 1$
52. $\cos \theta \sec \theta = 1$
53. $\tan \alpha \cos \alpha = \sin \alpha$
54. $\cot \alpha \sin \alpha = \cos \alpha$
55. $(1 + \cos \theta)(1 - \cos \theta) = \sin^2 \theta$
56. $(1 + \sin \theta)(1 - \sin \theta) = \cos^2 \theta$
57. $(\sec \theta + \tan \theta)(\sec \theta - \tan \theta) = 1$
58. $\sin^2 \theta - \cos^2 \theta = 2 \sin^2 \theta - 1$
59. $\dfrac{\sin \theta}{\cos \theta} + \dfrac{\cos \theta}{\sin \theta} = \csc \theta \sec \theta$
60. $\dfrac{\tan \beta + \cot \beta}{\tan \beta} = \csc^2 \beta$

61. **Height** A 6-foot person walks from the base of a broadcasting tower directly toward the tip of the shadow cast by the tower. When the person is 132 feet from the tower and 3 feet from the tip of the shadow, the person's shadow starts to appear beyond the tower's shadow.

    (a) Draw a right triangle that gives a visual representation of the problem. Show the known quantities of the triangle and use a variable to indicate the height of the tower.
    (b) Use a trigonometric function to write an equation involving the unknown quantity.
    (c) What is the height of the tower?

62. **Height** A 6-foot person standing 20 feet from a streetlight casts a 10-foot shadow (see figure). What is the height of the streetlight?

63. *Length* A 20-meter line is used to tether a helium-filled balloon. Because of a breeze, the line makes an angle of approximately 85° with the ground.
   (a) Draw a right triangle that gives a visual representation of the problem. Show the known quantities on the triangle and use a variable to indicate the height of the balloon.
   (b) Use a trigonometric function to write an equation involving the unknown quantity.
   (c) What is the height of the balloon?

64. *Width of a River* A biologist wants to know the width w of a river in order to set instruments for studying the pollutants in the water. From point A, the biologist walks downstream 100 feet and sights to point C (see figure). From this sighting, it is determined that $\theta = 54°$. How wide is the river?

65. *Distance* From a 40-foot observation tower on the coast, a Coast Guard officer sights a boat in difficulty. The angle of depression of the boat is 4° (see figure). How far is the boat from the shoreline?

(Not drawn to scale)

66. *Angle of Elevation* A ramp 20 feet in length rises to a loading platform that is $3\frac{1}{3}$ feet off the ground.
   (a) Draw a right triangle that gives a visual representation of the problem. Show the known quantities on the triangle and use a variable to indicate the angle of elevation of the ramp.
   (b) Use a trigonometric function to write an equation involving the unknown quantity.
   (c) What is the angle of elevation of the ramp?

67. *Machine Shop Calculations* A steel plate has the form of one-fourth of a circle with a radius of 60 centimeters. Two 2-centimeter holes are to be drilled in the plate positioned as shown in the figure. Find the coordinates of the center of each hole.

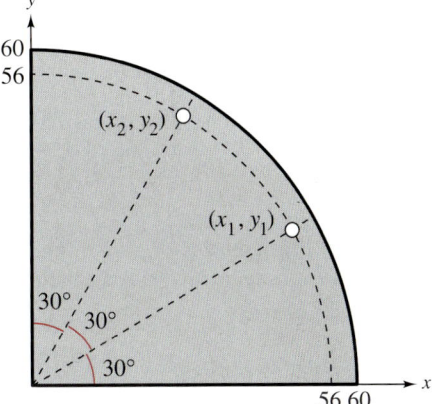

68. *Machine Shop Calculations* A tapered shaft has a diameter of 5 centimeters at the small end and is 15 centimeters long (see figure). If the taper is 3°, find the diameter d of the large end of the shaft.

148 Chapter 1 ▶ Trigonometry

69. *Geometry* Use a compass to sketch a quarter of a circle of radius 10 centimeters. Using a protractor, construct an angle of 20° in standard position (see figure). Drop a perpendicular from the point of intersection of the terminal side of the angle and the arc of the circle. By actual measurement, calculate the coordinates (x, y) of the point of intersection and use these measurements to approximate the six trigonometric functions of a 20° angle.

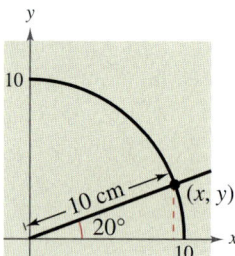

70. *Geometry* Repeat Exercise 69 using a 75° angle.

Synthesis

True or False? In Exercises 71–76, determine whether the statement is true or false. Justify your answer.

71. $\sin 60° \csc 60° = 1$

72. $\sec 30° = \csc 60°$

73. $\sin 45° + \cos 45° = 1$

74. $\cot^2 10° - \csc^2 10° = -1$

75. $\dfrac{\sin 60°}{\sin 30°} = \sin 2°$

76. $\tan[(5°)^2] = \tan^2(5°)$

77. In right triangle trigonometry, $\sin 30° = \tfrac{1}{2}$ regardless of the size of the triangle. Explain.

78. You are given only the value $\tan \theta$. Is it possible to find the value of $\sec \theta$ without finding the measure of θ? Explain.

79. *Exploration*

(a) Complete the table below.

θ	0.1	0.2	0.3	0.4	0.5
$\sin \theta$					

(b) As θ approaches 0, how do θ and $\sin \theta$ compare? Explain.

80. *Exploration*

(a) Complete the table.

θ	0°	18°	36°	54°	72°	90°
$\sin \theta$						
$\cos \theta$						

(b) Discuss the behavior of the sine function for θ in the range from 0° to 90°.

(c) Discuss the behavior of the cosine function for θ in the range from 0° to 90°.

(d) Use the definitions of the sine and cosine functions to explain the results of parts (b) and (c).

Review

In Exercises 81–84, perform the operations and simplify.

81. $\dfrac{x^2 - 6x}{x^2 + 4x - 12} \cdot \dfrac{x^2 + 12x + 36}{x^2 - 36}$

82. $\dfrac{2t^2 + 5t - 12}{9 - 4t^2} \div \dfrac{t^2 - 16}{4t^2 + 12t + 9}$

83. $\dfrac{3}{x + 2} - \dfrac{2}{x - 2} + \dfrac{x}{x^2 + 4x + 4}$

84. $\dfrac{\left(\dfrac{3}{x} - \dfrac{1}{4}\right)}{\left(\dfrac{12}{x} - 1\right)}$

In Exercises 85–88, solve for *x*.

85. $\dfrac{4}{x - 4} = \dfrac{12x}{24 - x}$

86. $\dfrac{6}{x} = \dfrac{72}{9x^2 - 6x}$

87. $\dfrac{2}{x + 3} + \dfrac{4}{x - 2} = \dfrac{12}{x^2 + x - 6}$

88. $\dfrac{3x + 2}{x^2 + x - 2} = \dfrac{4}{x + 2} - \dfrac{2}{1 - x}$

1.4 Trigonometric Functions of Any Angle

What you should learn

- How to evaluate trigonometric functions of any angle
- How to use reference angles to evaluate trigonometric functions
- How to evaluate trigonometric functions of real numbers

Why you should learn it

You can use trigonometric functions to model and solve real-life problems. For instance, Exercise 91 on page 156 shows how trigonometric functions can be used to model the average daily temperature in a city.

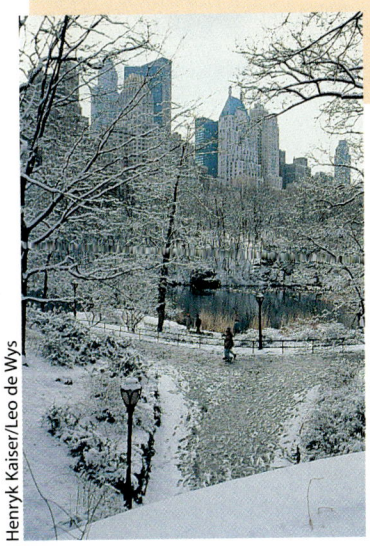

Introduction

In Section 1.3, the definitions of trigonometric functions were restricted to acute angles. In this section, the definitions are extended to cover *any* angle. If θ is an *acute* angle, these definitions coincide with those given in the preceding section.

Definitions of Trigonometric Functions of Any Angle

Let θ be an angle in standard position with (x, y) a point on the terminal side of θ and $r = \sqrt{x^2 + y^2} \neq 0$.

$$\sin \theta = \frac{y}{r} \qquad \cos \theta = \frac{x}{r}$$

$$\tan \theta = \frac{y}{x}, \quad x \neq 0 \qquad \cot \theta = \frac{x}{y}, \quad y \neq 0$$

$$\sec \theta = \frac{r}{x}, \quad x \neq 0 \qquad \csc \theta = \frac{r}{y}, \quad y \neq 0$$

Because $r = \sqrt{x^2 + y^2}$ cannot be zero, it follows that the sine and cosine functions are defined for any real value of θ. However, if $x = 0$, the tangent and secant of θ are undefined. For example, the tangent of 90° is undefined. Similarly, if $y = 0$, the cotangent and cosecant of θ are undefined.

Example 1 ▶ Evaluating Trigonometric Functions

Let $(-3, 4)$ be a point on the terminal side of θ. Find the sine, cosine, and tangent of θ.

Solution

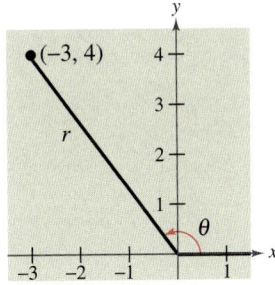

FIGURE 1.30

Referring to Figure 1.30, you see that $x = -3$, $y = 4$, and

$$r = \sqrt{x^2 + y^2} = \sqrt{(-3)^2 + 4^2} = \sqrt{25} = 5.$$

So, you have the following.

$$\sin \theta = y/r = 4/5, \qquad \cos \theta = x/r = -3/5, \qquad \tan \theta = y/x = -4/3$$

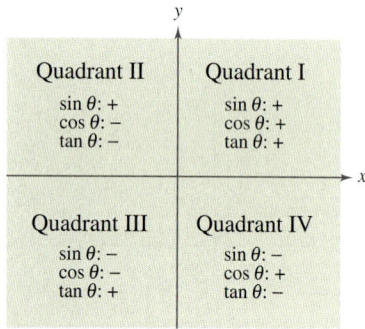

FIGURE 1.31

The *signs* of the trigonometric functions in the four quadrants can be determined easily from the definitions of the functions. For instance, because $\cos \theta = x/r$, it follows that $\cos \theta$ is positive wherever $x > 0$, which is in Quadrants I and IV. (Remember, r is always positive.) In a similar manner, you can verify the results shown in Figure 1.31.

Example 2 ▶ Evaluating Trigonometric Functions

Given $\tan \theta = -\frac{5}{4}$ and $\cos \theta > 0$, find $\sin \theta$ and $\sec \theta$.

Solution

Note that θ lies in Quadrant IV because that is the only quadrant in which the tangent is negative and the cosine is positive. Moreover, using

$$\tan \theta = \frac{y}{x}$$

$$= -\frac{5}{4}$$

and the fact that y is negative in Quadrant IV, you can let $y = -5$ and $x = 4$. So, $r = \sqrt{16 + 25} = \sqrt{41}$ and you have

$$\sin \theta = \frac{y}{r} = \frac{-5}{\sqrt{41}} \approx -0.7809$$

$$\sec \theta = \frac{r}{x} = \frac{\sqrt{41}}{4} \approx 1.6008.$$

Example 3 ▶ Trigonometric Functions of Quadrant Angles

Evaluate the sine function at the four quadrant angles 0, $\dfrac{\pi}{2}$, π, and $\dfrac{3\pi}{2}$.

Solution

To begin, choose a point on the terminal side of each angle, as shown in Figure 1.32. For each of the four given points, $r = 1$, and you have

$$\sin 0 = \frac{y}{r} = \frac{0}{1} = 0 \qquad (x, y) = (1, 0)$$

$$\sin \frac{\pi}{2} = \frac{y}{r} = \frac{1}{1} = 1 \qquad (x, y) = (0, 1)$$

$$\sin \pi = \frac{y}{r} = \frac{0}{1} = 0 \qquad (x, y) = (-1, 0)$$

$$\sin \frac{3\pi}{2} = \frac{y}{r} = \frac{-1}{1} = -1. \qquad (x, y) = (0, -1)$$

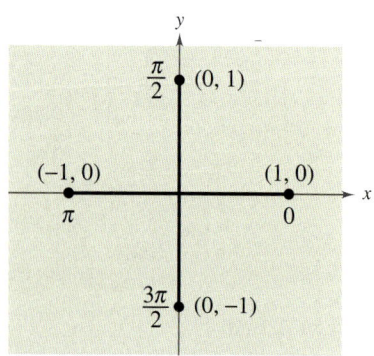

FIGURE 1.32

Reference Angles

The values of the trigonometric functions of angles greater than 90° (or less than 0°) can be determined from their values at corresponding acute angles called **reference angles.**

Definition of Reference Angle

Let θ be an angle in standard position. Its **reference angle** is the acute angle θ' formed by the terminal side of θ and the horizontal axis.

Figure 1.33 shows the reference angles for θ in Quadrants II, III, and IV.

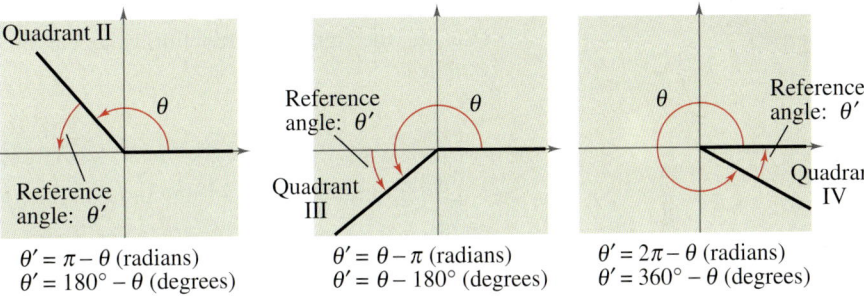

FIGURE 1.33

Example 4 ▶ Finding Reference Angles

Find the reference angle θ'.

a. $\theta = 300°$ **b.** $\theta = 2.3$ **c.** $\theta = -135°$

Solution

a. Because 300° lies in Quadrant IV, the angle it makes with the x-axis is

$$\theta' = 360° - 300° = 60°. \qquad \text{Degrees}$$

Figure 1.34(a) shows the angle $\theta = 300°$ and its reference angle $\theta' = 60°$.

b. Because 2.3 lies between $\pi/2 \approx 1.5708$ and $\pi \approx 3.1416$, it follows that it is in Quadrant II and its reference angle is

$$\theta' = \pi - 2.3 \approx 0.8416. \qquad \text{Radians}$$

Figure 1.34(b) shows the angle $\theta = 2.3$ and its reference angle $\theta' = \pi - 2.3$.

c. First, determine that $-135°$ is coterminal with 225°, which lies in Quadrant III. Hence, the reference angle is

$$\theta' = 225° - 180° = 45°. \qquad \text{Degrees}$$

Figure 1.34(c) shows the angle $\theta = -135°$ and its reference angle $\theta' = 45°$.

(a)

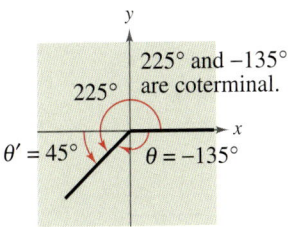

(b)

(c)

FIGURE 1.34

Trigonometric Functions of Real Numbers

To see how a reference angle is used to evaluate a trigonometric function, consider the point (x, y) on the terminal side of θ, as shown in Figure 1.35. By definition, you know that

$$\sin \theta = \frac{y}{r} \quad \text{and} \quad \tan \theta = \frac{y}{x}.$$

For the right triangle with acute angle θ' and sides of lengths $|x|$ and $|y|$, you have

$$\sin \theta' = \frac{\text{opp}}{\text{hyp}} = \frac{|y|}{r} \quad \text{and} \quad \tan \theta' = \frac{\text{opp}}{\text{adj}} = \frac{|y|}{|x|}.$$

So, it follows that $\sin \theta$ and $\sin \theta'$ are equal, *except possibly in sign*. The same is true for $\tan \theta$ and $\tan \theta'$ *and* for the other four trigonometric functions. In all cases, the sign of the function value can be determined by the quadrant in which θ lies.

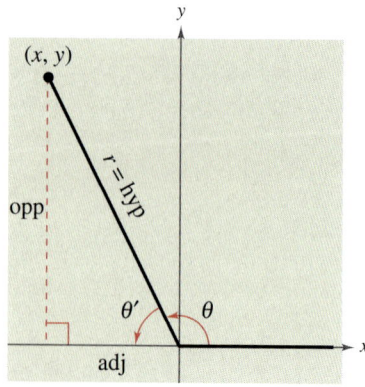

opp = $|y|$, adj = $|x|$
FIGURE 1.35

Evaluating Trigonometric Functions of Any Angle

To find the value of a trigonometric function of any angle θ,

1. determine the function value for the associated reference angle θ';
2. depending on the quadrant in which θ lies, affix the appropriate sign to the function value.

By using reference angles and the special angles discussed in the preceding section, you can greatly extend the scope of *exact* trigonometric values. For instance, knowing the function values of 30° means that you know the function values of all angles for which 30° is a reference angle. For convenience, the table below gives the exact values of the trigonometric functions of special angles and quadrant angles.

STUDY TIP

Learning the table of values at the right is worth the effort. Doing so will increase both your efficiency and your confidence. Here is a pattern for the sine function that may help you remember the values.

θ	0°	30°	45°	60°	90°
$\sin \theta$	$\frac{\sqrt{0}}{2}$	$\frac{\sqrt{1}}{2}$	$\frac{\sqrt{2}}{2}$	$\frac{\sqrt{3}}{2}$	$\frac{\sqrt{4}}{2}$

Reverse the order to get cosine values of the same angles.

Trigonometric Values of Common Angles

θ (degrees)	0°	30°	45°	60°	90°	180°	270°
θ (radians)	0	$\frac{\pi}{6}$	$\frac{\pi}{4}$	$\frac{\pi}{3}$	$\frac{\pi}{2}$	π	$\frac{3\pi}{2}$
$\sin \theta$	0	$\frac{1}{2}$	$\frac{\sqrt{2}}{2}$	$\frac{\sqrt{3}}{2}$	1	0	-1
$\cos \theta$	1	$\frac{\sqrt{3}}{2}$	$\frac{\sqrt{2}}{2}$	$\frac{1}{2}$	0	-1	0
$\tan \theta$	0	$\frac{\sqrt{3}}{3}$	1	$\sqrt{3}$	Undef.	0	Undef.

Example 5 Trigonometric Functions of Nonacute Angles

Evaluate the trigonometric functions.

a. $\cos \dfrac{4\pi}{3}$ **b.** $\tan(-210°)$ **c.** $\csc \dfrac{11\pi}{4}$

Solution

a. Because $\theta = 4\pi/3$ lies in Quadrant III, the reference angle is $\theta' = (4\pi/3) - \pi = \pi/3$, as shown in Figure 1.36(a). Moreover, the cosine is negative in Quadrant III, so

$$\cos \frac{4\pi}{3} = (-)\cos \frac{\pi}{3}$$

$$= -\frac{1}{2}.$$

b. Because $-210° + 360° = 150°$, it follows that $-210°$ is coterminal with the second-quadrant angle $150°$. Therefore, the reference angle is $\theta' = 180° - 150° = 30°$, as shown in Figure 1.36(b). Finally, because the tangent is negative in Quadrant II, you have

$$\tan(-210°) = (-)\tan 30°$$

$$= -\frac{\sqrt{3}}{3}.$$

c. Because $(11\pi/4) - 2\pi = 3\pi/4$, it follows that $11\pi/4$ is coterminal with the second-quadrant angle $3\pi/4$. Therefore, the reference angle is $\theta' = \pi - (3\pi/4) = \pi/4$, as shown in Figure 1.36(c). Because the cosecant is positive in Quadrant II, you have

$$\csc \frac{11\pi}{4} = (+)\csc \frac{\pi}{4}$$

$$= \frac{1}{\sin(\pi/4)}$$

$$= \sqrt{2}.$$

(a)

(b)

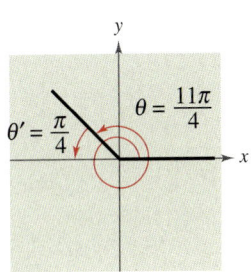
(c)

FIGURE 1.36

Example 6 ▶ Using Trigonometric Identities

Let θ be an angle in Quadrant II such that $\sin \theta = \frac{1}{3}$. Find (a) $\cos \theta$ and (b) $\tan \theta$ by using trigonometric identities.

Solution

a. Using the Pythagorean identity $\sin^2 \theta + \cos^2 \theta = 1$, you obtain

$$\left(\frac{1}{3}\right)^2 + \cos^2 \theta = 1 \qquad \text{Substitute } \tfrac{1}{3} \text{ for } \sin \theta.$$

$$\cos^2 \theta = 1 - \frac{1}{9} = \frac{8}{9}.$$

Because $\cos \theta < 0$ in Quadrant II, you can use the negative root to obtain

$$\cos \theta = -\frac{\sqrt{8}}{\sqrt{9}}$$

$$= -\frac{2\sqrt{2}}{3}.$$

b. Using the trigonometric identity $\tan \theta = \sin \theta / \cos \theta$, you obtain

$$\tan \theta = \frac{1/3}{-2\sqrt{2}/3}$$

$$= -\frac{1}{2\sqrt{2}}$$

$$= -\frac{\sqrt{2}}{4}.$$

Example 7 ▶ Using a Calculator

a. Use a calculator to evaluate $\cot 410°$ and $\sin(-7)$.

b. Use a calculator to solve $\tan \theta = 4.812$, $0 \leq \theta < 2\pi$.

Solution

Function	Mode	Calculator Keystrokes	Display
a. $\cot 410°$	Degree	(TAN 410) x^{-1} ENTER	0.8390996
$\sin(-7)$	Radian	SIN ((−) 7) ENTER	−0.6569866

b. To solve the equation $\tan \theta = 4.812$, you can use the inverse tangent key, as follows.

Equation	Mode	Calculator Keystrokes	Display
$\tan \theta = 4.812$	Radian	TAN^{-1} 4.812 ENTER	1.365898912

The angle $\theta \approx 1.366$ lies in Quadrant I. A second value of θ lies in Quadrant III (tangent is positive) and is

$$\theta = \pi + 1.366 \approx 4.508.$$

1.4 Exercises

In Exercises 1–4, determine the exact values of the six trigonometric functions of the angle θ.

1. (a) (b)

2. (a) (b)

3. (a) (b)

4. (a) (b)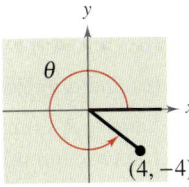

In Exercises 5–10, the point is on the terminal side of an angle in standard position. Determine the exact values of the six trigonometric functions of the angle.

5. $(7, 24)$
6. $(8, 15)$
7. $(-4, 10)$
8. $(-5, -2)$
9. $(-3.5, 6.8)$
10. $(3\tfrac{1}{2}, -7\tfrac{3}{4})$

In Exercises 11–14, state the quadrant in which θ lies.

11. $\sin \theta < 0$ and $\cos \theta < 0$
12. $\sin \theta > 0$ and $\cos \theta > 0$
13. $\sin \theta > 0$ and $\tan \theta < 0$
14. $\sec \theta > 0$ and $\cot \theta < 0$

In Exercises 15–24, find the values of the six trigonometric functions of θ.

Function Value	Constraint
15. $\sin \theta = \tfrac{3}{5}$	θ lies in Quadrant II.
16. $\cos \theta = -\tfrac{4}{5}$	θ lies in Quadrant III.
17. $\tan \theta = -\tfrac{15}{8}$	$\sin \theta < 0$
18. $\cos \theta = \tfrac{8}{17}$	$\tan \theta < 0$
19. $\cot \theta = -3$	$\cos \theta > 0$
20. $\csc \theta = 4$	$\cot \theta < 0$
21. $\sec \theta = -2$	$\sin \theta > 0$
22. $\sin \theta = 0$	$\sec \theta = -1$
23. $\cot \theta$ is undefined.	$\pi/2 \leq \theta \leq 3\pi/2$
24. $\tan \theta$ is undefined.	$\pi \leq \theta \leq 2\pi$

In Exercises 25–28, the terminal side of θ lies on the given line in the specified quadrant. Find the values of the six trigonometric functions of θ.

Line	Quadrant
25. $y = -x$	II
26. $y = \tfrac{1}{3}x$	III
27. $2x - y = 0$	III
28. $4x + 3y = 0$	IV

In Exercises 29–36, evaluate the trigonometric function of the quadrant angle.

29. $\cos \pi$
30. $\cos \dfrac{3\pi}{2}$
31. $\sec \pi$
32. $\sec \dfrac{3\pi}{2}$
33. $\tan \dfrac{\pi}{2}$
34. $\tan \pi$
35. $\cot \dfrac{\pi}{2}$
36. $\csc \pi$

In Exercises 37–44, find the reference angle θ′, and sketch θ and θ′ in standard position.

37. $\theta = 203°$
38. $\theta = 309°$
39. $\theta = -245°$
40. $\theta = -145°$
41. $\theta = \dfrac{2\pi}{3}$
42. $\theta = \dfrac{7\pi}{4}$

43. $\theta = 3.5$ **44.** $\theta = \dfrac{11\pi}{3}$

In Exercises 45–58, evaluate the sine, cosine, and tangent of the angle without using a calculator.

45. $225°$ **46.** $300°$
47. $750°$ **48.** $-405°$
49. $-150°$ **50.** $-840°$
51. $\dfrac{4\pi}{3}$ **52.** $\dfrac{\pi}{4}$
53. $-\dfrac{\pi}{6}$ **54.** $-\dfrac{\pi}{2}$
55. $\dfrac{11\pi}{4}$ **56.** $\dfrac{10\pi}{3}$
57. $-\dfrac{3\pi}{2}$ **58.** $-\dfrac{25\pi}{4}$

In Exercises 59–68, use a calculator to evaluate the trigonometric function to four decimal places. (Be sure the calculator is set in the correct mode.)

59. $\sin 10°$ **60.** $\sec 225°$
61. $\cos(-110°)$ **62.** $\csc(-330°)$
63. $\tan 4.5$ **64.** $\cot 1.35$
65. $\tan \dfrac{\pi}{9}$ **66.** $\tan\left(-\dfrac{\pi}{9}\right)$
67. $\sin(-0.65)$ **68.** $\sin 0.65$

In Exercises 69–74, find two solutions of the equation. Give your answers in degrees $(0° \le \theta < 360°)$ and radians $(0 \le \theta < 2\pi)$. Do not use a calculator.

69. (a) $\sin \theta = \dfrac{1}{2}$ (b) $\sin \theta = -\dfrac{1}{2}$
70. (a) $\cos \theta = \dfrac{\sqrt{2}}{2}$ (b) $\cos \theta = -\dfrac{\sqrt{2}}{2}$
71. (a) $\csc \theta = \dfrac{2\sqrt{3}}{3}$ (b) $\cot \theta = -1$
72. (a) $\sec \theta = 2$ (b) $\sec \theta = -2$
73. (a) $\tan \theta = 1$ (b) $\cot \theta = -\sqrt{3}$
74. (a) $\sin \theta = \dfrac{\sqrt{3}}{2}$ (b) $\sin \theta = -\dfrac{\sqrt{3}}{2}$

In Exercises 75–78, use a calculator to approximate two values of $\theta (0° \le \theta < 360°)$ that satisfy the equation. Round the values to two decimal places.

75. $\sin \theta = 0.8191$ **76.** $\cos \theta = 0.8746$
77. $\cos \theta = -0.4367$ **78.** $\sin \theta = -0.6514$

In Exercises 79–84, use a calculator to approximate two values of $\theta (0 \le \theta < 2\pi)$ that satisfy the equation. Round the values to three decimal places.

79. $\cos \theta = 0.9848$ **80.** $\sin \theta = 0.0175$
81. $\tan \theta = 1.192$ **82.** $\cot \theta = 5.671$
83. $\sec \theta = -2.6667$ **84.** $\cos \theta = -0.3214$

In Exercises 85–90, find the indicated trigonometric value in the specified quadrant.

	Function	Quadrant	Trigonometric Value
85.	$\sin \theta = -\dfrac{3}{5}$	IV	$\cos \theta$
86.	$\cot \theta = -3$	II	$\sin \theta$
87.	$\tan \theta = \dfrac{3}{2}$	III	$\sec \theta$
88.	$\csc \theta = -2$	IV	$\cot \theta$
89.	$\cos \theta = \dfrac{5}{8}$	I	$\sec \theta$
90.	$\sec \theta = -\dfrac{9}{4}$	III	$\tan \theta$

91. *Average Temperature* The average daily temperature T (in degrees Fahrenheit) for a certain city is

$$T = 45 - 23 \cos\left[\dfrac{2\pi}{365}(t - 32)\right]$$

where t is the time in days, with $t = 1$ corresponding to January 1. Find the average daily temperatures on the following days.

(a) January 1
(b) July 4 ($t = 185$)
(c) October 18 ($t = 291$)

92. *Sales* A company that produces a seasonal product forecasts monthly sales over the next 2 years to be

$$S = 23.1 + 0.442t + 4.3 \sin \dfrac{\pi t}{6}$$

where S is measured in thousands of units and t is the time in months, with $t = 1$ representing January 2001. Predict sales for each of the following months.

(a) February 2001 (b) February 2002
(c) September 2001 (d) September 2002

93. *Harmonic Motion* The displacement from equilibrium of an oscillating weight suspended by a spring is

$$y(t) = 2 \cos 6t$$

where y is the displacement in centimeters and t is the time in seconds (see figure). Find the displacement when (a) $t = 0$, (b) $t = \dfrac{1}{4}$, and (c) $t = \dfrac{1}{2}$.

Synthesis

True or False? In Exercises 97 and 98, determine whether the statement is true or false. Justify your answer.

97. In each of the four quadrants, the sign of the secant function and sine function will be the same.

98. To find the reference angle for an angle θ (given in degrees), find the integer n such that $0 \le 360n - \theta \le 360$. The difference $360n - \theta$ is the reference angle.

99. *Writing* Consider an angle in standard position with $r = 12$ centimeters, as shown in the figure. Write a short paragraph describing the changes in the magnitudes of x, y, $\sin \theta$, $\cos \theta$, and $\tan \theta$ as θ increases continually from $0°$ to $90°$.

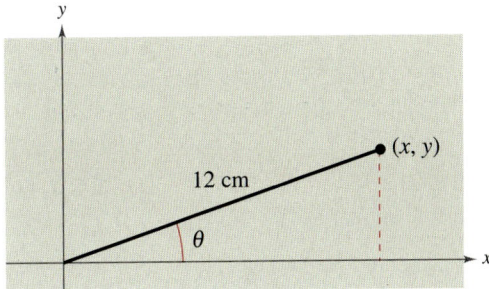

100. Explain how reference angles are used to find the trigonometric functions of obtuse angles.

101. Explain how reference angles are used to find the trigonometric functions of angles with negative measure.

Review

In Exercises 102–105, solve the equation.

102. $\dfrac{t}{3} + \dfrac{t}{5} = 1$

103. $\dfrac{3}{t} + \dfrac{5}{t} = 1$

104. $\dfrac{3}{x(x+1)} - \dfrac{4}{x} = \dfrac{1}{x+1}$

105. $\dfrac{12}{x} - 3 = \dfrac{4}{x} + 9$

In Exercises 106–109, find the equation of the line passing through the given pair of points.

106. $(-2, -5), (4, -1)$

107. $(10, 0), (1, 9)$

108. $(-6, 5), (3, -5)$

109. $\left(-\frac{1}{2}, 3\right), \left(\frac{11}{2}, -\frac{1}{3}\right)$

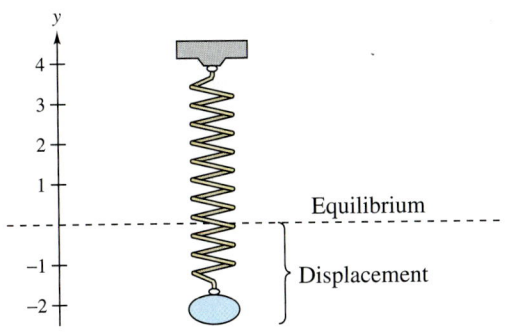

FIGURE FOR 93

94. *Distance* An airplane, flying at an altitude of 6 miles, is on a flight path that passes directly over an observer (see figure). If θ is the angle of elevation from the observer to the plane, find the distance d from the observer to the plane when (a) $\theta = 30°$, (b) $\theta = 90°$, and (c) $\theta = 120°$.

Path of a Projectile In Exercises 95 and 96, use the following information. The horizontal distance traveled by a projectile (in feet) with an initial speed of v feet per second is

$$d = \dfrac{v^2}{32} \sin \theta$$

where θ is the angle at which the projectile is launched.

95. Find the horizontal distance traveled by a golf ball that is hit with an initial speed of 100 feet per second when the ball is hit at an angle of (a) $\theta = 30°$, (b) $\theta = 50°$, and (c) $\theta = 60°$.

96. Find the horizontal distance traveled by a model rocket launched with an initial speed of 120 feet per second when the rocket is launched at an angle of (a) $\theta = 60°$, (b) $\theta = 70°$, and (c) $\theta = 80°$.

1.5 Graphs of Sine and Cosine Functions

▶ **What you should learn**
- How to sketch the graphs of basic sine and cosine functions
- How to use amplitude and period to help sketch the graphs of sine and cosine functions
- How to sketch translations of the graphs of sine and cosine functions
- How to use sine and cosine functions to model real-life data

▶ **Why you should learn it**

Many biological rhythms can be modeled with sine and cosine functions. For instance, Exercises 73 and 74 on pages 166 and 167 use a sine function to model the respiration of a person at rest and while exercising.

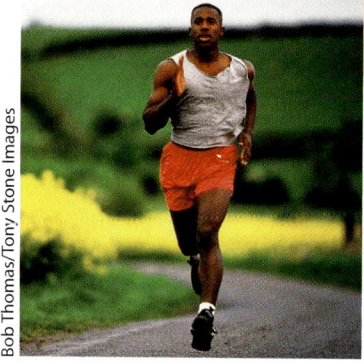

Basic Sine and Cosine Curves

In this section you will study techniques for sketching the graphs of the sine and cosine functions. The graph of the sine function is a **sine curve.** In Figure 1.37, the black portion of the graph represents one period of the function and is called **one cycle** of the sine curve. The gray portion of the graph indicates that the basic sine wave repeats indefinitely to the right and left. The graph of the cosine function is shown in Figure 1.38.

Recall from Section 1.2 that the domain of the sine and cosine functions is the set of all real numbers. Moreover, the range of each function is the interval $[-1, 1]$, and each function has a period of 2π. Do you see how this information is consistent with the basic graphs given in Figures 1.37 and 1.38?

FIGURE 1.37

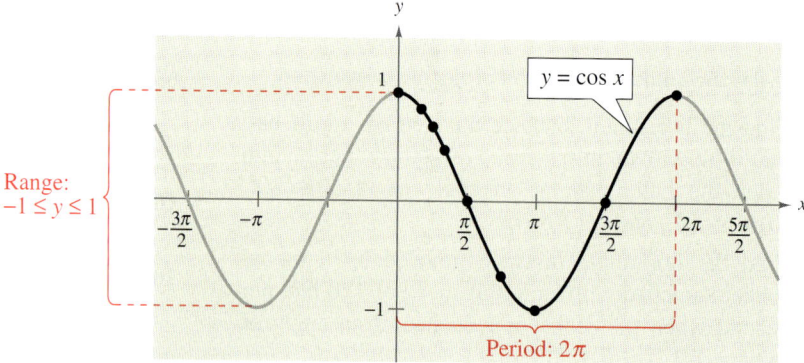

FIGURE 1.38

Note in Figures 1.37 and 1.38 that the sine curve is symmetric with respect to the *origin*, whereas the cosine curve is symmetric with respect to the *y-axis*. These properties of symmetry occur because the sine function is odd and the cosine function is even.

To sketch the graphs of the basic sine and cosine functions by hand, it helps to note five **key points** in one period of each graph: the *intercepts*, *maximum points*, and *minimum points* (see Figure 1.39).

 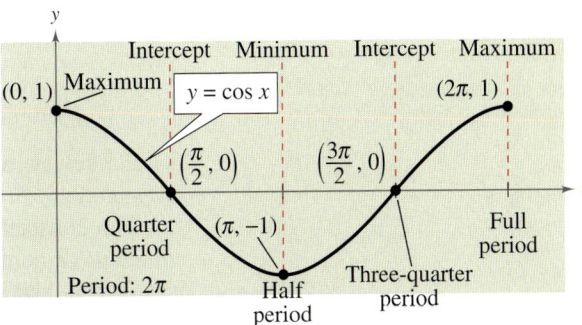

FIGURE 1.39

Example 1 ▶ Using Key Points to Sketch a Sine Curve

Sketch the graph of $y = 2 \sin x$ on the interval $[-\pi, 4\pi]$.

Solution

Note that

$$y = 2 \sin x = 2(\sin x)$$

indicates that the y-values for the key points will have twice the magnitude of those on the graph of $y = \sin x$. Divide the period 2π into four equal parts to get the key points for $y = 2 \sin x$.

$$(0, 0), \quad \left(\frac{\pi}{2}, 2\right), \quad (\pi, 0), \quad \left(\frac{3\pi}{2}, 2\right), \quad \text{and} \quad (2\pi, 0)$$

By connecting these key points with a smooth curve and extending the curve in both directions over the interval $[-\pi, 4\pi]$, you obtain the graph shown in Figure 1.40.

> **Technology**
>
> When using a graphing utility to graph trigonometric functions, pay special attention to the viewing window you use. For instance, try graphing $y = [\sin(10x)]/10$ in the standard viewing window in radian mode. What do you observe? Use the zoom feature to find a viewing window that displays a good view of the graph.

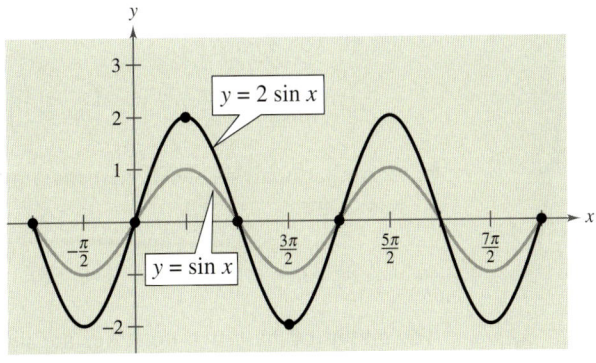

FIGURE 1.40

Amplitude and Period

In the remainder of this section you will study the graphic effect of each of the constants a, b, c, and d in equations of the forms

$$y = d + a \sin(bx - c)$$

and

$$y = d + a \cos(bx - c).$$

A quick review of the transformations studied in Section P.7 should help in this investigation.

The constant factor a in $y = a \sin x$ acts as a *scaling factor*—a *vertical stretch* or *vertical shrink* of the basic sine curve. If $|a| > 1$, the basic sine curve is stretched, and if $|a| < 1$, the basic sine curve is shrunk. The result is that the graph of $y = a \sin x$ ranges between $-a$ and a instead of between -1 and 1. The absolute value of a is the **amplitude** of the function $y = a \sin x$. The range of the function $y = a \sin x$ is $-a \le y \le a$.

> ### Definition of Amplitude of Sine and Cosine Curves
> The **amplitude** of $y = a \sin x$ and $y = a \cos x$ represents half the distance between the maximum and minimum values of the function and is given by
>
> $$\text{Amplitude} = |a|.$$

Example 2 ▶ Scaling: Vertical Shrinking and Stretching

On the same coordinate axes, sketch the graphs of the functions.

a. $y = \dfrac{1}{2} \cos x$

b. $y = 3 \cos x$

Solution

a. Because the amplitude of $y = \frac{1}{2} \cos x$ is $\frac{1}{2}$, the maximum value is $\frac{1}{2}$ and the minimum value is $-\frac{1}{2}$. Divide one cycle, $0 \le x \le 2\pi$, into four equal parts to get the key points

$$\left(0, \frac{1}{2}\right), \quad \left(\frac{\pi}{2}, 0\right), \quad \left(\pi, -\frac{1}{2}\right), \quad \left(\frac{3\pi}{2}, 0\right), \quad \text{and} \quad \left(2\pi, \frac{1}{2}\right).$$

b. A similar analysis shows that the amplitude of $y = 3 \cos x$ is 3, and the key points are

$$(0, 3), \quad \left(\frac{\pi}{2}, 0\right), \quad (\pi, -3), \quad \left(\frac{3\pi}{2}, 0\right), \quad \text{and} \quad (2\pi, 3).$$

The graphs of these two functions are shown in Figure 1.41.

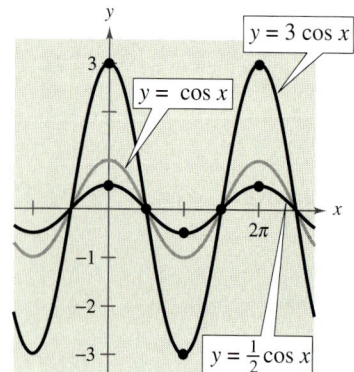

FIGURE 1.41

Section 1.5 ▶ Graphs of Sine and Cosine Functions 161

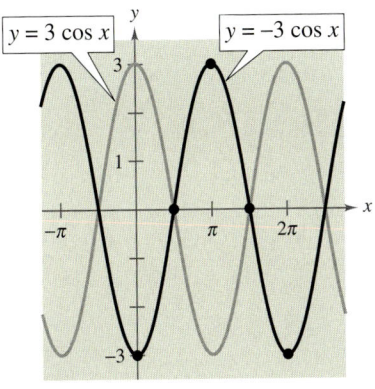

FIGURE 1.42

You know from Section P.7 that the graph of $y = -f(x)$ is a **reflection** in the x-axis of the graph of $y = f(x)$. For instance, the graph of

$$y = -3 \cos x$$

is a reflection of the graph of

$$y = 3 \cos x,$$

as shown in Figure 1.42.

Because $y = a \sin x$ completes one cycle from $x = 0$ to $x = 2\pi$, it follows that $y = a \sin bx$ completes one cycle from $x = 0$ to $x = 2\pi/b$.

Period of Sine and Cosine Functions

Let b be a positive real number. The **period** of $y = a \sin bx$ and $y = a \cos bx$ is $2\pi/b$.

Note that if $0 < b < 1$, the period of $y = a \sin bx$ is greater than 2π and represents a *horizontal stretching* of the graph of $y = a \sin x$. Similarly, if $b > 1$, the period of $y = a \sin bx$ is less than 2π and represents a *horizontal shrinking* of the graph of $y = a \sin x$. If b is negative, the identities $\sin(-x) = -\sin x$ and $\cos(-x) = \cos x$ are used to rewrite the function.

◀ Exploration ▶

Sketch the graph of $y = \cos bx$ for $b = \frac{1}{2}$, 2, and 3. How does the value of b affect the graph? How many complete cycles occur between 0 and 2π for each value of b?

Example 3 ▶ Scaling: Horizontal Stretching

Sketch the graph of $y = \sin \dfrac{x}{2}$.

Solution

The amplitude is 1. Moreover, because $b = \frac{1}{2}$, the period is

$$\frac{2\pi}{b} = \frac{2\pi}{\frac{1}{2}} = 4\pi. \qquad \text{Substitute for } b.$$

Now, divide the period-interval $[0, 4\pi]$ into four equal parts with the values π, 2π, and 3π to obtain the key points on the graph.

$(0, 0), \qquad (\pi, 1), \qquad (2\pi, 0), \qquad (3\pi, -1), \qquad \text{and} \qquad (4\pi, 0)$

The graph is shown in Figure 1.43.

STUDY T!P

In general, to divide a period-interval into four equal parts, successively add "period/4," starting with the left endpoint of the interval. For instance, for the period-interval $[-\pi/6, \pi/2]$ of length $2\pi/3$, you would successively add

$$\frac{2\pi/3}{4} = \frac{\pi}{6}$$

to get $-\pi/6, 0, \pi/6, \pi/3,$ and $\pi/2$.

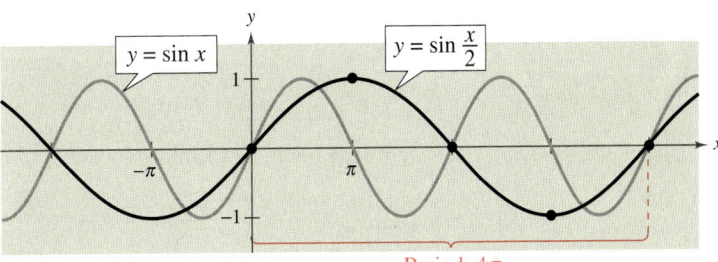

FIGURE 1.43

Translations of Sine and Cosine Curves

The constant c in the general equations

$$y = a\sin(bx - c) \quad \text{and} \quad y = a\cos(bx - c)$$

creates a *horizontal translation* (shift) of the basic sine and cosine curves. Comparing $y = a\sin bx$ with $y = a\sin(bx - c)$, you find that the graph of $y = a\sin(bx - c)$ completes one cycle from $bx - c = 0$ to $bx - c = 2\pi$. By solving for x, you can find the interval for one cycle to be

$$\underbrace{\dfrac{c}{b}}_{\text{Left endpoint}} \leq x \leq \underbrace{\dfrac{c}{b} + \dfrac{2\pi}{b}}_{\text{Right endpoint}}.$$

$$\underbrace{\phantom{\dfrac{c}{b} \leq x \leq \dfrac{c}{b} + \dfrac{2\pi}{b}}}_{\text{Period}}$$

This implies that the period of $y = a\sin(bx - c)$ is $2\pi/b$, and the graph of $y = a\sin bx$ is shifted by an amount c/b. The number c/b is the **phase shift**.

▶ Exploration

Sketch the graph of

$$y = \sin(x + c)$$

where $c = -\pi/4$, 0, and $\pi/4$. How does the value of c affect the graph?

Graphs of Sine and Cosine Functions

The graphs of $y = a\sin(bx - c)$ and $y = a\cos(bx - c)$ have the following characteristics. (Assume $b > 0$.)

$$\text{Amplitude} = |a| \qquad \text{Period} = \dfrac{2\pi}{b}$$

The left and right endpoints of a one-cycle interval can be determined by solving the equations $bx - c = 0$ and $bx - c = 2\pi$.

Example 4 ▶ Horizontal Translation

Sketch the graph of $y = \tfrac{1}{2}\sin(x - \pi/3)$.

Solution

The amplitude is $\tfrac{1}{2}$ and the period is 2π. By solving the equations

$$x - \dfrac{\pi}{3} = 0 \quad \Longrightarrow \quad x = \dfrac{\pi}{3}$$

and

$$x - \dfrac{\pi}{3} = 2\pi \quad \Longrightarrow \quad x = \dfrac{7\pi}{3}$$

you see that the interval $[\pi/3, 7\pi/3]$ corresponds to one cycle of the graph. Dividing this interval into four equal parts produces the key points

$$\left(\dfrac{\pi}{3}, 0\right), \quad \left(\dfrac{5\pi}{6}, \dfrac{1}{2}\right), \quad \left(\dfrac{4\pi}{3}, 0\right), \quad \left(\dfrac{11\pi}{6}, -\dfrac{1}{2}\right), \quad \text{and} \quad \left(\dfrac{7\pi}{3}, 0\right).$$

The graph is shown in Figure 1.44.

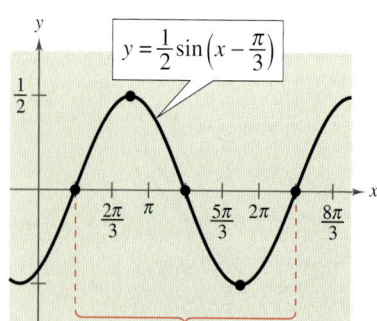

FIGURE 1.44

Section 1.5 ▶ Graphs of Sine and Cosine Functions 163

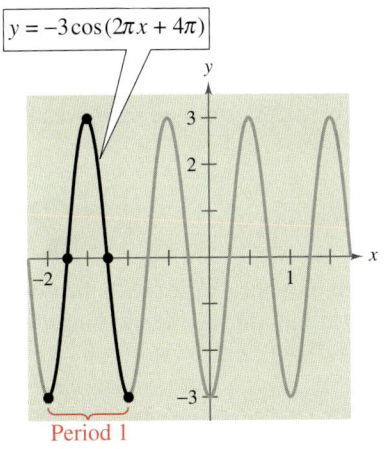

FIGURE 1.45

A computer animation of this example appears in the *Interactive* CD-ROM and *Internet* versions of this text.

Example 5 ▶ Horizontal Translation

Sketch the graph of

$$y = -3\cos(2\pi x + 4\pi).$$

Solution

The amplitude is 3 and the period is $2\pi/2\pi = 1$. By solving the equations

$$2\pi x + 4\pi = 0$$
$$2\pi x = -4\pi$$
$$x = -2$$

and

$$2\pi x + 4\pi = 2\pi$$
$$2\pi x = -2\pi$$
$$x = -1$$

you see that the interval $[-2, -1]$ corresponds to one cycle of the graph. Dividing this interval into four equal parts produces the key points

$$(-2, -3), \quad \left(-\frac{7}{4}, 0\right), \quad \left(-\frac{3}{2}, 3\right), \quad \left(-\frac{5}{4}, 0\right), \quad \text{and} \quad (-1, -3).$$

The graph is shown in Figure 1.45.

The final type of transformation is the *vertical translation* caused by the constant d in the equations

$$y = d + a\sin(bx - c)$$

and

$$y = d + a\cos(bx - c).$$

The shift is d units upward for $d > 0$ and downward for $d < 0$. In other words, the graph oscillates about the horizontal line $y = d$ instead of the x-axis.

Example 6 ▶ Vertical Translation

Sketch the graph of

$$y = 2 + 3\cos 2x.$$

Solution

The amplitude is 3 and the period is π. The key points over the interval $[0, \pi]$ are

$$(0, 5), \quad \left(\frac{\pi}{4}, 2\right), \quad \left(\frac{\pi}{2}, -1\right), \quad \left(\frac{3\pi}{4}, 2\right), \quad \text{and} \quad (\pi, 5).$$

The graph is shown in Figure 1.46.

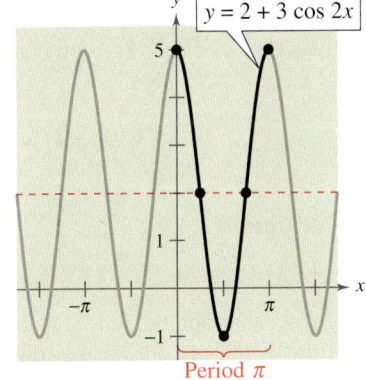

FIGURE 1.46

Mathematical Modeling

Sine and cosine functions can be used to model many real-life situations, including electric currents, musical tones, radio waves, tides, sunrises, and weather patterns.

Example 7 ▶ Finding a Trigonometric Model

Throughout the day, the depth of water at the end of a dock varies with the tides. The table shows the depths (in meters) at various times during the morning.

t (time)	Midnight	2 A.M.	4 A.M.	6 A.M.	8 A.M.	10 A.M.	Noon
y (depth)	2.55	3.80	4.40	3.80	2.55	1.80	2.27

a. Use a trigonometric function to model this data.
b. Find the depths at 9 A.M. and 3 P.M.
c. A boat needs at least 3 meters of water to moor at the dock. During what times in the afternoon can it safely dock?

Solution

a. Begin by graphing the data, as shown in Figure 1.47. You can use either a sine or cosine model. Suppose you use a cosine model of the form

$$y = a\cos(bt - c) + d.$$

The amplitude is given by

$$a = \frac{1}{2}[(\text{high}) - (\text{low})] = \frac{1}{2}(4.4 - 1.8) = 1.3.$$

The period is

$$p = 2[(\text{low time}) - (\text{high time})] = 2(10 - 4) = 12$$

which implies that $b = 2\pi/p \approx 0.524$. Because high tide occurs 4 hours after midnight, you can conclude that $c/b = 4$, so $c \approx 2.094$. Moreover, because the average depth is $\frac{1}{2}(4.4 + 1.8) = 3.1$, it follows that $d = 3.1$. So, you can model the depth with the function

$$y = 1.3\cos(0.524t - 2.094) + 3.1.$$

b. The depths at 9 A.M. and 3 P.M. are as follows.

$y = 1.3\cos(0.524 \cdot 9 - 2.094) + 3.1 \approx 1.97$ meters   9 A.M.

$y = 1.3\cos(0.524 \cdot 15 - 2.094) + 3.1 \approx 4.23$ meters   3 P.M.

c. To find out when the depth y is at least 3 meters, you can graph the model with the line $y = 3$, as shown in Figure 1.48. From the graph, it follows that the depth is at least 3 meters between 12:54 P.M. ($t \approx 12.9$) and 7:06 P.M. ($t \approx 19.1$).

FIGURE 1.47

FIGURE 1.48

1.5 Exercises

In Exercises 1–14, find the period and amplitude.

1. $y = 3 \sin 2x$

2. $y = 2 \cos 3x$

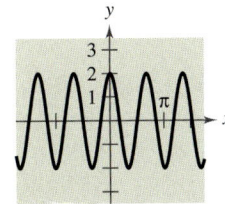

3. $y = \dfrac{5}{2} \cos \dfrac{x}{2}$

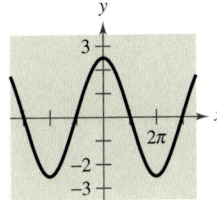

4. $y = -3 \sin \dfrac{x}{3}$

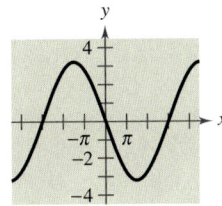

5. $y = \dfrac{1}{2} \sin \dfrac{\pi x}{3}$

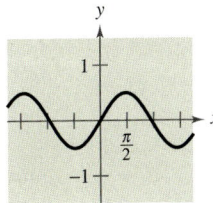

6. $y = \dfrac{3}{2} \cos \dfrac{\pi x}{2}$

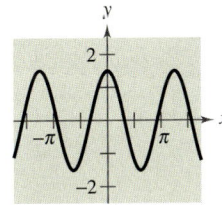

7. $y = -2 \sin x$

8. $y = -\cos \dfrac{2x}{3}$

9. $y = 3 \sin 10x$

10. $y = \dfrac{1}{3} \sin 8x$

11. $y = \dfrac{1}{2} \cos \dfrac{2x}{3}$

12. $y = \dfrac{5}{2} \cos \dfrac{x}{4}$

13. $y = \dfrac{1}{4} \sin 2\pi x$

14. $y = \dfrac{2}{3} \cos \dfrac{\pi x}{10}$

In Exercises 15–22, describe the relationship between the graphs of f and g.

15. $f(x) = \sin x$
    $g(x) = \sin(x - \pi)$

16. $f(x) = \cos x$
    $g(x) = \cos(x + \pi)$

17. $f(x) = \cos 2x$
    $g(x) = -\cos 2x$

18. $f(x) = \sin 3x$
    $g(x) = \sin(-3x)$

19. $f(x) = \cos x$
    $g(x) = \cos 2x$

20. $f(x) = \sin x$
    $g(x) = \sin 3x$

21. $f(x) = \sin 2x$
    $g(x) = 3 + \sin 2x$

22. $f(x) = \cos 4x$
    $g(x) = -2 + \cos 4x$

In Exercises 23–26, describe the relationship between the graphs of f and g.

23.

24.

25.

26.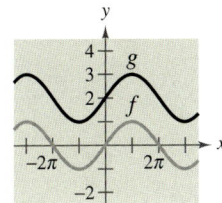

In Exercises 27–34, graph f and g on the same set of coordinate axes. (Include two full periods.)

27. $f(x) = -2 \sin x$
    $g(x) = 4 \sin x$

28. $f(x) = \sin x$
    $g(x) = \sin \dfrac{x}{3}$

29. $f(x) = \cos x$
    $g(x) = 1 + \cos x$

30. $f(x) = 2 \cos 2x$
    $g(x) = -\cos 4x$

31. $f(x) = -\dfrac{1}{2} \sin \dfrac{x}{2}$
    $g(x) = 3 - \dfrac{1}{2} \sin \dfrac{x}{2}$

32. $f(x) = 4 \sin \pi x$
    $g(x) = 4 \sin \pi x - 3$

33. $f(x) = 2 \cos x$
    $g(x) = 2 \cos(x + \pi)$

34. $f(x) = -\cos x$
    $g(x) = -\cos(x - \pi)$

In Exercises 35–52, sketch the graph of the function. (Include two full periods.)

35. $y = -2 \sin 6x$

36. $y = -3 \cos 4x$

37. $y = \cos 2\pi x$ **38.** $y = \sin \dfrac{\pi x}{4}$

39. $y = -\sin \dfrac{2\pi x}{3}$ **40.** $y = -10 \cos \dfrac{\pi x}{6}$

41. $y = \sin\left(x - \dfrac{\pi}{4}\right)$ **42.** $y = \sin(x - \pi)$

43. $y = 3\cos(x + \pi)$ **44.** $y = 4\cos\left(x + \dfrac{\pi}{4}\right)$

45. $y = 2 - \sin \dfrac{2\pi x}{3}$ **46.** $y = -3 + 5\cos \dfrac{\pi t}{12}$

47. $y = 2 + \tfrac{1}{10}\cos 60\pi x$ **48.** $y = 2\cos x - 3$

49. $y = 3\cos(x + \pi) - 3$ **50.** $y = 4\cos\left(x + \dfrac{\pi}{4}\right) + 4$

51. $y = \dfrac{2}{3}\cos\left(\dfrac{x}{2} - \dfrac{\pi}{4}\right)$ **52.** $y = -3\cos(6x + \pi)$

In Exercises 53–60, use a graphing utility to graph the function. Include two full periods. Be sure to choose an appropriate viewing window.

53. $y = -2\sin(4x + \pi)$

54. $y = -4\sin\left(\dfrac{2}{3}x - \dfrac{\pi}{3}\right)$

55. $y = \cos\left(2\pi x - \dfrac{\pi}{2}\right) + 1$

56. $y = 3\cos\left(\dfrac{\pi x}{2} + \dfrac{\pi}{2}\right) - 2$

57. $y = 5\sin(\pi - 2x) + 10$

58. $y = 5\cos(\pi - 2x) + 2$

59. $y = -0.1\sin\left(\dfrac{\pi x}{10} + \pi\right)$

60. $y = \tfrac{1}{100}\sin 120\pi t$

Graphical Reasoning In Exercises 61–64, find a and d for the function $f(x) = a\cos x + d$ such that the graph of f matches the figure.

61.

62.

63.

64.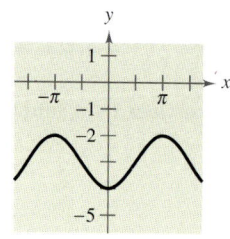

Graphical Reasoning In Exercises 65–68, find a, b, and c for the function $f(x) = a\sin(bx - c)$ such that the graph of f matches the figure.

65.

66.

67.

68.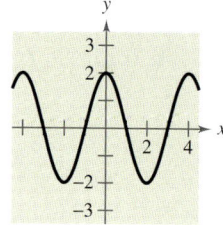

In Exercises 69–72, use a graphing utility to graph y_1 and y_2 in the interval $[-2\pi, 2\pi]$. Use the graphs to find real numbers x such that $y_1 = y_2$.

69. $y_1 = \sin x$
$y_2 = -\dfrac{1}{2}$

70. $y_1 = \cos x$
$y_2 = -1$

71. $y_1 = \cos x$
$y_2 = \dfrac{\sqrt{2}}{2}$

72. $y_1 = \sin x$
$y_2 = \dfrac{\sqrt{3}}{2}$

73. Respiratory Cycle For a person at rest, the velocity v (in liters per second) of air flow during a respiratory cycle (the time from the beginning of one breath to the beginning of the next) is

$$v = 0.85 \sin \dfrac{\pi t}{3}$$

where t is the time in seconds. (Inhalation occurs when $v > 0$, and exhalation occurs when $v < 0$.)

(a) Find the time for one full respiratory cycle.

(b) Find the number of cycles per minute.

(c) Sketch the graph of the velocity function.

The symbol indicates an exercise or part of an exercise in which you are instructed to use a graphing utility.

74. Respiratory Cycle After exercising for a few minutes, a person has a respiratory cycle for which the velocity of air flow is approximated by $v = 1.75 \sin(\pi t/2)$, where t is the time in seconds. (Inhalation occurs when $v > 0$, and exhalation occurs when $v < 0$.)

(a) Find the time for one full respiratory cycle.

(b) Find the number of cycles per minute.

(c) Sketch the graph of the velocity function.

75. Piano Tuning When tuning a piano, a technician strikes a tuning fork for the A above middle C and sets up a wave motion that can be approximated by $y = 0.001 \sin 880\pi t$, where t is the time in seconds.

(a) What is the period of the function?

(b) The frequency f is given by $f = 1/p$. What is the frequency of the note?

76. Biology The function

$$P = 100 - 20 \cos \frac{5\pi t}{3}$$

approximates the blood pressure P in millimeters of mercury at time t in seconds for a person at rest.

(a) Find the period of the function.

(b) Find the number of heartbeats per minute.

77. Data Analysis The table gives the normal daily high temperatures for Honolulu H and Chicago C (in degrees Fahrenheit) for month t, with $t = 1$ corresponding to January. (Source: National Oceanic and Atmospheric Administration)

t	1	2	3	4	5	6
H	80.1	80.5	81.6	82.8	84.7	86.5
C	29.0	33.5	45.8	58.6	70.1	79.6

t	7	8	9	10	11	12
H	87.5	88.7	88.5	86.9	84.1	81.2
C	83.7	81.8	74.8	63.3	48.4	34.0

(a) A model for the temperature in Honolulu is

$$H(t) = 84.40 + 4.28 \sin\left(\frac{\pi t}{6} + 3.86\right).$$

Find a trigonometric model for Chicago.

(b) Use a graphing utility to graph the data points and the model for the temperatures in Honolulu. How well does the model fit the data?

(c) Use a graphing utility to graph the data points and the model for the temperatures in Chicago. How well does the model fit the data?

(d) Use the models to estimate the average annual temperature in each city. Which term of the models did you use? Explain.

(e) What is the period of each model? Are the periods what you expected? Explain.

(f) Which city has the greater variability in temperature throughout the year? Which factor of the models determines this variability? Explain.

Sales In Exercises 78 and 79, use a graphing utility to graph the sales function over 1 year where S is the sales in thousands of units and t is the time in months, with $t = 1$ corresponding to January.

78. $S = 22.3 - 3.4 \cos \dfrac{\pi t}{6}$

79. $S = 74.50 + 43.75 \sin \dfrac{\pi t}{6}$

80. Fuel Consumption The daily consumption C (in gallons) of diesel fuel on a farm is modeled by

$$C = 30.3 + 21.6 \sin\left(\frac{2\pi t}{365} + 10.9\right)$$

where t is the time in days, with $t = 1$ corresponding to January 1.

(a) What is the period of the model? Is it what you expected? Explain.

(b) What is the average daily fuel consumption? Which term of the model did you use? Explain.

(c) Use a graphing utility to graph the model. Use the graph to approximate the time of the year when consumption exceeds 40 gallons per day.

81. Data Analysis The percent y of the moon's face that is illuminated on day x of the year 2005, where $x = 70$ represents March 11, is given in the table. (Source: U.S. Naval Observatory)

x	76	84	91	98	106	114
y	0.5	1.0	0.5	0.0	0.5	1.0

(a) Create a scatter plot of the data.

(b) Find a trigonometric model that fits the data.

(c) Add the graph of your model in part (b) to the scatter plot. How well does the model fit the data?

(d) Estimate the moon's percent illumination for May 8, 2005.

Synthesis

True or False? In Exercises 82 and 83, determine whether the statement is true or false. Justify your answer.

82. The graph of the function $f(x) = \sin(x + 2\pi)$ translates the graph of $f(x) = \sin x$ exactly one period to the right so that the two graphs look identical.

83. The function $y = \frac{1}{2}\cos 2x$ has an amplitude that is twice that of the function $y = \cos x$.

84. *Writing* Use a graphing utility to graph the function $y = a \sin x$ for $a = \frac{1}{2}$, $a = \frac{3}{2}$, and $a = -3$. Write a paragraph describing the changes in the graph corresponding to the specified changes in a.

85. *Writing* Use a graphing utility to graph the function $y = d + \sin x$ for $d = 2$, $d = 3.5$, and $d = -2$. Write a paragraph describing the changes in the graph corresponding to the specified changes in d.

86. *Writing* Use a graphing utility to graph the function $y = \sin bx$ for $b = \frac{1}{2}$, $b = \frac{3}{2}$, and $b = 4$. Write a paragraph describing the changes in the graph corresponding to the specified changes in b.

87. *Writing* Use a graphing utility to graph the function $y = \sin(x - c)$ for $c = 1$, $c = 3$, and $c = -2$. Write a paragraph describing the changes in the graph corresponding to the specified changes in c.

Conjecture In Exercises 88–91, graph f and g on the same set of coordinate axes. Include two full periods. Make a conjecture about the functions.

88. $f(x) = \sin x$, $g(x) = \cos\left(x - \dfrac{\pi}{2}\right)$

89. $f(x) = \sin x$, $g(x) = -\cos\left(x + \dfrac{\pi}{2}\right)$

90. $f(x) = \cos x$, $g(x) = -\sin\left(x - \dfrac{\pi}{2}\right)$

91. $f(x) = \cos x$, $g(x) = -\cos(x - \pi)$

92. *Exploration* Using calculus, it can be shown that the sine and cosine functions can be approximated by the polynomials

$$\sin x \approx x - \frac{x^3}{3!} + \frac{x^5}{5!} \quad \text{and} \quad \cos x \approx 1 - \frac{x^2}{2!} + \frac{x^4}{4!}$$

where x is in radians.

(a) Use a graphing utility to graph the sine function and its polynomial approximation in the same viewing window. How do the graphs compare?

(b) Use a graphing utility to graph the cosine function and its polynomial approximation in the same viewing window. How do the graphs compare?

(c) Study the patterns in the polynomial approximations of the sine and cosine functions and guess the next term in each. Then repeat parts (a) and (b). How did the accuracy of the approximations change when additional terms were added?

93. *Exploration* Use the polynomial approximations for the sine and cosine functions given in Exercise 92 to approximate the following functional values. Compare the results with those given by a calculator. Is the error in the approximation the same in each case? Explain.

(a) $\sin \dfrac{1}{2}$ (b) $\sin 1$ (c) $\sin \dfrac{\pi}{6}$

(d) $\cos(-0.5)$ (e) $\cos 1$ (f) $\cos \dfrac{\pi}{4}$

94. *Exploration* Use a graphing utility to graph h, and use the graph to decide whether h is even, odd, or neither.

(a) $h(x) = \cos^2 x$ (b) $h(x) = \sin^2 x$

95. *Conjecture* If f is an even function and g is an odd function, use the results of Exercise 94 to make a conjecture about h where

(a) $h(x) = [f(x)]^2$.   (b) $h(x) = [g(x)]^2$.

Review

In Exercises 96–99, perform the operations and simplify.

96. $\dfrac{4}{x} + \dfrac{4}{1-x}$

97. $\dfrac{2}{x+5} - \dfrac{2}{x-5}$

98. $\dfrac{3}{x-1} - \dfrac{2}{x(x-1)}$

99. $\dfrac{x}{x-5} + \dfrac{1}{2}$

In Exercises 100–103, find the domain of the function.

100. $f(x) = \dfrac{2}{11-x}$

101. $f(x) = \dfrac{\sqrt{x-3}}{x-8}$

102. $f(x) = \sqrt{81-x^2}$

103. $f(x) = \sqrt[3]{4-x^2}$

The symbol indicates an example or exercise that highlights algebraic techniques specifically used in calculus.

1.6 Graphs of Other Trigonometric Functions

▶ What you should learn

- How to sketch the graphs of tangent functions
- How to sketch the graphs of cotangent functions
- How to sketch the graphs of secant and cosecant functions
- How to sketch the graphs of damped trigonometric functions

▶ Why you should learn it

You can use tangent, cotangent, secant, and cosecant functions to model real-life data. For instance, Exercise 74 on page 178 shows you how a tangent function can be used to model and analyze the distance between a television camera and a parade unit.

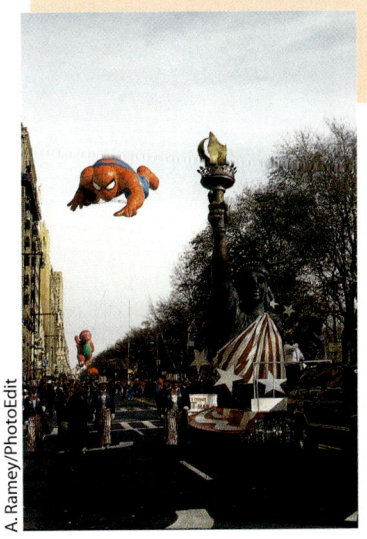

Graph of the Tangent Function

Recall that the tangent function is odd. That is, $\tan(-x) = -\tan x$. Consequently, the graph of $y = \tan x$ is symmetric with respect to the origin. You also know from the identity $\tan x = \sin x / \cos x$ that the tangent is undefined for values at which $\cos x = 0$. Two such values are $x = \pm \pi/2 \approx \pm 1.5708$.

x	$-\dfrac{\pi}{2}$	-1.57	-1.5	-1	0	1	1.5	1.57	$\dfrac{\pi}{2}$
$\tan x$	Undef.	-1255.8	-14.1	-1.56	0	1.56	14.1	1255.8	Undef.

As indicated in the table, $\tan x$ increases without bound as x approaches $\pi/2$ from the left, and decreases without bound as x approaches $-\pi/2$ from the right. So, the graph of $y = \tan x$ has *vertical asymptotes* at $x = \pi/2$ and $x = -\pi/2$, as shown in Figure 1.49. Moreover, because the period of the tangent function is π, vertical asymptotes also occur when $x = \pi/2 + n\pi$, where n is an integer. The domain of the tangent function is the set of all real numbers other than $x = \pi/2 + n\pi$, and the range is the set of all real numbers.

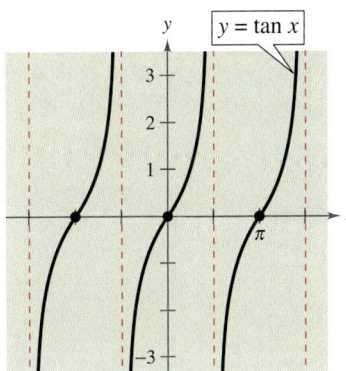

Period: π
Domain: all $x \neq \dfrac{\pi}{2} + n\pi$
Range: $(-\infty, \infty)$
Vertical asymptotes: $x = \dfrac{\pi}{2} + n\pi$

FIGURE 1.49

Sketching the graph of a function of the form $y = a \tan(bx - c)$ is similar to sketching the graph of $y = a \sin(bx - c)$ in that you locate key points that identify the intercepts and asymptotes. Two consecutive asymptotes can be found by solving the equations

$$bx - c = -\frac{\pi}{2} \quad \text{and} \quad bx - c = \frac{\pi}{2}.$$

The midpoint between two consecutive asymptotes is an x-intercept of the graph. The period of the function $y = a \tan(bx - c)$ is the distance between two consecutive asymptotes. The amplitude of a tangent function is not defined. After plotting the asymptotes and the x-intercept, plot a few additional points between the two asymptotes and sketch one cycle. Finally, sketch one or two additional cycles to the left and right.

Example 1 ▶ Sketching the Graph of a Tangent Function

Sketch the graph of $y = \tan \dfrac{x}{2}$.

Solution

By solving the equations

$$\dfrac{x}{2} = -\dfrac{\pi}{2} \quad \text{and} \quad \dfrac{x}{2} = \dfrac{\pi}{2}$$

$$x = -\pi \quad\quad\quad\quad x = \pi$$

you can see that two consecutive asymptotes occur at $x = -\pi$ and $x = \pi$. Between these two asymptotes, plot a few points, including the x-intercept, as shown in the table. Three cycles of the graph are shown in Figure 1.50.

x	$-\dfrac{\pi}{2}$	0	$\dfrac{\pi}{2}$
$\tan \dfrac{x}{2}$	-1	0	1

FIGURE 1.50

A computer animation of this example appears in the *Interactive* CD-ROM and *Internet* versions of this text.

Example 2 ▶ Sketching the Graph of a Tangent Function

Sketch the graph of $y = -3 \tan 2x$.

Solution

By solving the equations

$$2x = -\dfrac{\pi}{2} \quad \text{and} \quad 2x = \dfrac{\pi}{2}$$

$$x = -\dfrac{\pi}{4} \quad\quad\quad\quad x = \dfrac{\pi}{4}$$

you can see that two consecutive asymptotes occur at $x = -\pi/4$ and $x = \pi/4$. Between these two asymptotes, plot a few points, including the x-intercept, as shown in the table. Three cycles of the graph are shown in Figure 1.51.

x	$-\dfrac{\pi}{8}$	0	$\dfrac{\pi}{8}$
$-3 \tan 2x$	3	0	-3

FIGURE 1.51

By comparing the graphs in Examples 1 and 2, you can see that the graph of $y = a \tan(bx - c)$ is increasing between consecutive vertical asymptotes if $a > 0$, and decreasing between consecutive vertical asymptotes if $a < 0$. In other words, the graph for $a < 0$ is a reflection in the x-axis of the graph for $a > 0$.

Graph of the Cotangent Function

The graph of the cotangent function is similar to the graph of the tangent function. It also has a period of π. However, from the identity

$$y = \cot x = \frac{\cos x}{\sin x}$$

you can see that the cotangent function has vertical asymptotes at $x = n\pi$, where n is an integer, because $\sin x$ is zero at these x-values. The graph of the cotangent function is shown in Figure 1.52.

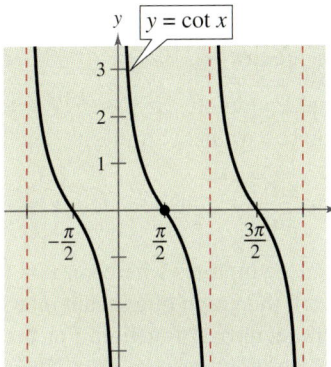

Period: π
Domain: all $x \neq n\pi$
Range: $(-\infty, \infty)$
Vertical asymptotes: $x = n\pi$

FIGURE **1.52**

Example 3 ▶ Sketching the Graph of a Cotangent Function

Sketch the graph of $y = 2 \cot \dfrac{x}{3}$.

Solution

To locate two consecutive vertical asymptotes of the graph, solve the equations $x/3 = 0$ and $x/3 = \pi$, as follows.

$$\frac{x}{3} = 0 \quad \text{and} \quad \frac{x}{3} = \pi$$

$$x = 0 \quad\quad\quad\quad x = 3\pi$$

Then, between these two asymptotes, plot a few points, including the x-intercept, as shown in the table. Three cycles of the graph are shown in Figure 1.53. (Note that the period is 3π, the distance between consecutive asymptotes.)

FIGURE **1.53**

x	$\dfrac{3\pi}{4}$	$\dfrac{3\pi}{2}$	$\dfrac{9\pi}{4}$
$2\cot\dfrac{x}{3}$	2	0	-2

Graphs of the Reciprocal Functions

The graphs of the two remaining trigonometric functions can be obtained from the graphs of the sine and cosine functions using the reciprocal identities

$$\csc x = \frac{1}{\sin x} \quad \text{and} \quad \sec x = \frac{1}{\cos x}.$$

For instance, at a given value of x, the y-coordinate of $\sec x$ is the reciprocal of the y-coordinate of $\cos x$. Of course, when $\cos x = 0$, the reciprocal does not exist. Near such values of x, the behavior of the secant function is similar to that of the tangent function. In other words, the graphs of

$$\tan x = \frac{\sin x}{\cos x} \quad \text{and} \quad \sec x = \frac{1}{\cos x}$$

have vertical asymptotes at $x = \pi/2 + n\pi$, where n is an integer, and the cosine is zero at these x-values. Similarly,

$$\cot x = \frac{\cos x}{\sin x} \quad \text{and} \quad \csc x = \frac{1}{\sin x}$$

have vertical asymptotes where $\sin x = 0$—that is, at $x = n\pi$.

To sketch the graph of a secant or cosecant function, you should first make a sketch of its reciprocal function. For instance, to sketch the graph of $y = \csc x$, first sketch the graph of $y = \sin x$. Then take reciprocals of the y-coordinates to obtain points on the graph of $y = \csc x$. This procedure is used to obtain the graphs shown in Figure 1.54.

Period: 2π
Domain: all $x \neq n\pi$
Range: $(-\infty, -1]$ and $[1, \infty)$
Vertical asymptotes: $x = n\pi$
Symmetry: origin

Period: 2π
Domain: all $x \neq \frac{\pi}{2} + n\pi$
Range: $(-\infty, -1]$ and $[1, \infty)$
Vertical asymptotes: $x = \frac{\pi}{2} + n\pi$
Symmetry: y-axis

FIGURE 1.54

In comparing the graphs of the secant and cosecant functions with those of the sine and cosine functions, note that the "hills" and "valleys" are interchanged. For example, a hill (or maximum point) on the sine curve corresponds to a valley (a local minimum) on the cosecant curve. Similarly, a valley (or minimum point) on the sine curve corresponds to a hill (a local maximum) on the cosecant curve. Additionally, x-intercepts of the sine and cosine functions become vertical asymptotes of the cosecant and secant functions, as shown in Figure 1.55.

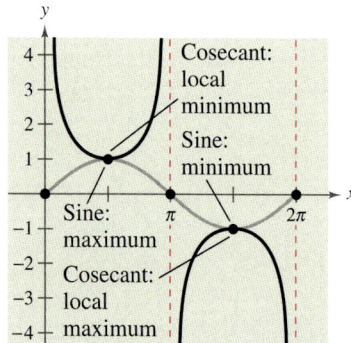

FIGURE 1.55

Section 1.6 ▶ Graphs of Other Trigonometric Functions 173

Example 4 ▶ Sketching the Graph of a Cosecant Function

Sketch the graph of $y = 2 \csc\left(x + \dfrac{\pi}{4}\right)$.

Solution

Begin by sketching the graph of

$$y = 2 \sin\left(x + \dfrac{\pi}{4}\right).$$

For this function, the amplitude is 2 and the period is 2π. By solving the equations

$$x + \dfrac{\pi}{4} = 0 \qquad \text{and} \qquad x + \dfrac{\pi}{4} = 2\pi$$

$$x = -\dfrac{\pi}{4} \qquad\qquad\qquad x = \dfrac{7\pi}{4}$$

you can see that one cycle of the sine function corresponds to the interval from $x = -\pi/4$ to $x = 7\pi/4$. The graph of this sine function is represented by the gray curve in Figure 1.56. Because the sine function is zero at the midpoint and endpoints of this interval, the corresponding cosecant function

$$y = 2 \csc\left(x + \dfrac{\pi}{4}\right)$$

$$= 2\left(\dfrac{1}{\sin[x + (\pi/4)]}\right)$$

has vertical asymptotes at $x = -\pi/4, x = 3\pi/4, x = 7\pi/4$, etc. The graph of the cosecant function is represented by the black curve in Figure 1.56.

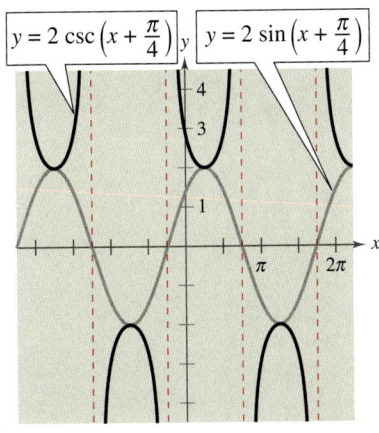

FIGURE 1.56

Example 5 ▶ Sketching the Graph of a Secant Function

Sketch the graph of $y = \sec 2x$.

Solution

Begin by sketching the graph of $y = \cos 2x$, as indicated by the gray curve in Figure 1.57. Then, form the graph of $y = \sec 2x$ as the black curve in the figure. Note that the x-intercepts of $y = \cos 2x$

$$\left(-\dfrac{\pi}{4}, 0\right), \quad \left(\dfrac{\pi}{4}, 0\right), \quad \left(\dfrac{3\pi}{4}, 0\right), \ldots$$

correspond to the vertical asymptotes

$$x = -\dfrac{\pi}{4}, \quad x = \dfrac{\pi}{4}, \quad x = \dfrac{3\pi}{4}, \ldots$$

of the graph of $y = \sec 2x$.

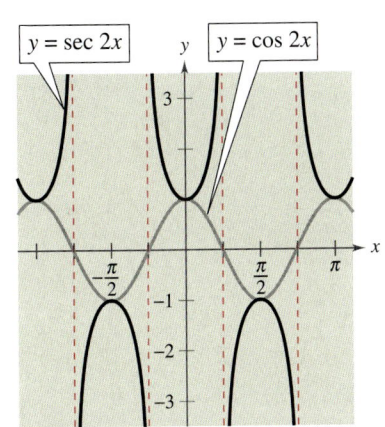

FIGURE 1.57

Damped Trigonometric Graphs

A *product* of two functions can be graphed using properties of the individual functions. For instance, consider the function

$$f(x) = x \sin x$$

as the product of the functions $y = x$ and $y = \sin x$. Using properties of absolute value and the fact that $|\sin x| \leq 1$, you have $0 \leq |x||\sin x| \leq |x|$. Consequently,

$$-|x| \leq x \sin x \leq |x|$$

which means that the graph of $f(x) = x \sin x$ lies between the lines $y = -x$ and $y = x$. Furthermore, because

$$f(x) = x \sin x = \pm x \quad \text{at} \quad x = \frac{\pi}{2} + n\pi$$

and

$$f(x) = x \sin x = 0 \quad \text{at} \quad x = n\pi$$

the graph of f touches the line $y = -x$ or the line $y = x$ at $x = \pi/2 + n\pi$ and has x-intercepts at $x = n\pi$. A sketch of f is shown in Figure 1.58. In the function $f(x) = x \sin x$, the factor x is called the **damping factor.**

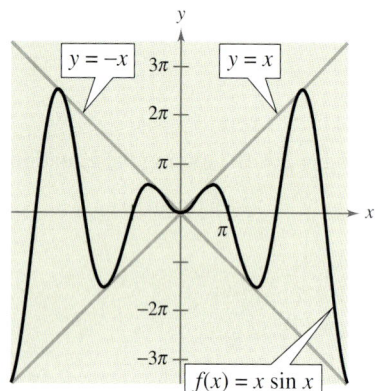

FIGURE 1.58

Example 6 ▶ Damped Sine Wave

Sketch the graph of

$$f(x) = x^2 \sin 3x.$$

Solution

Consider $f(x)$ as the product of the two functions

$$y = x^2 \quad \text{and} \quad y = \sin 3x$$

each of which has the set of real numbers as its domain. For any real number x, $x^2 \geq 0$ and $|\sin 3x| \leq 1$. Therefore, $x^2|\sin 3x| \leq x^2$, which means that

$$-x^2 \leq x^2 \sin 3x \leq x^2.$$

Furthermore, because

$$f(x) = x^2 \sin 3x = \pm x^2 \quad \text{at} \quad x = \frac{\pi}{6} + \frac{n\pi}{3}$$

and

$$f(x) = x^2 \sin 3x = 0 \quad \text{at} \quad x = \frac{n\pi}{3}$$

the graph of f touches the curves $y = -x^2$ and $y = x^2$ at $x = \pi/6 + n\pi/3$ and has intercepts at $x = n\pi/3$. A sketch is shown in Figure 1.59.

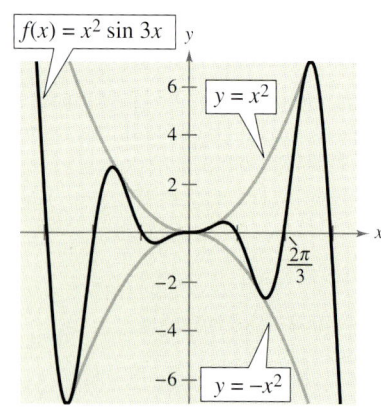

FIGURE 1.59

Figure 1.60 summarizes the six basic trigonometric functions.

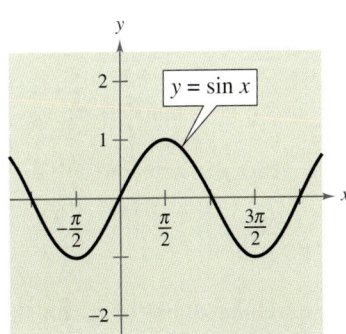

Domain: all reals
Range: $[-1, 1]$
Period: 2π

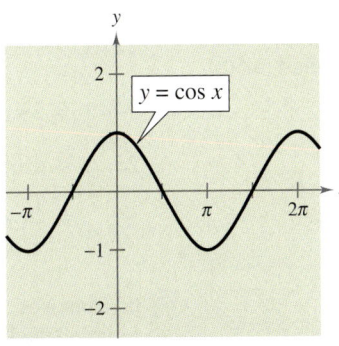

Domain: all reals
Range: $[-1, 1]$
Period: 2π

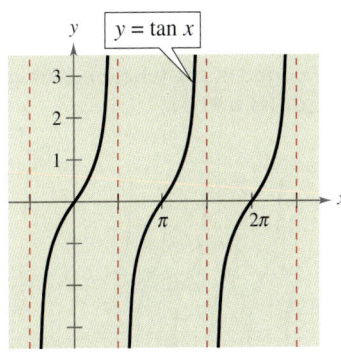

Domain: all $x \neq \frac{\pi}{2} + n\pi$
Range: $(-\infty, \infty)$
Period: π

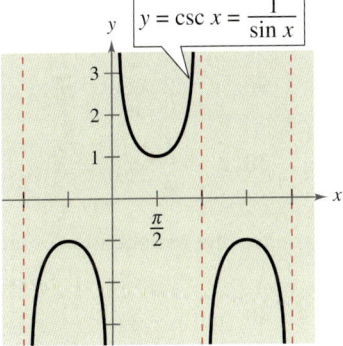

Domain: all $x \neq n\pi$
Range: $(-\infty, -1]$ and $[1, \infty)$
Period: 2π

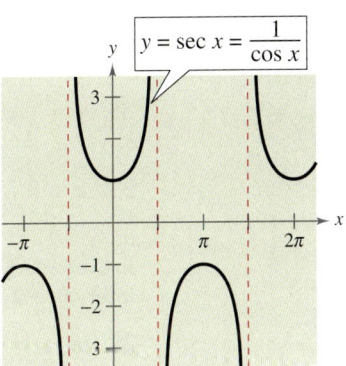

Domain: all $x \neq \frac{\pi}{2} + n\pi$
Range: $(-\infty, -1]$ and $[1, \infty)$
Period: 2π

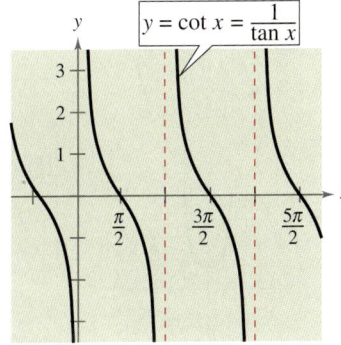

Domain: all $x \neq n\pi$
Range: $(-\infty, \infty)$
Period: π

FIGURE 1.60

Writing ABOUT MATHEMATICS

Combining Trigonometric Functions Recall from Section P.8 that functions can be combined arithmetically. This also applies to trigonometric functions. For each of the functions

$$h(x) = x + \sin x \quad \text{and} \quad h(x) = \cos x - \sin 3x$$

(a) identify two simpler functions f and g that comprise the combination, (b) use a table to show how to obtain the numerical values of $h(x)$ from the numerical values of $f(x)$ and $g(x)$, and (c) use a graph of f and g to show how h may be formed.

Can you find functions

$$f(x) = d + a\sin(bx + c) \quad \text{and} \quad g(x) = d + a\cos(bx + c)$$

such that $f(x) + g(x) = 0$ for all x?

1.6 Exercises

In Exercises 1–6, match the function with its graph. State the period of the function. [The graphs are labeled (a), (b), (c), (d), (e), and (f).]

(a)

(b)

(c)

(d)

(e)

(f)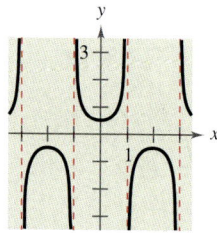

1. $y = \sec 2x$
2. $y = \tan \dfrac{x}{2}$
3. $y = \dfrac{1}{2} \cot \pi x$
4. $y = \dfrac{1}{2} \sec \dfrac{\pi x}{2}$
5. $y = -\csc x$
6. $y = -2 \sec \dfrac{\pi x}{2}$

In Exercises 7–28, sketch the graph of the function. Include two full periods.

7. $y = \dfrac{1}{3} \tan x$
8. $y = \dfrac{1}{4} \tan x$
9. $y = \tan 3x$
10. $y = -3 \tan \pi x$
11. $y = -\dfrac{1}{2} \sec x$
12. $y = \dfrac{1}{4} \sec x$
13. $y = \csc \pi x$
14. $y = 3 \csc 4x$
15. $y = \sec \pi x - 1$
16. $y = -2 \sec 4x + 2$
17. $y = \csc \dfrac{x}{2}$
18. $y = \csc \dfrac{x}{3}$
19. $y = \cot \dfrac{x}{2}$
20. $y = 3 \cot \dfrac{\pi x}{2}$
21. $y = \dfrac{1}{2} \sec 2x$
22. $y = -\dfrac{1}{2} \tan x$
23. $y = \tan \dfrac{\pi x}{4}$
24. $y = \tan(x + \pi)$
25. $y = \csc(\pi - x)$
26. $y = \sec(\pi - x)$
27. $y = \dfrac{1}{4} \csc\left(x + \dfrac{\pi}{4}\right)$
28. $y = 2 \cot\left(x + \dfrac{\pi}{2}\right)$

In Exercises 29–38, use a graphing utility to graph the function. Include two full periods.

29. $y = \tan \dfrac{x}{3}$
30. $y = -\tan 2x$
31. $y = -2 \sec 4x$
32. $y = \sec \pi x$
33. $y = \tan\left(x - \dfrac{\pi}{4}\right)$
34. $y = -\csc(4x - \pi)$
35. $y = \dfrac{1}{4} \cot\left(x - \dfrac{\pi}{2}\right)$
36. $y = 2 \sec(2x - \pi)$
37. $y = 0.1 \tan\left(\dfrac{\pi x}{4} + \dfrac{\pi}{4}\right)$
38. $y = \dfrac{1}{3} \sec\left(\dfrac{\pi x}{2} + \dfrac{\pi}{2}\right)$

In Exercises 39–46, use a graph to solve the equation on the interval $[-2\pi, 2\pi]$.

39. $\tan x = 1$
40. $\tan x = \sqrt{3}$
41. $\cot x = -\dfrac{\sqrt{3}}{3}$
42. $\cot x = 1$
43. $\sec x = -2$
44. $\sec x = 2$
45. $\csc x = \sqrt{2}$
46. $\csc x = -\dfrac{2\sqrt{3}}{3}$

In Exercises 47 and 48, use the graph of the function to determine whether the function is even, odd, or neither.

47. $f(x) = \sec x$
48. $f(x) = \tan x$

49. **Graphical Reasoning** Consider the functions

$$f(x) = 2 \sin x \quad \text{and} \quad g(x) = \dfrac{1}{2} \csc x$$

on the interval $(0, \pi)$.

(a) Graph f and g in the same coordinate plane.

(b) Approximate the interval where $f > g$.

(c) Describe the behavior of each of the functions as x approaches π. How is the behavior of g related to the behavior of f as x approaches π?

50. *Graphical Reasoning* Consider the functions

$$f(x) = \tan \frac{\pi x}{2} \quad \text{and} \quad g(x) = \frac{1}{2}\sec \frac{\pi x}{2}$$

on the interval $(-1, 1)$.

(a) Use a graphing utility to graph f and g in the same viewing window.

(b) Approximate the interval where $f < g$.

(c) Approximate the interval where $2f < 2g$. How does the result compare with that of part (b)? Explain.

In Exercises 51–54, use a graphing utility to graph the two equations in the same viewing window. Determine analytically whether the expressions are equivalent.

51. $y_1 = \sin x \csc x, \quad y_2 = 1$

52. $y_1 = \sin x \sec x, \quad y_2 = \tan x$

53. $y_1 = \dfrac{\cos x}{\sin x}, \quad y_2 = \cot x$

54. $y_1 = \sec^2 x - 1, \quad y_2 = \tan^2 x$

In Exercises 55–58, match the function with its graph. Describe the behavior of the function as x approaches zero. [The graphs are labeled (a), (b), (c), and (d).]

(a)

(b)

(c)

(d)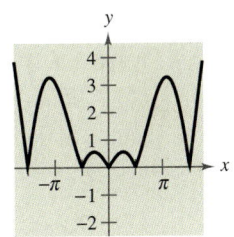

55. $f(x) = |x \cos x|$

56. $f(x) = x \sin x$

57. $g(x) = |x| \sin x$

58. $g(x) = |x| \cos x$

Conjecture In Exercises 59–62, graph the functions f and g. Use the graphs to make a conjecture about the relationship between the functions.

59. $f(x) = \sin x + \cos\left(x + \dfrac{\pi}{2}\right), \quad g(x) = 0$

60. $f(x) = \sin x - \cos\left(x + \dfrac{\pi}{2}\right), \quad g(x) = 2\sin x$

61. $f(x) = \sin^2 x, \quad g(x) = \tfrac{1}{2}(1 - \cos 2x)$

62. $f(x) = \cos^2 \dfrac{\pi x}{2}, \quad g(x) = \dfrac{1}{2}(1 + \cos \pi x)$

In Exercises 63–66, use a graphing utility to graph the function and the damping factor of the function in the same viewing window. Describe the behavior of the function as x increases without bound.

63. $f(x) = x \cos \pi x$

64. $f(x) = x^2 \cos x$

65. $g(x) = x^3 \sin x$

66. $h(x) = x^3 \cos x$

Exploration In Exercises 67–72, use a graphing utility to graph the function. Describe the behavior of the function as x approaches zero.

67. $y = \dfrac{6}{x} + \cos x, \quad x > 0$

68. $y = \dfrac{4}{x} + \sin 2x, \quad x > 0$

69. $g(x) = \dfrac{\sin x}{x}$

70. $f(x) = \dfrac{1 - \cos x}{x}$

71. $f(x) = \sin \dfrac{1}{x}$

72. $h(x) = x \sin \dfrac{1}{x}$

73. *Distance* A plane flying at an altitude of 7 miles above a radar antenna will pass directly over the radar antenna (see figure). Let d be the ground distance from the antenna to the point directly under the plane and let x be the angle of elevation to the plane from the antenna. (d is positive as the plane approaches the antenna.) Write d as a function of x and graph the function over the interval $0 < x < \pi$.

74. *Television Coverage* A television camera is on a reviewing platform 27 meters from the street on which a parade will be passing from left to right (see figure). Express the distance d from the camera to a particular unit in the parade as a function of the angle x, and graph the function over the interval $-\pi/2 < x < \pi/2$. (Consider x as negative when a unit in the parade approaches from the left.)

75. *Sales* The projected monthly sales S (in thousands of units) of a seasonal product are modeled by

$$S = 74 + 3t + 40 \sin \frac{\pi t}{6}$$

where t is the time in months, with $t = 1$ corresponding to January. Graph the sales function over 1 year.

76. *Predator-Prey Model* Suppose the population of a certain predator at time t (in months) in a given region is estimated to be

$$P = 10{,}000 + 3000 \sin \frac{2\pi t}{24}$$

and the population of its primary food source (its prey) is estimated to be

$$p = 15{,}000 + 5000 \cos \frac{2\pi t}{24}.$$

Use the graphs of the models to explain the oscillations in the size of each population.

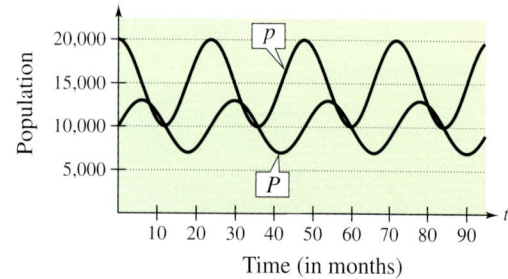

77. *Normal Temperatures* The normal monthly high temperatures in degrees Fahrenheit for Erie, Pennsylvania are approximated by

$$H(t) = 54.33 - 20.38 \cos \frac{\pi t}{6} - 15.69 \sin \frac{\pi t}{6}$$

and the normal monthly low temperatures are approximated by

$$L(t) = 39.36 - 15.70 \cos \frac{\pi t}{6} - 14.16 \sin \frac{\pi t}{6}$$

where t is the time in months, with $t = 1$ corresponding to January (see figure). (Source: National Oceanic and Atmospheric Administration)

(a) What is the period of each function?

(b) During what part of the year is the difference between the normal high and low temperatures greatest? When is it smallest?

(c) The sun is northernmost in the sky around June 21, but the graph shows the warmest temperatures at a later date. Approximate the lag time of the temperatures relative to the position of the sun.

78. *Harmonic Motion* An object weighing W pounds is suspended from the ceiling by a steel spring (see figure). The weight is pulled downward (positive direction) from its equilibrium position and released. The resulting motion of the weight is described by the function

$$y = \frac{1}{2} t \cos 4t, \quad t > 0$$

where y is the distance in feet and t is the time in seconds.

(a) Use a graphing utility to graph the function.

(b) Describe the behavior of the displacement function for increasing values of time t.

FIGURE FOR 78

Synthesis

True or False? In Exercises 79 and 80, determine whether the statement is true or false. Justify your answer.

79. The graph of $y = \csc x$ can be obtained on a calculator by graphing the reciprocal of $y = \sin x$.

80. The graph of $y = \sec x$ can be obtained on a calculator by graphing a translation of the reciprocal of $y = \sin x$.

81. *Writing* Describe the behavior of $f(x) = \tan x$ as x approaches $\pi/2$ from the left and from the right.

82. *Writing* Describe the behavior of $f(x) = \csc x$ as x approaches π from the left and from the right.

83. *Exploration* Consider the function
$$f(x) = x - \cos x.$$
   (a) Use a graphing utility to graph the function and verify that there exists a zero between 0 and 1. Use the graph to approximate the zero.
   (b) Starting with $x_0 = 1$, generate a sequence $x_1, x_2, x_3, \ldots$ where $x_n = \cos(x_{n-1})$. For example,
   $$x_0 = 1$$
   $$x_1 = \cos(x_0)$$
   $$x_2 = \cos(x_1)$$
   $$x_3 = \cos(x_2)$$
   $$\vdots$$
   Verify that the sequence approaches the zero of f.

84. *Approximation* Using calculus, it can be shown that the tangent function can be approximated by the polynomial
$$\tan x \approx x + \frac{2x^3}{3!} + \frac{16x^5}{5!}$$
where x is in radians. Use a graphing utility to graph the tangent function and its polynomial approximation in the same viewing window. How do the graphs compare?

85. *Approximation* Using calculus, it can be shown that the secant function can be approximated by the polynomial
$$\sec x \approx 1 + \frac{x^2}{2!} + \frac{5x^4}{4!}$$
where x is in radians. Use a graphing utility to graph the secant function and its polynomial approximation in the same viewing window. How do the graphs compare?

86. *Pattern Recognition*
   (a) Use a graphing utility to graph each function.
   $$y_1 = \frac{4}{\pi}\left(\sin \pi x + \frac{1}{3}\sin 3\pi x\right)$$
   $$y_2 = \frac{4}{\pi}\left(\sin \pi x + \frac{1}{3}\sin 3\pi x + \frac{1}{5}\sin 5\pi x\right)$$
   (b) Identify the pattern started in part (a) and find a function y_3 that continues the pattern one more term. Use a graphing utility to graph y_3.
   (c) The graphs in parts (a) and (b) approximate the periodic function in the figure. Find a function y_4 that is a better approximation.

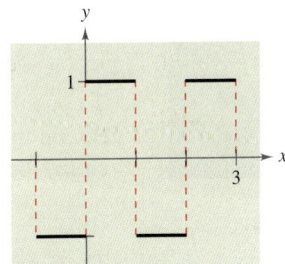

Review

In Exercises 87–94, solve the equation by any convenient method.

87. $x^2 = 64$
88. $(x - 5)^2 = 8$
89. $4x^2 - 12x + 9 = 0$
90. $9x^2 + 12x + 3 = 0$
91. $x^2 - 6x + 4 = 0$
92. $2x^2 - 4x - 6 = 0$
93. $50 + 5x = 3x^2$
94. $2x^2 + 4x - 9 = 2(x - 1)^2$

1.7 Inverse Trigonometric Functions

▶ **What you should learn**
- How to evaluate the inverse sine function
- How to evaluate the other inverse trigonometric functions
- How to evaluate the compositions of trigonometric functions

▶ **Why you should learn it**
You can use inverse trigonometric functions to model and solve real-life problems. For instance, Exercise 91 on page 187 shows how an inverse trigonometric function can be used to model the angle of elevation from a boat to a winch and the length of the rope joining them.

Inverse Sine Function

Recall from Section P.9 that, for a function to have an inverse, it must pass the Horizontal Line Test. From Figure 1.61 you can see that $y = \sin x$ does not pass the test because different values of x yield the same y-value.

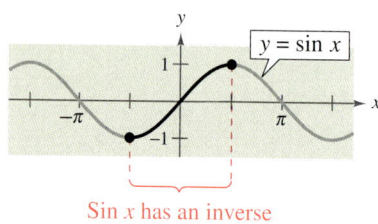

Sin x has an inverse on this interval.

FIGURE 1.61

However, if you restrict the domain to the interval $-\pi/2 \leq x \leq \pi/2$ (corresponding to the black portion of the graph in Figure 1.61), the following properties hold.

1. On the interval $[-\pi/2, \pi/2]$, the function $y = \sin x$ is increasing.
2. On the interval $[-\pi/2, \pi/2]$, $y = \sin x$ takes on its full range of values, $-1 \leq \sin x \leq 1$.
3. On the interval $[-\pi/2, \pi/2]$, $y = \sin x$ passes the Horizontal Line Test.

So, on the restricted domain $-\pi/2 \leq x \leq \pi/2$, $y = \sin x$ has a unique inverse called the **inverse sine function.** It is denoted by

$$y = \arcsin x \quad \text{or} \quad y = \sin^{-1} x.$$

The notation $\sin^{-1} x$ is consistent with the inverse function notation $f^{-1}(x)$. The arcsin x notation (read as "the arcsine of x") comes from the association of a central angle with its subtended *arc length* on a unit circle. So, arcsin x means the angle (or arc) whose sine is x. Both notations, arcsin x and $\sin^{-1} x$, are commonly used in mathematics, so remember that $\sin^{-1} x$ denotes the *inverse sine function* rather than $1/\sin x$. The values of arcsin x lie in the interval $-\pi/2 \leq \arcsin x \leq \pi/2$. The graph of $y = \arcsin x$ is shown in Figure 1.62.

> **Definition of Inverse Sine Function**
>
> The **inverse sine function** is defined by
>
> $$y = \arcsin x \quad \text{if and only if} \quad \sin y = x$$
>
> where $-1 \leq x \leq 1$ and $-\pi/2 \leq y \leq \pi/2$. The domain of $y = \arcsin x$ is $[-1, 1]$, and the range is $[-\pi/2, \pi/2]$.

When evaluating the inverse sine function, it helps to remember the phrase "the arcsine of x is the angle (or number) whose sine is x."

STUDY TIP

As with the trigonometric functions, much of the work with the inverse trigonometric functions can be done by *exact* calculations rather than by calculator approximations. Exact calculations help to increase your understanding of the inverse functions by relating them to the triangle definitions of the trigonometric functions.

Example 1 ▶ Evaluating the Inverse Sine Function

If possible, find the exact value.

a. $\arcsin\left(-\dfrac{1}{2}\right)$ **b.** $\sin^{-1}\dfrac{\sqrt{3}}{2}$ **c.** $\sin^{-1} 2$

Solution

a. Because $\sin(-\pi/6) = -\tfrac{1}{2}$ for $-\pi/2 \le y \le \pi/2$, it follows that

$$\arcsin\left(-\dfrac{1}{2}\right) = -\dfrac{\pi}{6}. \qquad \text{Angle whose sine is } -\tfrac{1}{2}$$

b. Because $\sin(\pi/3) = \sqrt{3}/2$ for $-\pi/2 \le y \le \pi/2$, it follows that

$$\sin^{-1}\dfrac{\sqrt{3}}{2} = \dfrac{\pi}{3}. \qquad \text{Angle whose sine is } \sqrt{3}/2$$

c. It is not possible to evaluate $y = \sin^{-1} x$ when $x = 2$ because there is no angle whose sine is 2. Remember that the domain of the inverse sine function is $[-1, 1]$.

Example 2 ▶ Graphing the Arcsine Function

Sketch a graph of $y = \arcsin x$.

Solution

By definition, the equations

$$y = \arcsin x \quad \text{and} \quad \sin y = x$$

are equivalent for $-\pi/2 \le y \le \pi/2$. So, their graphs are the same. From the interval $[-\pi/2, \pi/2]$, you can assign values to y in the second equation to make a table of values.

y	$-\dfrac{\pi}{2}$	$-\dfrac{\pi}{4}$	$-\dfrac{\pi}{6}$	0	$\dfrac{\pi}{6}$	$\dfrac{\pi}{4}$	$\dfrac{\pi}{2}$
$x = \sin y$	-1	$-\dfrac{\sqrt{2}}{2}$	$-\dfrac{1}{2}$	0	$\dfrac{1}{2}$	$\dfrac{\sqrt{2}}{2}$	1

The resulting graph for

$$y = \arcsin x$$

is shown in Figure 1.62. Note that it is the reflection (in the line $y = x$) of the black portion of the graph in Figure 1.61. Be sure you see that Figure 1.62 shows the *entire* graph of the inverse sine function. Remember that the range of $y = \arcsin x$ is the closed interval $[-\pi/2, \pi/2]$.

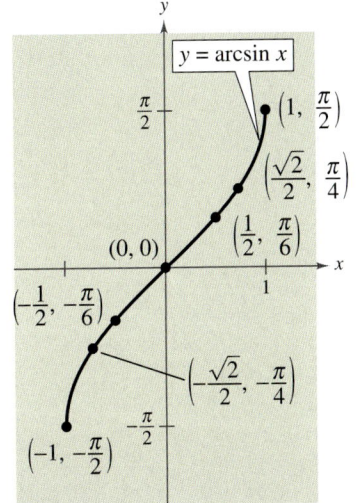

FIGURE 1.62

Other Inverse Trigonometric Functions

The cosine function is decreasing on the interval $0 \leq x \leq \pi$, as shown in Figure 1.63.

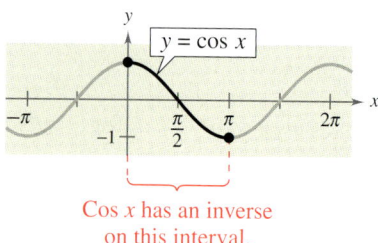

Cos x has an inverse on this interval.

FIGURE 1.63

Consequently, on this interval the cosine function has an inverse function—the **inverse cosine function**—denoted by

$$y = \arccos x \quad \text{or} \quad y = \cos^{-1} x.$$

Similarly, you can define an **inverse tangent function** by restricting the domain of $y = \tan x$ to the interval $(-\pi/2, \pi/2)$. The following list summarizes the definitions of the three most common inverse trigonometric functions. The remaining three are defined in Exercises 99–101.

Definitions of the Inverse Trigonometric Functions

Function	Domain	Range
$y = \arcsin x$ if and only if $\sin y = x$	$-1 \leq x \leq 1$	$-\dfrac{\pi}{2} \leq y \leq \dfrac{\pi}{2}$
$y = \arccos x$ if and only if $\cos y = x$	$-1 \leq x \leq 1$	$0 \leq y \leq \pi$
$y = \arctan x$ if and only if $\tan y = x$	$-\infty < x < \infty$	$-\dfrac{\pi}{2} < y < \dfrac{\pi}{2}$

The graphs of these three inverse trigonometric functions are shown in Figure 1.64.

A computer animation of this concept appears in the *Interactive* CD-ROM and *Internet* versions of this text.

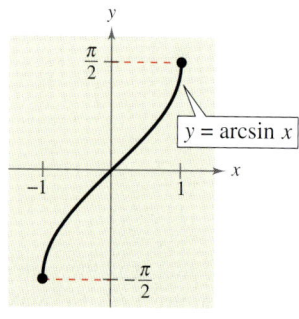

Domain: $[-1, 1]$
Range: $\left[-\dfrac{\pi}{2}, \dfrac{\pi}{2}\right]$

FIGURE 1.64

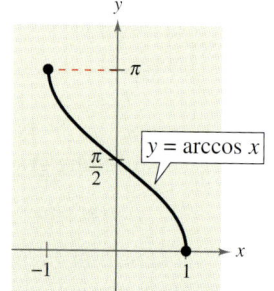

Domain: $[-1, 1]$
Range: $[0, \pi]$

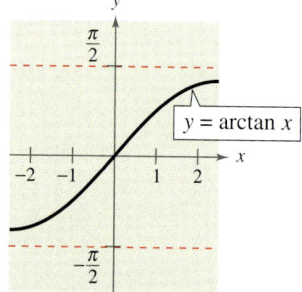

Domain: $(-\infty, \infty)$
Range: $\left(-\dfrac{\pi}{2}, \dfrac{\pi}{2}\right)$

Example 3 Evaluating Inverse Trigonometric Functions

Find the exact value.

a. $\arccos \dfrac{\sqrt{2}}{2}$ b. $\arccos(-1)$ c. $\arctan 0$ d. $\arctan(-1)$

Solution

a. Because $\cos(\pi/4) = \sqrt{2}/2$, and $\pi/4$ lies in $[0, \pi]$, it follows that

$$\arccos \dfrac{\sqrt{2}}{2} = \dfrac{\pi}{4}.$$ *Angle whose cosine is $\sqrt{2}/2$*

b. Because $\cos \pi = -1$, and π lies in $[0, \pi]$, it follows that

$$\arccos(-1) = \pi.$$ *Angle whose cosine is -1*

c. Because $\tan 0 = 0$, and 0 lies in $(-\pi/2, \pi/2)$, it follows that

$$\arctan 0 = 0.$$ *Angle whose tangent is 0*

d. Because $\tan(-\pi/4) = -1$, and $-\pi/4$ lies in $(-\pi/2, \pi/2)$, it follows that

$$\arctan(-1) = -\dfrac{\pi}{4}.$$ *Angle whose tangent is -1*

Example 4 Calculators and Inverse Trigonometric Functions

Use a calculator to approximate the value (if possible).

a. $\arctan(-8.45)$

b. $\arcsin 0.2447$

c. $\arccos 2$

Solution

Function	Mode	Calculator Keystrokes
a. $\arctan(-8.45)$	Radian	TAN⁻¹ ((−) 8.45) ENTER

From the display, it follows that $\arctan(-8.45) \approx -1.453001$.

b. $\arcsin 0.2447$	Radian	SIN⁻¹ 0.2447 ENTER

From the display, it follows that $\arcsin 0.2447 \approx 0.2472103$.

c. $\arccos 2$	Radian	COS⁻¹ 2 ENTER

In real number mode, the calculator should display an *error message* because the domain of the inverse cosine function is $[-1, 1]$.

In Example 4, if you had set the calculator to degree mode, the display would have been in degrees rather than radians. This convention is peculiar to calculators. By definition, the values of inverse trigonometric functions are *always in radians*.

> **STUDY TIP**
>
> It is important to remember that the domain of the inverse sine function and the inverse cosine function is $[-1, 1]$, as shown in Example 4(c).

Compositions of Functions

Recall from Section P.8 that inverse functions have the properties

$$f(f^{-1}(x)) = x \quad \text{and} \quad f^{-1}(f(x)) = x$$

for all x in the domain of f and f^{-1}.

> **Inverse Properties of Trigonometric Functions**
>
> If $-1 \le x \le 1$ and $-\pi/2 \le y \le \pi/2$, then
>
> $\sin(\arcsin x) = x \quad$ and $\quad \arcsin(\sin y) = y.$
>
> If $-1 \le x \le 1$ and $0 \le y \le \pi$, then
>
> $\cos(\arccos x) = x \quad$ and $\quad \arccos(\cos y) = y.$
>
> If x is a real number and $-\pi/2 < y < \pi/2$, then
>
> $\tan(\arctan x) = x \quad$ and $\quad \arctan(\tan y) = y.$

Keep in mind that these inverse properties do not apply for arbitrary values of x and y. For instance,

$$\arcsin\left(\sin \frac{3\pi}{2}\right) = \arcsin(-1) = -\frac{\pi}{2} \ne \frac{3\pi}{2}.$$

In other words, the property

$$\arcsin(\sin y) = y$$

is not valid for values of y outside the interval $[-\pi/2, \pi/2]$.

Example 5 ▶ Using Inverse Properties

If possible, find the exact value.

a. $\tan[\arctan(-5)]$ **b.** $\arcsin\left(\sin \frac{5\pi}{3}\right)$ **c.** $\cos(\cos^{-1} \pi)$

Solution

a. Because -5 lies in the domain of the arctan function, the inverse property applies, and you have

$$\tan[\arctan(-5)] = -5.$$

b. In this case, $5\pi/3$ does not lie within the range of the arcsine function, $-\pi/2 \le y \le \pi/2$. However, $5\pi/3$ is coterminal with

$$\frac{5\pi}{3} - 2\pi = -\frac{\pi}{3}$$

which does lie in the range of the arcsine function, and you have

$$\arcsin\left(\sin \frac{5\pi}{3}\right) = \arcsin\left[\sin\left(-\frac{\pi}{3}\right)\right] = -\frac{\pi}{3}.$$

c. The expression $\cos(\cos^{-1} \pi)$ is not defined because $\cos^{-1} \pi$ is not defined. Remember that the domain of the inverse cosine function is $[-1, 1]$.

Example 6 shows how to use right triangles to find exact values of compositions of inverse functions. Then, Example 7 shows how to use triangles to convert a trigonometric expression into an algebraic expression. This conversion technique is used frequently in calculus.

Example 6 ▶ Evaluating Compositions of Functions

Find the exact value.

a. $\tan\left(\arccos \dfrac{2}{3}\right)$ **b.** $\cos\left[\arcsin\left(-\dfrac{3}{5}\right)\right]$

Solution

a. If you let $u = \arccos \dfrac{2}{3}$, then $\cos u = \dfrac{2}{3}$. Because $\cos u$ is positive, u is a *first*-quadrant angle. You can sketch and label angle u as shown in Figure 1.65(a). Consequently,

$$\tan\left(\arccos \dfrac{2}{3}\right) = \tan u = \dfrac{\text{opp}}{\text{adj}} = \dfrac{\sqrt{5}}{2}.$$

b. If you let $u = \arcsin\left(-\dfrac{3}{5}\right)$, then $\sin u = -\dfrac{3}{5}$. Because $\sin u$ is negative, u is a *fourth*-quadrant angle. You can sketch and label angle u as shown in Figure 1.65(b). Consequently,

$$\cos\left[\arcsin\left(-\dfrac{3}{5}\right)\right] = \cos u = \dfrac{\text{adj}}{\text{hyp}} = \dfrac{4}{5}.$$

(a)

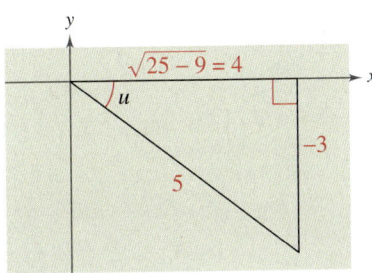

(b)

FIGURE 1.65

Example 7 ▶ Some Problems from Calculus

Write each of the following as an algebraic expression in x.

a. $\sin(\arccos 3x), \quad 0 \le x \le \dfrac{1}{3}$ **b.** $\cot(\arccos 3x), \quad 0 \le x \le \dfrac{1}{3}$

Solution

If you let $u = \arccos 3x$, then $\cos u = 3x$. Because

$$\cos u = \dfrac{3x}{1} = \dfrac{\text{adj}}{\text{hyp}}$$

you can sketch a right triangle with acute angle u, as shown in Figure 1.66. From this triangle, you can easily convert each expression to algebraic form.

a. $\sin(\arccos 3x) = \sin u = \dfrac{\text{opp}}{\text{hyp}} = \sqrt{1 - 9x^2}, \quad 0 \le x \le \dfrac{1}{3}$

b. $\cot(\arccos 3x) = \cot u = \dfrac{\text{adj}}{\text{opp}} = \dfrac{3x}{\sqrt{1 - 9x^2}}, \quad 0 \le x \le \dfrac{1}{3}$

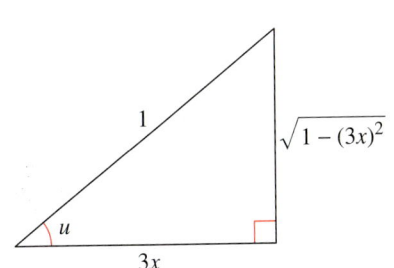

FIGURE 1.66

In Example 7, similar arguments can be made for x-values lying in the interval $\left[-\dfrac{1}{3}, 0\right]$.

The symbol indicates an example or exercise that highlights algebraic techniques specifically used in calculus.

1.7 Exercises

In Exercises 1–16, evaluate the expression without the aid of a calculator.

1. $\arcsin \frac{1}{2}$
2. $\arcsin 0$
3. $\arccos \frac{1}{2}$
4. $\arccos 0$
5. $\arctan \frac{\sqrt{3}}{3}$
6. $\arctan(-1)$
7. $\arccos\left(-\frac{\sqrt{3}}{2}\right)$
8. $\arcsin\left(-\frac{\sqrt{2}}{2}\right)$
9. $\arctan(-\sqrt{3})$
10. $\arctan \sqrt{3}$
11. $\arccos\left(-\frac{1}{2}\right)$
12. $\arcsin \frac{\sqrt{2}}{2}$
13. $\arcsin \frac{\sqrt{3}}{2}$
14. $\arctan\left(-\frac{\sqrt{3}}{3}\right)$
15. $\arctan 0$
16. $\arccos 1$

In Exercises 17–32, use a calculator to approximate the expression. Round your result to two decimal places.

17. $\arccos 0.28$
18. $\arcsin 0.45$
19. $\arcsin(-0.75)$
20. $\arccos(-0.7)$
21. $\arctan(-3)$
22. $\arctan 15$
23. $\arcsin 0.31$
24. $\arccos 0.26$
25. $\arccos(-0.41)$
26. $\arcsin(-0.125)$
27. $\arctan 0.92$
28. $\arctan 2.8$
29. $\arcsin \frac{3}{4}$
30. $\arccos\left(-\frac{1}{3}\right)$
31. $\arctan \frac{7}{2}$
32. $\arctan\left(-\frac{95}{7}\right)$

In Exercises 33 and 34, determine the missing coordinates of the points on the graph of the function.

33.
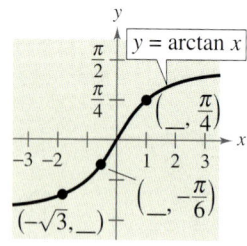

34.

In Exercises 35 and 36, use a graphing utility to graph f, g, and $y = x$ in the same viewing window to verify geometrically that g is the inverse of f. (Be sure to restrict the domain of f properly.)

35. $f(x) = \tan x$, $g(x) = \arctan x$
36. $f(x) = \sin x$, $g(x) = \arcsin x$

In Exercises 37–42, use an inverse trigonometric function to write θ as a function of x.

37.

38.

39.

40.

41.

42.
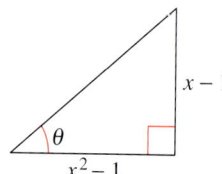

In Exercises 43–48, use the properties of inverse functions to evaluate the expression.

43. $\sin(\arcsin 0.3)$
44. $\tan(\arctan 25)$
45. $\cos[\arccos(-0.1)]$
46. $\sin[\arcsin(-0.2)]$
47. $\arcsin(\sin 3\pi)$
48. $\arccos\left(\cos \frac{7\pi}{2}\right)$

In Exercises 49–58, find the exact value of the expression. (Hint: Make a sketch of a right triangle.)

49. $\sin\left(\arctan \frac{3}{4}\right)$
50. $\sec\left(\arcsin \frac{4}{5}\right)$
51. $\cos(\arctan 2)$
52. $\sin\left(\arccos \frac{\sqrt{5}}{5}\right)$
53. $\cos\left(\arcsin \frac{5}{13}\right)$
54. $\csc\left[\arctan\left(-\frac{5}{12}\right)\right]$
55. $\sec\left[\arctan\left(-\frac{3}{5}\right)\right]$
56. $\tan\left[\arcsin\left(-\frac{3}{4}\right)\right]$
57. $\sin\left[\arccos\left(-\frac{2}{3}\right)\right]$
58. $\cot\left(\arctan \frac{5}{8}\right)$

In Exercises 59–68, write an algebraic expression that is equivalent to the expression. (*Hint:* Sketch a right triangle, as demonstrated in Example 7.)

59. $\cot(\arctan x)$
60. $\sin(\arctan x)$
61. $\cos(\arcsin 2x)$
62. $\sec(\arctan 3x)$
63. $\sin(\arccos x)$
64. $\sec[\arcsin(x - 1)]$
65. $\tan\left(\arccos \dfrac{x}{3}\right)$
66. $\cot\left(\arctan \dfrac{1}{x}\right)$
67. $\csc\left(\arctan \dfrac{x}{\sqrt{2}}\right)$
68. $\cos\left(\arcsin \dfrac{x - h}{r}\right)$

In Exercises 69 and 70, use a graphing utility to graph f and g in the same viewing window to verify that the two are equal. Explain why they are equal. Identify any asymptotes of the graphs.

69. $f(x) = \sin(\arctan 2x)$, $g(x) = \dfrac{2x}{\sqrt{1 + 4x^2}}$
70. $f(x) = \tan\left(\arccos \dfrac{x}{2}\right)$, $g(x) = \dfrac{\sqrt{4 - x^2}}{x}$

In Exercises 71–74, fill in the blank.

71. $\arctan \dfrac{9}{x} = \arcsin(\quad)$, $x \neq 0$
72. $\arcsin \dfrac{\sqrt{36 - x^2}}{6} = \arccos(\quad)$, $0 \leq x \leq 6$
73. $\arccos \dfrac{3}{\sqrt{x^2 - 2x + 10}} = \arcsin(\quad)$
74. $\arccos \dfrac{x - 2}{2} = \arctan(\quad)$, $|x - 2| \leq 2$

In Exercises 75–82, sketch a graph of the function.

75. $y = 2 \arccos x$
76. $y = \arcsin \dfrac{x}{2}$
77. $f(x) = \arcsin(x - 1)$
78. $g(t) = \arccos(t + 2)$
79. $f(x) = \arctan 2x$
80. $f(x) = \dfrac{\pi}{2} + \arctan x$
81. $h(v) = \tan(\arccos v)$
82. $f(x) = \arccos \dfrac{x}{4}$

In Exercises 83–88, use a graphing utility to sketch a graph of the function.

83. $f(x) = 2 \arccos(2x)$
84. $f(x) = \pi \arcsin(4x)$
85. $f(x) = \arctan(2x - 3)$
86. $f(x) = -3 + \arctan(\pi x)$
87. $f(x) = \pi - \arcsin\left(\dfrac{2}{3}\right)$
88. $f(x) = \dfrac{\pi}{2} + \arccos\left(\dfrac{1}{\pi}\right)$

In Exercises 89 and 90, write the given function in terms of the sine function by using the identity

$$A \cos \omega t + B \sin \omega t = \sqrt{A^2 + B^2} \sin\left(\omega t + \arctan \dfrac{A}{B}\right).$$

Use a graphing utility to graph both forms of the function. What does the graph imply?

89. $f(t) = 3 \cos 2t + 3 \sin 2t$
90. $f(t) = 4 \cos \pi t + 3 \sin \pi t$

91. *Docking a Boat* A boat is pulled in by means of a winch located on a dock 5 feet above the deck of the boat (see figure). Let θ be the angle of elevation from the boat to the winch and let s be the length of the rope from the winch to the boat.

(a) Write θ as a function of s.
(b) Find θ when $s = 40$ feet and $s = 20$ feet.

92. *Photography* A television camera at ground level is filming the lift-off of a space shuttle at a point 750 meters from the launch pad (see figure). Let θ be the angle of elevation to the shuttle and let s be the height of the shuttle.

(a) Write θ as a function of s.
(b) Find θ when $s = 300$ meters and $s = 1200$ meters.

93. *Photography* A photographer is taking a picture of a 2-foot painting hung in an art gallery. The camera lens is 1 foot below the lower edge of the painting (see figure). The angle β subtended by the camera lens x feet from the painting is

$$\beta = \arctan \frac{3x}{x^2 + 4}, \quad x > 0.$$

(a) Use a graphing utility to graph β as a function of x.

(b) Move the cursor along the graph to approximate the distance from the picture when β is maximum.

(c) Identify the asymptote of the graph and discuss its meaning in the context of the problem.

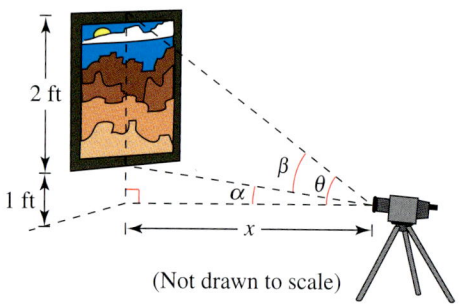

(Not drawn to scale)

94. *Angle of Elevation* An airplane flies at an altitude of 6 miles toward a point directly over an observer. Consider θ and x as shown in the figure.

(a) Write θ as a function of x.

(b) Find θ when $x = 7$ miles and $x = 1$ mile.

95. *Security Patrol* A security car with its spotlight on is parked 20 meters from a warehouse. Consider θ and x as shown in the figure.

(a) Write θ as a function of x.

(b) Find θ when $x = 5$ meters and $x = 12$ meters.

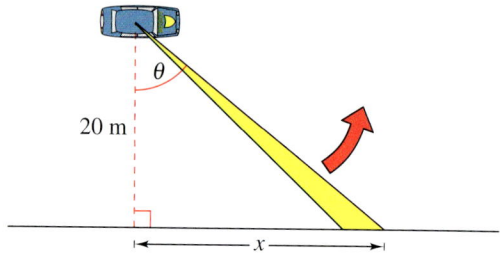

FIGURE FOR 95

Synthesis

True or False? In Exercises 96 and 97, determine whether the statement is true or false. Justify your answer.

96. $\sin \dfrac{5\pi}{6} = \dfrac{1}{2} \implies \arcsin \dfrac{1}{2} = \dfrac{5\pi}{6}$

97. $\tan \dfrac{5\pi}{4} = 1 \implies \arctan 1 = \dfrac{5\pi}{4}$

98. *Area* In calculus, it is shown that the area of the region bounded by the graphs of $y = 0$, $y = 1/(x^2 + 1)$, $x = a$, and $x = b$ is

Area $= \arctan b - \arctan a$

(see figure). Find the area for the following values of a and b.

(a) $a = 0, b = 1$ (b) $a = -1, b = 1$

(c) $a = 0, b = 3$ (d) $a = -1, b = 3$

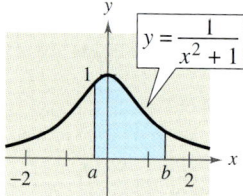

99. Define the inverse cotangent function by restricting the domain of the cotangent function to the interval $(0, \pi)$, and sketch its graph.

100. Define the inverse secant function by restricting the domain of the secant function to the intervals $[0, \pi/2)$ and $(\pi/2, \pi]$, and sketch its graph.

101. Define the inverse cosecant function by restricting the domain of the cosecant function to the intervals $[-\pi/2, 0)$ and $(0, \pi/2]$, and sketch its graph.

102. Use the results of Exercises 99–101 to evaluate the following without using a calculator.
  (a) $\text{arcsec } \sqrt{2}$ (b) $\text{arcsec } 1$
  (c) $\text{arccot}(-\sqrt{3})$ (d) $\text{arccsc } 2$

103. *Think About It* Use a graphing utility to graph the functions $f(x) = \sqrt{x}$ and $g(x) = 6 \arctan x$. For $x > 0$, it appears that $g > f$. Explain why you know that there exists a positive real number a such that $g < f$ for $x > a$. Approximate the number a.

104. *Think About It* Consider the functions
$$f(x) = \sin x \quad \text{and} \quad f^{-1}(x) = \arcsin x.$$
  (a) Use a graphing utility to graph the composite functions $f \circ f^{-1}$ and $f^{-1} \circ f$.
  (b) Explain why the graphs in part (a) are not the graph of the line $y = x$. Why do the graphs of $f \circ f^{-1}$ and $f^{-1} \circ f$ differ?

In Exercises 105–110, prove the identity.

105. $\arcsin(-x) = -\arcsin x$

106. $\arctan(-x) = -\arctan x$

107. $\arccos(-x) = \pi - \arccos x$

108. $\arctan x + \arctan \dfrac{1}{x} = \dfrac{\pi}{2}, \quad x > 0$

109. $\arcsin x + \arccos x = \dfrac{\pi}{2}$

110. $\arcsin x = \arctan \dfrac{x}{\sqrt{1 - x^2}}$

Review

In Exercises 111–114, evaluate the six trigonometric functions of θ.

111.   **112.**

113.   **114.**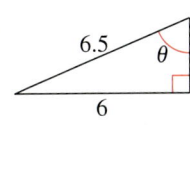

In Exercises 115–118, sketch a right triangle corresponding to the trigonometric function of the acute angle θ. Use the Pythagorean Theorem to determine the third side.

115. $\sin \theta = \dfrac{3}{4}$ **116.** $\tan \theta = 2$

117. $\cos \theta = \dfrac{5}{6}$ **118.** $\sec \theta = 3$

119. *Partnership Costs* A group of people agree to share equally in the cost of a $250,000 endowment to a college. If they could find two more people to join the group, each person's share of the cost would decrease by $6250. How many people are presently in the group?

120. *Speed* A boat travels at a speed of 18 miles per hour in still water. It travels 35 miles upstream and then returns to the starting point in a total of 4 hours. Find the speed of the current.

1.8 Applications and Models

▶ **What you should learn**
- How to solve real-life problems involving right triangles
- How to solve real-life problems involving directional bearings
- How to solve real-life problems involving harmonic motion

▶ **Why you should learn it**

Trigonometric functions frequently model real-life problems involving wave motion. For instance, Exercise 60 on page 199 shows how a trigonometric function can be used to model the harmonic motion of a buoy.

Applications Involving Right Triangles

In this section the three angles of a right triangle are denoted by the letters A, B, and C (where C is the right angle), and the lengths of the sides opposite these angles by the letters a, b, and c (where c is the hypotenuse).

Example 1 ▶ Solving a Right Triangle

Solve the right triangle shown in Figure 1.67 for all unknown sides and angles.

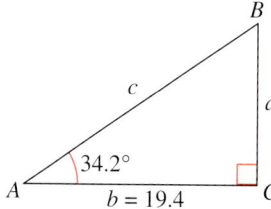

FIGURE 1.67

Solution

Because $C = 90°$, it follows that $A + B = 90°$ and $B = 90° - 34.2° = 55.8°$. To solve for a, use the fact that

$$\tan A = \frac{\text{opp}}{\text{adj}} = \frac{a}{b} \quad \Longrightarrow \quad a = b \tan A.$$

So, $a = 19.4 \tan 34.2° \approx 13.18$. Similarly, to solve for c, use the fact that

$$\cos A = \frac{\text{adj}}{\text{hyp}} = \frac{b}{c} \quad \Longrightarrow \quad c = \frac{b}{\cos A}.$$

So, $c = \dfrac{19.4}{\cos 34.2°} \approx 23.46.$

In the next three examples, the term **angle of elevation** represents the angle from the horizontal upward to an object. For objects that lie below the horizontal, it is common to use the term **angle of depression**, as shown in Figure 1.68.

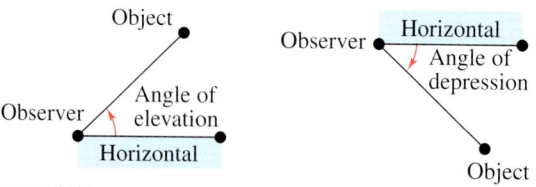

FIGURE 1.68

Section 1.8 ▶ Applications and Models   191

Example 2 ▶ Finding a Side of a Right Triangle

A safety regulation states that the maximum angle of elevation for a rescue ladder is 72°. If a fire department's longest ladder is 110 feet, what is the maximum safe rescue height?

Solution

A sketch is shown in Figure 1.69. From the equation $\sin A = a/c$, it follows that

$$a = c \sin A = 110 \sin 72° \approx 104.6.$$

So, the maximum safe rescue height is about 104.6 feet above the height of the fire truck.

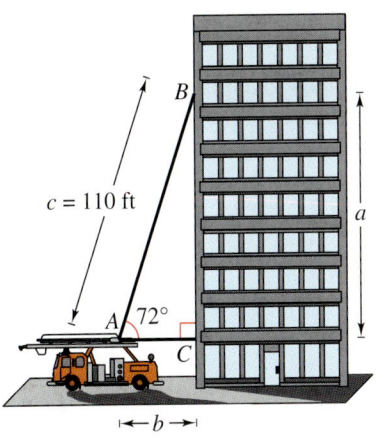

FIGURE 1.69

Example 3 ▶ Finding a Side of a Right Triangle

At a point 200 feet from the base of a building, the angle of elevation to the *bottom* of a smokestack is 35°, whereas the angle of elevation to the *top* is 53°, as shown in Figure 1.70. Find the height s of the smokestack alone.

Solution

Note from Figure 1.70 that this problem involves two right triangles. In the smaller right triangle, use the fact that $\tan 35° = a/200$ to conclude that the height of the building is

$$a = 200 \tan 35°.$$

In the larger right triangle, use the equation

$$\tan 53° = \frac{a + s}{200}$$

to conclude that $a + s = 200 \tan 53°$. So, the height of the smokestack is

$$s = 200 \tan 53° - a$$
$$= 200 \tan 53° - 200 \tan 35°$$
$$\approx 125.4 \text{ feet.}$$

FIGURE 1.70

Example 4 ▶ Finding an Acute Angle of a Right Triangle

A swimming pool is 20 meters long and 12 meters wide. The bottom of the pool is slanted so that the water depth is 1.3 meters at the shallow end and 4 meters at the deep end, as shown in Figure 1.71. Find the angle of depression of the bottom of the pool.

Solution

Using the tangent function, you see that

$$\tan A = \frac{\text{opp}}{\text{adj}} = \frac{2.7}{20} = 0.135.$$

So, the angle of depression is

$$A = \arctan 0.135 \approx 0.13419 \text{ radian} \approx 7.69°.$$

FIGURE 1.71

Trigonometry and Bearings

In surveying and navigation, directions are generally given in terms of **bearings**. A bearing measures the acute angle a path or line of sight makes with a fixed north-south line, as shown in Figure 1.72. For instance, the bearing S 35° E in Figure 1.72 means 35 degrees east of south.

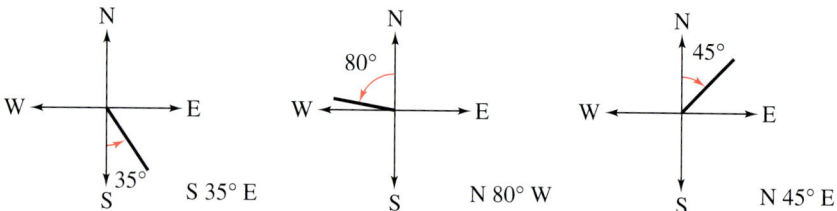

FIGURE 1.72

Example 5 ▶ Finding Directions in Terms of Bearings

A ship leaves port at noon and heads due west at 20 knots, or 20 nautical miles (nm) per hour. At 2 P.M. the ship changes course to N 54° W, as shown in Figure 1.73. Find the ship's bearing and distance from the port of departure at 3 P.M.

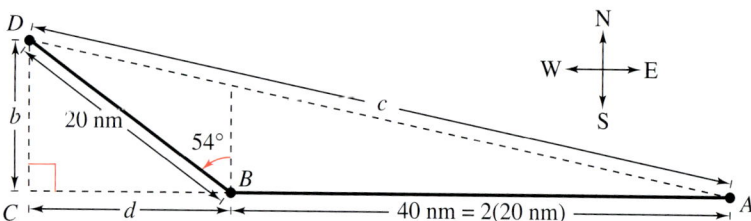

FIGURE 1.73

Solution

In triangle BCD, you have $B = 90° - 54° = 36°$. The two sides of this triangle can be determined to be

$$b = 20 \sin 36° \quad \text{and} \quad d = 20 \cos 36°.$$

In triangle ACD, you find angle A as follows.

$$\tan A = \frac{b}{d + 40} = \frac{20 \sin 36°}{20 \cos 36° + 40} \approx 0.2092494$$

$$A \approx \arctan 0.2092494 \approx 0.2062732 \text{ radian} \approx 11.82°$$

The angle with the north-south line is $90° - 11.82° = 78.18°$. Therefore, the bearing of the ship is

N 78.18° W.   *Bearing*

Finally, from triangle ACD, you have $\sin A = b/c$, which yields

$$c = \frac{b}{\sin A} = \frac{20 \sin 36°}{\sin 11.82°}$$

$$\approx 57.4 \text{ nautical miles.} \quad \textit{Distance from port}$$

Harmonic Motion

The periodic nature of the trigonometric functions is useful for describing the motion of a point on an object that vibrates, oscillates, rotates, or is moved by wave motion.

For example, consider a ball that is bobbing up and down on the end of a spring, as shown in Figure 1.74. Suppose that 10 centimeters is the maximum distance the ball moves vertically upward or downward from its equilibrium (at rest) position. Suppose further that the time it takes for the ball to move from its maximum displacement above zero to its maximum displacement below zero and back again is $t = 4$ seconds. Assuming the ideal conditions of perfect elasticity and no friction or air resistance, the ball would continue to move up and down in a uniform and regular manner.

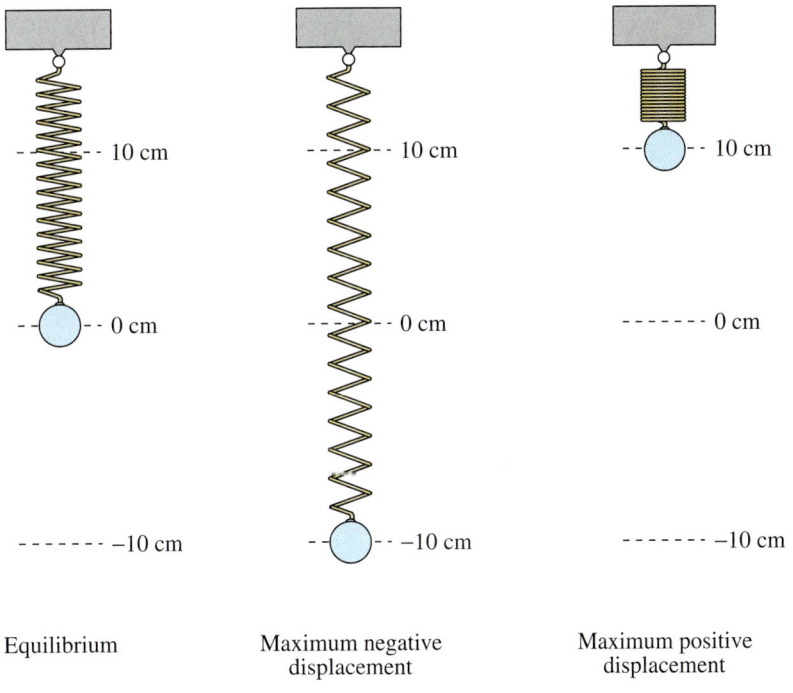

Equilibrium   Maximum negative   Maximum positive
               displacement   displacement

FIGURE 1.74

From this spring you can conclude that the period (time for one complete cycle) of the motion is

Period = 4 seconds

and that its amplitude (maximum displacement from equilibrium) is

Amplitude = 10 centimeters.

Motion of this nature can be described by a sine or cosine function, and is called **simple harmonic motion.**

Definition of Simple Harmonic Motion

A point that moves on a coordinate line is said to be in **simple harmonic motion** if its distance d from the origin at time t is given by either

$$d = a \sin \omega t \quad \text{or} \quad d = a \cos \omega t$$

where a and ω are real numbers such that $\omega > 0$. The motion has amplitude $|a|$, period $2\pi/\omega$, and frequency $\omega/2\pi$.

Example 6 ▶ Simple Harmonic Motion

Write the equation for the simple harmonic motion of the ball described in Figure 1.74, where the period is 4 seconds. What is the frequency of this harmonic motion?

Solution

Because the spring is at equilibrium ($d = 0$) when $t = 0$, you use the equation

$$d = a \sin \omega t.$$

Moreover, because the maximum displacement from zero is 10 and the period is 4, you have

$$\text{Amplitude} = |a| = 10$$

$$\text{Period} = \frac{2\pi}{\omega} = 4 \quad \Longrightarrow \quad \omega = \frac{\pi}{2}.$$

Consequently, the equation of motion is

$$d = 10 \sin \frac{\pi}{2} t.$$

Note that the choice of $a = 10$ or $a = -10$ depends on whether the ball initially moves up or down. The frequency is

$$\text{Frequency} = \frac{\omega}{2\pi} = \frac{\pi/2}{2\pi} = \frac{1}{4} \text{ cycle per second.}$$

One illustration of the relationship between sine waves and harmonic motion is seen in the wave motion resulting when a stone is dropped into a calm pool of water. The waves move outward in roughly the shape of sine (or cosine) waves, as shown in Figure 1.75. As an example, suppose you are fishing and your fishing bob is attached so that it does not move horizontally. As the waves move outward from the dropped stone, your fishing bob will move up and down in simple harmonic motion, as shown in Figure 1.76.

FIGURE 1.75

FIGURE 1.76

Example 7 ▶ Simple Harmonic Motion

Given the equation for simple harmonic motion

$$d = 6 \cos \frac{3\pi}{4} t$$

find (a) the maximum displacement, (b) the frequency, (c) the value of d when $t = 4$, and (d) the least positive value of t for which $d = 0$.

Solution

The given equation has the form $d = a \cos \omega t$, with $a = 6$ and $\omega = 3\pi/4$.

a. The maximum displacement (from the point of equilibrium) is given by the amplitude. So, the maximum displacement is 6.

b. Frequency $= \dfrac{\omega}{2\pi}$

$= \dfrac{3\pi/4}{2\pi}$

$= \dfrac{3}{8}$ cycle per unit of time

c. $d = 6 \cos\left[\dfrac{3\pi}{4}(4)\right]$

$= 6 \cos 3\pi$

$= 6(-1)$

$= -6$

d. To find the least positive value of t for which $d = 0$, solve the equation

$$d = 6 \cos \frac{3\pi}{4} t = 0$$

to obtain

$$\frac{3\pi}{4} t = \frac{\pi}{2}, \frac{3\pi}{2}, \frac{5\pi}{2}, \ldots$$ $$t = \frac{2}{3}, 2, \frac{10}{3}, \ldots.$$

So, the least positive value of t is $t = \frac{2}{3}$.

Writing ABOUT MATHEMATICS

Radio Waves Many different physical phenomena can be characterized by wave motion. These phenomena include electromagnetic waves such as radio waves, television waves, and microwaves. Radio waves transmit sound in two different ways. For an AM station, the *amplitude* of the wave is modified to carry sound. The letters AM stand for "amplitude modulation." An FM radio signal has its *frequency* modified in order to carry sound, hence the term "frequency modulation." Of the two graphs in Figure 1.77, one shows an AM wave and the other shows an FM wave. Which is which? Explain your reasoning.

(a)

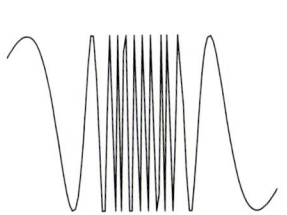
(b)

FIGURE 1.77

1.8 Exercises

In Exercises 1–10, solve the right triangle shown in the figure. Round your answer to two decimal places.

1. $A = 20°$, $b = 10$
2. $B = 54°$, $c = 15$
3. $B = 71°$, $b = 24$
4. $A = 8.4°$, $a = 40.5$
5. $a = 6$, $b = 10$
6. $a = 25$, $c = 35$
7. $b = 16$, $c = 52$
8. $b = 1.32$, $c = 9.45$
9. $A = 12°15'$, $c = 430.5$
10. $B = 65°12'$, $a = 14.2$

FIGURE FOR 1–10

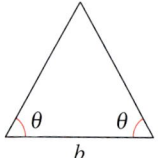

FIGURE FOR 11–14

In Exercises 11–14, find the altitude of the isosceles triangle shown in the figure. Round your answer to two decimal places.

11. $\theta = 52°$, $b = 4$ inches
12. $\theta = 18°$, $b = 10$ meters
13. $\theta = 41°$, $b = 46$ inches
14. $\theta = 27°$, $b = 11$ feet

15. *Length of a Shadow* If the sun is 25° above the horizon, find the length of a shadow cast by a silo that is 50 feet tall (see figure).

16. *Length of a Shadow* If the sun is 20° above the horizon, find the length of a shadow cast by a building that is 600 feet tall.

17. *Height* A ladder 20 feet long leans against the side of a house. Find the height h from the top of the ladder to the ground if the angle of elevation of the ladder is 80°.

18. *Height* The length of a shadow of a tree is 125 feet when the angle of elevation of the sun is 33°. Approximate the height h of the tree.

19. *Height* From a point 50 feet in front of a church, the angles of elevation to the base of the steeple and the top of the steeple are 35° and 47° 40', respectively.
    (a) Draw right triangles that represent the problem. Label the known and unknown quantities.
    (b) Use a trigonometric function to write an equation involving the unknown quantity.
    (c) Find the height of the steeple.

20. *Height* You are standing 100 feet from the base of a platform from which people are bungee jumping. The angle of elevation from your position to the top of the platform from which they jump is 51°. From what height are the people jumping?

21. *Depth of a Submarine* The sonar of a navy cruiser detects a submarine that is 4000 feet from the cruiser. The angle between the water line and the submarine is 34° (see figure). How deep is the submarine?

22. *Height of a Kite* A 75-foot line is attached to a kite. When the kite has pulled the line taut, the angle of elevation to the kite is approximately 60°. Approximate the height of the kite.

23. *Angle of Elevation* An amateur radio operator erects a 75-foot vertical tower for an antenna. Find the angle of elevation to the top of the tower at a point on level ground 50 feet from its base.

24. *Angle of Elevation* The height of an outdoor basketball backboard is $12\frac{1}{2}$ feet, and the backboard casts a shadow $17\frac{1}{3}$ feet long.
    (a) Draw a right triangle that represents the problem. Label the known and unknown quantities.
    (b) Use a trigonometric function to write an equation involving the unknown quantity.
    (c) Find the angle of elevation of the sun.

25. *Angle of Depression* A Global Positioning System satellite orbits 10,900 feet above earth's surface. Find the angle of depression from the satellite to the horizon. Assume the radius of earth is 4000 miles.

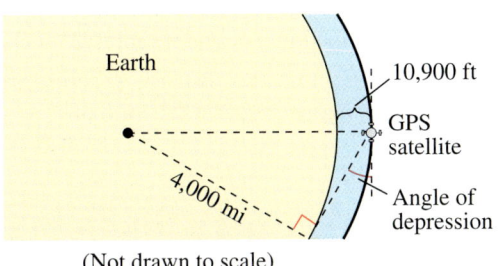

(Not drawn to scale)

26. *Angle of Depression* A cellular telephone tower that is 150 feet tall is placed on top of a mountain that is 1200 feet above sea level. What is the angle of depression from the top of the tower to a cell phone user who is 5 horizontal miles away and 400 feet above sea level?

27. *Airplane Ascent* During takeoff, an airplane's angle of climb is 18° and its speed is 275 feet per second. Find the plane's altitude after 1 minute.

28. *Airplane Ascent* How long will it take the plane in Exercise 27 to climb to an altitude of 10,000 feet?

29. *Mountain Descent* A sign on a roadway at the top of a mountain indicates that for the next 4 miles the grade is 10.5° (see figure). Find the change in elevation for a car descending the mountain.

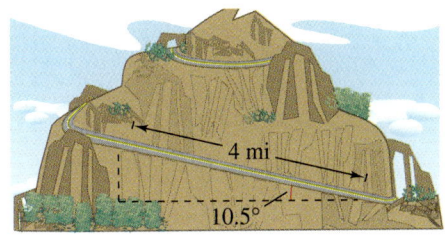

30. *Mountain Descent* A roadway sign at the top of a mountain indicates that for the next 4 miles the grade is 12%. Find the angle of the grade and the change in elevation for a car descending the mountain.

31. *Navigation* An airplane flying at 600 miles per hour has a bearing of N 52° E. After flying for 1.5 hours, how far north and how far east will the plane have traveled from its point of departure?

32. *Navigation* A ship leaves port at noon and has a bearing of S 27° W. If its speed is 20 knots, how many nautical miles south and how many nautical miles west will the ship have traveled by 6:00 P.M.?

33. *Surveying* A surveyor wishes to find the distance across a swamp (see figure). The bearing from A to B is N 32° W. The surveyor walks 50 meters from A, and at the point C the bearing to B is N 68° W. Find (a) the bearing from A to C and (b) the distance from A to B.

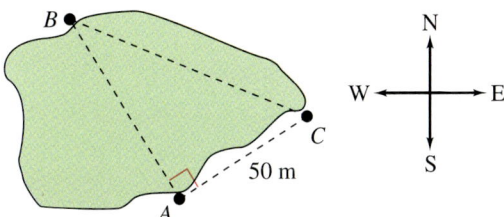

34. *Location of a Fire* Two fire towers are 30 kilometers apart, tower A being due west of tower B. A fire is spotted from the towers, and the bearings from A and B are E 14° N and W 34° N, respectively (see figure). Find the distance d of the fire from the line segment AB.

35. *Navigation* A ship is 45 miles east and 30 miles south of port. If the captain wants to sail directly to port, what bearing should be taken?

36. *Navigation* A plane is 120 miles north and 85 miles east of an airport. If the pilot wants to fly directly to the airport, what bearing should be taken?

37. *Distance Between Ships* An observer in a lighthouse 350 feet above sea level observes two ships directly offshore. The angles of depression to the ships are 4° and 6.5° (see figure). How far apart are the ships?

(Not drawn to scale)

38. *Distance Between Towns* A passenger in an airplane at an altitude of 10 kilometers sees two towns directly to the east of the plane. The angles of depression to the towns are 28° and 55° (see figure). How far apart are the towns?

39. *Altitude of a Plane* A plane is observed approaching your home and you assume that its speed is 550 miles per hour. If the angle of elevation of the plane is 16° at one time and 57° one minute later, approximate the altitude of the plane.

40. *Height of a Mountain* While traveling across flat land, you notice a mountain directly in front of you. The angle of elevation to the peak is 2.5°. After you drive 17 miles closer to the mountain, the angle of elevation is 9°. Approximate the height of the mountain.

Geometry In Exercises 41 and 42, find the angle α between two nonvertical lines L_1 and L_2. The angle α satisfies the equation

$$\tan \alpha = \left| \frac{m_2 - m_1}{1 + m_2 m_1} \right|$$

where m_1 and m_2 are the slopes of L_1 and L_2, respectively. (Assume that $m_1 m_2 \neq -1$.)

41. L_1: $3x - 2y = 5$
    L_2: $x + y = 1$

42. L_1: $2x - y = 8$
    L_2: $x - 5y = -4$

43. *Geometry* Determine the angle between the diagonal of the cube and the diagonal of its base, as shown in the figure.

FIGURE FOR 43

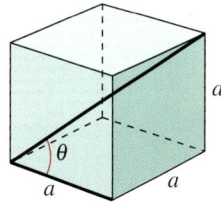

FIGURE FOR 44

44. *Geometry* Determine the angle between the diagonal of the cube and its edge, as shown in the figure.

45. *Geometry* Find the length of the sides of a regular pentagon inscribed in a circle of radius 25 inches.

46. *Geometry* Find the length of the sides of a regular hexagon inscribed in a circle of radius 25 inches.

47. *Hardware* Express the distance y across the flat sides of a hexagonal nut as a function of r, as shown in the figure.

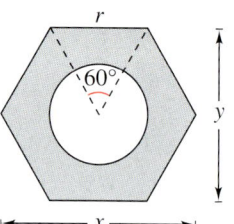

48. *Bolt Holes* The figure shows a circular piece of sheet metal that has a diameter of 40 centimeters and contains 12 equally spaced bolt holes. Determine the straight-line distance between the centers of consecutive bolt holes.

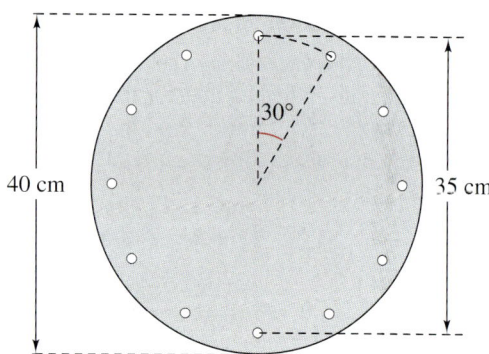

Trusses In Exercises 49 and 50, find the lengths of all the unknown members of the truss.

49.

50.

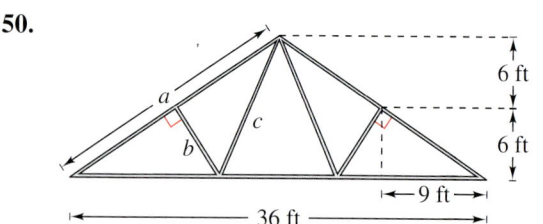

Section 1.8 Applications and Models

Harmonic Motion In Exercises 51–54, for the simple harmonic motion described by the trigonometric function, find (a) the maximum displacement, (b) the frequency, and (c) the least positive value of t for which $d = 0$.

51. $d = 4 \cos 8\pi t$ **52.** $d = \frac{1}{2} \cos 20\pi t$

53. $d = \frac{1}{16} \sin 120\pi t$ **54.** $d = \frac{1}{64} \sin 792\pi t$

Harmonic Motion In Exercises 55–58, find a model for simple harmonic motion satisfying the specified conditions.

	Displacement ($t = 0$)	Amplitude	Period
55.	0	4 cm	2 sec
56.	0	3 m	6 sec
57.	3 in.	3 in.	1.5 sec
58.	2 ft	2 ft	10 sec

59. *Tuning Fork* A point on the end of a tuning fork moves in simple harmonic motion described by $d = a \sin \omega t$. Find ω given that the tuning fork for middle C has a frequency of 264 vibrations per second.

60. *Wave Motion* A buoy oscillates in simple harmonic motion as waves go past. It is noted that the buoy moves a total of 3.5 feet from its low point to its high point (see figure), and that it returns to its high point every 10 seconds. Write an equation that describes the motion of the buoy if its high point is at $t = 0$.

61. *Springs* A weight stretches a spring 1.5 inches. The weight is pushed 3 inches above its equilibrium position and released. Its motion is modeled by

$$y = \frac{1}{4} \cos 16t, \quad t > 0$$

where y is in feet and t is in seconds.

(a) Graph the function.

(b) What is the period of the oscillations?

(c) Determine the first time the weight passes the point of equilibrium ($y = 0$).

Synthesis

True or False? In Exercises 62 and 63, determine whether the statement is true or false. Justify your answer.

62. A building that is famous for not being perfectly vertical is the Leaning Tower of Pisa. If you know the exact angle of elevation θ to the 191-foot tower when you stand near it, then you can determine the exact distance to the tower d by using the formula

$$\tan \theta = \frac{191}{d}.$$

63. For the harmonic motion of a ball bobbing up and down on the end of a spring, one period can be described as the length of one coil of the spring.

64. *Numerical and Graphical Analysis* A 2-meter-high fence is 3 meters from the side of a grain storage bin. A grain elevator must reach from ground level outside the fence to the storage bin (see figure). The objective is to determine the shortest elevator that meets the constraints.

(a) Complete four rows of the table.

θ	L_1	L_2	$L_1 + L_2$
0.1	$\dfrac{2}{\sin 0.1}$	$\dfrac{3}{\cos 0.1}$	23.0
0.2	$\dfrac{2}{\sin 0.2}$	$\dfrac{3}{\cos 0.2}$	13.1

(b) Use a graphing utility to generate additional rows of the table. Use the table to estimate the minimum length of the elevator.

(c) Write the length $L_1 + L_2$ as a function of θ.

(d) Use a graphing utility to graph the function. Use the graph to estimate the minimum length. How does your estimate compare with that of part (b)?

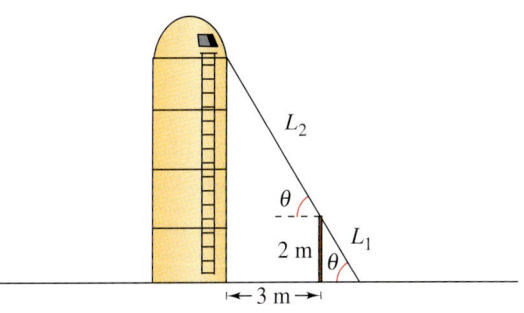

65. *Numerical and Graphical Analysis* The cross sections of an irrigation canal are isosceles trapezoids where the length of three of the sides is 8 feet (see figure). The objective is to find the angle θ that maximizes the area of the cross sections. [*Hint:* The area of a trapezoid is $(h/2)(b_1 + b_2)$.]

(a) Complete seven rows of the table.

Base 1	Base 2	Altitude	Area
8	$8 + 16 \cos 10°$	$8 \sin 10°$	22.1
8	$8 + 16 \cos 20°$	$8 \sin 20°$	42.5

(b) Use a graphing utility to generate additional rows of the table. Use the table to estimate the maximum cross-sectional area.

(c) Write the area A as a function of θ.

(d) Use a graphing utility to graph the function. Use the graph to estimate the maximum cross-sectional area. How does your estimate compare with that of part (b)?

66. *Data Analysis* The times S of sunset (Greenwich Mean Time) at 40° north latitude on the 15th of each month are: 1(16:59), 2(17:35), 3(18:06), 4(18:38), 5(19:08), 6(19:30), 7(19:28), 8(18:57), 9(18:09), 10(17:21), 11(16:44), 12(16:36). The month is represented by t, with $t = 1$ corresponding to January. A model (where minutes have been converted to the decimal parts of an hour) for this data is

$$S(t) = 18.09 + 1.41 \sin\left(\frac{\pi t}{6} + 4.60\right).$$

(a) Use a graphing utility to graph the data points and the model in the same viewing window.

(b) What is the period of the model? Is it what you expected? Explain.

(c) What is the amplitude of the function? What does it represent in the model? Explain.

67. *Data Analysis* The table gives the average sales S (in millions of dollars) of an outerwear manufacturer for each month t, where $t = 1$ represents January.

t	1	2	3	4	5	6
S	13.46	11.15	8.00	4.85	2.54	1.70

t	7	8	9	10	11	12
S	2.54	4.85	8.00	11.15	13.46	14.30

(a) Create a scatter plot of the data.

(b) Find a trigonometric model that fits the data. Graph the model on your scatter plot. How well does the model fit the data?

(c) What is the period of the model? Do you think it is reasonable given the context? Explain your reasoning.

(d) Interpret the meaning of the model's amplitude in the context of the problem.

Review

In Exercises 68–77, graph the equation by hand.

68. $3x - 2y = 4$

69. $5y - 3x = 12$

70. $(y - 2)^2 = 8(x + 2)$

71. $(x + 3)^2 = 5y - 8$

72. $\dfrac{x^2}{4} + y^2 = 1$

73. $2x^2 + y^2 - 4 = 0$

74. $\dfrac{x^2}{4} - y^2 = 1$

75. $\dfrac{y^2}{4} - \dfrac{(x + 2)^2}{25} - 1 = 0$

76. $\dfrac{x^2}{4} + \dfrac{y^2}{4} = 1$

77. $(x - 2)^2 + y^2 = 25$

Chapter Summary

What did you learn?

	Review Exercises
Section 1.1	
☐ How to describe angles	1–4
☐ How to use radian and degree measure	5–20
☐ How to use angles to model and solve real-life problems	21, 22
Section 1.2	
☐ How to identify a unit circle and its relationship to real numbers	23–26
☐ How to evaluate trigonometric functions using the unit circle	27–30
☐ How to use the domain and period to evaluate sine and cosine functions	31–34
☐ How to use a calculator to evaluate trigonometric functions	35–38
Section 1.3	
☐ How to evaluate trigonometric functions of acute angles	39–42
☐ How to use the fundamental trigonometric identities	43–46
☐ How to use a calculator to evaluate trigonometric functions	47–52
☐ How to use trigonometric functions to model and solve real-life problems	53, 54
Section 1.4	
☐ How to evaluate trigonometric functions of any angle	55–68
☐ How to use reference angles to evaluate trigonometric functions	69–74
☐ How to evaluate trigonometric functions of real numbers	75–82
Section 1.5	
☐ How to use amplitude and period to sketch the graphs of sine and cosine functions	83–86
☐ How to sketch translations of graphs of sine and cosine functions	87–90
☐ How to use sine and cosine functions to model real-life data	91, 92
Section 1.6	
☐ How to sketch the graphs of tangent and cotangent functions	93–96
☐ How to sketch the graphs of secant and cosecant functions	97–100
☐ How to sketch the graphs of damped trigonometric functions	101, 102
Section 1.7	
☐ How to evaluate the inverse sine function	103–108
☐ How to evaluate the other inverse trigonometric functions	109–120
☐ How to evaluate the compositions of trigonometric functions	121–128
Section 1.8	
☐ How to solve real-life problems involving right triangles	129, 130
☐ How to solve real-life problems involving directional bearings	131
☐ How to solve real-life problems involving harmonic motion	132

Review Exercises

1.1 In Exercises 1–4, estimate the angle to the nearest one-half radian.

1.
2.
3.
4.

In Exercises 5–12, sketch the angle in standard position. List one positive and one negative coterminal angle.

5. $\dfrac{11\pi}{4}$
6. $\dfrac{2\pi}{9}$
7. $-\dfrac{4\pi}{3}$
8. $-\dfrac{23\pi}{3}$
9. $70°$
10. $280°$
11. $-110°$
12. $-405°$

In Exercises 13–16, convert the measure from radians to degrees. Round to two decimal places.

13. $\dfrac{5\pi}{7}$
14. $-\dfrac{11\pi}{6}$
15. -3.5
16. 5.7

In Exercises 17–20, convert the measure from degrees to radians. Round to four decimal places.

17. $480°$
18. $-127.5°$
19. $-33°\ 45'$
20. $196°\ 77'$

21. **Phonograph** Compact discs have all but replaced phonograph records. Phonograph records are vinyl discs that rotate on a turntable. A typical record album is 12 inches in diameter and plays at $33\tfrac{1}{3}$ revolutions per minute.
    (a) What is the angular speed of a record album?
    (b) What is the linear speed of the outer edge of a record album?

22. **Bicycle** At what speed is a bicyclist traveling if his 27-inch-diameter tires are rotating at an angular speed of 5π radians per second?

1.2 In Exercises 23–26, find the point (x, y) on the unit circle that corresponds to the real number t.

23. $t = \dfrac{2\pi}{3}$
24. $t = \dfrac{3\pi}{4}$
25. $t = \dfrac{5\pi}{6}$
26. $t = -\dfrac{4\pi}{3}$

In Exercises 27–30, evaluate (if possible) the six trigonometric functions of the real number.

27. $t = \dfrac{7\pi}{6}$
28. $t = \dfrac{\pi}{4}$
29. $t = -\dfrac{2\pi}{3}$
30. $t = 2\pi$

In Exercises 31–34, evaluate the trigonometric function using its period as an aid.

31. $\sin \dfrac{11\pi}{4}$
32. $\cos 4\pi$
33. $\sin\left(-\dfrac{17\pi}{6}\right)$
34. $\cos\left(-\dfrac{13\pi}{3}\right)$

In Exercises 35–38, use a calculator to evaluate the trigonometric function. Round to two decimal places.

35. $\tan 33$
36. $\csc 10.5$
37. $\sec \dfrac{12\pi}{5}$
38. $\sin\left(-\dfrac{\pi}{9}\right)$

1.3 In Exercises 39–42, find the values of the six trigonometric functions of the angle θ in the figure.

39.
40.
41.
42.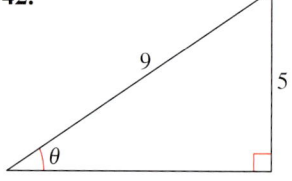

In Exercises 43–46, use the given function value and trigonometric identities (including the cofunction identities) to find the indicated trigonometric functions.

43. $\sin \theta = \frac{1}{3}$
  (a) $\csc \theta$
  (b) $\cos \theta$
  (c) $\sec \theta$
  (d) $\tan \theta$

44. $\tan \theta = 4$
  (a) $\cot \theta$
  (b) $\sec \theta$
  (c) $\cos \theta$
  (d) $\csc \theta$

45. $\csc \theta = 4$
  (a) $\sin \theta$
  (b) $\cos \theta$
  (c) $\sec \theta$
  (d) $\tan \theta$

46. $\csc \theta = 5$
  (a) $\sin \theta$
  (b) $\cot \theta$
  (c) $\tan \theta$
  (d) $\sec(90° - \theta)$

In Exercises 47–52, use a calculator to evaluate the trigonometric function. Round your answer to two decimal places.

47. $\tan 33°$
48. $\csc 11°$
49. $\sin 34.2°$
50. $\sec 79.3°$
51. $\cot 15° \, 14'$
52. $\cos 78° \, 11' \, 58''$

53. *Railroad Grade* A train travels 3.5 kilometers on a straight track with a grade of $1° \, 10'$ (see figure). What is the vertical rise of the train in that distance?

54. *Guy Wire* A guy wire runs from the ground to the top of a 25-foot telephone pole. The angle formed between the wire and the ground is 52°. How far from the base of the pole is the wire attached to the ground?

1.4 **In Exercises 55–62, find the six trigonometric functions of the angle θ (in standard position) whose terminal side passes through the point.**

55. $(12, 16)$
56. $(3, -4)$
57. $\left(\frac{2}{3}, \frac{5}{2}\right)$
58. $\left(-\frac{10}{3}, -\frac{2}{3}\right)$
59. $(-0.5, 4.5)$
60. $(0.3, 0.4)$
61. $(x, 4x), \, x > 0$
62. $(-2x, -3x), \, x > 0$

In Exercises 63–68, find the remaining five trigonometric functions of θ satisfying the condition.

63. $\sec \theta = \frac{6}{5}, \, \tan \theta < 0$
64. $\csc \theta = \frac{3}{2}, \, \cos \theta < 0$
65. $\sin \theta = \frac{3}{8}, \, \cos \theta < 0$
66. $\tan \theta = \frac{5}{4}, \, \cos \theta < 0$
67. $\cos \theta = -\frac{2}{5}, \, \sin \theta > 0$
68. $\sin \theta = -\frac{2}{4}, \, \cos \theta > 0$

In Exercises 69–74, evaluate the trigonometric function without using a calculator.

69. $\tan \pi/3$
70. $\sec \pi/4$
71. $\cos(-7\pi/3)$
72. $\cot(-5\pi/4)$
73. $\cos 495°$
74. $\sin(-150°)$

In Exercises 75–82, evaluate the trigonometric function of the real number. Round your answer to two decimal places.

75. $\sin 4$
76. $\tan 3$
77. $\sin(-3.2)$
78. $\cot(-4.8)$
79. $\sin 3\pi$
80. $\cot 1.5\pi$
81. $\sec 12\pi/5$
82. $\tan(-25\pi/7)$

1.5 **In Exercises 83–90, sketch a graph of the function. Include two full periods.**

83. $y = \sin x$
84. $y = \cos x$
85. $f(x) = 5 \sin \dfrac{2x}{5}$
86. $f(x) = 8 \cos\left(-\dfrac{x}{4}\right)$
87. $y = 2 + \sin x$
88. $y = -4 - \cos \pi x$
89. $g(t) = \frac{5}{2} \sin(t - \pi)$
90. $g(t) = 3 \cos(t + \pi)$

91. *Sound Waves* Sound waves can be modeled by sine functions of the form $y = a \sin bx$, where x is measured in seconds.
  (a) Write an equation of a sound wave whose amplitude is 2 and whose period is $\frac{1}{264}$ second.
  (b) What is the frequency of the sound wave described in part (a)?

92. *Sound Waves* Use the cosine function $y = a \cos bx$ to model the sound wave described in Exercise 91.

1.6 In Exercises 93–102, sketch a graph of the function. Include two full periods.

93. $f(x) = \tan x$
94. $f(t) = \tan\left(t - \dfrac{\pi}{4}\right)$
95. $f(x) = \cot x$
96. $g(t) = 2\cot 2t$
97. $f(x) = \sec x$
98. $h(t) = \sec\left(t - \dfrac{\pi}{4}\right)$
99. $f(x) = \csc x$
100. $f(t) = 3\csc\left(2t + \dfrac{\pi}{4}\right)$
101. $f(x) = x^4 \cos x$
102. $g(x) = x^4 \sin x$

1.7 In Exercises 103–108, evaluate the expression. If necessary, round your answer to two decimal places.

103. $\arcsin\left(-\dfrac{1}{2}\right)$
104. $\arcsin(-1)$
105. $\arcsin 0.4$
106. $\arcsin 0.213$
107. $\sin^{-1}(-0.44)$
108. $\sin^{-1} 0.89$

In Exercises 109–112, evaluate the expression without the aid of a calculator.

109. $\arccos \dfrac{\sqrt{3}}{2}$
110. $\arccos \dfrac{\sqrt{2}}{2}$
111. $\cos^{-1}(-1)$
112. $\cos^{-1} \dfrac{\sqrt{3}}{2}$

In Exercises 113–120, use a calculator to approximate the value of the expression. Round your answer to two decimal places.

113. $\arccos 0.324$
114. $\arccos(-0.888)$
115. $\arctan 0.123$
116. $\arctan 2.34$
117. $\arctan 5.783$
118. $\arctan 99.1$
119. $\tan^{-1}(-1.5)$
120. $\tan^{-1} 8.2$

In Exercises 121–128, find the exact value of the expression.

121. $\sin(\arcsin 0.72)$
122. $\cos(\arccos 0.25)$
123. $\arctan(\tan \pi)$
124. $\arccos[\cos(-5\pi)]$
125. $\cos\left(\arctan \dfrac{3}{4}\right)$
126. $\tan\left(\arccos \dfrac{3}{5}\right)$
127. $\sec\left(\arctan \dfrac{12}{5}\right)$
128. $\cot\left[\arcsin\left(-\dfrac{12}{13}\right)\right]$

1.8 129. *Angle of Elevation* The height of a radio transmission tower is 70 meters, and it casts a shadow of length 30 meters (see figure). Find the angle of elevation of the sun.

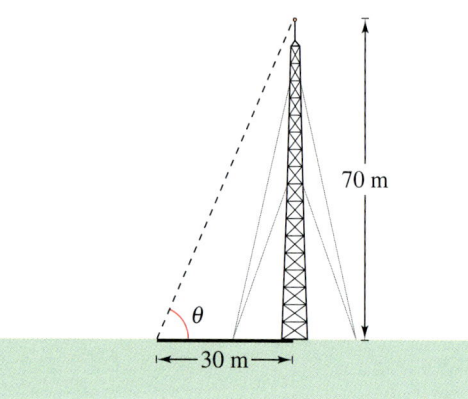

FIGURE FOR 129

130. *Lost Ball* Your football has landed at the edge of the roof of your school building. When you are 25 feet from the base of the building, the angle of elevation to your football is 21°. How high off the ground is your football?

131. *Distance* From city A to city B, a plane flies 650 miles at a bearing of N 48° E. From city B to city C, the plane flies 810 miles at a bearing S 65° E. Find the distance from A to C and the bearing from A to C.

132. *Wave Motion* Your fishing bobber oscillates in simple harmonic motion from the waves in the lake where you fish. Your bobber moves a total of 1.5 inches from its high point to its low point and returns to its high point every 3 seconds. Write an equation modeling the motion of your bobber if it is at its high point at time $t = 0$.

Synthesis

True or False? In Exercises 133–136, determine whether the statement is true or false. Justify your answer.

133. The tangent function is often useful for modeling simple harmonic motion.

134. The inverse sine function $y = \arcsin x$ cannot be defined as a function over any interval that is greater than the interval defined as $-\pi/2 \leq y \leq \pi/2$.

135. $y = \sin \theta$ is not a function because $\sin 30° = \sin 150°$.

136. Because $\tan 3\pi/4 = -1$, $\arctan(-1) = 3\pi/4$.

In Exercises 137–140, match the function $y = a \sin bx$ with its graph. Base your selection solely on your interpretation of the constants a and b. Explain your reasoning. [The graphs are labeled (a), (b), (c), and (d).]

(a)

(b)

(c)

(d)
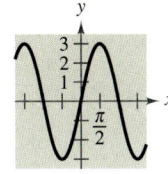

137. $y = 3 \sin x$

138. $y = -3 \sin x$

139. $y = 2 \sin \pi x$

140. $y = 2 \sin \dfrac{x}{2}$

141. Describe the behavior of $f(\theta) = \sec \theta$ at the zeros of $g(\theta) = \cos \theta$. Explain.

142. The function f is periodic, with period c. Therefore, $f(t + c) = f(t)$. Are the following equal? Explain.
(a) $f(t - 2c) \stackrel{?}{=} f(t)$
(b) $f(t + \tfrac{1}{2}c) \stackrel{?}{=} f(\tfrac{1}{2}t)$
(c) $f(\tfrac{1}{2}(t + c)) \stackrel{?}{=} f(\tfrac{1}{2}t)$

143. When graphing the sine and cosine functions, determining the amplitude is part of the analysis. Why is this not true for the other four trigonometric functions?

144. *Exploration* The base of the triangle in the figure is also the radius of a circular arc.

(a) Find the area A of the shaded region as a function of θ for $0 < \theta < \pi/2$.

(b) Use a graphing utility to graph the area function over the given domain. Interpret the graph in the context of the problem.

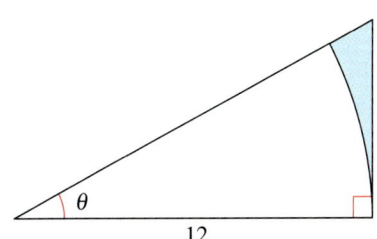

145. *Exploration* In calculus it can be shown that the arcsine and arctangent functions can be approximated by the polynomials

$$\arcsin x \approx x + \frac{x^3}{6} + \frac{3x^5}{40} + \frac{5x^7}{112}$$

$$\arctan x \approx x - \frac{x^3}{3} + \frac{x^5}{5} - \frac{x^7}{7}$$

where x is in radians.

(a) Use a graphing utility to graph the arcsine function and its polynomial approximation in the same viewing window. How do the graphs compare?

(b) Use a graphing utility to graph the arctangent function and its polynomial approximation in the same viewing window. How do the graphs compare?

(c) Study the pattern in the polynomial approximation of the arctangent function and guess the next term. Then repeat part (b). How did the accuracy of the approximation change when additional terms were added?

146. Describe a real-life application that can be represented by a simple harmonic motion model and is different from any that you've seen in this chapter. Explain which function you would use to model your application, and why. Explain how you would determine the amplitude, period, and frequency of the model for your application.

Chapter Project ▶ Analyzing a Graph

Graphs of functions that are combinations of algebraic functions and trigonometric functions can be difficult to sketch by hand. For such graphs, a graphing utility is helpful.

Example ▶ Sketching the Graph of a Function

Since 1974, the Mauna Loa Climate Observatory in Hawaii has been collecting data on the carbon dioxide level of the earth's atmosphere. A model that closely represents the data is

$$y = 323 + 1.5t + 0.001t^2 + 2.5 \sin 2\pi t$$

where y represents the monthly average of carbon dioxide concentration (in parts per million) and $t = 4$ represents January 1974, $t = 5$ represents January 1975, and so on. Sketch the graph of this function and explain the oscillations in the graph. *(Source: National Oceanic and Atmospheric Administration, Climate Monitoring and Diagnostic Laboratory, Carbon Cycle-Greenhouse Gases)*

Solution

The graph of the function is shown below. From the graph, you can see that the carbon dioxide level fluctuates each year. The low level each year, which occurs toward the end of the summer in the northern hemisphere, is caused by the intake of carbon dioxide in growing plants.

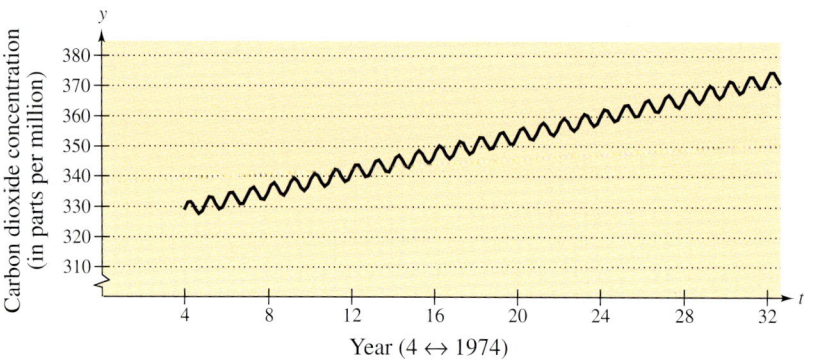

Chapter Project Investigations

1. Sketch the graph of the model given in the example for $27 \leq t \leq 29$. Between January 1997 and January 1999, what were the highest and lowest levels of carbon dioxide? When did each occur?

In Questions 2–4, use a graphing utility to graph the function. Choose a viewing window that you think produces a good representation of the important features of the graph.

2. $y = x^2 + \sin x$
3. $y = x^2 \sin x$
4. $y = \dfrac{\sin x}{x}$

Chapter Test

Take this test as you would take a test in class. After you are done, check your work against the answers given in the back of the book.

1. Consider the angle of magnitude $5\pi/4$ radians.
   (a) Sketch the angle in standard position.
   (b) Determine two coterminal angles (one positive and one negative).
   (c) Convert the angle to degree measure.

2. A truck is moving at a rate of 90 kilometers per hour, and the diameter of its wheels is 1 meter. Find the angular speed of the wheels in radians per minute.

3. Find the exact values of the six trigonometric functions of the angle θ shown in the figure.

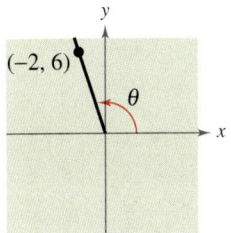

FIGURE FOR 3

4. Given that $\tan \theta = \frac{3}{2}$, find the other five trigonometric functions of θ.

5. Determine the reference angle θ' of the angle $\theta = 290°$ and sketch θ and θ' in standard position.

6. Determine the quadrant in which θ lies if $\sec \theta < 0$ and $\tan \theta > 0$.

7. Find two values of θ in degrees $(0 \le \theta < 360°)$ if $\cos \theta = -\sqrt{3}/2$. (Do not use a calculator.)

8. Use a calculator to approximate two values of θ in radians $(0 \le \theta < 2\pi)$ if $\csc \theta = 1.030$. Round the result to two decimal places.

In Exercises 9 and 10, find the remaining five trigonometric functions of θ satisfying the conditions.

9. $\cos \theta = \frac{3}{5}$, $\tan \theta < 0$ 10. $\sec \theta = -\frac{17}{8}$, $\sin \theta > 0$

In Exercises 11 and 12, graph the function through two full periods without the aid of a graphing utility.

11. $g(x) = -2 \sin\left(x - \dfrac{\pi}{4}\right)$ 12. $f(\alpha) = \dfrac{1}{2} \tan 2\alpha$

In Exercises 13 and 14, use a graphing utility to graph the function. If the function is periodic, find its period.

13. $y = \sin 2\pi x + 2 \cos \pi x$ 14. $y = 6t \cos(0.25t)$, $0 \le t \le 32$

15. Find a, b, and c for the function $f(x) = a \sin(bx + c)$ such that the graph of f matches the figure.

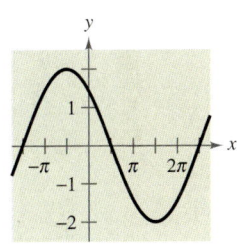

FIGURE FOR 15

16. Find the exact value of $\tan\left(\arccos \frac{2}{3}\right)$ without the aid of a calculator.

17. Graph the function $f(x) = 2 \arcsin\left(\frac{1}{2}x\right)$.

18. A plane is 80 miles south and 95 miles east of an airport. What bearing should be taken to fly directly to the airport?

19. Write the equation for the simple harmonic motion of a ball on a spring that starts at its lowest point of 6 inches below equilibrium, bounces to its maximum height of 6 inches above equilibrium, and returns to its lowest point in a total of 2 seconds.

2.1 ▶ **Using Fundamental Identities**
2.2 ▶ **Verifying Trigonometric Identities**
2.3 ▶ **Solving Trigonometric Equations**
2.4 ▶ **Sum and Difference Formulas**
2.5 ▶ **Multiple-Angle and Product-to-Sum Formulas**

Andy Lyons/Allsport

A football thrown by a quarterback follows a parabolic path. The horizontal distance the football travels depends not only on the speed of the throw, but on the angle at which the ball is thrown.

2 Analytic Trigonometry

▶ How to Study This Chapter

The Big Picture
In this chapter you will learn the following skills and concepts.

- ▶ How to use fundamental trigonometric identities to evaluate trigonometric functions and simplify trigonometric expressions
- ▶ How to verify trigonometric identities
- ▶ How to use standard algebraic techniques and inverse trigonometric functions to solve trigonometric equations
- ▶ How to use sum and difference formulas, multiple-angle formulas, power-reducing formulas, half-angle formulas, and product-to-sum formulas to rewrite and evaluate trigonometric functions

Important Vocabulary
As you encounter each new vocabulary term in this chapter, add the term and its definition to your notebook glossary.

Sum and difference formulas (p. 236)
Reduction formulas (p. 238)
Double-angle formulas (p. 243)
Power-reducing formulas (p. 245)
Half-angle formulas (p. 246)
Product-to-sum formulas (p. 247)
Sum-to-product formulas (p. 248)

Study Tools
- Learning objectives at the beginning of each section
- Chapter Summary (p. 254)
- Review Exercises (pp. 255–257)
- Chapter Test (p. 259)

Additional Resources
- Study and Solutions Guide
- Interactive Trigonometry
- Videotapes for Chapter 2
- Trigonometry Website
- Student Success Organizer

STUDY T!P
Avoid getting frustrated or spending too much time on one problem. Ask for help, take a break to clear your thoughts, sleep on it, rework the problem, or reread the section in the text.

2.1 Using Fundamental Identities

▶ What you should learn

- How to recognize and write the fundamental trigonometric identities
- How to use the fundamental trigonometric identities to evaluate trigonometric functions, simplify trigonometric expressions, and rewrite trigonometric expressions

▶ Why you should learn it

Fundamental trigonometric identities can be used to simplify trigonometric expressions. For instance, in Exercise 99 on page 217 you can use trigonometric identities to simplify an expression for the coefficient of friction.

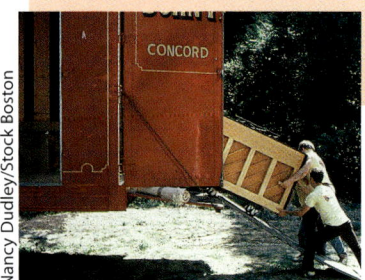

Nancy Dudley/Stock Boston

Introduction

In Chapter 1, you studied the basic definitions, properties, graphs, and applications of the individual trigonometric functions. In this chapter, you will learn how to use the fundamental identities to

1. evaluate trigonometric functions.
2. simplify trigonometric expressions.
3. develop additional trigonometric identities.
4. solve trigonometric equations.

Fundamental Trigonometric Identities

Reciprocal Identities

$$\sin u = \frac{1}{\csc u} \qquad \cos u = \frac{1}{\sec u} \qquad \tan u = \frac{1}{\cot u}$$

$$\csc u = \frac{1}{\sin u} \qquad \sec u = \frac{1}{\cos u} \qquad \cot u = \frac{1}{\tan u}$$

Quotient Identities

$$\tan u = \frac{\sin u}{\cos u} \qquad \cot u = \frac{\cos u}{\sin u}$$

Pythagorean Identities

$$\sin^2 u + \cos^2 u = 1 \qquad 1 + \tan^2 u = \sec^2 u \qquad 1 + \cot^2 u = \csc^2 u$$

Cofunction Identities

$$\sin\left(\frac{\pi}{2} - u\right) = \cos u \qquad \cos\left(\frac{\pi}{2} - u\right) = \sin u$$

$$\tan\left(\frac{\pi}{2} - u\right) = \cot u \qquad \cot\left(\frac{\pi}{2} - u\right) = \tan u$$

$$\sec\left(\frac{\pi}{2} - u\right) = \csc u \qquad \csc\left(\frac{\pi}{2} - u\right) = \sec u$$

Even/Odd Identities

$$\sin(-u) = -\sin u, \qquad \cos(-u) = \cos u, \qquad \tan(-u) = -\tan u$$

$$\csc(-u) = -\csc u, \qquad \sec(-u) = \sec u, \qquad \cot(-u) = -\cot u$$

Pythagorean identities are sometimes used in radical form such as

$$\sin u = \pm\sqrt{1 - \cos^2 u}$$

or

$$\tan u = \pm\sqrt{\sec^2 u - 1}$$

where the sign depends on the choice of u.

Using the Fundamental Identities

The *Interactive* CD-ROM and *Internet* versions of this text offer a built-in graphing calculator, which can be used in the Examples, Explorations, Technology notes, and Exercises.

One common use of trigonometric identities is to use given values of trigonometric functions to evaluate other trigonometric functions.

Example 1 ▶ Using Identities to Evaluate a Function

Use the values $\sec u = -\frac{3}{2}$ and $\tan u > 0$ to find the values of all six trigonometric functions.

Solution

Using a reciprocal identity, you have

$$\cos u = \frac{1}{\sec u} = \frac{1}{-3/2} = -\frac{2}{3}.$$

Using a Pythagorean identity, you have

$$\sin^2 u = 1 - \cos^2 u$$
$$= 1 - \left(-\frac{2}{3}\right)^2$$
$$= 1 - \frac{4}{9}$$
$$= \frac{5}{9}.$$

Because $\sec u < 0$ and $\tan u > 0$, it follows that u lies in Quadrant III. Moreover, because $\sin u$ is negative when u is in Quadrant III, you can choose the negative root and obtain $\sin u = -\sqrt{5}/3$. Now, knowing the values of the sine and cosine, you can find the values of all six trigonometric functions.

$$\sin u = -\frac{\sqrt{5}}{3} \qquad\qquad \csc u = \frac{1}{\sin u} = -\frac{3}{\sqrt{5}}$$

$$\cos u = -\frac{2}{3} \qquad\qquad \sec u = \frac{1}{\cos u} = -\frac{3}{2}$$

$$\tan u = \frac{\sin u}{\cos u} = \frac{-\sqrt{5}/3}{-2/3} = \frac{\sqrt{5}}{2} \qquad\qquad \cot u = \frac{1}{\tan u} = \frac{2}{\sqrt{5}}$$

Technology

You can use a graphing utility to check the result of Example 2. To do this, graph

$$y = \sin x \cos^2 x - \sin x$$

and

$$y = -\sin^3 x$$

in the same viewing window, as shown below. Because Example 2 shows the equivalence algebraically and the two graphs appear to coincide, you can conclude that the expressions are equivalent.

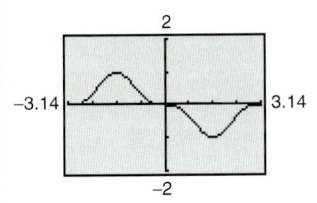

Example 2 ▶ Simplifying a Trigonometric Expression

Simplify $\sin x \cos^2 x - \sin x$.

Solution

Factor the expression and then use a fundamental identity.

$$\sin x \cos^2 x - \sin x = \sin x (\cos^2 x - 1) \qquad \text{Monomial factor}$$
$$= -\sin x (1 - \cos^2 x) \qquad \text{Factor out } -1.$$
$$= -\sin x (\sin^2 x) \qquad \text{Pythagorean identity}$$
$$= -\sin^3 x \qquad \text{Multiply.}$$

The *Interactive* CD-ROM and *Internet* versions of this text show every example with its solution; clicking on the *Try It!* button brings up similar problems. Guided Examples and Integrated Examples show step-by-step solutions to additional examples. Integrated Examples are related to several concepts in the section.

Example 3 ▶ Factoring Trigonometric Expressions

Factor each expression.

a. $\sec^2 \theta - 1$ **b.** $4\tan^2 \theta + \tan \theta - 3$

Solution

a. Here you have the difference of two squares, which factors as

$$\sec^2 \theta - 1 = (\sec \theta - 1)(\sec \theta + 1).$$

b. This expression has the polynomial form $ax^2 + bx + c$, and it factors as

$$4\tan^2 \theta + \tan \theta - 3 = (4\tan \theta - 3)(\tan \theta + 1).$$

On occasion, factoring or simplifying can best be done by first rewriting the expression in terms of just *one* trigonometric function or in terms of *sine and cosine only*. These strategies are illustrated in Examples 4 and 5, respectively.

Example 4 ▶ Factoring a Trigonometric Expression

Factor $\csc^2 x - \cot x - 3$.

Solution

You can use the identity $\csc^2 x = 1 + \cot^2 x$ to rewrite the expression in terms of the cotangent.

$$\csc^2 x - \cot x - 3 = (1 + \cot^2 x) - \cot x - 3 \quad \text{Pythagorean identity}$$

$$= \cot^2 x - \cot x - 2 \quad \text{Combine like terms.}$$

$$= (\cot x - 2)(\cot x + 1) \quad \text{Factor.}$$

Example 5 ▶ Simplifying a Trigonometric Expression

Simplify $\sin t + \cot t \cos t$.

Solution

Begin by rewriting $\cot t$ in terms of sine and cosine.

$$\sin t + \cot t \cos t = \sin t + \left(\frac{\cos t}{\sin t}\right)\cos t \quad \text{Quotient identity}$$

$$= \frac{\sin^2 t + \cos^2 t}{\sin t} \quad \text{Add fractions.}$$

$$= \frac{1}{\sin t} \quad \text{Pythagorean identity}$$

$$= \csc t \quad \text{Reciprocal identity}$$

Example 6   Adding Trigonometric Expressions

Perform the addition and simplify.

$$\frac{\sin \theta}{1 + \cos \theta} + \frac{\cos \theta}{\sin \theta}$$

Solution

$$\frac{\sin \theta}{1 + \cos \theta} + \frac{\cos \theta}{\sin \theta} = \frac{(\sin \theta)(\sin \theta) + (\cos \theta)(1 + \cos \theta)}{(1 + \cos \theta)(\sin \theta)}$$

$$= \frac{\sin^2 \theta + \cos^2 \theta + \cos \theta}{(1 + \cos \theta)(\sin \theta)} \qquad \text{Multiply.}$$

$$= \frac{1 + \cos \theta}{(1 + \cos \theta)(\sin \theta)} \qquad \text{Pythagorean identity}$$

$$= \frac{1}{\sin \theta} \qquad \text{Divide out common factor.}$$

$$= \csc \theta \qquad \text{Reciprocal identity}$$

The last two examples in this section involve techniques for rewriting expressions in forms that are used in calculus.

Example 7   Rewriting a Trigonometric Expression

Rewrite $\dfrac{1}{1 + \sin x}$ so that it is *not* in fractional form.

Solution

From the Pythagorean identity $\cos^2 x = 1 - \sin^2 x = (1 - \sin x)(1 + \sin x)$, you can see that by multiplying both the numerator and the denominator by $(1 - \sin x)$ you produce a monomial denominator.

$$\frac{1}{1 + \sin x} = \frac{1}{1 + \sin x} \cdot \frac{1 - \sin x}{1 - \sin x} \qquad \text{Multiply numerator and denominator by } (1 - \sin x).$$

$$= \frac{1 - \sin x}{1 - \sin^2 x} \qquad \text{Multiply.}$$

$$= \frac{1 - \sin x}{\cos^2 x} \qquad \text{Pythagorean identity}$$

$$= \frac{1}{\cos^2 x} - \frac{\sin x}{\cos^2 x} \qquad \text{Separate fractions.}$$

$$= \frac{1}{\cos^2 x} - \frac{\sin x}{\cos x} \cdot \frac{1}{\cos x} \qquad \text{Separate fractions.}$$

$$= \sec^2 x - \tan x \sec x \qquad \text{Identities}$$

Example 8 ▶ Trigonometric Substitution

Use the substitution $x = 2\tan\theta$, $0 < \theta < \pi/2$, to express

$$\sqrt{4 + x^2}$$

as a trigonometric function of θ.

Solution

Begin by letting $x = 2\tan\theta$. Then, you can obtain

$$\begin{aligned}
\sqrt{4 + x^2} &= \sqrt{4 + (2\tan\theta)^2} && \text{Substitute } 2\tan\theta \text{ for } x. \\
&= \sqrt{4 + 4\tan^2\theta} && \text{Rule of exponents} \\
&= \sqrt{4(1 + \tan^2\theta)} && \text{Factor.} \\
&= \sqrt{4\sec^2\theta} && \text{Pythagorean identity} \\
&= 2\sec\theta. && \sec\theta > 0 \text{ for } 0 < \theta < \pi/2
\end{aligned}$$

Figure 2.1 shows the right triangle illustration of the trigonometric substitution in Example 8. For $0 < \theta < \pi/2$, you have

$$\text{opp} = x, \quad \text{adj} = 2, \quad \text{and} \quad \text{hyp} = \sqrt{4 + x^2}.$$

With these expressions, you can write the following.

$$\sec\theta = \frac{\sqrt{4 + x^2}}{2} \quad \Longrightarrow \quad 2\sec\theta = \sqrt{4 + x^2}$$

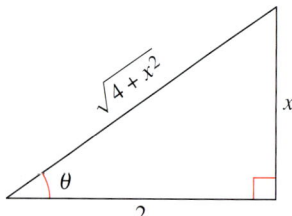

FIGURE 2.1

Writing ABOUT MATHEMATICS

Remembering Trigonometric Identities Most people find the Pythagorean identity involving sine and cosine to be fairly easy to remember: $\sin^2 u + \cos^2 u = 1$. The one involving tangent and secant, however, tends to give some people trouble. They can't remember if the identity is

$$1 + \tan^2 u \stackrel{?}{=} \sec^2 u \quad \text{or} \quad 1 + \sec^2 u \stackrel{?}{=} \tan^2 u.$$

Which of these two is the correct Pythagorean identity involving tangent and secant? Discuss how to remember (or derive) this identity. Can you think of easy ways to remember other fundamental trigonometric identities?

2.1 Exercises

In Exercises 1–14, use the given values to evaluate (if possible) the remaining trigonometric functions.

1. $\sin x = \dfrac{\sqrt{3}}{2}, \quad \cos x = -\dfrac{1}{2}$
2. $\tan x = \dfrac{\sqrt{3}}{3}, \quad \cos x = -\dfrac{\sqrt{3}}{2}$
3. $\sec \theta = \sqrt{2}, \quad \sin \theta = -\dfrac{\sqrt{2}}{2}$
4. $\csc \theta = \tfrac{5}{3}, \quad \tan \theta = \tfrac{3}{4}$
5. $\tan x = \tfrac{5}{12}, \quad \sec x = -\tfrac{13}{12}$
6. $\cot \phi = -3, \quad \sin \phi = \dfrac{\sqrt{10}}{10}$
7. $\sec \phi = \dfrac{3}{2}, \quad \csc \phi = -\dfrac{3\sqrt{5}}{5}$
8. $\cos\left(\dfrac{\pi}{2} - x\right) = \dfrac{3}{5}, \quad \cos x = \dfrac{4}{5}$
9. $\sin(-x) = -\dfrac{1}{3}, \quad \tan x = -\dfrac{\sqrt{2}}{4}$
10. $\sec x = 4, \quad \sin x > 0$
11. $\tan \theta = 2, \quad \sin \theta < 0$
12. $\csc \theta = -5, \quad \cos \theta < 0$
13. $\sin \theta = -1, \quad \cot \theta = 0$
14. $\tan \theta$ is undefined, $\quad \sin \theta > 0$

In Exercises 15–20, match the trigonometric expression with one of the following.

(a) $\sec x$ (b) -1 (c) $\cot x$
(d) 1 (e) $-\tan x$ (f) $\sin x$

15. $\sec x \cos x$
16. $\tan x \csc x$
17. $\cot^2 x - \csc^2 x$
18. $(1 - \cos^2 x)(\csc x)$
19. $\dfrac{\sin(-x)}{\cos(-x)}$
20. $\dfrac{\sin[(\pi/2) - x]}{\cos[(\pi/2) - x]}$

In Exercises 21–26, match the trigonometric expression with one of the following.

(a) $\csc x$ (b) $\tan x$ (c) $\sin^2 x$
(d) $\sin x \tan x$ (e) $\sec^2 x$ (f) $\sec^2 x + \tan^2 x$

21. $\sin x \sec x$
22. $\cos^2 x (\sec^2 x - 1)$
23. $\sec^4 x - \tan^4 x$
24. $\cot x \sec x$
25. $\dfrac{\sec^2 x - 1}{\sin^2 x}$
26. $\dfrac{\cos^2[(\pi/2) - x]}{\cos x}$

In Exercises 27–44, use the fundamental identities to simplify the expression. There is more than one correct form of each answer.

27. $\cot \theta \sec \theta$
28. $\cos \beta \tan \beta$
29. $\sin \phi (\csc \phi - \sin \phi)$
30. $\sec^2 x (1 - \sin^2 x)$
31. $\dfrac{\cot x}{\csc x}$
32. $\dfrac{\csc \theta}{\sec \theta}$
33. $\dfrac{1 - \sin^2 x}{\csc^2 x - 1}$
34. $\dfrac{1}{\tan^2 x + 1}$
35. $\sec \alpha \cdot \dfrac{\sin \alpha}{\tan \alpha}$
36. $\dfrac{\tan^2 \theta}{\sec^2 \theta}$
37. $\cos\left(\dfrac{\pi}{2} - x\right) \sec x$
38. $\cot\left(\dfrac{\pi}{2} - x\right) \cos x$
39. $\dfrac{\cos^2 y}{1 - \sin y}$
40. $\cos t (1 + \tan^2 t)$
41. $\sin \beta \tan \beta + \cos \beta$
42. $\csc \phi \tan \phi + \sec \phi$
43. $\cot u \sin u + \tan u \cos u$
44. $\sin \theta \sec \theta + \cos \theta \csc \theta$

In Exercises 45–56, factor the expression and use the fundamental identities to simplify. There is more than one correct form of each answer.

45. $\tan^2 x - \tan^2 x \sin^2 x$
46. $\sin^2 x \csc^2 x - \sin^2 x$
47. $\sin^2 x \sec^2 x - \sin^2 x$
48. $\cos^2 x + \cos^2 x \tan^2 x$
49. $\dfrac{\sec^2 x - 1}{\sec x - 1}$
50. $\dfrac{\cos^2 x - 4}{\cos x - 2}$
51. $\tan^4 x + 2\tan^2 x + 1$
52. $1 - 2\cos^2 x + \cos^4 x$
53. $\sin^4 x - \cos^4 x$
54. $\sec^4 x - \tan^4 x$
55. $\csc^3 x - \csc^2 x - \csc x + 1$
56. $\sec^3 x - \sec^2 x - \sec x + 1$

In Exercises 57–60, perform the multiplication and use the fundamental identities to simplify. There is more than one correct form of each answer.

57. $(\sin x + \cos x)^2$
58. $(\cot x + \csc x)(\cot x - \csc x)$
59. $(2 \csc x + 2)(2 \csc x - 2)$
60. $(3 - 3 \sin x)(3 + 3 \sin x)$

In Exercises 61–64, perform the addition or subtraction and use the fundamental identities to simplify. There is more than one correct form of each answer.

61. $\dfrac{1}{1 + \cos x} + \dfrac{1}{1 - \cos x}$
62. $\dfrac{1}{\sec x + 1} - \dfrac{1}{\sec x - 1}$
63. $\dfrac{\cos x}{1 + \sin x} + \dfrac{1 + \sin x}{\cos x}$
64. $\tan x - \dfrac{\sec^2 x}{\tan x}$

In Exercises 65–68, rewrite the expression so that it is not in fractional form. There is more than one correct form of each answer.

65. $\dfrac{\sin^2 y}{1 - \cos y}$
66. $\dfrac{5}{\tan x + \sec x}$
67. $\dfrac{3}{\sec x - \tan x}$
68. $\dfrac{\tan^2 x}{\csc x + 1}$

Numerical and Graphical Analysis In Exercises 69–72, use a graphing utility to complete the table and graph the functions. Make a conjecture about y_1 and y_2.

x	0.2	0.4	0.6	0.8	1.0	1.2	1.4
y_1							
y_2							

69. $y_1 = \cos\left(\dfrac{\pi}{2} - x\right), \quad y_2 = \sin x$
70. $y_1 = \sec x - \cos x, \quad y_2 = \sin x \tan x$
71. $y_1 = \dfrac{\cos x}{1 - \sin x}, \quad y_2 = \dfrac{1 + \sin x}{\cos x}$
72. $y_1 = \sec^4 x - \sec^2 x, \quad y_2 = \tan^2 x + \tan^4 x$

In Exercises 73–76, use a graphing utility to determine which of the six trigonometric functions is equal to the expression. Verify your answer algebraically.

73. $\cos x \cot x + \sin x$
74. $\sec x \csc x - \tan x$
75. $\dfrac{1}{\sin x}\left(\dfrac{1}{\cos x} - \cos x\right)$
76. $\dfrac{1}{2}\left(\dfrac{1 + \sin \theta}{\cos \theta} + \dfrac{\cos \theta}{1 + \sin \theta}\right)$

In Exercises 77–82, use the trigonometric substitution to write the algebraic expression as a trigonometric function of θ, where $0 < \theta < \pi/2$.

77. $\sqrt{9 - x^2}, \quad x = 3 \cos \theta$
78. $\sqrt{64 - 16x^2}, \quad x = 2 \cos \theta$
79. $\sqrt{x^2 - 9}, \quad x = 3 \sec \theta$
80. $\sqrt{x^2 - 4}, \quad x = 2 \sec \theta$
81. $\sqrt{x^2 + 25}, \quad x = 5 \tan \theta$
82. $\sqrt{x^2 + 100}, \quad x = 10 \tan \theta$

In Exercises 83–86, use the trigonometric substitution to write the algebraic equation as a trigonometric function of θ, where $-\pi/2 < \theta < \pi/2$. Then find $\sin \theta$ and $\cos \theta$.

83. $3 = \sqrt{9 - x^2}, \quad x = 3 \sin \theta$
84. $3 = \sqrt{36 - x^2}, \quad x = 6 \sin \theta$
85. $2\sqrt{2} = \sqrt{16 - 4x^2}, \quad x = 2 \cos \theta$
86. $-5\sqrt{3} = \sqrt{100 - x^2}, \quad x = 10 \cos \theta$

In Exercises 87–90, use a graphing utility to solve the equation for θ, where $0 \leq \theta < 2\pi$.

87. $\sin \theta = \sqrt{1 - \cos^2 \theta}$
88. $\cos \theta = -\sqrt{1 - \sin^2 \theta}$
89. $\sec \theta = \sqrt{1 + \tan^2 \theta}$
90. $\csc \theta = \sqrt{1 + \cot^2 \theta}$

In Exercises 91–98, use a calculator to demonstrate the identity for the given values of θ.

91. $\csc^2 \theta - \cot^2 \theta = 1$
    (a) $\theta = 132°$ (b) $\theta = \dfrac{2\pi}{7}$

92. $\tan^2 \theta + 1 = \sec^2 \theta$
    (a) $\theta = 346°$ (b) $\theta = 3.1$

93. $\cos\left(\dfrac{\pi}{2} - \theta\right) = \sin \theta$
    (a) $\theta = 80°$ (b) $\theta = 0.8$

94. $\sin(-\theta) = -\sin \theta$
    (a) $\theta = 250°$ (b) $\theta = \dfrac{1}{2}$

95. $\cot \theta = \dfrac{\cos \theta}{\sin \theta}$
    (a) $\theta = 25°$ (b) $\theta = \dfrac{\pi}{8}$

96. $1 + \cot^2 \theta = \csc^2 \theta$
    (a) $\theta = 240°$ (b) $\theta = 2.2$

97. $\tan\left(\dfrac{\pi}{2} - \theta\right) = \cot \theta$
    (a) $\theta = 5°$ (b) $\theta = \dfrac{11\pi}{12}$

98. $\cos(-\theta) = \cos \theta$
    (a) $\theta = 125°$ (b) $\theta = \dfrac{5\pi}{6}$

99. *Friction* The forces acting on an object weighing W units on an inclined plane positioned at an angle of θ with the horizontal (see figure) are modeled by $\mu W \cos \theta = W \sin \theta$, where μ is the coefficient of friction. Solve the equation for μ and simplify the result.

Synthesis

True or False? In Exercises 100 and 101, determine whether the statement is true or false. Justify your answer.

100. The even and odd trigonometric identities are helpful for determining whether the value of a trigonometric function is positive or negative.

101. A cofunction identity can be used to transform a tangent function so that it can be represented by a cosecant function.

Calculus In Exercises 102–105, fill in the blanks. (*Note:* The notation $x \to c^+$ indicates that x approaches c from the right and the notion $x \to c^-$ indicates that x approaches c from the left.)

102. As $x \to \dfrac{\pi^-}{2}$, $\sin x \to$ ▢ and $\csc x \to$ ▢.

103. As $x \to 0^+$, $\cos x \to$ ▢ and $\sec x \to$ ▢.

104. As $x \to \dfrac{\pi^-}{2}$, $\tan x \to$ ▢ and $\cot x \to$ ▢.

105. As $x \to \pi^+$, $\sin x \to$ ▢ and $\csc x \to$ ▢.

In Exercises 106–111, determine whether or not the equation is an identity, and give a reason for your answer.

106. $\cos \theta = \sqrt{1 - \sin^2 \theta}$
107. $\cot \theta = \sqrt{\csc^2 \theta + 1}$
108. $(\sin k\theta)/(\cos k\theta) = \tan \theta$, k is a constant.
109. $1/(5 \cos \theta) = 5 \sec \theta$
110. $\sin \theta \csc \theta = 1$
111. $\sin \theta \csc \phi = 1$

112. Express each of the other trigonometric functions of θ in terms of $\sin \theta$.

113. Express each of the other trigonometric functions of θ in terms of $\cos \theta$.

Review

In Exercises 114–117, perform the operation and simplify.

114. $(\sqrt{x} + 5)(\sqrt{x} - 5)$
115. $\sqrt{v}(\sqrt{20} - \sqrt{5})$
116. $(2\sqrt{z} + 3)^2$
117. $50x/(\sqrt{30} - 5)$

In Exercises 118–121, sketch the graph of the function. Include two full periods.

118. $y = -4 \sin(2x - 2\pi)$
119. $y = 3 \sec(\pi x - \pi)$
120. $y = 2 - \dfrac{1}{2} \tan 2x$
121. $y = \dfrac{1}{4} \sec 2x$

2.2 Verifying Trigonometric Identities

▶ **What you should learn**
- How to plan a strategy for verifying trigonometric identities
- How to verify trigonometric identities

▶ **Why you should learn it**

You can use trigonometric identities to rewrite trigonometric expressions. For instance, in Exercise 65 on page 224 you can use trigonometric identities to simplify the expression for a rate of change.

A computer animation of this concept appears in the *Interactive* CD-ROM and *Internet* versions of this text.

Introduction

In this section, you will study techniques for verifying trigonometric identities. In the next section, you will study techniques for solving trigonometric equations. The key to verifying identities *and* solving equations is the ability to use the fundamental identities and the rules of algebra to rewrite trigonometric expressions.

Remember that a *conditional equation* is an equation that is true for only some of the values in its domain. For example, the conditional equation

$$\sin x = 0 \qquad \text{Conditional equation}$$

is true only for $x = n\pi$, where n is an integer. When you find these values, you are *solving* the equation.

On the other hand, an equation that is true for all real values in the domain of the variable is an *identity*. For example, the familiar equation

$$\sin^2 x = 1 - \cos^2 x \qquad \text{Identity}$$

is true for all real numbers x. So, it is an identity.

Although there are similarities, verifying that a trigonometric equation is an identity is quite different from solving an equation. There is no well-defined set of rules to follow in verifying trigonometric identities, and the process is best learned by practice.

For instance, to verify that the trigonometric equation $\tan\theta \cos\theta = \sin\theta$ is an identity, begin by working with the more complicated, left side of the equation.

$$\tan\theta \cos\theta = \left(\frac{\sin\theta}{\cos\theta}\right)\cos\theta \qquad \text{Rewrite } \tan\theta \text{ as } \frac{\sin\theta}{\cos\theta}.$$

$$= \sin\theta \qquad \text{Simplify.}$$

The result shows that the left side of the equation is equal to the right side. Therefore, the identity has been verified.

Guidelines for Verifying Trigonometric Identities

1. Work with one side of the equation at a time. It is often better to work with the more complicated side first.
2. Look for opportunities to factor an expression, add fractions, square a binomial, or create a monomial denominator.
3. Look for opportunities to use the fundamental identities. Note which functions are in the final expression you want. Sines and cosines pair up well, as do secants and tangents, and cosecants and cotangents.
4. If the preceding guidelines do not help, try converting all terms to sines and cosines.
5. Always try *something*. Even paths that lead to dead ends give you insights.

Verifying Trigonometric Identities

Example 1 ▶ Verifying a Trigonometric Identity

Verify the identity $\dfrac{\sec^2 \theta - 1}{\sec^2 \theta} = \sin^2 \theta$.

Solution

Because the left side is more complicated, start with it.

$\dfrac{\sec^2 \theta - 1}{\sec^2 \theta} = \dfrac{(\tan^2 \theta + 1) - 1}{\sec^2 \theta}$ Pythagorean identity

$\phantom{\dfrac{\sec^2 \theta - 1}{\sec^2 \theta}} = \dfrac{\tan^2 \theta}{\sec^2 \theta}$ Simplify.

$\phantom{\dfrac{\sec^2 \theta - 1}{\sec^2 \theta}} = \tan^2 \theta (\cos^2 \theta)$ Reciprocal identity

$\phantom{\dfrac{\sec^2 \theta - 1}{\sec^2 \theta}} = \dfrac{\sin^2 \theta}{(\cos^2 \theta)} (\cos^2 \theta)$ Quotient identity

$\phantom{\dfrac{\sec^2 \theta - 1}{\sec^2 \theta}} = \sin^2 \theta$ Simplify.

Here is another way to verify the identity in Example 1.

$\dfrac{\sec^2 \theta - 1}{\sec^2 \theta} = \dfrac{\sec^2 \theta}{\sec^2 \theta} - \dfrac{1}{\sec^2 \theta}$ Rewrite as the difference of fractions.

$\phantom{\dfrac{\sec^2 \theta - 1}{\sec^2 \theta}} = 1 - \cos^2 \theta$ Reciprocal identity

$\phantom{\dfrac{\sec^2 \theta - 1}{\sec^2 \theta}} = \sin^2 \theta$ Pythagorean identity

As you can see, there can be more than one way to verify an identity. Your method may differ from that used by your instructor or fellow students. Here is a good chance to be creative and establish your own style, but try to be as efficient as possible.

Example 2 ▶ Combining Fractions Before Using Identities

Verify the identity $\dfrac{1}{1 - \sin \alpha} + \dfrac{1}{1 + \sin \alpha} = 2 \sec^2 \alpha$.

Solution

$\dfrac{1}{1 - \sin \alpha} + \dfrac{1}{1 + \sin \alpha} = \dfrac{1 + \sin \alpha + 1 - \sin \alpha}{(1 - \sin \alpha)(1 + \sin \alpha)}$ Add fractions.

$\phantom{\dfrac{1}{1 - \sin \alpha} + \dfrac{1}{1 + \sin \alpha}} = \dfrac{2}{1 - \sin^2 \alpha}$ Simplify.

$\phantom{\dfrac{1}{1 - \sin \alpha} + \dfrac{1}{1 + \sin \alpha}} = \dfrac{2}{\cos^2 \alpha}$ Pythagorean identity

$\phantom{\dfrac{1}{1 - \sin \alpha} + \dfrac{1}{1 + \sin \alpha}} = 2 \sec^2 \alpha$ Reciprocal identity

Example 3 ▶ Verifying a Trigonometric Identity

Verify the identity

$$(\tan^2 x + 1)(\cos^2 x - 1) = -\tan^2 x.$$

Solution

By applying identities before multiplying, you obtain the following.

$$(\tan^2 x + 1)(\cos^2 x - 1) = (\sec^2 x)(-\sin^2 x) \quad \text{Pythagorean identities}$$

$$= -\frac{\sin^2 x}{\cos^2 x} \quad \text{Reciprocal identity}$$

$$= -\left(\frac{\sin x}{\cos x}\right)^2 \quad \text{Rule of exponents}$$

$$= -\tan^2 x \quad \text{Quotient identity}$$

Example 4 ▶ Converting to Sines and Cosines

Verify the identity

$$\tan x + \cot x = \sec x \csc x.$$

Solution

In this case there appear to be no fractions to add, no products to find, and no opportunities to use the Pythagorean identities. So, try converting the left side into sines and cosines to see what happens.

$$\tan x + \cot x = \frac{\sin x}{\cos x} + \frac{\cos x}{\sin x} \quad \text{Quotient identities}$$

$$= \frac{\sin^2 x + \cos^2 x}{\cos x \sin x} \quad \text{Add fractions.}$$

$$= \frac{1}{\cos x \sin x} \quad \text{Pythagorean identity}$$

$$= \frac{1}{\cos x} \cdot \frac{1}{\sin x} \quad \text{Product of fractions}$$

$$= \sec x \csc x \quad \text{Reciprocal identities}$$

Recall from algebra that *rationalizing the denominator* using conjugates is, on occasion, a powerful simplification technique. A related form of this technique works for simplifying trigonometric expressions as well. For instance, to simplify $1/(1 - \cos x)$, multiply the numerator and the denominator by $1 + \cos x$.

$$\frac{1}{1 - \cos x} = \frac{1}{1 - \cos x}\left(\frac{1 + \cos x}{1 + \cos x}\right) = \frac{1 + \cos x}{1 - \cos^2 x} = \frac{1 + \cos x}{\sin^2 x}$$

$$= \csc^2 x (1 + \cos x)$$

This technique is demonstrated in the next example.

Example 5 Verifying Trigonometric Identities

Verify the identity $\sec y + \tan y = \dfrac{\cos y}{1 - \sin y}$.

Solution

Begin with the *right* side. Note that you can create a monomial denominator by multiplying the numerator and denominator by $(1 + \sin y)$.

$$\dfrac{\cos y}{1 - \sin y} = \dfrac{\cos y}{1 - \sin y}\left(\dfrac{1 + \sin y}{1 + \sin y}\right) \qquad \text{Multiply numerator and denominator by } (1 + \sin y).$$

$$= \dfrac{\cos y + \cos y \sin y}{1 - \sin^2 y} \qquad \text{Multiply.}$$

$$= \dfrac{\cos y + \cos y \sin y}{\cos^2 y} \qquad \text{Pythagorean identity}$$

$$= \dfrac{\cos y}{\cos^2 y} + \dfrac{\cos y \sin y}{\cos^2 y} \qquad \text{Separate fractions.}$$

$$= \dfrac{1}{\cos y} + \dfrac{\sin y}{\cos y} \qquad \text{Simplify.}$$

$$= \sec y + \tan y \qquad \text{Identities}$$

In Examples 1 through 5, you have been verifying trigonometric identities by working with one side of the equation and converting to the form given on the other side. On occasion it is practical to work with each side *separately*, to obtain one common form equivalent to both sides. This is illustrated in Example 6.

Example 6 Working with Each Side Separately

Verify the identity $\dfrac{\cot^2 \theta}{1 + \csc \theta} = \dfrac{1 - \sin \theta}{\sin \theta}$.

Solution

Working with the left side, you have

$$\dfrac{\cot^2 \theta}{1 + \csc \theta} = \dfrac{\csc^2 \theta - 1}{1 + \csc \theta} \qquad \text{Pythagorean identity}$$

$$= \dfrac{(\csc \theta - 1)(\csc \theta + 1)}{1 + \csc \theta} \qquad \text{Factor.}$$

$$= \csc \theta - 1. \qquad \text{Simplify.}$$

Now, simplifying the right side, you have

$$\dfrac{1 - \sin \theta}{\sin \theta} = \dfrac{1}{\sin \theta} - \dfrac{\sin \theta}{\sin \theta} \qquad \text{Separate fractions.}$$

$$= \csc \theta - 1. \qquad \text{Reciprocal identity}$$

The identity is verified because both sides are equal to $\csc \theta - 1$.

STUDY TIP

The technique of cross multiplication is not used when verifying trigonometric identities because you do not know that the expressions are equal. Cross multiplication is used in solving an equation, when you know that the left side equals the right side.

In Example 7, powers of trigonometric functions are rewritten as more complicated sums of products of trigonometric functions. This is a common procedure used in calculus.

Example 7 ▶ Three Examples from Calculus

Verify each identity.

a. $\tan^4 x = \tan^2 x \sec^2 x - \tan^2 x$
b. $\sin^3 x \cos^4 x = (\cos^4 x - \cos^6 x) \sin x$
c. $\csc^4 x \cot x = \csc^2 x(\cot x + \cot^3 x)$

Solution

a. $\tan^4 x = (\tan^2 x)(\tan^2 x)$ Separate factors.
$= \tan^2 x(\sec^2 x - 1)$ Pythagorean identity
$= \tan^2 x \sec^2 x - \tan^2 x$ Multiply.

b. $\sin^3 x \cos^4 x = \sin^2 x \cos^4 x \sin x$ Separate factors.
$= (1 - \cos^2 x)\cos^4 x \sin x$ Pythagorean identity
$= (\cos^4 x - \cos^6 x) \sin x$ Multiply.

c. $\csc^4 x \cot x = \csc^2 x \csc^2 x \cot x$ Separate factors.
$= \csc^2 x(1 + \cot^2 x) \cot x$ Pythagorean identity
$= \csc^2 x(\cot x + \cot^3 x)$ Multiply.

Writing ABOUT MATHEMATICS

Error Analysis Suppose you are tutoring a student in trigonometry. One of the homework problems your student encounters asks whether the following statement is an identity.

$$\tan^2 x \sin^2 x \stackrel{?}{=} \frac{5}{6} \tan^2 x$$

Your student does not attempt to verify the equivalence algebraically, but mistakenly uses only a graphical approach. Using range settings of

Xmin $= -3\pi$

Xmax $= 3\pi$

Xscl $= \pi/2$

Ymin $= -20$

Ymax $= 20$

Yscl $= 1$

your student graphs both sides of the expression on a graphing utility and concludes that the statement is an identity.

What is wrong with your student's reasoning? Explain. Discuss the limitations of verifying identities graphically.

2.2 Exercises

In Exercises 1–44, verify the identity.

1. $\sin t \csc t = 1$
2. $\sec y \cos y = 1$
3. $(1 + \sin \alpha)(1 - \sin \alpha) = \cos^2 \alpha$
4. $\cot^2 y(\sec^2 y - 1) = 1$
5. $\cos^2 \beta - \sin^2 \beta = 1 - 2\sin^2 \beta$
6. $\cos^2 \beta - \sin^2 \beta = 2\cos^2 \beta - 1$
7. $\tan^2 \theta + 4 = \sec^2 \theta + 3$
8. $2 - \sec^2 z = 1 - \tan^2 z$
9. $\sin^2 \alpha - \sin^4 \alpha = \cos^2 \alpha - \cos^4 \alpha$
10. $\cos x + \sin x \tan x = \sec x$
11. $\dfrac{\csc^2 \theta}{\cot \theta} = \csc \theta \sec \theta$
12. $\dfrac{\cot^3 t}{\csc t} = \cos t(\csc^2 t - 1)$
13. $\dfrac{\cot^2 t}{\csc t} = \csc t - \sin t$
14. $\dfrac{1}{\tan \beta} + \tan \beta = \dfrac{\sec^2 \beta}{\tan \beta}$
15. $\sin^{1/2} x \cos x - \sin^{5/2} x \cos x = \cos^3 x \sqrt{\sin x}$
16. $\sec^6 x(\sec x \tan x) - \sec^4 x(\sec x \tan x) = \sec^5 x \tan^3 x$
17. $\dfrac{1}{\sec x \tan x} = \csc x - \sin x$
18. $\dfrac{\sec \theta - 1}{1 - \cos \theta} = \sec \theta$
19. $\cot \alpha + \tan \alpha = \csc \alpha \sec \alpha$
20. $\sec x - \cos x = \sin x \tan x$
21. $\sin x \cos x + \sin^3 x \sec x = \tan x$
22. $\dfrac{\sec x + \tan x}{\sec x - \tan x} = (\sec x + \tan x)^2$
23. $\dfrac{1}{\tan x} + \dfrac{1}{\cot x} = \tan x + \cot x$
24. $\dfrac{1}{\sin x} - \dfrac{1}{\csc x} = \csc x - \sin x$
25. $\dfrac{\cos \theta \cot \theta}{1 - \sin \theta} - 1 = \csc \theta$
26. $\dfrac{1 + \sin \theta}{\cos \theta} + \dfrac{\cos \theta}{1 + \sin \theta} = 2\sec \theta$
27. $\dfrac{1}{\sin x + 1} + \dfrac{1}{\csc x + 1} = 1$
28. $\cos x - \dfrac{\cos x}{1 - \tan x} = \dfrac{\sin x \cos x}{\sin x - \cos x}$
29. $\tan\left(\dfrac{\pi}{2} - \theta\right) \tan \theta = 1$
30. $\dfrac{\cos[(\pi/2) - x]}{\sin[(\pi/2) - x]} = \tan x$
31. $\dfrac{\csc(-x)}{\sec(-x)} = -\cot x$
32. $(1 + \sin y)[1 + \sin(-y)] = \cos^2 y$
33. $\dfrac{\cos(-\theta)}{1 + \sin(-\theta)} = \sec \theta + \tan \theta$
34. $\dfrac{\csc(-\theta) + 1}{\cos(-\theta) + \cot(-\theta)} = \sec \theta$
35. $\dfrac{\sin x \cos y + \cos x \sin y}{\cos x \cos y - \sin x \sin y} = \dfrac{\tan x + \tan y}{1 - \tan x \tan y}$
36. $\dfrac{\tan x + \tan y}{1 - \tan x \tan y} = \dfrac{\cot x + \cot y}{\cot x \cot y - 1}$
37. $\dfrac{\tan x + \cot y}{\tan x \cot y} = \tan y + \cot x$
38. $\dfrac{\cos x - \cos y}{\sin x + \sin y} + \dfrac{\sin x - \sin y}{\cos x + \cos y} = 0$
39. $\sqrt{\dfrac{1 + \sin \theta}{1 - \sin \theta}} = \dfrac{1 + \sin \theta}{|\cos \theta|}$
40. $\sqrt{\dfrac{1 - \cos \theta}{1 + \cos \theta}} = \dfrac{1 - \cos \theta}{|\sin \theta|}$
41. $\cos^2 \beta + \cos^2\left(\dfrac{\pi}{2} - \beta\right) = 1$
42. $\sec^2 y - \cot^2\left(\dfrac{\pi}{2} - y\right) = 1$
43. $\sin t \csc\left(\dfrac{\pi}{2} - t\right) = \tan t$
44. $\sec^2\left(\dfrac{\pi}{2} - x\right) - 1 = \cot^2 x$

In Exercises 45–56, use a graphing utility to verify the identity graphically, then confirm it algebraically.

45. $2\sec^2 x - 2\sec^2 x \sin^2 x - \sin^2 x - \cos^2 x = 1$
46. $\csc x(\csc x - \sin x) + \dfrac{\sin x - \cos x}{\sin x} + \cot x = \csc^2 x$
47. $2 + \cos^2 x - 3\cos^4 x = \sin^2 x(2 + 3\cos^2 x)$
48. $4\tan^4 x + \tan^2 x - 3 = \sec^2 x(4\tan^2 x - 3)$
49. $\csc^4 x - 2\csc^2 x + 1 = \cot^4 x$
50. $(\sin^4 \beta - 2\sin^2 \beta + 1)\cos \beta = \cos^5 \beta$
51. $\sec^4 \theta - \tan^4 \theta = 1 + 2\tan^2 \theta$
52. $\csc^4 \theta - \cot^4 \theta = 2\csc^2 \theta - 1$
53. $\dfrac{\cos x}{1 + \sin x} = \dfrac{1 - \sin x}{\cos x}$
54. $\dfrac{\cot \alpha}{\csc \alpha - 1} = \dfrac{\csc \alpha + 1}{\cot \alpha}$
55. $\dfrac{\tan^3 \alpha - 1}{\tan \alpha - 1} = \tan^2 \alpha + \tan \alpha + 1$
56. $\dfrac{\sin^3 \beta + \cos^3 \beta}{\sin \beta + \cos \beta} = 1 - \sin \beta \cos \beta$

In Exercises 57–60, use the cofunction identities to evaluate the expression without the aid of a calculator.

57. $\sin^2 25° + \sin^2 65°$
58. $\cos^2 55° + \cos^2 35°$
59. $\cos^2 20° + \cos^2 52° + \cos^2 38° + \cos^2 70°$
60. $\sin^2 12° + \sin^2 40° + \sin^2 50° + \sin^2 78°$

61. *Rate of Change*   The rate of change of the function
$$f(x) = \sin x + \csc x$$
with respect to change in the variable x is given by the expression
$$\cos x - \csc x \cot x.$$
Show that the expression for the rate of change can also be
$$-\cos x \cot^2 x.$$

Synthesis

True or False?   **In Exercises 62 and 63, determine whether the statement is true or false. Justify your answer.**

62. The equation $\sin^2 \theta + \cos^2 \theta = 1 + \tan^2 \theta$ is an identity, because $\sin^2(0) + \cos^2(0) = 1$ and $1 + \tan^2(0) = 1$.
63. The equation $1 + \tan^2 \theta = 1 + \cot^2 \theta$ is *not* an identity, because it is true that $1 + \tan^2(\pi/6) = 1\tfrac{1}{3}$, and $1 + \cot^2(\pi/6) = 4$.

Think About It   **In Exercises 64–67, explain why the equation is *not* an identity and find one value of the variable for which the equation is not true.**

64. $\sin \theta = \sqrt{1 - \cos^2 \theta}$
65. $\tan \theta = \sqrt{\sec^2 \theta - 1}$
66. $\sqrt{\tan^2 x} = \tan x$
67. $\sqrt{\sin^2 x + \cos^2 x} = \sin x + \cos x$

68. Verify that for all integers n, $\cos\left[\dfrac{(2n + 1)\pi}{2}\right] = 0$.
69. Verify that for all integers n, $\sin\left[\dfrac{(12n + 1)\pi}{6}\right] = \dfrac{1}{2}$.

Review

In Exercises 70–75, solve the equation by factoring.

70. $x^2 - 5x - 14 = 0$
71. $x^2 + 5x - 36 = 0$
72. $x^2 - 19x + 88 = 0$
73. $2x^2 - 7x - 15 = 0$
74. $36x^2 - 16 = 0$
75. $9x^2 - 1 = 0$

In Exercises 76–81, evaluate the expression without using a calculator. Give your answer in both radians and degrees.

76. $\cos^{-1}\dfrac{\sqrt{2}}{2}$
77. $\tan^{-1}\sqrt{3}$
78. $\sin^{-1}\left(-\dfrac{\sqrt{3}}{2}\right)$
79. $\sin^{-1}\dfrac{1}{2}$
80. $\tan^{-1}(-1)$
81. $\cos^{-1}\left(-\tfrac{1}{2}\right)$

2.3 Solving Trigonometric Equations

What you should learn
- How to use standard algebraic techniques to solve trigonometric equations
- How to solve trigonometric equations of quadratic type
- How to solve trigonometric equations involving multiple angles
- How to use inverse trigonometric functions to solve trigonometric equations

Why you should learn it
You can use trigonometric equations to solve a variety of real-life problems. For instance, in Exercise 76 on page 235 you solve a trigonometric equation to help answer questions about the unemployment rate in the United States.

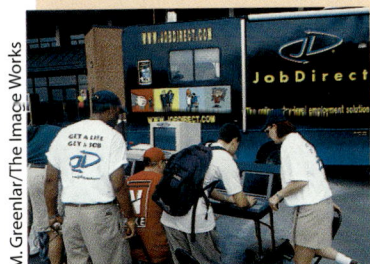

Introduction

To solve a trigonometric equation, use standard algebraic techniques such as collecting like terms and factoring. Your preliminary goal in solving trigonometric equations is to isolate the trigonometric function involved in the equation.

Example 1 ▶ Solving a Trigonometric Equation

$2 \sin x - 1 = 0$ Write original equation.

$2 \sin x = 1$ Add 1 to each side.

$\sin x = \dfrac{1}{2}$ Divide each side by 2.

To solve for x, note in Figure 2.2 that the equation $\sin x = \frac{1}{2}$ has solutions $x = \pi/6$ and $x = 5\pi/6$ in the interval $[0, 2\pi)$. Moreover, because $\sin x$ has a period of 2π, there are infinitely many other solutions, which can be written as

$x = \pi/6 + 2n\pi$ and $x = 5\pi/6 + 2n\pi$ General solution

where n is an integer, as shown in Figure 2.2.

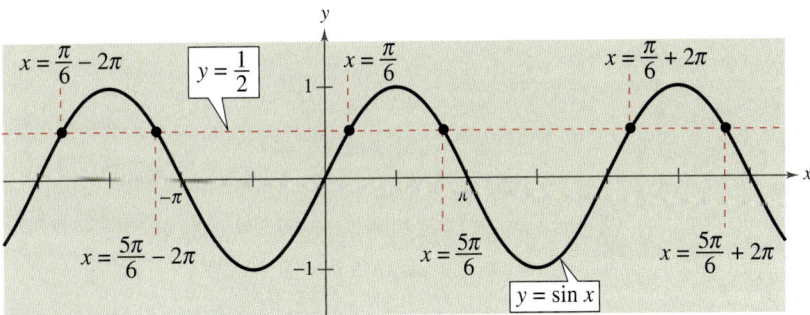

FIGURE 2.2

Another way to see that the equation $\sin x = \frac{1}{2}$ has infinitely many solutions is indicated in Figure 2.3. Any angles that are coterminal with $\pi/6$ or $5\pi/6$ will also be solutions of the equation.

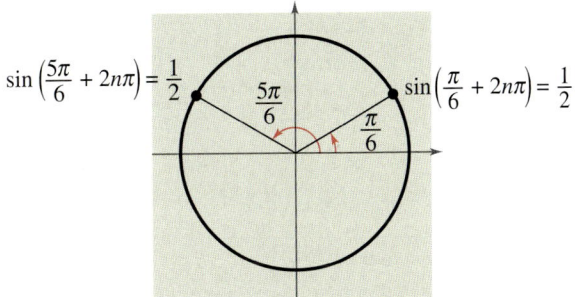

FIGURE 2.3

Example 2 ▶ Collecting Like Terms

Solve $\sin x + \sqrt{2} = -\sin x$.

Solution

Begin by rewriting the equation so that $\sin x$ is isolated on one side of the equation.

$$\sin x + \sqrt{2} = -\sin x \quad \text{Write original equation.}$$

$$\sin x + \sin x + \sqrt{2} = 0 \quad \text{Add } \sin x \text{ to each side.}$$

$$\sin x + \sin x = -\sqrt{2} \quad \text{Subtract } \sqrt{2} \text{ from each side.}$$

$$2 \sin x = -\sqrt{2} \quad \text{Combine like terms.}$$

$$\sin x = -\frac{\sqrt{2}}{2} \quad \text{Divide each side by 2.}$$

Because $\sin x$ has a period of 2π, first find all solutions in the interval $[0, 2\pi)$. These are $x = 5\pi/4$ and $x = 7\pi/4$. Finally, add $2n\pi$ to each of these solutions to get the general form

$$x = \frac{5\pi}{4} + 2n\pi \quad \text{and} \quad x = \frac{7\pi}{4} + 2n\pi \quad \text{General solution}$$

where n is an integer.

Example 3 ▶ Extracting Square Roots

Solve $3 \tan^2 x - 1 = 0$.

Solution

Begin by rewriting the equation so that $\tan x$ is isolated on one side of the equation.

$$3 \tan^2 x - 1 = 0 \quad \text{Write original equation.}$$

$$3 \tan^2 x = 1 \quad \text{Add 1 to each side.}$$

$$\tan^2 x = \frac{1}{3} \quad \text{Divide each side by 3.}$$

$$\tan x = \pm \frac{1}{\sqrt{3}} \quad \text{Extract square roots.}$$

Because $\tan x$ has a period of π, first find all solutions in the interval $[0, \pi)$. These are $x = \pi/6$ and $x = 5\pi/6$. Finally, add $n\pi$ to each of these solutions to get the general form

$$x = \frac{\pi}{6} + n\pi \quad \text{and} \quad x = \frac{5\pi}{6} + n\pi \quad \text{General solution}$$

where n is an integer.

Technology

The solutions in Examples 2 and 3 are obtained analytically. You can use a graphing utility to confirm the solutions graphically. For instance, to confirm the solutions found in Example 3, sketch the graph of

$$y = 3 \tan^2 x - 1$$

as shown below.

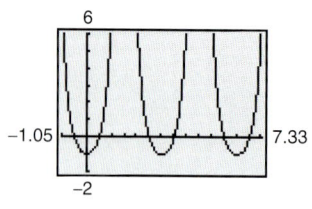

The equations in Examples 1, 2, and 3 involved only one trigonometric function. When two or more functions occur in the same equation, collect all terms on one side and try to separate the functions by factoring or by using appropriate identities. This may produce factors that yield no solutions, as illustrated in Example 4.

Example 4 ▶ Factoring

Solve $\cot x \cos^2 x = 2 \cot x$.

Solution

Begin by rewriting the equation so that all terms are collected on one side of the equation.

$$\cot x \cos^2 x = 2 \cot x \qquad \text{Write original equation.}$$

$$\cot x \cos^2 x - 2 \cot x = 0 \qquad \text{Subtract } 2 \cot x \text{ from each side.}$$

$$\cot x (\cos^2 x - 2) = 0 \qquad \text{Factor.}$$

By setting each of these factors equal to zero, you obtain

$$\cot x = 0 \quad \text{and} \quad \cos^2 x - 2 = 0$$

$$x = \frac{\pi}{2} \qquad\qquad \cos^2 x = 2$$

$$\cos x = \pm\sqrt{2}.$$

The equation $\cot x = 0$ has the solution $x = \pi/2$. No solution is obtained from $\cos x = \pm\sqrt{2}$ because $\pm\sqrt{2}$ are outside the range of the cosine function. Therefore, the general form of the solution is obtained by adding multiples of π to $x = \pi/2$, to get

$$x = \frac{\pi}{2} + n\pi \qquad \text{General solution}$$

where n is an integer. You can confirm this graphically by sketching the graph of $y = \cot x \cos^2 x - 2 \cot x$, as shown in Figure 2.4.

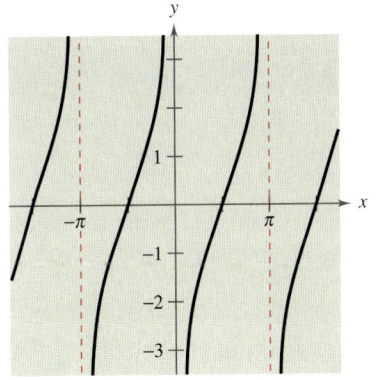

FIGURE 2.4

In Example 4, don't make the mistake of dividing each side of the equation by $\cot x$. If you do this, you lose the solutions. Can you see why?

Equations of Quadratic Type

Many trigonometric equations are of quadratic type. Here are a couple of examples.

Quadratic in $\sin x$	*Quadratic in* $\sec x$
$2 \sin^2 x - \sin x - 1 = 0$	$\sec^2 x - 3 \sec x - 2 = 0$
$2(\sin x)^2 - (\sin x) - 1 = 0$	$(\sec x)^2 - 3(\sec x) - 2 = 0$

To solve equations of this type, factor the quadratic or, if this is not possible, use the Quadratic Formula.

Example 5 ▶ Factoring an Equation of Quadratic Type

Find all solutions of $2\sin^2 x - \sin x - 1 = 0$ in the interval $[0, 2\pi)$.

Solution

Begin by treating the equation as a quadratic in $\sin x$ and factoring.

$2\sin^2 x - \sin x - 1 = 0$ Write original equation.

$(2\sin x + 1)(\sin x - 1) = 0$ Factor.

Setting each factor equal to zero, you can find the solutions in the interval $[0, 2\pi)$.

$2\sin x + 1 = 0$ and $\sin x - 1 = 0$

$\sin x = -\dfrac{1}{2}$ $\sin x = 1$

$x = \dfrac{7\pi}{6}, \dfrac{11\pi}{6}$ $x = \dfrac{\pi}{2}$

When working with an equation of quadratic type, be sure that the equation involves a *single* trigonometric function, as shown in the next example.

Example 6 ▶ Rewriting with a Single Trigonometric Function

Solve $2\sin^2 x + 3\cos x - 3 = 0$.

Solution

This equation contains both sine and cosine functions. You can rewrite the equation so that it has only cosine functions by using the identity $\sin^2 x = 1 - \cos^2 x$.

$2\sin^2 x + 3\cos x - 3 = 0$ Write original equation.

$2(1 - \cos^2 x) + 3\cos x - 3 = 0$ Pythagorean identity

$2\cos^2 x - 3\cos x + 1 = 0$ Multiply each side by -1.

$(2\cos x - 1)(\cos x - 1) = 0$ Factor.

By setting each factor equal to zero, you can find the solutions in the interval $[0, 2\pi)$.

$2\cos x - 1 = 0$ and $\cos x - 1 = 0$

$\cos x = \dfrac{1}{2}$ $\cos x = 1$

                    $x = 0$

$x = \dfrac{\pi}{3}, \dfrac{5\pi}{3}$

The general solution is therefore

$$x = 2n\pi, \quad x = \dfrac{\pi}{3} + 2n\pi, \quad x = \dfrac{5\pi}{3} + 2n\pi \qquad \text{General solution}$$

where n is an integer.

Sometimes you must square both sides of an equation to obtain a quadratic, as demonstrated in the next example. Because this procedure can introduce extraneous solutions, you should check any solutions in the original equation to see whether they are valid or extraneous.

Example 7 ▶ Squaring and Converting to Quadratic Type

Find all solutions of $\cos x + 1 = \sin x$ in the interval $[0, 2\pi)$.

Solution

It is not clear how to rewrite this equation in terms of a single trigonometric function. See what happens when you square both sides of the equation.

$\cos x + 1 = \sin x$	Write original equation.
$\cos^2 x + 2\cos x + 1 = \sin^2 x$	Square each side.
$\cos^2 x + 2\cos x + 1 = 1 - \cos^2 x$	Pythagorean identity
$\cos^2 x + \cos^2 x + 2\cos x + 1 - 1 = 0$	Rewrite equation.
$2\cos^2 x + 2\cos x = 0$	Combine like terms.
$2\cos x(\cos x + 1) = 0$	Factor.

Setting each factor equal to zero produces

$$2\cos x = 0 \quad \text{and} \quad \cos x + 1 = 0$$
$$\cos x = 0 \quad\quad\quad\quad\quad \cos x = -1$$
$$x = \frac{\pi}{2}, \frac{3\pi}{2} \quad\quad\quad\quad x = \pi.$$

Because you squared the original equation, check for extraneous solutions.

Check for $x = \pi/2$

$\cos \dfrac{\pi}{2} + 1 \stackrel{?}{=} \sin \dfrac{\pi}{2}$ Substitute $\pi/2$ for x.

$0 + 1 = 1$ Solution checks. ✓

Check for $x = 3\pi/2$

$\cos \dfrac{3\pi}{2} + 1 \stackrel{?}{=} \sin \dfrac{3\pi}{2}$ Substitute $3\pi/2$ for x.

$0 + 1 = -1$ Solution does not check. ✗

Check for $x = \pi$

$\cos \pi + 1 \stackrel{?}{=} \sin \pi$ Substitute π for x.

$-1 + 1 = 0$ Solution checks. ✓

Of the three possible solutions, $x = 3\pi/2$ is extraneous. So, in the interval $[0, 2\pi)$, the only two solutions are $x = \pi/2$ and $x = \pi$.

◀ Exploration ▶

Use a graphing utility to confirm the solutions found in Example 7 in two different ways. Do both methods produce the same x-values? Which method do you prefer? Why?

1. Graph both sides of the equation and find the x-coordinates of the points at which the graphs intersect.

   Left side: $y = \cos x + 1$

   Right side: $y = \sin x$

2. Graph the equation

   $y = \cos x + 1 - \sin x$

   and find the x-intercepts of the graph.

Functions Involving Multiple Angles

The next two examples involve trigonometric functions of multiple angles of the forms sin ku or cos ku.

Example 8 ▶ Functions of Multiple Angles

Find all solutions of $2 \cos 3t - 1 = 0$.

Solution

$$2 \cos 3t - 1 = 0 \quad \text{Write original equation.}$$

$$2 \cos 3t = 1 \quad \text{Add 1 to each side.}$$

$$\cos 3t = \frac{1}{2} \quad \text{Divide each side by 2.}$$

In the interval $[0, 2\pi)$, you know that $3t = \pi/3$ and $3t = 5\pi/3$ are the only solutions so that, in general, you have

$$3t = \frac{\pi}{3} + 2n\pi \quad \text{and} \quad 3t = \frac{5\pi}{3} + 2n\pi.$$

Dividing these results by 3, you obtain the general solution

$$t = \frac{\pi}{9} + \frac{2n\pi}{3} \quad \text{and} \quad t = \frac{5\pi}{9} + \frac{2n\pi}{3} \quad \text{General solution}$$

where n is an integer.

Example 9 ▶ Functions of Multiple Angles

Find all solutions of $3 \tan(x/2) + 3 = 0$.

Solution

$$3 \tan \frac{x}{2} + 3 = 0 \quad \text{Write original equation.}$$

$$3 \tan \frac{x}{2} = -3 \quad \text{Subtract 3 from each side.}$$

$$\tan \frac{x}{2} = -1 \quad \text{Divide each side by 3.}$$

In the interval $[0, \pi)$, you know that $x/2 = 3\pi/4$ is the only solution so that, in general, you have

$$\frac{x}{2} = \frac{3\pi}{4} + n\pi.$$

Multiplying this result by 2, you obtain the general solution

$$x = \frac{3\pi}{2} + 2n\pi \quad \text{General solution}$$

where n is an integer.

Using Inverse Functions

In the next example, you will see how inverse trigonometric functions are used to solve an equation.

Example 10 ▶ Using Inverse Functions

Find all solutions of $\sec^2 x - 2 \tan x = 4$.

Solution

$$\sec^2 x - 2 \tan x = 4 \quad \text{Write original equation.}$$
$$1 + \tan^2 x - 2 \tan x - 4 = 0 \quad \text{Pythagorean identity}$$
$$\tan^2 x - 2 \tan x - 3 = 0 \quad \text{Combine like terms.}$$
$$(\tan x - 3)(\tan x + 1) = 0 \quad \text{Factor.}$$

Setting each factor equal to zero, you obtain two solutions in the interval $(-\pi/2, \pi/2)$. [Recall that the range of the inverse tangent function is $(-\pi/2, \pi/2)$.]

$$\tan x - 3 = 0 \quad \text{and} \quad \tan x + 1 = 0$$
$$\tan x = 3 \quad \quad \tan x = -1$$
$$x = \arctan 3 \quad \quad x = -\frac{\pi}{4}$$

Finally, by adding multiples of π, you obtain the general solution

$$x = \arctan 3 + n\pi \quad \text{and} \quad x = -\frac{\pi}{4} + n\pi \quad \text{General solution}$$

where n is an integer.

Writing ABOUT MATHEMATICS

Equations with No Solutions One of the following equations has solutions and the other two don't. Which two equations do not have solutions?

a. $\sin^2 x - 5 \sin x + 6 = 0$
b. $\sin^2 x - 4 \sin x + 6 = 0$
c. $\sin^2 x - 5 \sin x - 6 = 0$

Can you find conditions involving the constants b and c that will guarantee that the equation

$$\sin^2 x + b \sin x + c = 0$$

has at least one solution on some interval of length 2π?

2.3 Exercises

In Exercises 1–6, verify that the x-values are solutions.

1. $2\cos x - 1 = 0$
   (a) $x = \dfrac{\pi}{3}$ (b) $x = \dfrac{5\pi}{3}$

2. $\sec x - 2 = 0$
   (a) $x = \dfrac{\pi}{3}$ (b) $x = \dfrac{5\pi}{3}$

3. $3\tan^2 2x - 1 = 0$
   (a) $x = \dfrac{\pi}{12}$ (b) $x = \dfrac{5\pi}{12}$

4. $2\cos^2 4x - 1 = 0$
   (a) $x = \dfrac{\pi}{16}$ (b) $x = \dfrac{3\pi}{16}$

5. $2\sin^2 x - \sin x - 1 = 0$
   (a) $x = \dfrac{\pi}{2}$ (b) $x = \dfrac{7\pi}{6}$

6. $\csc^4 x - 4\csc^2 x = 0$
   (a) $x = \dfrac{\pi}{6}$ (b) $x = \dfrac{5\pi}{6}$

In Exercises 7–20, solve the equation.

7. $2\cos x + 1 = 0$
8. $2\sin x + 1 = 0$
9. $\sqrt{3}\csc x - 2 = 0$
10. $\tan x + \sqrt{3} = 0$
11. $3\sec^2 x - 4 = 0$
12. $3\cot^2 x - 1 = 0$
13. $\sin x(\sin x + 1) = 0$
14. $(3\tan^2 x - 1)(\tan^2 x - 3) = 0$
15. $4\cos^2 x - 1 = 0$
16. $\sin^2 x = 3\cos^2 x$
17. $2\sin^2 2x = 1$
18. $\tan^2 3x = 3$
19. $\tan 3x(\tan x - 1) = 0$
20. $\cos 2x(2\cos x + 1) = 0$

In Exercises 21–32, find all solutions of the equation in the interval $[0, 2\pi)$.

21. $\cos^3 x = \cos x$
22. $\sec^2 x - 1 = 0$
23. $3\tan^3 x = \tan x$
24. $2\sin^2 x = 2 + \cos x$
25. $\sec^2 x - \sec x = 2$
26. $\sec x \csc x = 2\csc x$
27. $2\sin x + \csc x = 0$
28. $\sec x + \tan x = 1$
29. $2\cos^2 x + \cos x - 1 = 0$
30. $2\sin^2 x + 3\sin x + 1 = 0$
31. $2\sec^2 x + \tan^2 x - 3 = 0$
32. $\cos x + \sin x \tan x = 2$

In Exercises 33–38, find all solutions of the equation.

33. $\cos 2x = \dfrac{1}{2}$
34. $\sin 2x = -\dfrac{\sqrt{3}}{2}$
35. $\tan 3x = 1$
36. $\sec 4x = 2$
37. $\cos \dfrac{x}{2} = \dfrac{\sqrt{2}}{2}$
38. $\sin \dfrac{x}{2} = -\dfrac{\sqrt{3}}{2}$

In Exercises 39–42, find the x-intercepts of the graph.

39. $y = \sin \dfrac{\pi x}{2} + 1$
40. $y = \sin \pi x + \cos \pi x$

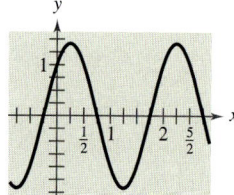

41. $y = \tan^2\left(\dfrac{\pi x}{6}\right) - 3$
42. $y = \sec^4\left(\dfrac{\pi x}{8}\right) - 4$

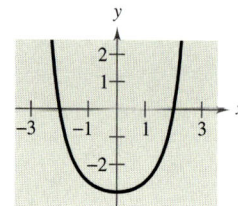

In Exercises 43 and 44, solve both equations. How do the solutions of the algebraic equation compare with the solutions of the trigonometric equation?

43. $6y^2 - 13y + 6 = 0$
    $6\cos^2 x - 13\cos x + 6 = 0$
44. $y^2 + y - 20 = 0$
    $\sin^2 x + \sin x - 20 = 0$

In Exercises 45–54, use a graphing utility to approximate the solutions of the equation in the interval $[0, 2\pi)$.

45. $2\sin x + \cos x = 0$
46. $4\sin^3 x + 2\sin^2 x - 2\sin x - 1 = 0$

47. $\dfrac{1 + \sin x}{\cos x} + \dfrac{\cos x}{1 + \sin x} = 4$ 48. $\dfrac{\cos x \cot x}{1 - \sin x} = 3$

49. $x \tan x - 1 = 0$ 50. $x \cos x - 1 = 0$

51. $\sec^2 x + 0.5 \tan x - 1 = 0$

52. $\csc^2 x + 0.5 \cot x - 5 = 0$

53. $2 \tan^2 x + 7 \tan x - 15 = 0$

54. $6 \sin^2 x - 7 \sin x + 2 = 0$

In Exercises 55–58, use the Quadratic Formula to solve the equation in the interval $[0, 2\pi)$. Then use a graphing utility to approximate the angle x.

55. $12 \sin^2 x - 13 \sin x + 3 = 0$
56. $3 \tan^2 x + 4 \tan x - 4 = 0$
57. $\tan^2 x + 3 \tan x + 1 = 0$
58. $4 \cos^2 x - 4 \cos x - 1 = 0$

In Exercises 59–62, use inverse functions where needed to find all solutions of the equation in the interval $[0, 2\pi)$.

59. $\tan^2 x - 6 \tan x + 5 = 0$
60. $\sec^2 x + \tan x - 3 = 0$
61. $2 \cos^2 x - 5 \cos x + 2 = 0$
62. $2 \sin^2 x - 7 \sin x + 3 = 0$

In Exercises 63 and 64, (a) use a graphing utility to graph the function and approximate the maximum and minimum points on the graph in the interval $[0, 2\pi)$, and (b) solve the trigonometric equation and demonstrate that its solutions are the x-coordinates of the maximum and minimum points of f. (Calculus is required to find the trigonometric equation.)

Function	Trigonometric Equation
63. $f(x) = \sin x + \cos x$	$\cos x - \sin x = 0$
64. $f(x) = 2 \sin x + \cos 2x$	$2 \cos x - 4 \sin x \cos x = 0$

Fixed Point In Exercises 65 and 66, find the smallest positive fixed point of the function f. [A fixed point of a function f is a real number c such that $f(c) = c$.]

65. $f(x) = \tan \dfrac{\pi x}{4}$ 66. $f(x) = \cos x$

67. *Graphical Reasoning* Consider the function

$$f(x) = \cos \dfrac{1}{x}$$

and its graph shown in the figure.

(a) What is the domain of the function?

(b) Identify any symmetry or asymptotes of the graph.

(c) Describe the behavior of the function as $x \to 0$.

(d) How many solutions does the equation

$$\cos \dfrac{1}{x} = 0$$

have in the interval $[-1, 1]$?

(e) Does the equation $\cos(1/x) = 0$ have a greatest solution? If so, approximate the solution. If not, explain.

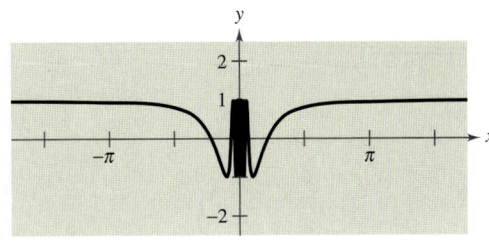

FIGURE FOR 67

68. *Graphical Reasoning* Consider the function

$$f(x) = \dfrac{\sin x}{x}$$

and its graph shown in the figure.

(a) What is the domain of the function?

(b) Identify any symmetry or asymptotes of the graph.

(c) Describe the behavior of the function as $x \to 0$.

(d) How many solutions does the equation

$$\dfrac{\sin x}{x} = 0$$

have in the interval $[-8, 8]$? Find the solutions.

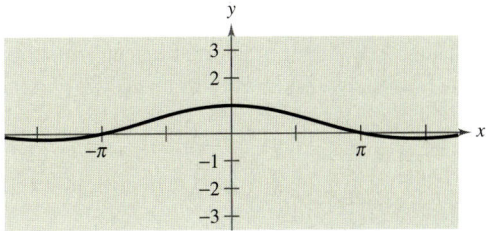

69. *Harmonic Motion* A weight is oscillating on the end of a spring (see figure). The position of the weight relative to the point of equilibrium is

$$y = \frac{1}{12}(\cos 8t - 3\sin 8t)$$

where y is the displacement in meters and t is the time in seconds. Find the times when the weight is at the point of equilibrium ($y = 0$) for $0 \leq t \leq 1$.

70. *Damped Harmonic Motion* The displacement from equilibrium of a weight oscillating on the end of a spring is

$$y = 1.56t^{-1/2}\cos 1.9t$$

where y is the displacement in feet and t is the time in seconds. Use a graphing utility to graph the displacement function for $0 \leq t \leq 10$. Find the time beyond which the displacement does not exceed 1 foot from equilibrium.

71. *Sales* The monthly sales (in thousands of units) of a seasonal product are approximated by

$$S = 74.50 + 43.75 \sin \frac{\pi t}{6}$$

where t is the time in months, with $t = 1$ corresponding to January. Determine the months when sales exceed 100,000 units.

72. *Projectile Motion* A batted baseball leaves the bat at an angle of θ with the horizontal and an initial velocity of $v_0 = 100$ feet per second. The ball is caught by an outfielder 300 feet from home plate (see figure). Find θ if the range r of a projectile is

$$r = \frac{1}{32}v_0^2 \sin 2\theta.$$

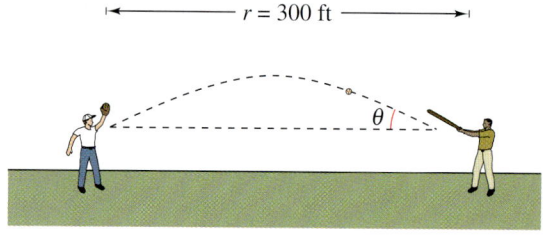

FIGURE FOR 72

73. *Projectile Motion* A sharpshooter intends to hit a target at a distance of 1000 yards with a gun that has a muzzle velocity of 1200 feet per second (see figure). Neglecting air resistance, determine the gun's minimum angle of elevation θ if the range r is

$$r = \frac{1}{32}v_0^2 \sin 2\theta.$$

74. *Area* The area of a rectangle (see figure) inscribed in one arc of the graph of $y = \cos x$ is

$$A = 2x \cos x, \quad 0 < x < \frac{\pi}{2}.$$

(a) Use a graphing utility to graph the area function, and approximate the area of the largest inscribed rectangle.

(b) Determine the values of x for which $A \geq 1$.

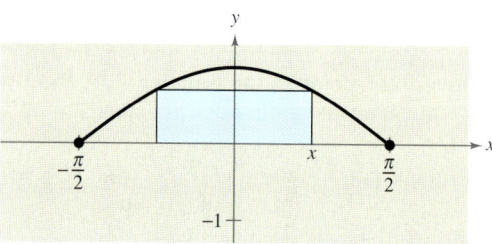

75. *Quadratic Approximation* Consider the function

$$f(x) = 3 \sin(0.6x - 2).$$

(a) Approximate the zero of the function in the interval $[0, 6]$.

(b) A quadratic approximation agreeing with f at $x = 5$ is $g(x) = -0.45x^2 + 5.52x - 13.70$. Use a graphing utility to graph f and g in the same viewing window. Describe the result.

(c) Use the Quadratic Formula to find the zeros of g. Compare the zero in the interval $[0, 6]$ with the result of part (a).

76. *Data Analysis* The table gives the unemployment rate r for the years 1985 through 1996 in the United States. The time t is measured in years, with $t = 0$ corresponding to 1990. (Source: U.S. Bureau of Labor Statistics)

t	-5	-4	-3	-2	-1	0
r	7.2	7.0	6.2	5.5	5.3	5.6

t	1	2	3	4	5	6
r	6.8	7.5	6.9	6.1	5.6	5.4

(a) Create a scatter plot of the data.

(b) Which of the following models best represents the data? Explain your reasoning.

 (1) $r = 1.5 \cos(t + 3.9) + 6.37$
 (2) $r = 1.03 \sin(0.9t + 0.44) + 6.19$
 (3) $r = \sin[0.91(t + 6.44)] + 6.26$
 (4) $r = 1.5 \sin[0.5(t + 2.8)] + 6.25$

(c) What term in the model gives the average unemployment rate? What is the rate?

(d) Economists study the lengths of business cycles such as unemployment rates. Based on this short span of time, use the model to give the length of this cycle.

(e) Use the model to estimate the next time the unemployment rate will be 6% or less.

Synthesis

True or False? **In Exercises 77 and 78, determine whether the statement is true or false. Justify your answer.**

77. The equation $2 \sin 4t - 1 = 0$ has four times the number of solutions in the interval $[0, 2\pi)$ as the equation $2 \sin t - 1 = 0$.

78. If you correctly solve a trigonometric equation down to the statement $\sin x = 3.4$, then you can finish solving the equation by using an inverse function.

In Exercises 79 and 80, use the graph to approximate the number of points of intersection of the graphs of y_1 and y_2.

79. $y_1 = 2 \sin x$
 $y_2 = 3x + 1$

80. $y_1 = 2 \sin x$
 $y_2 = \frac{1}{2}x + 1$

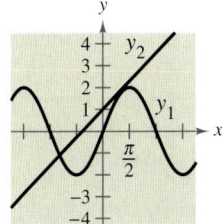

Review

In Exercises 81–84, solve triangle ABC by finding all missing angle measures and side lengths.

81.
82.

83.
84.

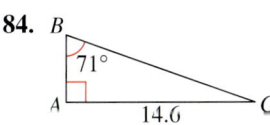

In Exercises 85–90, use reference angles to find the sine, cosine, and tangent of an angle with the given measure.

85. $390°$
86. $570°$
87. $495°$
88. $855°$
89. $-1845°$
90. $-1410°$

91. *Height* From a point 100 feet in front of the public library, the angles of elevation to the base of the flagpole and the top of the pole are $28°$ and $39° \, 45'$, respectively. The flagpole is mounted on the front of the library's roof. Find the height of the pole.

92. *Angle of Depression* Find the angle of depression from the top of a lighthouse 250 feet above water level to the water line of a ship 2 miles offshore.

2.4 Sum and Difference Formulas

What you should learn
- How to use sum and difference formulas to evaluate trigonometric functions
- How to use sum and difference formulas to verify identities and solve trigonometric equations

Why you should learn it
You can use sum and difference formulas to rewrite trigonometric expressions. For instance, in Exercise 76 on page 241 you use sum and difference formulas to rewrite a trigonometric expression in a form that helps you find the equation of a standing wave.

Richard Megna/Fundamental Photographs

Using Sum and Difference Formulas

In this and the following section, you will study the uses of several trigonometric identities and formulas. (Proofs of these formulas are given in Appendix A.)

Sum and Difference Formulas

$$\sin(u + v) = \sin u \cos v + \cos u \sin v$$
$$\sin(u - v) = \sin u \cos v - \cos u \sin v$$
$$\cos(u + v) = \cos u \cos v - \sin u \sin v$$
$$\cos(u - v) = \cos u \cos v + \sin u \sin v$$
$$\tan(u + v) = \frac{\tan u + \tan v}{1 - \tan u \tan v}$$
$$\tan(u - v) = \frac{\tan u - \tan v}{1 + \tan u \tan v}$$

◀ Exploration ▶

Use a graphing utility to graph $y = \cos(x + 2)$ and $y = \cos x + \cos 2$ in the same viewing window. What can you conclude about the graphs? Is it true that $\cos(x + 2) = \cos x + \cos 2$?

Use a graphing utility to graph $y = \sin(x + 4)$ and $y = \sin x + \sin 4$ in the same viewing window. What can you conclude about the graphs? Is it true that $\sin(x + 4) = \sin x + \sin 4$?

Examples 1 and 2 show how **sum and difference formulas** can be used to find exact values of trigonometric functions involving sums or differences of special angles.

Example 1 ▶ Evaluating a Trigonometric Function

Find the exact value of $\cos 75°$.

Solution

To find the *exact* value of $\cos 75°$, use the fact that $75° = 30° + 45°$. Consequently, the formula for $\cos(u + v)$ yields

$$\cos 75° = \cos(30° + 45°)$$
$$= \cos 30° \cos 45° - \sin 30° \sin 45°$$
$$= \frac{\sqrt{3}}{2}\left(\frac{\sqrt{2}}{2}\right) - \frac{1}{2}\left(\frac{\sqrt{2}}{2}\right)$$
$$= \frac{\sqrt{6} - \sqrt{2}}{4}.$$

Try checking this result on your calculator. You will find that $\cos 75° \approx 0.259$.

Section 2.4 ▶ Sum and Difference Formulas 237

Historical Note
Hipparchus, considered the most eminent of Greek astronomers, was born about 160 B.C. in Nicaea. He was credited with the invention of trigonometry. He also derived the sum and difference formulas for $\sin(A \pm B)$ and $\cos(A \pm B)$.

Example 2 ▶ Evaluating a Trigonometric Function

Find the exact value of $\sin \dfrac{\pi}{12}$.

Solution
Using the fact that

$$\frac{\pi}{12} = \frac{\pi}{3} - \frac{\pi}{4}$$

together with the formula for $\sin(u - v)$, you obtain

$$\sin \frac{\pi}{12} = \sin\left(\frac{\pi}{3} - \frac{\pi}{4}\right)$$

$$= \sin \frac{\pi}{3} \cos \frac{\pi}{4} - \cos \frac{\pi}{3} \sin \frac{\pi}{4}$$

$$= \frac{\sqrt{3}}{2}\left(\frac{\sqrt{2}}{2}\right) - \frac{1}{2}\left(\frac{\sqrt{2}}{2}\right)$$

$$= \frac{\sqrt{6} - \sqrt{2}}{4}.$$

Example 3 ▶ Evaluating a Trigonometric Expression

Find the exact value of $\sin 42° \cos 12° - \cos 42° \sin 12°$.

Solution
Recognizing that this expression fits the formula for $\sin(u - v)$, you can write

$$\sin 42° \cos 12° - \cos 42° \sin 12° = \sin(42° - 12°)$$

$$= \sin 30°$$

$$= \frac{1}{2}.$$

Example 4 ▶ An Application of a Sum Formula

Write $\cos(\arctan 1 + \arccos x)$ as an algebraic expression.

Solution
This expression fits the formula for $\cos(u + v)$. Angles $u = \arctan 1$ and $v = \arccos x$ are shown in Figure 2.5. So

$$\cos(u + v) = \cos(\arctan 1)\cos(\arccos x) - \sin(\arctan 1)\sin(\arccos x)$$

$$= \frac{1}{\sqrt{2}} \cdot x - \frac{1}{\sqrt{2}} \cdot \sqrt{1 - x^2}$$

$$= \frac{x - \sqrt{1 - x^2}}{\sqrt{2}}.$$

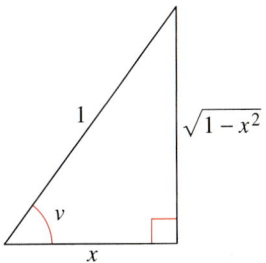

FIGURE 2.5

Example 5 ▶ Proving a Cofunction Identity

Prove the cofunction identity $\cos\left(\dfrac{\pi}{2} - x\right) = \sin x$.

Solution

Using the formula for $\cos(u - v)$, you have

$$\cos\left(\dfrac{\pi}{2} - x\right) = \cos\dfrac{\pi}{2}\cos x + \sin\dfrac{\pi}{2}\sin x$$

$$= (0)(\cos x) + (1)(\sin x)$$

$$= \sin x.$$

Sum and difference formulas can be used to derive **reduction formulas** involving expressions such as

$$\sin\left(\theta + \dfrac{n\pi}{2}\right) \quad \text{and} \quad \cos\left(\theta + \dfrac{n\pi}{2}\right), \quad \text{where } n \text{ is an integer.}$$

Example 6 ▶ Deriving Reduction Formulas

Simplify each expression.

a. $\cos\left(\theta - \dfrac{3\pi}{2}\right)$ **b.** $\tan(\theta + 3\pi)$

Solution

a. Using the formula for $\cos(u - v)$, you have

$$\cos\left(\theta - \dfrac{3\pi}{2}\right) = \cos\theta\cos\dfrac{3\pi}{2} + \sin\theta\sin\dfrac{3\pi}{2}$$

$$= (\cos\theta)(0) + (\sin\theta)(-1)$$

$$= -\sin\theta.$$

b. Using the formula for $\tan(u + v)$, you have

$$\tan(\theta + 3\pi) = \dfrac{\tan\theta + \tan 3\pi}{1 - \tan\theta\tan 3\pi}$$

$$= \dfrac{\tan\theta + 0}{1 - (\tan\theta)(0)}$$

$$= \tan\theta.$$

The next example was taken from calculus. It is used to derive the derivative of the sine function.

Section 2.4 ▶ Sum and Difference Formulas   239

Example 7 ▶ An Application from Calculus

Verify that

$$\frac{\sin(x+h) - \sin x}{h} = (\cos x)\left(\frac{\sin h}{h}\right) - (\sin x)\left(\frac{1 - \cos h}{h}\right)$$

where $h \neq 0$.

Solution

Using the formula for $\sin(u + v)$, you have

$$\frac{\sin(x+h) - \sin x}{h} = \frac{\sin x \cos h + \cos x \sin h - \sin x}{h}$$

$$= \frac{\cos x \sin h - \sin x(1 - \cos h)}{h}$$

$$= (\cos x)\left(\frac{\sin h}{h}\right) - (\sin x)\left(\frac{1 - \cos h}{h}\right).$$

Example 8 ▶ Solving a Trigonometric Equation

Find all solutions of

$$\sin\left(x + \frac{\pi}{4}\right) + \sin\left(x - \frac{\pi}{4}\right) = -1$$

in the interval $[0, 2\pi)$.

Solution

Using sum and difference formulas, rewrite the equation as

$$\sin x \cos \frac{\pi}{4} + \cos x \sin \frac{\pi}{4} + \sin x \cos \frac{\pi}{4} - \cos x \sin \frac{\pi}{4} = -1$$

$$2 \sin x \cos \frac{\pi}{4} = -1$$

$$2(\sin x)\left(\frac{\sqrt{2}}{2}\right) = -1$$

$$\sin x = -\frac{1}{\sqrt{2}}$$

$$\sin x = -\frac{\sqrt{2}}{2}.$$

Therefore, the only solutions in the interval $[0, 2\pi)$ are

$$x = \frac{5\pi}{4} \quad \text{and} \quad x = \frac{7\pi}{4}.$$

These solutions are checked graphically in Figure 2.6.

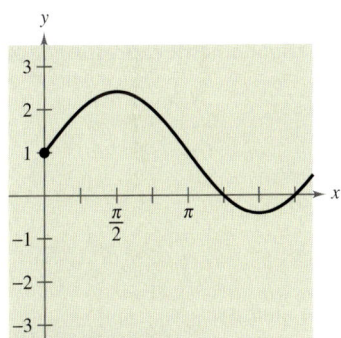

FIGURE 2.6

2.4 Exercises

In Exercises 1–6, find the exact value of each expression.

1. (a) $\cos\left(\dfrac{\pi}{4} + \dfrac{\pi}{3}\right)$ (b) $\cos\dfrac{\pi}{4} + \cos\dfrac{\pi}{3}$
2. (a) $\sin\left(\dfrac{3\pi}{4} + \dfrac{5\pi}{6}\right)$ (b) $\sin\dfrac{3\pi}{4} + \sin\dfrac{5\pi}{6}$
3. (a) $\sin\left(\dfrac{7\pi}{6} - \dfrac{\pi}{3}\right)$ (b) $\sin\dfrac{7\pi}{6} - \sin\dfrac{\pi}{3}$
4. (a) $\cos\left(\dfrac{2\pi}{3} - \dfrac{\pi}{6}\right)$ (b) $\cos\dfrac{2\pi}{3} + \cos\dfrac{\pi}{6}$
5. (a) $\cos(120° + 45°)$ (b) $\cos 120° + \cos 45°$
6. (a) $\sin(135° - 30°)$ (b) $\sin 135° - \cos 30°$

In Exercises 7–14, find the exact values of the sine, cosine, and tangent of the angle.

7. $105° = 60° + 45°$
8. $165° = 135° + 30°$
9. $195° = 225° - 30°$
10. $255° = 300° - 45°$
11. $\dfrac{11\pi}{12} = \dfrac{3\pi}{4} + \dfrac{\pi}{6}$
12. $\dfrac{7\pi}{12} = \dfrac{\pi}{3} + \dfrac{\pi}{4}$
13. $\dfrac{17\pi}{12} = \dfrac{9\pi}{4} - \dfrac{5\pi}{6}$
14. $-\dfrac{\pi}{12} = \dfrac{\pi}{6} - \dfrac{\pi}{4}$

In Exercises 15–22, find the exact values of the sine, cosine, and tangent of the angle.

15. $285°$
16. $-105°$
17. $-165°$
18. $15°$
19. $\dfrac{13\pi}{12}$
20. $-\dfrac{7\pi}{12}$
21. $-\dfrac{13\pi}{12}$
22. $\dfrac{5\pi}{12}$

In Exercises 23–30, write the expression as the sine, cosine, or tangent of an angle.

23. $\cos 25° \cos 15° - \sin 25° \sin 15°$
24. $\sin 140° \cos 50° + \cos 140° \sin 50°$
25. $\dfrac{\tan 325° - \tan 86°}{1 + \tan 325° \tan 86°}$
26. $\dfrac{\tan 140° - \tan 60°}{1 + \tan 140° \tan 60°}$
27. $\sin 3 \cos 1.2 - \cos 3 \sin 1.2$
28. $\cos\dfrac{\pi}{7}\cos\dfrac{\pi}{5} - \sin\dfrac{\pi}{7}\sin\dfrac{\pi}{5}$
29. $\dfrac{\tan 2x + \tan x}{1 - \tan 2x \tan x}$
30. $\cos 3x \cos 2y + \sin 3x \sin 2y$

In Exercises 31–36, find the exact value of the expression.

31. $\sin 330° \cos 30° - \cos 330° \sin 30°$
32. $\cos 15° \cos 60° + \sin 15° \sin 60°$
33. $\sin\dfrac{\pi}{12}\cos\dfrac{\pi}{4} + \cos\dfrac{\pi}{12}\sin\dfrac{\pi}{4}$
34. $\cos\dfrac{\pi}{16}\cos\dfrac{3\pi}{16} - \sin\dfrac{\pi}{16}\sin\dfrac{3\pi}{16}$
35. $\dfrac{\tan 25° + \tan 110°}{1 - \tan 25° \tan 110°}$
36. $\dfrac{\tan(5\pi/4) - \tan(\pi/12)}{1 + \tan(5\pi/4)\tan(\pi/12)}$

In Exercises 37–44, find the exact value of the trigonometric function given that $\sin u = \tfrac{5}{13}$ and $\cos v = -\tfrac{3}{5}$. (Both u and v are in Quadrant II.)

37. $\sin(u + v)$
38. $\cos(u - v)$
39. $\cos(u + v)$
40. $\sin(v - u)$
41. $\tan(u + v)$
42. $\csc(u - v)$
43. $\sec(v - u)$
44. $\cot(u + v)$

In Exercises 45–50, find the exact value of the trigonometric function given that $\sin u = -\tfrac{7}{25}$ and $\cos v = -\tfrac{4}{5}$. (Both u and v are in Quadrant III.)

45. $\cos(u + v)$
46. $\sin(u + v)$
47. $\tan(u - v)$
48. $\cot(v - u)$
49. $\sec(u + v)$
50. $\csc(u - v)$

In Exercises 51–54, write the trigonometric expression as an algebraic expression.

51. $\sin(\arcsin x + \arccos x)$
52. $\sin(\arctan 2x - \arccos x)$
53. $\cos(\arccos x + \arcsin x)$
54. $\cos(\arccos x - \arctan x)$

In Exercises 55–64, verify the identity.

55. $\sin(3\pi - x) = \sin x$
56. $\sin\left(\dfrac{\pi}{2} + x\right) = \cos x$

57. $\sin\left(\dfrac{\pi}{6} + x\right) = \dfrac{1}{2}(\cos x + \sqrt{3} \sin x)$

58. $\cos\left(\dfrac{5\pi}{4} - x\right) = -\dfrac{\sqrt{2}}{2}(\cos x + \sin x)$

59. $\cos(\pi - \theta) + \sin\left(\dfrac{\pi}{2} + \theta\right) = 0$

60. $\tan\left(\dfrac{\pi}{4} - \theta\right) = \dfrac{1 - \tan \theta}{1 + \tan \theta}$

61. $\cos(x + y)\cos(x - y) = \cos^2 x - \sin^2 y$

62. $\sin(x + y)\sin(x - y) = \sin^2 x - \sin^2 y$

63. $\sin(x + y) + \sin(x - y) = 2 \sin x \cos y$

64. $\cos(x + y) + \cos(x - y) = 2 \cos x \cos y$

In Exercises 65–68, simplify the expression algebraically and use a graphing utility to confirm it graphically.

65. $\cos\left(\dfrac{3\pi}{2} - x\right)$

66. $\cos(\pi + x)$

67. $\sin\left(\dfrac{3\pi}{2} + \theta\right)$

68. $\tan(\pi + \theta)$

In Exercises 69–72, find all solutions of the equation in the interval $[0, 2\pi)$.

69. $\sin\left(x + \dfrac{\pi}{3}\right) + \sin\left(x - \dfrac{\pi}{3}\right) = 1$

70. $\sin\left(x + \dfrac{\pi}{6}\right) - \sin\left(x - \dfrac{\pi}{6}\right) = \dfrac{1}{2}$

71. $\cos\left(x + \dfrac{\pi}{4}\right) - \cos\left(x - \dfrac{\pi}{4}\right) = 1$

72. $\tan(x + \pi) + 2 \sin(x + \pi) = 0$

In Exercises 73 and 74, use a graphing utility to approximate the solutions in the interval $[0, 2\pi)$.

73. $\cos\left(x + \dfrac{\pi}{4}\right) + \cos\left(x - \dfrac{\pi}{4}\right) = 1$

74. $\tan(x + \pi) - \cos\left(x + \dfrac{\pi}{2}\right) = 0$

75. *Harmonic Motion* A weight is attached to a spring suspended vertically from a ceiling. When a driving force is applied to the system, the weight moves vertically from its equilibrium position, and this motion is modeled by

$$y = \dfrac{1}{3} \sin 2t + \dfrac{1}{4} \cos 2t$$

where y is the distance from equilibrium measured in feet and t is the time in seconds.

(a) Use the identity

$$a \sin B\theta + b \cos B\theta = \sqrt{a^2 + b^2} \sin(B\theta + C)$$

where $C = \arctan(b/a)$, $a > 0$, to write the model in the form

$$y = \sqrt{a^2 + b^2} \sin(Bt + C).$$

(b) Find the amplitude of the oscillations of the weight.

(c) Find the frequency of the oscillations of the weight.

76. *Standing Waves* The equation of a standing wave is obtained by adding the displacements of two waves traveling in opposite directions (see figure). Assume that each of the waves has amplitude A, period T, and wavelength λ. If the models for these waves are

$$y_1 = A \cos 2\pi\left(\dfrac{t}{T} - \dfrac{x}{\lambda}\right) \quad \text{and}$$

$$y_2 = A \cos 2\pi\left(\dfrac{t}{T} + \dfrac{x}{\lambda}\right)$$

show that

$$y_1 + y_2 = 2A \cos \dfrac{2\pi t}{T} \cos \dfrac{2\pi x}{\lambda}.$$

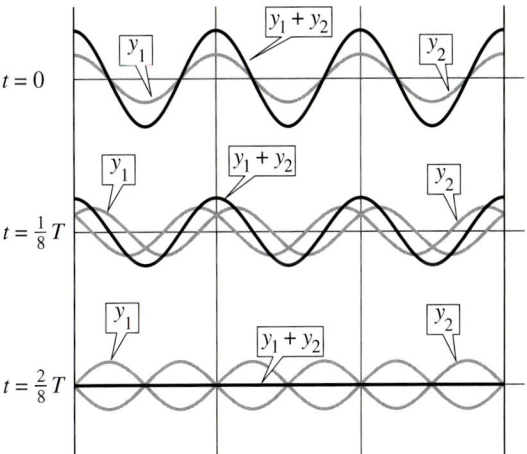

Synthesis

True or False? In Exercises 77–80, determine whether the statement is true or false. Justify your answer.

77. $\sin(u \pm v) = \sin u \pm \sin v$
78. $\cos(u \pm v) = \cos u \pm \cos v$
79. $\cos\left(x - \dfrac{\pi}{2}\right) = -\sin x$
80. $\sin\left(x - \dfrac{\pi}{2}\right) = -\cos x$

In Exercises 81–84, verify the identity.

81. $\cos(n\pi + \theta) = (-1)^n \cos \theta, \quad n$ is an integer.
82. $\sin(n\pi + \theta) = (-1)^n \sin \theta, \quad n$ is an integer.
83. $a \sin B\theta + b \cos B\theta = \sqrt{a^2 + b^2} \sin(B\theta + C)$, where $C = \arctan(b/a)$ and $a > 0$.
84. $a \sin B\theta + b \cos B\theta = \sqrt{a^2 + b^2} \cos(B\theta - C)$, where $C = \arctan(a/b)$ and $b > 0$.

In Exercises 85–88, use the formulas given in Exercises 83 and 84 to write the trigonometric expression in the following forms.

(a) $\sqrt{a^2 + b^2} \sin(B\theta + C)$
(b) $\sqrt{a^2 + b^2} \cos(B\theta - C)$

85. $\sin \theta + \cos \theta$
86. $3 \sin 2\theta + 4 \cos 2\theta$
87. $12 \sin 3\theta + 5 \cos 3\theta$
88. $\sin 2\theta - \cos 2\theta$

In Exercises 89 and 90, use the formulas given in Exercises 83 and 84 to write the trigonometric expression in the form $a \sin B\theta + b \cos B\theta$.

89. $2 \sin\left(\theta + \dfrac{\pi}{2}\right)$
90. $5 \cos\left(\theta + \dfrac{3\pi}{4}\right)$

In Exercises 91 and 92, use the figure, which shows two lines whose equations are

$y_1 = m_1 x + b_1$ and $y_2 = m_2 x + b_2$.

Assume that both lines have positive slopes. Derive a formula for the angle between the two lines. Then use your formula to find the angle between the given pair of lines.

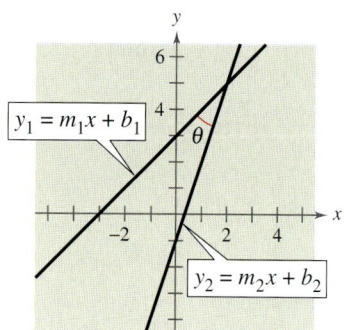

91. $y = x$ and $y = \sqrt{3}x$
92. $y = x$ and $y = \dfrac{1}{\sqrt{3}}x$

93. ***Conjecture*** Consider the function

$$f(\theta) = \sin^2\left(\theta + \dfrac{\pi}{4}\right) + \sin^2\left(\theta - \dfrac{\pi}{4}\right).$$

Use a graphing utility to graph the function and use the graph to create an identity. Prove your conjecture.

94. ***Conjecture*** Three squares of side s are placed side by side (see figure). Make a conjecture about the relationship between the sum $u + v$ and w. Prove your conjecture by using the identity for the tangent of the sum of two angles.

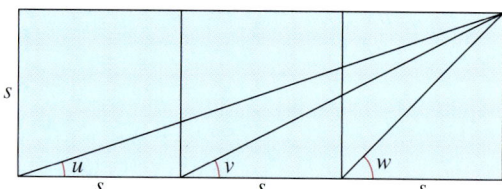

95. ***Proof*** Write a proof for the formula for $\sin(u + v)$.
96. ***Proof*** Write a proof for the formula for $\sin(u - v)$.

Review

In Exercises 97–100, find the inverse of f. Verify that $f(f^{-1}(x)) = x$ and $f^{-1}(f(x)) = x$.

97. $f(x) = 5(x - 3)$
98. $f(x) = \dfrac{7 - x}{8}$
99. $f(x) = x^2 - 8$
100. $f(x) = \sqrt{x - 16}$

2.5 Multiple-Angle and Product-to-Sum Formulas

▶ **What you should learn**
- How to use multiple-angle formulas to rewrite and evaluate trigonometric functions
- How to use power-reducing formulas to rewrite and evaluate trigonometric functions
- How to use half-angle formulas to rewrite and evaluate trigonometric functions
- How to use product-to-sum formulas to rewrite and evaluate trigonometric functions

▶ **Why you should learn it**
You can use a variety of trigonometric formulas to rewrite trigonometric functions in more convenient forms. For instance, in Exercise 121 on page 252 you use a half-angle formula to determine the apex angle of a sound wave cone from the speed of an airplane.

NASA-Liaison Agency

A computer animation of this concept appears in the *Interactive* CD-ROM and *Internet* versions of this text.

Multiple-Angle Formulas

In this section you will study four other categories of trigonometric identities.

1. The first category involves functions of multiple angles such as $\sin ku$ and $\cos ku$.
2. The second category involves squares of trigonometric functions such as $\sin^2 u$.
3. The third category involves functions of half-angles such as $\sin(u/2)$.
4. The fourth category involves products of trigonometric functions such as $\sin u \cos v$.

The most commonly used multiple-angle formulas are the **double-angle formulas.** They are used often, so you should learn them. (Proofs of the double-angle formulas are given in Appendix A.)

Double-Angle Formulas

$\sin 2u = 2 \sin u \cos u$ $\qquad \tan 2u = \dfrac{2 \tan u}{1 - \tan^2 u}$

$\cos 2u = \cos^2 u - \sin^2 u$

$\qquad = 2 \cos^2 u - 1$

$\qquad = 1 - 2 \sin^2 u$

Example 1 ▶ Solving a Multiple-Angle Equation

Find all solutions of $2 \cos x + \sin 2x = 0$.

Solution

Begin by rewriting the equation so that it involves functions of x (rather than $2x$). Then factor and solve as usual.

$2 \cos x + \sin 2x = 0$ Write original equation.

$2 \cos x + 2 \sin x \cos x = 0$ Double-angle formula

$2 \cos x (1 + \sin x) = 0$ Factor.

$\cos x = 0$ and $1 + \sin x = 0$ Set factors equal to zero.

$x = \dfrac{\pi}{2}, \dfrac{3\pi}{2}$ $\qquad x = \dfrac{3\pi}{2}$ Solutions in $[0, 2\pi)$

Therefore, the general solution is

$x = \dfrac{\pi}{2} + 2n\pi$ and $x = \dfrac{3\pi}{2} + 2n\pi$

where n is an integer. Try verifying these solutions graphically.

Example 2 ▶ Using Double-Angle Formulas in Sketching Graphs

Use a double-angle formula to rewrite the equation

$$y = 4\cos^2 x - 2.$$

Then sketch the graph of the equation over the interval $[0, 2\pi]$.

Solution

Using a double-angle formula, you can rewrite the given function as

$$y = 4\cos^2 x - 2$$
$$= 2(2\cos^2 x - 1)$$
$$= 2\cos 2x.$$

Using the techniques discussed in Section 1.5, you can recognize that the graph of this function has an amplitude of 2 and a period of π. The key points in the interval $[0, \pi]$ are as follows.

Maximum	Intercept	Minimum	Intercept	Maximum
$(0, 2)$	$\left(\dfrac{\pi}{4}, 0\right)$	$\left(\dfrac{\pi}{2}, -2\right)$	$\left(\dfrac{3\pi}{4}, 0\right)$	$(\pi, 2)$

Two cycles of the graph are shown in Figure 2.7.

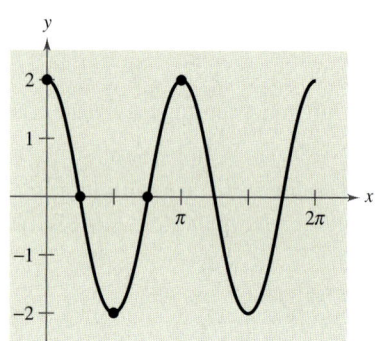

FIGURE 2.7

Example 3 ▶ Evaluating Functions Involving Double Angles

Use the following to find $\sin 2\theta$, $\cos 2\theta$, and $\tan 2\theta$.

$$\cos \theta = \frac{5}{13}, \quad \frac{3\pi}{2} < \theta < 2\pi$$

Solution

From Figure 2.8, you can see that $\sin \theta = y/r = -12/13$. Consequently, you can write

$$\sin 2\theta = 2\sin\theta\cos\theta = 2\left(\frac{-12}{13}\right)\left(\frac{5}{13}\right) = -\frac{120}{169}$$

$$\cos 2\theta = 2\cos^2\theta - 1 = 2\left(\frac{25}{169}\right) - 1 = -\frac{119}{169}$$

$$\tan 2\theta = \frac{\sin 2\theta}{\cos 2\theta} = \frac{120}{119}.$$

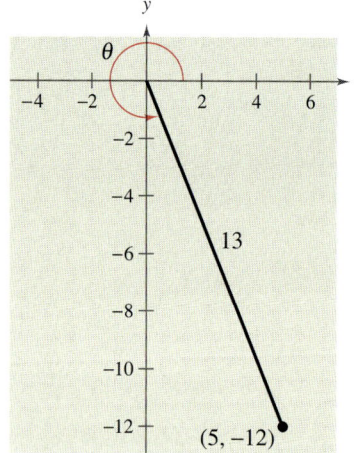

FIGURE 2.8

The double-angle formulas are not restricted to angles 2θ and θ. Other *double* combinations, such as 4θ and 2θ or 6θ and 3θ, are also valid. Here are two examples.

$$\sin 4\theta = 2\sin 2\theta \cos 2\theta \quad \text{and} \quad \cos 6\theta = \cos^2 3\theta - \sin^2 3\theta$$

By using double-angle formulas together with the sum formulas given in the preceding section, you can form other multiple-angle formulas.

Example 4 ▶ Deriving a Triple-Angle Formula

Express $\sin 3x$ in terms of $\sin x$.

Solution

$$\begin{aligned}
\sin 3x &= \sin(2x + x) \\
&= \sin 2x \cos x + \cos 2x \sin x \\
&= 2 \sin x \cos x \cos x + (1 - 2 \sin^2 x)\sin x \\
&= 2 \sin x \cos^2 x + \sin x - 2 \sin^3 x \\
&= 2 \sin x(1 - \sin^2 x) + \sin x - 2 \sin^3 x \\
&= 2 \sin x - 2 \sin^3 x + \sin x - 2 \sin^3 x \\
&= 3 \sin x - 4 \sin^3 x
\end{aligned}$$

Power-Reducing Formulas

The double-angle formulas can be used to obtain the following **power-reducing formulas**. (Proofs of the power-reducing formulas are given in Appendix A.) Example 5 shows a typical power reduction that is used in calculus.

Power-Reducing Formulas

$$\sin^2 u = \frac{1 - \cos 2u}{2} \qquad \cos^2 u = \frac{1 + \cos 2u}{2} \qquad \tan^2 u = \frac{1 - \cos 2u}{1 + \cos 2u}$$

A computer animation of this concept appears in the *Interactive* CD-ROM and *Internet* versions of this text.

Example 5 ▶ Reducing a Power

Rewrite $\sin^4 x$ as a sum of first powers of the cosines of multiple angles.

Solution

Note the repeated use of power-reducing formulas.

$$\begin{aligned}
\sin^4 x &= (\sin^2 x)^2 \\
&= \left(\frac{1 - \cos 2x}{2}\right)^2 \\
&= \frac{1}{4}(1 - 2 \cos 2x + \cos^2 2x) \\
&= \frac{1}{4}\left(1 - 2 \cos 2x + \frac{1 + \cos 4x}{2}\right) \\
&= \frac{1}{4} - \frac{1}{2}\cos 2x + \frac{1}{8} + \frac{1}{8}\cos 4x \\
&= \frac{1}{8}(3 - 4 \cos 2x + \cos 4x)
\end{aligned}$$

Half-Angle Formulas

You can derive some useful alternative forms of the power-reducing formulas by replacing u with $u/2$. The results are called **half-angle formulas.**

A computer animation of this concept appears in the *Interactive* CD-ROM and *Internet* versions of this text.

Half-Angle Formulas

$$\sin \frac{u}{2} = \pm \sqrt{\frac{1 - \cos u}{2}}$$

$$\cos \frac{u}{2} = \pm \sqrt{\frac{1 + \cos u}{2}}$$

$$\tan \frac{u}{2} = \frac{1 - \cos u}{\sin u} = \frac{\sin u}{1 + \cos u}$$

The signs of $\sin(u/2)$ and $\cos(u/2)$ depend on the quadrant in which $u/2$ lies.

Example 6 ▶ Using a Half-Angle Formula

Find the exact value of $\sin 105°$.

Solution

Begin by noting that $105°$ is half of $210°$. Then, using the half-angle formula for $\sin(u/2)$ and the fact that $105°$ lies in Quadrant II, you have

$$\sin 105° = \sqrt{\frac{1 - \cos 210°}{2}}$$

$$= \sqrt{\frac{1 - (-\cos 30°)}{2}}$$

$$= \sqrt{\frac{1 + (\sqrt{3}/2)}{2}}$$

$$= \frac{\sqrt{2 + \sqrt{3}}}{2}.$$

The positive square root is chosen because $\sin \theta$ is positive in Quadrant II.

Use your calculator to verify the result obtained in Example 6. That is, evaluate $\sin 105°$ and $(\sqrt{2 + \sqrt{3}})/2$.

$$\sin 105° \approx 0.9659258$$

$$\frac{\sqrt{2 + \sqrt{3}}}{2} \approx 0.9659258$$

You can see that both values are approximately 0.9659258.

Example 7 Solving a Trigonometric Equation

Find all solutions of

$$2 - \sin^2 x = 2\cos^2 \frac{x}{2}$$

in the interval $[0, 2\pi)$.

Solution

$$2 - \sin^2 x = 2\cos^2 \frac{x}{2} \qquad \text{Write original equation.}$$

$$2 - \sin^2 x = 2\left(\pm\sqrt{\frac{1+\cos x}{2}}\right)^2 \qquad \text{Half-angle formula}$$

$$2 - \sin^2 x = 2\left(\frac{1+\cos x}{2}\right) \qquad \text{Simplify.}$$

$$2 - \sin^2 x = 1 + \cos x \qquad \text{Simplify.}$$

$$2 - (1 - \cos^2 x) = 1 + \cos x \qquad \text{Pythagorean identity}$$

$$\cos^2 x - \cos x = 0 \qquad \text{Simplify.}$$

$$\cos x(\cos x - 1) = 0 \qquad \text{Factor.}$$

By setting the factors $\cos x$ and $(\cos x - 1)$ equal to zero, you find that the solutions in the interval $[0, 2\pi)$ are

$$x = \frac{\pi}{2}, \quad x = \frac{3\pi}{2}, \quad \text{and} \quad x = 0.$$

Product-to-Sum Formulas

Each of the following **product-to-sum formulas** is easily verified using the sum and difference formulas discussed in the preceding section.

Product-to-Sum Formulas

$$\sin u \sin v = \frac{1}{2}[\cos(u - v) - \cos(u + v)]$$

$$\cos u \cos v = \frac{1}{2}[\cos(u - v) + \cos(u + v)]$$

$$\sin u \cos v = \frac{1}{2}[\sin(u + v) + \sin(u - v)]$$

$$\cos u \sin v = \frac{1}{2}[\sin(u + v) - \sin(u - v)]$$

Example 8 ▶ Writing Products as Sums

Rewrite the product as a sum or difference.

$$\cos 5x \sin 4x$$

Solution

$$\cos 5x \sin 4x = \frac{1}{2}[\sin(5x + 4x) - \sin(5x - 4x)]$$

$$= \frac{1}{2}\sin 9x - \frac{1}{2}\sin x$$

Occasionally, it is useful to reverse the procedure and write a sum of trigonometric functions as a product. This can be accomplished with the following **sum-to-product formulas.** (A proof of the first formula is given in Appendix A.)

Sum-to-Product Formulas

$$\sin x + \sin y = 2 \sin\left(\frac{x+y}{2}\right) \cos\left(\frac{x-y}{2}\right)$$

$$\sin x - \sin y = 2 \cos\left(\frac{x+y}{2}\right) \sin\left(\frac{x-y}{2}\right)$$

$$\cos x + \cos y = 2 \cos\left(\frac{x+y}{2}\right) \cos\left(\frac{x-y}{2}\right)$$

$$\cos x - \cos y = -2 \sin\left(\frac{x+y}{2}\right) \sin\left(\frac{x-y}{2}\right)$$

Example 9 ▶ Using a Sum-to-Product Formula

Find the exact value of

$$\cos 195° + \cos 105°.$$

Solution

Using the appropriate sum-to-product formula, you obtain

$$\cos 195° + \cos 105° = 2 \cos\left(\frac{195° + 105°}{2}\right) \cos\left(\frac{195° - 105°}{2}\right)$$

$$= 2 \cos 150° \cos 45°$$

$$= 2\left(-\frac{\sqrt{3}}{2}\right)\left(\frac{\sqrt{2}}{2}\right)$$

$$= -\frac{\sqrt{6}}{2}.$$

Section 2.5 ▶ Multiple-Angle and Product-to-Sum Formulas   249

Example 10 ▶ **Solving a Trigonometric Equation**

Find all solutions of $\sin 5x + \sin 3x = 0$.

Solution

$$\sin 5x + \sin 3x = 0 \qquad \text{Write original equation.}$$

$$2 \sin\left(\frac{5x + 3x}{2}\right) \cos\left(\frac{5x - 3x}{2}\right) = 0 \qquad \text{Sum-to-product formula}$$

$$2 \sin 4x \cos x = 0 \qquad \text{Simplify.}$$

By setting the factor $\sin 4x$ equal to zero, you can find that the solutions in the interval $[0, 2\pi)$ are

$$x = 0, \frac{\pi}{4}, \frac{\pi}{2}, \frac{3\pi}{4}, \pi, \frac{5\pi}{4}, \frac{3\pi}{2}, \frac{7\pi}{4}.$$

The equation $\cos x = 0$ yields no additional solutions, and you can conclude that the solutions are of the form

$$x = \frac{n\pi}{4}$$

where n is an integer. These solutions are verified graphically in Figure 2.9.

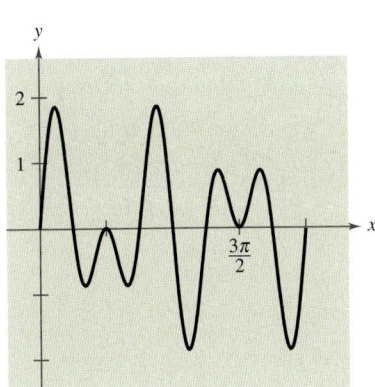

FIGURE 2.9

Example 11 ▶ **Verifying a Trigonometric Identity**

Verify the identity

$$\frac{\sin t + \sin 3t}{\cos t + \cos 3t} = \tan 2t.$$

Solution

Using appropriate sum-to-product formulas, you have

$$\frac{\sin t + \sin 3t}{\cos t + \cos 3t} = \frac{2 \sin 2t \cos(-t)}{2 \cos 2t \cos(-t)}$$

$$= \frac{\sin 2t}{\cos 2t}$$

$$= \tan 2t.$$

Writing ABOUT MATHEMATICS

Deriving an Area Formula   Describe how you can use a double-angle formula or a half-angle formula to derive a formula for the area of an isosceles triangle. Use a labeled sketch to illustrate your derivation. Then write two examples that show how your formula can be used.

2.5 Exercises

In Exercises 1–8, use the figure to find the exact value of the trigonometric function.

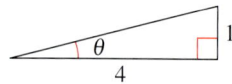

1. $\sin \theta$
2. $\tan \theta$
3. $\cos 2\theta$
4. $\sin 2\theta$
5. $\tan 2\theta$
6. $\sec 2\theta$
7. $\csc 2\theta$
8. $\cot 2\theta$

In Exercises 9–18, find the exact solutions of the equation in the interval $[0, 2\pi)$.

9. $\sin 2x - \sin x = 0$
10. $\sin 2x + \cos x = 0$
11. $4 \sin x \cos x = 1$
12. $\sin 2x \sin x = \cos x$
13. $\cos 2x - \cos x = 0$
14. $\cos 2x + \sin x = 0$
15. $\tan 2x - \cot x = 0$
16. $\tan 2x - 2 \cos x = 0$
17. $\sin 4x = -2 \sin 2x$
18. $(\sin 2x + \cos 2x)^2 = 1$

In Exercises 19–22, use a double-angle formula to rewrite the expression.

19. $6 \sin x \cos x$
20. $6 \cos^2 x - 3$
21. $4 - 8 \sin^2 x$
22. $(\cos x + \sin x)(\cos x - \sin x)$

In Exercises 23–28, find the exact values of $\sin 2u$, $\cos 2u$, and $\tan 2u$ using the double-angle formulas.

23. $\sin u = -\dfrac{4}{5}, \quad \pi < u < \dfrac{3\pi}{2}$
24. $\cos u = -\dfrac{2}{3}, \quad \dfrac{\pi}{2} < u < \pi$
25. $\tan u = \dfrac{3}{4}, \quad 0 < u < \dfrac{\pi}{2}$
26. $\cot u = -4, \quad \dfrac{3\pi}{2} < u < 2\pi$
27. $\sec u = -\dfrac{5}{2}, \quad \dfrac{\pi}{2} < u < \pi$
28. $\csc u = 3, \quad \dfrac{\pi}{2} < u < \pi$

In Exercises 29–34, use the power-reducing formulas to rewrite the expression in terms of the first power of the cosine.

29. $\cos^4 x$
30. $\sin^8 x$
31. $\sin^2 x \cos^2 x$
32. $\sin^4 x \cos^4 x$
33. $\sin^2 x \cos^4 x$
34. $\sin^4 x \cos^2 x$

In Exercises 35–40, use the figure to find the exact value of the trigonometric function.

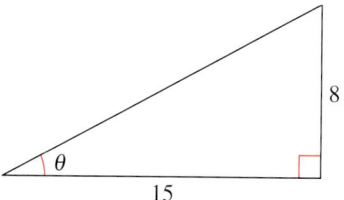

35. $\cos \dfrac{\theta}{2}$
36. $\sin \dfrac{\theta}{2}$
37. $\tan \dfrac{\theta}{2}$
38. $\sec \dfrac{\theta}{2}$
39. $\csc \dfrac{\theta}{2}$
40. $\cot \dfrac{\theta}{2}$

In Exercises 41–48, use the half-angle formulas to determine the exact values of the sine, cosine, and tangent of the angle.

41. $75°$
42. $165°$
43. $112° \, 30'$
44. $67° \, 30'$
45. $\dfrac{\pi}{8}$
46. $\dfrac{\pi}{12}$
47. $\dfrac{3\pi}{8}$
48. $\dfrac{7\pi}{12}$

In Exercises 49–54, find the exact values of $\sin(u/2)$, $\cos(u/2)$, and $\tan(u/2)$ using the half-angle formulas.

49. $\sin u = \dfrac{5}{13}, \quad \dfrac{\pi}{2} < u < \pi$
50. $\cos u = \dfrac{3}{5}, \quad 0 < u < \dfrac{\pi}{2}$

51. $\tan u = -\dfrac{5}{8}, \quad \dfrac{3\pi}{2} < u < 2\pi$

52. $\cot u = 3, \quad \pi < u < \dfrac{3\pi}{2}$

53. $\csc u = -\dfrac{5}{3}, \quad \pi < u < \dfrac{3\pi}{2}$

54. $\sec u = -\dfrac{7}{2}, \quad \dfrac{\pi}{2} < u < \pi$

In Exercises 55–58, use the half-angle formulas to simplify the expression.

55. $\sqrt{\dfrac{1 - \cos 6x}{2}}$

56. $\sqrt{\dfrac{1 + \cos 4x}{2}}$

57. $-\sqrt{\dfrac{1 - \cos 8x}{1 + \cos 8x}}$

58. $-\sqrt{\dfrac{1 - \cos(x - 1)}{2}}$

In Exercises 59–62, find all solutions in the interval $[0, 2\pi)$. Use a graphing utility to graph the function and verify the solutions.

59. $\sin \dfrac{x}{2} + \cos x = 0$

60. $\sin \dfrac{x}{2} + \cos x - 1 = 0$

61. $\cos \dfrac{x}{2} - \sin x = 0$

62. $\tan \dfrac{x}{2} - \sin x = 0$

In Exercises 63–74, use the product-to-sum formulas to write the product as a sum or difference.

63. $6 \sin \dfrac{\pi}{4} \cos \dfrac{\pi}{4}$

64. $4 \cos \dfrac{\pi}{3} \sin \dfrac{5\pi}{6}$

65. $\cos 4\theta \sin 6\theta$

66. $3 \sin 2\alpha \sin 3\alpha$

67. $5 \cos(-5\beta) \cos 3\beta$

68. $\cos 2\theta \cos 4\theta$

69. $\sin(x + y) \sin(x - y)$

70. $\sin(x + y) \cos(x - y)$

71. $\cos(\theta - \pi) \sin(\theta + \pi)$

72. $\sin(\theta + \pi) \sin(\theta - \pi)$

73. $10 \cos 75° \cos 15°$

74. $6 \sin 45° \cos 15°$

In Exercises 75–86, use the sum-to-product formulas to write the sum or difference as a product.

75. $\sin 60° + \sin 30°$

76. $\cos 120° + \cos 30°$

77. $\cos \dfrac{3\pi}{4} - \cos \dfrac{\pi}{4}$

78. $\sin \dfrac{5\pi}{4} - \sin \dfrac{3\pi}{4}$

79. $\sin 5\theta - \sin 3\theta$

80. $\sin 3\theta + \sin \theta$

81. $\cos 6x + \cos 2x$

82. $\sin x + \sin 5x$

83. $\sin(\alpha + \beta) - \sin(\alpha - \beta)$

84. $\cos(\phi + 2\pi) + \cos \phi$

85. $\cos\left(\theta + \dfrac{\pi}{2}\right) - \cos\left(\theta - \dfrac{\pi}{2}\right)$

86. $\sin\left(x + \dfrac{\pi}{2}\right) + \sin\left(x - \dfrac{\pi}{2}\right)$

In Exercises 87–90, find all solutions in the interval $[0, 2\pi)$. Use a graphing utility to graph the function and verify the solutions.

87. $\sin 6x + \sin 2x = 0$

88. $\cos 2x - \cos 6x = 0$

89. $\dfrac{\cos 2x}{\sin 3x - \sin x} - 1 = 0$

90. $\sin^2 3x - \sin^2 x = 0$

In Exercises 91–94, use the figure to find the exact value of the trigonometric function in two ways.

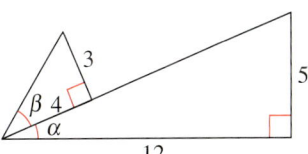

91. $\sin^2 \alpha$

92. $\cos^2 \alpha$

93. $\sin \alpha \cos \beta$

94. $\cos \alpha \sin \beta$

In Exercises 95–110, verify the identity.

95. $\csc 2\theta = \dfrac{\csc \theta}{2 \cos \theta}$

96. $\sec 2\theta = \dfrac{\sec^2 \theta}{2 - \sec^2 \theta}$

97. $\cos^2 2\alpha - \sin^2 2\alpha = \cos 4\alpha$

98. $\cos^4 x - \sin^4 x = \cos 2x$

99. $(\sin x + \cos x)^2 = 1 + \sin 2x$

100. $\sin \dfrac{\alpha}{3} \cos \dfrac{\alpha}{3} = \dfrac{1}{2} \sin \dfrac{2\alpha}{3}$

101. $1 + \cos 10y = 2 \cos^2 5y$

102. $\dfrac{\cos 3\beta}{\cos \beta} = 1 - 4 \sin^2 \beta$

103. $\sec \dfrac{u}{2} = \pm \sqrt{\dfrac{2 \tan u}{\tan u + \sin u}}$

104. $\tan \dfrac{u}{2} = \csc u - \cot u$

105. $\dfrac{\sin x \pm \sin y}{\cos x + \cos y} = \tan \dfrac{x \pm y}{2}$

106. $\dfrac{\sin x + \sin y}{\cos x - \cos y} = -\cot \dfrac{x - y}{2}$

107. $\dfrac{\cos 4x + \cos 2x}{\sin 4x + \sin 2x} = \cot 3x$

108. $\dfrac{\cos t + \cos 3t}{\sin 3t - \sin t} = \cot t$

109. $\sin\left(\dfrac{\pi}{6} + x\right) + \sin\left(\dfrac{\pi}{6} - x\right) = \cos x$

110. $\cos\left(\dfrac{\pi}{3} + x\right) + \cos\left(\dfrac{\pi}{3} - x\right) = \cos x$

In Exercises 111–114, use a graphing utility to verify the identity. Confirm that it is an identity algebraically.

111. $\cos 3\beta = \cos^3 \beta - 3 \sin^2 \beta \cos \beta$

112. $\sin 4\beta = 4 \sin \beta \cos \beta (1 - 2 \sin^2 \beta)$

113. $(\cos 4x - \cos 2x)/(2 \sin 3x) = -\sin x$

114. $(\cos 3x - \cos x)/(\sin 3x - \sin x) = -\tan 2x$

In Exercises 115 and 116, graph the function by hand in the interval $[0, 2\pi)$ by using the power-reducing formulas.

115. $f(x) = \sin^2 x$

116. $f(x) = \cos^2 x$

In Exercises 117 and 118, write the trigonometric expression as an algebraic expression.

117. $\sin(2 \arcsin x)$

118. $\cos(2 \arccos x)$

119. ***Area*** The length of each of the two equal sides of an isosceles triangle is 10 meters (see figure). The angle between the two sides is θ.

(a) Express the area of the triangle as a function of $\theta/2$.

(b) Express the area of the triangle as a function of θ. Determine the value of θ such that the area is a maximum.

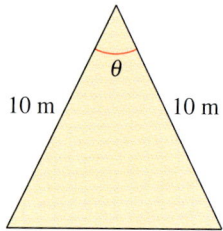

120. ***Projectile Motion*** The range of a projectile fired at an angle θ with the horizontal and with an initial velocity of v_0 feet per second is

$$r = \dfrac{1}{32} v_0^2 \sin 2\theta$$

where r is measured in feet. Determine the expression for the range in terms of θ.

121. ***Mach Number*** The mach number M of an airplane is the ratio of its speed to the speed of sound. When an airplane travels faster than the speed of sound, the sound waves form a cone behind the airplane. The mach number is related to the apex angle θ of the cone by

$$\sin \dfrac{\theta}{2} = \dfrac{1}{M}.$$

Find the angle θ that corresponds to a mach number of 4.5.

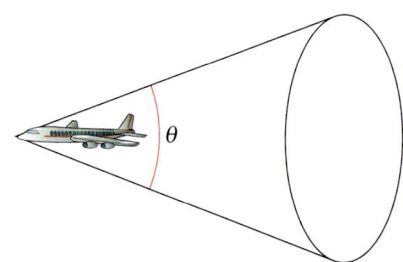

122. *Railroad Track* When two railroad tracks merge, the overlapping portions of the tracks are in the shapes of circular arcs. The radius of each arc r (in feet) and the angle θ are related by

$$\frac{x}{2} = 2r \sin^2 \frac{\theta}{2}.$$

Write a formula for x in terms of $\cos \theta$.

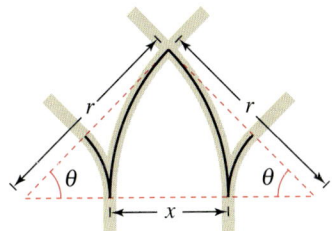

Synthesis

True or False? In Exercises 123 and 124, determine whether the statement is true or false. Justify your answer.

123. Because the sine function is an odd function, for a negative number u, $\sin 2u = -2 \sin u \cos u$.

124. $\sin \dfrac{u}{2} = -\sqrt{\dfrac{1 - \cos u}{2}}$ when u is in the second quadrant.

In Exercises 125 and 126, (a) use a graphing utility to graph the function and approximate the maximum and minimum points on the graph in the interval $[0, 2\pi)$ and (b) solve the trigonometric equation and verify that its solutions are the x-coordinates of the maximum and minimum points of f. (Calculus is required to find the trigonometric equation.)

	Function	Trigonometric Equation
125.	$f(x) = 4 \sin \dfrac{x}{2} + \cos x$	$2 \cos \dfrac{x}{2} - \sin x = 0$
126.	$f(x) = \cos 2x - 2 \sin x$	$-2 \cos x(2 \sin x + 1) = 0$

127. *Exploration* Consider the function

$$f(x) = \sin^4 x + \cos^4 x.$$

(a) Use the power-reducing formulas to write the function in terms of cosine to the first power.

(b) Determine another way of rewriting the function. Use a graphing utility to rule out incorrectly rewritten functions.

(c) Add a trigonometric term to the function so that it becomes a perfect square trinomial. Rewrite the function as a perfect square trinomial minus the term that you added. Use a graphing utility to rule out incorrectly rewritten functions.

(d) Rewrite the result of part (c) in terms of the sine of a double angle. Use a graphing utility to rule out incorrectly rewritten functions.

(e) When you rewrite a trigonometric expression, the result may not be the same as a friend's. Does this mean that one of you is wrong? Explain.

128. *Conjecture* Consider the function

$$f(x) = 2 \sin x [2 \cos^2(x/2) - 1].$$

(a) Use a graphing utility to graph the function.

(b) Make a conjecture about the function that is an identity with f.

(c) Verify your conjecture analytically.

Review

In Exercises 129–132, write the equation of the parabola in standard form and identify its vertex.

129. $y = x^2 - 6x + 13$ **130.** $y = x^2 + 10x + 23$

131. $y = 2x^2 - 4x + 3$ **132.** $y = \frac{1}{2}x^2 + 4x + 2$

In Exercises 133 and 134, find the vertex of the parabola and sketch its graph.

133. $(x - 5)^2 + y + 8 = 0$

134. $x^2 - 2x + 12y + 13 = 0$

135. *Profit* The total profit for a company in October was 16% higher than it was in September. The total profit for the two months was $507,600. Find the profit for each month.

136. *Average Speed* Two cars start at a given point and travel in the same direction at average speeds of 48 miles per hour and 56 miles per hour. How much time must elapse before the two cars are 12 miles apart?

137. *Mixture Problem* A 55-gallon barrel contains a mixture with a concentration of 30%. How much of this mixture must be withdrawn and replaced by 100% concentrate to bring the mixture up to 50% concentration?

138. *Baseball Diamond* A baseball diamond has the shape of a square where the distance between each of the consecutive bases is 90 feet. Approximate the distance from home plate to second base.

Chapter Summary

What did you learn?

Section 2.1   **Review Exercises**
- ☐ How to recognize and write the fundamental trigonometric identities — 1–6
- ☐ How to use the fundamental trigonometric identities to evaluate trigonometric functions, simplify trigonometric expressions, and rewrite trigonometric expressions — 7–23

Section 2.2
- ☐ How to plan a strategy for verifying trigonometric identities — 24–31
- ☐ How to verify trigonometric identities — 24–31

Section 2.3
- ☐ How to use standard algebraic techniques to solve trigonometric equations — 32–37
- ☐ How to solve trigonometric equations of quadratic type — 38–41
- ☐ How to solve trigonometric equations involving multiple angles — 42–45
- ☐ How to use inverse trigonometric functions to solve trigonometric equations — 46–49

Section 2.4
- ☐ How to use sum and difference formulas to evaluate trigonometric functions — 50–63
- ☐ How to use sum and difference formulas to verify identities and solve trigonometric equations — 64–67

Section 2.5
- ☐ How to use multiple-angle formulas to rewrite and evaluate trigonometric functions — 68–72
- ☐ How to use power-reducing formulas to rewrite and evaluate trigonometric functions — 73–76
- ☐ How to use half-angle formulas to rewrite and evaluate trigonometric functions — 77–83
- ☐ How to use product-to-sum formulas to rewrite and evaluate trigonometric functions — 84–92

Review Exercises

2.1 In Exercises 1–6, name the trigonometric function that is equivalent to the expression.

1. $\dfrac{1}{\cos x}$
2. $\dfrac{1}{\sin x}$
3. $\dfrac{1}{\sec x}$
4. $\dfrac{1}{\tan x}$
5. $\dfrac{\cos x}{\sin x}$
6. $\sqrt{1 + \tan^2 x}$

In Exercises 7–10, use the given values and trigonometric identities to evaluate (if possible) the other trigonometric functions of the angle.

7. $\sin x = \tfrac{3}{5}, \quad \cos x = \tfrac{4}{5}$
8. $\tan \theta = \tfrac{2}{3}, \quad \sec \theta = \dfrac{\sqrt{13}}{3}$
9. $\sin\left(\dfrac{\pi}{2} - x\right) = \dfrac{\sqrt{2}}{2}, \quad \sin x = -\dfrac{\sqrt{2}}{2}$
10. $\csc\left(\dfrac{\pi}{2} - \theta\right) = 9, \quad \sin \theta = \dfrac{4\sqrt{5}}{9}$

In Exercises 11–22, use the fundamental trigonometric Identities to simplify the trigonometric expression.

11. $\dfrac{1}{\cot^2 x + 1}$
12. $\dfrac{\tan \theta}{1 - \cos^2 \theta}$
13. $\tan^2 x(\csc^2 x - 1)$
14. $\cot^2 x(\sin^2 x)$
15. $\dfrac{\sin\left(\dfrac{\pi}{2} - \theta\right)}{\sin \theta}$
16. $\dfrac{\cot\left(\dfrac{\pi}{2} - u\right)}{\cos u}$
17. $\cos^2 x + \cos^2 x \cot^2 x$
18. $\tan^2 \theta \csc^2 \theta - \tan^2 \theta$
19. $(\tan x + 1)^2 \cos x$
20. $(\sec x - \tan x)^2$
21. $\dfrac{1}{\csc \theta + 1} - \dfrac{1}{\csc \theta - 1}$
22. $\dfrac{\cos^2 x}{1 - \sin x}$

23. *Rate of Change* The rate of change of the function $f(x) = 2\sqrt{\sin x}$ is the expression $\sin^{-1/2} x \cos x$. Show that this expression can also be written as $\cot x \sqrt{\sin x}$.

2.2 In Exercises 24–31, verify the identity.

24. $\cos x(\tan^2 x + 1) = \sec x$
25. $\sec^2 x \cot x - \cot x = \tan x$
26. $\cos\left(x + \dfrac{\pi}{2}\right) = -\sin x$
27. $\cot\left(\dfrac{\pi}{2} - x\right) = \tan x$
28. $\dfrac{1}{\tan \theta \csc \theta} = \cos \theta$
29. $\dfrac{1}{\tan x \csc x \sin x} = \cot x$
30. $\sin^5 x \cos^2 x = (\cos^2 x - 2\cos^4 x + \cos^6 x)\sin x$
31. $\cos^3 x \sin^2 x = (\sin^2 x - \sin^4 x) \cos x$

2.3 In Exercises 32–37, solve the equation.

32. $\sin x = \sqrt{3} - \sin x$
33. $4 \cos \theta = 1 + 2\cos \theta$
34. $3\sqrt{3} \tan u = 3$
35. $\tfrac{1}{2} \sec x - 1 = 0$
36. $3 \csc^2 x = 4$
37. $4 \tan^2 u - 1 = \tan^2 u$

In Exercises 38–45, find all solutions of the equation in the interval $[0, 2\pi)$.

38. $2\cos^2 x - \cos x - 1$
39. $2 \sin^2 x - 3 \sin x = -1$
40. $\cos^2 x + \sin x = 1$
41. $\sin^2 x + 2 \cos x = 2$
42. $2 \sin 2x - \sqrt{2} = 0$
43. $\sqrt{3} \tan 3x = 0$
44. $\cos 4x(\cos x - 1) = 0$
45. $3 \csc^2 5x = -4$

In Exercises 46–49, use inverse functions where needed to find all solutions of the equation in the interval $[0, 2\pi)$.

46. $\sin^2 x - 2 \sin x = 0$
47. $2\cos^2 x + 3 \cos x = 0$
48. $\tan^2 \theta + \tan \theta - 12 = 0$
49. $\sec^2 x + 6 \tan x + 4 = 0$

2.4 In Exercises 50–53, find the exact values of the sine, cosine, and tangent of the angle by using a sum or difference formula.

50. $285° = 315° - 30°$
51. $345° = 300° + 45°$
52. $\dfrac{25\pi}{12} = \dfrac{11\pi}{6} + \dfrac{\pi}{4}$
53. $\dfrac{19\pi}{12} = \dfrac{11\pi}{6} - \dfrac{\pi}{4}$

In Exercises 54–57, write the expression as the sine, cosine, or tangent of an angle.

54. $\sin 60° \cos 45° - \cos 60° \sin 45°$
55. $\cos 45° \cos 120° - \sin 45° \sin 120°$
56. $\dfrac{\tan 25° + \tan 10°}{1 - \tan 25° \tan 10°}$
57. $\dfrac{\tan 68° - \tan 115°}{1 + \tan 68° \tan 115°}$

In Exercises 58–63, find the exact value of the trigonometric function given that $\sin u = \tfrac{3}{4}$, $\cos v = -\tfrac{5}{13}$, and u and v are in Quadrant II.

58. $\sin(u + v)$
59. $\tan(u + v)$
60. $\cos(u - v)$
61. $\sin(u - v)$
62. $\cos(u + v)$
63. $\tan(u - v)$

In Exercises 64–67, find all solutions of the equation in the interval $[0, 2\pi)$.

64. $\sin\left(x + \dfrac{\pi}{4}\right) - \sin\left(x - \dfrac{\pi}{4}\right) = 1$
65. $\cos\left(x + \dfrac{\pi}{6}\right) - \cos\left(x - \dfrac{\pi}{6}\right) = 1$
66. $\sin\left(x + \dfrac{\pi}{2}\right) - \sin\left(x - \dfrac{\pi}{2}\right) = \sqrt{3}$
67. $\cos\left(x + \dfrac{3\pi}{4}\right) - \cos\left(x - \dfrac{3\pi}{4}\right) = 0$

2.5 In Exercises 68 and 69, use double-angle formulas to verify the identity algebraically and use a graphing utility to confirm it graphically.

68. $\sin 4x = 8\cos^3 x \sin x - 4\cos x \sin x$
69. $\tan^2 x = \dfrac{1 - \cos 2x}{1 + \cos 2x}$

In Exercises 70 and 71, find the exact values of $\sin 2u$, $\cos 2u$, and $\tan 2u$ using the double-angle formulas.

70. $\sin u = -\dfrac{4}{5}$, $\pi < u < \dfrac{3\pi}{2}$

71. $\cos u = -\dfrac{2}{\sqrt{5}}$, $\dfrac{\pi}{2} < u < \pi$

72. *Projectile Motion* A baseball leaves the hand of the person at first base at an angle of θ with the horizontal and an initial velocity of $v_0 = 80$ feet per second. The ball is caught by the person at second base 100 feet away. Find θ if the range r of a projectile is

$$r = \dfrac{1}{32} v_0^2 \sin 2\theta.$$

In Exercises 73–76, use the power-reducing formulas to rewrite the expression in terms of the first power of the cosine.

73. $\tan^2 2x$
74. $\cos^2 3x$
75. $\sin^2 x \tan^2 x$
76. $\cos^2 x \tan^2 x$

In Exercises 77–80, use the half-angle formulas to determine the exact values of the sine, cosine, and tangent of the angle.

77. $-75°$
78. $15°$
79. $\dfrac{19\pi}{12}$
80. $-\dfrac{17\pi}{12}$

In Exercises 81 and 82, use the half-angle formulas to simplify the expression.

81. $-\sqrt{\dfrac{1 + \cos 10x}{2}}$
82. $\dfrac{\sin 6x}{1 + \cos 6x}$

83. *Volume* A trough for feeding cattle is 4 meters long and its cross sections are isosceles triangles with the two equal sides being $\tfrac{1}{2}$ meter (see figure). The angle between the two sides is θ.

(a) Express the trough's volume as a function of $\dfrac{\theta}{2}$.

(b) Express the volume of the trough as a function of θ and determine the value of θ such that the volume is maximum.

In Exercises 84–87, use the product-to-sum formulas to write the product as a sum or difference.

84. $\cos \dfrac{\pi}{6} \sin \dfrac{\pi}{6}$

85. $6 \sin 15° \sin 45°$

86. $\cos 5\theta \cos 3\theta$

87. $4 \sin 3\alpha \cos 2\alpha$

In Exercises 88–91, use the sum-to-product formulas to write the sum or difference as a product.

88. $\sin 60° + \sin 90°$

89. $\cos 3\theta + \cos 2\theta$

90. $\cos\left(x + \dfrac{\pi}{6}\right) - \cos\left(x - \dfrac{\pi}{6}\right)$

91. $\sin\left(x + \dfrac{\pi}{4}\right) - \sin\left(x - \dfrac{\pi}{4}\right)$

92. *Harmonic Motion* A weight is attached to a spring suspended vertically from a ceiling. When a driving force is applied to the system, the weight moves vertically from its equilibrium position, and this motion is described by the model

$$y = 1.5 \sin 8t - 0.5 \cos 8t$$

where y is the distance from equilibrium measured in feet and t is the time in seconds.

(a) Write the model in the form

$$y = \sqrt{a^2 + b^2} \sin(Bt + C).$$

(b) Find the amplitude of the oscillations of the weight.

(c) Find the frequency of the oscillations of the weight.

Synthesis

True or False? In Exercises 93–96, determine whether the statement is true or false. Justify your answer.

93. If $\dfrac{\pi}{2} < \theta < \pi$, then $\cos \dfrac{\theta}{2} < 0$.

94. $\sin(x + y) = \sin x + \sin y$

95. $4 \sin(-x) \cos(-x) = -2 \sin 2x$

96. $4 \sin 45° \cos 15° = 1 + \sqrt{3}$

97. List the reciprocal identities, quotient identities, and Pythagorean identities from memory.

98. Name the trigonometric functions that are considered to be odd.

99. *Think About It* If a trigonometric equation has an infinite number of solutions, is it true that the equation is an identity? Explain.

100. *Think About It* Explain why you know from observation that the equation $a \sin x - b = 0$ has no solution if $|a| < |b|$.

In Exercises 101 and 102, use the graphs of y_1 and y_2 to determine how to change one function to form the identity $y_1 = y_2$.

101. $y_1 = \sec^2\left(\dfrac{\pi}{2} - x\right)$

$y_2 = \cot^2 x$

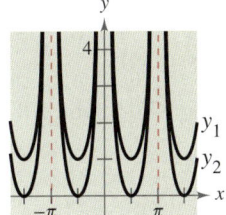

102. $y_1 = \dfrac{\cos 3x}{\cos x}$

$y_2 = (2 \sin x)^2$

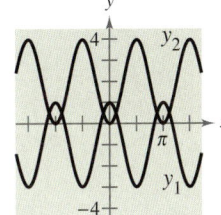

In Exercises 103 and 104, use a graphing utility with root- or zero-finding capabilities to approximate the zeros of the function.

103. $y = \sqrt{x + 3} + 4 \cos x$

104. $y = 2 - \dfrac{1}{2}x^2 + 3 \sin \dfrac{\pi x}{2}$

Chapter Project ▶ Projectile Motion

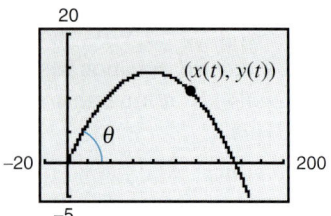

```
Tmin = 0
Tmax = 5
Tscl = .05
Xmin = -20
Xmax = 200
Xscl = 20
Ymin = -5
Ymax = 20
Yscl = 5
```

In this project, you will use parametric equations to model the path of a projectile. Parametric equations use a third variable t to represent time. The variables x and y are written as functions of t. For any time t, the horizontal position $x(t)$ and vertical position $y(t)$ of a projectile (ignoring air resistance) launched at ground level are given by the equations

$$x(t) = (v_0 \cos \theta)t$$
$$y(t) = (v_0 \sin \theta)t - 16t^2.$$

In these equations, θ is the angle with the horizontal and v_0 is the initial velocity in feet per second, as indicated in the figure at the left.

Set your graphing utility to parametric and degree modes, and use the viewing window shown at the left. Let $v_0 = 88$ feet per second and $\theta = 20°$ and graph the parametric equations

$$x(t) = (88 \cos 20°)t$$
$$y(t) = (88 \sin 20°)t - 16t^2.$$

Use the zoom and trace features to (a) find the maximum height attained by the projectile, (b) find the time at which the maximum height occurs, (c) determine the length of time that the projectile is in the air, and (d) determine the range of the projectile.

Chapter Project Investigations

1. Verify analytically the range of the projectile by solving the equation $(88 \sin 20°)t - 16t^2 = 0$ for t and then evaluating $x(t)$ at this t-value.

2. Use a graphing utility to find the maximum height and range of the projectile when $\theta = 30°$ and $v_0 = 132$ feet per second. What viewing window did you use?

3. Let $v_0 = 60$ feet per second and find the maximum range for the angles $\theta = 20°, 30°, 40°, 50°,$ and $60°$. In general, what angle should you use to produce the maximum range?

4. What is the relationship between the time the projectile reaches its maximum height and the time it takes for the projectile to return to the ground? Explain.

5. Eliminate t from the parametric equations

$$x(t) = (v_0 \cos \theta)t \quad \text{and} \quad y(t) = (v_0 \sin \theta)t - 16t^2$$

by solving for t in the first equation and substituting this value into the equation for y. Show in this case that the height of the projectile is given by the equation

$$y = (\tan \theta)x - \frac{16 \sec^2 \theta}{v_0^2}x^2.$$

Use this equation to find the angle θ corresponding to a maximum range of 200 feet and initial velocity of $v_0 = 80$ feet per second.

Chapter Test

The *Interactive* CD-ROM and *Internet* versions of this text provide answers to the Chapter Tests and Cumulative Tests. They also offer Chapter Pre-Tests (which test key skills and concepts covered in previous chapters) and Chapter Post-Tests, both of which have randomly generated exercises with diagnostic capabilities.

Take this test as you would take a test in class. After you are done, check your work against the answers given in the back of the book.

1. If $\tan \theta = \frac{3}{2}$ and $\cos \theta < 0$, use the fundamental identities to evaluate the other five trigonometric functions of θ.

2. Use the fundamental identities to simplify $\csc^2 \beta (1 - \cos^2 \beta)$.

3. Factor and simplify $\dfrac{\sec^4 x - \tan^4 x}{\sec^2 x + \tan^2 x}$.

4. Add and simplify $\dfrac{\cos \theta}{\sin \theta} + \dfrac{\sin \theta}{\cos \theta}$.

5. Determine the values of θ, $0 \leq \theta < 2\pi$, for which $\tan \theta = -\sqrt{\sec^2 \theta - 1}$ is true.

6. Use a graphing utility to graph the functions $y_1 = \cos x + \sin x \tan x$ and $y_2 = \sec x$. Make a conjecture about y_1 and y_2. Verify the result analytically.

In Exercises 7–12, verify the identity.

7. $\sin \theta \sec \theta = \tan \theta$

8. $\sec^2 x \tan^2 x + \sec^2 x = \sec^4 x$

9. $\dfrac{\csc \alpha + \sec \alpha}{\sin \alpha + \cos \alpha} = \cot \alpha + \tan \alpha$

10. $\cos\left(x + \dfrac{\pi}{2}\right) = -\sin x$

11. $\sin(n\pi + \theta) = (-1)^n \sin \theta$, n is an integer.

12. $(\sin x + \cos x)^2 = 1 + \sin 2x$

13. Rewrite $\sin^4 x \tan^2 x$ in terms of the first power of the cosine.

14. Use a half-angle formula to simplify the expression $\dfrac{\sin 4\theta}{1 + \cos 4\theta}$.

15. Write $4 \cos 2\theta \sin 4\theta$ as a sum or difference.

16. Write $\sin 3\theta - \sin 4\theta$ as a product.

In Exercises 17–20, find all solutions of the equation in the interval $[0, 2\pi)$.

17. $\tan^2 x + \tan x = 0$

18. $\sin 2\alpha - \cos \alpha = 0$

19. $4 \cos^2 x - 3 = 0$

20. $\csc^2 x - \csc x - 2 = 0$

21. Use a graphing utility to approximate the solutions of the equation $3 \cos x - x = 0$ accurate to three decimal places.

22. Explain why the equation $\cos^2 x + \cos x - 6 = 0$ has no solution.

23. Find the exact value of $\cos 105°$ using the fact that $105° = 135° - 30°$.

24. Use the figure to find the exact values of $\sin 2u$ and $\tan 2u$.

25. The index of refraction n of a transparent material is the ratio of the speed of light in a vacuum to the speed of light in the material. For the glass triangular prism in the figure, $n = 1.5$ and $\alpha = 60°$. Find the angle θ for the glass prism.

$$n = \dfrac{\sin(\theta/2 + \alpha/2)}{\sin(\theta/2)}$$

FIGURE FOR 24

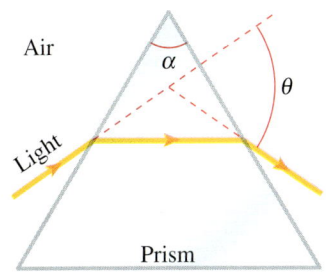

FIGURE FOR 25

3.1 ▶ **Law of Sines**
3.2 ▶ **Law of Cosines**
3.3 ▶ **Vectors in the Plane**
3.4 ▶ **Vectors and Dot Products**

In November of 1999, 81.4% of U.S. airline flights arrived on time. Factors such as severe weather, aircraft maintenance, and air traffic control decisions can cause flight delays.
(Source: U.S. Department of Transportation)

Superstock

3 Additional Topics in Trigonometry

▶ How to Study This Chapter

The Big Picture
In this chapter you will learn the following skills and concepts.

- How to use the Law of Sines and the Law of Cosines to solve oblique triangles
- How to find the areas of oblique triangles
- How to write the component forms of vectors and perform basic vector operations
- How to find the direction angles of vectors and the angle between two vectors

Important Vocabulary
As you encounter each new vocabulary term in this chapter, add the term and its definition to your notebook glossary.

Oblique triangle (p. 262)
Law of Sines (p. 262)
Law of Cosines (p. 271)
Directed line segment (p. 279)
Initial point (p. 279)
Terminal point (p. 279)
Magnitude of directed line segment (p. 279)
Vector **v** in the plane (p. 279)
Standard position (p. 280)
Component form of a vector **v** (p. 280)
Zero vector (p. 280)
Magnitude of **v** (p. 280)
Unit vector (p. 280)
Parallelogram law (p. 281)
Resultant (p. 281)
Standard unit vectors (p. 284)
Horizontal and vertical components of **v** (p. 284)

Linear combination of vectors (p. 284)
Direction angle (p. 285)
Dot product (p. 292)
Angle between two nonzero vectors (p. 293)
Orthogonal (p. 294)
Work (p. 297)

Study Tools
- Learning objectives at the beginning of each section
- Chapter Summary (p. 300)
- Review Exercises (pp. 301–303)
- Chapter Test (p. 305)
- Cumulative Test for Chapters 1–3 (pp. 306–307)

Additional Resources
- Study and Solutions Guide
- Interactive Trigonometry
- Videotapes for Chapter 3
- Trigonometry Website
- Student Success Organizer

STUDY T!P
When your test is returned, review it carefully. Rework the problems you answered incorrectly. Discovering your mistakes will help you to avoid repeating them.

3.1 Law of Sines

What you should learn

- How to use the Law of Sines to solve oblique triangles (AAS, ASA, or SSA)
- How to find the areas of oblique triangles
- How to use the Law of Sines to model and solve real-life problems

Why you should learn it

You can use the Law of Sines to solve real-life problems involving oblique triangles. For instance, in Exercise 44 on page 269 you use the Law of Sines to determine the distance from a ranger station to a forest fire.

Introduction

In Chapter 1 you looked at techniques for solving right triangles. In this section and the next, you will solve **oblique triangles**—triangles that have no right angles. As standard notation, the angles of a triangle are labeled as A, B, and C, and their opposite sides as a, b, and c, as shown in Figure 3.1.

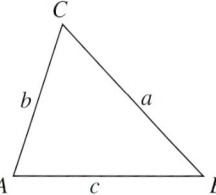

FIGURE 3.1

To solve an oblique triangle, you need to know the measure of at least one side and any two other parts of the triangle—either two sides, two angles, or one angle and one side. This breaks down into the following four cases.

1. Two angles and any side (AAS or ASA)
2. Two sides and an angle opposite one of them (SSA)
3. Three sides (SSS)
4. Two sides and their included angle (SAS)

The first two cases can be solved using the **Law of Sines,** whereas the last two cases require the Law of Cosines (Section 3.2). A proof of the Law of Sines is given in Appendix A.

Law of Sines

If ABC is a triangle with sides a, b, and c, then

$$\frac{a}{\sin A} = \frac{b}{\sin B} = \frac{c}{\sin C}.$$

A is acute.

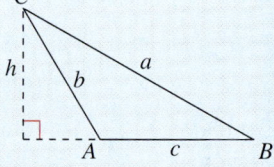

A is obtuse.

The Law of Sines can also be written in the reciprocal form

$$\frac{\sin A}{a} = \frac{\sin B}{b} = \frac{\sin C}{c}.$$

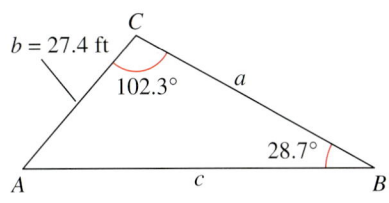

FIGURE 3.2

Example 1 ▶ Given Two Angles and One Side—AAS

For the triangle in Figure 3.2, $C = 102.3°$, $B = 28.7°$, and $b = 27.4$ feet. Find the remaining angle and sides.

Solution

The third angle of the triangle is

$$A = 180° - B - C$$
$$= 180° - 28.7° - 102.3°$$
$$= 49.0°.$$

By the Law of Sines, you have

$$\frac{a}{\sin 49°} = \frac{b}{\sin 28.7°} = \frac{c}{\sin 102.3°}.$$

Using $b = 27.4$ produces

$$a = \frac{b}{\sin B}(\sin A) = \frac{27.4}{\sin 28.7°}(\sin 49°) \approx 43.06 \text{ feet}$$

and

$$c = \frac{b}{\sin B}(\sin C) = \frac{27.4}{\sin 28.7°}(\sin 102.3°) \approx 55.75 \text{ feet}.$$

Example 2 ▶ Given Two Angles and One Side—ASA

A pole tilts *toward* the sun at an 8° angle from the vertical, and it casts a 22-foot shadow. The angle of elevation from the tip of the shadow to the top of the pole is 43°. How tall is the pole?

Solution

From Figure 3.3, note that $A = 43°$ and $B = 90° + 8° = 98°$. So, the third angle is

$$C = 180° - A - B$$
$$= 180° - 43° - 98°$$
$$= 39°.$$

By the Law of Sines, you have

$$\frac{a}{\sin 43°} = \frac{c}{\sin 39°}.$$

Because $c = 22$ feet, the length of the pole is

$$a = \frac{c}{\sin C}(\sin A) = \frac{22}{\sin 39°}(\sin 43°) \approx 23.84 \text{ feet}.$$

For practice, try reworking Example 2 for a pole that tilts *away from* the sun under the same conditions.

> **STUDY TIP**
>
> When solving triangles, a careful sketch is useful as a quick test for the feasibility of an answer. Remember that the longest side lies opposite the largest angle, and the shortest side lies opposite the smallest angle.

FIGURE 3.3

A computer simulation to accompany this concept appears in the *Interactive CD-ROM* and *Internet* versions of this text.

The Ambiguous Case (SSA)

In Examples 1 and 2 you saw that two angles and one side determine a unique triangle. However, if two sides and one opposite angle are given, three possible situations can occur: (1) no such triangle exists, (2) one such triangle exists, or (3) two distinct triangles may satisfy the conditions.

The Ambiguous Case (SSA)

Consider a triangle in which you are given a, b, and A.   ($h = b \sin A$)

	A is acute.	A is acute.	A is acute.	A is acute.	A is obtuse.	A is obtuse.
Sketch						
Necessary condition	$a < h$	$a = h$	$a > b$	$h < a < b$	$a \leq b$	$a > b$
Triangles possible	None	One	One	Two	None	One

Example 3 ▶ Single-Solution Case—SSA

For the triangle in Figure 3.4, $a = 22$ inches, $b = 12$ inches, and $A = 42°$. Find the remaining side and angles.

Solution

By the Law of Sines, you have

$$\frac{22}{\sin 42°} = \frac{12}{\sin B}$$

$$\sin B = b\left(\frac{\sin A}{a}\right) = 12\left(\frac{\sin 42°}{22}\right) \approx 0.3649803$$

$$B \approx 21.41°. \qquad \text{B is acute.}$$

Now, you can determine that

$$C \approx 180° - 42° - 21.41°$$
$$= 116.59°.$$

Then the remaining side is

$$\frac{c}{\sin 116.59°} = \frac{22}{\sin 42°}$$

$$c = \sin C\left(\frac{a}{\sin A}\right) = \sin 116.59°\left(\frac{22}{\sin 42°}\right) \approx 29.40 \text{ inches.}$$

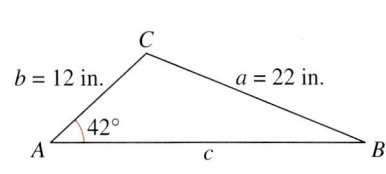

One solution: $a > b$
FIGURE 3.4

No solution: a < h
FIGURE 3.5

Example 4 ▶ No-Solution Case—SSA

Show that there is no triangle for which $a = 15$, $b = 25$, and $A = 85°$.

Solution

Begin by making the sketch shown in Figure 3.5. From this figure it appears that no triangle is formed. You can verify this using the Law of Sines.

$$\frac{a}{\sin A} = \frac{b}{\sin B}$$

$$\frac{15}{\sin 85°} = \frac{25}{\sin B}$$

$$\sin B = b\left(\frac{\sin A}{a}\right) = 25\left(\frac{\sin 85°}{15}\right) \approx 1.660 > 1$$

This contradicts the fact that $|\sin B| \leq 1$. So, no triangle can be formed having sides $a = 15$ and $b = 25$ and an angle of $A = 85°$.

Example 5 ▶ Two-Solution Case—SSA

Find two triangles for which $a = 12$ meters, $b = 31$ meters, and $A = 20.5°$.

Solution

By the Law of Sines, you have

$$\frac{a}{\sin A} = \frac{b}{\sin B}$$

$$\sin B = b\left(\frac{\sin A}{a}\right) = 31\left(\frac{\sin 20.5°}{12}\right) \approx 0.9047.$$

There are two angles $B_1 \approx 64.8°$ and $B_2 \approx 180° - 64.8° \approx 115.2°$ between $0°$ and $180°$ whose sine is 0.9047. For $B_1 \approx 64.8°$, you obtain

$$C \approx 180° - 20.5° - 64.8° = 94.7°$$

$$c = \frac{a}{\sin A}(\sin C) = \frac{12}{\sin 20.5°}(\sin 94.7°) \approx 34.15 \text{ meters.}$$

For $B_2 \approx 115.2°$, you obtain

$$C \approx 180° - 20.5° - 115.2° = 44.3°$$

$$c = \frac{a}{\sin A}(\sin C) = \frac{12}{\sin 20.5°}(\sin 44.3°) \approx 23.93 \text{ meters.}$$

The resulting triangles are shown in Figure 3.6.

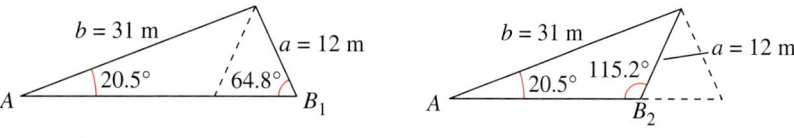

FIGURE 3.6

Area of an Oblique Triangle

The procedure used to prove the Law of Sines leads to a simple formula for the area of an oblique triangle. Referring to Figure 3.7, note that each triangle has a height of $h = b \sin A$. Consequently, the area of each triangle is

$$\text{Area} = \frac{1}{2}(\text{base})(\text{height}) = \frac{1}{2}(c)(b \sin A) = \frac{1}{2}bc \sin A.$$

By similar arguments, you can develop the formulas

$$\text{Area} = \frac{1}{2}ab \sin C = \frac{1}{2}ac \sin B.$$

A is acute.

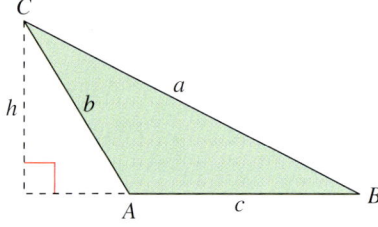
A is obtuse.

FIGURE 3.7

Area of an Oblique Triangle

The area of any triangle is one-half the product of the lengths of two sides times the sine of their included angle. That is,

$$\text{Area} = \frac{1}{2}bc \sin A = \frac{1}{2}ab \sin C = \frac{1}{2}ac \sin B.$$

Note that if angle A is $90°$, the formula gives the area for a right triangle:

$$\text{Area} = \frac{1}{2}bc = \frac{1}{2}(\text{base})(\text{height}).$$

Similar results are obtained for angles C and B equal to $90°$.

Example 6 ▶ Finding the Area of a Triangular Lot

Find the area of a triangular lot having two sides of lengths 90 meters and 52 meters and an included angle of $102°$.

Solution

Consider $a = 90$ m, $b = 52$ m, and angle $C = 102°$, as shown in Figure 3.8. Then the area of the triangle is

$$\text{Area} = \frac{1}{2}ab \sin C = \frac{1}{2}(90)(52)(\sin 102°) \approx 2289 \text{ square meters}.$$

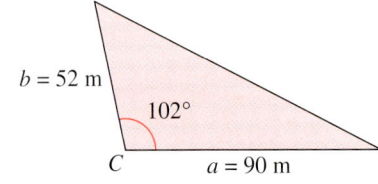

FIGURE 3.8

Section 3.1 ▶ Law of Sines 267

FIGURE 3.9

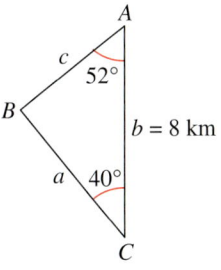

FIGURE 3.10

Application

Example 7 ▶ An Application of the Law of Sines

The course for a boat race starts at point A and proceeds in the direction S 52° W to point B, then in the direction S 40° E to point C, and finally back to A, as shown in Figure 3.9. Point C lies 8 kilometers directly south of point A. Approximate the total distance of the race course.

Solution

Because lines BD and AC are parallel, it follows that $\angle BCA \cong \angle DBC$. Consequently, triangle ABC has the measures shown in Figure 3.10. For angle B, you have $B = 180° - 52° - 40° = 88°$. Using the Law of Sines

$$\frac{a}{\sin 52°} = \frac{b}{\sin 88°} = \frac{c}{\sin 40°}$$

you can let $b = 8$ and obtain

$$a = \frac{8}{\sin 88°}(\sin 52°) \approx 6.308$$

and

$$c = \frac{8}{\sin 88°}(\sin 40°) \approx 5.145.$$

The total length of the course is approximately

Length $\approx 8 + 6.308 + 5.145 = 19.453$ kilometers.

Writing ABOUT MATHEMATICS

Using the Law of Sines In this section, you have been using the Law of Sines to solve *oblique* triangles. Can the Law of Sines also be used to solve a right triangle? If so, write a short paragraph explaining how to use the Law of Sines to solve the following two triangles. Is there an easier way to solve these triangles?

a. (AAS)

b. (ASA)

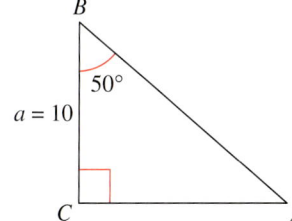

3.1 Exercises

In Exercises 1–18, use the information to solve the triangle.

1.

2.

3.

4.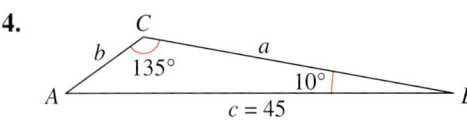

5. $A = 36°$, $a = 8$, $b = 5$
6. $A = 60°$, $a = 9$, $c = 10$
7. $A = 102.4°$, $C = 16.7°$, $a = 21.6$
8. $A = 24.3°$, $C = 54.6°$, $c = 2.68$
9. $A = 83° 20'$, $C = 54.6°$, $c = 18.1$
10. $A = 5° 40'$, $B = 8° 15'$, $b = 4.8$
11. $B = 15° 30'$, $a = 4.5$, $b = 6.8$
12. $B = 2° 45'$, $b = 6.2$, $c = 5.8$
13. $C = 145°$, $b = 4$, $c = 14$
14. $A = 100°$, $a = 125$, $c = 10$
15. $A = 110° 15'$, $a = 48$, $b = 16$
16. $C = 85° 20'$, $a = 35$, $c = 50$
17. $A = 55°$, $B = 42°$, $c = \frac{3}{4}$
18. $B = 28°$, $C = 104°$, $a = 3\frac{5}{8}$

In Exercises 19–26, use the information to solve (if possible) the triangle. If two solutions exist, find both.

19. $A = 58°$, $a = 4.5$, $b = 12.8$
20. $A = 58°$, $a = 11.4$, $b = 12.8$
21. $A = 76°$, $a = 18$, $b = 20$
22. $A = 76°$, $a = 34$, $b = 21$
23. $A = 110°$, $a = 125$, $b = 200$
24. $A = 110°$, $a = 125$, $b = 100$
25. $A = 22°$, $a = \frac{5}{12}$, $b = 1\frac{3}{8}$
26. $A = 22°$, $a = \frac{5}{7}$, $b = \frac{5}{7}$

In Exercises 27–30, find a value for b such that the triangle has (a) one solution, (b) two solutions, and (c) no solution.

27. $A = 36°$, $a = 5$
28. $A = 60°$, $a = 10$
29. $A = 10°$, $a = 10.8$
30. $A = 88°$, $a = 315.6$

In Exercises 31–36, find the area of the triangle having the indicated sides and angle.

31. $C = 120°$, $a = 4$, $b = 6$
32. $B = 130°$, $a = 62$, $c = 20$
33. $A = 43° 45'$, $b = 57$, $c = 85$
34. $A = 5° 15'$, $b = 4.5$, $c = 22$
35. $B = 72° 30'$, $a = 105$, $c = 64$
36. $C = 84° 30'$, $a = 16$, $b = 20$

37. **Height** Because of prevailing winds, a tree grew so that it was leaning 4° from the vertical. At a point 35 meters from the tree, the angle of elevation to the top of the tree is 23° (see figure). Find the height h of the tree.

38. **Height** A flagpole at a right angle to the horizontal is located on a slope that makes an angle of 12° with the horizontal. The pole's shadow is 16 meters long and points directly up the slope. The angle of elevation from the tip of the shadow to the sun is 20°.

   (a) Draw a triangle that represents the problem. Show the known quantities on the triangle and use a variable to indicate the height of the flagpole.

   (b) Write an equation involving the unknown quantity.

   (c) Find the height of the flagpole.

39. *Angle of Elevation* A 10-meter telephone pole casts a 17-meter shadow directly down a slope when the angle of elevation of the sun is 42° (see figure). Find θ, the angle of elevation of the ground.

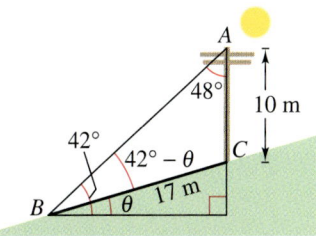

40. *Flight Path* A plane flies 500 kilometers with a bearing of N 44° W from Naples to Elgin (see figure). The plane then flies 720 kilometers from Elgin to Canton. Find the bearing of the flight from Elgin to Canton.

41. *Bridge Design* A bridge is to be built across a small lake from a gazebo to a dock (see figure). The bearing from the gazebo to the dock is S 41° W. From a tree 100 meters from the gazebo, the bearings to the gazebo and the dock are S 74° E and S 28° E, respectively. Find the distance from the gazebo to the dock.

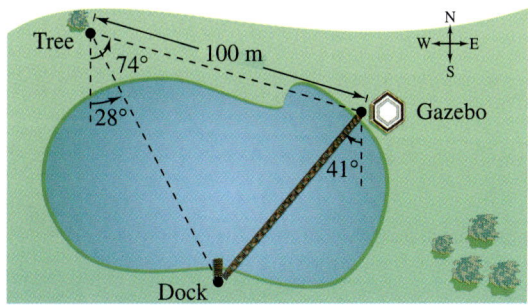

42. *Railroad Track Design* The circular arc of a railroad curve has a chord of length 3000 feet and a central angle of 40°.

(a) Draw a figure that visually represents the problem. Show the known quantities on the figure and use variables r and s to represent the radius of the arc and the length of the arc, respectively.

(b) Find the radius r of the circular arc.

(c) Find the length s of the circular arc.

43. *Glide Path* A pilot has just started on the glide path for landing at an airport where the length of the runway is 9000 feet. The angles of depression from the plane to the ends of the runway are 17.5° and 18.8°.

(a) Draw a figure that visually represents the problem.

(b) Find the air distance the plane must travel until touching down on the near end of the runway.

(c) Find the ground distance the plane must travel until touching down.

(d) Find the altitude of the plane when the pilot begins the descent.

44. *Locating a Fire* The bearing from the Pine Knob fire tower to the Colt Station fire tower is N 65° E and the two towers are 30 kilometers apart. A fire spotted by rangers in each tower has a bearing of N 80° E from Pine Knob and S 70° E from Colt Station. Find the distance of the fire from each tower.

45. *Distance* A boat is sailing due east parallel to the shoreline at a speed of 10 miles per hour. At a given time the bearing to the lighthouse is S 70° E, and 15 minutes later the bearing is S 63° E (see figure). Find the distance from the boat to the shoreline if the lighthouse is at the shoreline.

46. *Distance* A family is traveling due west on a road that passes a famous landmark. At a given time the bearing to the landmark is N 62° W, and after the family travels 5 miles farther the bearing is N 38° W. What is the closest the family will come to the landmark while on the road?

47. *Distance* The angles of elevation θ and ϕ to an airplane from the airport control tower and from an observation post 2 miles away are being continuously monitored (see figure). Write an equation giving the distance d between the plane and observation post in terms of θ and ϕ.

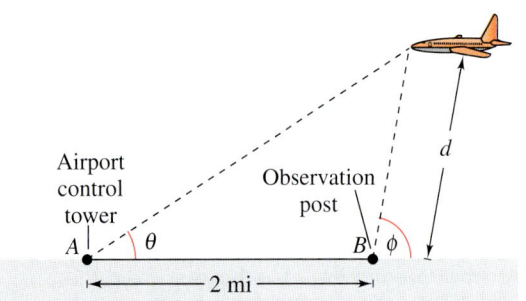

Synthesis

True or False? In Exercises 48 and 49, determine whether the statement is true or false. Justify your answer.

48. It is not possible to create an obtuse triangle whose longest side is one of the sides that forms its obtuse angle.

49. Two angles and one side of a triangle do not necessarily determine a unique triangle.

50. *Graphical and Numerical Analysis* In the figure, α and β are positive angles.

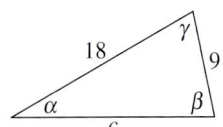

(a) Write α as a function of β.

(b) Use a graphing utility to graph the function. Determine its domain and range.

(c) Use the result of part (a) to write c as a function of β.

(d) Use a graphing utility to graph the function in part (c). Determine its domain and range.

(e) Complete the table. What can you infer?

β	0.4	0.8	1.2	1.6	2.0	2.4	2.8
α							
c							

51. *Graphical Analysis*

(a) Write the area A of the shaded region in the figure as a function of θ.

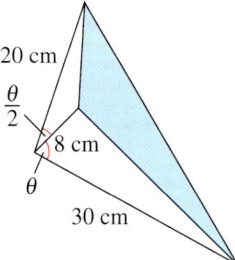

(b) Use a graphing utility to graph the area function.

(c) Determine the domain of the area function. Explain how the area of the region and the domain of the function would change if the 8-centimeter line segment were decreased in length.

Review

In Exercises 52–55, use the given values to evaluate the other five trigonometric functions.

52. $\sin x = \dfrac{2}{5}, \quad \dfrac{\pi}{2} < x < \pi$

53. $\cos x = \dfrac{1}{5}, \quad \dfrac{3\pi}{2} < x < 2\pi$

54. $\sec x = -4, \quad \pi < x < \dfrac{3\pi}{2}$

55. $\tan x = -5, \quad \dfrac{\pi}{2} < x < \pi$

In Exercises 56–59, use the fundamental trigonometric identities to simplify the expression.

56. $\sin x \cot x$ **57.** $\tan x \cos x \sec x$

58. $1 - \sin^2\left(\dfrac{\pi}{2} - x\right)$ **59.** $1 + \cot^2\left(\dfrac{\pi}{2} - x\right)$

3.2 Law of Cosines

▶ What you should learn

- How to use the Law of Cosines to solve oblique triangles (SSS or SAS)
- How to use the Law of Cosines to model and solve real-life problems
- How to use Heron's Area Formula to find the area of a triangle

▶ Why you should learn it

You can use the Law of Cosines to solve real-life problems involving oblique triangles. For instance, in Exercise 40 on page 276 you use the Law of Cosines to approximate the number of feet a baseball player must run to catch a ball.

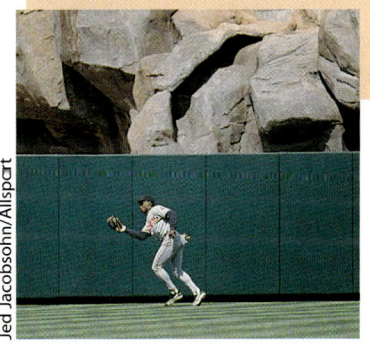

Introduction

Two cases remain in the list of conditions needed to solve an oblique triangle—SSS and SAS. To use the Law of Sines, you must know at least one side and its opposite angle. If you are given three sides (SSS), or two sides and their included angle (SAS), none of the ratios would be complete. In such cases you can use the **Law of Cosines**. A proof of the Law of Cosines is given in Appendix A.

Law of Cosines

Standard Form	Alternative Form
$a^2 = b^2 + c^2 - 2bc \cos A$	$\cos A = \dfrac{b^2 + c^2 - a^2}{2bc}$
$b^2 = a^2 + c^2 - 2ac \cos B$	$\cos B = \dfrac{a^2 + c^2 - b^2}{2ac}$
$c^2 = a^2 + b^2 - 2ab \cos C$	$\cos C = \dfrac{a^2 + b^2 - c^2}{2ab}$

Example 1 ▶ Three Sides of a Triangle—SSS

Find the three angles of the triangle in Figure 3.11.

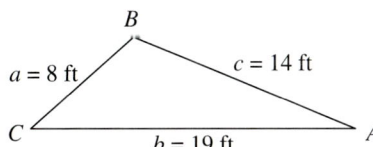

FIGURE 3.11

Solution

It is a good idea first to find the angle opposite the longest side—side b in this case (see Figure 3.11). Using the Law of Cosines, you find that

$$\cos B = \frac{a^2 + c^2 - b^2}{2ac} = \frac{8^2 + 14^2 - 19^2}{2(8)(14)} \approx -0.45089.$$

Because $\cos B$ is negative, you know that B is an *obtuse* angle given by $B \approx 116.80°$. At this point, knowing that $B \approx 116.80°$, it is simpler to use the Law of Sines to determine A.

$$\sin A = a\left(\frac{\sin B}{b}\right) \approx 8\left(\frac{\sin 116.80°}{19}\right) \approx 0.37582$$

Because B is obtuse, you know that A must be acute, because a triangle can have, at most, one obtuse angle. So, $A \approx 22.08°$ and

$$C \approx 180° - 22.08° - 116.80° = 41.12°.$$

Exploration

What familiar formula do you obtain when you use the third form of the Law of Cosines

$$c^2 = a^2 + b^2 - 2ab \cos C$$

and you let $C = 90°$? What is the relationship between the Law of Cosines and this formula?

Do you see why it was wise to find the largest angle *first* in Example 1? Knowing the cosine of an angle, you can determine whether the angle is acute or obtuse. That is,

$\cos \theta > 0$ for $0° < \theta < 90°$ Acute

$\cos \theta < 0$ for $90° < \theta < 180°$.   Obtuse

So, in Example 1, once you found that angle B was obtuse, you knew that angles A and C were both acute. If the largest angle is acute, the remaining two angles are acute also.

Example 2 ▶ Two Sides and the Included Angle—SAS

Find the remaining angles and side of the triangle in Figure 3.12.

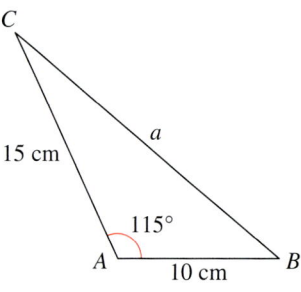

FIGURE 3.12

The *Interactive* CD-ROM and *Internet* versions of this text show every example with its solution; clicking on the *Try It!* button brings up similar problems. Guided Examples and Integrated Examples show step-by-step solutions to additional examples. Integrated Examples are related to several concepts in the section.

Solution

Use the Law of Cosines to find the unknown side a in the figure.

$$a^2 = b^2 + c^2 - 2bc \cos A$$

$$a^2 = 15^2 + 10^2 - 2(15)(10) \cos 115°$$

$$a^2 \approx 451.79$$

$$a \approx 21.26$$

Because $a \approx 21.26$ cm, you now know the ratio $a/\sin A$ and you can use the Law of Sines

$$\frac{a}{\sin A} = \frac{b}{\sin B}$$

to solve for B.

$$\sin B = b\left(\frac{\sin A}{a}\right)$$

$$= 15\left(\frac{\sin 115°}{21.26}\right)$$

$$\approx 0.63945$$

So, $B = \arcsin 0.63945 \approx 39.75°$ and $C \approx 180° - 115° - 39.75° = 25.25°$.

Section 3.2 ▶ Law of Cosines   273

Applications

Example 3 ▶ An Application of the Law of Cosines

The pitcher's mound on a softball field is 46 feet from home plate and the distance between the bases is 60 feet, as shown in Figure 3.13. (The pitcher's mound is not halfway between home plate and second base.) How far is the pitcher's mound from first base?

Solution

In triangle HPF, $H = 45°$ (line HP bisects the right angle at H), $f = 46$, and $p = 60$. Using the Law of Cosines for this SAS case, you have

$$h^2 = f^2 + p^2 - 2fp \cos H$$
$$= 46^2 + 60^2 - 2(46)(60) \cos 45°$$
$$\approx 1812.8.$$

Therefore, the approximate distance from the pitcher's mound to first base is

$$h \approx \sqrt{1812.8}$$
$$\approx 42.58 \text{ feet}.$$

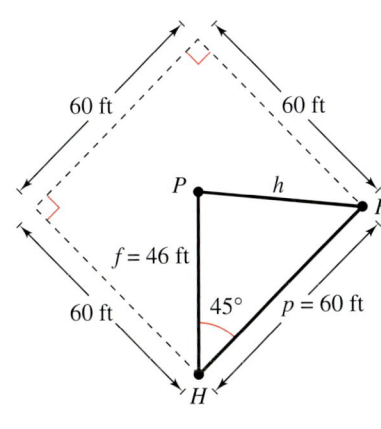

FIGURE 3.13

Example 4 ▶ An Application of the Law of Cosines

A ship travels 60 miles due east, then adjusts its course northward, as shown in Figure 3.14. After traveling 80 miles in that direction, the ship is 139 miles from its point of departure. Describe the bearing from point B to point C.

 The *Interactive* CD-ROM and *Internet* versions of this text offer a built-in graphing calculator, which can be used in the Examples, Explorations, Technology notes, and Exercises.

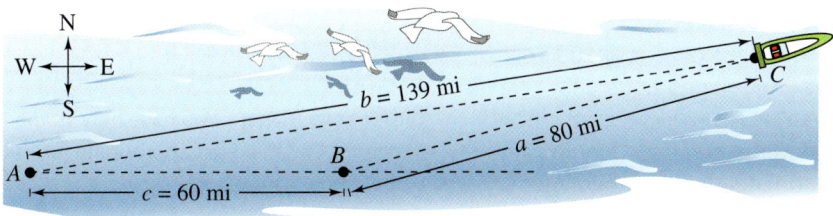

FIGURE 3.14

Solution

You have $a = 80$, $b = 139$, and $c = 60$; so, using the alternative form of the Law of Cosines, you have

$$\cos B = \frac{a^2 + c^2 - b^2}{2ac}$$
$$= \frac{80^2 + 60^2 - 139^2}{2(80)(60)}$$
$$\approx -0.97094.$$

Therefore, $B \approx \arccos(-0.97094) \approx 166.15°$. So, the bearing measured from due north from point B to point C is N 76.15° E.

Heron's Area Formula

The Law of Cosines can be used to establish the following formula for the area of a triangle. This formula is credited to the Greek mathematician Heron (c. 100 B.C.). A proof of this formula is given in Appendix A.

Heron's Area Formula

Given any triangle with sides of lengths a, b, and c, the area of the triangle is

$$\text{Area} = \sqrt{s(s-a)(s-b)(s-c)}$$

where $s = (a + b + c)/2$.

Example 5 ▶ Using Heron's Area Formula

Find the area of the triangular region having sides of lengths $a = 43$ meters, $b = 53$ meters, and $c = 72$ meters.

Solution

Because $s = (a + b + c)/2 = 168/2 = 84$, Heron's Area Formula yields

$$\text{Area} = \sqrt{s(s-a)(s-b)(s-c)}$$
$$= \sqrt{84(41)(31)(12)}$$
$$\approx 1131.89 \text{ square meters.}$$

Historical Note

Heron of Alexandria (c. 100 B.C.) was a Greek geometer and inventor. His works describe how to find the areas of triangles, quadrilaterals, regular polygons having 3 to 12 sides, and circles as well as the surface areas and volumes of three-dimensional objects.

Writing ABOUT MATHEMATICS

The Area of a Triangle You have now studied three different formulas for the area of a triangle. Use the most appropriate formula to find the area of each triangle below. Show your work and give your reasons for choosing each formula.

Standard Formula   Area $= \frac{1}{2}bh$

Oblique Triangle   Area $= \frac{1}{2}bc \sin A = \frac{1}{2}ab \sin C = \frac{1}{2}ac \sin B$

Heron's Area Formula   Area $= \sqrt{s(s-a)(s-b)(s-c)}$

a.

b.

c.

d.
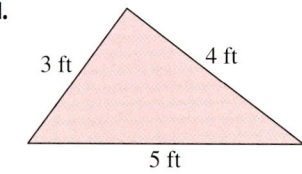

3.2 Exercises

In Exercises 1–16, use the Law of Cosines to solve the triangle.

1.
2.
3.
4.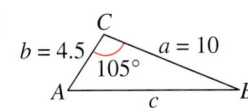

5. $a = 11$, $b = 14$, $c = 20$
6. $a = 55$, $b = 25$, $c = 72$
7. $a = 75.4$, $b = 52$, $c = 52$
8. $a = 1.42$, $b = 0.75$, $c = 1.25$
9. $A = 135°$, $b = 4$, $c = 9$
10. $A = 55°$, $b = 3$, $c = 10$
11. $B = 10° \, 35'$, $a = 40$, $c = 30$
12. $B = 75° \, 20'$, $a = 6.2$, $c = 9.5$
13. $B = 125° \, 40'$, $a = 32$, $c = 32$
14. $C = 15° \, 15'$, $a = 6.25$, $b = 2.15$
15. $C = 43°$, $a = \frac{4}{9}$, $b = \frac{7}{9}$
16. $C = 103°$, $a = \frac{3}{8}$, $b = \frac{3}{4}$

In Exercises 17–22, complete the table by solving the parallelogram shown in the figure. (The lengths of the diagonals are given by c and d.)

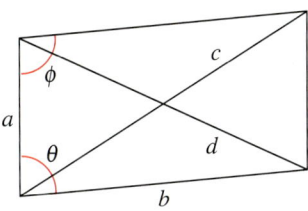

	a	b	c	d	θ	φ
17.	5	8			45°	
18.	25	35				120°
19.	10	14	20			
20.	40	60		80		
21.	15		25	20		
22.		25	50	35		

In Exercises 23–28, use Heron's Area Formula to find the area of the triangle.

23. $a = 5$, $b = 7$, $c = 10$
24. $a = 12$, $b = 15$, $c = 9$
25. $a = 2.5$, $b = 10.2$, $c = 9$
26. $a = 75.4$, $b = 52$, $c = 52$
27. $a = 12.32$, $b = 8.46$, $c = 15.05$
28. $a = 3.05$, $b = 0.75$, $c = 2.45$

29. *Navigation* A boat race runs along a triangular course marked by buoys A, B, and C. The race starts with the boats headed west for 3700 meters. The other two sides of the course lie to the north of the first side, and their lengths are 1700 meters and 3000 meters. Draw a figure that gives a visual representation of the problem, and find the bearings for the last two legs of the race.

30. *Navigation* A plane flies 810 miles from Niagara to Cuyahoga with a bearing of N 75° E. Then it flies 648 miles from Cuyahoga to Rosemount with a bearing of N 32° E. Draw a figure that visually represents the problem, and find the straight-line distance and bearing from Niagara to Rosemount.

31. *Surveying* To approximate the length of a marsh, a surveyor walks 250 meters from point A to point B, then turns 75° and walks 220 meters to point C (see figure). Approximate the length AC of the marsh.

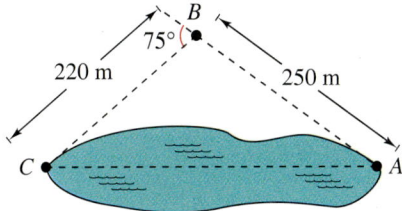

32. *Surveying* A triangular parcel of land has 115 meters of frontage, and the other boundaries have lengths of 76 meters and 92 meters. What angles does the frontage make with the two other boundaries?

33. *Surveying* A triangular parcel of ground has sides of lengths 725 feet, 650 feet, and 575 feet. Find the measure of the largest angle.

34. *Streetlight Design* Determine the angle θ in the design of the streetlight shown in the figure.

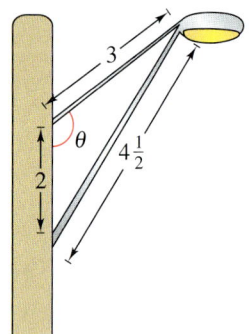

35. *Distance* Two ships leave a port at 9 A.M. One travels at a bearing of N 53° W at 12 miles per hour and the other travels at a bearing of S 67° W at 16 miles per hour. Approximate how far apart they are at noon that day.

36. *Distance* A 100-foot vertical tower is to be erected on the side of a hill that makes a 6° angle with the horizontal (see figure). Find the length of each of the two guy wires that will be anchored 75 feet uphill and downhill from the base of the tower.

37. *Navigation* On a map, Orlando is 178 millimeters due south of Niagara Falls, Denver is 273 millimeters from Orlando, and Denver is 235 millimeters from Niagara Falls (see figure).
(a) Find the bearing of Denver from Orlando.
(b) Find the bearing of Denver from Niagara Falls.

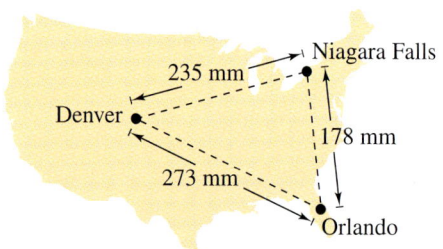

38. *Navigation* On a map, Minneapolis is 165 millimeters due west of Albany, Phoenix is 216 millimeters from Minneapolis, and Phoenix is 368 millimeters from Albany (see figure).
(a) Find the bearing of Minneapolis from Phoenix.
(b) Find the bearing of Albany from Phoenix.

39. *Baseball* On a baseball diamond with 90-foot sides, the pitcher's mound is 60.5 feet from home plate. How far is it from the pitcher's mound to third base?

40. *Baseball* The baseball player in center field is playing approximately 330 feet from the television camera that is behind home plate. A batter hits a fly ball that goes to the wall 420 feet from the camera (see figure). Approximate the number of feet that the center fielder has to run to make the catch if the camera turns 8° to follow the play.

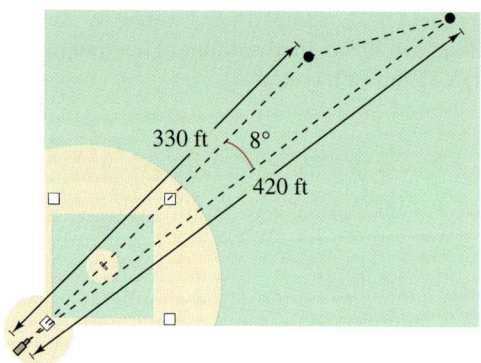

41. *Aircraft Tracking* To determine the distance between two aircraft, a tracking station continuously determines the distance to each aircraft and the angle A between them (see figure). Determine the distance a between the planes when $A = 42°$, $b = 35$ miles, and $c = 20$ miles.

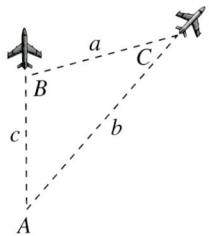

FIGURE FOR 41

with the paper on the 4-inch roller is s inches. Complete the following table.

d (inches)	9	10	12	13	14	15	16
θ (degrees)							
s (inches)							

42. *Aircraft Tracking* Use the figure for Exercise 41 to determine the distance a between the planes when $A = 11°$, $b = 20$ miles, and $c = 20$ miles.

43. *Engineering* If Q is the midpoint of the line segment $\overline{PR}$ in the truss rafter shown in the figure, find the lengths of the line segments $\overline{PQ}$, $\overline{QS}$, and $\overline{RS}$.

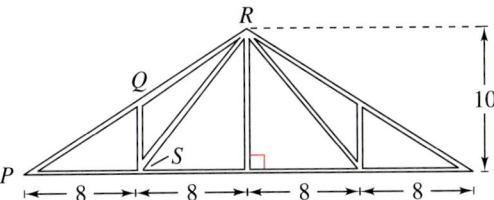

44. *Engine Design* An engine has a 7-inch connecting rod fastened to a crank (see figure).

(a) Use the Law of Cosines to write an equation giving the relationship between x and θ.

(b) Write x as a function of θ. (Select the sign that yields positive values of x.)

(c) Use a graphing utility to graph the function in part (b).

(d) Use the graph in part (c) to determine the maximum distance the piston moves in one cycle.

45. *Paper Manufacturing* In a certain process with continuous paper, the paper passes across three rollers of radii 3 inches, 4 inches, and 6 inches (see figure). The centers of the 3-inch and 6-inch rollers are d inches apart, and the length of the arc in contact

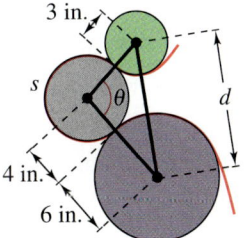

FIGURE FOR 45

46. *Awning Design* A retractable awning lowers at an angle of 50° from the top of a patio door that is 7 feet tall (see figure). Find the length x of the awning if no direct sunlight is to enter the door when the angle of elevation of the sun is greater than 70°.

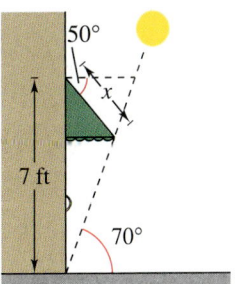

47. *Area* The lengths of the sides of a triangular parcel of land are approximately 200 feet, 500 feet, and 600 feet. Approximate the area of the parcel.

48. *Area* A parking lot has the shape of a parallelogram (see figure). The lengths of two adjacent sides are 70 meters and 100 meters. Find the area of the parking lot if the angle between the two sides is 70°.

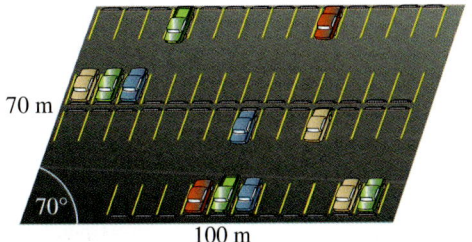

Synthesis

True or False? In Exercises 49 and 50, determine whether the statement is true or false. Justify your answer.

49. In Heron's Area Formula

    Area $= \sqrt{s(s-a)(s-b)(s-c)}$

    s is the average of the lengths of the three sides of the triangle.

50. In addition to SSS and SAS, the Law of Cosines can be used to solve triangles with SSA conditions.

51. ***Circumscribed and Inscribed Circles*** Let R and r be the radii of the circumscribed and inscribed circles of a triangle ABC, respectively (see figure), and let $s = (a + b + c)/2$.

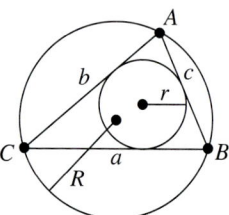

   (a) Prove that $2R = \dfrac{a}{\sin A} = \dfrac{b}{\sin B} = \dfrac{c}{\sin C}$.

   (b) Prove that $r = \sqrt{\dfrac{(s-a)(s-b)(s-c)}{s}}$.

Circumscribed and Inscribed Circles In Exercises 52 and 53, use the results of Exercise 51.

52. Given a triangle with $a = 25$, $b = 55$, and $c = 72$, find the area of (a) the triangle, (b) the circumscribed circle, and (c) the inscribed circle.

53. Find the length of the largest circular track that can be built on a triangular piece of property with sides of lengths 200 feet, 250 feet, and 325 feet.

54. Use the Law of Cosines to prove that

$$\frac{1}{2}bc(1 + \cos A) = \frac{a+b+c}{2} \cdot \frac{-a+b+c}{2}.$$

55. Use the Law of Cosines to prove that

$$\frac{1}{2}bc(1 - \cos A) = \frac{a-b+c}{2} \cdot \frac{a+b-c}{2}.$$

Review

In Exercises 56–61, evaluate the expression without the aid of a calculator.

56. $\arcsin(-1)$
57. $\arccos 0$
58. $\arctan \sqrt{3}$
59. $\arctan -\sqrt{3}$
60. $\arcsin\left(-\dfrac{\sqrt{3}}{2}\right)$
61. $\arccos\left(-\dfrac{\sqrt{3}}{2}\right)$

In Exercises 62–65, write an algebraic expression that is equivalent to the expression.

62. $\sec(\arcsin 2x)$
63. $\tan(\arccos 3x)$
64. $\cot[\arctan(x-2)]$
65. $\cos\left(\arcsin \dfrac{x-1}{2}\right)$

In Exercises 66–69, use trigonometric substitution to write the algebraic equation as a trigonometric function of θ, where $-\pi/2 < \theta < \pi/2$. Then find $\sec \theta$ and $\csc \theta$.

66. $5 = \sqrt{25 - x^2}$, $\quad x = 5 \sin \theta$
67. $-\sqrt{2} = \sqrt{4 - x^2}$, $\quad x = 2 \cos \theta$
68. $-\sqrt{3} = \sqrt{x^2 - 9}$, $\quad x = 3 \sec \theta$
69. $12 = \sqrt{36 + x^2}$, $\quad x = 6 \tan \theta$

3.3 Vectors in the Plane

▶ **What you should learn**
- How to represent vectors as directed line segments
- How to write the component forms of vectors
- How to perform basic vector operations and represent them graphically
- How to write vectors as linear combinations of unit vectors
- How to find the direction angles of vectors
- How to use vectors to model and solve real-life problems

▶ **Why you should learn it**

You can use vectors to model and solve real-life problems involving magnitude and direction. For instance, in Exercise 77 on page 290 you use vectors to determine the tension in the tow lines between a tugboat and a barge.

Introduction

Quantities such as force and velocity involve both *magnitude* and *direction* and cannot be completely characterized by a single real number. To represent such a quantity, you can use a **directed line segment,** as shown in Figure 3.15. The directed line segment $\overrightarrow{PQ}$ has **initial point** P and **terminal point** Q. Its **magnitude** (or length) is denoted by $\|\overrightarrow{PQ}\|$ and can be found using the distance formula.

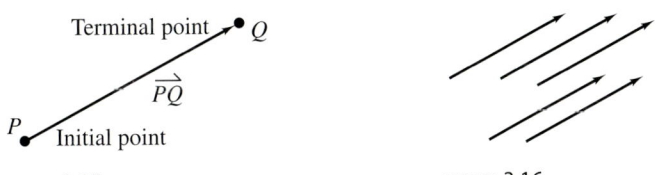

FIGURE 3.15   FIGURE 3.16

Two directed line segments that have the same magnitude and direction are equivalent. For example, the directed line segments in Figure 3.16 are all equivalent. The set of all directed line segments that are equivalent to given directed line segment $\overrightarrow{PQ}$ is a **vector v in the plane,** written $\mathbf{v} = \overrightarrow{PQ}$. Vectors are denoted by lowercase, boldface letters such as **u**, **v**, and **w**.

Example 1 ▶ Vector Representation by Directed Line Segments

Let **u** be represented by the directed line segment from $P = (0, 0)$ to $Q = (3, 2)$, and let **v** be represented by the directed line segment from $R = (1, 2)$ to $S = (4, 4)$, as shown in Figure 3.17. Show that $\mathbf{u} = \mathbf{v}$.

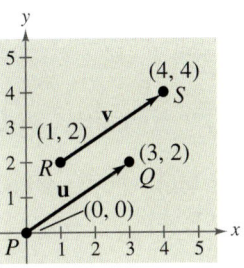

FIGURE 3.17

Solution

From the distance formula, it follows that $\overrightarrow{PQ}$ and $\overrightarrow{RS}$ have the *same magnitude*.

$$\|\overrightarrow{PQ}\| = \sqrt{(3-0)^2 + (2-0)^2} = \sqrt{13}$$
$$\|\overrightarrow{RS}\| = \sqrt{(4-1)^2 + (4-2)^2} = \sqrt{13}$$

Moreover, both line segments have the *same direction* because they are both directed toward the upper right on lines having a slope of $\frac{2}{3}$. So, $\overrightarrow{PQ}$ and $\overrightarrow{RS}$ have the same magnitude and direction, and it follows that $\mathbf{u} = \mathbf{v}$.

Component Form of a Vector

The directed line segment whose initial point is the origin is often the most convenient representative of a set of equivalent directed line segments. This representative of the vector **v** is in **standard position.**

A vector whose initial point is at the origin $(0, 0)$ can be uniquely represented by the coordinates of its terminal point (v_1, v_2). This is the **component form of a vector v,** written

$$\mathbf{v} = \langle v_1, v_2 \rangle.$$

The coordinates v_1 and v_2 are the *components* of **v**. If both the initial point and the terminal point lie at the origin, **v** is the **zero vector** and is denoted by $\mathbf{0} = \langle 0, 0 \rangle$.

> **Component Form of a Vector**
>
> The component form of the vector with initial point $P = (p_1, p_2)$ and terminal point $Q = (q_1, q_2)$ is
>
> $$\overrightarrow{PQ} = \langle q_1 - p_1, q_2 - p_2 \rangle = \langle v_1, v_2 \rangle = \mathbf{v}.$$
>
> The **magnitude** (or length) of **v** is
>
> $$\|\mathbf{v}\| = \sqrt{(q_1 - p_1)^2 + (q_2 - p_2)^2} = \sqrt{v_1^2 + v_2^2}.$$
>
> If $\|\mathbf{v}\| = 1$, **v** is a **unit vector.** Moreover, $\|\mathbf{v}\| = 0$ if and only if **v** is the zero vector **0**.

Two vectors $\mathbf{u} = \langle u_1, u_2 \rangle$ and $\mathbf{v} = \langle v_1, v_2 \rangle$ are *equal* if and only if $u_1 = v_1$ and $u_2 = v_2$. For instance, in Example 1, the vector **u** from $P = (0, 0)$ to $Q = (3, 2)$ is

$$\mathbf{u} = \overrightarrow{PQ} = \langle 3 - 0, 2 - 0 \rangle = \langle 3, 2 \rangle$$

and the vector **v** from $R = (1, 2)$ to $S = (4, 4)$ is

$$\mathbf{v} = \overrightarrow{RS} = \langle 4 - 1, 4 - 2 \rangle = \langle 3, 2 \rangle.$$

Example 2 ▶ Finding the Component Form of a Vector

Find the component form and magnitude of the vector **v** that has initial point $(4, -7)$ and terminal point $(-1, 5)$.

Solution

Let $P = (4, -7) = (p_1, p_2)$ and let $Q = (-1, 5) = (q_1, q_2)$, as shown in Figure 3.18. Then, the components of $\mathbf{v} = \langle v_1, v_2 \rangle$ are

$$v_1 = q_1 - p_1 = -1 - 4 = -5$$

$$v_2 = q_2 - p_2 = 5 - (-7) = 12.$$

So, $\mathbf{v} = \langle -5, 12 \rangle$ and the magnitude of **v** is

$$\|\mathbf{v}\| = \sqrt{(-5)^2 + 12^2} = \sqrt{169} = 13.$$

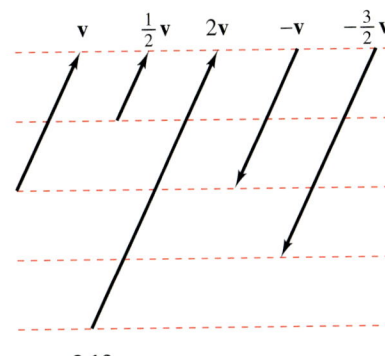

FIGURE 3.19

Vector Operations

The two basic vector operations are **scalar multiplication** and **vector addition.** In operations with vectors, numbers are usually referred to as **scalars.** In this text, scalars will always be real numbers. Geometrically, the product of a vector **v** and a scalar k is the vector that is $|k|$ times as long as **v**. If k is positive, $k\mathbf{v}$ has the same direction as **v**, and if k is negative, $k\mathbf{v}$ has the direction opposite that of **v**, as shown in Figure 3.19.

To add two vectors geometrically, position them (without changing length or direction) so that the initial point of one coincides with the terminal point of the other. The sum $\mathbf{u} + \mathbf{v}$ is formed by joining the initial point of the second vector **v** with the terminal point of the first vector **u**, as shown in Figure 3.20. This technique is called the **parallelogram law** for vector addition because the vector $\mathbf{u} + \mathbf{v}$, often called the **resultant** of vector addition, is the diagonal of a parallelogram having **u** and **v** as its adjacent sides.

A computer animation of this concept appears in the *Interactive* CD-ROM and *Internet* versions of this text.

 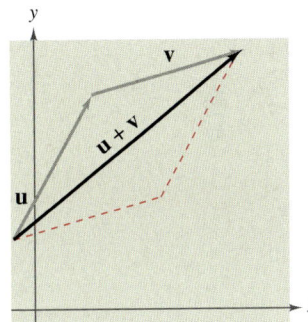

FIGURE 3.20

Definitions of Vector Addition and Scalar Multiplication

Let $\mathbf{u} = \langle u_1, u_2 \rangle$ and $\mathbf{v} = \langle v_1, v_2 \rangle$ be vectors and let k be a scalar (a real number). Then the *sum* of **u** and **v** is the vector

$$\mathbf{u} + \mathbf{v} = \langle u_1 + v_1, u_2 + v_2 \rangle \qquad \text{Sum}$$

and the *scalar multiple* of k times **u** is the vector

$$k\mathbf{u} = k\langle u_1, u_2 \rangle = \langle ku_1, ku_2 \rangle. \qquad \text{Scalar multiple}$$

The *negative* of $\mathbf{v} = \langle v_1, v_2 \rangle$ is

$$-\mathbf{v} = (-1)\mathbf{v}$$
$$= \langle -v_1, -v_2 \rangle \qquad \text{Negative}$$

and the *difference* of **u** and **v** is

$$\mathbf{u} - \mathbf{v} = \mathbf{u} + (-\mathbf{v})$$
$$= \langle u_1 - v_1, u_2 - v_2 \rangle. \qquad \text{Difference}$$

To represent $\mathbf{u} - \mathbf{v}$ geometrically, you can use directed line segments with the *same* initial point. The difference $\mathbf{u} - \mathbf{v}$ is the vector from the terminal point of **v** to the terminal point of **u**, as shown in Figure 3.21.

FIGURE 3.21

The component definitions of vector addition and scalar multiplication are illustrated in Example 3. In this example, notice that each of the vector operations can be interpreted geometrically.

A computer animation of this example appears in the *Interactive* CD-ROM and *Internet* versions of this text.

Example 3 ▶ Vector Operations

Let $\mathbf{v} = \langle -2, 5 \rangle$ and $\mathbf{w} = \langle 3, 4 \rangle$, and find each of the following vectors.

a. $2\mathbf{v}$ **b.** $\mathbf{w} - \mathbf{v}$ **c.** $\mathbf{v} + 2\mathbf{w}$

Solution
a. Because $\mathbf{v} = \langle -2, 5 \rangle$, you have

$$2\mathbf{v} = 2\langle -2, 5 \rangle$$
$$= \langle 2(-2), 2(5) \rangle$$
$$= \langle -4, 10 \rangle.$$

A sketch of $2\mathbf{v}$ is shown in Figure 3.22(a).

b. The difference of $\mathbf{w}$ and $\mathbf{v}$ is

$$\mathbf{w} - \mathbf{v} = \langle 3 - (-2), 4 - 5 \rangle$$
$$= \langle 5, -1 \rangle.$$

A sketch of $\mathbf{w} - \mathbf{v}$ is shown in Figure 3.22(b).

c. The sum of $\mathbf{v}$ and $2\mathbf{w}$ is

$$\mathbf{v} + 2\mathbf{w} = \langle -2, 5 \rangle + 2\langle 3, 4 \rangle$$
$$= \langle -2, 5 \rangle + \langle 2(3), 2(4) \rangle$$
$$= \langle -2, 5 \rangle + \langle 6, 8 \rangle$$
$$= \langle -2 + 6, 5 + 8 \rangle$$
$$= \langle 4, 13 \rangle.$$

A sketch of $\mathbf{v} + 2\mathbf{w}$ is shown in Figure 3.22(c).

(a)

(b)

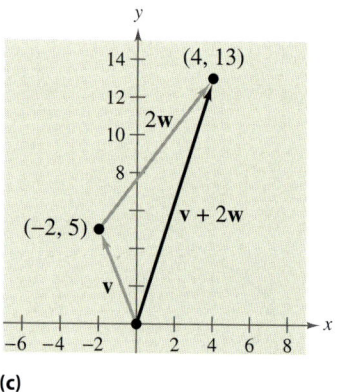
(c)

FIGURE 3.22

Vector addition and scalar multiplication share many of the properties of ordinary arithmetic.

Properties of Vector Addition and Scalar Multiplication

Let **u**, **v**, and **w** be vectors and let c and d be scalars. Then the following properties are true.

1. $\mathbf{u} + \mathbf{v} = \mathbf{v} + \mathbf{u}$
2. $(\mathbf{u} + \mathbf{v}) + \mathbf{w} = \mathbf{u} + (\mathbf{v} + \mathbf{w})$
3. $\mathbf{u} + \mathbf{0} = \mathbf{u}$
4. $\mathbf{u} + (-\mathbf{u}) = \mathbf{0}$
5. $c(d\mathbf{u}) = (cd)\mathbf{u}$
6. $(c + d)\mathbf{u} = c\mathbf{u} + d\mathbf{u}$
7. $c(\mathbf{u} + \mathbf{v}) = c\mathbf{u} + c\mathbf{v}$
8. $1(\mathbf{u}) = \mathbf{u}, 0(\mathbf{u}) = \mathbf{0}$
9. $\|c\mathbf{v}\| = |c|\|\mathbf{v}\|$

Property 9 can be stated as follows: the magnitude of the vector $c\mathbf{v}$ is the absolute value of c times the magnitude of **v**.

Unit Vectors

In many applications of vectors it is useful to find a unit vector that has the same direction as a given nonzero vector **v**. To do this, you can divide **v** by its magnitude to obtain

$$\mathbf{u} = \text{unit vector} = \frac{\mathbf{v}}{\|\mathbf{v}\|} = \left(\frac{1}{\|\mathbf{v}\|}\right)\mathbf{v}.$$

Note that **u** is a scalar multiple of **v**. The vector **u** has magnitude 1 and the same direction as **v**. The vector **u** is called a **unit vector in the direction of v.**

Example 4 ▶ Finding a Unit Vector

Find a unit vector in the direction of $\mathbf{v} = \langle -2, 5 \rangle$ and verify that the result has magnitude 1.

Solution

The unit vector in the direction of **v** is

$$\frac{\mathbf{v}}{\|\mathbf{v}\|} = \frac{\langle -2, 5 \rangle}{\sqrt{(-2)^2 + (5)^2}}$$

$$= \frac{1}{\sqrt{29}}\langle -2, 5 \rangle$$

$$= \left\langle \frac{-2}{\sqrt{29}}, \frac{5}{\sqrt{29}} \right\rangle.$$

This vector has magnitude 1 because

$$\sqrt{\left(\frac{-2}{\sqrt{29}}\right)^2 + \left(\frac{5}{\sqrt{29}}\right)^2} = \sqrt{\frac{4}{29} + \frac{25}{29}} = \sqrt{\frac{29}{29}} = 1.$$

Historical Note
William Rowan Hamilton (1805–1865), an Irish mathematician, did some of the earliest work with vectors. Hamilton spent many years developing a system of vector-like quantities called quaternions. Although Hamilton was convinced of the benefits of quaternions, the operations he defined did not produce good models for physical phenomena. It wasn't until the latter half of the nineteenth century that the Scottish physicist James Maxwell (1831–1879) restructured Hamilton's quaternions in a form useful for representing physical quantities such as force, velocity, and acceleration.

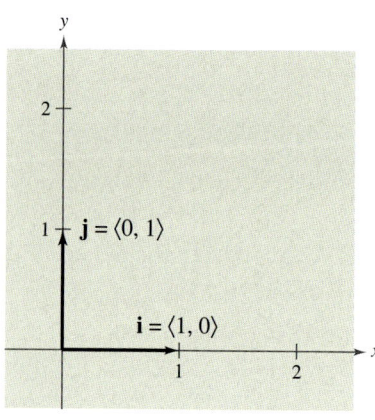

FIGURE 3.23

The unit vectors $\langle 1, 0 \rangle$ and $\langle 0, 1 \rangle$ are called the **standard unit vectors** and are denoted by

$$\mathbf{i} = \langle 1, 0 \rangle \quad \text{and} \quad \mathbf{j} = \langle 0, 1 \rangle,$$

as shown in Figure 3.23. (Note that the lowercase letter **i** is written in boldface to distinguish it from the imaginary number $i = \sqrt{-1}$.) These vectors can be used to represent any vector $\mathbf{v} = \langle v_1, v_2 \rangle$ as follows.

$$\begin{aligned} \mathbf{v} &= \langle v_1, v_2 \rangle \\ &= v_1 \langle 1, 0 \rangle + v_2 \langle 0, 1 \rangle \\ &= v_1 \mathbf{i} + v_2 \mathbf{j} \end{aligned}$$

The scalars v_1 and v_2 are called the **horizontal** and **vertical components of v**, respectively. The vector sum

$$v_1 \mathbf{i} + v_2 \mathbf{j}$$

is called a **linear combination** of the vectors **i** and **j**. Any vector in the plane can be expressed as a linear combination of the standard unit vectors **i** and **j**.

Example 5 ▶ Writing a Linear Combination of Unit Vectors

Let **u** be the vector with initial point $(2, -5)$ and terminal point $(-1, 3)$. Write **u** as a linear combination of the standard unit vectors **i** and **j**.

Solution

Begin by writing the component form of the vector **u**.

$$\begin{aligned} \mathbf{u} &= \langle -1 - 2, 3 + 5 \rangle \\ &= \langle -3, 8 \rangle \\ &= -3\mathbf{i} + 8\mathbf{j} \end{aligned}$$

This result is shown graphically in Figure 3.24.

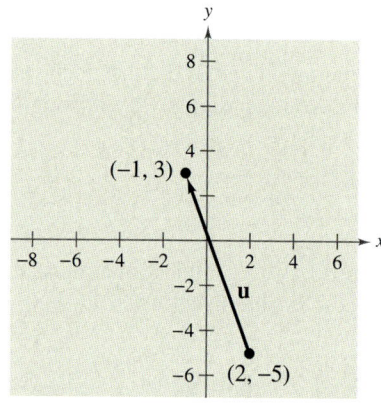

FIGURE 3.24

Example 6 ▶ Vector Operations

Let $\mathbf{u} = -3\mathbf{i} + 8\mathbf{j}$ and let $\mathbf{v} = 2\mathbf{i} - \mathbf{j}$. Find $2\mathbf{u} - 3\mathbf{v}$.

Solution

You could solve this problem by converting **u** and **v** to component form. This, however, is not necessary. It is just as easy to perform the operations in unit vector form.

$$\begin{aligned} 2\mathbf{u} - 3\mathbf{v} &= 2(-3\mathbf{i} + 8\mathbf{j}) - 3(2\mathbf{i} - \mathbf{j}) \\ &= -6\mathbf{i} + 16\mathbf{j} - 6\mathbf{i} + 3\mathbf{j} \\ &= -12\mathbf{i} + 19\mathbf{j} \end{aligned}$$

Direction Angles

If **u** is a *unit vector* such that θ is the angle (measured counterclockwise) from the positive x-axis to **u**, the terminal point of **u** lies on the unit circle and you have

$$\mathbf{u} = \langle x, y \rangle = \langle \cos\theta, \sin\theta \rangle = (\cos\theta)\mathbf{i} + (\sin\theta)\mathbf{j}$$

as shown in Figure 3.25. The angle θ is the **direction angle** of the vector **u**.

Suppose that **u** is a unit vector with direction angle θ. If **v** is any vector that makes an angle θ with the positive x-axis, it has the same direction as **u** and you can write

$$\mathbf{v} = \|\mathbf{v}\| \langle \cos\theta, \sin\theta \rangle$$
$$= \|\mathbf{v}\|(\cos\theta)\mathbf{i} + \|\mathbf{v}\|(\sin\theta)\mathbf{j}.$$

Because $\mathbf{v} = a\mathbf{i} + b\mathbf{j} = \|\mathbf{v}\|(\cos\theta)\mathbf{i} + \|\mathbf{v}\|(\sin\theta)\mathbf{j}$, it follows that the direction angle θ for **v** is determined from

$$\tan\theta = \frac{\sin\theta}{\cos\theta}$$
$$= \frac{\|\mathbf{v}\|\sin\theta}{\|\mathbf{v}\|\cos\theta}$$
$$= \frac{b}{a}.$$

FIGURE 3.25

Example 7 ▶ Finding Direction Angles of Vectors

Find the direction angle of each vector.

a. $\mathbf{u} = 3\mathbf{i} + 3\mathbf{j}$ **b.** $\mathbf{v} = 3\mathbf{i} - 4\mathbf{j}$

Solution

a. The direction angle is

$$\tan\theta = \frac{b}{a} = \frac{3}{3} = 1.$$

Therefore, $\theta = 45°$, as shown in Figure 3.26(a).

b. The direction angle is

$$\tan\theta = \frac{b}{a} = \frac{-4}{3}.$$

Moreover, because $\mathbf{v} = 3\mathbf{i} - 4\mathbf{j}$ lies in Quadrant IV, θ lies in Quadrant IV and its reference angle is

$$\theta = \left| \arctan\left(-\frac{4}{3}\right) \right| \approx |-53.13°| = 53.13°.$$

Therefore, it follows that $\theta \approx 360° - 53.13° = 306.87°$, as shown in Figure 3.26(b).

(a)

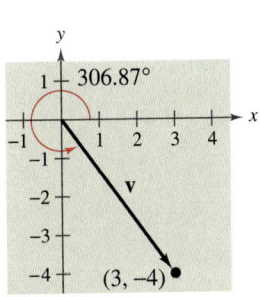

(b)

FIGURE 3.26

Applications of Vectors

Example 8 ▶ Finding the Component Form of a Vector

Find the component form of the vector that represents the velocity of an airplane descending at a speed of 100 miles per hour at an angle 30° below the horizontal, as shown in Figure 3.27.

Solution

The velocity vector **v** has a magnitude of 100 and a direction angle of $\theta = 210°$.

$$\mathbf{v} = \|\mathbf{v}\|(\cos\theta)\mathbf{i} + \|\mathbf{v}\|(\sin\theta)\mathbf{j}$$
$$= 100(\cos 210°)\mathbf{i} + 100(\sin 210°)\mathbf{j}$$
$$= 100\left(\frac{-\sqrt{3}}{2}\right)\mathbf{i} + 100\left(\frac{-1}{2}\right)\mathbf{j}$$
$$= -50\sqrt{3}\,\mathbf{i} - 50\mathbf{j}$$
$$= \langle -50\sqrt{3}, -50 \rangle$$

You can check that **v** has a magnitude of 100.

$$\|\mathbf{v}\| = \sqrt{\left(-50\sqrt{3}\right)^2 + (50)^2}$$
$$= \sqrt{7500 + 2500}$$
$$= \sqrt{10{,}000}$$
$$= 100$$

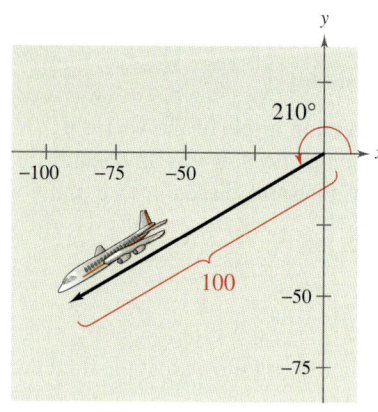

FIGURE 3.27

Example 9 ▶ Using Vectors to Determine Weight

A force of 600 pounds is required to pull a boat and trailer up a ramp inclined at 15° from the horizontal. Find the combined weight of the boat and trailer.

Solution

Based on Figure 3.28, you can make the following observations.

$\|\overrightarrow{BA}\|$ = force of gravity = combined weight of boat and trailer

$\|\overrightarrow{BC}\|$ = force against ramp

$\|\overrightarrow{AC}\|$ = force required to move boat up ramp = 600 pounds

By construction, triangles BWD and ABC are similar. So, angle ABC is 15°. Therefore, in triangle ABC you have

$$\sin 15° = \frac{\|\overrightarrow{AC}\|}{\|\overrightarrow{BA}\|} = \frac{600}{\|\overrightarrow{BA}\|}$$

$$\|BA\| = \frac{600}{\sin 15°} \approx 2318.$$

Consequently, the combined weight is approximately 2318 pounds. (In Figure 3.28, note that $\overrightarrow{AC}$ is parallel to the ramp.)

FIGURE 3.28

Example 10 ▶ Using Vectors to Find Speed and Direction

An airplane is traveling at a fixed altitude with a negligible wind velocity. The airplane is headed N 30° W at a speed of 500 miles per hour, as shown in Figure 3.29. As the airplane reaches a certain point, it encounters a wind with a velocity of 70 miles per hour in the direction N 45° E. What are the resultant speed and direction of the airplane?

 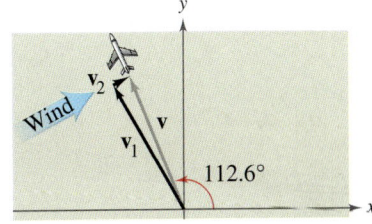

FIGURE 3.29

Solution

Using Figure 3.29, the velocity of the airplane (alone) is

$$\mathbf{v}_1 = 500 \langle \cos 120°, \sin 120° \rangle$$
$$= \langle -250, 250\sqrt{3} \rangle$$

and the velocity of the wind is

$$\mathbf{v}_2 = 70 \langle \cos 45°, \sin 45° \rangle$$
$$= \langle 35\sqrt{2}, 35\sqrt{2} \rangle.$$

So, the velocity of the airplane (in the wind) is

$$\mathbf{v} = \mathbf{v}_1 + \mathbf{v}_2$$
$$= \langle -250 + 35\sqrt{2}, 250\sqrt{3} + 35\sqrt{2} \rangle$$
$$\approx \langle -200.5, 482.5 \rangle$$

and the speed of the airplane is

$$\|\mathbf{v}\| = \sqrt{(-200.5)^2 + (482.5)^2}$$
$$\approx 522.5 \text{ miles per hour.}$$

Finally, if θ is the direction angle of the flight path, you have

$$\tan \theta = \frac{482.5}{-200.5}$$
$$\approx -2.4065$$

which implies that

$$\theta \approx 180° + \arctan(-2.4065)$$
$$\approx 180° - 67.4°$$
$$= 112.6°.$$

3.3 Exercises

In Exercises 1–12, find the component form and the magnitude of the vector **v**.

1.
2.
3.
4.
5.
6.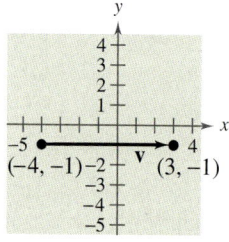

	Initial Point	Terminal Point
7.	$(-1, 5)$	$(15, 12)$
8.	$(1, 11)$	$(9, 3)$
9.	$(-3, -5)$	$(5, 1)$
10.	$(-3, 11)$	$(9, 40)$
11.	$(1, 3)$	$(-8, -9)$
12.	$(-2, 7)$	$(5, -17)$

In Exercises 13–18, use the figure to sketch a graph of the specified vector.

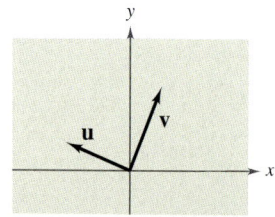

13. $-\mathbf{v}$
14. $5\mathbf{v}$
15. $\mathbf{u} + \mathbf{v}$
16. $\mathbf{u} - \mathbf{v}$
17. $\mathbf{u} + 2\mathbf{v}$
18. $\mathbf{v} - \frac{1}{2}\mathbf{u}$

In Exercises 19–26, find (a) $\mathbf{u} + \mathbf{v}$, (b) $\mathbf{u} - \mathbf{v}$, and (c) $2\mathbf{u} - 3\mathbf{v}$. Then sketch your resultant vector.

19. $\mathbf{u} = \langle 2, 1 \rangle$, $\mathbf{v} = \langle 1, 3 \rangle$
20. $\mathbf{u} = \langle 2, 3 \rangle$, $\mathbf{v} = \langle 4, 0 \rangle$
21. $\mathbf{u} = \langle -5, 3 \rangle$, $\mathbf{v} = \langle 0, 0 \rangle$
22. $\mathbf{u} = \langle 0, 0 \rangle$, $\mathbf{v} = \langle 2, 1 \rangle$
23. $\mathbf{u} = \mathbf{i} + \mathbf{j}$, $\mathbf{v} = 2\mathbf{i} - 3\mathbf{j}$
24. $\mathbf{u} = -2\mathbf{i} + \mathbf{j}$, $\mathbf{v} = -\mathbf{i} + 2\mathbf{j}$
25. $\mathbf{u} = 2\mathbf{i}$, $\mathbf{v} = \mathbf{j}$
26. $\mathbf{u} = 3\mathbf{j}$, $\mathbf{v} = 2\mathbf{i}$

In Exercises 27–36, find a unit vector in the direction of the given vector.

27. $\mathbf{u} = \langle 3, 0 \rangle$
28. $\mathbf{u} = \langle 0, -2 \rangle$
29. $\mathbf{v} = \langle -2, 2 \rangle$
30. $\mathbf{v} = \langle 5, -12 \rangle$
31. $\mathbf{v} = 6\mathbf{i} - 2\mathbf{j}$
32. $\mathbf{v} = \mathbf{i} + \mathbf{j}$
33. $\mathbf{w} = 4\mathbf{j}$
34. $\mathbf{w} = -6\mathbf{i}$
35. $\mathbf{w} = \mathbf{i} - 2\mathbf{j}$
36. $\mathbf{w} = 7\mathbf{j} - 3\mathbf{i}$

In Exercises 37–40, find the vector **v** with the given magnitude and the same direction as **u**.

	Magnitude	Direction
37.	$\|\mathbf{v}\| = 5$	$\mathbf{u} = \langle 3, 3 \rangle$
38.	$\|\mathbf{v}\| = 6$	$\mathbf{u} = \langle -3, 3 \rangle$
39.	$\|\mathbf{v}\| = 9$	$\mathbf{u} = \langle 2, 5 \rangle$
40.	$\|\mathbf{v}\| = 10$	$\mathbf{u} = \langle -10, 0 \rangle$

In Exercises 41–46, find the component form of **v** and sketch the specified vector operations geometrically, where $\mathbf{u} = 2\mathbf{i} - \mathbf{j}$ and $\mathbf{w} = \mathbf{i} + 2\mathbf{j}$.

41. $\mathbf{v} = \frac{3}{2}\mathbf{u}$
42. $\mathbf{v} = \frac{3}{4}\mathbf{w}$
43. $\mathbf{v} = \mathbf{u} + 2\mathbf{w}$
44. $\mathbf{v} = -\mathbf{u} + \mathbf{w}$
45. $\mathbf{v} = \frac{1}{2}(3\mathbf{u} + \mathbf{w})$
46. $\mathbf{v} = \mathbf{u} - 2\mathbf{w}$

In Exercises 47–50, find the magnitude and direction angle of the vector **v**.

47. $\mathbf{v} = 3(\cos 60° \mathbf{i} + \sin 60° \mathbf{j})$

48. $\mathbf{v} = 8(\cos 135° \mathbf{i} + \sin 135° \mathbf{j})$
49. $\mathbf{v} = 6\mathbf{i} - 6\mathbf{j}$
50. $\mathbf{v} = -5\mathbf{i} + 4\mathbf{j}$

In Exercises 51–58, find the component form of v given its magnitude and the angle it makes with the positive x-axis. Sketch v.

Magnitude	Angle
51. $\|\mathbf{v}\| = 3$	$\theta = 0°$
52. $\|\mathbf{v}\| = 1$	$\theta = 45°$
53. $\|\mathbf{v}\| = \frac{7}{2}$	$\theta = 150°$
54. $\|\mathbf{v}\| = \frac{5}{2}$	$\theta = 45°$
55. $\|\mathbf{v}\| = 3\sqrt{2}$	$\theta = 150°$
56. $\|\mathbf{v}\| = 4\sqrt{3}$	$\theta = 90°$
57. $\|\mathbf{v}\| = 2$	**v** in the direction $\mathbf{i} + 3\mathbf{j}$
58. $\|\mathbf{v}\| = 3$	**v** in the direction $3\mathbf{i} + 4\mathbf{j}$

In Exercises 59–62, find the component form of the sum of u and v with direction angles $\theta_\mathbf{u}$ and $\theta_\mathbf{v}$.

Magnitude	Angle
59. $\|\mathbf{u}\| = 5$	$\theta_\mathbf{u} = 0°$
$\|\mathbf{v}\| = 5$	$\theta_\mathbf{v} = 90°$
60. $\|\mathbf{u}\| = 4$	$\theta_\mathbf{u} = 60°$
$\|\mathbf{v}\| = 4$	$\theta_\mathbf{v} = 90°$
61. $\|\mathbf{u}\| = 20$	$\theta_\mathbf{u} = 45°$
$\|\mathbf{v}\| = 50$	$\theta_\mathbf{v} = 180°$
62. $\|\mathbf{u}\| = 50$	$\theta_\mathbf{u} = 30°$
$\|\mathbf{v}\| = 30$	$\theta_\mathbf{v} = 110°$

In Exercises 63–66, use the Law of Cosines to find the angle α between the given vectors. (Assume $0° \leq \alpha \leq 180°$.)

63. $\mathbf{v} = \mathbf{i} + \mathbf{j}$, $\mathbf{w} = 2\mathbf{i} - 2\mathbf{j}$
64. $\mathbf{v} = 3\mathbf{i} - 2\mathbf{j}$, $\mathbf{w} = 2\mathbf{i} + 2\mathbf{j}$
65. $\mathbf{v} = \mathbf{i} + \mathbf{j}$, $\mathbf{w} = 3\mathbf{i} - \mathbf{j}$
66. $\mathbf{v} = \mathbf{i} + 2\mathbf{j}$, $\mathbf{w} = 2\mathbf{i} - \mathbf{j}$

In Exercises 67 and 68, find the angle between the forces given the magnitude of their resultant. (*Hint*: Write force 1 as a vector in the direction of the positive x-axis and force 2 as a vector at an angle θ with the positive x-axis.)

	Force 1	Force 2	Resultant Force
67.	45 pounds	60 pounds	90 pounds
68.	3000 pounds	1000 pounds	3750 pounds

69. *Resultant Force* Forces with magnitudes of 125 newtons and 300 newtons act on a hook (see figure). The angle between the two forces is 45°. Find the direction and magnitude of the resultant of these forces.

70. *Resultant Force* Forces with magnitudes of 2000 newtons and 900 newtons act on a machine part at angles of 30° and −45°, respectively, with the x-axis (see figure). Find the direction and magnitude of the resultant of these forces.

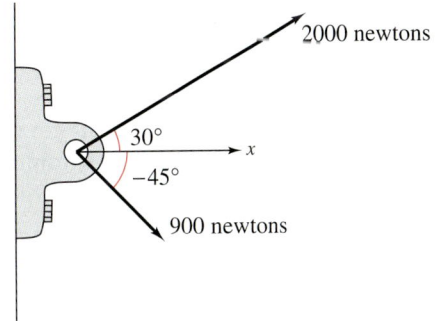

71. *Resultant Force* Three forces with magnitudes of 75 pounds, 100 pounds, and 125 pounds act on an object at angles of 30°, 45°, and 120°, respectively, with the positive x-axis. Find the direction and magnitude of the resultant of these forces.

72. *Resultant Force* Three forces with magnitudes of 70 pounds, 40 pounds, and 60 pounds act on an object at angles of −30°, 45°, and 135°, respectively, with the positive x-axis. Find the direction and magnitude of the resultant of these forces.

73. *Horizontal and Vertical Components of Velocity* A ball is thrown with an initial velocity of 70 feet per second, at an angle of 35° with the horizontal (see figure). Find the vertical and horizontal components of the velocity.

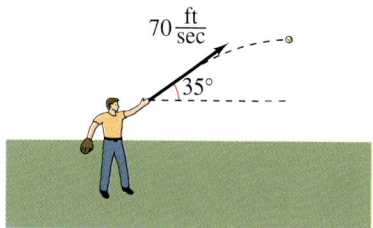

74. *Horizontal and Vertical Components of Velocity* A gun with a muzzle velocity of 1200 feet per second is fired at an angle of 6° with the horizontal. Find the vertical and horizontal components of the velocity.

Cable Tension In Exercises 75 and 76, use the figure to determine the tension in each cable supporting the load.

75. 76.

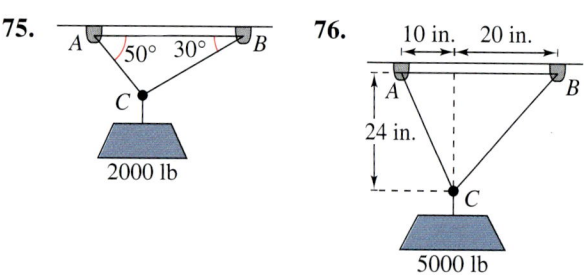

77. *Barge Towing* A loaded barge is being towed by two tugboats, and the magnitude of the resultant is 6000 pounds directed along the axis of the barge (see figure). Find the tension in the tow lines if they each make an 18° angle with the axis of the barge.

78. *Shared Load* To carry a 100-pound cylindrical weight, two people lift on the ends of short ropes that are tied to an eyelet on the top center of the cylinder. Each rope makes a 20° angle with the vertical. Draw a figure that gives a visual representation of the problem, and find the tension in the ropes.

79. *Navigation* An airplane is flying in the direction S 32° E, with an airspeed of 875 kilometers per hour. Because of the wind, its groundspeed and direction are 800 kilometers per hour and S 40° E, respectively (see figure). Find the direction and speed of the wind.

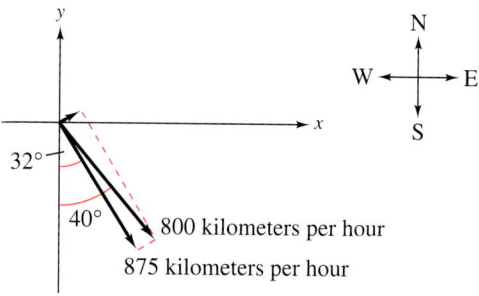

80. *Navigation* An airplane's velocity with respect to the air is 580 miles per hour, and it is heading N 60° W. The wind, at the altitude of the plane, is from the southwest and has a velocity of 60 miles per hour. Draw a figure that gives a visual representation of the problem. What is the true direction of the plane, and what is its speed with respect to the ground?

81. *Work* A heavy implement is pulled 30 feet across a floor, using a force of 100 pounds. Find the work done if the direction of the force is 50° above the horizontal (see figure). (Use the formula for work, $W = FD$, where F is the component of the force in the direction of motion and D is the distance.)

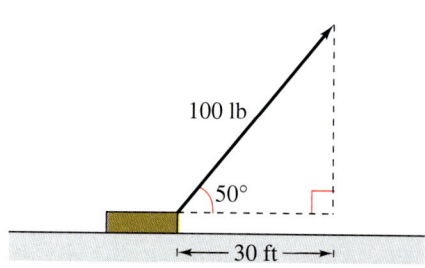

82. Tetherball A tetherball weighing 1 pound is pulled outward from the pole by a horizontal force **u** until the rope makes a 45° angle with the pole (see figure). Determine the resulting tension in the rope and the magnitude of **u**.

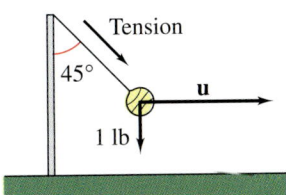

Synthesis

True or False? In Exercises 83–86, decide whether the statement is true or false. Justify your answer.

83. If **u** and **v** have the same magnitude and direction, then **u** = **v**.

84. If **u** is a unit vector in the direction of **v**, then $\mathbf{v} = \|\mathbf{v}\|\mathbf{u}$.

85. If $\mathbf{v} = a\mathbf{i} + b\mathbf{j} = \mathbf{0}$, then $a = -b$.

86. If $\mathbf{u} = a\mathbf{i} + b\mathbf{j}$ is a unit vector, then $a^2 + b^2 = 1$.

87. *Think About It* Consider two forces of equal magnitude acting on a point.

(a) If the magnitude of the resultant is the sum of the magnitudes of the two forces, make a conjecture about the angle between the forces.

(b) If the resultant of the forces is **0**, make a conjecture about the angle between the forces.

(c) Can the magnitude of the resultant be greater than the sum of the magnitudes of the two forces? Explain.

88. *Graphical Reasoning* Consider two forces $\mathbf{F}_1 = \langle 10, 0 \rangle$ and $\mathbf{F}_2 = 5\langle \cos\theta, \sin\theta \rangle$.

(a) Find $\|\mathbf{F}_1 + \mathbf{F}_2\|$ as a function of θ.

(b) Use a graphing utility to graph the function in part (a) for $0 \leq \theta < 2\pi$.

(c) Use the graph in part (b) to determine the range of the function. What is its maximum, and for what value of θ does it occur? What is its minimum, and for what value of θ does it occur?

(d) Explain why the magnitude of the resultant is never 0.

89. Prove that $(\cos\theta)\mathbf{i} + (\sin\theta)\mathbf{j}$ is a unit vector for any value of θ.

90. *Technology* Write a program for your graphing utility that graphs two vectors and their difference given the vectors in component form.

In Exercises 91 and 92, use the program in Exercise 90 to find the difference of the vectors shown in the figure.

91.

92.

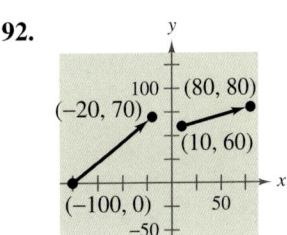

Review

In Exercises 93–96, use the specified trigonometric substitution to write the algebraic expression as a trigonometric function of θ, where $0 < \theta < \pi/2$.

93. $\sqrt{x^2 - 64}$, $x = 8\sec\theta$

94. $\sqrt{64 - x^2}$, $x = 8\sin\theta$

95. $\sqrt{x^2 + 36}$, $x = 6\tan\theta$

96. $\sqrt{(x^2 - 25)^3}$, $x = 5\sec\theta$

In Exercises 97–102, solve the equation.

97. $\cos x(\cos x + 1) = 0$

98. $\sin x(2\sin x + \sqrt{2}) = 0$

99. $3\sec x \sin x - 2\sqrt{3}\sin x = 0$

100. $\cos x \csc x + \cos x\sqrt{2} = 0$

101. $(\sin^2 x - 1)\sin^2 x = 0$

102. $(2\cos^2 x - 3)(\cos^2 x - 1) = 0$

3.4 Vectors and Dot Products

What you should learn

- How to find the dot product of two vectors and use the Properties of the Dot Product
- How to find the angle between two vectors
- How to determine whether two vectors are orthogonal
- How to write a vector as the sum of two vector components
- How to use vectors to find the work done by a force

Why you should learn it

You can use the dot product of two vectors to solve real-life problems involving two vector quantities. For instance, in Exercise 51 on page 299 you use the dot product to find the force necessary to keep a truck from rolling down a hill.

The Dot Product of Two Vectors

So far you have studied two vector operations—vector addition and multiplication by a scalar—each of which yields another vector. In this section you will study a third vector operation, the **dot product**. This product yields a scalar, rather than a vector.

Definition of Dot Product

The **dot product** of $\mathbf{u} = \langle u_1, u_2 \rangle$ and $\mathbf{v} = \langle v_1, v_2 \rangle$ is

$$\mathbf{u} \cdot \mathbf{v} = u_1 v_1 + u_2 v_2.$$

Properties of the Dot Product

Let $\mathbf{u}$, $\mathbf{v}$, and $\mathbf{w}$ be vectors in the plane or in space and let c be a scalar.

1. $\mathbf{u} \cdot \mathbf{v} = \mathbf{v} \cdot \mathbf{u}$
2. $\mathbf{0} \cdot \mathbf{v} = 0$
3. $\mathbf{u} \cdot (\mathbf{v} + \mathbf{w}) = \mathbf{u} \cdot \mathbf{v} + \mathbf{u} \cdot \mathbf{w}$
4. $\mathbf{v} \cdot \mathbf{v} = \|\mathbf{v}\|^2$
5. $c(\mathbf{u} \cdot \mathbf{v}) = c\mathbf{u} \cdot \mathbf{v} = \mathbf{u} \cdot c\mathbf{v}$

Proofs of Properties 1, 4, and 5 are given in Appendix A.

Example 1 ▶ Finding Dot Products

Find each dot product.

a. $\langle 4, 5 \rangle \cdot \langle 2, 3 \rangle$
b. $\langle 2, -1 \rangle \cdot \langle 1, 2 \rangle$
c. $\langle 0, 3 \rangle \cdot \langle 4, -2 \rangle$

Solution

a. $\langle 4, 5 \rangle \cdot \langle 2, 3 \rangle = 4(2) + 5(3)$
$= 8 + 15$
$= 23$
b. $\langle 2, -1 \rangle \cdot \langle 1, 2 \rangle = 2(1) + (-1)(2) = 2 - 2 = 0$
c. $\langle 0, 3 \rangle \cdot \langle 4, -2 \rangle = 0(4) + 3(-2) = 0 - 6 = -6$

In Example 1, be sure you see that the dot product of two vectors is a scalar (a real number), not a vector. Moreover, notice that the dot product can be positive, zero, or negative.

Example 2 ▶ Using Properties of Dot Products

Let $\mathbf{u} = \langle -1, 3 \rangle$, $\mathbf{v} = \langle 2, -4 \rangle$, and $\mathbf{w} = \langle 1, -2 \rangle$. Find each dot product.

a. $(\mathbf{u} \cdot \mathbf{v})\mathbf{w}$ **b.** $\mathbf{u} \cdot 2\mathbf{v}$

Solution

Begin by finding the dot product of $\mathbf{u}$ and $\mathbf{v}$.

$$\mathbf{u} \cdot \mathbf{v} = \langle -1, 3 \rangle \cdot \langle 2, -4 \rangle$$
$$= (-1)(2) + 3(-4)$$
$$= -14$$

a. $(\mathbf{u} \cdot \mathbf{v})\mathbf{w} = -14 \langle 1, -2 \rangle$
$$= \langle -14, 28 \rangle$$

b. $\mathbf{u} \cdot 2\mathbf{v} = 2(\mathbf{u} \cdot \mathbf{v})$
$$= 2(-14)$$
$$= -28$$

Notice that the first product is a vector, whereas the second is a scalar. Can you see why?

Example 3 ▶ Dot Product and Magnitude

The dot product of $\mathbf{u}$ with itself is 5. What is the magnitude of $\mathbf{u}$?

Solution

Because $\|\mathbf{u}\|^2 = \mathbf{u} \cdot \mathbf{u}$ and $\mathbf{u} \cdot \mathbf{u} = 5$, it follows that

$$\|\mathbf{u}\| = \sqrt{\mathbf{u} \cdot \mathbf{u}}$$
$$= \sqrt{5}.$$

The Angle Between Two Vectors

The **angle between two nonzero vectors** is the angle θ, $0 \leq \theta \leq \pi$, between their respective standard position vectors, as shown in Figure 3.30 on page 294. This angle can be found using the dot product. (Note that the angle between the zero vector and another vector is not defined.) A proof is given in Appendix A.

Angle Between Two Vectors

If θ is the angle between two nonzero vectors $\mathbf{u}$ and $\mathbf{v}$, then

$$\cos \theta = \frac{\mathbf{u} \cdot \mathbf{v}}{\|\mathbf{u}\| \, \|\mathbf{v}\|}.$$

Example 4 Finding the Angle Between Two Vectors

Find the angle between $\mathbf{u} = \langle 4, 3 \rangle$ and $\mathbf{v} = \langle 3, 5 \rangle$.

Solution

$$\cos \theta = \frac{\mathbf{u} \cdot \mathbf{v}}{\|\mathbf{u}\| \|\mathbf{v}\|}$$

$$= \frac{\langle 4, 3 \rangle \cdot \langle 3, 5 \rangle}{\|\langle 4, 3 \rangle\| \|\langle 3, 5 \rangle\|}$$

$$= \frac{27}{5\sqrt{34}}$$

This implies that the angle between the two vectors is

$$\theta = \arccos \frac{27}{5\sqrt{34}} \approx 22.2°$$

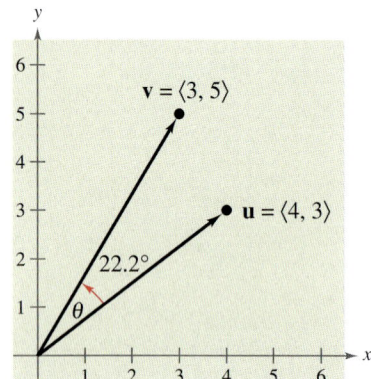

FIGURE 3.30

as shown in Figure 3.30.

Rewriting the expression for the angle between two vectors in the form

$$\mathbf{u} \cdot \mathbf{v} = \|\mathbf{u}\| \|\mathbf{v}\| \cos \theta \qquad \text{Alternative form of dot product}$$

produces an alternative way to calculate the dot product. From this form, you can see that because $\|\mathbf{u}\|$ and $\|\mathbf{v}\|$ are always positive, $\mathbf{u} \cdot \mathbf{v}$ and $\cos \theta$ will always have the same sign. Figure 3.31 shows the five possible orientations of two vectors.

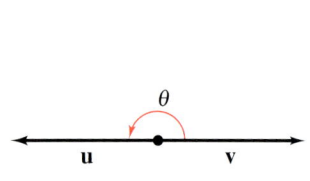
$\theta = \pi$
$\cos \theta = -1$
Opposite Direction

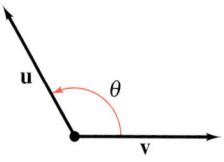
$\frac{\pi}{2} < \theta < \pi$
$-1 < \cos \theta < 0$
Obtuse Angle

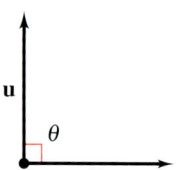
$\theta = \frac{\pi}{2}$
$\cos \theta = 0$
90° Angle

$0 < \theta < \frac{\pi}{2}$
$0 < \cos \theta < 1$
Acute Angle

$\theta = 0$
$\cos \theta = 1$
Same Direction

FIGURE 3.31

Definition of Orthogonal Vectors
The vectors $\mathbf{u}$ and $\mathbf{v}$ are **orthogonal** if $\mathbf{u} \cdot \mathbf{v} = 0$.

The terms "orthogonal" and "perpendicular" mean essentially the same thing—meeting at right angles. Even though the angle between the zero vector and another vector is not defined, it is convenient to extend the definition of orthogonality to include the zero vector. In other words, the zero vector is orthogonal to every vector $\mathbf{u}$, because $\mathbf{0} \cdot \mathbf{u} = 0$.

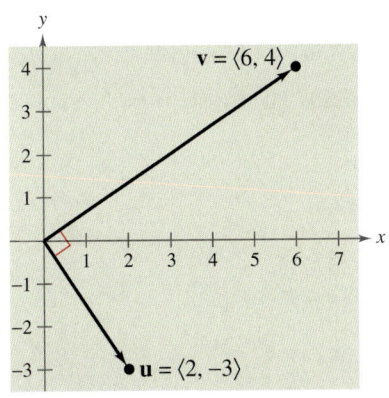

FIGURE 3.32

A computer simulation to accompany this concept appears in the *Interactive CD-ROM* and *Internet* versions of this text.

Example 5 ▶ Determining Orthogonal Vectors

Are the vectors $\mathbf{u} = \langle 2, -3 \rangle$ and $\mathbf{v} = \langle 6, 4 \rangle$ orthogonal?

Solution

Begin by finding the dot product of the two vectors.

$$\mathbf{u} \cdot \mathbf{v} = \langle 2, -3 \rangle \cdot \langle 6, 4 \rangle = 2(6) + (-3)(4) = 0$$

Because the dot product is 0, the two vectors are orthogonal, as shown in Figure 3.32.

Finding Vector Components

You have already seen applications in which two vectors are added to produce a resultant vector. Many applications in physics and engineering pose the reverse problem—decomposing a given vector into the sum of two **vector components.**

Consider a boat on an inclined ramp, as shown in Figure 3.33. The force $\mathbf{F}$ due to gravity pulls the boat *down* the ramp and *against* the ramp. These two orthogonal forces, $\mathbf{w}_1$ and $\mathbf{w}_2$, are vector components of $\mathbf{F}$. That is,

$$\mathbf{F} = \mathbf{w}_1 + \mathbf{w}_2. \quad \text{Vector components of } \mathbf{F}$$

The negative of component $\mathbf{w}_1$ represents the force needed to keep the boat from rolling down the ramp, whereas $\mathbf{w}_2$ represents the force that the tires must withstand against the ramp. A procedure for finding $\mathbf{w}_1$ and $\mathbf{w}_2$ is shown below.

FIGURE 3.33

Definition of Vector Components

Let $\mathbf{u}$ and $\mathbf{v}$ be nonzero vectors such that

$$\mathbf{u} = \mathbf{w}_1 + \mathbf{w}_2$$

where $\mathbf{w}_1$ and $\mathbf{w}_2$ are orthogonal and $\mathbf{w}_1$ is parallel to (or a scalar multiple of) $\mathbf{v}$, as shown in Figure 3.34. The vectors $\mathbf{w}_1$ and $\mathbf{w}_2$ are called vector components of $\mathbf{u}$. The vector $\mathbf{w}_1$ is the **projection** of $\mathbf{u}$ onto $\mathbf{v}$ and is denoted by

$$\mathbf{w}_1 = \text{proj}_\mathbf{v} \mathbf{u}.$$

The vector $\mathbf{w}_2$ is given by $\mathbf{w}_2 = \mathbf{u} - \mathbf{w}_1$.

θ is acute.

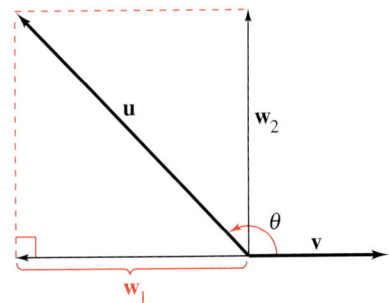

θ is obtuse.

FIGURE 3.34

From the definition of vector components, you can see that it is easy to find the component $\mathbf{w}_2$ once you have found the projection of $\mathbf{u}$ onto $\mathbf{v}$. To find the projection, you can use the dot product.

Projection of u onto v

Let $\mathbf{u}$ and $\mathbf{v}$ be nonzero vectors. The projection of $\mathbf{u}$ onto $\mathbf{v}$ is

$$\text{proj}_\mathbf{v} \mathbf{u} = \left(\frac{\mathbf{u} \cdot \mathbf{v}}{\|\mathbf{v}\|^2} \right) \mathbf{v}.$$

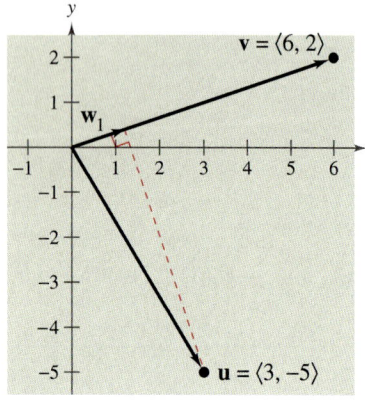

FIGURE 3.35

Example 6 ▶ Decomposing a Vector into Components

Find the projection of $\mathbf{u} = \langle 3, -5 \rangle$ onto $\mathbf{v} = \langle 6, 2 \rangle$. Then write $\mathbf{u}$ as the sum of two orthogonal vectors, one of which is $\text{proj}_{\mathbf{v}}\mathbf{u}$.

Solution

The projection of $\mathbf{u}$ onto $\mathbf{v}$ is

$$\mathbf{w}_1 = \text{proj}_{\mathbf{v}}\mathbf{u} = \left(\frac{\mathbf{u} \cdot \mathbf{v}}{\|\mathbf{v}\|^2}\right)\mathbf{v} = \left(\frac{8}{40}\right)\langle 6, 2 \rangle = \left\langle \frac{6}{5}, \frac{2}{5} \right\rangle$$

as shown in Figure 3.35. The other component, $\mathbf{w}_2$, is

$$\mathbf{w}_2 = \mathbf{u} - \mathbf{w}_1 = \langle 3, -5 \rangle - \left\langle \frac{6}{5}, \frac{2}{5} \right\rangle = \left\langle \frac{9}{5}, -\frac{27}{5} \right\rangle.$$

So,

$$\mathbf{u} = \mathbf{w}_1 + \mathbf{w}_2 = \left\langle \frac{6}{5}, \frac{2}{5} \right\rangle + \left\langle \frac{9}{5}, -\frac{27}{5} \right\rangle = \langle 3, -5 \rangle.$$

Example 7 ▶ Finding a Force

A 600-pound boat sits on a ramp inclined at 30°, as shown in Figure 3.36. What force is required to keep the boat from rolling down the ramp?

Solution

Because the force due to gravity is vertical and downward, you can represent the gravitational force by the vector

$$\mathbf{F} = -600\mathbf{j}. \qquad \text{Force due to gravity}$$

To find the force required to keep the boat from rolling down the ramp, project $\mathbf{F}$ onto a unit vector $\mathbf{v}$ in the direction of the ramp, as follows.

$$\mathbf{v} = (\cos 30°)\mathbf{i} + (\sin 30°)\mathbf{j} = \frac{\sqrt{3}}{2}\mathbf{i} + \frac{1}{2}\mathbf{j} \qquad \text{Unit vector along ramp}$$

FIGURE 3.36

Therefore, the projection of $\mathbf{F}$ onto $\mathbf{v}$ is

$$\mathbf{w}_1 = \text{proj}_{\mathbf{v}}\mathbf{F}$$
$$= \left(\frac{\mathbf{F} \cdot \mathbf{v}}{\|\mathbf{v}\|^2}\right)\mathbf{v}$$
$$= (\mathbf{F} \cdot \mathbf{v})\mathbf{v}$$
$$= (-600)\left(\frac{1}{2}\right)\mathbf{v}$$
$$= -300\left(\frac{\sqrt{3}}{2}\mathbf{i} + \frac{1}{2}\mathbf{j}\right).$$

The magnitude of this force is 300, and therefore a force of 300 pounds is required to keep the boat from rolling down the ramp.

Work

The work W done by a constant force $\mathbf{F}$ acting along the line of motion of an object is

$$W = (\text{magnitude of force})(\text{distance})$$
$$= \|\mathbf{F}\|\,\|\overrightarrow{PQ}\|$$

as shown in Figure 3.37(a). If the constant force $\mathbf{F}$ is not directed along the line of motion, as shown in Figure 3.37(b), the work W done by the force is

$$W = \|\text{proj}_{\overrightarrow{PQ}}\mathbf{F}\|\,\|\overrightarrow{PQ}\| = (\cos\theta)\|\mathbf{F}\|\,\|\overrightarrow{PQ}\| = \mathbf{F}\cdot\overrightarrow{PQ}.$$

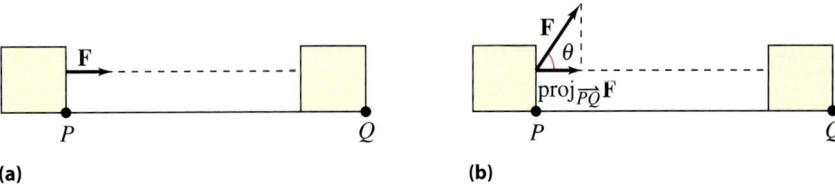

(a)   **(b)**

FIGURE 3.37

This notion of work is summarized in the following definition.

Definition of Work

The **work** W done by a constant force $\mathbf{F}$ as its point of application moves along the vector $\overrightarrow{PQ}$ is given by either of the following.

1. $W = \|\text{proj}_{\overrightarrow{PQ}}\mathbf{F}\|\,\|\overrightarrow{PQ}\|$ Projection form
2. $W = \mathbf{F}\cdot\overrightarrow{PQ}$ Dot product form

Example 8 ▶ Finding Work

To close a sliding door, a person pulls on a rope with a constant force of 50 pounds at a constant angle of 60°, as shown in Figure 3.38. Find the work done in moving the door 12 feet to its closed position.

Solution

Using a projection, you can calculate the work as follows.

$$W = \|\text{proj}_{\overrightarrow{PQ}}\mathbf{F}\|\,\|\overrightarrow{PQ}\|$$
$$= (\cos 60°)\|\mathbf{F}\|\,\|\overrightarrow{PQ}\|$$
$$= \frac{1}{2}(50)(12)$$
$$= 300 \text{ foot-pounds}$$

So, the work done is 300 foot-pounds. You can verify this result by finding the vectors $\mathbf{F}$ and $\overrightarrow{PQ}$ and calculating their dot product.

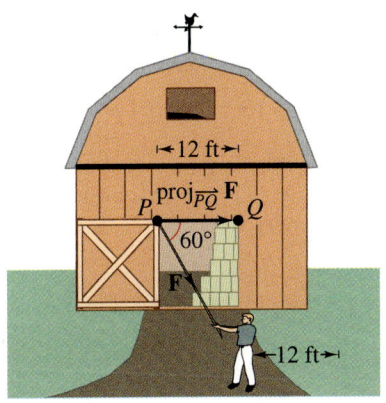

FIGURE 3.38

3.4 Exercises

In Exercises 1–4, find the dot product of u and v.

1. $\mathbf{u} = \langle 6, 1 \rangle$
   $\mathbf{v} = \langle -2, 3 \rangle$
2. $\mathbf{u} = \langle 5, 12 \rangle$
   $\mathbf{v} = \langle -3, 2 \rangle$
3. $\mathbf{u} = 4\mathbf{i} - 2\mathbf{j}$
   $\mathbf{v} = \mathbf{i} - \mathbf{j}$
4. $\mathbf{u} = 3\mathbf{i} + 4\mathbf{j}$
   $\mathbf{v} = 7\mathbf{i} - 2\mathbf{j}$

In Exercises 5–8, use the vectors $\mathbf{u} = \langle 2, 2 \rangle$ and $\mathbf{v} = \langle -3, 4 \rangle$ to find the indicated quantity. State whether the result is a vector or a scalar.

5. $\mathbf{u} \cdot \mathbf{u}$
6. $\|\mathbf{v}\| + 3$
7. $(\mathbf{u} \cdot \mathbf{v})\mathbf{v}$
8. $3\mathbf{u} \cdot \mathbf{v}$

In Exercises 9–14, use the dot product to find the magnitude of u.

9. $\mathbf{u} = \langle -5, 12 \rangle$
10. $\mathbf{u} = \langle 2, -4 \rangle$
11. $\mathbf{u} = 20\mathbf{i} + 25\mathbf{j}$
12. $\mathbf{u} = 12\mathbf{i} - 16\mathbf{j}$
13. $\mathbf{u} = 6\mathbf{j}$
14. $\mathbf{u} = -21\mathbf{i}$

In Exercises 15–24, find the angle θ between the vectors.

15. $\mathbf{u} = \langle 1, 0 \rangle$
    $\mathbf{v} = \langle 0, -2 \rangle$
16. $\mathbf{u} = \langle 3, 2 \rangle$
    $\mathbf{v} = \langle 4, 0 \rangle$
17. $\mathbf{u} = 3\mathbf{i} + 4\mathbf{j}$
    $\mathbf{v} = -2\mathbf{j}$
18. $\mathbf{u} = 2\mathbf{i} - 3\mathbf{j}$
    $\mathbf{v} = \mathbf{i} - 2\mathbf{j}$
19. $\mathbf{u} = 2\mathbf{i} - \mathbf{j}$
    $\mathbf{v} = 6\mathbf{i} + 4\mathbf{j}$
20. $\mathbf{u} = -6\mathbf{i} - 3\mathbf{j}$
    $\mathbf{v} = -8\mathbf{i} + 4\mathbf{j}$
21. $\mathbf{u} = 5\mathbf{i} + 5\mathbf{j}$
    $\mathbf{v} = -6\mathbf{i} + 6\mathbf{j}$
22. $\mathbf{u} = 2\mathbf{i} - 3\mathbf{j}$
    $\mathbf{v} = 4\mathbf{i} + 3\mathbf{j}$
23. $\mathbf{u} = \cos\left(\frac{\pi}{3}\right)\mathbf{i} + \sin\left(\frac{\pi}{3}\right)\mathbf{j}$
    $\mathbf{v} = \cos\left(\frac{3\pi}{4}\right)\mathbf{i} + \sin\left(\frac{3\pi}{4}\right)\mathbf{j}$
24. $\mathbf{u} = \cos\left(\frac{\pi}{4}\right)\mathbf{i} + \sin\left(\frac{\pi}{4}\right)\mathbf{j}$
    $\mathbf{v} = \cos\left(\frac{\pi}{2}\right)\mathbf{i} + \sin\left(\frac{\pi}{2}\right)\mathbf{j}$

In Exercises 25–28, use vectors to find the interior angles of the triangle with the given vertices.

25. $(1, 2), (3, 4), (2, 5)$
26. $(-3, -4), (1, 7), (8, 2)$
27. $(-3, 0), (2, 2), (0, 6)$
28. $(-3, 5), (-1, 9), (7, 9)$

In Exercises 29–32, find $\mathbf{u} \cdot \mathbf{v}$, where θ is the angle between u and v.

29. $\|\mathbf{u}\| = 4, \|\mathbf{v}\| = 10, \theta = \frac{2\pi}{3}$
30. $\|\mathbf{u}\| = 100, \|\mathbf{v}\| = 250, \theta = \frac{\pi}{6}$
31. $\|\mathbf{u}\| = 81, \|\mathbf{v}\| = 64, \theta = \frac{\pi}{4}$
32. $\|\mathbf{u}\| = 9, \|\mathbf{v}\| = 144, \theta = \frac{\pi}{2}$

In Exercises 33–38, determine whether u and v are orthogonal, parallel, or neither.

33. $\mathbf{u} = \langle -12, 30 \rangle$
    $\mathbf{v} = \langle \frac{1}{2}, -\frac{5}{4} \rangle$
34. $\mathbf{u} = \langle 3, 15 \rangle$
    $\mathbf{v} = \langle -1, 5 \rangle$
35. $\mathbf{u} = \frac{1}{4}(3\mathbf{i} - \mathbf{j})$
    $\mathbf{v} = 5\mathbf{i} + 6\mathbf{j}$
36. $\mathbf{u} = \mathbf{i}$
    $\mathbf{v} = -2\mathbf{i} + 2\mathbf{j}$
37. $\mathbf{u} = 2\mathbf{i} - 2\mathbf{j}$
    $\mathbf{v} = -\mathbf{i} - \mathbf{j}$
38. $\mathbf{u} = \langle \cos\theta, \sin\theta \rangle$
    $\mathbf{v} = \langle \sin\theta, -\cos\theta \rangle$

In Exercises 39–42, find the projection of u onto v and the vector component of u orthogonal to v.

39. $\mathbf{u} = \langle 2, 2 \rangle$
    $\mathbf{v} = \langle 6, 1 \rangle$
40. $\mathbf{u} = \langle 4, 2 \rangle$
    $\mathbf{v} = \langle 1, -2 \rangle$
41. $\mathbf{u} = \langle 0, 3 \rangle$
    $\mathbf{v} = \langle 2, 15 \rangle$
42. $\mathbf{u} = \langle -3, -2 \rangle$
    $\mathbf{v} = \langle -4, -1 \rangle$

In Exercises 43–46, find two vectors in opposite directions that are orthogonal to the vector u. (The answers are not unique.)

43. $\mathbf{u} = \langle 3, 5 \rangle$
44. $\mathbf{u} = \langle -8, 3 \rangle$
45. $\mathbf{u} = \frac{1}{2}\mathbf{i} - \frac{2}{3}\mathbf{j}$
46. $\mathbf{u} = -\frac{5}{2}\mathbf{i} - 3\mathbf{j}$

Work In Exercises 47 and 48, find the work done in moving a particle from P to Q if the magnitude and direction of the force are given by v.

47. $P = (0, 0), \quad Q = (4, 7), \quad \mathbf{v} = \langle 1, 4 \rangle$
48. $P = (1, 3), \quad Q = (-3, 5), \quad \mathbf{v} = -2\mathbf{i} + 3\mathbf{j}$

49. *Revenue* The vector $\mathbf{u} = \langle 1650, 3200 \rangle$ gives the numbers of units of two products produced by a company. The vector $\mathbf{v} = \langle 15.25, 10.50 \rangle$ gives the price (in dollars) of each unit, respectively. Find the dot product $\mathbf{u} \cdot \mathbf{v}$ and explain what information it gives.

50. *Revenue* Repeat Exercise 49 after increasing the prices by 5%. Identify the vector operation used to increase the prices by 5%.

51. *Braking Load* A truck with a gross weight of 30,000 pounds is parked on a 5° slope (see figure). Assume that the only force to overcome is the force of gravity.

(a) Find the force required to keep the truck from rolling down the hill.

(b) Find the force perpendicular to the hill.

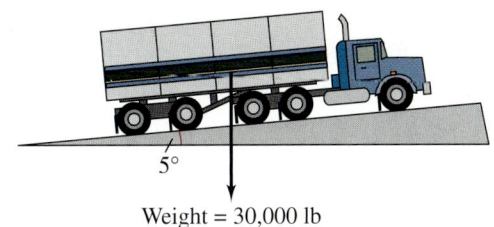

Weight = 30,000 lb

52. *Braking Load* Rework Exercise 51 for a truck that is parked on an 8° slope.

53. *Work* A 25-kilogram (245-newton) bag of sugar is lifted 3 meters. Determine the work done.

54. *Work* Determine the work done by a crane lifting a 2400-pound car 5 feet.

55. *Work* A force of 45 pounds in the direction of 30° above the horizontal is required to slide an implement across a floor (see figure). Find the work done if the implement is dragged 20 feet.

56. *Work* A tractor pulls a log 800 meters and the tension in the cable connecting the tractor and log is approximately 1600 kilograms (15,691 newtons). Approximate the work done if the direction of the force is 35° above the horizontal.

Synthesis

True or False? In Exercises 57 and 58, determine whether the statement is true or false. Justify your answer.

57. The work W done by a constant force $\mathbf{F}$ acting along the line of motion of an object is represented by a vector.

58. A sliding door moves along the line of vector $\overrightarrow{PQ}$. If a force is applied to the door along a vector that is orthogonal to $\overrightarrow{PQ}$, then no work is done.

59. *Think About It* What is known about θ, the angle between two nonzero vectors $\mathbf{u}$ and $\mathbf{v}$, under the following conditions?

(a) $\mathbf{u} \cdot \mathbf{v} = 0$ (b) $\mathbf{u} \cdot \mathbf{v} > 0$ (c) $\mathbf{u} \cdot \mathbf{v} < 0$

60. *Think About It* What can be said about the vectors $\mathbf{u}$ and $\mathbf{v}$ under the following conditions?

(a) The projection of $\mathbf{u}$ onto $\mathbf{v}$ equals $\mathbf{u}$.

(b) The projection of $\mathbf{u}$ onto $\mathbf{v}$ equals $\mathbf{0}$.

61. Use vectors to prove that the diagonals of a rhombus are perpendicular.

62. Prove the following.

$$\|\mathbf{u} - \mathbf{v}\|^2 = \|\mathbf{u}\|^2 + \|\mathbf{v}\|^2 - 2\mathbf{u} \cdot \mathbf{v}$$

63. Prove the following Properties of the Dot Product.

(a) $\mathbf{0} \cdot \mathbf{v} = 0$

(b) $\mathbf{u} \cdot (\mathbf{v} + \mathbf{w}) = \mathbf{u} \cdot \mathbf{v} + \mathbf{u} \cdot \mathbf{w}$

(c) $c(\mathbf{u} \cdot \mathbf{v}) = \mathbf{u} \cdot c\mathbf{v}$

64. Prove that if $\mathbf{u}$ is orthogonal to $\mathbf{v}$ and $\mathbf{w}$, then $\mathbf{u}$ is orthogonal to $c\mathbf{v} + d\mathbf{w}$ for any scalars c and d.

Review

In Exercises 65–68, find the exact solutions of the equation in the interval $[0, 2\pi)$.

65. $\sin 2x - \sqrt{3} \sin x = 0$ **66.** $\sin 2x + \sqrt{2} \cos x = 0$

67. $2 \tan x = \tan 2x$ **68.** $\cos 2x - 3 \sin x = 2$

In Exercises 69–72, find the exact value of the trigonometric function given that $\sin u = -\frac{12}{13}$ and $\cos v = \frac{24}{25}$. (Both u and v are in Quadrant IV.)

69. $\sin(u - v)$ **70.** $\sin(u + v)$

71. $\cos(v - u)$ **72.** $\tan(u - v)$

Chapter Summary

What did you learn?

Section 3.1 | **Review Exercises**
- How to use the Law of Sines to solve oblique triangles (AAS, ASA, or SSA) — 1–12
- How to find the areas of oblique triangles — 13–16
- How to use the Law of Sines to model and solve real-life problems — 17–19

Section 3.2
- How to use the Law of Cosines to solve oblique triangles (SSS or SAS) — 20–27
- How to use the Law of Cosines to model and solve real-life problems — 28, 29
- How to use Heron's Area Formula to find the area of a triangle — 30–33

Section 3.3
- How to represent vectors as directed line segments — 34–37
- How to write the component forms of vectors — 38–43
- How to perform basic vector operations and represent them graphically — 44–47
- How to write vectors as linear combinations of unit vectors — 48–53
- How to find the direction angles of vectors — 54–59
- How to use vectors to model and solve real-life problems — 60–62

Section 3.4
- How to find the dot product of two vectors and use the Properties of the Dot Product — 63–70
- How to find the angle between two vectors — 71–74
- How to determine whether two vectors are orthogonal — 75–78
- How to write a vector as the sum of two vector components — 79–82
- How to use vectors to find the work done by a force — 83, 84

Review Exercises

3.1 In Exercises 1–12, use the Law of Sines to solve (if possible) the triangle. If two solutions exist, list both. Round your answers to two decimal places.

1.
2.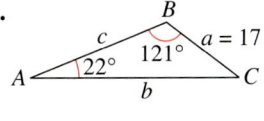

3. $B = 72°$, $C = 82°$, $b = 54$
4. $B = 10°$, $C = 20°$, $c = 33$
5. $A = 16°$, $B = 98°$, $c = 8.4$
6. $A = 95°$, $B = 45°$, $c = 104.8$
7. $A = 24°$, $C = 48°$, $b = 27.5$
8. $B = 64°$, $C = 36°$, $a = 367$
9. $B = 150°$, $b = 30$, $c = 10$
10. $B = 150°$, $a = 10$, $b = 3$
11. $A = 75°$, $a = 51.2$, $b = 33.7$
12. $B = 25°$, $a = 6.2$, $b = 4$

In Exercises 13–16, use the information to find the area of the triangle.

13. $A = 27°$, $b = 5$, $c = 7$
14. $B = 80°$, $a = 4$, $c = 8$
15. $C = 123°$, $a = 16$, $b = 5$
16. $A = 11°$, $b = 22$, $c = 21$

17. *Height* From a certain distance, the angle of elevation to the top of a building is 17°. At a point 50 meters closer to the building, the angle of elevation is 31° (see figure). Approximate the height of the building.

18. *Geometry* Find the length of the side w of the parallelogram.

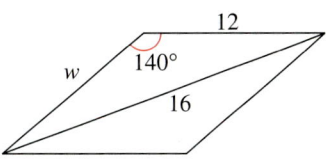

FIGURE FOR 18

19. *Height of a Tree* Find the height of a tree that stands on a hillside of slope 28° (from the horizontal) if from a point 75 feet down the hill the angle of elevation to the top of the tree is 45° (see figure).

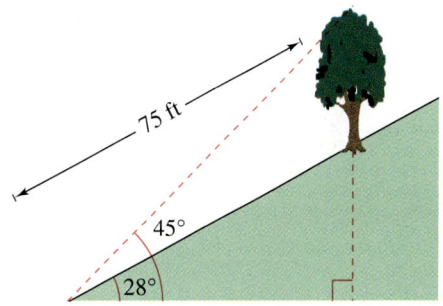

3.2 In Exercises 20–27, use the Law of Cosines to solve the triangle.

20. $a = 5$, $b = 8$, $c = 10$
21. $a = 80$, $b = 60$, $c = 100$
22. $a = 2.5$, $b = 5.0$, $c = 4.5$
23. $a = 16.4$, $b = 8.8$, $c = 12.2$
24. $B = 110°$, $a = 4$, $c = 4$
25. $B = 150°$, $a = 10$, $c = 20$
26. $C = 43°$, $a = 22.5$, $b = 31.4$
27. $A = 62°$, $b = 11.34$, $c = 19.52$

28. *Surveying* To approximate the length of a marsh, a surveyor walks 425 meters from point A to point B. Then the surveyor turns 65° and walks 300 meters to point C. Approximate the length AC of the marsh (see figure).

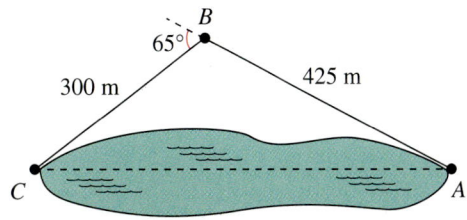

29. *Navigation* Two planes leave an airport at approximately the same time. One is flying 425 miles per hour at a bearing of N 5° W, and the other is flying 530 miles per hour at a bearing of N 67° E. Draw a figure that gives a visual representation of the problem, and determine the distance between the planes after they have flown for 2 hours.

In Exercises 30–33, use Heron's Area Formula to find the area of the triangle.

30. $a = 4$, $b = 5$, $c = 7$
31. $a = 15$, $b = 8$, $c = 10$
32. $a = 12.3$, $b = 15.8$, $c = 3.7$
33. $a = 38.1$, $b = 26.7$, $c = 19.4$

3.3 **In Exercises 34–37, graph the vector with the specified initial point and terminal point.**

	Initial Point	Terminal Point
34.	$(0, 0)$	$(8, 7)$
35.	$(3, 4)$	$(-5, -7)$
36.	$(-3, 9)$	$(8, -4)$
37.	$(-6, -8)$	$(8, 3)$

In Exercises 38–43, find the component form of the vector v satisfying the conditions.

38.

39.

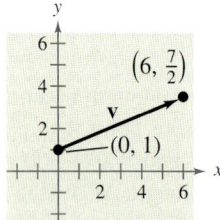

40. Initial point: $(0, 10)$; Terminal point: $(7, 3)$
41. Initial point: $(1, 5)$; Terminal point: $(15, 9)$
42. $\|\mathbf{v}\| = 8$, $\theta = 120°$
43. $\|\mathbf{v}\| = \frac{1}{2}$, $\theta = 225°$

In Exercises 44–47, find the component form of the specified vector given that $\mathbf{u} = 6\mathbf{i} - 5\mathbf{j}$ and $\mathbf{v} = 10\mathbf{i} + 3\mathbf{j}$. Then sketch your result.

44. $2\mathbf{u} + \mathbf{v}$
45. $4\mathbf{u} - 5\mathbf{v}$
46. $3\mathbf{v}$
47. $\frac{1}{2}\mathbf{v}$

In Exercises 48–51, write vector u as a linear combination of the standard unit vectors i and j.

48. $\mathbf{u} = \langle -3, 4 \rangle$
49. $\mathbf{u} = \langle -6, -8 \rangle$
50. $\mathbf{u}$ has initial point $(3, 4)$ and terminal point $(9, 8)$.
51. $\mathbf{u}$ has initial point $(-2, 7)$ and terminal point $(5, -9)$.

In Exercises 52 and 53, write the vector v in the form $\|\mathbf{v}\|(\mathbf{i} \cos \theta + \mathbf{j} \sin \theta)$.

52. $\mathbf{v} = -10\mathbf{i} + 10\mathbf{j}$
53. $\mathbf{v} = 4\mathbf{i} - \mathbf{j}$

In Exercises 54–59, find the magnitude and the direction angle of the vector v.

54. $\mathbf{v} = 7(\cos 60°\mathbf{i} + \sin 60°\mathbf{j})$
55. $\mathbf{v} = 3(\cos 150°\mathbf{i} + \sin 150°\mathbf{j})$
56. $\mathbf{v} = 5\mathbf{i} + 4\mathbf{j}$
57. $\mathbf{v} = -4\mathbf{i} + 7\mathbf{j}$
58. $\mathbf{v} = -3\mathbf{i} - 3\mathbf{j}$
59. $\mathbf{v} = 8\mathbf{i} - \mathbf{j}$

60. *Resultant Force* Forces of 85 pounds and 50 pounds act on a single point. The angle between the forces is 15°. Describe the resultant force.

61. *Rope Tension* A 180-pound weight is supported by two ropes, as shown in the figure. Find the tension exerted on each rope.

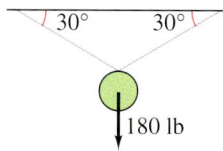

62. *Navigation* An airplane has an airspeed of 724 kilometers per hour at a bearing of N 30° E. If the wind velocity is 32 kilometers per hour from the west, find the groundspeed and the direction of the plane.

3.4 **In Exercises 63–66, find the dot product of u and v.**

63. $\mathbf{u} = \langle 6, 7 \rangle$
$\mathbf{v} = \langle -3, 9 \rangle$

64. $\mathbf{u} = \langle -7, 12 \rangle$
$\mathbf{v} = \langle -4, -14 \rangle$

65. $\mathbf{u} = 3\mathbf{i} + 7\mathbf{j}$
$\mathbf{v} = 11\mathbf{i} - 5\mathbf{j}$

66. $\mathbf{u} = -7\mathbf{i} + 2\mathbf{j}$
$\mathbf{v} = 16\mathbf{i} - 12\mathbf{j}$

In Exercises 67–70, use the vectors $\mathbf{u} = \langle -3, 4 \rangle$ and $\mathbf{v} = \langle 2, 1 \rangle$ to find the indicated quantity. State whether the result is a vector or a scalar.

67. $2\mathbf{u} \cdot \mathbf{u}$

68. $\|\mathbf{v}\|^2$

69. $\mathbf{u}(\mathbf{u} \cdot \mathbf{v})$

70. $3\mathbf{u} \cdot \mathbf{v}$

In Exercises 71–74, find the angle between $\mathbf{u}$ and $\mathbf{v}$.

71. $\mathbf{u} = \cos\dfrac{7\pi}{4}\mathbf{i} + \sin\dfrac{7\pi}{4}\mathbf{j}$
$\mathbf{v} = \cos\dfrac{5\pi}{6}\mathbf{i} + \sin\dfrac{5\pi}{6}\mathbf{j}$

72. $\mathbf{u} = \cos 45°\mathbf{i} + \sin 45°\mathbf{j}$
$\mathbf{v} = \cos 300°\mathbf{i} + \sin 300°\mathbf{j}$

73. $\mathbf{u} = \langle 2\sqrt{2}, -4 \rangle, \quad \mathbf{v} = \langle -\sqrt{2}, 1 \rangle$

74. $\mathbf{u} = \langle 3, \sqrt{3} \rangle, \quad \mathbf{v} = \langle 4, 3\sqrt{3} \rangle$

In Exercises 75–78, determine whether $\mathbf{u}$ and $\mathbf{v}$ are orthogonal, parallel, or neither.

75. $\mathbf{u} = \langle -3, 8 \rangle$
$\mathbf{v} = \langle 8, 3 \rangle$

76. $\mathbf{u} = \langle \tfrac{1}{4}, \tfrac{1}{2} \rangle$
$\mathbf{v} = \langle -2, 4 \rangle$

77. $\mathbf{u} = -\mathbf{i}$
$\mathbf{v} = \mathbf{i} + 2\mathbf{j}$

78. $\mathbf{u} = -2\mathbf{i} + \mathbf{j}$
$\mathbf{v} = 3\mathbf{i} + 6\mathbf{j}$

In Exercises 79–82, find $\text{proj}_\mathbf{v}\mathbf{u}$ and the vector component of $\mathbf{u}$ orthogonal to $\mathbf{v}$.

79. $\mathbf{u} = \langle -4, 3 \rangle, \ \mathbf{v} = \langle -8, -2 \rangle$

80. $\mathbf{u} = \langle 5, 6 \rangle, \ \mathbf{v} = \langle 10, 0 \rangle$

81. $\mathbf{u} = \langle 2, 7 \rangle, \ \mathbf{v} = \langle 1, -1 \rangle$

82. $\mathbf{u} = \langle -3, 5 \rangle, \ \mathbf{v} = \langle -5, 2 \rangle$

In Exercises 83 and 84, find the work done in moving a particle from P to Q if the magnitude and direction of the force are given by $\mathbf{v}$.

83. $P = (5, 3), Q = (8, 9), \mathbf{v} = \langle 2, 7 \rangle$

84. $P = (-2, -9), Q = (-12, 8), \mathbf{v} = 3\mathbf{i} - 6\mathbf{j}$

Synthesis

True or False? In Exercises 85 and 86, determine whether the statement is true or false. Justify your answer.

85. The Law of Sines is true if one of the angles in the triangle is a right angle.

86. When the Law of Sines is used, the solution is always unique.

87. State the Law of Sines from memory.

88. State the Law of Cosines from memory.

89. If one of the angles in the triangle is a right angle, the Law of Cosines simplifies to what famous theorem?

90. What characterizes a vector in the plane?

91. Which vectors in the figure appear to be equivalent?

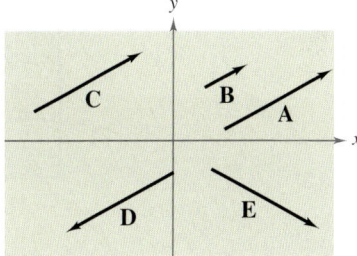

92. The vectors $\mathbf{u}$ and $\mathbf{v}$ have the same magnitudes in the two figures. In which figure will the magnitude of the resultant be greater? Give a reason for your answer.

(a)

(b)

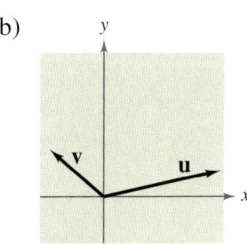

93. Give a geometric description of the scalar multiple $k\mathbf{u}$ of the vector $\mathbf{u}$.

94. Give a geometric description of the sum of the vectors $\mathbf{u}$ and $\mathbf{v}$.

Chapter Project ▶ Adding Vectors Graphically

Program
- Input *a*
- Input *b*
- Input *c*
- Input *d*
- Draw a line from $(0, 0)$ to (a, b).
- Draw a line from $(0, 0)$ to (c, d).
- Add $a + c$ and store in *e*.
- Add $b + d$ and store in *f*.
- Draw a line from $(0, 0)$ to (e, f).
- Draw a line from (a, b) to (c, d).
- Draw a line from (c, d) to (e, f).
- Pause to view graph.
- End program

The psuedo-code at the left can be translated into a program for a graphing utility. (The program for several models of graphing calculators can be found at our website *college.hmco.com*.) The program sketches two vectors

$$\mathbf{u} = a\mathbf{i} + b\mathbf{j} \quad \text{and} \quad \mathbf{v} = c\mathbf{i} + d\mathbf{j}$$

in standard position. Then, using the parallelogram law for vector addition, the program also sketches the vector sum $\mathbf{u} + \mathbf{v}$. *Before* running the program, you should set values that produce an appropriate viewing window.

(a) Use the program to sketch the sum of the vectors $\mathbf{u} = 5\mathbf{i} + 2\mathbf{j}$ and $\mathbf{v} = -4\mathbf{i} + 3\mathbf{j}$. Set your viewing window as indicated in the figure below. Identify the vectors $\mathbf{u}$, $\mathbf{v}$, and $\mathbf{u} + \mathbf{v}$ in the graph.

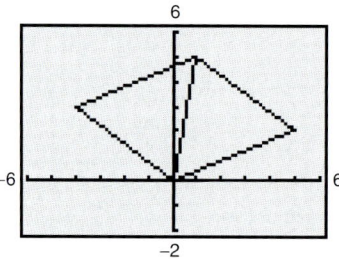

(b) An airplane is headed N 60° W at a speed of 400 miles per hour. The airplane encounters wind of velocity 75 miles per hour in the direction N 40° E. Use the program to find the resultant speed and direction of the airplane.

Chapter Project Investigations

In Questions 1–4, use the program above to sketch the sum of the vectors. Use the result to estimate graphically the components of the sum. Then check your result analytically. (Use $-9 \leq x \leq 9$ and $-6 \leq y \leq 6$.)

1. $\mathbf{u} = 3\mathbf{i} + 4\mathbf{j}$, $\mathbf{v} = -5\mathbf{i} + \mathbf{j}$
2. $\mathbf{u} = 5\mathbf{i} - 4\mathbf{j}$, $\mathbf{v} = 3\mathbf{i} + 2\mathbf{j}$
3. $\mathbf{u} = -4\mathbf{i} + 4\mathbf{j}$, $\mathbf{v} = -2\mathbf{i} - 6\mathbf{j}$
4. $\mathbf{u} = 7\mathbf{i} + 3\mathbf{j}$, $\mathbf{v} = -2\mathbf{i} - 6\mathbf{j}$

5. After encountering the wind, is the airplane in part (b) above traveling at a higher speed or a lower speed? Explain.

6. Consider the airplane described in part (b), headed N 60° W at a speed of 400 miles per hour. What wind velocity, in the direction of N 40° E, will produce a resultant direction of N 50° W? Explain how to use the program above to obtain the answer *experimentally*. Then explain how to obtain the answer analytically.

7. Consider the airplane described in part (b), headed N 60° W at a speed of 400 miles per hour. What wind direction, at a speed of 75 miles per hour, will produce a resultant direction of N 50° W? Explain how to use the program above to obtain the answer *experimentally*. Then explain how to obtain the answer analytically.

Chapter Test

The *Interactive* CD-ROM and *Internet* versions of this text provide answers to the Chapter Tests and Cumulative Tests. They also offer Chapter Pre-Tests (which test key skills and concepts covered in previous chapters) and Chapter Post-Tests, both of which have randomly generated exercises with diagnostic capabilities.

Take this test as you would take a test in class. After you are done, check your work against the answers given in the back of the book.

In Exercises 1–6, use the given information to solve the triangle. If two solutions exist, find both.

1. $A = 24°$, $B = 68°$, $a = 12.2$
2. $B = 104°$, $C = 33°$, $a = 18.1$
3. $A = 24°$, $a = 11.2$, $b = 13.4$
4. $a = 4.0$, $b = 7.3$, $c = 12.4$
5. $B = 100°$, $a = 15$, $b = 23$
6. $C = 123°$, $a = 41$, $b = 57$

7. A triangular parcel of land has borders of 60 meters, 70 meters, and 82 meters. Find the area of the parcel of land.

8. A plane flies 370 miles from point A to point B with a bearing of N 24° E. It then flies 240 miles from point B to point C with a bearing of N 37° E (see figure). Find the distance and bearing from point A to point C.

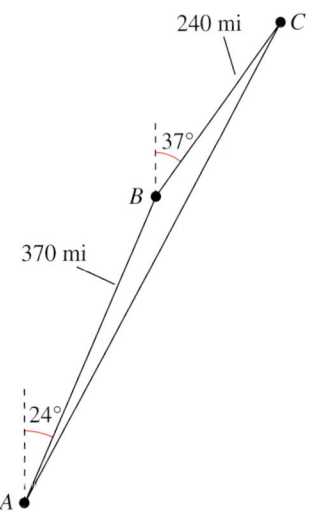

FIGURE FOR 8

In Exercises 9 and 10, find the component form of the vector **v** with the given components.

9. Initial point of **v**: $(-3, 7)$; Terminal point of **v**: $(11, -16)$
10. Magnitude of **v**: $\|\mathbf{v}\| = 12$; Direction of **v**: $\mathbf{u} = \langle 3, -5 \rangle$

In Exercises 11–13, $\mathbf{u} = \langle 3, 5 \rangle$ and $\mathbf{v} = \langle -7, 1 \rangle$. Find the resultant vector and sketch its graph.

11. $\mathbf{u} + \mathbf{v}$
12. $\mathbf{u} - \mathbf{v}$
13. $5\mathbf{u} - 3\mathbf{v}$

14. Find a unit vector in the direction of $\mathbf{u} = \langle 4, -3 \rangle$.

15. Forces with magnitudes of 250 pounds and 130 pounds act on an object at angles of 45° and $-60°$, respectively, with the x-axis. Find the direction and magnitude of the resultant of these forces.

16. Find the angle between the vectors $\mathbf{u} = \langle -1, 5 \rangle$ and $\mathbf{v} = \langle 3, -2 \rangle$.

17. Are the vectors $\mathbf{u} = \langle 6, 10 \rangle$ and $\mathbf{v} = \langle 2, 3 \rangle$ orthogonal?

18. Find the projection of $\mathbf{u} = \langle 6, 7 \rangle$ onto $\mathbf{v} = \langle -5, -1 \rangle$ and the vector component of **u** orthogonal to **v**.

Cumulative Test for Chapters 1–3

Take this test to review the material from earlier chapters. After you are done, check your work against the answers given in the back of the book.

1. Consider the angle $\theta = -120°$.
   (a) Sketch the angle in standard position.
   (b) Determine a coterminal angle in the interval $[0°, 360°)$.
   (c) Convert the angle to radian measure.
   (d) Find the reference angle θ'.
   (e) Find the exact values of the six trigonometric functions of θ.

2. Convert the angle of measure 2.35 radians to degrees. Round the answer to one decimal place.

3. Find $\cos\theta$ if $\tan\theta = -\frac{4}{3}$ and $\sin\theta < 0$.

In Exercises 4 and 5, find the period and amplitude, and sketch the graph of the trigonometric function.

4. $f(x) = 3 - 2\sin \pi x$

5. $g(x) = \dfrac{1}{2}\tan\left(x - \dfrac{\pi}{2}\right)$

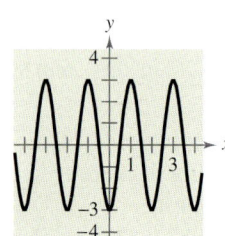

FIGURE FOR 6

6. Find a, b, and c such that the graph of the function $h(x) = a\cos(bx + c)$ matches the graph in the figure.

7. Sketch the graph of the function $f(x) = \dfrac{x}{2}\sin x$ over the interval $-3\pi \leq x \leq 3\pi$.

In Exercises 8 and 9, find the exact value of the expression without the aid of a calculator.

8. $\tan(\arctan 6.7)$

9. $\tan\left(\arcsin \frac{3}{5}\right)$

10. Write an algebraic expression equivalent to $\sin(\arccos 2x)$.

11. Use the fundamental identities to simplify: $\cos\left(\dfrac{\pi}{2} - x\right)\csc x$.

12. Subtract and simplify: $\dfrac{\sin\theta - 1}{\cos\theta} - \dfrac{\cos\theta}{\sin\theta - 1}$.

In Exercises 13–15, prove the identity.

13. $\cot^2\alpha(\sec^2\alpha - 1) = 1$
14. $\sin(x + y)\sin(x - y) = \sin^2 x - \sin^2 y$
15. $\sin^2 x \cos^2 x = \frac{1}{8}(1 - \cos 4x)$

In Exercises 16 and 17, find all solutions of the equation in the interval $[0, 2\pi)$.

16. $2\cos^2\beta - \cos\beta = 0$

17. $3\tan\theta - \cot\theta = 0$

Cumulative Test for Chapters 1–3   **307**

18. Use the Quadratic Formula to solve the equation in the interval $[0, 2\pi)$: $\sin^2 x + 2 \sin x + 1 = 0$.

19. Given that $\sin u = \frac{12}{13}$, $\cos v = \frac{3}{5}$, and angles u and v are both in Quadrant I, find $\tan(u - v)$.

20. If $\tan \theta = \frac{1}{2}$, find the exact value of $\tan(2\theta)$.

21. If $\tan \theta = \frac{4}{3}$, find the exact value of $\sin \frac{\theta}{2}$.

22. Write the product $5 \sin \frac{3\pi}{4} \cdot \cos \frac{7\pi}{4}$ as a sum or difference.

In Exercises 23–26, use the information to solve the triangle shown in the figure.

23. $A = 30°$, $a = 9$, $b = 8$
24. $A = 30°$, $b = 8$, $c = 10$
25. $A = 30°$, $C = 90°$, $b = 10$
26. $a = 4$, $b = 8$, $c = 9$

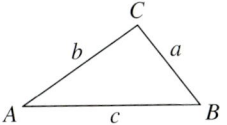

FIGURE FOR 23–26

27. Two sides of a triangle have lengths 7 and 12 inches. Their included angle measures 60°. Find the area of the triangle.

28. Find the area of a triangle with sides of lengths 11, 16, and 17 inches.

29. Write vector $\mathbf{u} = \langle 3, 5 \rangle$ as a linear combination of the standard unit vectors $\mathbf{i}$ and $\mathbf{j}$.

30. Find $\mathbf{u} \cdot \mathbf{v}$ for $\mathbf{u} = 3\mathbf{i} + 4\mathbf{j}$ and $\mathbf{v} = \mathbf{i} - 2\mathbf{j}$.

31. Find the projection of $\mathbf{u} = \langle 8, -2 \rangle$ onto $\mathbf{v} = \langle 1, 5 \rangle$ and the vector component of $\mathbf{u}$ orthogonal to $\mathbf{v}$.

32. From a point 200 feet from a flagpole, the angles of elevation to the bottom and top of the flag are 16° 45′ and 18°, respectively. Approximate the height of the flag to the nearest foot.

33. A record single rotates on a turntable at 45 revolutions per minute. Find the angular speed of the record. Then find the speed of the groove that the needle is in when the needle is 3 inches from the center of the record.

34. To determine the angle of elevation of a star in the sky, you get the star in your line of vision with the backboard of a basketball hoop that is 5 feet higher than your eyes (see figure). Your horizontal distance from the backboard is 12 feet. What is the angle of elevation of the star?

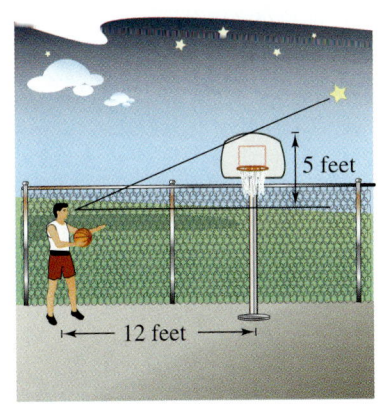

FIGURE FOR 34

35. Write a model for a particle in simple harmonic motion with a displacement of 4 inches and a period of 8 seconds.

36. An airplane's velocity with respect to the air is 500 kilometers per hour, with a bearing of N 30° E. The wind at the altitude of the plane has a velocity of 50 kilometers per hour with a bearing of N 60° E. What is the true direction of the plane, and what is its speed relative to the ground?

4.1 ▶ **Complex Numbers**
4.2 ▶ **Complex Solutions of Equations**
4.3 ▶ **Trigonometric Form of a Complex Number**
4.4 ▶ **DeMoivre's Theorem**

Gregory Sams/SPL/Photo Researchers, Inc.

To add more interest to the picture of a fractal, computer scientists discovered that points not in the set could be assigned a variety of colors, depending on how quickly their sequences diverge.

4 Complex Numbers

▶ How to Study This Chapter

The Big Picture
In this chapter you will learn the following skills and concepts.

- ▶ How to perform operations with complex numbers
- ▶ How to determine the number of zeros of polynomial functions
- ▶ How to find the zeros of polynomial functions
- ▶ How to multiply and divide complex numbers written in trigonometric form
- ▶ How to find powers and nth roots of complex numbers

Important Vocabulary
As you encounter each new vocabulary term in this chapter, add the term and its definition to your notebook glossary.

Imaginary unit i (p. 310)
Complex numbers (p. 310)
Standard form of a complex number (p. 310)
Real part (p. 310)
Imaginary part (p. 310)
Additive identity (p. 311)
Additive inverse (p. 311)
Complex conjugates (p. 313)
Discriminant (p. 318)
Complex plane (p. 324)
Real axis (p. 324)

Imaginary axis (p. 324)
Absolute value of a complex number (p. 324)
Trigonometric form of a complex number (p. 325)
Modulus (p. 325)
Argument (p. 325)
nth root of a complex number (p. 332)
nth roots of unity (p. 334)

Study Tools
- Learning objectives at the beginning of each section
- Chapter Summary (p. 337)
- Review Exercises (pp. 338–339)
- Chapter Test (p. 341)

Additional Resources
- Study and Solutions Guide
- Interactive Trigonometry
- Videotapes for Chapter 4
- Trigonometry Website
- Student Success Organizer

STUDY T!P
The best time to do homework is right after class, when the concepts are still fresh in your mind. This increases your chances of retaining the information in long-term memory.

4.1 Complex Numbers

▶ **What you should learn**
- How to use the imaginary unit i to write complex numbers
- How to add, subtract, and multiply complex numbers
- How to use complex conjugates to divide complex numbers
- How to use the Quadratic Formula to find complex solutions of quadratic equations

▶ **Why you should learn it**

You can use complex numbers to model and solve real-life problems in electronics. For instance, in Exercise 84 on page 316, you will learn how to use complex numbers to find the impedance of an electrical circuit.

The Imaginary Unit i

Some quadratic equations have no real solutions. For instance, the quadratic equation

$$x^2 + 1 = 0 \qquad \text{Equation with no real solution}$$

has no real solution because there is no real number x that can be squared to produce -1. To overcome this deficiency, mathematicians created an expanded system of numbers using the **imaginary unit i,** defined as

$$i = \sqrt{-1} \qquad \text{Imaginary unit}$$

where $i^2 = -1$. By adding real numbers to real multiples of this imaginary unit, the set of **complex numbers** is obtained. Each complex number can be written in the **standard form $a + bi$.** The real number a is called the **real part** of the complex number $a + bi$, and the number bi (where b is a real number) is called the **imaginary part** of the complex number.

Definition of a Complex Number

If a and b are real numbers, the number $a + bi$ is a **complex number,** and it is said to be written in **standard form.** If $b = 0$, the number $a + bi = a$ is a real number. If $b \neq 0$, the number $a + bi$ is called an **imaginary number.** A number of the form bi, where $b \neq 0$, is called a **pure imaginary number.**

The set of real numbers is a subset of the set of complex numbers, as shown in Figure 4.1. This is true because every real number a can be written as a complex number using $b = 0$. That is, for every real number a, you can write $a = a + 0i$.

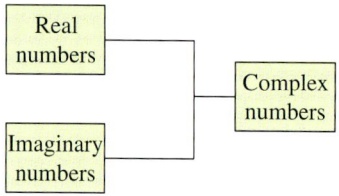

FIGURE 4.1

Equality of Complex Numbers

Two complex numbers $a + bi$ and $c + di$, written in standard form, are equal to each other

$$a + bi = c + di \qquad \text{Equality of two complex numbers}$$

if and only if $a = c$ and $b = d$.

Operations with Complex Numbers

To add (or subtract) two complex numbers, you add (or subtract) the real and imaginary parts of the numbers separately.

> **Addition and Subtraction of Complex Numbers**
>
> If $a + bi$ and $c + di$ are two complex numbers written in standard form, their sum and difference are defined as follows.
>
> *Sum:* $(a + bi) + (c + di) = (a + c) + (b + d)i$
>
> *Difference:* $(a + bi) - (c + di) = (a - c) + (b - d)i$

The **additive identity** in the complex number system is zero (the same as in the real number system). Furthermore, the **additive inverse** of the complex number $a + bi$ is

$$-(a + bi) = -a - bi. \quad \text{Additive inverse}$$

So, you have

$$(a + bi) + (-a - bi) = 0 + 0i = 0.$$

The *Interactive* CD-ROM and *Internet* versions of this text show every example with its solution; clicking on the *Try It!* button brings up similar problems. Guided Examples and Integrated Examples show step-by-step solutions to additional examples. Integrated Examples are related to several concepts in the section.

Example 1 ▶ Adding and Subtracting Complex Numbers

a. $(3 - i) + (2 + 3i) = 3 - i + 2 + 3i$ Remove parentheses.
$ = 3 + 2 - i + 3i$ Group like terms.
$ = (3 + 2) + (-1 + 3)i$
$ = 5 + 2i$ Write in standard form.

b. $2i + (-4 - 2i) = 2i - 4 - 2i$ Remove parentheses.
$ = -4 + 2i - 2i$ Group like terms.
$ = -4$ Write in standard form.

c. $3 - (-2 + 3i) + (-5 + i) = 3 + 2 - 3i - 5 + i$
$ = 3 + 2 - 5 - 3i + i$
$ = 0 - 2i$
$ = -2i$

d. $(3 + 2i) + (4 - i) - (7 + i) = 3 + 2i + 4 - i - 7 - i$
$ = 3 + 4 - 7 + 2i - i - i$
$ = 0 + 0i$
$ = 0$

Note in Example 1(b) that the sum of two complex numbers can be a real number.

Exploration

Complete the table:

$i^1 = i$ $i^7 = $
$i^2 = -1$ $i^8 = $
$i^3 = -i$ $i^9 = $
$i^4 = 1$ $i^{10} = $
$i^5 = $ $i^{11} = $
$i^6 = $ $i^{12} = $

What pattern do you see? Write a brief description of how you would find i raised to any positive integer power.

Many of the properties of real numbers are valid for complex numbers as well. Here are some examples.

Associative Properties of Addition and Multiplication

Commutative Properties of Addition and Multiplication

Distributive Property of Multiplication Over Addition

Notice below how these properties are used when two complex numbers are multiplied.

$(a + bi)(c + di) = a(c + di) + bi(c + di)$ Distributive Property
$ = ac + (ad)i + (bc)i + (bd)i^2$ Distributive Property
$ = ac + (ad)i + (bc)i + (bd)(-1)$ $i^2 = -1$
$ = ac - bd + (ad)i + (bc)i$ Commutative Property
$ = (ac - bd) + (ad + bc)i$ Associative Property

Rather than trying to memorize this multiplication rule, you should simply remember how the Distributive Property is used to multiply two complex numbers. The procedure is similar to multiplying two polynomials and combining like terms (as in the FOIL Method).

The *Interactive* CD-ROM and *Internet* versions of this text offer a built-in graphing calculator, which can be used in the Examples, Explorations, Technology notes, and Exercises.

Example 2 ▶ Multiplying Complex Numbers

a. $4(-2 + 3i) = 4(-2) + 4(3i)$ Distributive Property
$ = -8 + 12i$ Simplify.

b. $(i)(-3i) = -3i^2$ Multiply.
$ = -3(-1)$ $i^2 = -1$
$ = 3$ Simplify.

c. $(2 - i)(4 + 3i) = 8 + 6i - 4i - 3i^2$ Product of binomials
$ = 8 + 6i - 4i - 3(-1)$ $i^2 = -1$
$ = (8 + 3) + (6i - 4i)$ Group like terms.
$ = 11 + 2i$ Write in standard form.

d. $(3 + 2i)(3 - 2i) = 9 - 6i + 6i - 4i^2$ Product of binomials
$ = 9 - 4(-1)$ $i^2 = -1$
$ = 9 + 4$ Simplify.
$ = 13$ Write in standard form.

e. $(3 + 2i)^2 = 9 + 6i + 6i + 4i^2$ Product of binomials
$ = 9 + 4(-1) + 12i$ $i^2 = -1$
$ = 9 - 4 + 12i$ Simplify.
$ = 5 + 12i$ Write in standard form.

Complex Conjugates and Division

Notice in Example 2(d) that the product of two complex numbers can be a real number. This occurs with pairs of complex numbers of the form $a + bi$ and $a - bi$, called **complex conjugates.**

$$(a + bi)(a - bi) = a^2 - abi + abi - b^2i^2$$
$$= a^2 - b^2(-1)$$
$$= a^2 + b^2$$

To find the quotient of $a + bi$ and $c + di$ where c and d are not both zero, multiply the numerator and denominator by the conjugate of the *denominator* to obtain

$$\frac{a + bi}{c + di} = \frac{a + bi}{c + di}\left(\frac{c - di}{c - di}\right)$$
$$= \frac{(ac + bd) + (bc - ad)i}{c^2 + d^2}.$$

Example 3 ▶ Dividing Complex Numbers

$$\frac{1}{1 + i} = \frac{1}{1 + i}\left(\frac{1 - i}{1 - i}\right) \quad \text{Multiply numerator and denominator by conjugate of denominator.}$$

$$= \frac{1 - i}{1^2 - i^2} \quad \text{Expand.}$$

$$= \frac{1 - i}{1 - (-1)} \quad i^2 = -1$$

$$= \frac{1 - i}{2} \quad \text{Simplify.}$$

$$= \frac{1}{2} - \frac{1}{2}i \quad \text{Write in standard form.}$$

Example 4 ▶ Dividing Complex Numbers

$$\frac{2 + 3i}{4 - 2i} = \frac{2 + 3i}{4 - 2i}\left(\frac{4 + 2i}{4 + 2i}\right) \quad \text{Multiply numerator and denominator by conjugate of denominator.}$$

$$= \frac{8 + 4i + 12i + 6i^2}{16 - 4i^2} \quad \text{Expand.}$$

$$= \frac{8 - 6 + 16i}{16 + 4} \quad i^2 = -1$$

$$= \frac{2 + 16i}{20} \quad \text{Simplify.}$$

$$= \frac{1}{10} + \frac{4}{5}i \quad \text{Write in standard form.}$$

Complex Solutions of Quadratic Equations

When using the Quadratic Formula to solve a quadratic equation, you often obtain a result such as $\sqrt{-3}$, which you know is not a real number. By factoring out $i = \sqrt{-1}$, you can write this number in standard form.

$$\sqrt{-3} = \sqrt{3(-1)} = \sqrt{3}\sqrt{-1} = \sqrt{3}i$$

The number $\sqrt{3}i$ is called the *principal square root* of -3.

Principal Square Root of a Negative Number

If a is a positive number, the **principal square root** of the negative number $-a$ is defined as

$$\sqrt{-a} = \sqrt{a}\,i.$$

> **STUDY TIP**
>
> The definition of principal square root uses the rule
>
> $$\sqrt{ab} = \sqrt{a}\sqrt{b}$$
>
> for $a > 0$ and $b < 0$. This rule is not valid if *both* a and b are negative. For example,
>
> $$\sqrt{-5}\sqrt{-5} = \sqrt{5(-1)}\sqrt{5(-1)}$$
> $$= \sqrt{5}i\sqrt{5}i$$
> $$= \sqrt{25}i^2$$
> $$= 5i^2 = -5$$
>
> whereas
>
> $$\sqrt{(-5)(-5)} = \sqrt{25} = 5.$$
>
> To avoid problems with multiplying square roots of negative numbers, be sure to convert to standard form *before* multiplying.

Example 5 ▶ Writing Complex Numbers in Standard Form

a. $\sqrt{-3}\sqrt{-12} = \sqrt{3}i\sqrt{12}i = \sqrt{36}i^2 = 6(-1) = -6$

b. $\sqrt{-48} - \sqrt{-27} = \sqrt{48}i - \sqrt{27}i = 4\sqrt{3}i - 3\sqrt{3}i = \sqrt{3}i$

c. $(-1 + \sqrt{-3})^2 = (-1 + \sqrt{3}i)^2$
$\phantom{(-1 + \sqrt{-3})^2} = (-1)^2 - 2\sqrt{3}i + (\sqrt{3})^2(i^2)$
$\phantom{(-1 + \sqrt{-3})^2} = 1 - 2\sqrt{3}i + 3(-1)$
$\phantom{(-1 + \sqrt{-3})^2} = -2 - 2\sqrt{3}i$

Example 6 ▶ Complex Solutions of a Quadratic Equation

Solve (a) $x^2 + 4 = 0$ and (b) $3x^2 - 2x + 5 = 0$.

Solution

a. $x^2 + 4 = 0$ Write original equation.

$ x^2 = -4$ Subtract 4 from each side.

$ x = \pm 2i$ Extract square roots.

b. $3x^2 - 2x + 5 = 0$ Write original equation.

$ x = \dfrac{-(-2) \pm \sqrt{(-2)^2 - 4(3)(5)}}{2(3)}$ Quadratic Formula

$ = \dfrac{2 \pm \sqrt{-56}}{6}$ Simplify.

$ = \dfrac{2 \pm 2\sqrt{14}i}{6}$ Write $\sqrt{-56}$ in standard form.

$ = \dfrac{1}{3} \pm \dfrac{\sqrt{14}}{3}i$ Write in standard form.

4.1 Exercises

In Exercises 1–4, find real numbers a and b such that the equation is true.

1. $a + bi = -10 + 6i$
2. $a + bi = 13 + 4i$
3. $(a - 1) + (b + 3)i = 5 + 8i$
4. $(a + 6) + 2bi = 6 - 5i$

In Exercises 5–16, write the complex number in standard form.

5. $4 + \sqrt{-9}$
6. $3 + \sqrt{-16}$
7. $2 - \sqrt{-27}$
8. $1 + \sqrt{-8}$
9. $\sqrt{-75}$
10. $\sqrt{-4}$
11. 8
12. 45
13. $-6i + i^2$
14. $-4i^2 + 2i$
15. $\sqrt{-0.09}$
16. $\sqrt{-0.0004}$

In Exercises 17–26, perform the addition or subtraction and write the result in standard form.

17. $(5 + i) + (6 - 2i)$
18. $(13 - 2i) + (-5 + 6i)$
19. $(8 - i) - (4 - i)$
20. $(3 + 2i) - (6 + 13i)$
21. $\left(-2 + \sqrt{-8}\right) + \left(5 - \sqrt{-50}\right)$
22. $\left(8 + \sqrt{-18}\right) - \left(4 + 3\sqrt{2}i\right)$
23. $13i - (14 - 7i)$
24. $22 + (-5 + 8i) + 10i$
25. $-\left(\frac{3}{2} + \frac{5}{2}i\right) + \left(\frac{5}{3} + \frac{11}{3}i\right)$
26. $(1.6 + 3.2i) + (-5.8 + 4.3i)$

In Exercises 27–40, perform the operation and write the result in standard form.

27. $\sqrt{-6} \cdot \sqrt{-2}$
28. $\sqrt{-5} \cdot \sqrt{-10}$
29. $\left(\sqrt{-10}\right)^2$
30. $\left(\sqrt{-75}\right)^2$
31. $(1 + i)(3 - 2i)$
32. $(6 - 2i)(2 - 3i)$
33. $6i(5 - 2i)$
34. $-8i(9 + 4i)$
35. $\left(\sqrt{14} + \sqrt{10}i\right)\left(\sqrt{14} - \sqrt{10}i\right)$
36. $\left(3 + \sqrt{-5}\right)\left(7 - \sqrt{-10}\right)$
37. $(4 + 5i)^2$
38. $(2 - 3i)^2$
39. $(2 + 3i)^2 + (2 - 3i)^2$
40. $(1 - 2i)^2 - (1 + 2i)^2$

In Exercises 41–48, write the conjugate of the complex number. Then multiply the number and its conjugate.

41. $6 + 3i$
42. $7 - 12i$
43. $-1 - \sqrt{5}i$
44. $-3 + \sqrt{2}i$
45. $\sqrt{-20}$
46. $\sqrt{-15}$
47. $\sqrt{8}$
48. $1 + \sqrt{8}$

In Exercises 49–62, perform the operation and write the result in standard form.

49. $\dfrac{5}{i}$
50. $-\dfrac{14}{2i}$
51. $\dfrac{2}{4 - 5i}$
52. $\dfrac{5}{1 - i}$
53. $\dfrac{3 + i}{3 - i}$
54. $\dfrac{6 - 7i}{1 - 2i}$
55. $\dfrac{6 - 5i}{i}$
56. $\dfrac{8 + 16i}{2i}$
57. $\dfrac{3i}{(4 - 5i)^2}$
58. $\dfrac{5i}{(2 + 3i)^2}$
59. $\dfrac{2}{1 + i} - \dfrac{3}{1 - i}$
60. $\dfrac{2i}{2 + i} + \dfrac{5}{2 - i}$
61. $\dfrac{i}{3 - 2i} + \dfrac{2i}{3 + 8i}$
62. $\dfrac{1 + i}{i} - \dfrac{3}{4 - i}$

In Exercises 63–72, use the Quadratic Formula to solve the quadratic equation.

63. $x^2 - 2x + 2 = 0$
64. $x^2 + 6x + 10 = 0$
65. $4x^2 + 16x + 17 = 0$
66. $9x^2 - 6x + 37 = 0$
67. $4x^2 + 16x + 15 = 0$
68. $16t^2 - 4t + 3 = 0$
69. $\frac{3}{2}x^2 - 6x + 9 = 0$
70. $\frac{7}{8}x^2 - \frac{3}{4}x + \frac{5}{16} = 0$
71. $1.4x^2 - 2x - 10 = 0$
72. $4.5x^2 - 3x + 12 = 0$

73. Express each of the following powers of i as i, $-i$, 1, or -1.
(a) i^{40} (b) i^{25} (c) i^{50} (d) i^{67}

In Exercises 74–81, simplify the complex number and write it in standard form.

74. $-6i^3 + i^2$
75. $4i^2 - 2i^3$
76. $-5i^5$
77. $(-i)^3$
78. $(\sqrt{-75})^3$
79. $(\sqrt{-2})^6$
80. $\dfrac{1}{i^3}$
81. $\dfrac{1}{(2i)^3}$

82. Cube each complex number.
(a) 2 (b) $-1 + \sqrt{3}i$ (c) $-1 - \sqrt{3}i$

83. Raise each complex number to the fourth power.
(a) 2 (b) -2 (c) $2i$ (d) $-2i$

84. *Impedance* The opposition to current in an electrical circuit is called its impedance. The impedance in a parallel circuit with two pathways satisfies the equation
$$\dfrac{1}{z} = \dfrac{1}{z_1} + \dfrac{1}{z_2}$$
where z_1 is the impedance (in ohms) of pathway 1 and z_2 is the impedance of pathway 2. Use the table to determine the impedance of each parallel circuit. The impedance of each pathway is found by adding the impedance of each component in the pathway.

	Resistor	Inductor	Capacitor
Symbol	$a\Omega$	$b\Omega$	$c\Omega$
Impedance	a	bi	$-ci$

(a)
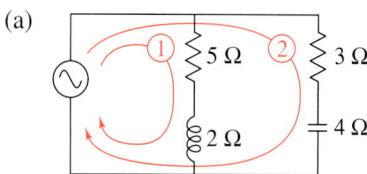
$z_1 = 5 + 2i,\ z_2 = 3 - 4i$

(b)
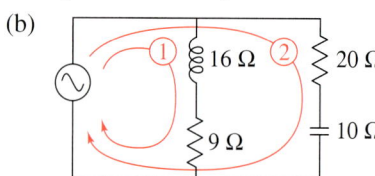
$z_1 = 9 + 16i,\ z_2 = 20 - 10i$

Synthesis

True or False? In Exercises 85–87, determine whether the statement is true or false. Justify your answer.

85. There is no complex number that is equal to its conjugate.

86. $-i\sqrt{6}$ is a solution of $x^4 - x^2 + 14 = 56$.

87. $i^{44} + i^{150} - i^{74} - i^{109} + i^{61} = -1$

88. *Error Analysis* Describe the error.
$$\sqrt{-6}\sqrt{-6} = \sqrt{(-6)(-6)} = \sqrt{36} = 6 \quad \times$$

89. Prove that the conjugate of the product of two complex numbers $a_1 + b_1 i$ and $a_2 + b_2 i$ is the product of their conjugates.

90. Prove that the conjugate of the sum of two complex numbers $a_1 + b_1 i$ and $a_2 + b_2 i$ is the sum of their conjugates.

Review

In Exercises 91–94, perform the operations and write the result in standard form.

91. $(4 + 3x) + (8 - 6x - x^2)$
92. $(x^3 - 3x^2) - (6 - 2x - 4x^2)$
93. $\left(3x - \tfrac{1}{2}\right)(x + 4)$
94. $(2x - 5)^2$

In Exercises 95–98, solve the equation and check your solution.

95. $-x - 12 = 19$
96. $8 - 3x = -34$
97. $4(5x - 6) - 3(6x + 1) = 0$
98. $5[x - (3x + 11)] = 20x - 15$

99. *Volume of an Oblate Spheroid*
Solve for a: $V = \tfrac{4}{3}\pi a^2 b$

100. *Newton's Law of Universal Gravitation*
Solve for r: $F = \alpha \dfrac{m_1 m_2}{r^2}$

101. *Mixture Problem* A 5-liter container contains a mixture with a concentration of 50%. How much of this mixture must be withdrawn and replaced by 100% concentrate to bring the mixture up to 60% concentration?

102. *Travel* A business executive drove at an average speed of 100 kilometers per hour on a 200-kilometer trip. Because of heavy traffic, the average speed on the return trip was 80 kilometers per hour. Find the average speed for the round trip.

4.2 Complex Solutions of Equations

▶ **What you should learn**
- How to determine the numbers of solutions of polynomial equations
- How to find the zeros of a polynomial function
- How to find a polynomial function given the zeros of the function

▶ **Why you should learn it**

Finding zeros of polynomial functions is an important part of solving real-life problems. For instance, Exercise 62 on page 323 shows how the zeros of a polynomial function can help you analyze the profit function for a business.

The Number of Solutions of a Polynomial Equation

The Fundamental Theorem of Algebra implies that a polynomial equation of degree n has precisely n solutions in the complex number system. These solutions can be real or complex and may be repeated.

Example 1 ▶ Zeros of Polynomial Functions

a. The first-degree equation $x - 2 = 0$ has exactly one solution: $x = 2$.

b. The second-degree equation

$$x^2 - 6x + 9 = 0 \qquad \text{Second-degree equation}$$

$$(x - 3)(x - 3) = 0 \qquad \text{Factor.}$$

has exactly two solutions: $x = 3$ and $x = 3$. (This is called a repeated zero.)

c. The third-degree equation

$$x^3 + 4x = 0 \qquad \text{Third-degree equation}$$

$$x(x - 2i)(x + 2i) = 0 \qquad \text{Factor.}$$

has exactly three solutions: $x = 0$, $x = 2i$, and $x = -2i$.

d. The fourth-degree equation

$$x^4 - 1 = 0 \qquad \text{Fourth-degree equation}$$

$$(x - 1)(x + 1)(x - i)(x + i) = 0 \qquad \text{Factor.}$$

has exactly four solutions: $x = 1$, $x = -1$, $x = i$, and $x = -i$.

You can use a graph to check the number of *real* solutions of an equation. For instance, to check the number of real solutions of $x^4 - 1 = 0$, sketch the graph of $f(x) = x^4 - 1$. As shown in Figure 4.2, the graph has two x-intercepts, which implies that the equation has two real solutions.

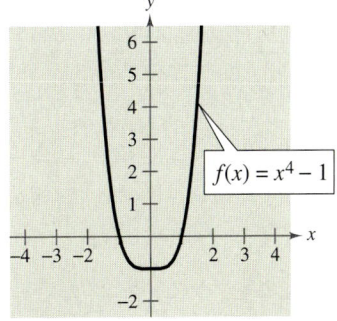

FIGURE 4.2

Every second-degree equation, $ax^2 + bx + c = 0$, has precisely two solutions given by the Quadratic Formula.

$$x = \frac{-b \pm \sqrt{b^2 - 4ac}}{2a}$$

The expression inside the radical, $b^2 - 4ac$, is called the **discriminant,** and can be used to determine whether the solutions are real, repeated, or complex.

1. If $b^2 - 4ac < 0$, the equation has two complex solutions.
2. If $b^2 - 4ac = 0$, the equation has one repeated real solution.
3. If $b^2 - 4ac > 0$, the equation has two real solutions.

Example 2 ▶ Using the Discriminant

Use the discriminant to find the number of real solutions of each equation.

a. $4x^2 - 20x + 25 = 0$ **b.** $13x^2 + 7x + 2 = 0$ **c.** $5x^2 - 8x = 0$

Solution

a. For this equation, $a = 4$, $b = -20$, and $c = 25$. So, the discriminant is

$$b^2 - 4ac = (-20)^2 - 4(4)(25) = 400 - 400 = 0.$$

Because the discriminant is zero, there is one repeated real solution.

b. For this equation, $a = 13$, $b = 7$, and $c = 2$. So, the discriminant is

$$b^2 - 4ac = 7^2 - 4(13)(2) = 49 - 104 = -55.$$

Because the discriminant is negative, there are two complex solutions.

c. For this equation, $a = 5$, $b = -8$, and $c = 0$. So, the discriminant is

$$b^2 - 4ac = (-8)^2 - 4(5)(0) = 64 - 0 = 64.$$

Because the discriminant is positive, there are two real solutions.

Figure 4.3 shows the graphs of the functions corresponding to the equations in Example 2. Notice that with one repeated solution, the graph touches the x-axis at its x-intercept. With two complex solutions, the graph has no x-intercepts. With two real solutions, the graph crosses the x-axis at its x-intercepts.

(a)

(b)

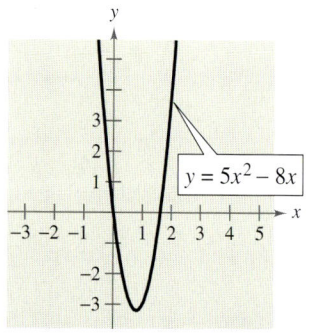

(c)

FIGURE 4.3

Finding Solutions of Polynomial Equations

Example 3 ▶ **Solving a Quadratic Equation**

Solve $x^2 + 2x + 2 = 0$. List complex solutions in standard form.

Solution

Using $a = 1$, $b = 2$, and $c = 2$, you can apply the Quadratic Formula as follows.

$$x = \frac{-b \pm \sqrt{b^2 - 4ac}}{2a} \qquad \text{Quadratic Formula}$$

$$= \frac{-2 \pm \sqrt{2^2 - 4(1)(2)}}{2(1)} \qquad \text{Substitute 1 for } a\text{, 2 for } b\text{, and 2 for } c.$$

$$= \frac{-2 \pm \sqrt{-4}}{2} \qquad \text{Simplify.}$$

$$= \frac{-2 \pm 2i}{2} \qquad \text{Simplify.}$$

$$= -1 \pm i \qquad \text{Standard form}$$

In Example 3, the two complex solutions are **conjugates.** That is, they are of the form $a \pm bi$. This is not a coincidence, as indicated by the following theorem.

Complex Solutions Occur in Conjugate Pairs

If $a + bi$, $b \neq 0$ is a solution of a polynomial equation with real coefficients, the conjugate $a - bi$ is also a solution of the equation.

Be sure you see that this result is true only if the polynomial has *real* coefficients. For instance, the result applies to the equation $x^2 + 1 = 0$, but not to the equation $x - i = 0$.

Example 4 ▶ **Solving a Polynomial Equation**

Solve $x^4 - x^2 - 20 = 0$.

Solution

$$x^4 - x^2 - 20 = 0$$

$$(x^2 - 5)(x^2 + 4) = 0$$

$$\left(x + \sqrt{5}\right)\left(x - \sqrt{5}\right)(x + 2i)(x - 2i) = 0$$

Setting each factor equal to zero yields the solutions $x = -\sqrt{5}, \sqrt{5}, -2i,$ and $2i$.

Finding Zeros of Polynomial Functions

The problem of finding the zeros of a polynomial function is essentially the same problem as finding the solutions of a polynomial equation. For instance, the zeros of the polynomial function

$$f(x) = 3x^2 - 4x + 5$$

are simply the solutions of the polynomial equation

$$3x^2 - 4x + 5 = 0.$$

Example 5 ▶ Finding the Zeros of a Polynomial Function

Find all the zeros of

$$f(x) = x^4 - 3x^3 + 6x^2 + 2x - 60$$

given that $1 + 3i$ is a zero of f.

Solution

Because complex zeros occur in conjugate pairs, you know that $1 - 3i$ is also a zero of f. This means that both

$$[x - (1 + 3i)] \quad \text{and} \quad [x - (1 - 3i)]$$

are factors of f. Multiplying these two factors produces

$$[x - (1 + 3i)][x - (1 - 3i)] = [(x - 1) - 3i][(x - 1) + 3i]$$
$$= (x - 1)^2 - 9i^2$$
$$= x^2 - 2x + 1 - 9(-1)$$
$$= x^2 - 2x + 10.$$

Using long division, you can divide $x^2 - 2x + 10$ into f to obtain the following.

$$\begin{array}{r}
x^2 - x - 6 \\
x^2 - 2x + 10 \overline{\smash{\big)}\, x^4 - 3x^3 + 6x^2 + 2x - 60} \\
\underline{x^4 - 2x^3 + 10x^2} \\
-x^3 - 4x^2 + 2x \\
\underline{-x^3 + 2x^2 - 10x} \\
-6x^2 + 12x - 60 \\
\underline{-6x^2 + 12x - 60} \\
0
\end{array}$$

Therefore, you have

$$f(x) = (x^2 - 2x + 10)(x^2 - x - 6)$$
$$= (x^2 - 2x + 10)(x - 3)(x + 2)$$

and you can conclude that the zeros of f are $1 + 3i$, $1 - 3i$, 3, and -2.

Example 6 ▶ Finding a Polynomial with Given Zeros

Find a fourth-degree polynomial function with real coefficients that has -1, -1, and $3i$ as zeros.

Solution

Because $3i$ is a zero *and* the polynomial is stated to have real coefficients, you know that the conjugate $-3i$ must also be a zero. So, $f(x)$ can be written as

$$f(x) = a(x + 1)(x + 1)(x - 3i)(x + 3i).$$

For simplicity, let $a = 1$ to obtain

$$f(x) = (x^2 + 2x + 1)(x^2 + 9)$$
$$= x^4 + 2x^3 + 10x^2 + 18x + 9.$$

Example 7 ▶ Finding a Polynomial with Given Zeros

Find a cubic polynomial function f with real coefficients that has 2 and $1 - i$ as zeros, such that $f(1) = 3$.

Solution

Because $1 - i$ is a zero of f, so is $1 + i$. Therefore,

$$f(x) = a(x - 2)[x - (1 - i)][x - (1 + i)]$$
$$= a(x - 2)[(x - 1) + i][(x - 1) - i]$$
$$= a(x - 2)[(x - 1)^2 - i^2]$$
$$= a(x - 2)(x^2 - 2x + 2)$$
$$= a(x^3 - 4x^2 + 6x - 4).$$

To find the value of a, use the fact that $f(1) = 3$ and obtain

$$f(1) = a(1 - 4 + 6 - 4) = 3.$$

So, $a = -3$ and it follows that

$$f(x) = -3(x^3 - 4x^2 + 6x - 4)$$
$$= -3x^3 + 12x^2 - 18x + 12.$$

Writing ABOUT MATHEMATICS

Solutions, Zeros, and Intercepts Write a paragraph explaining the relationships among the solutions of a polynomial equation, the zeros of a polynomial function, and the *x*-intercepts of the graph of a polynomial function. Include examples in your paragraph.

4.2 Exercises

In Exercises 1–4, determine the number of solutions of the equation in the complex number system.

1. $2x^3 + 3x + 1 = 0$
2. $x^6 + 4x^2 + 12 = 0$
3. $50 - 2x^4 = 0$
4. $14 - x + 4x^2 - 7x^5 = 0$

In Exercises 5–12, use the discriminant to determine the number of real solutions of the quadratic equation.

5. $2x^2 - 5x + 5 = 0$
6. $2x^2 - x - 1 = 0$
7. $\frac{1}{5}x^2 + \frac{6}{5}x - 8 = 0$
8. $\frac{1}{3}x^2 - 5x + 25 = 0$
9. $2x^2 - x - 15 = 0$
10. $-2x^2 + 11x - 2 = 0$
11. $x^2 + 2x + 10 = 0$
12. $x^2 - 4x + 53 = 0$

In Exercises 13–26, solve the equation. List complex solutions in standard form.

13. $x^2 - 5 = 0$
14. $3x^2 - 1 = 0$
15. $(x + 5)^2 - 6 = 0$
16. $16 - (x - 1)^2 = 0$
17. $x^2 - 8x + 16 = 0$
18. $4x^2 + 4x + 1 = 0$
19. $x^2 + 2x + 5 = 0$
20. $54 + 16x - x^2 = 0$
21. $4x^2 - 4x + 5 = 0$
22. $4x^2 - 4x + 21 = 0$
23. $230 + 20x - 0.5x^2 = 0$
24. $125 - 30x + 0.4x^2 = 0$
25. $8 + (x + 3)^2 = 0$
26. $6 - (x - 1)^2 = 0$

Graphical and Analytical Analysis In Exercises 27–30, find all the zeros of the function. Is there a relationship between the number of real zeros and the number of x-intercepts of the graph? Explain.

27. $f(x) = x^3 - 4x^2 + x - 4$
28. $f(x) = x^3 - 4x^2 - 4x + 16$

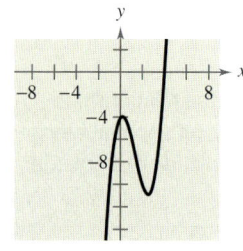

FIGURE FOR 27   FIGURE FOR 28

29. $f(x) = x^4 + 4x^2 + 4$
30. $f(x) = x^4 - 3x^2 - 4$

 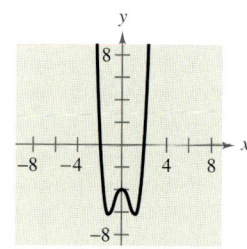

In Exercises 31–46, find all the zeros of the function and write the polynomial as a product of linear factors.

31. $f(x) = x^2 + 25$
32. $f(x) = x^2 - x + 56$
33. $h(x) = x^2 - 4x + 1$
34. $g(x) = x^2 + 10x + 23$
35. $f(x) = x^4 - 81$
36. $f(y) = y^4 - 625$
37. $g(x) = x^3 - 6x^2 + 13x - 10$
38. $f(x) = x^3 - 2x^2 - 11x + 52$
39. $h(x) = x^3 - x + 6$
40. $h(x) = x^3 + 9x^2 + 27x + 35$
41. $f(x) = 5x^3 - 9x^2 + 28x + 6$
42. $g(x) = 3x^3 - 4x^2 + 8x + 8$
43. $g(x) = x^4 - 4x^3 + 8x^2 - 16x + 16$
44. $h(x) = x^4 + 6x^3 + 10x^2 + 6x + 9$
45. $f(x) = x^4 + 10x^2 + 9$
46. $f(x) = x^4 + 29x^2 + 100$

In Exercises 47–54, use the given zero to find all the zeros of the function.

Function	Zero
47. $f(x) = 2x^3 + 3x^2 + 50x + 75$	$5i$
48. $f(x) = x^3 + x^2 + 9x + 9$	$3i$
49. $f(x) = 2x^4 - x^3 + 7x^2 - 4x - 4$	$2i$
50. $g(x) = x^3 - 7x^2 - x + 87$	$5 + 2i$
51. $g(x) = 4x^3 + 23x^2 + 34x - 10$	$-3 + i$
52. $h(x) = 3x^3 - 4x^2 + 8x + 8$	$1 - \sqrt{3}i$
53. $f(x) = x^4 + 3x^3 - 5x^2 - 21x + 22$	$-3 + \sqrt{2}i$
54. $f(x) = x^3 + 4x^2 + 14x + 20$	$-1 - 3i$

In Exercises 55–60, find a polynomial function with integer coefficients that has the given zeros. (There are many correct answers.)

55. $1, 5i, -5i$
56. $4, 3i, -3i$

57. $6, -5 + 2i, -5 - 2i$ 58. $2, 4 + i, 4 - i$
59. $\frac{2}{3}, -1, 3 + \sqrt{2}i$ 60. $-5, -5, 1 + \sqrt{3}i$

61. **Physics** A baseball is thrown upward from ground level with an initial velocity of 48 feet per second, and its height h (in feet) is

$$h = -16t^2 + 48t, \quad 0 \le t \le 3$$

where t is the time (in seconds). Suppose you are told the ball reaches a height of 64 feet. Is this possible?

62. **Economics** The demand equation for a certain product is $p = 140 - 0.0001x$, where p is the unit price (in dollars) of the product and x is the number of units produced and sold. The cost equation for the product is $C = 80x + 150,000$, where C is the total cost (in dollars) and x is the number of units produced. The total profit obtained by producing and selling x units is $P = R - C = xp - C$. You are working in the marketing department of the company that produces this product, and you are asked to determine a price p that will yield a profit of 9 million dollars. Is this possible? Explain.

Synthesis

True or False? In Exercises 63 and 64, determine whether the statement is true or false. Justify your answer.

63. It is possible for a third-degree polynomial function with integer coefficients to have no real zeros.

64. If $x = -i$ is a zero of the function $f(x) = x^3 + ix^2 + ix - 1$, then $x = i$ must also be a zero of f.

Think About It In Exercises 65–70, determine (if possible) the zeros of the function g if the function f has zeros at $x = r_1, x = r_2,$ and $x = r_3$.

65. $g(x) = -f(x)$ 66. $g(x) = 3f(x)$
67. $g(x) = f(x - 5)$ 68. $g(x) = f(2x)$
69. $g(x) = 3 + f(x)$ 70. $g(x) = f(-x)$

71. **Exploration** Use a graphing utility to graph the function $f(x) = x^4 - 4x^2 + k$ for different values of k. Find values of k such that the zeros of f satisfy the specified characteristics. (Some parts do not have unique answers.)

    (a) Four real zeros
    (b) Two real zeros, each of multiplicity 2
    (c) Two real zeros and two complex roots
    (d) Four complex zeros

72. **Think About It** Will the answers to Exercise 71 change for the function g?

    (a) $g(x) = f(x - 2)$
    (b) $g(x) = f(2x)$

73. (a) Find a quadratic function f (with integer coefficients) that has $\pm\sqrt{b}i$ as zeros. Assume that b is a positive integer.
    (b) Find a quadratic function f (with integer coefficients) that has $a \pm bi$ as zeros. Assume that b is a positive integer.

74. **Graphical Reasoning** The graph of one of the following functions is shown below. Identify the function shown in the graph. Explain why each of the others is not the correct function. Use a graphing utility to verify your result.

    (a) $f(x) = x^2(x + 2)(x - 3.5)$
    (b) $g(x) = (x + 2)(x - 3.5)$
    (c) $h(x) = (x + 2)(x - 3.5)(x^2 + 1)$
    (d) $k(x) = (x + 1)(x + 2)(x - 3.5)$

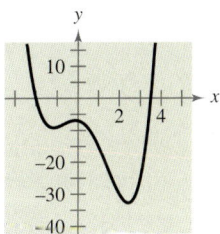

Review

In Exercises 75–80, perform the operation and simplify.

75. $(-3 + 6i) - (8 - 3i)$ 76. $(12 - 5i) + 16i$
77. $(6 - 2i)(1 + 7i)$ 78. $(9 - 5i)(9 + 5i)$
79. $\dfrac{1 + i}{1 - i}$ 80. $(3 + i)^3$

In Exercises 81–86, use the graph of f to graph the function g.

81. $g(x) = f(x - 2)$
82. $g(x) = f(x) - 2$
83. $g(x) = 2f(x)$
84. $g(x) = f(-x)$
85. $g(x) = f(2x)$
86. $g(x) = f(\frac{1}{2}x)$

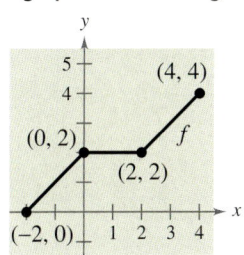

4.3 Trigonometric Form of a Complex Number

▶ **What you should learn**

- How to plot complex numbers in the complex plane
- How to write the trigonometric forms of complex numbers
- How to multiply and divide complex numbers written in trigonometric form

▶ **Why you should learn it**

You can perform the operations of multiplication and division on complex numbers by learning to write complex numbers in trigonometric form. For instance, in Exercises 61–68 on page 330, you will multiply and divide complex numbers in trigonometric form and standard form.

The Complex Plane

Just as real numbers can be represented by points on the real number line, you can represent a complex number

$$z = a + bi$$

as the point (a, b) in a coordinate plane (the **complex plane**). The horizontal axis is called the **real axis** and the vertical axis is called the **imaginary axis,** as shown in Figure 4.4.

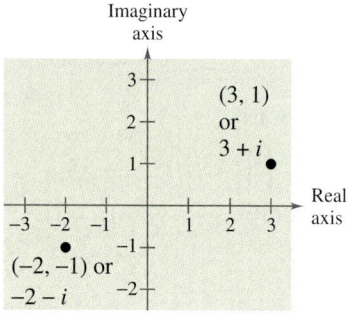

FIGURE 4.4

The **absolute value** of the complex number $a + bi$ is defined as the distance between the origin $(0, 0)$ and the point (a, b).

Definition of the Absolute Value of a Complex Number

The **absolute value** of the complex number $z = a + bi$ is

$$|a + bi| = \sqrt{a^2 + b^2}.$$

If the complex number $a + bi$ is a real number (that is, if $b = 0$), then this definition agrees with that given for the absolute value of a real number

$$|a + 0i| = \sqrt{a^2 + 0^2} = |a|.$$

Example 1 ▶ Finding the Absolute Value of a Complex Number

Plot $z = -2 + 5i$ and find its absolute value.

Solution

The number is plotted in Figure 4.5. It has an absolute value of

$$|z| = \sqrt{(-2)^2 + 5^2}$$

$$= \sqrt{29}.$$

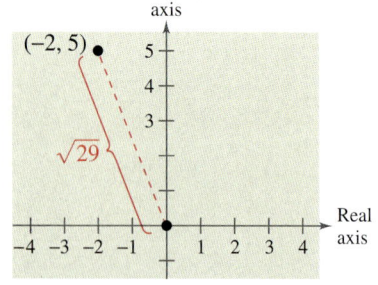

FIGURE 4.5

Trigonometric Form of a Complex Number

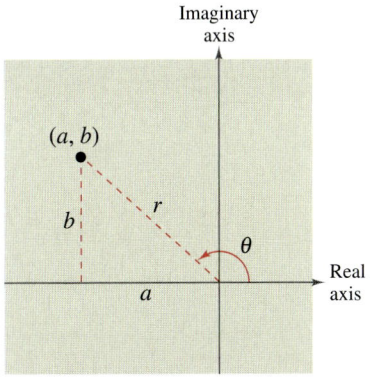

FIGURE 4.6

In Section 4.1 you learned how to add, subtract, multiply, and divide complex numbers. To work effectively with *powers* and *roots* of complex numbers, it is helpful to write complex numbers in trigonometric form. In Figure 4.6, consider the nonzero complex number $a + bi$. By letting θ be the angle from the positive real axis (measured counterclockwise) to the line segment connecting the origin and the point (a, b), you can write

$$a = r \cos \theta \quad \text{and} \quad b = r \sin \theta$$

where $r = \sqrt{a^2 + b^2}$. Consequently, you have

$$a + bi = (r \cos \theta) + (r \sin \theta)i$$

from which you can obtain the **trigonometric form of a complex number.**

> **Trigonometric Form of a Complex Number**
>
> The **trigonometric form** of the complex number $z = a + bi$ is
>
> $$z = r(\cos \theta + i \sin \theta)$$
>
> where $a = r \cos \theta$, $b = r \sin \theta$, $r = \sqrt{a^2 + b^2}$, and $\tan \theta = b/a$. The number r is the **modulus** of z, and θ is called an **argument** of z.

The trigonometric form of a complex number is also called the *polar form*. Because there are infinitely many choices for θ, the trigonometric form of a complex number is not unique. Normally, θ is restricted to the interval $0 \leq \theta < 2\pi$, although on occasion it is convenient to use $\theta < 0$.

Example 2 ▶ Writing a Complex Number in Trigonometric Form

Write the complex number $z = -2 - 2\sqrt{3}i$ in trigonometric form.

Solution

The absolute value of z is

$$r = \left| -2 - 2\sqrt{3}i \right| = \sqrt{(-2)^2 + \left(-2\sqrt{3}\right)^2} = \sqrt{16} = 4$$

and the angle θ is

$$\tan \theta = \frac{b}{a} = \frac{-2\sqrt{3}}{-2} = \sqrt{3}.$$

Because $\tan(\pi/3) = \sqrt{3}$ and $z = -2 - 2\sqrt{3}i$ lies in Quadrant III, you choose θ to be $\theta = \pi + \pi/3 = 4\pi/3$. So, the trigonometric form is

$$z = r(\cos \theta + i \sin \theta)$$
$$= 4\left(\cos \frac{4\pi}{3} + i \sin \frac{4\pi}{3}\right).$$

See Figure 4.7.

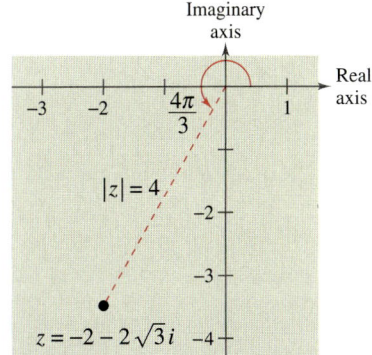

FIGURE 4.7

Example 3 ▶ Writing a Complex Number in Trigonometric Form

Write the complex number $z = 6 + 2i$ in trigonometric form.

Solution

The absolute value of z is

$$r = |6 + 2i|$$
$$= \sqrt{6^2 + 2^2}$$
$$= \sqrt{40}$$
$$= 2\sqrt{10}$$

and the angle θ is

$$\tan \theta = \frac{b}{a} = \frac{2}{6} = \frac{1}{3}.$$

Because θ is in Quadrant I, you can conclude that

$$\theta = \arctan \frac{1}{3} \approx 0.32175 \text{ radian} \approx 18.4°.$$

Therefore, the trigonometric form of z is

$$z = r(\cos \theta + i \sin \theta)$$
$$= 2\sqrt{10}\left[\cos\left(\arctan \frac{1}{3}\right) + i \sin\left(\arctan \frac{1}{3}\right)\right]$$
$$\approx 2\sqrt{10}(\cos 18.4° + i \sin 18.4°).$$

This result is illustrated graphically in Figure 4.8.

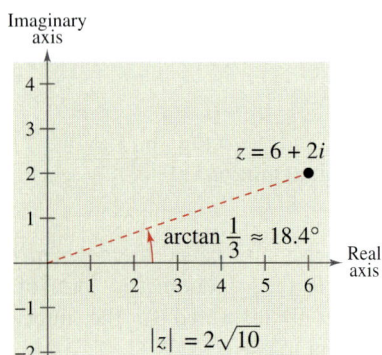

FIGURE 4.8

Example 4 ▶ Writing a Complex Number in Standard Form

Write the complex number in standard form $a + bi$.

$$z = \sqrt{8}\left[\cos\left(-\frac{\pi}{3}\right) + i \sin\left(-\frac{\pi}{3}\right)\right]$$

Solution

Because $\cos(-\pi/3) = 1/2$ and $\sin(-\pi/3) = -\sqrt{3}/2$, you can write

$$z = \sqrt{8}\left[\cos\left(-\frac{\pi}{3}\right) + i \sin\left(-\frac{\pi}{3}\right)\right]$$
$$= \sqrt{8}\left(\frac{1}{2} - \frac{\sqrt{3}}{2}i\right)$$
$$= 2\sqrt{2}\left(\frac{1}{2} - \frac{\sqrt{3}}{2}i\right)$$
$$= \sqrt{2} - \sqrt{6}i.$$

Technology

A graphing utility can be used to convert a complex number in trigonometric (or polar) form to rectangular form and vice versa. For specific keystrokes, see the user's manual for your graphing utility.

Multiplication and Division of Complex Numbers

The trigonometric form adapts nicely to multiplication and division of complex numbers. Suppose you are given two complex numbers

$$z_1 = r_1(\cos\theta_1 + i\sin\theta_1) \quad \text{and} \quad z_2 = r_2(\cos\theta_2 + i\sin\theta_2).$$

The product of z_1 and z_2 is

$$z_1 z_2 = r_1 r_2(\cos\theta_1 + i\sin\theta_1)(\cos\theta_2 + i\sin\theta_2)$$
$$= r_1 r_2[(\cos\theta_1 \cos\theta_2 - \sin\theta_1 \sin\theta_2) + i(\sin\theta_1 \cos\theta_2 + \cos\theta_1 \sin\theta_2)].$$

Using the sum and difference formulas for cosine and sine, you can rewrite this equation as

$$z_1 z_2 = r_1 r_2 [\cos(\theta_1 + \theta_2) + i\sin(\theta_1 + \theta_2)].$$

This establishes the first part of the following rule. The second part is left to you (see Exercise 75).

Product and Quotient of Two Complex Numbers

Let $z_1 = r_1(\cos\theta_1 + i\sin\theta_1)$ and $z_2 = r_2(\cos\theta_2 + i\sin\theta_2)$ be complex numbers.

$$z_1 z_2 = r_1 r_2 [\cos(\theta_1 + \theta_2) + i\sin(\theta_1 + \theta_2)] \qquad \text{Product}$$

$$\frac{z_1}{z_2} = \frac{r_1}{r_2}[\cos(\theta_1 - \theta_2) + i\sin(\theta_1 - \theta_2)], \quad z_2 \neq 0 \qquad \text{Quotient}$$

Note that this rule says that to *multiply* two complex numbers you multiply moduli and add arguments, whereas to *divide* two complex numbers you divide moduli and subtract arguments.

Example 5 ▶ Dividing Complex Numbers

Find the quotient, z_1/z_2, of the complex numbers.

$$z_1 = 24(\cos 300° + i\sin 300°) \qquad z_2 = 8(\cos 75° + i\sin 75°)$$

Solution

$$\frac{z_1}{z_2} = \frac{24(\cos 300° + i\sin 300°)}{8(\cos 75° + i\sin 75°)}$$

$$= \frac{24}{8}[\cos(300° - 75°) + i\sin(300° - 75°)]$$

$$= 3(\cos 225° + i\sin 225°)$$

$$= 3\left[\left(-\frac{\sqrt{2}}{2}\right) + i\left(-\frac{\sqrt{2}}{2}\right)\right]$$

$$= -\frac{3\sqrt{2}}{2} - \frac{3\sqrt{2}}{2}i$$

Example 6 Multiplying Complex Numbers

Find the product of the complex numbers.

$$z_1 = 2\left(\cos\frac{2\pi}{3} + i\sin\frac{2\pi}{3}\right) \qquad z_2 = 8\left(\cos\frac{11\pi}{6} + i\sin\frac{11\pi}{6}\right)$$

Solution

$$\begin{aligned}
z_1 z_2 &= 2\left(\cos\frac{2\pi}{3} + i\sin\frac{2\pi}{3}\right) \times 8\left(\cos\frac{11\pi}{6} + i\sin\frac{11\pi}{6}\right) \\
&= 16\left[\cos\left(\frac{2\pi}{3} + \frac{11\pi}{6}\right) + i\sin\left(\frac{2\pi}{3} + \frac{11\pi}{6}\right)\right] \\
&= 16\left(\cos\frac{5\pi}{2} + i\sin\frac{5\pi}{2}\right) \\
&= 16\left(\cos\frac{\pi}{2} + i\sin\frac{\pi}{2}\right) \\
&= 16[0 + i(1)] \\
&= 16i
\end{aligned}$$

You can check this result by first converting the complex numbers to the standard forms $z_1 = -1 + \sqrt{3}i$ and $z_2 = 4\sqrt{3} - 4i$ and then multiplying algebraically, as in Section 4.1.

$$\begin{aligned}
z_1 z_2 &= (-1 + \sqrt{3}i)(4\sqrt{3} - 4i) \\
&= -4\sqrt{3} + 4i + 12i + 4\sqrt{3} \\
&= 16i
\end{aligned}$$

Writing ABOUT MATHEMATICS

Multiplying Complex Numbers Graphically Discuss how you can graphically approximate the product of the complex numbers. Then, approximate the values of the products and check your answers analytically.

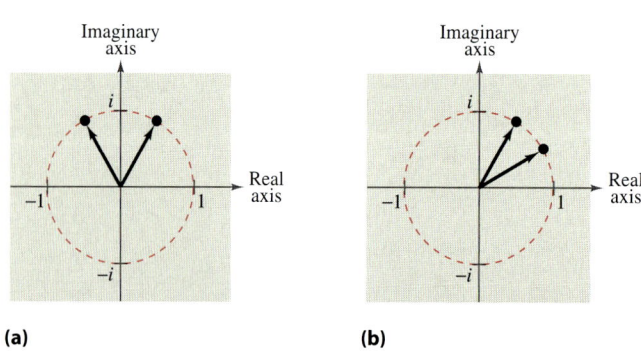

(a)   (b)

4.3 Exercises

In Exercises 1–6, plot the complex number and find its absolute value.

1. $-7i$
2. -7
3. $-4 + 4i$
4. $5 - 12i$
5. $6 - 7i$
6. $-8 + 3i$

In Exercises 7–10, write the complex number in trigonometric form.

7.
8.
9.
10.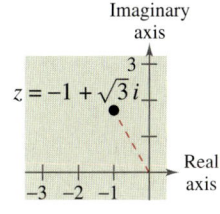

In Exercises 11–26, represent the complex number graphically, and find the trigonometric form of the number.

11. $3 - 3i$
12. $2 + 2i$
13. $\sqrt{3} + i$
14. $4 - 4\sqrt{3}i$
15. $-2(1 + \sqrt{3}i)$
16. $\frac{5}{2}(\sqrt{3} - i)$
17. $-5i$
18. $4i$
19. $-7 + 4i$
20. $3 - i$
21. 7
22. 4
23. $3 + \sqrt{3}i$
24. $2\sqrt{2} - i$
25. $-3 - i$
26. $1 + 3i$

In Exercises 27–34, use a graphing utility to represent the complex number in trigonometric form.

27. $5 + 2i$
28. $8 + 3i$
29. $-3 + i$
30. $-5 - i$
31. $3\sqrt{2} - 7i$
32. $4\sqrt{5} - 4i$
33. $-8 - 5\sqrt{3}i$
34. $-9 - 2\sqrt{10}i$

In Exercises 35–44, represent the complex number graphically, and find the standard form of the number.

35. $3(\cos 120° + i \sin 120°)$
36. $5(\cos 135° + i \sin 135°)$
37. $\frac{3}{2}(\cos 300° + i \sin 300°)$
38. $\frac{1}{4}(\cos 225° + i \sin 225°)$
39. $3.75\left(\cos \frac{3\pi}{4} + i \sin \frac{3\pi}{4}\right)$
40. $6\left(\cos \frac{5\pi}{12} + i \sin \frac{5\pi}{12}\right)$
41. $8\left(\cos \frac{\pi}{2} + i \sin \frac{\pi}{2}\right)$
42. $7(\cos 0 + i \sin 0)$
43. $3[\cos(18° \ 45') + i \sin(18° \ 45')]$
44. $6[\cos(230° \ 30') + i \sin(230° \ 30')]$

In Exercises 45–48, use a graphing utility to represent the complex number in standard form.

45. $5\left(\cos \frac{\pi}{9} + i \sin \frac{\pi}{9}\right)$
46. $10\left(\cos \frac{2\pi}{5} + i \sin \frac{2\pi}{5}\right)$
47. $3(\cos 165.5° + i \sin 165.5°)$
48. $9(\cos 58° + i \sin 58°)$

In Exercises 49–60, perform the operation and leave the result in trigonometric form.

49. $\left[2\left(\cos \frac{\pi}{4} + i \sin \frac{\pi}{4}\right)\right]\left[6\left(\cos \frac{\pi}{12} + i \sin \frac{\pi}{12}\right)\right]$
50. $\left[\frac{3}{4}\left(\cos \frac{\pi}{3} + i \sin \frac{\pi}{3}\right)\right]\left[4\left(\cos \frac{3\pi}{4} + i \sin \frac{3\pi}{4}\right)\right]$
51. $\left[\frac{5}{3}(\cos 140° + i \sin 140°)\right]\left[\frac{2}{3}(\cos 60° + i \sin 60°)\right]$
52. $[0.5(\cos 100° + i \sin 100°)] \times$
    $[0.8(\cos 300° + i \sin 300°)]$
53. $[0.45(\cos 310° + i \sin 310°)] \times$
    $[0.60(\cos 200° + i \sin 200°)]$
54. $(\cos 5° + i \sin 5°)(\cos 20° + i \sin 20°)$
55. $\dfrac{\cos 50° + i \sin 50°}{\cos 20° + i \sin 20°}$

56. $\dfrac{2(\cos 120° + i \sin 120°)}{4(\cos 40° + i \sin 40°)}$

57. $\dfrac{\cos(5\pi/3) + i \sin(5\pi/3)}{\cos \pi + i \sin \pi}$

58. $\dfrac{5(\cos 4.3 + i \sin 4.3)}{4(\cos 2.1 + i \sin 2.1)}$

59. $\dfrac{12(\cos 52° + i \sin 52°)}{3(\cos 110° + i \sin 110°)}$

60. $\dfrac{6(\cos 40° + i \sin 40°)}{7(\cos 100° + i \sin 100°)}$

In Exercises 61–68, (a) give the trigonometric form of the complex numbers, (b) perform the indicated operation using the trigonometric form, and (c) perform the indicated operation using the standard form, and check your result with that of part (b).

61. $(2 + 2i)(1 - i)$
62. $(\sqrt{3} + i)(1 + i)$
63. $-2i(1 + i)$
64. $4(1 - \sqrt{3}i)$
65. $\dfrac{3 + 4i}{1 - \sqrt{3}i}$
66. $\dfrac{1 + \sqrt{3}i}{6 - 3i}$
67. $\dfrac{5}{2 + 3i}$
68. $\dfrac{4i}{-4 + 2i}$

In Exercises 69–72, sketch the graphs of all complex numbers z satisfying the given condition.

69. $|z| = 2$
70. $|z| = 3$
71. $\theta = \dfrac{\pi}{6}$
72. $\theta = \dfrac{5\pi}{4}$

Synthesis

True or False? In Exercises 73 and 74, determine whether the statement is true or false. Justify your answer.

73. Although the square of the complex number bi is given by $(bi)^2 = -b^2$, the absolute value of the complex number $z = a + bi$ is defined as $|a + bi| = \sqrt{a^2 + b^2}$.

74. The product of two complex numbers, $z = r_1(\cos \theta_1 + i \sin \theta_1)$ and $z_2 = r_2(\cos \theta_2 + i \sin \theta_2)$, is zero only when $r_1 = 0$ and/or $r_2 = 0$.

75. Given two complex numbers $z_1 = r_1(\cos \theta_1 + i \sin \theta_1)$ and $z_2 = r_2(\cos \theta_2 + i \sin \theta_2)$, $z_2 \neq 0$, show that
$$\dfrac{z_1}{z_2} = \dfrac{r_1}{r_2}[\cos(\theta_1 - \theta_2) + i \sin(\theta_1 - \theta_2)].$$

76. Show that $\bar{z} = r[\cos(-\theta) + i \sin(-\theta)]$ is the complex conjugate of $z = r(\cos \theta + i \sin \theta)$.

77. Use the trigonometric forms of z and $\bar{z}$ in Exercise 76 to find (a) $z\bar{z}$ and (b) $z/\bar{z}, \bar{z} \neq 0$.

78. Show that the negative of $z = r(\cos \theta + i \sin \theta)$ is $-z = r[\cos(\theta + \pi) + i \sin(\theta + \pi)]$.

Review

In Exercises 79–84, solve the right triangle shown in the figure. Approximate the result to two decimal places.

79. $A = 22°, \quad a = 8$
80. $B = 66°, \quad a = 33.5$
81. $A = 30°, \quad b = 112.6$
82. $B = 6°, \quad b = 211.2$
83. $A = 42° 15', \quad c = 11.2$
84. $B = 81° 30', \quad c = 6.8$

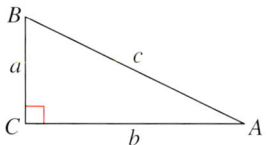

In Exercises 85–88, for the simple harmonic motion described by the trigonometric function, find the maximum displacement and the lowest possible value of t for which $d = 0$.

85. $d = 16 \cos \dfrac{\pi}{4} t$
86. $d = \tfrac{1}{8} \cos 12\pi t$
87. $d = \tfrac{1}{16} \sin \tfrac{5}{4} \pi t$
88. $d = \tfrac{1}{12} \sin 60\pi t$

4.4 DeMoivre's Theorem

▶ **What you should learn**
- How to use DeMoivre's Theorem to find powers of complex numbers
- How to find nth roots of complex numbers

▶ **Why you should learn it**

The trigonometric form of a complex number is useful when finding powers and roots of complex numbers. For instance, Exercises 45–60 on pages 335–336 show you how the trigonometric forms of complex numbers can help you solve a polynomial equation.

Powers of Complex Numbers

To raise a complex number to a power, consider repeated use of the multiplication rule.

$$z = r(\cos\theta + i\sin\theta)$$
$$z^2 = r(\cos\theta + i\sin\theta)r(\cos\theta + i\sin\theta) = r^2(\cos 2\theta + i\sin 2\theta)$$
$$z^3 = r^2(\cos 2\theta + i\sin 2\theta)r(\cos\theta + i\sin\theta) = r^3(\cos 3\theta + i\sin 3\theta)$$
$$z^4 = r^4(\cos 4\theta + i\sin 4\theta)$$
$$z^5 = r^5(\cos 5\theta + i\sin 5\theta)$$
$$\vdots$$

This pattern leads to the following important theorem, which is named after the French mathematician Abraham DeMoivre (1667–1754).

> **DeMoivre's Theorem**
>
> If $z = r(\cos\theta + i\sin\theta)$ is a complex number and n is a positive integer, then
>
> $$z^n = [r(\cos\theta + i\sin\theta)]^n = r^n(\cos n\theta + i\sin n\theta).$$

Example 1 ▶ **Finding Powers of a Complex Number**

Use DeMoivre's Theorem to find $(-1 + \sqrt{3}i)^{12}$.

Solution

First convert the complex number to trigonometric form using

$$r = \sqrt{(-1)^2 + (\sqrt{3})^2} = 2 \text{ and } \theta = \arctan\frac{\sqrt{3}}{-1} = \frac{2\pi}{3}.$$

$$-1 + \sqrt{3}i = 2\left(\cos\frac{2\pi}{3} + i\sin\frac{2\pi}{3}\right)$$

Then, by DeMoivre's Theorem, you have

$$(-1 + \sqrt{3}i)^{12} = \left[2\left(\cos\frac{2\pi}{3} + i\sin\frac{2\pi}{3}\right)\right]^{12}$$

$$= 2^{12}\left[\cos(12)\frac{2\pi}{3} + i\sin(12)\frac{2\pi}{3}\right]$$

$$= 4096(\cos 8\pi + i\sin 8\pi)$$

$$= 4096(1 + 0) = 4096.$$

Roots of Complex Numbers

Recall that a consequence of the Fundamental Theorem of Algebra is that a polynomial equation of degree n has n solutions in the complex number system. So, the equation $x^6 = 1$ has six solutions, and in this particular case you can find the six solutions by factoring and using the Quadratic Formula.

$$x^6 - 1 = (x^3 - 1)(x^3 + 1)$$
$$= (x - 1)(x^2 + x + 1)(x + 1)(x^2 - x + 1) = 0$$

Consequently, the solutions are

$$x = \pm 1, \quad x = \frac{-1 \pm \sqrt{3}i}{2}, \quad \text{and} \quad x = \frac{1 \pm \sqrt{3}i}{2}.$$

Each of these numbers is a sixth root of 1. In general, the **nth root** of a complex number is defined as follows.

Definition of nth Root of a Complex Number

The complex number $u = a + bi$ is an **nth root** of the complex number z if

$$z = u^n = (a + bi)^n.$$

To find a formula for an nth root of a complex number, let u be an nth root of z, where

$$u = s(\cos \beta + i \sin \beta)$$

and

$$z = r(\cos \theta + i \sin \theta).$$

By DeMoivre's Theorem and the fact that $u^n = z$, you have

$$s^n(\cos n\beta + i \sin n\beta) = r(\cos \theta + i \sin \theta).$$

Taking the absolute values of both sides of this equation, it follows that $s^n = r$. Substituting back into the previous equation and dividing by r, you get

$$\cos n\beta + i \sin n\beta = \cos \theta + i \sin \theta.$$

So, it follows that

$$\cos n\beta = \cos \theta$$

and

$$\sin n\beta = \sin \theta.$$

Because both sine and cosine have a period of 2π, these last two equations have solutions if and only if the angles differ by a multiple of 2π. Consequently, there must exist an integer k such that

$$n\beta = \theta + 2\pi k$$

$$\beta = \frac{\theta + 2\pi k}{n}.$$

By substituting this value of β into the trigonometric form of u, you get the result stated on the following page.

Historical Note
Abraham DeMoivre (1667–1754) is remembered for his work in probability theory and DeMoivre's Theorem. His *The Doctrine of Chances* (published in 1718) includes the theory of recurring series and the theory of partial fractions.

Exploration

The nth roots of a complex number are useful for solving some polynomial equations. For instance, explain how you can use DeMoivre's Theorem to solve the polynomial equation

$$x^4 + 16 = 0.$$

[*Hint:* Write -16 as $16(\cos \pi + i \sin \pi)$.]

nth Roots of a Complex Number

For a positive integer n, the complex number $z = r(\cos \theta + i \sin \theta)$ has exactly n distinct nth roots given by

$$\sqrt[n]{r}\left(\cos \frac{\theta + 2\pi k}{n} + i \sin \frac{\theta + 2\pi k}{n}\right)$$

where $k = 0, 1, 2, \ldots, n - 1$.

When k exceeds $n - 1$, the roots begin to repeat. For instance, if $k = n$, the angle

$$\frac{\theta + 2\pi n}{n} = \frac{\theta}{n} + 2\pi$$

is coterminal with θ/n, which is also obtained when $k = 0$.

The formula for the nth roots of a complex number z has a nice geometrical interpretation, as shown in Figure 4.9. Note that because the nth roots of z all have the same magnitude $\sqrt[n]{r}$, they all lie on a circle of radius $\sqrt[n]{r}$ with center at the origin. Furthermore, because successive nth roots have arguments that differ by $2\pi/n$, the n roots are equally spaced around the circle.

You have already found the sixth roots of 1 by factoring and by using the Quadratic Formula. Example 2 shows how you can solve the same problem with the formula for nth roots.

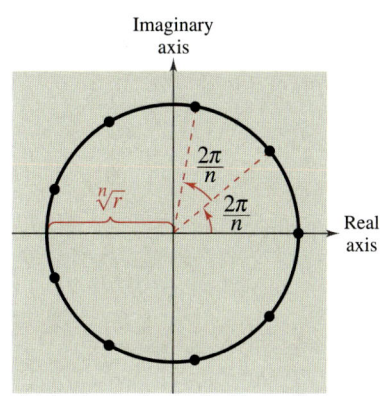

FIGURE 4.9

Example 2 ▶ Finding the nth Roots of a Real Number

Find all the sixth roots of 1.

Solution

First write 1 in the trigonometric form $1 = 1(\cos 0 + i \sin 0)$. Then, by the nth root formula, with $n = 6$ and $r = 1$, the roots have the form

$$\sqrt[6]{1}\left(\cos \frac{0 + 2\pi k}{6} + i \sin \frac{0 + 2\pi k}{6}\right)$$

or simply $\cos(\pi k/3) + i \sin(\pi k/3)$. So, for $k = 0, 1, 2, 3, 4,$ and 5, the sixth roots are as follows. (See Figure 4.10.)

$\cos 0 + i \sin 0 = 1$

$\cos \dfrac{\pi}{3} + i \sin \dfrac{\pi}{3} = \dfrac{1}{2} + \dfrac{\sqrt{3}}{2} i$ Increment by $\dfrac{2\pi}{n} = \dfrac{2\pi}{6} = \dfrac{\pi}{3}$.

$\cos \dfrac{2\pi}{3} + i \sin \dfrac{2\pi}{3} = -\dfrac{1}{2} + \dfrac{\sqrt{3}}{2} i$

$\cos \pi + i \sin \pi = -1$

$\cos \dfrac{4\pi}{3} + i \sin \dfrac{4\pi}{3} = -\dfrac{1}{2} - \dfrac{\sqrt{3}}{2} i$

$\cos \dfrac{5\pi}{3} + i \sin \dfrac{5\pi}{3} = \dfrac{1}{2} - \dfrac{\sqrt{3}}{2} i$

FIGURE 4.10

In Figure 4.10, notice that the roots obtained in Example 2 all have a magnitude of 1 and are equally spaced around the unit circle. Also notice that the complex roots occur in conjugate pairs, as discussed in Section 4.2. The n distinct nth roots of 1 are called the **nth roots of unity.**

Example 3 ▶ Finding the nth Roots of a Complex Number

Find the three cube roots of
$$z = -2 + 2i.$$

Solution

Because z lies in Quadrant II, the trigonometric form of z is
$$z = -2 + 2i$$
$$= \sqrt{8}\,(\cos 135° + i \sin 135°).$$

By the formula for nth roots, the cube roots have the form
$$\sqrt[6]{8}\left(\cos \frac{135° + 360°k}{3} + i \sin \frac{135° + 360°k}{3}\right).$$

Finally, for $k = 0, 1,$ and 2, you obtain the roots

$$\sqrt[6]{8}\left(\cos \frac{135° + 360°(0)}{3} + i \sin \frac{135° + 360°(0)}{3}\right) = \sqrt{2}(\cos 45° + i \sin 45°)$$
$$= 1 + i$$

$$\sqrt[6]{8}\left(\cos \frac{135° + 360°(1)}{3} + i \sin \frac{135° + 360°(1)}{3}\right) = \sqrt{2}(\cos 165° + i \sin 165°)$$
$$\approx -1.3660 + 0.3660i$$

$$\sqrt[6]{8}\left(\cos \frac{135° + 360°(2)}{3} + i \sin \frac{135° + 360°(2)}{3}\right) = \sqrt{2}(\cos 285° + i \sin 285°)$$
$$\approx 0.3660 - 1.3660i.$$

Writing ABOUT MATHEMATICS

A Famous Mathematical Formula The famous formula

$$e^{a+bi} = e^a(\cos b + i \sin b)$$

is called Euler's Formula, after the German mathematician Leonhard Euler (1707–1783). Although the interpretation of this formula is beyond the scope of this text, we decided to include it because it gives rise to one of the most wonderful equations in mathematics.

$$e^{\pi i} + 1 = 0$$

This elegant equation relates the five most famous numbers in mathematics—0, 1, π, e, and i—in a single equation. Show how Euler's Formula can be used to derive this equation.

4.4 Exercises

In Exercises 1–24, use DeMoivre's Theorem to find the indicated power of the complex number. Express the result in standard form.

1. $(1 + i)^5$
2. $(2 + 2i)^6$
3. $(-1 + i)^{10}$
4. $(3 - 2i)^8$
5. $2(\sqrt{3} + i)^7$
6. $4(1 - \sqrt{3}i)^3$
7. $[5(\cos 20° + i \sin 20°)]^3$
8. $[3(\cos 150° + i \sin 150°)]^4$
9. $\left(\cos \dfrac{\pi}{4} + i \sin \dfrac{\pi}{4}\right)^{12}$
10. $\left[2\left(\cos \dfrac{\pi}{2} + i \sin \dfrac{\pi}{2}\right)\right]^8$
11. $[5(\cos 3.2 + i \sin 3.2)]^4$
12. $(\cos 0 + i \sin 0)^{20}$
13. $(3 - 2i)^5$
14. $(2 + 5i)^6$
15. $(\sqrt{5} - 4i)^3$
16. $(\sqrt{3} + 2i)^4$
17. $[3(\cos 15° + i \sin 15°)]^4$
18. $[2(\cos 10° + i \sin 10°)]^8$
19. $[5(\cos 95° + i \sin 95°)]^3$
20. $[4(\cos 110° + i \sin 110°)]^4$
21. $\left[2\left(\cos \dfrac{\pi}{10} + i \sin \dfrac{\pi}{10}\right)\right]^5$
22. $\left[2\left(\cos \dfrac{\pi}{8} + i \sin \dfrac{\pi}{8}\right)\right]^6$
23. $\left[3\left(\cos \dfrac{2\pi}{3} + i \sin \dfrac{2\pi}{3}\right)\right]^3$
24. $\left[3\left(\cos \dfrac{\pi}{12} + i \sin \dfrac{\pi}{12}\right)\right]^5$

In Exercises 25–44, (a) use the theorem on page 333 to find the indicated roots of the complex number, (b) represent each of the roots graphically, and (c) express each of the roots in standard form.

25. Square roots of $5(\cos 120° + i \sin 120°)$
26. Square roots of $16(\cos 60° + i \sin 60°)$
27. Cube roots of $8\left(\cos \dfrac{2\pi}{3} + i \sin \dfrac{2\pi}{3}\right)$
28. Cube roots of $64\left(\cos \dfrac{\pi}{3} + i \sin \dfrac{\pi}{3}\right)$
29. Fifth roots of $243\left(\cos \dfrac{\pi}{6} + i \sin \dfrac{\pi}{6}\right)$
30. Fifth roots of $32\left(\cos \dfrac{5\pi}{6} + i \sin \dfrac{5\pi}{6}\right)$
31. Square roots of $-25i$
32. Square roots of $-36i$
33. Fourth roots of $81i$
34. Fourth roots of $625i$
35. Cube roots of $-\dfrac{125}{2}(1 + \sqrt{3}i)$
36. Cube roots of $-4\sqrt{2}(1 - i)$
37. Fourth roots of 16
38. Fourth roots of i
39. Fifth roots of 1
40. Cube roots of 1000
41. Cube roots of -125
42. Fourth roots of -4
43. Fifth roots of $128(-1 + i)$
44. Sixth roots of $64i$

In Exercises 45–60, use the theorem on page 333 to find all the solutions of the equation and represent the solutions graphically.

45. $x^4 - i = 0$
46. $x^3 - i = 0$
47. $x^6 + 1 = 0$
48. $x^3 + 1 = 0$
49. $x^5 + 243 = 0$
50. $x^3 + 125 = 0$
51. $x^5 - 32 = 0$
52. $x^3 - 27 = 0$
53. $x^4 + 16i = 0$
54. $x^3 + 27i = 0$
55. $x^4 - 16i = 0$
56. $x^6 - 64i = 0$

336 Chapter 4 ▶ Complex Numbers

57. $x^3 - (1 - i) = 0$
58. $x^5 - (1 - i) = 0$
59. $x^6 + (1 + i) = 0$
60. $x^4 + (1 + i) = 0$

Synthesis

True or False? In Exercises 61 and 62, determine whether the statement is true or false. Justify your answer.

61. Geometrically, the *n*th roots of any complex number *z* are all equally spaced around the unit circle centered at the origin.

62. By DeMoivre's Theorem,
$$(4 + \sqrt{6}i)^8 = \cos(32) + i\sin(8\sqrt{6}).$$

63. Show that $-\frac{1}{2}(1 + \sqrt{3}i)$ is a sixth root of 1.
64. Show that $2^{-1/4}(1 - i)$ is a fourth root of -2.

Graphical Reasoning In Exercises 65 and 66, use the graph of the roots of a complex number.

(a) Write each of the roots in trigonometric form.
(b) Identify the complex number whose roots are given.
(c) Use a graphing utility to verify the results of part (b).

65.

66.

Review

In Exercises 67–76, use the figure and trigonometric identities to find the exact value of the trigonometric function.

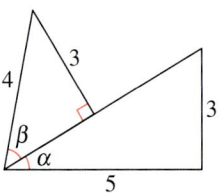

67. $\cos(\alpha + \beta)$
68. $\sin(\alpha + \beta)$
69. $\sin(\alpha - \beta)$
70. $\cos(\beta - \alpha)$
71. $\tan(\alpha + \beta)$
72. $\tan(\beta - \alpha)$
73. $\tan 2\alpha$
74. $\sin 2\beta$
75. $\sin \dfrac{\beta}{2}$
76. $\cos \dfrac{\alpha}{2}$

In Exercises 77–84, use the dot product to find the length of **u**.

77. $\mathbf{u} = \langle -3, 4 \rangle$
78. $\mathbf{u} = \langle -5, 7 \rangle$
79. $\mathbf{u} = \langle -9, 40 \rangle$
80. $\mathbf{u} = \langle -5, -12 \rangle$
81. $\mathbf{u} = 22\mathbf{i} - 3\mathbf{j}$
82. $\mathbf{u} = 16\mathbf{i} + 4\mathbf{j}$
83. $\mathbf{u} = -13\mathbf{i} + 6\mathbf{j}$
84. $\mathbf{u} = -24\mathbf{i} - 16\mathbf{j}$

Chapter Summary

What did you learn?

Section 4.1	Review Exercises
☐ How to use the imaginary unit *i* to write complex numbers	1–4
☐ How to add, subtract, and multiply complex numbers	5–10
☐ How to use complex conjugates to divide complex numbers	11–14
☐ How to use the Quadratic Formula to find complex solutions of quadratic equations	15–18

Section 4.2	
☐ How to determine the numbers of solutions of polynomial equations	19–26
☐ How to find the zeros of a polynomial function	27–44, 53
☐ How to find a polynomial function given the zeros of the function	45–52

Section 4.3	
☐ How to plot complex numbers in the complex plane	54–57
☐ How to write the trigonometric forms of complex numbers	58–61
☐ How to multiply and divide complex numbers written in trigonometric form	62, 63

Section 4.4	
☐ How to use DeMoivre's Theorem to find powers of complex numbers	64–67
☐ How to find *n*th roots of complex numbers	68–75

Review Exercises

4.1 In Exercises 1–4, write the complex number in standard form.

1. $6 + \sqrt{-4}$
2. $3 - \sqrt{-25}$
3. $i^2 + 3i$
4. $-5i + i^2$

In Exercises 5–14, perform the operations and write the result in standard form.

5. $(7 + 5i) + (-4 + 2i)$
6. $\left(\dfrac{\sqrt{2}}{2} - \dfrac{\sqrt{2}}{2}i\right) - \left(\dfrac{\sqrt{2}}{2} + \dfrac{\sqrt{2}}{2}i\right)$
7. $5i(13 - 8i)$
8. $(1 + 6i)(5 - 2i)$
9. $(10 - 8i)(2 - 3i)$
10. $i(6 + i)(3 - 2i)$
11. $\dfrac{6 + i}{4 - i}$
12. $\dfrac{3 + 2i}{5 + i}$
13. $\dfrac{4}{2 - 3i} + \dfrac{2}{1 + i}$
14. $\dfrac{1}{2 + i} - \dfrac{5}{1 + 4i}$

In Exercises 15–18, find all solutions of the equation.

15. $3x^2 + 1 = 0$
16. $2 + 8x^2 = 0$
17. $x^2 - 2x + 10 = 0$
18. $6x^2 + 3x + 27 = 0$

4.2 In Exercises 19–22, determine the number of solutions of the equation in the complex number system.

19. $x^5 - 2x^4 + 3x^2 - 5 = 0$
20. $-2x^6 + 7x^3 + x^2 + 4x - 19 = 0$
21. $\tfrac{1}{2}x^4 + \tfrac{2}{3}x^3 - x^2 + \tfrac{3}{10} = 0$
22. $\tfrac{3}{4}x^3 + \tfrac{1}{2}x^2 + \tfrac{3}{2}x + 2 = 0$

In Exercises 23–26, use the discriminant to determine the number of real solutions of the equation.

23. $6x^2 + x - 2 = 0$
24. $9x^2 - 12x + 4 = 0$
25. $0.13x^2 - 0.45x + 0.65 = 0$
26. $4x^2 + \tfrac{4}{3}x + \tfrac{1}{9} = 0$

In Exercises 27–36, find all the zeros of the function.

27. $g(x) = x^2 - 2x$
28. $f(x) = 6x - x^2$
29. $f(x) = x^2 + 8x + 10$
30. $h(x) = 3 + 4x - x^2$
31. $r(x) = 2x^2 + 2x + 3$
32. $s(x) = 2x^2 + 5x + 4$
33. $f(x) = 4x^3 - 11x^2 + 10x - 3$
34. $f(x) = x^3 - 1.3x^2 - 1.7x + 0.6$
35. $f(x) = 6x^4 - 25x^3 + 14x^2 + 27x - 18$
36. $f(x) = 5x^4 + 126x^2 + 25$

In Exercises 37–44, use the given zero of the function as an aid in finding all its zeros. Write the polynomial as a product of linear factors.

Function	Zero
37. $f(x) = x^3 + 3x^2 - 24x + 28$	2
38. $f(x) = 10x^3 + 21x^2 - x - 6$	-2
39. $f(x) = x^3 + 3x^2 - 5x + 25$	-5
40. $g(x) = x^3 - 8x^2 + 29x - 52$	4
41. $h(x) = 2x^3 - 19x^2 + 58x + 34$	$5 + 3i$
42. $f(x) = 5x^3 - 4x^2 + 20x - 16$	$2i$
43. $f(x) = x^4 + 5x^3 + 2x^2 - 50x - 84$	$-3 + \sqrt{5}i$
44. $g(x) = x^4 - 6x^3 + 18x^2 - 26x + 21$	$2 + \sqrt{3}i$

In Exercises 45–52, find a polynomial function with integer coefficients that has the given zeros.

45. $1, 1, \tfrac{1}{4}, -\tfrac{2}{3}$
46. $-2, 2, 3, 3$
47. $3, 2 - \sqrt{3}, 2 + \sqrt{3}$
48. $5, 1 - \sqrt{2}, 1 + \sqrt{2}$
49. $\tfrac{2}{3}, 4, \sqrt{3}i, -\sqrt{3}i$
50. $2, -3, 1 - 2i, 1 + 2i$
51. $-\sqrt{2}i, \sqrt{2}i, -5i, 5i$
52. $-2i, 2i, -4i, 4i$

53. **Profit** The demand equation for a certain product is $p = 140 - 0.0001x$, where p is the unit price (in dollars) of the product and x is the number of units produced and sold. The cost equation for the product is $C = 75x + 100{,}000$, where C is the total cost (in dollars) and x is the number of units produced. The total profit obtained by producing and selling x units is

$$P = xp - C.$$

Suppose you work in the marketing department of the company that produces this product and are asked to determine a price p that would yield a profit of 9 million dollars. Is this possible? Explain.

4.3 In Exercises 54–57, plot the complex number and find its absolute value.

54. $7i$
55. $-6i$
56. $5 + 3i$
57. $-10 - 4i$

In Exercises 58–61, write the trigonometric form of the complex number.

58. $5 - 5i$
59. $5 + 12i$
60. $-3\sqrt{3} + 3i$
61. -7

In Exercises 62 and 63, (a) express the two complex numbers in trigonometric form, and (b) use the trigonometric form to find $z_1 z_2$ and z_1/z_2.

62. $z_1 = 2\sqrt{3} - 2i, \quad z_2 = -10i$
63. $z_1 = -3(1 + i), \quad z_2 = 2(\sqrt{3} + i)$

4.4 In Exercises 64–67, use DeMoivre's Theorem to find the indicated power of the complex number. Express the result in standard form.

64. $\left[5\left(\cos\dfrac{\pi}{12} + i\sin\dfrac{\pi}{12}\right)\right]^4$
65. $\left[2\left(\cos\dfrac{4\pi}{15} + i\sin\dfrac{4\pi}{15}\right)\right]^5$
66. $(2 + 3i)^6$
67. $(1 - i)^8$

Graphical Reasoning In Exercises 68 and 69, use the graph of the roots of a complex number.

(a) Write each of the roots in trigonometric form.
(b) Identify the complex number whose roots are given.
(c) Use a graphing utility to verify the results of part (b).

68.

69.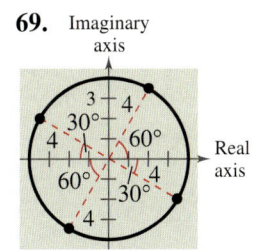

In Exercises 70 and 71, use the theorem on page 333 to find the roots of the complex number.

70. Sixth roots of $-729i$
71. Fourth roots of 256

In Exercises 72–75, find all solutions of the equation and represent the solutions graphically.

72. $x^4 + 81 = 0$
73. $x^5 - 32 = 0$
74. $x^3 + 8i = 0$
75. $(x^3 - 1)(x^2 + 1) = 0$

Synthesis

True or False? In Exercises 76–78, determine whether the statement is true or false. Justify your answer.

76. $\sqrt{-18}\sqrt{-2} = \sqrt{(-18)(-2)}$
77. The equation $325x^2 - 717x + 398 = 0$ has no solution.
78. A fourth-degree polynomial can have $-5, 128i, 4i,$ and 5 as its zeros.
79. Write quadratic equations that have (a) two distinct real solutions, (b) two complex solutions, and (c) no real solution.
80. The figure shows z_1 and z_2. Describe $z_1 z_2$ and z_1/z_2.

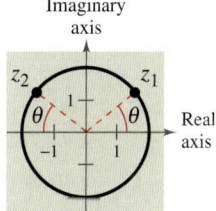

81. One of the fourth roots of a complex number z is shown in the figure.

    (a) How many roots are not shown?
    (b) Describe the other roots.

Chapter Project ▶ Using a Graphing Utility to Create Fractals

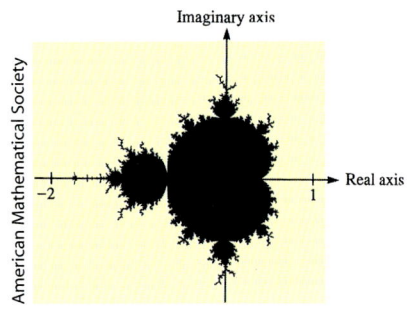

Graphing utilities can be used to draw pictures of **fractals** in the complex plane. The most famous fractal is called the **Mandelbrot Set,** after the Polish-born mathematician Benoit Mandelbrot. To construct the Mandelbrot Set, consider the following sequence of numbers.

$$c, c^2 + c, (c^2 + c)^2 + c, [(c^2 + c)^2 + c]^2 + c, \ldots$$

The behavior of this sequence depends on the value of the complex number c. For some values of c this sequence is **bounded,** and for other values it is **unbounded.** If the sequence is bounded, the complex number c is in the Mandelbrot Set. If the sequence is unbounded, the complex number c is not in the Mandelbrot Set.

Example 1 ▶ Members of the Mandelbrot Set

a. The complex number i is in the Mandelbrot Set because for $c = i$, the Mandelbrot sequence is $i, -1 + i, -i, -1 + i, -i, -1 + i, \ldots$, which is bounded. (No number in the sequence has an absolute value greater than $\sqrt{2}$.)

b. The complex number $1 + i$ is *not* in the Mandelbrot Set because for $c = 1 + i$, the Mandelbrot sequence is $1 + i, 1 + 3i, -7 + 7i, 1 - 97i, -9407 - 193i, \ldots$, which is unbounded. (The absolute values of the numbers in the sequence become arbitrarily large.)

Example 2 ▶ Writing a Program to Determine Mandelbrot Points

The pseudo-code at the left can be translated into a program for a graphing utility. (The programs for several models of graphing calculators can be found at our website *college.hmco.com.*) The program determines whether the complex number c is in the Mandelbrot Set. Run the program for $c = -1 + 0.2i$.

Program
- Enter the real part A
- Enter the imaginary part B
- Store A in C
- Store B in D
- Store 0 in N (number of term)
- Label 1
- Increment N
- Display N
- Display A
- Display B
- Store A in F
- Store B in G
- Store $F^2 - G^2 + C$ in A
- Store $2FG + D$ in B
- Goto Label 1

Solution

Run the program, entering -1 for A and 0.2 for B. Press [ENTER] to see the first term of the sequence. Press [ENTER] again to see the second term of the sequence. Continue pressing [ENTER]. If the terms begin to become large, the sequence is unbounded. For the number $c = -1 + 0.2i$, the terms are $-1 + 0.2i$, $-0.04 - 0.2i$, $-1.038 + 0.216i$, $0.032 - 0.249i, \ldots$, and so the sequence is bounded. Therefore, $c = -1 + 0.2i$ is in the Mandelbrot Set.

Chapter Project Investigation

1. Let $c = C + Di$ and let $c^2 + c = F + Gi$. Explain why $(c^2 + c)^2 + c$ is equal to $(F^2 - G^2 + C) + (2FG + D)i$.

In Questions 2–4, use a graphing calculator program or a computer program to determine whether the complex number is in the Mandelbrot Set.

2. $c = 1$ 3. $c = -1 + 0.5i$ 4. $c = 0.1 + i$

Chapter Test

 The *Interactive* CD-ROM and *Internet* versions of this text provide answers to the Chapter Tests and Cumulative Tests. They also offer Chapter Pre-Tests (which test key skills and concepts covered in previous chapters) and Chapter Post-Tests, both of which have randomly generated exercises with diagnostic capabilities.

Take this test as you would take a test in class. After you are done, check your work against the answers given in the back of the book.

1. Write the complex number $-3 + \sqrt{-81}$ in standard form.

In Exercises 2–5, perform the operations and write the result in standard form.

2. $10i - (3 + \sqrt{-25})$

3. $(2 + 6i)^2$

4. $(2 + \sqrt{3}i)(2 - \sqrt{3}i)$

5. $\dfrac{5}{2 + i}$

6. Use the Quadratic Formula to solve the equation $2x^2 - 2x + 3 = 0$.

In Exercises 7 and 8, determine the number of solutions of the equation in the complex number system.

7. $x^5 + x^3 - x + 1 = 0$

8. $x^4 - 3x^3 + 2x^2 - 4x - 5 = 0$

In Exercises 9 and 10, find all the zeros of the function.

9. $x^3 - 6x^2 + 5x - 30 = 0$

10. $x^4 - 2x^2 - 24 = 0$

In Exercises 11 and 12, use the zero(s) of the function as an aid in finding all its zeros. Write the polynomial as a product of linear factors.

	Function	Zero(s)
11.	$h(x) = x^4 - 2x^2 - 8$	$-2, 2$
12.	$g(v) = 2v^3 - 11v^2 + 22v - 15$	$\frac{3}{2}$

In Exercises 13 and 14, find a polynomial function with integer coefficients that has the given zeros.

13. $0, 3, 3 + i, 3 - i$

14. $1 + \sqrt{6}i, 1 - \sqrt{6}i, 3, 3$

15. Is it possible for a polynomial function with integer coefficients to have exactly one complex zero? Explain.

16. Write the complex number $z = 5 - 5i$ in trigonometric form.

17. Write the complex number $z = 6(\cos 120° + i \sin 120°)$ in standard form.

In Exercises 18 and 19, use DeMoivre's Theorem to find the indicated power of the complex number.

18. $\left[3\left(\cos \dfrac{7\pi}{6} + i \sin \dfrac{7\pi}{6}\right)\right]^8$

19. $(3 - 3i)^6$

20. Find the fourth roots of $256(1 + \sqrt{3}i)$.

21. Find all solutions of the equation $x^3 - 27i = 0$ and represent the solutions graphically.

- **5.1 ▶ Exponential Functions and Their Graphs**
- **5.2 ▶ Logarithmic Functions and Their Graphs**
- **5.3 ▶ Properties of Logarithms**
- **5.4 ▶ Exponential and Logarithmic Equations**
- **5.5 ▶ Exponential and Logarithmic Models**

Bill Aron/PhotoEdit

In the United States in 1997, personal savings as a percent of disposable income was 3.9% and the disposable per capita income was $21,969. So, the per capita personal savings in 1997 was $856.79. (Source: U.S. Bureau of Economic Analysis)

5 Exponential and Logarithmic Functions

▶ How to Study This Chapter

The Big Picture

In this chapter you will learn the following skills and concepts.

- ▶ How to recognize and evaluate exponential and logarithmic functions
- ▶ How to graph exponential and logarithmic functions
- ▶ How to rewrite logarithmic functions with a different base
- ▶ How to use properties of logarithms to evaluate, rewrite, expand, or condense logarithmic expressions
- ▶ How to solve exponential and logarithmic equations
- ▶ How to use exponential growth models, exponential decay models, Gaussian models, logistic growth models, and logarithmic models to solve real-life problems

Important Vocabulary

As you encounter each new vocabulary term in this chapter, add the term and its definition to your notebook glossary.

Algebraic functions (p. 344)
Transcendental functions (p. 344)
Exponential function f with base a (p. 344)
Natural base e (p. 348)
Natural exponential function (p. 348)
Continuous compounding (p. 349)
Logarithmic function with base a (p. 356)
Common logarithmic function (p. 357)

Natural logarithmic function (p. 360)
Exponential growth model (p. 384)
Exponential decay model (p. 384)
Gaussian model (p. 384)
Logistic growth model (p. 384)
Logarithmic models (p. 384)
Bell-shaped curve (p. 388)
Logistic curve (p. 389)
Sigmoidal curve (p. 389)

Study Tools

- Learning objectives at the beginning of each section
- Chapter Summary (p. 397)
- Review Exercises (pp. 398–401)
- Chapter Test (p. 403)

Additional Resources

- Study and Solutions Guide
- Interactive Trigonometry
- Videotapes for Chapter 5
- Trigonometry Website
- Student Success Organizer

STUDY T!P

Attend every class. Arrive on time with your text, a pen or pencil and paper for notes, and your calculator. If you must miss class, get the notes from another student, go to your tutor for help, or view the appropriate mathematics videotape.

5.1 Exponential Functions and Their Graphs

▶ **What you should learn**
- How to recognize and evaluate exponential functions with base a
- How to graph exponential functions
- How to recognize and evaluate exponential functions with base e
- How to use exponential functions to model and solve real-life applications

▶ **Why you should learn it**

Exponential functions can be used to model and solve real-life problems. For instance, Exercise 74 on page 354 shows how to use an exponential function to model the amount of defoliation caused by the gypsy moth.

Jenny Hager/The Image Works

Exponential Functions

So far, this book has dealt only with **algebraic functions,** which include polynomial functions and rational functions. In this chapter you will study two types of nonalgebraic functions—*exponential* functions and *logarithmic* functions. These functions are examples of **transcendental functions.**

Definition of Exponential Function
The **exponential function** f **with base** a is denoted by
$$f(x) = a^x$$
where $a > 0$, $a \neq 1$, and x is any real number.

The base $a = 1$ is excluded because it yields $f(x) = 1^x = 1$. This is a constant function, not an exponential function.

You already know how to evaluate a^x for integer and rational values of x. For example, you know that $4^3 = 64$ and $4^{1/2} = 2$. However, to evaluate 4^x for any real number x, you need to interpret forms with *irrational* exponents. For the purposes of this book, it is sufficient to think of

$$a^{\sqrt{2}} \quad (\text{where } \sqrt{2} \approx 1.41421356)$$

as the number that has the successively closer approximations

$$a^{1.4}, a^{1.41}, a^{1.414}, a^{1.4142}, a^{1.41421}, \ldots.$$

Example 1 shows how to use a calculator to evaluate exponential expressions.

Example 1 ▶ Evaluating Exponential Expressions

Use a calculator to evaluate each expression.

a. $2^{-3.1}$
b. $2^{-\pi}$
c. $12^{5/7}$
d. $(0.6)^{3/2}$

Solution

Number	Graphing Calculator Keystrokes	Display
a. $2^{-3.1}$	2 ^ (−) 3.1 ENTER	0.1166291
b. $2^{-\pi}$	2 ^ (−) π ENTER	0.1133147
c. $12^{5/7}$	12 ^ (5 ÷ 7) ENTER	5.8998877
d. $(0.6)^{3/2}$	.6 ^ (3 ÷ 2) ENTER	0.4647580

Graphs of Exponential Functions

The graphs of all exponential functions have similar characteristics, as shown in Examples 2, 3, and 4.

Example 2 ▶ Graphs of $y = a^x$

In the same coordinate plane, sketch the graph of each function.

a. $f(x) = 2^x$ **b.** $g(x) = 4^x$

Solution

The table below lists some values for each function, and Figure 5.1 shows their graphs. Note that both graphs are increasing. Moreover, the graph of $g(x) = 4^x$ is increasing more rapidly than the graph of $f(x) = 2^x$.

x	-2	-1	0	1	2	3
2^x	$\frac{1}{4}$	$\frac{1}{2}$	1	2	4	8
4^x	$\frac{1}{16}$	$\frac{1}{4}$	1	4	16	64

FIGURE 5.1

The table in Example 2 was evaluated by hand. You could, of course, use a graphing utility to construct tables with even more values.

Example 3 ▶ Graphs of $y = a^{-x}$

In the same coordinate plane, sketch the graph of each function.

a. $F(x) = 2^{-x}$ **b.** $G(x) = 4^{-x}$

Solution

The table below lists some values for each function, and Figure 5.2 shows their graphs. Note that both graphs are decreasing. Moreover, the graph of $G(x) = 4^{-x}$ is decreasing more rapidly than the graph of $F(x) = 2^{-x}$.

x	-3	-2	-1	0	1	2
2^{-x}	8	4	2	1	$\frac{1}{2}$	$\frac{1}{4}$
4^{-x}	64	16	4	1	$\frac{1}{4}$	$\frac{1}{16}$

FIGURE 5.2

In Example 3, note that the functions $F(x) = 2^{-x}$ and $G(x) = 4^{-x}$ can be rewritten with positive exponents.

$$F(x) = 2^{-x} = \left(\frac{1}{2}\right)^x \quad \text{and} \quad G(x) = 4^{-x} = \left(\frac{1}{4}\right)^x$$

Comparing the functions in Examples 2 and 3, observe that

$$F(x) = 2^{-x} = f(-x) \quad \text{and} \quad G(x) = 4^{-x} = g(-x).$$

Consequently, the graph of F is a reflection (in the y-axis) of the graph of f. The graphs of G and g have the same relationship. The graphs in Figures 5.1 and 5.2 are typical of the exponential functions a^x and a^{-x}. They have one y-intercept and one horizontal asymptote (the x-axis), and they are continuous. The basic characteristics of these exponential functions are summarized in Figures 5.3 and 5.4.

STUDY TIP

Notice that the range of an exponential function is $(0, \infty)$, which means that $a^x > 0$ for all values of x.

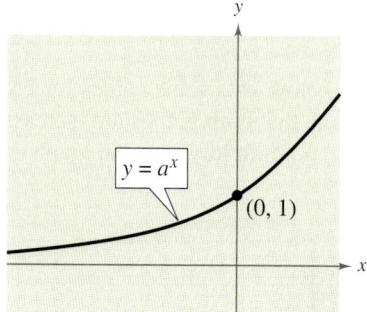

FIGURE 5.3

Graph of $y = a^x, a > 1$
- Domain: $(-\infty, \infty)$
- Range: $(0, \infty)$
- Intercept: $(0, 1)$
- Increasing
- x-Axis is a horizontal asymptote $(a^x \to 0$ as $x \to -\infty)$.
- Continuous

Graph of $y = a^{-x}, a > 1$
- Domain: $(-\infty, \infty)$
- Range: $(0, \infty)$
- Intercept: $(0, 1)$
- Decreasing
- x-Axis is a horizontal asymptote $(a^{-x} \to 0$ as $x \to \infty)$.
- Continuous

FIGURE 5.4

 The *Interactive* CD-ROM and *Internet* versions of this text offer a built-in graphing calculator, which can be used in the Examples, Explorations, Technology notes, and Exercises.

Exploration

Use a graphing utility to graph

$$y = a^x$$

for $a = 3, 5,$ and 7 in the same viewing window. (Use a viewing window in which $-2 \leq x \leq 1$ and $0 \leq y \leq 2$.) For instance, the graph of

$$y = 3^x$$

is shown in Figure 5.5. How do the graphs compare with each other? Which graph is on the top in the interval $(-\infty, 0)$? Which is on the bottom? Which graph is on the top in the interval $(0, \infty)$? Which is on the bottom?
  Repeat this experiment with the graphs of $y = b^x$ for $b = \frac{1}{3}, \frac{1}{5},$ and $\frac{1}{7}$. (Use a viewing window in which $-1 \leq x \leq 2$ and $0 \leq y \leq 2$.) What can you conclude about the shape of the graph of $y = b^x$ and the value of b?

FIGURE 5.5

Section 5.1 ▶ Exponential Functions and Their Graphs 347

In the following example, notice how the graph of $y = a^x$ can be used to sketch the graphs of functions of the form $f(x) = b \pm a^{x+c}$.

Example 4 ▶ Transformations of Graphs of Exponential Functions

Each of the following graphs is a transformation of the graph of $f(x) = 3^x$, as shown in Figure 5.6.

a. Because $g(x) = 3^{x+1} = f(x + 1)$, the graph of g can be obtained by shifting the graph of f 1 unit to the left.

b. Because $h(x) = 3^x - 2 = f(x) - 2$, the graph of h can be obtained by shifting the graph of f down 2 units.

c. Because $k(x) = -3^x = -f(x)$, the graph of k can be obtained by reflecting the graph of f in the x-axis.

d. Because $j(x) = 3^{-x} = f(-x)$, the graph of j can be obtained by reflecting the graph of f in the y-axis.

A computer animation of this example appears in the *Interactive* CD-ROM and *Internet* versions of this text.

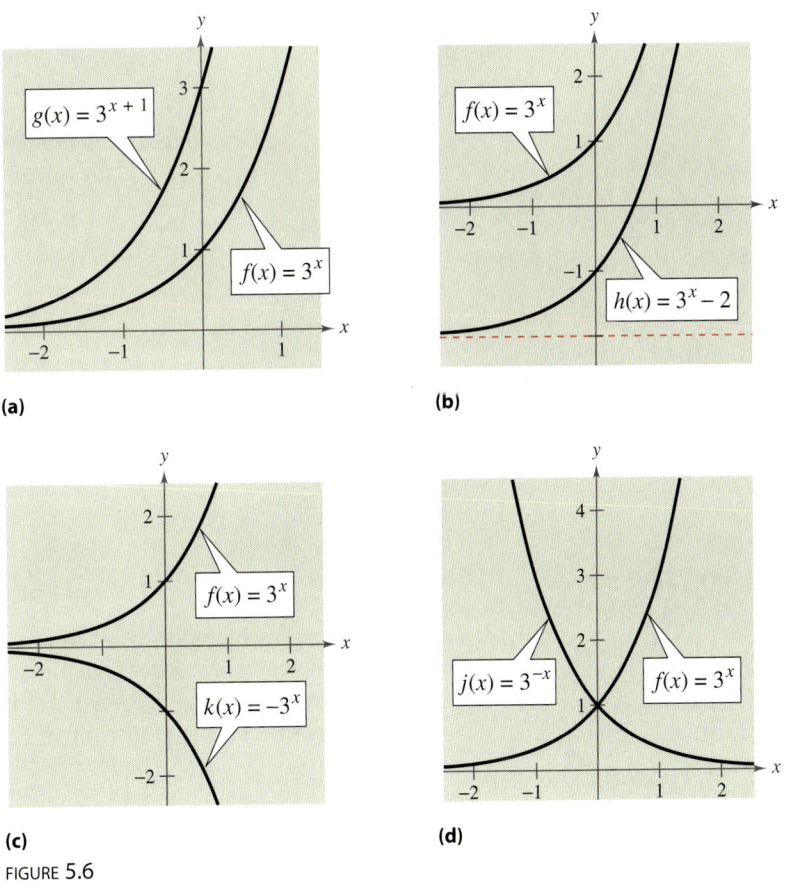

FIGURE 5.6

In Figure 5.6, notice that the transformations in parts (a), (c), and (d) keep the x-axis as a horizontal asymptote, but the transformation in part (b) yields a new horizontal asymptote of $y = -2$. Also, be sure to note how the y-intercept is affected by each transformation.

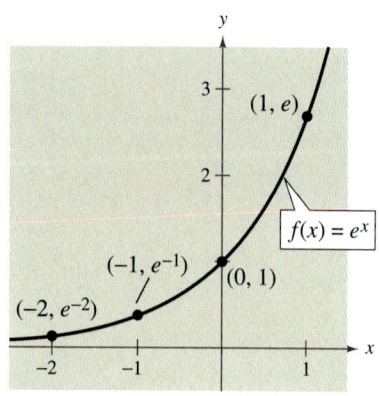

FIGURE 5.7

The Natural Base e

In many applications, the most convenient choice for a base is the irrational number

$$e \approx 2.718281828 \ldots.$$

This number is called the **natural base.** The function $f(x) = e^x$ is called the **natural exponential function.** Its graph is shown in Figure 5.7. Be sure you see that for the exponential function $f(x) = e^x$, e is the constant 2.71828183 . . . , whereas x is the variable.

Example 5 ▶ Evaluating the Natural Exponential Function

Use a calculator to evaluate each expression.

a. e^{-2}
b. e^{-1}
c. e^1
d. e^2

Solution

Number	Graphing Calculator Keystrokes	Display
a. e^{-2}	e^x (−) 2 ENTER	0.1353353
b. e^{-1}	e^x (−) 1 ENTER	0.3678794
c. e^1	e^x 1 ENTER	2.7182818
d. e^2	e^x 2 ENTER	7.3890561

Example 6 ▶ Graphing Natural Exponential Functions

Sketch the graph of each natural exponential function.

a. $f(x) = 2e^{0.24x}$
b. $g(x) = \frac{1}{2}e^{-0.58x}$

Solution

To sketch these two graphs, you can use a graphing utility to construct a table of values, as shown below. After constructing the table, plot the points and connect them with smooth curves, as shown in Figure 5.8. Note that the graph in part (a) is increasing whereas the graph in part (b) is decreasing.

x	−3	−2	−1	0	1	2	3
$f(x)$	0.974	1.238	1.573	2.000	2.542	3.232	4.109
$g(x)$	2.849	1.595	0.893	0.500	0.280	0.157	0.088

(a)

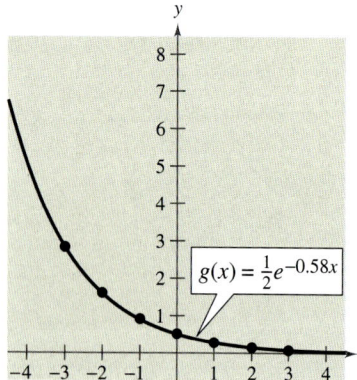

(b)

FIGURE 5.8

Applications

One of the most familiar examples of exponential growth is that of an investment earning *continuously compounded interest.* Using exponential functions, you can *develop* a formula for the balance in an account that pays compound interest, and show how it leads to continuous compounding.

Suppose a principal P is invested at an annual interest rate r, compounded once a year. If the interest is added to the principal at the end of the year, the new balance P_1 is

$$P_1 = P + Pr$$
$$= P(1 + r).$$

This pattern of multiplying the previous principal by $1 + r$ is then repeated each successive year, as shown below.

Year	Balance After Each Compounding
0	$P = P$
1	$P_1 = P(1 + r)$
2	$P_2 = P_1(1 + r) = P(1 + r)(1 + r) = P(1 + r)^2$
3	$P_3 = P_2(1 + r) = P(1 + r)^2(1 + r) = P(1 + r)^3$
⋮	
t	$P_t = P(1 + r)^t$

To accommodate more frequent (quarterly, monthly, or daily) compounding of interest, let n be the number of compoundings per year and let t be the number of years. Then the rate per compounding is r/n and the account balance after t years is

$$A = P\left(1 + \frac{r}{n}\right)^{nt}. \qquad \text{Amount (balance) with } n \text{ compoundings per year}$$

If you let the number of compoundings n increase without bound, the process approaches what is called **continuous compounding.** In the formula for n compoundings per year, let $m = n/r$. This produces

$$A = P\left(1 + \frac{r}{n}\right)^{nt} \qquad \text{Amount with } n \text{ compoundings per year}$$
$$= P\left(1 + \frac{r}{mr}\right)^{mrt} \qquad \text{Substitute } mr \text{ for } n.$$
$$= P\left(1 + \frac{1}{m}\right)^{mrt} \qquad \text{Simplify.}$$
$$= P\left[\left(1 + \frac{1}{m}\right)^m\right]^{rt}. \qquad \text{Property of exponents}$$

As m increases without bound, it can be shown that $[1 + (1/m)]^m$ approaches e. (Try the values $m = 10$, $10{,}000$, and $10{,}000{,}000$.) From this, you can conclude that the formula for continuous compounding is

$$A = Pe^{rt}. \qquad \text{Substitute } e \text{ for } (1 + 1/m)^m.$$

Technology

Use the formula

$$A = P\left(1 + \frac{r}{n}\right)^{nt}$$

to calculate the amount in an account when $P = \$3000$, $r = 6\%$, $t = 10$ years, and compounding is done (1) by the day, (2) by the hour, (3) by the minute, and (4) by the second. Use these results to present an argument that increasing the number of compoundings does not mean unlimited growth of the amount in the account.

A computer simulation of this concept appears in the *Interactive* CD-ROM and *Internet* versions of this text.

STUDY TIP

Be sure you see that the annual interest rate must be expressed in decimal form. For instance, 6% should be expressed as 0.06.

Formulas for Compound Interest

After t years, the balance A in an account with principal P and annual interest rate r (in decimal form) is given by the following formulas.

1. For n compoundings per year: $A = P\left(1 + \dfrac{r}{n}\right)^{nt}$

2. For continuous compounding: $A = Pe^{rt}$

Example 7 ▶ Compound Interest

A total of $12,000 is invested at an annual interest rate of 9%. Find the balance after 5 years if it is compounded

a. quarterly.
b. monthly.
c. continuously.

Solution

a. For quarterly compoundings, you have $n = 4$. So, in 5 years at 9%, the balance is

$A = P\left(1 + \dfrac{r}{n}\right)^{nt}$ Formula for compound interest

$= 12{,}000\left(1 + \dfrac{0.09}{4}\right)^{4(5)}$ Substitute for P, r, n, and t.

$\approx \$18{,}726.11.$ Use a calculator.

b. For monthly compoundings, you have $n = 12$. So, in 5 years at 9%, the balance is

$A = P\left(1 + \dfrac{r}{n}\right)^{nt}$ Formula for compound interest

$= 12{,}000\left(1 + \dfrac{0.09}{12}\right)^{12(5)}$ Substitute for P, r, n, and t.

$\approx \$18{,}788.17.$ Use a calculator.

c. For continuous compounding, the balance is

$A = Pe^{rt}$ Formula for continuous compounding

$= 12{,}000e^{0.09(5)}$ Substitute for P, r, n, and t.

$\approx \$18{,}819.75.$ Use a calculator.

In Example 7, note that continuous compounding yields more than quarterly or monthly compounding. This is typical of the two types of compounding. That is, for a given principal, interest rate, and time, continuous compounding will always yield a larger balance than compounding n times a year.

Example 8 ▶ Radioactive Decay

In 1986, a nuclear reactor accident occurred in Chernobyl in what was then the Soviet Union. The explosion spread highly toxic radioactive chemicals, such as plutonium, over hundreds of square miles, and the government evacuated the city and the surrounding area. To see why the city is now uninhabited, consider the model

$$P = 10e^{-0.00002845t}.$$

This model represents the amount of plutonium that remains (from an initial amount of 10 pounds) after t years. Sketch the graph of this function over the interval from $t = 0$ to $t = 100{,}000$. How much of the 10 pounds will remain in the year 2002? How much of the 10 pounds will remain after 100,000 years?

Solution

The graph of this function is shown in Figure 5.9. Note from this graph that plutonium has a *half-life* of about 24,360 years. That is, after 24,360 years, *half* of the original amount will remain. After another 24,360 years, one-quarter of the original amount will remain, and so on. In the year 2002 ($t = 16$), there will still be

$$P = 10e^{-0.00002845(16)}$$
$$= 10e^{-0.0004552}$$
$$\approx 9.995 \text{ pounds}$$

of plutonium remaining. After 100,000 years, there will still be

$$P = 10e^{-0.00002845(100{,}000)}$$
$$= 10e^{-2.845}$$
$$\approx 0.58 \text{ pound}$$

of plutonium remaining.

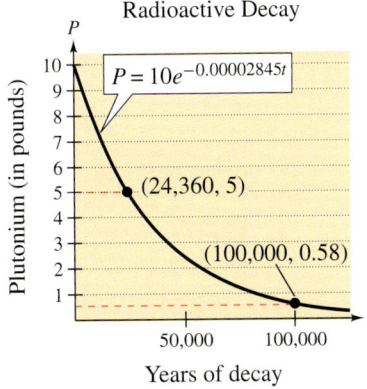

FIGURE 5.9

Writing ABOUT MATHEMATICS

Identifying Exponential Functions Which of the following functions generated the two tables below? Discuss how you were able to decide. What do these functions have in common? Are any the same? If so, explain why.

a. $f_1(x) = 2^{(x+3)}$ **b.** $f_2(x) = 8\left(\frac{1}{2}\right)^x$ **c.** $f_3(x) = \left(\frac{1}{2}\right)^{(x-3)}$

d. $f_4(x) = \left(\frac{1}{2}\right)^x + 7$ **e.** $f_5(x) = 7 + 2^x$ **f.** $f_6(x) = (8)2^x$

x	-1	0	1	2	3
$g(x)$	7.5	8	9	11	15

x	-2	-1	0	1	2
$h(x)$	32	16	8	4	2

Create two different exponential functions of the forms $y = a(b)^x$ and $y = c^x + d$ with y-intercepts of $(0, -3)$.

5.1 Exercises

In Exercises 1–10, evaluate the expression. Round your result to three decimal places.

1. $(3.4)^{5.6}$
2. $5000(2^{-1.5})$
3. $(1.005)^{400}$
4. $8^{2\pi}$
5. $5^{-\pi}$
6. $\sqrt[3]{4395}$
7. $100^{\sqrt{2}}$
8. $e^{1/2}$
9. $e^{-3/4}$
10. $e^{3.2}$

In Exercises 11–14, match the exponential function with its graph. [The graphs are labeled (a), (b), (c), and (d).]

(a)

(b)

(c)

(d)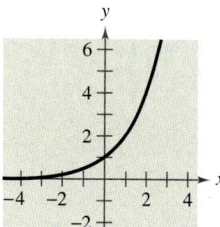

11. $f(x) = 2^x$
12. $f(x) = 2^x + 1$
13. $f(x) = 2^{-x}$
14. $f(x) = 2^{x-2}$

In Exercises 15–22, use the graph of f to describe the transformation that yields the graph of g.

15. $f(x) = 3^x$, $g(x) = 3^{x-4}$
16. $f(x) = 4^x$, $g(x) = 4^x + 1$
17. $f(x) = -2^x$, $g(x) = 5 - 2^x$
18. $f(x) = 10^x$, $g(x) = 10^{-x+3}$
19. $f(x) = \left(\frac{3}{5}\right)^x$, $g(x) = -\left(\frac{3}{5}\right)^{x+4}$
20. $f(x) = \left(\frac{7}{2}\right)^x$, $g(x) = -\left(\frac{7}{2}\right)^{-x+6}$
21. $f(x) = 0.3^x$, $g(x) = -0.3^x + 5$
22. $f(x) = 3.6^x$, $g(x) = -3.6^{-x} + 8$

In Exercises 23–36, use a graphing utility to construct a table of values for each function. Then sketch the graph of the function.

23. $f(x) = \left(\frac{1}{2}\right)^x$
24. $f(x) = 6^x$
25. $f(x) = \left(\frac{1}{2}\right)^{-x}$
26. $f(x) = 6^{-x}$
27. $f(x) = 2^{x-1}$
28. $f(x) = 3^{x+2}$
29. $f(x) = e^x$
30. $f(x) = e^{-x}$
31. $f(x) = 3e^{x+4}$
32. $f(x) = 2e^{-0.5x}$
33. $f(x) = 2e^{x-2} + 4$
34. $f(x) = 2 + e^{x-5}$
35. $f(x) = 4^{x-3} + 3$
36. $f(x) = -4^{x-3} - 3$

In Exercises 37–54, use a graphing utility to graph the exponential function.

37. $g(x) = 5^x$
38. $f(x) = \left(\frac{3}{2}\right)^x$
39. $f(x) = \left(\frac{1}{5}\right)^x = 5^{-x}$
40. $h(x) = \left(\frac{3}{2}\right)^{-x}$
41. $h(x) = 5^{x-2}$
42. $g(x) = \left(\frac{3}{2}\right)^{x+2}$
43. $g(x) = 5^{-x} - 3$
44. $f(x) = \left(\frac{3}{2}\right)^{-x} + 2$
45. $y = 2^{-x^2}$
46. $y = 3^{-|x|}$
47. $y = 3^{x-2} + 1$
48. $y = 4^{x+1} - 2$
49. $y = 1.08^{-5x}$
50. $y = 1.08^{5x}$
51. $s(t) = 2e^{0.12t}$
52. $s(t) = 3e^{-0.2t}$
53. $g(x) = 1 + e^{-x}$
54. $h(x) = e^{x-2}$

Finance In Exercises 55–58, complete the table to determine the balance A for P dollars invested at rate r for t years and compounded n times per year.

n	1	2	4	12	365	Continuous
A						

55. $P = \$2500$, $r = 8\%$, $t = 10$ years
56. $P = \$1000$, $r = 6\%$, $t = 10$ years
57. $P = \$2500$, $r = 8\%$, $t = 20$ years
58. $P = \$1000$, $r = 6\%$, $t = 40$ years

Finance In Exercises 59–62, complete the table to determine the balance A for $12,000 invested at rate r for t years, compounded continuously.

t	1	10	20	30	40	50
A						

59. $r = 8\%$

60. $r = 6\%$

61. $r = 6.5\%$

62. $r = 7.5\%$

63. *Finance* On the day of a child's birth, a deposit of $25,000 is made in a trust fund that pays 8.75% interest, compounded continuously. Determine the balance in this account on the child's 25th birthday.

64. *Finance* A deposit of $5000 is made in a trust fund that pays 7.5% interest, compounded continuously. It is specified that the balance will be given to the college from which the donor graduated after the money has earned interest for 50 years. How much will the college receive?

65. *Graphical Reasoning* There are two options for investing $500. The first earns 7% compounded annually and the second earns 7% simple interest. The figure shows the growth of each investment over a 30-year period.

(a) Identify which graph represents each type of investment in the figure. Explain your reasoning.

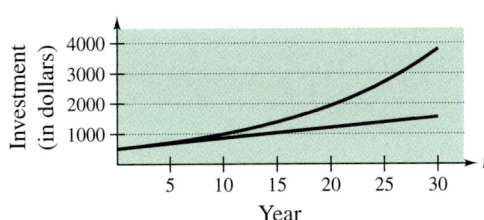

(b) Verify your answer in part (a) by finding the equations that model the investment growth and graphing the models.

66. *Depreciation* After t years, the value of a car that cost $20,000 is modeled by

$$V(t) = 20{,}000\left(\frac{3}{4}\right)^t.$$

Graph the function and determine the value of the car 2 years after it was purchased.

67. *Inflation* If the annual rate of inflation averages 4% over the next 10 years, the approximate cost C of goods or services during any year in that decade will be modeled by

$$C(t) = P(1.04)^t$$

where t is the time in years and P is the present cost. If the price of an oil change for your car is presently $23.95, estimate the price 10 years from now.

68. *Economics* The demand equation for a certain product is

$$p = 5000\left(1 - \frac{4}{4 + e^{-0.002x}}\right).$$

(a) Use a graphing utility to graph the demand function for $x > 0$ and $p > 0$.

(b) Find the price p for a demand of $x = 500$ units.

(c) Use the graph in part (a) to approximate the greatest price that will still yield a demand of at least 600 units.

69. *Population Growth* A certain type of bacterium increases according to the model $P(t) = 100e^{0.2197t}$, where t is the time in hours. Find (a) $P(0)$, (b) $P(5)$, and (c) $P(10)$.

70. *Population Growth* The population of a town increases according to the model $P(t) = 2500e^{0.0293t}$, where t is the time in years, with $t = 0$ corresponding to 1990. Use the model to estimate the population in (a) 2000 and (b) 2010.

71. *Radioactive Decay* Let Q (in grams) represent a mass of radioactive radium (^{226}Ra), whose half-life is 1620 years. The quantity of radium present after t years is

$$Q = 25\left(\frac{1}{2}\right)^{t/1620}.$$

(a) Determine the initial quantity (when $t = 0$).

(b) Determine the quantity present after 1000 years.

(c) Use a graphing utility to graph the function over the interval $t = 0$ to $t = 5000$.

72. *Radioactive Decay* Let Q (in grams) represent a mass of carbon 14 (^{14}C), whose half-life is 5730 years. The quantity of carbon 14 present after t years is

$$Q = 10\left(\frac{1}{2}\right)^{t/5730}.$$

(a) Determine the initial quantity (when $t = 0$).

(b) Determine the quantity present after 2000 years.

(c) Sketch the graph of this function over the interval $t = 0$ to $t = 10{,}000$.

73. *Data Analysis* A meteorologist measures the atmospheric pressure P (in pascals) at altitude h (in kilometers). The data is shown in the table.

h	0	5	10	15	20
P	101,293	54,735	23,294	12,157	5069

A model for the data is given by

$P = 102{,}303 e^{-0.137h}$.

(a) Sketch a scatter plot of the data and graph the model on the same set of axes.

(b) Create a table that compares the model with the sample data.

(c) Estimate the atmospheric pressure at a height of 8 kilometers.

(d) Use the graph in part (a) to estimate the altitude at which the atmospheric pressure is 21,000 pascals.

74. *Data Analysis* To estimate the amount of defoliation caused by the gypsy moth during a given year, a forester counts the number x of egg masses on $\frac{1}{40}$ of an acre (circle of radius 18.6 feet) in the fall. The percent of defoliation y the next spring is shown in the table. (Source: USDA, Forest Service)

x	0	25	50	75	100
y	12	44	81	96	99

(a) A model for the data is

$$y = \frac{300}{3 + 17e^{-0.065x}}.$$

Use a graphing utility to create a scatter plot of the data and graph the model in the same viewing window.

(b) Create a table that compares the model with the sample data.

(c) Estimate the percent of defoliation if 36 egg masses are counted on $\frac{1}{40}$ acre.

(d) Use the graph in part (a) to estimate the number of egg masses per $\frac{1}{40}$ acre if you observe that $\frac{2}{3}$ of a forest is defoliated the following spring.

Synthesis

True or False? In Exercises 75 and 76, determine whether the statement is true or false. Justify your answer.

75. The x-axis is an asymptote for the graph of $f(x) = 10^x$.

76. $e = \dfrac{271{,}801}{99{,}990}$

Think About It In Exercises 77–80, use properties of exponents to determine which functions (if any) are the same.

77. $f(x) = 3^{x-2}$
$g(x) = 3^x - 9$
$h(x) = \frac{1}{9}(3^x)$

78. $f(x) = 4^x + 12$
$g(x) = 2^{2x+6}$
$h(x) = 64(4^x)$

79. $f(x) = 16(4^{-x})$
$g(x) = \left(\frac{1}{4}\right)^{x-2}$
$h(x) = 16(2^{-2x})$

80. $f(x) = 5^{-x} + 3$
$g(x) = 5^{3-x}$
$h(x) = -5^{x-3}$

81. Graph the functions $y = 3^x$ and $y = 4^x$ and use the graphs to solve the inequalities.

(a) $4^x < 3^x$

(b) $4^x > 3^x$

82. Graph the functions $y = \left(\frac{1}{2}\right)^x$ and $y = \left(\frac{1}{4}\right)^x$ and use the graphs to solve the inequalities.

(a) $\left(\frac{1}{4}\right)^x < \left(\frac{1}{2}\right)^x$

(b) $\left(\frac{1}{4}\right)^x > \left(\frac{1}{2}\right)^x$

83. Use a graphing utility to graph each function. Use the graph to find any asymptotes of the function.

(a) $f(x) = \dfrac{8}{1 + e^{-0.5x}}$

(b) $g(x) = \dfrac{8}{1 + e^{-0.5/x}}$

84. Use a graphing utility to graph each function. Use the graph to find where the function is increasing and decreasing, and approximate any relative maximum or minimum values.

(a) $f(x) = x^2 e^{-x}$

(b) $g(x) = x 2^{3-x}$

85. Use a graphing utility to graph $y_1 = e^x$ and each of the functions $y_2 = x^2$, $y_3 = x^3$, $y_4 = \sqrt{x}$, and $y_5 = |x|$. Which function increases at the fastest rate as x approaches $+\infty$?

86. *Conjecture* Use the result of Exercise 85 to make a conjecture about the rate of growth of $y_1 = e^x$ and $y = x^n$, where n is a natural number and x approaches $+\infty$.

87. *Writing* Use the results of Exercises 85 and 86 to describe what is implied when it is stated that a quantity is growing exponentially.

88. *Graphical Analysis* Use a graphing utility to graph

$$f(x) = \left(1 + \frac{0.5}{x}\right)^x \quad \text{and} \quad g(x) = e^{0.5}$$

in the same viewing window. What is the relationship between f and g as x increases without bound?

89. *Conjecture* Use the result of Exercise 88 to make a conjecture about the value of $[1 + (r/x)]^x$ as x increases without bound. Create a table that illustrates your conjecture for $r = 1$.

90. *Think About It* Which functions are exponential?

(a) $3x$ (b) $3x^2$
(c) 3^x (d) 2^{-x}

91. *Think About It* Without using a calculator, why do you know that $2^{\sqrt{2}}$ is greater than 2, but less than 4?

92. *Exploration* Use a graphing utility to compare the graph of the function $y = e^x$ with the graph of each given function. [$n!$ (read "n factorial") is defined as $n! = 1 \cdot 2 \cdot 3 \cdots (n-1) \cdot n$.]

(a) $y_1 = 1 + \dfrac{x}{1!}$

(b) $y_2 = 1 + \dfrac{x}{1!} + \dfrac{x^2}{2!}$

(c) $y_3 = 1 + \dfrac{x}{1!} + \dfrac{x^2}{2!} + \dfrac{x^3}{3!}$

93. *Pattern Recognition* Identify the pattern of successive polynomials given in Exercise 92. Extend the pattern one more term and compare the graph of the resulting polynomial function with the graph of $y = e^x$. What do you think this pattern implies?

94. Given the exponential function

$$f(x) = a^x,$$

show that

(a) $f(u + v) = f(u) \cdot f(v)$.
(b) $f(2x) = [f(x)]^2$.

Review

In Exercises 95–98, solve for y.

95. $2x - 7y + 14 = 0$
96. $x^2 + 3y = 4$
97. $x^2 + y^2 = 25$
98. $x - |y| = 2$

In Exercises 99–102, find a polynomial function with integer coefficients that has the given zeros. (There are many correct answers.)

99. $1, 8i, -8i$
100. $i, -i, 4i, -4i$
101. $4, 6 + i, 6 - i$
102. $0, 0, 2, 1 - i$

5.2 Logarithmic Functions and Their Graphs

What you should learn

- How to recognize and evaluate logarithmic functions with base a
- How to graph logarithmic functions
- How to recognize and evaluate natural logarithmic functions
- How to use logarithmic functions to model and solve real-life applications

Why you should learn it

You can use logarithmic functions to model and solve real-life problems. For instance, Exercises 73 and 74 on page 364 show how to use a logarithmic function to model the minimum required ventilation rates in public school classrooms.

Mary Kate Kenny

Logarithmic Functions

In Section P.9, you studied the concept of the inverse of a function. There, you learned that if a function has the property that no horizontal line intersects the graph of the function more than once, the function must have an inverse. By looking back at the graphs of the exponential functions introduced in Section 5.1, you will see that every function of the form

$$f(x) = a^x$$

passes the Horizontal Line Test and therefore must have an inverse. This inverse function is called the **logarithmic function with base** a.

Definition of Logarithmic Function with Base a

For $x > 0$ and $0 < a \neq 1$,

$$y = \log_a x \text{ if and only if } x = a^y.$$

The function given by

$$f(x) = \log_a x$$

is called the **logarithmic function with base a.**

The equations

$$y = \log_a x \quad \text{and} \quad x = a^y$$

are equivalent. The first equation is in logarithmic form and the second is in exponential form.

When evaluating logarithms, remember that *a logarithm is an exponent*. This means that $\log_a x$ is the exponent to which a must be raised to obtain x. For instance, $\log_2 8 = 3$ because 2 must be raised to the third power to get 8.

Example 1 ▶ Evaluating Logarithms

Use the definition of logarithmic function to evaluate the logarithms.

a. $\log_2 32$ **b.** $\log_3 27$ **c.** $\log_4 2$
d. $\log_{10} \frac{1}{100}$ **e.** $\log_3 1$ **f.** $\log_2 2$

Solution

a. $\log_2 32 = 5$ because $2^5 = 32$.
b. $\log_3 27 = 3$ because $3^3 = 27$.
c. $\log_4 2 = \frac{1}{2}$ because $4^{1/2} = \sqrt{4} = 2$.
d. $\log_{10} \frac{1}{100} = -2$ because $10^{-2} = \frac{1}{10^2} = \frac{1}{100}$.
e. $\log_3 1 = 0$ because $3^0 = 1$.
f. $\log_2 2 = 1$ because $2^1 = 2$.

Exploration

Complete the table for $f(x) = 10^x$.

x	-2	-1	0	1	2
$f(x)$					

Complete the table for $f(x) = \log_{10} x$.

x	$\frac{1}{100}$	$\frac{1}{10}$	1	10	100
$f(x)$					

Compare the two tables. What is the relationship between $f(x) = 10^x$ and $f(x) = \log_{10} x$?

The logarithmic function with base 10 is called the **common logarithmic function.** On most calculators, this function is denoted by [LOG]. Because $\log_a x$ is the inverse function of a^x, it follows that the domain of $\log_a x$ is the range of a^x, $(0, \infty)$. In other words, $\log_a x$ is defined only if x is positive.

Example 2 ▶ Evaluating Logarithms on a Calculator

Use a calculator to evaluate each expression.

a. $\log_{10} 10$
b. $2 \log_{10} 2.5$
c. $\log_{10}(-2)$

Solution

Number	Graphing Calculator Keystrokes	Display
a. $\log_{10} 10$	[LOG] 10 [ENTER]	1
b. $2 \log_{10} 2.5$	2 [×] [LOG] 2.5 [ENTER]	0.7958800
c. $\log_{10}(-2)$	[LOG] [(-)] 2 [ENTER]	ERROR

Note that the calculator displays an error message (or a complex number) when you try to evaluate $\log_{10}(-2)$. The reason for this is that the domain of every logarithmic function is the set of *positive real numbers*.

The following properties follow directly from the definition of the logarithmic function with base a.

Properties of Logarithms

1. $\log_a 1 = 0$ because $a^0 = 1$.
2. $\log_a a = 1$ because $a^1 = a$.
3. $\log_a a^x = x$ and $a^{\log_a x} = x$ Inverse Properties
4. If $\log_a x = \log_a y$, then $x = y$.   One-to-One Property

Example 3 ▶ Using Properties of Logarithms

Solve each equation for x.

a. $\log_2 x = \log_2 3$
b. $\log_4 4 = x$

Solution

a. Using the One-to-One Property (Property 4), you can conclude that $x = 3$.
b. Using Property 2, you can conclude that $x = 1$.

Graphs of Logarithmic Functions

To sketch the graph of

$$y = \log_a x$$

you can use the fact that the graphs of inverse functions are reflections of each other in the line $y = x$.

Example 4 ▶ Graphs of Exponential and Logarithmic Functions

In the same coordinate plane, sketch the graph of each function.

a. $f(x) = 2^x$ **b.** $g(x) = \log_2 x$

Solution

a. For $f(x) = 2^x$, construct a table of values.

x	-2	-1	0	1	2	3
$f(x) = 2^x$	$\frac{1}{4}$	$\frac{1}{2}$	1	2	4	8

By plotting these points and connecting them with a smooth curve, you obtain the graph shown in Figure 5.10.

b. Because $g(x) = \log_2 x$ is the inverse of $f(x) = 2^x$, the graph of g is obtained by plotting the points $(f(x), x)$ and connecting them with a smooth curve. The graph of g is a reflection of the graph of f in the line $y = x$, as shown in Figure 5.10.

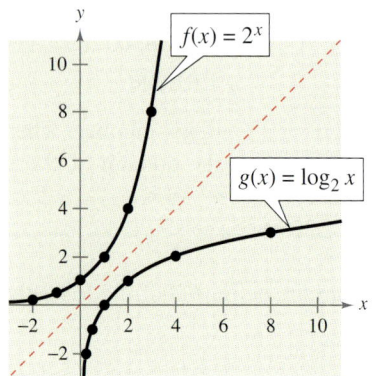

FIGURE 5.10

A computer animation of this example appears in the *Interactive* CD-ROM and *Internet* versions of this text.

Example 5 ▶ Sketching the Graph of a Logarithmic Function

Sketch the graph of the common logarithmic function

$$f(x) = \log_{10} x.$$

Identify the x-intercept and the vertical asymptote.

Solution

Begin by constructing a table of values. Note that some of the values can be obtained without a calculator by using the Inverse Property of Logarithms. Others require a calculator. Next, plot the points and connect them with a smooth curve, as shown in Figure 5.11. The x-intercept of the graph is $(1, 0)$ and the vertical asymptote is $x = 0$ (y-axis).

	Without calculator				With calculator		
x	$\frac{1}{100}$	$\frac{1}{10}$	1	10	2	5	8
$\log_{10} x$	-2	-1	0	1	0.301	0.699	0.903

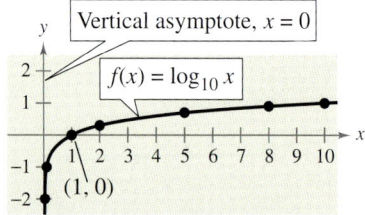

FIGURE 5.11

The nature of the graph in Figure 5.11 is typical of functions of the form $f(x) = \log_a x, a > 1$. They have one x-intercept and one vertical asymptote. Notice how slowly the graph rises for $x > 1$. In Figure 5.11 you would need to move out to $x = 1000$ before the graph rose to $y = 3$. The basic characteristics of logarithmic graphs are summarized in Figure 5.12.

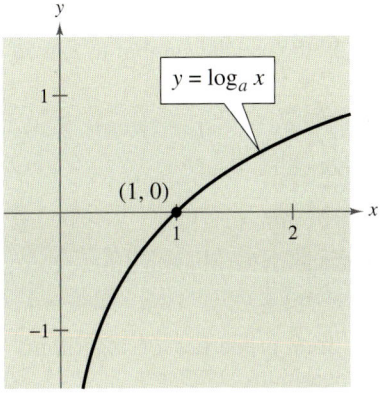

Graph of $y = \log_a x, a > 1$
- Domain: $(0, \infty)$
- Range: $(-\infty, \infty)$
- Intercept: $(1, 0)$
- Increasing
- y-axis is a vertical asymptote $(\log_a x \to -\infty \text{ as } x \to 0^+)$
- Continuous
- Reflection of graph of $y = a^x$ about the line $y = x$

FIGURE 5.12

In the next example, the graph of $\log_a x$ is used to sketch the graphs of functions of the form $y = b \pm \log_a(x + c)$. Notice how a horizontal shift of the graph results in a horizontal shift of the vertical asymptote.

STUDY T!P

You can use your understanding of transformations to identify vertical asymptotes of logarithmic functions. For instance, in Example 6(a) the graph of $g(x) = f(x - 1)$ shifts the graph of $f(x)$ 1 unit to the right. So the vertical asymptote of $g(x)$ is $x = 1$, 1 unit to the right of the asymptote of the graph of $f(x)$.

A computer animation of this example appears in the *Interactive* CD-ROM and *Internet* versions of this text.

Example 6 ▶ Shifting Graphs of Logarithmic Functions

The graph of each of the following functions is similar to the graph of $f(x) = \log_{10} x$, as shown in Figure 5.13.

a. Because

$$g(x) = \log_{10}(x - 1) = f(x - 1),$$

the graph of g can be obtained by shifting the graph of f 1 unit to the right.

b. Because

$$h(x) = 2 + \log_{10} x = 2 + f(x),$$

the graph of h can be obtained by shifting the graph of f 2 units up.

(a)

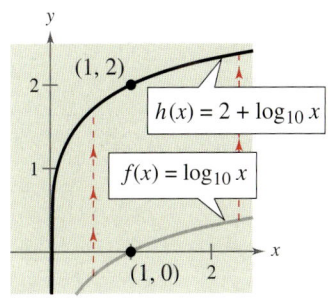

(b)

FIGURE 5.13

The Natural Logarithmic Function

As with exponential functions, the most widely used base for logarithmic functions is the number e, where

$$e \approx 2.718281828\ldots.$$

The logarithmic function with base e is the **natural logarithmic function** and is denoted by the special symbol $\ln x$, read as "the natural log of x" or "el en of x."

The Natural Logarithmic Function
The function defined by
$$f(x) = \log_e x = \ln x, \quad x > 0$$
is called the **natural logarithmic function.**

The four properties of logarithms listed on page 357 are also valid for natural logarithms.

Properties of Natural Logarithms
1. $\ln 1 = 0$ because $e^0 = 1$.
2. $\ln e = 1$ because $e^1 = e$.
3. $\ln e^x = x$ and $e^{\ln x} = x$ Inverse Properties
4. If $\ln x = \ln y$, then $x = y$.   One-to-One Property

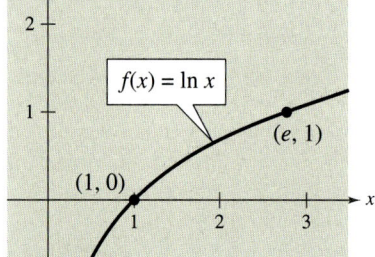

FIGURE 5.14

The graph of the natural logarithmic function is shown in Figure 5.14. Try using a graphing utility to confirm this graph. What is the domain of the natural logarithmic function?

Example 7 ▶ Using Properties of Natural Logarithms

Use the properties of natural logarithms to simplify the expression.

a. $\ln \dfrac{1}{e}$ **b.** $e^{\ln 5}$

c. $\dfrac{\ln 1}{3}$ **d.** $2 \ln e$

Solution
a. $\ln \dfrac{1}{e} = \ln e^{-1} = -1$ Inverse Property

b. $e^{\ln 5} = 5$ Inverse Property

c. $\dfrac{\ln 1}{3} = \dfrac{0}{3} = 0$ Property 1

d. $2 \ln e = 2(1) = 2$ Property 2

STUDY TIP

Some graphing utilities display a complex number instead of an ERROR message when evaluating an expression such as $\ln(-1)$.

On most calculators, the natural logarithm is denoted by [LN], as illustrated in Example 8.

Example 8 ▶ Evaluating the Natural Logarithmic Function

Use a calculator to evaluate each expression.

a. $\ln 2$ **b.** $\ln 0.3$ **c.** $\ln e^2$ **d.** $\ln(-1)$ **e.** $\ln(1 + \sqrt{2})$

Solution

Number	Graphing Calculator Keystrokes	Display
a. $\ln 2$	[LN] 2 [ENTER]	0.6931472
b. $\ln 0.3$	[LN] .3 [ENTER]	−1.2039728
c. $\ln e^2$	[LN] [e^x] 2 [ENTER]	2
d. $\ln(-1)$	[LN] [(−)] 1 [ENTER]	ERROR
e. $\ln(1 + \sqrt{2})$	[LN] [(] 1 [+] [√] 2 [)] [ENTER]	0.8813736

In Example 8, be sure you see that $\ln(-1)$ gives an error message on most calculators. This occurs because the domain of $\ln x$ is the set of positive real numbers (see Figure 5.14). So, $\ln(-1)$ is undefined.

Example 9 ▶ Finding the Domains of Logarithmic Functions

Find the domain of each function.

a. $f(x) = \ln(x - 2)$ **b.** $g(x) = \ln(2 - x)$ **c.** $h(x) = \ln x^2$

Solution

a. Because $\ln(x - 2)$ is defined only if

$$x - 2 > 0,$$

it follows that the domain of f is $(2, \infty)$.

b. Because $\ln(2 - x)$ is defined only if

$$2 - x > 0,$$

it follows that the domain of g is $(-\infty, 2)$. The graph of g is shown in Figure 5.15.

c. Because $\ln x^2$ is defined only if

$$x^2 > 0,$$

it follows that the domain of h is all real numbers except $x = 0$.

FIGURE 5.15

In Example 9, suppose you had been asked to analyze the function given by $h(x) = \ln|x - 2|$. How would the domain of this function compare with the domains of the functions given in parts (a) and (b) of the example?

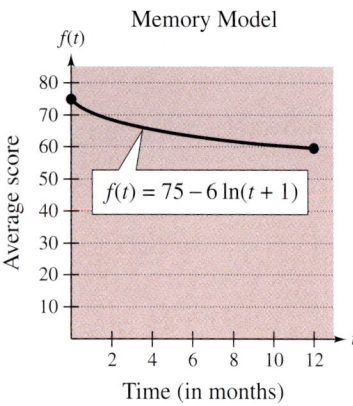

FIGURE 5.16

Application

Example 10 ▶ Human Memory Model

Students participating in a psychological experiment attended several lectures on a subject and were given an exam. Every month for a year after the exam, the students were retested to see how much of the material they remembered. The average scores for the group are given by the *human memory model*

$$f(t) = 75 - 6\ln(t + 1), \quad 0 \leq t \leq 12$$

where t is the time in months. The graph of f is shown in Figure 5.16.

a. What was the average score on the original ($t = 0$) exam?
b. What was the average score at the end of $t = 2$ months?
c. What was the average score at the end of $t = 6$ months?

Solution

a. The original average score was

$$f(0) = 75 - 6\ln(0 + 1) \quad \text{Substitute 0 for } t.$$
$$= 75 - 6\ln 1 \quad \text{Simplify.}$$
$$= 75 - 6(0) \quad \text{Property of natural logarithms}$$
$$= 75. \quad \text{Solution}$$

b. After 2 months, the average score was

$$f(2) = 75 - 6\ln(2 + 1)$$
$$= 75 - 6\ln 3$$
$$\approx 75 - 6(1.0986)$$
$$\approx 68.4.$$

c. After 6 months, the average score was

$$f(6) = 75 - 6\ln(6 + 1)$$
$$= 75 - 6\ln 7$$
$$\approx 75 - 6(1.9459)$$
$$\approx 63.3.$$

Writing ABOUT MATHEMATICS

Analyzing a Human Memory Model Use a graphing utility to determine the time in months when the average score in Example 10 was 60. Explain your method of solving the problem. Describe another way that you can use a graphing utility to determine the answer.

5.2 Exercises

In Exercises 1–8, write the logarithmic equation in exponential form. For example, the exponential form of $\log_5 25 = 2$ is $5^2 = 25$.

1. $\log_4 64 = 3$
2. $\log_3 81 = 4$
3. $\log_7 \frac{1}{49} = -2$
4. $\log_{10} \frac{1}{1000} = -3$
5. $\log_{32} 4 = \frac{2}{5}$
6. $\log_{16} 8 = \frac{3}{4}$
7. $\ln 1 = 0$
8. $\ln 4 = 1.386 \ldots$

In Exercises 9–18, write the exponential equation in logarithmic form. For example, the logarithmic form of $2^3 = 8$ is $\log_2 8 = 3$.

9. $5^3 = 125$
10. $8^2 = 64$
11. $81^{1/4} = 3$
12. $9^{3/2} = 27$
13. $6^{-2} = \frac{1}{36}$
14. $10^{-3} = 0.001$
15. $e^3 = 20.0855 \ldots$
16. $e^x = 4$
17. $e^0 = 1$
18. $u^v = w$

In Exercises 19–32, evaluate the expression without using a calculator.

19. $\log_2 16$
20. $\log_2 \frac{1}{8}$
21. $\log_{16} 4$
22. $\log_{27} 9$
23. $\log_7 1$
24. $\log_{10} 1000$
25. $\log_{10} 0.01$
26. $\log_{10} 10$
27. $\log_8 32$
28. $\log_9 243$
29. $\ln e^3$
30. $\ln e^{-2}$
31. $\log_a a^2$
32. $\log_b b^{-3}$

In Exercises 33–44, use a calculator to evaluate the logarithm. Round to three decimal places.

33. $\log_{10} 345$
34. $\log_{10} 145$
35. $\log_{10} \frac{4}{5}$
36. $\log_{10} 12.5$
37. $\ln 18.42$
38. $\ln \sqrt{42}$
39. $3 \ln 0.32$
40. $2 \ln 0.75$
41. $\ln(1 + \sqrt{3})$
42. $\ln(\sqrt{5} - 2)$
43. $\ln \frac{2}{3}$
44. $\ln \frac{1}{2}$

In Exercises 45–50, use the graph of $y = \log_3 x$ to match the given function with its graph. [The graphs are labeled (a), (b), (c), (d), (e), and (f).]

(a)
(b)
(c)
(d)
(e)
(f)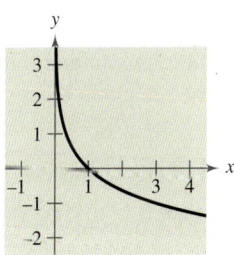

45. $f(x) = \log_3 x + 2$
46. $f(x) = -\log_3 x$
47. $f(x) = -\log_3(x + 2)$
48. $f(x) = \log_3(x - 1)$
49. $f(x) = \log_3(1 - x)$
50. $f(x) = -\log_3(-x)$

In Exercises 51–62, find the domain, x-intercept, and vertical asymptote of the logarithmic function and sketch its graph.

51. $f(x) = \log_4 x$
52. $g(x) = \log_6 x$
53. $y = -\log_3 x + 2$
54. $h(x) = \log_4(x - 3)$
55. $f(x) = -\log_6(x + 2)$
56. $y = \log_5(x - 1) + 4$
57. $y = \log_{10}\left(\frac{x}{5}\right)$
58. $y = \log_{10}(-x)$
59. $f(x) = \ln(x - 2)$
60. $h(x) = \ln(x + 1)$
61. $g(x) = \ln(-x)$
62. $f(x) = \ln(3 - x)$

In Exercises 63–68, use a graphing utility to graph the function. Be sure to use an appropriate viewing window.

63. $f(x) = \log_{10}(x + 1)$
64. $f(x) = \log_{10}(x - 1)$
65. $f(x) = \ln(x - 1)$
66. $f(x) = \ln(x + 2)$
67. $f(x) = \ln x + 2$
68. $f(x) = 3 \ln x - 1$

69. *Human Memory Model* Students in a mathematics class were given an exam and then retested monthly with an equivalent exam. The average scores for the class are given by the human memory model

$$f(t) = 80 - 17 \log_{10}(t + 1), \quad 0 \leq t \leq 12$$

where t is the time in months.

(a) What was the average score on the original exam $(t = 0)$?
(b) What was the average score after 4 months?
(c) What was the average score after 10 months?

70. *Population Growth* The population of a town will double in

$$t = \frac{10 \ln 2}{\ln 67 - \ln 50} \text{ years.}$$

Find t.

71. *Population* The time t in years for the world population to double if it is increasing at a continuous rate of r is given by

$$t = \frac{\ln 2}{r}.$$

(a) Complete the table.

r	0.005	0.01	0.015	0.02	0.025	0.03
t						

(b) Use a reference source to decide which value of r best approximates the actual rate of growth for the world population.

72. *Finance* A principal P, invested at $9\frac{1}{2}\%$ and compounded continuously, increases to an amount K times the original principal after t years, where t is given by

$$t = \frac{\ln K}{0.095}.$$

(a) Complete the table and interpret your results.

K	1	2	4	6	8	10	12
t							

(b) Sketch a graph of the function.

Ventilation In Exercises 73 and 74, use the model

$$y = 80.4 - 11 \ln x, \quad 100 \leq x \leq 1500$$

which approximates the minimum required ventilation rate in terms of the air space per child in a public school classroom. In the model, x is the air space per child in cubic feet and y is the ventilation rate in cubic feet per minute.

73. Use a graphing utility to graph the function and approximate the required ventilation rate if there is 300 cubic feet of air space per child.

74. A classroom is designed for 30 students. The air conditioning system in the room has the capacity of moving 450 cubic feet of air per minute.

(a) Determine the ventilation rate per child, assuming that the room is filled to capacity.
(b) Use the graph in Exercise 73 to estimate the air space required per child.
(c) Determine the minimum number of square feet of floor space required for the room if the ceiling height is 30 feet.

75. *Work* The work (in foot-pounds) done in compressing a volume of 9 cubic feet at a pressure of 15 pounds per square inch to a volume of 3 cubic feet is

$$W = 19{,}440(\ln 9 - \ln 3).$$

Find W.

76. *Sound Intensity* The relationship between the number of decibels β and the intensity of a sound I in watts per square meter is

$$\beta = 10 \log_{10}\left(\frac{I}{10^{-12}}\right).$$

(a) Determine the number of decibels of a sound with an intensity of 1 watt per square meter.
(b) Determine the number of decibels of a sound with an intensity of 10^{-2} watt per square meter.
(c) The intensity of the sound in part (a) is 100 times as great as that in part (b). Is the number of decibels 100 times as great? Explain.

Monthly Payment In Exercises 77–80, use the model

$$t = 12.542 \ln\left(\frac{x}{x - 1000}\right), \quad x > 1000$$

which approximates the length of a home mortgage of $150,000 at 8% in terms of the monthly payment. In the model, t is the length of the mortgage in years and x is the monthly payment in dollars (see figure).

77. Use the model to approximate the length of a $150,000 mortgage at 8% if the monthly payment is $1100.65.

78. Use the model to approximate the length of a $150,000 mortgage at 8% if the monthly payment is $1254.68.

79. Approximate the total amount paid over the term of the mortgage in Exercise 77 with a monthly payment of $1100.65. What is the total interest charge?

80. Approximate the total amount paid over the term of the mortgage in Exercise 78 with a monthly payment of $1254.68. What is the total interest charge?

81. (a) Complete the table for the function

$$f(x) = \frac{\ln x}{x}.$$

x	1	5	10	10^2	10^4	10^6
$f(x)$						

(b) Use the table in part (a) to determine what value $f(x)$ approaches as x increases without bound.

(c) Use a graphing utility to confirm the result of part (b).

Synthesis

True or False? In Exercises 82 and 83, determine whether the statement is true or false. Justify your answer.

82. You can determine the graph of $f(x) = \log_6 x$ by graphing $g(x) = 6^x$ and reflecting it about the x-axis.

83. The graph of $f(x) = \log_3 x$ contains the point $(27, 3)$.

In Exercises 84–87, describe the relationship between the graphs of f and g. What is the relationship between the functions f and g?

84. $f(x) = 3^x$
$g(x) = \log_3 x$

85. $f(x) = 5^x$
$g(x) = \log_5 x$

86. $f(x) = e^x$
$g(x) = \ln x$

87. $f(x) = 10^x$
$g(x) = \log_{10} x$

88. *Graphical Analysis* Use a graphing utility to graph f and g in the same viewing window and determine which is increasing at the greater rate as x approaches $+\infty$. What can you conclude about the rate of growth of the natural logarithmic function?

(a) $f(x) = \ln x, \quad g(x) = \sqrt{x}$

(b) $f(x) = \ln x, \quad g(x) = \sqrt[4]{x}$

89. *Exploration* The table of values was obtained by evaluating a function. Determine which of the statements may be true and which must be false.

x	1	2	8
y	0	1	3

(a) y is an exponential function of x.

(b) y is a logarithmic function of x.

(c) x is an exponential function of y.

(d) y is a linear function of x.

90. *Exploration* Use a graphing utility to compare the graph of the function $y = \ln x$ with the graph of each given function.

(a) $y_1 = x - 1$

(b) $y_2 = (x - 1) - \frac{1}{2}(x - 1)^2$

(c) $y_3 = (x - 1) - \frac{1}{2}(x - 1)^2 + \frac{1}{3}(x - 1)^3$

91. *Pattern Recognition* Identify the pattern of successive polynomials given in Exercise 90. Extend the pattern one more term and compare the graph of the resulting polynomial function with the graph of $y = \ln x$. What do you think the pattern implies?

92. *Exploration* Answer the following questions for the function $f(x) = \log_{10} x.$ Do not use a calculator.

(a) What is the domain of f?

(b) What is f^{-1}?

(c) If x is a real number between 1000 and 10,000, in which interval will $f(x)$ be found?

(d) In which interval will x be found if $f(x)$ is negative?

(e) If $f(x)$ is increased by 1 unit, x must have been increased by what factor?

(f) If $f(x_1) = 3n$ and $f(x_2) = n,$ what is the ratio of x_1 to x_2?

In Exercises 93–96, (a) use a graphing utility to graph the function, (b) use the graph to determine the intervals in which the function is increasing and decreasing, and (c) approximate any relative maximum or minimum values of the function.

93. $f(x) = |\ln x|$

94. $h(x) = \ln(x^2 + 1)$

95. $f(x) = \dfrac{x}{2} - \ln \dfrac{x}{4}$

96. $g(x) = \dfrac{12 \ln x}{x}$

In Exercises 101–104, evaluate the expression. Round your result to three decimal places.

101. e^6

102. $e^{3/2}$

103. e^{-4}

104. $4e^{-6}$

In exercises 105–108, sketch the graph of the function by hand.

105. $f(x) = 2^x + 3$

106. $f(x) = -2^{x-3}$

107. $f(x) = \left(\dfrac{3}{2}\right)^{x+1}$

108. $f(x) = 5^{-x} - 2$

Review

In Exercises 97–100, translate the statement into an algebraic expression.

97. The product of 8 and n is decreased by 3.

98. The total hourly wage for an employee is $9.25 per hour plus 75 cents for each of q units produced per hour.

99. The total cost for auto repairs if the cost of parts was $83.95 and there were t hours of labor at $37.50 per hour

100. The area of a rectangle if the length is 10 units more than the width w

5.3 Properties of Logarithms

What you should learn
- How to rewrite logarithmic functions with a different base
- How to use properties of logarithms to evaluate or rewrite logarithmic expressions
- How to use properties of logarithms to expand or condense logarithmic expressions
- How to use logarithmic functions to model and solve real-life applications

Why you should learn it
Logarithmic functions are often used to model scientific observations. For instance, Exercise 85 on page 372 shows how to use a logarithmic function to model human memory.

Change of Base

Most calculators have only two types of log keys, one for common logarithms (base 10) and one for natural logarithms (base e). Although common logs and natural logs are the most frequently used, you may occasionally need to evaluate logarithms to other bases. To do this, you can use the following *change-of-base formula*.

Change-of-Base Formula

Let a, b, and x be positive real numbers such that $a \neq 1$ and $b \neq 1$. Then $\log_a x$ can be converted to a different base as follows.

$$\text{Base } b \qquad \text{Base } 10 \qquad \text{Base } e$$

$$\log_a x = \frac{\log_b x}{\log_b a} \qquad \log_a x = \frac{\log_{10} x}{\log_{10} a} \qquad \log_a x = \frac{\ln x}{\ln a}$$

One way to look at the change-of-base formula is that logarithms to base a are simply *constant multiples* of logarithms to base b. The constant multiplier is $1/(\log_b a)$.

Example 1 ▶ Changing Bases Using Common Logarithms

a. $\log_4 30 = \dfrac{\log_{10} 30}{\log_{10} 4}$ $\qquad \log_a b = \dfrac{\log_{10} b}{\log_{10} a}$

$\qquad\quad \approx \dfrac{1.47712}{0.60206}$ $\qquad$ Use a calculator.

$\qquad\quad \approx 2.4534$ $\qquad$ Use a calculator.

b. $\log_2 14 = \dfrac{\log_{10} 14}{\log_{10} 2} \approx \dfrac{1.14613}{0.30103} \approx 3.8074$

Example 2 ▶ Changing Bases Using Natural Logarithms

a. $\log_4 30 = \dfrac{\ln 30}{\ln 4}$ $\qquad \log_a b = \dfrac{\ln b}{\ln a}$

$\qquad\quad \approx \dfrac{3.40120}{1.38629}$ $\qquad$ Use a calculator.

$\qquad\quad \approx 2.4534$ $\qquad$ Use a calculator.

b. $\log_2 14 = \dfrac{\ln 14}{\ln 2} \approx \dfrac{2.63906}{0.693147} \approx 3.8074$

Properties of Logarithms

You know from the preceding section that the logarithmic function with base a is the *inverse* of the exponential function with base a. So, it makes sense that the properties of exponents should have corresponding properties involving logarithms. For instance, the exponential property $a^0 = 1$ has the corresponding logarithmic property $\log_a 1 = 0$.

> **STUDY TIP**
>
> There is no general property that can be used to rewrite $\log_a(u \pm v)$. Specifically, $\log_a(x + y)$ is not equal to $\log_a x + \log_a y$.

Properties of Logarithms

Let a be a positive number such that $a \neq 1$, and let n be a real number. If u and v are positive real numbers, the following properties are true.

1. $\log_a(uv) = \log_a u + \log_a v$ 1. $\ln(uv) = \ln u + \ln v$

2. $\log_a \dfrac{u}{v} = \log_a u - \log_a v$ 2. $\ln \dfrac{u}{v} = \ln u - \ln v$

3. $\log_a u^n = n \log_a u$ 3. $\ln u^n = n \ln u$

A proof of the first property listed above is given in Appendix A.

Example 3 ▶ Using Properties of Logarithms

Write the logarithm in terms of $\ln 2$ and $\ln 3$.

a. $\ln 6$ **b.** $\ln \dfrac{2}{27}$

Solution

a. $\ln 6 = \ln(2 \cdot 3)$ Rewrite 6 as $2 \cdot 3$.

    $= \ln 2 + \ln 3$ Property 1

b. $\ln \dfrac{2}{27} = \ln 2 - \ln 27$ Property 2

    $= \ln 2 - \ln 3^3$ Rewrite 27 as 3^3.

    $= \ln 2 - 3 \ln 3$ Property 3

Example 4 ▶ Using Properties of Logarithms

Use the properties of logarithms to verify that $-\log_{10} \frac{1}{100} = \log_{10} 100$.

Solution

$-\log_{10} \frac{1}{100} = -\log_{10}(100^{-1})$ Rewrite $\frac{1}{100}$ as 100^{-1}.

    $= -(-1)\log_{10} 100$ Property 3

    $= \log_{10} 100$ Simplify.

Try checking this result on your calculator.

John Napier, a Scottish mathematician, developed logarithms as a way to simplify some of the tedious calculations of his day. Beginning in 1594, Napier worked about 20 years on the invention of logarithms. Napier was only partially successful in his quest to simplify tedious calculations. Nonetheless, the development of logarithms was a step forward and received immediate recognition.

Rewriting Logarithmic Expressions

The properties of logarithms are useful for rewriting logarithmic expressions in forms that simplify the operations of algebra. This is true because these properties convert complicated products, quotients, and exponential forms into simpler sums, differences, and products, respectively.

Example 5 ▶ Expanding Logarithmic Expressions

Expand the logarithmic expressions.

a. $\log_{10} 5x^3y$ **b.** $\ln \dfrac{\sqrt{3x-5}}{7}$

Solution

a. $\log_{10} 5x^3y = \log_{10} 5 + \log_{10} x^3y$ Property 1

$\qquad\qquad\;\; = \log_{10} 5 + \log_{10} x^3 + \log_{10} y$ Property 1

$\qquad\qquad\;\; = \log_{10} 5 + 3\log_{10} x + \log_{10} y$ Property 3

b. $\ln \dfrac{\sqrt{3x-5}}{7} = \ln \dfrac{(3x-5)^{1/2}}{7}$ Rewrite using rational exponent.

$\qquad\qquad\;\; = \ln(3x-5)^{1/2} - \ln 7$ Property 2

$\qquad\qquad\;\; = \dfrac{1}{2}\ln(3x-5) - \ln 7$ Property 3

In Example 5, the properties of logarithms were used to *expand* logarithmic expressions. In Example 6, this procedure is reversed and the properties of logarithms are used to *condense* logarithmic expressions.

Example 6 ▶ Condensing Logarithmic Expressions

Condense the logarithmic expressions.

a. $\frac{1}{2}\log_{10} x + 3\log_{10}(x+1)$

b. $2\ln(x+2) - \ln x$

Solution

a. $\frac{1}{2}\log_{10} x + 3\log_{10}(x+1) = \log_{10} x^{1/2} + \log_{10}(x+1)^3$ Property 3

$\qquad\qquad\qquad\qquad\qquad\;\; = \log_{10}\left[\sqrt{x} \cdot (x+1)^3\right]$ Property 1

b. $2\ln(x+2) - \ln x = \ln(x+2)^2 - \ln x$ Property 3

$\qquad\qquad\qquad\;\; = \ln\dfrac{(x+2)^2}{x}$ Property 2

◀ Exploration ▶

Use a graphing utility to graph the functions

$$y = \ln x - \ln(x-3)$$

and

$$y = \ln \dfrac{x}{x-3}$$

in the same viewing window. Does the graphing utility show the functions with the same domain? If so, should it? Explain your reasoning.

Application

One method of determining how the *x*- and *y*-values for a set of nonlinear data are related begins by taking the natural log of each of the *x*- and *y*-values. If the points are graphed and fall on a straight line, then you can determine that the *x*- and *y*-values are related by the equation

$$\ln y = m \ln x$$

where *m* is the slope of the straight line.

Example 7 ▶ Finding a Mathematical Model

The table gives the mean distance *x* and the period *y* of the six planets that are closest to the sun. In the table, the mean distance is given in terms of astronomical units (where the earth's mean distance is defined as 1.0), and the period is given in terms of years. Find an equation that expresses *y* as a function of *x*.

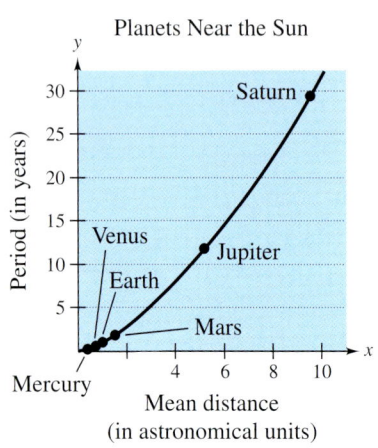

FIGURE 5.17

Planet	Mercury	Venus	Earth	Mars	Jupiter	Saturn
Period, *y*	0.241	0.615	1.0	1.881	11.862	29.458
Mean distance, *x*	0.387	0.723	1.0	1.524	5.203	9.539

Solution

The points in the table are plotted in Figure 5.17. From this figure it is not clear how to find an equation that relates *y* and *x*. To solve this problem, take the natural log of each of the *x*- and *y*-values given in the table. This produces the following results.

Planet	Mercury	Venus	Earth	Mars	Jupiter	Saturn
ln *y*	−1.423	−0.486	0.0	0.632	2.473	3.383
ln *x*	−0.949	−0.324	0.0	0.421	1.649	2.255

Now, by plotting the points in the second table, you can see that all six of the points appear to lie in a line (see Figure 5.18). You can use a graphical approach or the algebraic approach to find that the slope of this line is $\frac{3}{2}$. You can therefore conclude that

$$\ln y = \frac{3}{2} \ln x.$$

Try to convert this to $y = f(x)$ form. You will get a function of the form $y = ax^b$, which is a power model (the variable *x* is raised to a power, *b*).

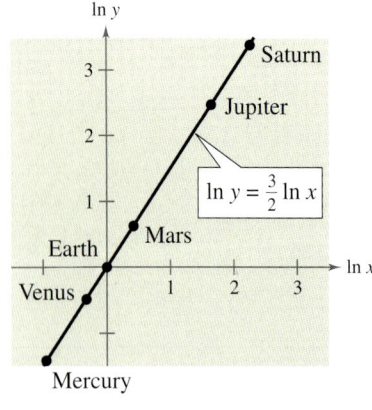

FIGURE 5.18

5.3 Exercises

In Exercises 1–8, evaluate the logarithm using the change-of-base formula. Round your result to three decimal places.

1. $\log_3 7$
2. $\log_7 4$
3. $\log_{1/2} 4$
4. $\log_{1/4} 5$
5. $\log_9 0.4$
6. $\log_{20} 0.125$
7. $\log_{15} 1250$
8. $\log_3 0.015$

In Exercises 9–16, rewrite the logarithm as a ratio of (a) common logarithms and (b) natural logarithms.

9. $\log_5 x$
10. $\log_3 x$
11. $\log_{1/5} x$
12. $\log_{1/3} x$
13. $\log_x \frac{3}{10}$
14. $\log_x \frac{3}{4}$
15. $\log_{2.6} x$
16. $\log_{7.1} x$

In Exercises 17–22, use the change-of-base formula to rewrite the logarithm as a ratio of logarithms. Then use a graphing utility to sketch the graph.

17. $f(x) = \log_2 x$
18. $f(x) = \log_4 x$
19. $f(x) = \log_{1/2} x$
20. $f(x) = \log_{1/4} x$
21. $f(x) = \log_{11.8} x$
22. $f(x) = \log_{12.4} x$

In Exercises 23–42, use the properties of logarithms to expand the expression as a sum, difference, and/or constant multiple of logarithms. (Assume all variables are positive.)

23. $\log_{10} 5x$
24. $\log_{10} 10z$
25. $\log_{10} \frac{5}{x}$
26. $\log_{10} \frac{y}{2}$
27. $\log_8 x^4$
28. $\log_6 z^{-3}$
29. $\ln \sqrt{z}$
30. $\ln \sqrt[3]{t}$
31. $\ln xyz$
32. $\ln \frac{xy}{z}$
33. $\ln \sqrt{a-1},\ a > 1$
34. $\ln\left(\frac{x^2-1}{x^3}\right),\ x > 1$
35. $\ln z(z-1)^2,\ z > 1$
36. $\ln \frac{x}{\sqrt{x^2+1}}$
37. $\ln \sqrt[3]{\frac{x}{y}}$
38. $\ln \sqrt{\frac{x^2}{y^3}}$
39. $\ln \frac{x^4 \sqrt{y}}{z^5}$
40. $\ln \sqrt{x^2(x+2)}$
41. $\log_b \frac{x^2}{y^2 z^3}$
42. $\log_b \frac{\sqrt{x}\, y^4}{z^4}$

In Exercises 43–62, condense the expression to the logarithm of a single quantity.

43. $\ln x + \ln 3$
44. $\ln y + \ln t$
45. $\log_4 z - \log_4 y$
46. $\log_5 8 - \log_5 t$
47. $2 \log_2(x+4)$
48. $-4 \log_6 2x$
49. $\frac{1}{4} \log_3 5x$
50. $\frac{2}{3} \log_7(z-2)$
51. $\ln x - 3\ln(x+1)$
52. $2\ln 8 + 5\ln z$
53. $\ln(x-2) - \ln(x+2)$
54. $3\ln x + 4\ln y - 4\ln z$
55. $\ln x - 4[\ln(x+2) + \ln(x-2)]$
56. $4[\ln z + \ln(z+5)] - 2\ln(z-5)$
57. $\frac{1}{3}[2\ln(x+3) + \ln x - \ln(x^2-1)]$
58. $2[\ln x - \ln(x+1) - \ln(x-1)]$
59. $\frac{1}{3}[\ln y + 2\ln(y+4)] - \ln(y-1)$
60. $\frac{1}{2}[\ln(x+1) + 2\ln(x-1)] + 6\ln x$
61. $2\ln 3 - \frac{1}{2}\ln(x^2+1)$
62. $\frac{3}{2}\ln 5t^6 - \frac{3}{4}\ln t^4$

In Exercises 63 and 64, compare the logarithmic quantities. If two are equal, explain why.

63. $\dfrac{\log_2 32}{\log_2 4},\quad \log_2 \dfrac{32}{4},\quad \log_2 32 - \log_2 4$
64. $\log_7 \sqrt{70},\quad \log_7 35,\quad \frac{1}{2} + \log_7 \sqrt{10}$

In Exercises 65–78, find the exact value of the logarithm without using a calculator. (If this is not possible, state the reason.)

65. $\log_3 9$
66. $\log_6 \sqrt[3]{6}$
67. $\log_4 16^{1.2}$
68. $\log_5 \frac{1}{125}$
69. $\log_3(-9)$
70. $\log_2(-16)$
71. $\log_5 75 - \log_5 3$
72. $\log_4 2 + \log_4 32$
73. $\ln e^2 - \ln e^5$
74. $3\ln e^4$
75. $\log_{10} 0$
76. $\ln 1$
77. $\ln e^{4.5}$
78. $\ln \sqrt[4]{e^3}$

In Exercises 79–84, use the properties of logarithms to rewrite and simplify the logarithmic expression.

79. $\log_4 8$

80. $\log_2(4^2 \cdot 3^4)$

81. $\log_5 \frac{1}{250}$

82. $\log_{10} \frac{9}{300}$

83. $\ln(5e^6)$

84. $\ln \frac{6}{e^2}$

85. *Human Memory Model* Students participating in a psychological experiment attended several lectures and were given an exam. Every month for a year after the exam, the students were retested to see how much of the material they remembered. The average scores for the group can be modeled by the memory model

$$f(t) = 90 - 15 \log_{10}(t + 1), \quad 0 \leq t \leq 12$$

where t is the time in months.

(a) What was the average score on the original exam ($t = 0$)?

(b) What was the average score after 6 months?

(c) What was the average score after 12 months?

(d) When will the average score decrease to 75?

(e) Use the properties of logarithms to write the function in another form.

(f) Sketch the graph of the function over the specified domain.

86. *Sound Intensity* The relationship between the number of decibels β and the intensity of a sound I in watts per square meter is

$$\beta = 10 \log_{10}\left(\frac{I}{10^{-12}}\right).$$

Use the properties of logarithms to write the formula in simpler form, and determine the number of decibels of a sound with an intensity of 10^{-6} watt per square meter.

Synthesis

True or False? **In Exercises 87–92, determine whether the statement is true or false given that $f(x) = \ln x$. Justify your answer.**

87. $f(0) = 0$

88. $f(ax) = f(a) + f(x), \quad a > 0, x > 0$

89. $f(x - 2) = f(x) - f(2), \quad x > 2$

90. $\sqrt{f(x)} = \frac{1}{2} f(x)$

91. If $f(u) = 2f(v)$, then $v = u^2$.

92. If $f(x) < 0$, then $0 < x < 1$.

93. Prove that $\log_b \frac{u}{v} = \log_b u - \log_b v$.

94. Prove that $\log_b u^n = n \log_b u$.

In Exercises 95 and 96, use a graphing utility to graph the two functions in the same viewing window. Use the graphs to verify that the expressions are equivalent.

95. $f(x) = \log_{10} x$

$g(x) = \dfrac{\ln x}{\ln 10}$

96. $f(x) = \ln x$

$g(x) = \dfrac{\log_{10} x}{\log_{10} e}$

97. *Think About It* Sketch the graphs of

$$f(x) = \ln \frac{x}{2}, \quad g(x) = \frac{\ln x}{\ln 2}, \quad h(x) = \ln x - \ln 2$$

on the same set of axes. Which two functions have identical graphs? Explain why.

98. *Exploration* Approximate the natural logarithms of as many integers as possible between 1 and 20 given that $\ln 2 \approx 0.6931$, $\ln 3 \approx 1.0986$, and $\ln 5 \approx 1.6094$. (Do not use a calculator.)

Review

In Exercises 99–102, simplify the expression.

99. $\dfrac{24xy^{-2}}{16x^{-3}y}$

100. $\left(\dfrac{2x^2}{3y}\right)^{-3}$

101. $(18x^3y^4)^{-3}(18x^3y^4)^3$

102. $xy(x^{-1} + y^{-1})^{-1}$

In Exercises 103–108, use a calculator to evaluate the expression. Round your result to three decimal places.

103. $(2.8)^{7.6}$

104. $40(6^{-3.2})$

105. $7^{-\pi}$

106. $1.4^{3\pi}$

107. $\sqrt[4]{350}$

108. $96^{\sqrt{3}}$

In Exercises 109–112, use a calculator to evaluate the logarithm. Round your result to three decimal places.

109. $\log_{10} 26$

110. $\log_{10} \frac{5}{8}$

111. $\ln 10.6$

112. $\ln(\sqrt{7} + 1)$

5.4 Exponential and Logarithmic Equations

What you should learn
- How to solve simple exponential and logarithmic equations
- How to solve more complicated exponential equations
- How to solve more complicated logarithmic equations
- How to use exponential and logarithmic equations to model and solve real-life applications

Why you should learn it
Applications of exponential and logarithmic equations are found in consumer safety testing. For instance, Exercise 122 on page 383 shows how to use a logarithmic function to model crumple zones for automobile crash tests.

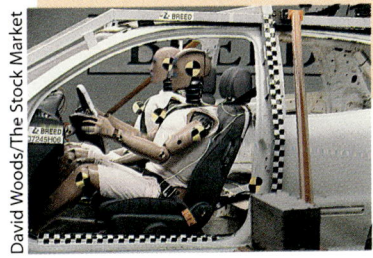

Introduction

So far in this chapter, you have studied the definitions, graphs, and properties of exponential and logarithmic functions. In this section, you will study procedures for *solving equations* involving these exponential and logarithmic functions.

There are two basic strategies for solving exponential or logarithmic equations. The first is based on the One-to-One Properties and the second is based on the Inverse Properties. For $a > 0$ and $a \neq 1$, the following properties are true for all x and y for which $\log_a x$ and $\log_a y$ are defined.

One-to-One Properties

$a^x = a^y$ if and only if $x = y$.

$\log_a x = \log_a y$ if and only if $x = y$.

Inverse Properties

$a^{\log_a x} = x$

$\log_a a^x = x$

Example 1 ▶ Solving Simple Equations

	Original Equation	Rewritten Equation	Solution	Property
a.	$2^x = 32$	$2^x = 2^5$	$x = 5$	One-to-One
b.	$\ln x - \ln 3 = 0$	$\ln x = \ln 3$	$x = 3$	One-to-One
c.	$\left(\frac{1}{3}\right)^x = 9$	$3^{-x} = 3^2$	$x = -2$	One-to-One
d.	$e^x = 7$	$\ln e^x = \ln 7$	$x = \ln 7$	Inverse
e.	$\ln x = -3$	$e^{\ln x} = e^{-3}$	$x = e^{-3}$	Inverse
f.	$\log_{10} x = -1$	$10^{\log_{10} x} = 10^{-1}$	$x = 10^{-1} = \frac{1}{10}$	Inverse

The strategies used in Example 1 are summarized as follows.

Strategies for Solving Exponential and Logarithmic Equations

1. Rewrite the given equation in a form that allows the use of the One-to-One Properties of exponential or logarithmic functions.
2. Rewrite an *exponential* equation in logarithmic form and apply the Inverse Property of logarithmic functions.
3. Rewrite a *logarithmic* equation in exponential form and apply the Inverse Property of exponential functions.

Solving Exponential Equations

Example 2 ▶ Solving Exponential Equations

Solve each equation and approximate the result to three decimal places.

a. $e^x = 72$ **b.** $3(2^x) = 42$

Solution

a.
$e^x = 72$	Write original equation.
$\ln e^x = \ln 72$	Take natural log of each side.
$x = \ln 72$	Inverse Property
$x \approx 4.277$	Use a calculator.

The solution is $\ln 72 \approx 4.277$. Check this in the original equation.

b.
$3(2^x) = 42$	Write original equation.
$2^x = 14$	Divide each side by 3.
$\log_2 2^x = \log_2 14$	Take log (base 2) of each side.
$x = \log_2 14$	Inverse Property
$x = \dfrac{\ln 14}{\ln 2}$	Change-of-base formula
$x \approx 3.807$	Use a calculator.

The solution is $\log_2 14 \approx 3.807$. Check this in the original equation.

In Example 2(a), the exact solution is $x = \ln 72$ and the approximate solution is $x \approx 4.277$. An exact answer is preferred when the solution is an intermediate step in a larger problem. For a final answer, an approximate solution is easier to comprehend.

Example 3 ▶ Solving an Exponential Equation

Solve $e^x + 5 = 60$ and approximate the result to three decimal places.

Solution

$e^x + 5 = 60$	Write original equation.
$e^x = 55$	Subtract 5 from each side.
$\ln e^x = \ln 55$	Take natural log of each side.
$x = \ln 55$	Inverse Property
$x \approx 4.007$	Use a calculator.

The solution is $\ln 55 \approx 4.007$. Check this in the original equation.

Technology

When solving an exponential or logarithmic equation, remember that you can check your solution graphically by "graphing the left and right sides separately" and using the intersect feature of your graphing utility to determine the point of intersection. For instance, to check the solution of the equation in Example 2(a), you can graph

$y = e^x$ and $y = 72$

in the same viewing window, as shown below. Using the intersect feature of your graphing utility, you can determine that the graphs intersect when $x \approx 4.277$, which confirms the solution found in Example 2(a).

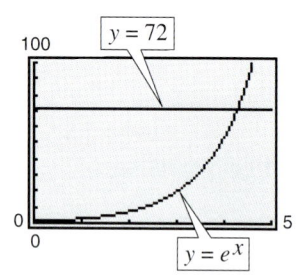

Example 4 ▶ Solving an Exponential Equation

Solve $2(3^{2t-5}) - 4 = 11$ and approximate the result to three decimal places.

Solution

$$2(3^{2t-5}) - 4 = 11 \qquad \text{Write original equation.}$$

$$2(3^{2t-5}) = 15 \qquad \text{Add 4 to each side.}$$

$$3^{2t-5} = \frac{15}{2} \qquad \text{Divide each side by 2.}$$

$$\log_3 3^{2t-5} = \log_3 \frac{15}{2} \qquad \text{Take log (base 3) of each side.}$$

$$2t - 5 = \log_3 \frac{15}{2} \qquad \text{Inverse Property}$$

$$2t = 5 + \log_3 7.5 \qquad \text{Add 5 to each side.}$$

$$t = \frac{5}{2} + \frac{1}{2} \log_3 7.5 \qquad \text{Divide each side by 2.}$$

$$t \approx 3.417 \qquad \text{Use a calculator.}$$

The solution is $\frac{5}{2} + \frac{1}{2} \log_3 7.5 \approx 3.417$. Check this in the original equation.

When an equation involves two or more exponential expressions, you can still use a procedure similar to that demonstrated in Examples 2, 3, and 4. However, the algebra is a bit more complicated.

Example 5 ▶ Solving an Exponential Equation of Quadratic Type

Solve $e^{2x} - 3e^x + 2 = 0$.

Solution

$$e^{2x} - 3e^x + 2 = 0 \qquad \text{Write original equation.}$$

$$(e^x)^2 - 3e^x + 2 = 0 \qquad \text{Write in quadratic form.}$$

$$(e^x - 2)(e^x - 1) = 0 \qquad \text{Factor.}$$

$$e^x - 2 = 0 \qquad \text{Set 1st factor equal to 0.}$$

$$x = \ln 2 \qquad \text{Solution}$$

$$e^x - 1 = 0 \qquad \text{Set 2nd factor equal to 0.}$$

$$x = 0 \qquad \text{Solution}$$

The solutions are $\ln 2$ and 0. Check these in the original equation.

In Example 5, use a graphing utility to graph $y = e^{2x} - 3e^x + 2$. The graph should have two *x*-intercepts: one at $x = \ln 2 \approx 0.693$ and one at $x = 0$.

Solving Logarithmic Equations

To solve a logarithmic equation such as

$$\ln x = 3 \qquad \text{Logarithmic form}$$

write the equation in exponential form as follows.

$$e^{\ln x} = e^3 \qquad \text{Exponentiate each side.}$$

$$x = e^3 \qquad \text{Exponential form}$$

This procedure is called *exponentiating* both sides of an equation.

Example 6 ▶ Solving a Logarithmic Equation

a. Solve $\ln x = 2$. **b.** Solve $\log_3(5x - 1) = \log_3(x + 7)$.

Solution

a. $\ln x = 2$ — Write original equation.

$e^{\ln x} = e^2$ — Exponentiate each side.

$x = e^2$ — Inverse Property

The solution is e^2. Check this in the original equation.

b. $\log_3(5x - 1) = \log_3(x + 7)$ — Write original equation.

$5x - 1 = x + 7$ — One-to-One Property

$4x = 8$ — Add $-x$ and 1 to each side.

$x = 2$ — Divide each side by 4.

The solution is 2. Check this in the original equation.

Example 7 ▶ Solving a Logarithmic Equation

Solve $5 + 2 \ln x = 4$ and approximate the result to three decimal places.

Solution

$5 + 2 \ln x = 4$ — Write original equation.

$2 \ln x = -1$ — Subtract 5 from each side.

$\ln x = -\dfrac{1}{2}$ — Divide each side by 2.

$e^{\ln x} = e^{-1/2}$ — Exponentiate each side.

$x = e^{-1/2}$ — Inverse Property

$x \approx 0.607$ — Use a calculator.

The solution is $e^{-1/2} \approx 0.607$. To check this result graphically, you can use a graphing utility to graph $y = 5 + 2 \ln x$ and $y = 4$ in the same viewing window, as shown in Figure 5.19.

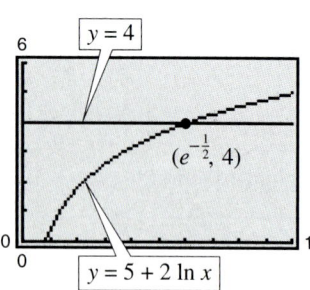

FIGURE 5.19

Example 8 ▶ Solving a Logarithmic Equation

Solve $2 \log_5 3x = 4$.

Solution

$2 \log_5 3x = 4$	Write original equation.
$\log_5 3x = 2$	Divide each side by 2.
$5^{\log_5 3x} = 5^2$	Exponentiate each side (base 5).
$3x = 25$	Inverse Property
$x = \dfrac{25}{3}$	Divide each side by 3.

The solution is $\frac{25}{3}$. Check this in the original equation. Or try performing a graphical check by graphing the functions

$$y = 2 \log_5 3x \quad \text{and} \quad y = 4$$

on the same screen. The two graphs should intersect when $x = \frac{25}{3}$ and $y = 4$.

Because the domain of a logarithmic function generally does not include all real numbers, you should be sure to check for extraneous solutions of logarithmic equations.

> **STUDY TIP**
>
> In Example 9 the domain of $\log_{10} 5x$ is $x > 0$ and the domain of $\log_{10}(x - 1)$ is $x > 1$, so the domain of the original equation is $x > 1$. Because the domain is all real numbers greater than 1, the solution $x = -4$ is extraneous.

Example 9 ▶ Checking for Extraneous Solutions

Solve $\log_{10} 5x + \log_{10}(x - 1) = 2$.

Solution

$\log_{10} 5x + \log_{10}(x - 1) = 2$	Write original equation.
$\log_{10}[5x(x - 1)] = 2$	Product Property of Logarithms
$10^{\log_{10}(5x^2 - 5x)} = 10^2$	Exponentiate each side (base 10).
$5x^2 - 5x = 100$	Inverse Property
$x^2 - x - 20 = 0$	Write in general form.
$(x - 5)(x + 4) = 0$	Factor.
$x - 5 = 0$	Set 1st factor equal to 0.
$x = 5$	Solution
$x + 4 = 0$	Set 2nd factor equal to 0.
$x = -4$	Solution

The solutions appear to be 5 and -4. However, when you check these in the original equation or use a graphical check, you can see that $x = 5$ is the only solution.

Applications

Example 10 ▶ Doubling an Investment

You have deposited $500 in an account that pays 6.75% interest, compounded continuously. How long will it take your money to double?

Solution

Using the formula for continuous compounding, you can find that the balance in the account is

$$A = Pe^{rt}$$
$$A = 500e^{0.0675t}.$$

To find the time required for the balance to double, let $A = 1000$ and solve the resulting equation for t.

$500e^{0.0675t} = 1000$	Let $A = 1000$.
$e^{0.0675t} = 2$	Divide each side by 500.
$\ln e^{0.0675t} = \ln 2$	Take natural log of each side.
$0.0675t = \ln 2$	Inverse Property
$t = \dfrac{\ln 2}{0.0675}$	Divide each side by 0.0675.
$t \approx 10.27$	Use a calculator.

The balance in the account will double after approximately 10.27 years. This result is demonstrated graphically in Figure 5.20.

Doubling an Investment

FIGURE 5.20

◀ Exploration ▶

The *effective yield* of a savings plan is the percent increase in the balance after 1 year. Find the effective yields for the following savings plans when $1000 is deposited in a savings account.

a. 7% annual interest rate, compounded annually

b. 7% annual interest rate, compounded continuously

c. 7% annual interest rate, compounded quarterly

d. 7.25% annual interest rate, compounded quarterly

Which savings plan has the greatest effective yield? Which savings plan will have the highest balance after 5 years?

In Example 10, an approximate answer of 10.27 years is given. The exact solution, $(\ln 2)/0.0675$ years, does not make sense as an answer.

Example 11 ▶ Consumer Price Index for Sugar

From 1970 to 1997, the Consumer Price Index (CPI) value y for a fixed amount of sugar for the year t can be modeled by the equation

$$y = -171.8 + 87.1 \ln t$$

where $t = 10$ represents 1970 (see Figure 5.21). During which year did the price of sugar reach 4.5 times its 1970 price of 30.5 on the CPI? *(Source: U.S. Bureau of Labor Statistics)*

Solution

$-171.8 + 87.1 \ln t = y$	Write original equation.
$-171.8 + 87.1 \ln t = 137.25$	Let $y = (4.5)(30.5) = 137.25$.
$87.1 \ln t = 309.05$	Add 171.8 to each side.
$\ln t \approx 3.548$	Divide each side by 87.1.
$e^{\ln t} \approx e^{3.548}$	Exponentiate each side.
$t \approx e^{3.548}$	Inverse Property
$t \approx 35$	Use a calculator.

The solution is $t \approx 35$ years. Because $t = 10$ represents 1970, it follows that the price of sugar reached 4.5 times its 1970 price in 1995.

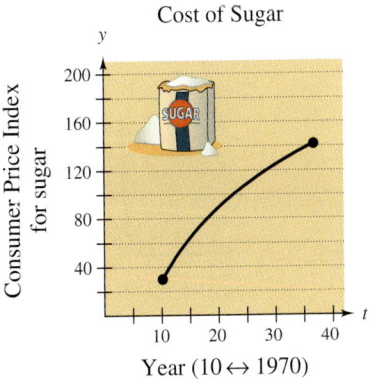

FIGURE 5.21

Writing ABOUT MATHEMATICS

Comparing Mathematical Models The table gives the numbers y (in millions) of single compact discs (CDs) shipped annually by manufacturers from 1993 through 1997, where $x = 3$ represents 1993. *(Source: Recording Industry Association of America)*

x	3	4	5	6	7
y	7.8	9.3	21.5	43.2	66.7

a. Create a scatter plot of the data. Find a linear model for the data, and add its graph to your scatter plot. According to this model, when will the annual shipment of single CDs reach 100 million?

b. Create a new table giving values for $\ln x$ and $\ln y$ and create a scatter plot of this transformed data. Use the method illustrated in Example 7 in Section 5.3 to find a model for the transformed data, and add its graph to your scatter plot. According to this model, when will the annual shipment of single CDs reach 100 million?

c. Solve the model in part (b) for y, and add its graph to your scatter plot in part (a). Which model better fits the original data? Which model will better predict future shipments? Explain.

5.4 Exercises

In Exercises 1–6, determine whether the x-values are solutions or approximate solutions of the equation.

1. $4^{2x-7} = 64$
   (a) $x = 5$
   (b) $x = 2$

2. $2^{3x+1} = 32$
   (a) $x = -1$
   (b) $x = 2$

3. $3e^{x+2} = 75$
   (a) $x = -2 + e^{25}$
   (b) $x = -2 + \ln 25$
   (c) $x \approx 1.219$

4. $5^{2x+3} = 812$
   (a) $x = -1.5 + \log_5 \sqrt{812}$
   (b) $x \approx 0.581$
   (c) $x = \dfrac{1}{2}\left(-3 + \dfrac{\ln 812}{\ln 5}\right)$

5. $\log_4(3x) = 3$
   (a) $x \approx 20.356$
   (b) $x = -4$
   (c) $x = \dfrac{64}{3}$

6. $\ln(x - 1) = 3.8$
   (a) $x = 1 + e^{3.8}$
   (b) $x \approx 45.701$
   (c) $x = 1 + \ln 3.8$

In Exercises 7–30, solve for x.

7. $4^x = 16$
8. $3^x = 243$
9. $5^x = 625$
10. $3^x = 729$
11. $7^x = \dfrac{1}{49}$
12. $8^x = 4$
13. $\left(\dfrac{1}{2}\right)^x = 32$
14. $\left(\dfrac{1}{4}\right)^x = 64$
15. $\left(\dfrac{3}{4}\right)^x = \dfrac{27}{64}$
16. $\left(\dfrac{2}{3}\right)^x = \dfrac{4}{9}$
17. $3^{x-1} = 27$
18. $2^{x-3} = 32$
19. $\ln x - \ln 2 = 0$
20. $\ln x - \ln 5 = 0$
21. $e^x = 2$
22. $e^x = 4$
23. $\ln x = -1$
24. $\ln x = -7$
25. $\log_4 x = 3$
26. $\log_x 625 = 4$
27. $\log_{10} x - 2 = 0$
28. $\log_{10} x + 3 = 0$
29. $\log_{10} x = -1$
30. $\ln(2x - 1) = 0$

In Exercises 31–34, approximate the point of intersection of the graphs of f and g. Then solve the equation $f(x) = g(x)$ algebraically.

31. $f(x) = 2^x$
    $g(x) = 8$

32. $f(x) = 27^x$
    $g(x) = 9$

33. $f(x) = \log_3 x$
    $g(x) = 2$

34. $f(x) = \ln(x - 4)$
    $g(x) = 0$

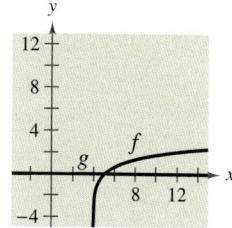

In Exercises 35–44, apply the inverse properties of ln x and e^x to simplify the expression.

35. $\log_{10} 10^{x^2}$
36. $\log_6 6^{2x-1}$
37. $8^{\log_8(x-2)}$
38. $4^{\log_4 x^3}$
39. $\ln e^{7x+2}$
40. $\ln e^{x^4}$
41. $e^{\ln(5x+2)}$
42. $e^{\ln x^2}$
43. $-1 + \ln e^{2x}$
44. $-8 + e^{\ln x^3}$

In Exercises 45–56, solve the exponential equation with base e algebraically. Approximate the result to three decimal places.

45. $e^x = 10$
46. $4e^x = 91$
47. $7 - 2e^x = 5$
48. $-14 + 3e^x = 11$
49. $e^{3x} = 12$
50. $e^{2x} = 50$
51. $500e^{-x} = 300$
52. $1000e^{-4x} = 75$
53. $e^{2x} - 4e^x - 5 = 0$
54. $e^{2x} - 5e^x + 6 = 0$
55. $20(100 - e^{x/2}) = 500$
56. $\dfrac{400}{1 + e^{-x}} = 350$

Section 5.4 Exponential and Logarithmic Equations

In Exercises 57–64, solve the exponential equation with base *a* algebraically. Approximate the result to three decimal places.

57. $10^x = 42$
58. $10^x = 570$
59. $3^{2x} = 80$
60. $6^{5x} = 3000$
61. $5^{-t/2} = 0.20$
62. $4^{-3t} = 0.10$
63. $2^{3-x} = 565$
64. $8^{-2-x} = 431$

In Exercises 65–72, use a graphing utility to graph the function. Approximate its zero to three decimal places.

65. $g(x) = 6e^{1-x} - 25$
66. $f(x) = -4e^{-x-1} + 15$
67. $f(x) = 3e^{3x/2} - 962$
68. $g(x) = 8e^{-2x/3} - 11$
69. $g(t) = e^{0.09t} - 3$
70. $f(x) = -e^{1.8x} + 7$
71. $h(t) = e^{0.125t} - 8$
72. $f(x) = e^{2.724x} - 29$

In Exercises 73–82, solve the exponential equation. Approximate the result to three decimal places.

73. $8(10^{3x}) = 12$
74. $5(10^{x-6}) = 7$
75. $3(5^{x-1}) = 21$
76. $8(3^{6-x}) = 40$
77. $\left(1 + \dfrac{0.065}{365}\right)^{365t} = 4$
78. $\left(4 - \dfrac{2.471}{40}\right)^{9t} = 21$
79. $\left(1 + \dfrac{0.10}{12}\right)^{12t} = 2$
80. $\left(16 - \dfrac{0.878}{26}\right)^{3t} = 30$
81. $\dfrac{3000}{2 + e^{2x}} = 2$
82. $\dfrac{119}{e^{6x} - 14} = 7$

In Exercises 83–96, solve the natural logarithmic equation algebraically. Approximate the result to three decimal places.

83. $\ln x = -3$
84. $\ln x = 2$
85. $\ln 2x = 2.4$
86. $\ln 4x = 1$
87. $3 \ln 5x = 10$
88. $2 \ln x = 7$
89. $\ln \sqrt{x + 2} = 1$
90. $\ln \sqrt{x - 8} = 5$
91. $\ln(x + 1)^2 = 2$
92. $\ln x + \ln(x + 1) = 1$
93. $\ln x + \ln(x - 2) = 1$
94. $\ln x + \ln(x + 3) = 1$
95. $\ln(x + 5) = \ln(x - 1) - \ln(x + 1)$
96. $\ln(x + 1) - \ln(x - 2) = \ln x^2$

In Exercises 97–106, solve the logarithmic equation algebraically. Approximate the result to three decimal places.

97. $\log_{10}(z - 3) = 2$
98. $\log_{10} x^2 = 6$
99. $6 \log_3(0.5x) = 11$
100. $5 \log_{10}(x - 2) = 11$
101. $\log_{10}(x + 4) - \log_{10} x = \log_{10}(x + 2)$
102. $\log_2 x + \log_2(x + 2) = \log_2(x + 6)$
103. $\log_4 x - \log_4(x - 1) = \dfrac{1}{2}$
104. $\log_3 x + \log_3(x - 8) = 2$
105. $\log_{10} 8x - \log_{10}(1 + \sqrt{x}) = 2$
106. $\log_{10} 4x - \log_{10}(12 + \sqrt{x}) = 2$

In Exercises 107–110, use a graphing utility to approximate the point of intersection of the graphs. Approximate the result to three decimal places.

107. $y_1 = 7$
 $y_2 = 2^x$
108. $y_1 = 500$
 $y_2 = 1500e^{-x/2}$
109. $y_1 = 3$
 $y_2 = \ln x$
110. $y_1 = 10$
 $y_2 = 4 \ln(x - 2)$

Finance In Exercises 111 and 112, find the time required for a $1000 investment to double at interest rate *r*, compounded continuously.

111. $r = 0.085$
112. $r = 0.12$

Finance In Exercises 113 and 114, find the time required for a $1000 investment to triple at interest rate *r*, compounded continuously.

113. $r = 0.085$
114. $r = 0.12$

115. *Economics* The demand equation for a certain product is
$$p = 500 - 0.5(e^{0.004x}).$$
Find the demand *x* for a price of (a) $p = \$350$ and (b) $p = \$300$.

116. *Economics* The demand equation for a certain product is
$$p = 5000\left(1 - \dfrac{4}{4 + e^{-0.002x}}\right).$$
Find the demand *x* for a price of (a) $p = \$600$ and (b) $p = \$400$.

117. *Forest Yield* The yield V (in millions of cubic feet per acre) for a forest at age t years is

$$V = 6.7e^{-48.1/t}.$$

(a) Use a graphing utility to graph the function.

(b) Determine the horizontal asymptote of the function. Interpret its meaning in the context of the problem.

(c) Find the time necessary to obtain a yield of 1.3 million cubic feet.

118. *Trees per Acre* The number of trees per acre N of a certain species is approximated by the model

$$N = 68(10^{-0.04x}), \quad 5 \leq x \leq 40$$

where x is the average diameter of the trees 3 feet above the ground. Use the model to approximate the average diameter of the trees in a test plot when $N = 21$.

119. *Average Heights* The percent of American males between the ages of 18 and 24 who are no more than x inches tall is

$$m(x) = \frac{100}{1 + e^{-0.6114(x - 69.71)}}.$$

The percent of American females between the ages of 18 and 24 who are no more than x inches tall is

$$f(x) = \frac{100}{1 + e^{-0.66607(x - 64.51)}}$$

where m and f are the percents and x is the height in inches. (Source: U.S. National Center for Health Statistics)

(a) Use the graph to determine any horizontal asymptotes of the functions. What do they mean?

(b) What is the average height of each sex?

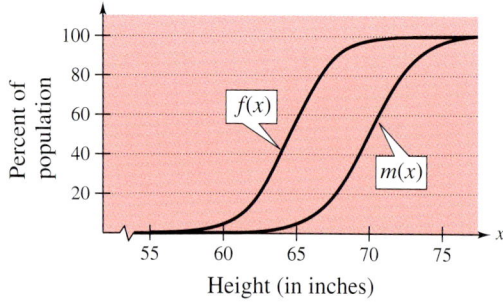

120. *Human Learning Model* In a group project in learning theory, a mathematical model for the proportion P of correct responses after n trials was found to be

$$P = \frac{0.83}{1 + e^{-0.2n}}.$$

(a) Use a graphing utility to graph the function.

(b) Use the graph to determine any horizontal asymptotes of the function. Interpret the meaning of the upper asymptote in the context of this problem.

(c) After how many trials will 60% of the responses be correct?

121. *Data Analysis* An object at a temperature of 160°C was removed from a furnace and placed in a room at 20°C. The temperature T of the object was measured each hour h and recorded in the table.

h	0	1	2	3	4	5
T	160°	90°	56°	38°	29°	24°

A model for this data is

$$T = 20[1 + 7(2^{-h})].$$

(a) The graph of this model is shown in the figure. Use the graph to identify the horizontal asymptote of the model and interpret the asymptote in the context of the problem.

(b) Use the model to approximate the time when the temperature of the object was 100°C.

122. *Automobiles* Automobiles are designed with crumple zones that help protect their occupants in crashes. The crumple zones allow the occupants to move short distances when the automobiles come to abrupt stops. The greater the distance moved, the fewer g's the crash victims experience. (One g is equal to the acceleration due to gravity. For very short periods of time, humans have withstood as much as 40 g's.) In crash tests with vehicles moving at 90 kilometers per hour, analysts measured the numbers y of g's experienced during deceleration by crash dummies that were permitted to move x meters during impact. The data is shown in the table.

x	0.2	0.4	0.6	0.8	1
y	158	80	53	40	32

A model for this data is

$$y = -3.00 + 11.88 \ln x + \frac{36.94}{x}.$$

(a) Use a graphing utility to graph the data points and the model in the same viewing window. How do they compare?

(b) Use the model to estimate the distance traveled during impact if the passenger deceleration must not exceed 30 g's.

(c) Do you think it is practical to lower the number of g's experienced during impact to fewer than 23? Explain your reasoning.

Synthesis

True or False? In Exercises 123–126, rewrite each verbal statement as an equation. Then decide whether the statement is true or false. Justify your answer.

123. The logarithm of the product of two numbers is equal to the sum of the logarithms of the numbers.

124. The logarithm of the sum of two numbers is equal to the product of the logarithms of the numbers.

125. The logarithm of the difference of two numbers is equal to the difference of the logarithms of the numbers.

126. The logarithm of the quotient of two numbers is equal to the difference of the logarithms of the numbers.

127. *Finance* You are investing P dollars at an annual interest rate of r, compounded continuously, for t years. Which of the following would result in the highest value of the investment? Explain your reasoning.

(a) Double the amount you invest.

(b) Double your interest rate.

(c) Double the number of years.

128. *Think About It* Are the times required for the investments in Exercises 111 and 112 to quadruple twice as long as the times for them to double? Give a reason for your answer and verify your answer algebraically.

129. *Writing* Write a paragraph explaining whether or not the time required for an investment to double is dependent on the size of the investment.

Review

In Exercises 130–133, simplify the expression.

130. $\sqrt{48x^2y^5}$

131. $\sqrt{32} - 2\sqrt{25}$

132. $\sqrt[3]{25} \cdot \sqrt[3]{15}$

133. $\dfrac{3}{\sqrt{10} - 2}$

In Exercises 134–138, find a mathematical model for the verbal statement.

134. M varies directly as the cube of p.

135. t varies inversely as the cube of s.

136. d varies jointly as a and b.

137. x is inversely proportional to $b - 3$.

138. c varies inversely as the square root of w.

In Exercises 139–142, evaluate the logarithm using the change-of-base formula. Approximate your result to three decimal places.

139. $\log_6 9$

140. $\log_3 4$

141. $\log_{3/4} 5$

142. $\log_8 22$

5.5 Exponential and Logarithmic Models

▶ What you should learn

- How to recognize the five most common types of models involving exponential and logarithmic functions
- How to use exponential growth and decay functions to model and solve real-life problems
- How to use Gaussian functions to model and solve real-life problems
- How to use logistic growth functions to model and solve real-life problems
- How to use logarithmic functions to model and solve real-life problems

▶ Why you should learn it

Real-life applications of mathematics often involve deciding which mathematical model to use for a given situation. For instance, Exercise 45 on page 393 compares an exponential decay model and a linear model for the depreciation of a computer over 3 years.

Introduction

The five most common types of mathematical models involving exponential functions and logarithmic functions are as follows.

1. **Exponential growth model:** $y = ae^{bx}, \quad b > 0$
2. **Exponential decay model:** $y = ae^{-bx}, \quad b > 0$
3. **Gaussian model:** $y = ae^{-(x-b)^2/c}$
4. **Logistic growth model:** $y = \dfrac{a}{1 + be^{-rx}}$
5. **Logarithmic models:** $y = a + b \ln x, \quad y = a + b \log_{10} x$

The graphs of the basic forms of these functions are shown in Figure 5.22.

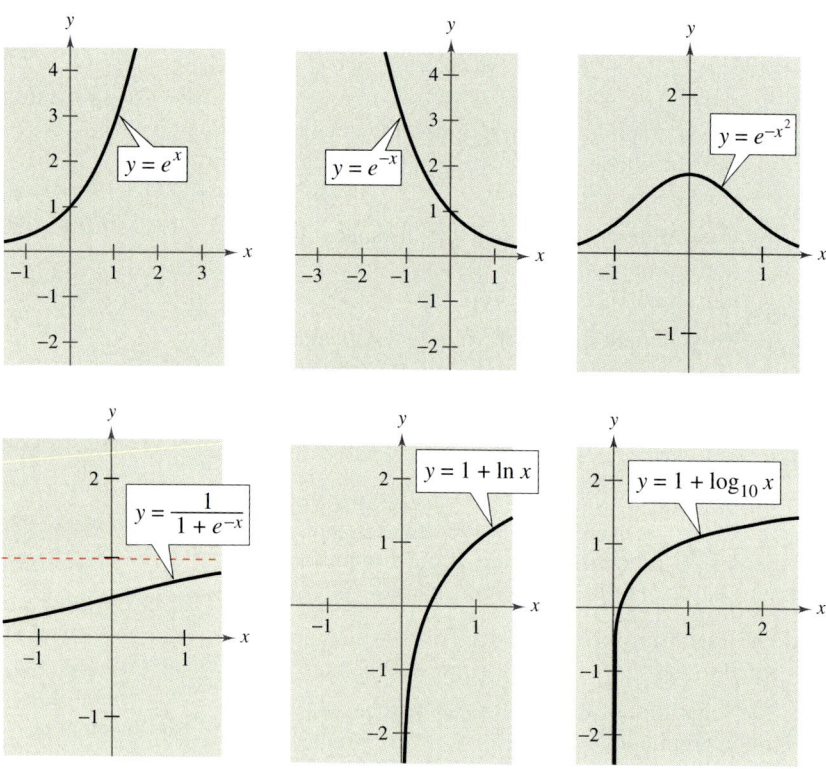

FIGURE 5.22

You can often gain quite a bit of insight into a situation modeled by an exponential or logarithmic function by identifying and interpreting the function's asymptotes. Use the graphs in Figure 5.22 to identify the asymptotes of each function.

Exponential Growth and Decay

Example 1 ▸ Population Increase

Estimates of the world population (in millions) from 1992 through 2000 are shown in the table. The scatter plot of the data is shown in Figure 5.23. (Source: U.S. Bureau of the Census, International Data Base)

Year	1992	1993	1994	1995	1996	1997	1998	1999	2000
Population	5445	5527	5607	5688	5767	5847	5926	6005	6083

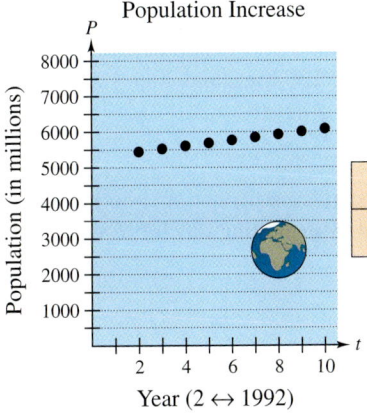

FIGURE 5.23

An exponential growth model that approximates this data is

$$P = 5304e^{0.013819t}, \quad 2 \le t \le 10$$

where P is the population (in millions) and $t = 2$ represents 1992. Compare the values given by the model with the estimates given by the U.S. Bureau of the Census. According to this model, when will the world population reach 6.5 billion?

Solution

The following table compares the two sets of population figures. The graph of the model is shown in Figure 5.24.

Year	1992	1993	1994	1995	1996	1997	1998	1999	2000
Population	5445	5527	5607	5688	5767	5847	5926	6005	6083
Model	5453	5529	5605	5683	5763	5843	5924	6006	6090

FIGURE 5.24

To find when the world population will reach 6.5 billion, let $P = 6500$ in the model and solve for t.

$5304e^{0.013819t} = P$ Write original model.

$5304e^{0.013819t} = 6500$ Let $P = 6500$.

$e^{0.013819t} \approx 1.22549$ Divide each side by 5304.

$\ln e^{0.013819t} \approx \ln 1.22549$ Take natural log of each side.

$0.013819t \approx 0.203341$ Inverse Property

$t \approx 14.71$ Divide each side by 0.013819.

According to the model, the world population will reach 6.5 billion in 2004.

An exponential model increases (or decreases) by the same percent each year. What is the annual percent increase for the model in Example 1?

Technology

Some graphing utilities have curve-fitting capabilities that can be used to find models that represent data. If you have such a graphing utility, try using it to find a model for the data given in Example 1. How does your model compare with the model given in Example 1?

In Example 1, you were given the exponential growth model. But suppose this model were not given; how could you find such a model? One technique for doing this is demonstrated in Example 2.

Example 2 ▶ Modeling Population Growth

In a research experiment, a population of fruit flies is increasing according to the law of exponential growth. After 2 days there are 100 flies, and after 4 days there are 300 flies. How many flies will there be after 5 days?

Solution

Let y be the number of flies at time t. From the given information, you know that $y = 100$ when $t = 2$ and $y = 300$ when $t = 4$. Substituting this information into the model $y = ae^{bt}$ produces

$$100 = ae^{2b} \quad \text{and} \quad 300 = ae^{4b}.$$

To solve for b, solve for a in the first equation.

$$100 = ae^{2b} \implies a = \frac{100}{e^{2b}} \qquad \text{Solve for } a \text{ in the first equation.}$$

Then substitute the result into the second equation.

$$300 = ae^{4b} \qquad \text{Write second equation.}$$

$$300 = \left(\frac{100}{e^{2b}}\right)e^{4b} \qquad \text{Substitute } 100/e^{2b} \text{ for } a.$$

$$\frac{300}{100} = e^{2b} \qquad \text{Divide each side by 100.}$$

$$\ln 3 = 2b \qquad \text{Take natural log of each side.}$$

$$\frac{1}{2}\ln 3 = b \qquad \text{Solve for } b.$$

Using $b = \frac{1}{2}\ln 3$ and the equation you found for a, you can determine that

$$a = \frac{100}{e^{2[(1/2)\ln 3]}} \qquad \text{Substitute } (1/2)\ln 3 \text{ for } b.$$

$$= \frac{100}{e^{\ln 3}} \qquad \text{Simplify.}$$

$$= \frac{100}{3} \qquad \text{Inverse Property}$$

$$\approx 33. \qquad \text{Simplify.}$$

So, with $a \approx 33$ and $b = \frac{1}{2}\ln 3 \approx 0.5493$, the exponential growth model is

$$y = 33e^{0.5493t}$$

as shown in Figure 5.25. This implies that, after 5 days, the population is

$$y = 33e^{0.5493(5)} \approx 514 \text{ flies}.$$

FIGURE 5.25

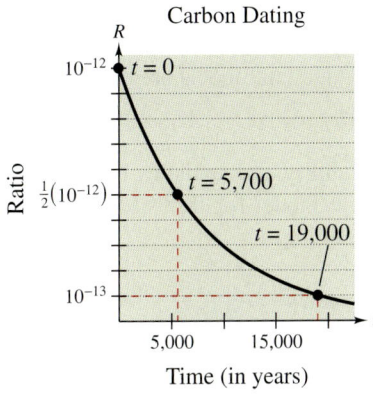

FIGURE 5.26

In living organic material, the ratio of the number of radioactive carbon isotopes (carbon 14) to the number of nonradioactive carbon isotopes (carbon 12) is about 1 to 10^{12}. When organic material dies, its carbon 12 content remains fixed, whereas its radioactive carbon 14 begins to decay with a half-life of about 5700 years. To estimate the age of dead organic material, scientists use the following formula, which denotes the ratio of carbon 14 to carbon 12 present at any time t (in years).

$$R = \frac{1}{10^{12}} e^{-t/8223} \qquad \text{Carbon dating model}$$

The graph of R is shown in Figure 5.26. Note that R decreases as t increases.

Example 3 ▶ Carbon Dating

The ratio of carbon 14 to carbon 12 in a newly discovered fossil is

$$R = \frac{1}{10^{13}}.$$

Estimate the age of the fossil.

Solution

In the carbon dating model, substitute the given value of R to obtain the following.

$$\frac{1}{10^{12}} e^{-t/8223} = R \qquad \text{Write original model.}$$

$$\frac{e^{-t/8223}}{10^{12}} = \frac{1}{10^{13}} \qquad \text{Let } R = \frac{1}{10^{13}}.$$

$$e^{-t/8223} = \frac{1}{10} \qquad \text{Multiply each side by } 10^{12}.$$

$$\ln e^{-t/8223} = \ln \frac{1}{10} \qquad \text{Take natural log of each side.}$$

$$-\frac{t}{8223} \approx -2.3026 \qquad \text{Inverse Property}$$

$$t \approx 18{,}934 \qquad \text{Multiply each side by } -8223.$$

So, to the nearest thousand years, you can estimate the age of the fossil to be 19,000 years.

The carbon dating model in Example 3 assumed that the carbon 14/carbon 12 ratio was one part in 10,000,000,000,000. Suppose an error in measurement occurred and the actual ratio was only one part in 8,000,000,000,000. The fossil age corresponding to the actual ratio would then be approximately 17,000 years. Try checking this result.

> **STUDY TIP**
>
> To find how much of an initial 10 grams of ^{226}Ra isotope with a half-life of 1620 years is left after 500 years, you would use the exponential decay model.
>
> $$y = ae^{-bt}$$
>
> $$\tfrac{1}{2}(10) = 10e^{-b(1620)}$$
>
> $$b = -\ln\!\left(\tfrac{1}{2}\right)/1620$$
>
> Using the value of b found above and $a = 10$, the amount left is
>
> $$y = 10e^{-[-\ln(1/2)/1620](500)}$$
>
> $$y \approx 8.07 \text{ grams}.$$

Gaussian Models

As mentioned at the beginning of this section, Gaussian models are of the form

$$y = ae^{-(x-b)^2/c}.$$

This type of model is commonly used in probability and statistics to represent populations that are **normally distributed.** One model for this situation takes the form

$$y = \frac{1}{\sigma\sqrt{2\pi}} e^{-x^2/(2\sigma^2)}$$

where σ is the standard deviation (σ is the lowercase Greek letter sigma). The graph of a Gaussian model is called a **bell-shaped curve.** Try assigning a value to σ and sketching a normal distribution curve with a graphing utility. Can you see why it is called a bell-shaped curve?

The average value for a population can be found from the bell-shaped curve by observing where the maximum y-value of the function occurs. The x-value corresponding to the maximum y-value of the function represents the average value of the independent variable—in this case, x.

Example 4 ▶ SAT Scores

In 1997, the Scholastic Aptitude Test (SAT) math scores for college-bound seniors roughly followed a normal distribution

$$y = 0.0036 e^{-(x-511)^2/25,088}, \quad 200 \leq x \leq 800$$

where x is the SAT score for mathematics. Sketch the graph of this function. From the graph, estimate the average SAT score.   (Source: College Board)

Solution

The graph of the function is given in Figure 5.27. From the graph, you can see that the average mathematics score for college-bound seniors in 1997 was 511.

FIGURE 5.27

Logistic Growth Models

Some populations initially have rapid growth, followed by a declining rate of growth, as indicated by the graph in Figure 5.28. One model for describing this type of growth pattern is the **logistic curve** given by the function

$$y = \frac{a}{1 + be^{-rx}}$$

where y is the population size and x is the time. An example is a bacteria culture that is initially allowed to grow under ideal conditions, and then under less favorable conditions that inhibit growth. A logistic growth curve is also called a **sigmoidal curve.**

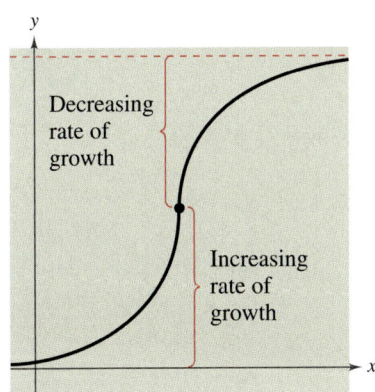

FIGURE 5.28

Example 5 ▶ Spread of a Virus

On a college campus of 5000 students, one student returns from vacation with a contagious and long-lasting flu virus. The spread of the virus is modeled by

$$y = \frac{5000}{1 + 4999e^{-0.8t}}, \quad 0 \leq t$$

where y is the total number of students infected after t days. The college will cancel classes when 40% or more of the students are infected.

a. How many students are infected after 5 days?

b. After how many days will the college cancel classes?

Solution

a. After 5 days, the number of students infected is

$$y = \frac{5000}{1 + 4999e^{-0.8(5)}} = \frac{5000}{1 + 4999e^{-4}} \approx 54.$$

b. Classes are canceled when the number infected is $(0.40)(5000) = 2000$.

$$2000 = \frac{5000}{1 + 4999e^{-0.8t}}$$

$$1 + 4999e^{-0.8t} = 2.5$$

$$e^{-0.8t} \approx \frac{1.5}{4999}$$

$$\ln e^{-0.8t} \approx \ln \frac{1.5}{4999}$$

$$-0.8t \approx -\ln \frac{1.5}{4999}$$

$$t = -\frac{1}{0.8} \ln \frac{1.5}{4999}$$

$$t \approx 10.1$$

So, after 10 days, at least 40% of the students will be infected, and classes will be canceled. The graph of the function is shown in Figure 5.29.

FIGURE 5.29

Twenty seconds of a 7.2 magnitude earthquake in Kobe, Japan on January 17, 1995 left damage approaching $60 billion.

t	Year	Population
1	1810	7.23
3	1830	12.87
5	1850	23.19
7	1870	39.82
9	1890	62.95
11	1910	91.97
13	1930	122.78
15	1950	151.33
17	1970	203.30
19	1990	250.00

Logarithmic Models

Example 6 ▶ Magnitude of Earthquakes

On the Richter scale, the magnitude R of an earthquake of intensity I is

$$R = \log_{10} \frac{I}{I_0}$$

where $I_0 = 1$ is the minimum intensity used for comparison. Find the intensities per unit of area for the following earthquakes. (Intensity is a measure of the wave energy of an earthquake.)

a. Tokyo and Yokohama, Japan in 1923: $R = 8.3$.
b. Kobe, Japan in 1995: $R = 7.2$.

Solution

a. Because $I_0 = 1$ and $R = 8.3$, you have

$$8.3 = \log_{10} \frac{I}{1} \qquad \text{Substitute 1 for } I_0 \text{ and 8.3 for } R.$$

$$10^{8.3} = 10^{\log_{10} I} \qquad \text{Exponentiate each side.}$$

$$I = 10^{8.3} \approx 199{,}526{,}000. \qquad \text{Inverse property of exponents and logs}$$

b. For $R = 7.2$, you have

$$7.2 = \log_{10} \frac{I}{1} \qquad \text{Substitute 1 for } I_0 \text{ and 7.2 for } R.$$

$$10^{7.2} = 10^{\log_{10} I} \qquad \text{Exponentiate each side.}$$

$$I = 10^{7.2} \approx 15{,}849{,}000. \qquad \text{Inverse property of exponents and logs}$$

Note that an increase of 1.1 units on the Richter scale (from 7.2 to 8.3) represents an increase in intensity by a factor of

$$\frac{199{,}526{,}000}{15{,}849{,}000} \approx 13.$$

In other words, the earthquake in 1923 had a magnitude about 13 times greater than that of the 1995 quake.

Writing ABOUT MATHEMATICS

Comparing Population Models The population (in millions) of the United States from 1810 to 1990 is given in the table. *(Source: U.S. Bureau of the Census)* Least squares regression analysis gives the best quadratic model for this data as $P = 0.6569t^2 + 0.305t + 6.12$ and the best exponential model for this data as $P = 8.325e^{0.195t}$. Which model better fits the data? Describe the method you used to reach your conclusion.

5.5 Exercises

In Exercises 1–6, match the function with its graph. [The graphs are labeled (a) through (f).]

(a)

(b)

(c)

(d)

(e)

(f)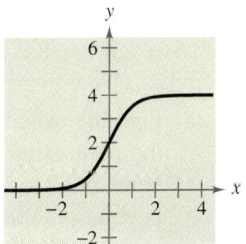

1. $y = 2e^{x/4}$
2. $y = 6e^{-x/4}$
3. $y = 6 + \log_{10}(x + 2)$
4. $y = 3e^{-(x-2)^2/5}$
5. $y = \ln(x + 1)$
6. $y = \dfrac{4}{1 + e^{-2x}}$

Finance In Exercises 7–14, complete the table for a savings account in which interest is compounded continuously.

Initial Investment	Annual % Rate	Time to Double	Amount after 10 Years
7. $1000	12%		
8. $20,000	$10\frac{1}{2}\%$		
9. $750		$7\frac{3}{4}$ yr	
10. $10,000		12 yr	
11. $500			$1505.00
12. $600			$19,205.00
13.	4.5%		$10,000.00
14.	8%		$20,000.00

Finance In Exercises 15 and 16, determine the principal P that must be invested at rate r, compounded monthly, so that $500,000 will be available for retirement in t years.

15. $r = 7\frac{1}{2}\%$, $t = 20$
16. $r = 12\%$, $t = 40$

Finance In Exercises 17 and 18, determine the time necessary for $1000 to double if it is invested at interest rate r compounded (a) annually, (b) monthly, (c) daily, and (d) continuously.

17. $r = 11\%$
18. $r = 10\frac{1}{2}\%$

19. **Finance** Complete the table for the time t necessary for P dollars to triple if interest is compounded continuously at rate r.

r	2%	4%	6%	8%	10%	12%
t						

20. **Modeling Data** Draw a scatter plot of the data in Exercise 19. Use the curve-fitting capabilities of a graphing utility to find a model for the data.

21. **Finance** Complete the table for the time t necessary for P dollars to triple if interest is compounded annually at rate r.

r	2%	4%	6%	8%	10%	12%
t						

22. **Modeling Data** Draw a scatter plot of the data in Exercise 21. Use the curve-fitting capabilities of a graphing utility to find a model for the data.

23. **Finance** If $1 is invested in an account over a 10-year period, the amount in the account, where t represents the time in years, is

$$A = 1 + 0.075[\![t]\!] \quad \text{or} \quad A = e^{0.07t}$$

depending on whether the account pays simple interest at $7\frac{1}{2}\%$ or continuous compound interest at 7%. Graph each function on the same set of axes. Which grows at the faster rate? (Remember that $[\![t]\!]$ is the greatest integer function discussed in Section P.6.)

24. *Finance* If $1 is invested in an account over a 10-year period, the amount in the account, where t represents the time in years, is

$$A = 1 + 0.06[\![t]\!] \quad \text{or} \quad A = \left(1 + \frac{0.055}{365}\right)^{[\![365t]\!]}$$

depending on whether the account pays simple interest at 6% or compound interest at $5\frac{1}{2}$% compounded daily. Use a graphing utility to graph each function in the same viewing window. Which grows at the faster rate?

In Exercises 25–30, complete the table for the radioactive isotope.

	Isotope	Half-life (years)	Initial Quantity	Amount after 1000 Years
25.	^{226}Ra	1620	10 g	
26.	^{226}Ra	1620		1.5 g
27.	^{14}C	5730		2 g
28.	^{14}C	5730	3 g	
29.	^{239}Pu	24,360		2.1 g
30.	^{239}Pu	24,360		0.4 g

In Exercises 31–34, find the exponential model $y = ae^{bx}$ that fits the points in the graph or table.

31.

32.
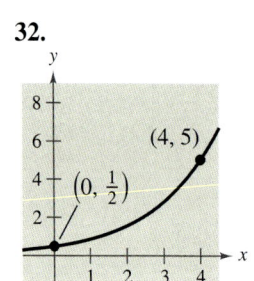

33.
x	0	4
y	5	1

34.
x	0	3
y	1	$\frac{1}{4}$

35. *Population* The population P of a city is

$$P = 105,300e^{0.015t}$$

where $t = 0$ represents the year 2000. According to this model, when will the population reach 150,000?

36. *Population* The population P of a city is

$$P = 240,360e^{0.012t}$$

where $t = 0$ represents the year 2000. According to this model, when will the population reach 275,000?

37. *Population* The population P of a city is

$$P = 2500e^{kt}$$

where $t = 0$ represents the year 2000. In 1945, the population was 1350. Find the value of k, and use this result to predict the population in the year 2010.

38. *Population* The population P of a city is

$$P = 140,500e^{kt}$$

where $t = 0$ represents the year 2000. In 1960, the population was 100,250. Find the value of k, and use this result to predict the population in the year 2020.

39. *Population* The table gives the population (in millions) of a country in 1997 and the projected population (in millions) for the year 2020. (Source: U.S. Bureau of the Census, International Data Base)

Country	1997	2020
Croatia	5.0	4.8
Mali	9.9	20.4
Singapore	3.5	4.3
Sweden	8.9	9.5

(a) Find the exponential growth model $y = ae^{bt}$ for the population in each country by letting $t = 0$ correspond to 1997. Use the model to predict the population of each country in 2030.

(b) You can see that the populations of Mali and Sweden are growing at different rates. What constant in the equation $y = ae^{bt}$ is determined by these different growth rates? Discuss the relationship between the different growth rates and the magnitude of the constant.

(c) You can see that the population of Singapore is increasing while the population of Croatia is decreasing. What constant in the equation $y = ae^{bt}$ reflects this difference? Explain.

40. *Bacteria Growth* The number of bacteria N in a culture is modeled by

$$N = 100e^{kt}$$

where t is the time in hours. If $N = 300$ when $t = 5$, estimate the time required for the population to double in size.

41. *Bacteria Growth* The number of bacteria N in a culture is modeled by

$$N = 250e^{kt}$$

where t is the time in hours. If $N = 280$ when $t = 10$, estimate the time required for the population to double in size.

42. *Radioactive Decay* The half-life of radioactive radium (^{226}Ra) is 1620 years. What percent of a present amount of radioactive radium will remain after 100 years?

43. *Radioactive Decay* Carbon 14 dating assumes that the carbon dioxide on earth today has the same radioactive content as it did centuries ago. If this is true, the amount of ^{14}C absorbed by a tree that grew several centuries ago should be the same as the amount of ^{14}C absorbed by a tree growing today. A piece of ancient charcoal contains only 15% as much radioactive carbon as a piece of modern charcoal. How long ago was the tree burned to make the ancient charcoal if the half-life of ^{14}C is 5730 years?

44. *Depreciation* A car that cost $22,000 new has a book value of $13,000 after 2 years.
(a) Find the straight-line model $V = mt + b$.
(b) Find the exponential model $V = ae^{kt}$.
(c) Use a graphing utility to graph the two models in the same viewing window. Which model depreciates faster in the first 2 years?
(d) Find the book values of the car after 1 year and after 3 years using each model.
(e) Interpret the slope of the straight-line model.

45. *Depreciation* A computer that costs $2000 new has a book value of $500 after 2 years.
(a) Find the straight-line model $V = mt + b$.
(b) Find the exponential model $V = ae^{kt}$.
(c) Use a graphing utility to graph the two models in the same viewing window. Which model depreciates faster in the first 2 years?
(d) Find the book values of the computer after 1 year and after 3 years using each model.
(e) Interpret the slope of the straight-line model.

46. *Business* The sales S (in thousands of units) of a new product after it has been on the market t years are modeled by

$$S(t) = 100(1 - e^{kt}).$$

Fifteen thousand units of the new product were sold the first year.

(a) Complete the model by solving for k.
(b) Sketch the graph of the model.
(c) Use the model to estimate the number of units sold after 5 years.

47. *Business* After discontinuing all advertising for a certain product in 1998, the manufacturer noted that sales began to drop according to the model

$$S = \frac{500{,}000}{1 + 0.6e^{kt}}$$

where S represents the number of units sold and $t = 0$ represents 1998. In 2000, the company sold 300,000 units.

(a) Complete the model by solving for k.
(b) Estimate sales in 2003.

48. *Business* The sales S (in thousands of units) of a product after x hundred dollars is spent on advertising are modeled by

$$S = 10(1 - e^{kx}).$$

When $500 is spent on advertising, 2500 units are sold.

(a) Complete the model by solving for k.
(b) Estimate the number of units that will be sold if advertising expenditures are raised to $700.

49. *Business* Because of a slump in the economy, a company finds that its annual profits have dropped from $742,000 in 1998 to $632,000 in 2000. If the profit follows an exponential pattern of decline, what is the expected profit for 2001? (Let $t = 0$ represent 1998.)

50. *Learning Curve* The management at a factory has found that the maximum number of units a worker can produce in a day is 30. The learning curve for the number of units N produced per day after a new employee has worked t days is

$$N = 30(1 - e^{kt}).$$

After 20 days on the job, a new employee produces 19 units.

(a) Find the learning curve for this employee (first, find the value of k).
(b) How many days should pass before this employee is producing 25 units per day?
(c) Is the employee's production increasing at a linear rate? Explain your reasoning.

51. *Population Growth* A conservation organization releases 100 animals of an endangered species into a game preserve. The organization believes that the preserve has a carrying capacity of 1000 animals and that the growth of the herd will be modeled by the logistic curve

$$p(t) = \frac{1000}{1 + 9e^{-0.1656t}}$$

where t is measured in months (see figure).

(a) Use a graphing utility to graph the function. Use the graph to determine the horizontal asymptotes, and interpret the meaning of the larger p-value in the context of the problem.

(b) Estimate the population after 5 months.

(c) After how many months will the population be 500?

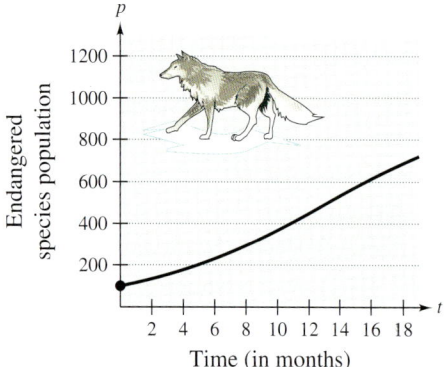

Geology In Exercises 52 and 53, use the Richter scale for measuring the magnitudes of earthquakes.

52. Find the magnitude R of an earthquake of intensity I (let $I_0 = 1$).

(a) $I = 80{,}500{,}000$

(b) $I = 48{,}275{,}000$

(c) $I = 251{,}200$

53. Find the intensity I of an earthquake measuring R on the Richter scale (let $I_0 = 1$).

(a) Chile in 1906, $R = 8.6$

(b) Los Angeles in 1971, $R = 6.7$

(c) Taiwan in 1999, $R = 7.7$

Physics In Exercises 54–57, use the following information for determining sound intensity. The level of sound β, in decibels, with an intensity of I is

$$\beta = 10 \log_{10} \frac{I}{I_0}$$

where I_0 is an intensity of 10^{-12} watt per square meter, corresponding roughly to the faintest sound that can be heard by the human ear. In Exercises 54 and 55, find the level of sound, β.

54. (a) $I = 10^{-10}$ watt per m² (faint whisper)

(b) $I = 10^{-5}$ watt per m² (busy street corner)

(c) $I = 10^{-2.5}$ watt per m² (air hammer)

(d) $I = 10^0$ watt per m² (threshold of pain)

55. (a) $I = 10^{-9}$ watt per m² (whisper)

(b) $I = 10^{-3.5}$ watt per m² (jet 4 miles from takeoff)

(c) $I = 10^{-3}$ watt per m² (diesel truck at 25 feet)

(d) $I = 10^{-0.5}$ watt per m² (auto horn at 3 feet)

56. Due to the installation of noise suppression materials, the noise level in an auditorium was reduced from 93 to 80 decibels. Find the percent decrease in the intensity level of the noise as a result of the installation of these materials.

57. Due to the installation of a muffler, the noise level of an engine was reduced from 88 to 72 decibels. Find the percent decrease in the intensity level of the noise as a result of the installation of the muffler.

Chemistry In Exercises 58–63, use the acidity model given by pH $= -\log_{10} [H^+]$, where acidity (pH) is a measure of the hydrogen ion concentration $[H^+]$ (measured in moles of hydrogen per liter) of a solution.

58. Find the pH if $[H^+] = 2.3 \times 10^{-5}$.

59. Find the pH if $[H^+] = 11.3 \times 10^{-6}$.

60. Compute $[H^+]$ for a solution in which pH $= 5.8$.

61. Compute $[H^+]$ for a solution in which pH $= 3.2$.

62. A certain fruit has a pH of 2.5 and an antacid tablet has a pH of 9.5. The hydrogen ion concentration of the fruit is how many times the concentration of the tablet?

63. If the pH of a solution is decreased by 1 unit, the hydrogen ion concentration is increased by what factor?

64. Finance A $120,000 home mortgage for 35 years at $7\frac{1}{2}\%$ has a monthly payment of $809.39. Part of the monthly payment goes for the interest charge on the unpaid balance, and the remainder of the payment is used to reduce the principal. The amount that goes for interest is

$$u = M - \left(M - \frac{Pr}{12}\right)\left(1 + \frac{r}{12}\right)^{12t}$$

and the amount that goes toward reduction of the principal is

$$v = \left(M - \frac{Pr}{12}\right)\left(1 + \frac{r}{12}\right)^{12t}.$$

In these formulas, P is the size of the mortgage, r is the interest rate, M is the monthly payment, and t is the time in years.

(a) Use a graphing utility to graph each function in the same viewing window. (The viewing window should show all 35 years of mortgage payments.)

(b) In the early years of the mortgage, the larger part of the monthly payment goes for what purpose? Approximate the time when the monthly payment is evenly divided between interest and principal reduction.

(c) Repeat parts (a) and (b) for a repayment period of 20 years ($M = \$966.71$). What can you conclude?

65. Finance The total interest u paid on a home mortgage of P dollars at interest rate r for t years is

$$u = P\left[\frac{rt}{1 - \left(\dfrac{1}{1 + r/12}\right)^{12t}} - 1\right].$$

Consider a $120,000 home mortgage at $7\frac{1}{2}\%$.

(a) Use a graphing utility to graph the total interest function.

(b) Approximate the length of the mortgage for which the total interest paid is the same as the size of the mortgage. Is it possible that some people are paying twice as much in interest charges as the size of the mortgage?

66. Data Analysis The table shows the time t (in seconds) required to attain a speed of s miles per hour from a standing start for a particular car.

s	30	40	50	60	70	80	90
t	3.4	5.0	7.0	9.3	12.0	15.8	20.0

Two models for this data are as follows.

$$t_1 = 40.757 + 0.556s - 15.817 \ln s$$
$$t_2 = 1.2259 + 0.0023s^2$$

(a) Use a graphing utility to fit a linear model t_3 and an exponential model t_4 to the data.

(b) Use a graphing utility to graph the data points and each model.

(c) Create a table comparing the data with estimates obtained from each model.

(d) Use the results of part (c) to find the sum of the absolute values of the differences between the data and estimated values given by each model. Based on the four sums, which model do you think better fits the data? Explain.

67. Forensics At 8:30 A.M., a coroner was called to the home of a person who had died during the night. In order to estimate the time of death, the coroner took the person's temperature twice. At 9:00 A.M. the temperature was 85.7°F, and at 9:30 A.M. the temperature was 82.8°F. From these two temperatures the coroner was able to determine that the time elapsed since death and the body temperature were related by the formula

$$t = -2.5 \ln \frac{T - 70}{98.6 - 70}$$

where t is the time in hours elapsed since the person died and T is the temperature (in degrees Fahrenheit) of the person's body. Assume that the person had a normal body temperature of 98.6°F at death, and that the room temperature was a constant 70°F. (This formula is derived from a general cooling principle called Newton's Law of Cooling.) Use the formula to estimate the time of death of the person.

Synthesis

True or False? In Exercises 68–70, determine whether the statement is true or false. Justify your answer.

68. The domain of a logistic growth function cannot be the set of real numbers.

69. A logistic growth function will always have an x-intercept.

70. The graph of $f(x) = \dfrac{4}{1 + 6e^{-2x}} + 5$ is the graph of $g(x) = \dfrac{4}{1 + 6e^{-2x}}$ shifted to the right 5 units.

71. Identify each model as linear, logarithmic, exponential, logistic, or none of the above. Explain your reasoning.

(a)

(b)

(c)

(d)

(e)

(f)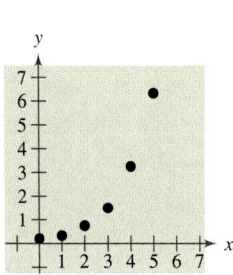

72. *Writing* Use your school's library or some other reference source to write a paper describing John Napier's work with logarithms.

73. *Writing* Before the development of electronic calculators and graphing utilities, some computations were done on slide rules. Use your school's library or some other reference source to write a paper describing the use of logarithmic scales on a slide rule.

Review

In Exercises 74–83, sketch the graph of the equation.

74. $y = 10 - 3x$
75. $y = -4x - 1$
76. $y = -2x^2 - 3$
77. $y = 2x^2 - 7x - 30$
78. $3x^2 - 4y = 0$
79. $-x^2 - 8y = 0$
80. $y = \dfrac{4}{1 - 3x}$
81. $y = \dfrac{x^2}{-x - 2}$
82. $x^2 + (y - 8)^2 = 25$
83. $(x - 4)^2 + (y + 7)^2 = 4$

In Exercises 84–87, graph the exponential function.

84. $f(x) = 2^{x-1} + 5$
85. $f(x) = -2^{-x-1} - 1$
86. $f(x) = 3^x - 4$
87. $f(x) = -3^x + 4$

Chapter Summary

What did you learn?

	Review Exercises
Section 5.1	
☐ How to recognize and evaluate exponential functions with base a	1–6
☐ How to graph exponential functions	7–18, 23–26
☐ How to recognize and evaluate exponential functions with base e	19–22
☐ How to use exponential functions to model and solve real-life applications	27–34
Section 5.2	
☐ How to recognize and evaluate logarithmic functions with base a	35–40
☐ How to graph logarithmic functions	41–46, 53–56
☐ How to recognize and evaluate natural logarithmic functions	47–52
☐ How to use logarithmic functions to model and solve real-life applications	57
Section 5.3	
☐ How to rewrite logarithmic functions with a different base	58–61
☐ How to use properties of logarithms to evaluate or rewrite logarithmic expressions	62–66
☐ How to use properties of logarithms to expand or condense logarithmic expressions	67–74
☐ How to use logarithmic functions to model and solve real-life applications	75
Section 5.4	
☐ How to solve simple exponential and logarithmic equations	76–81
☐ How to solve more complicated exponential equations	82–95
☐ How to solve more complicated logarithmic equations	96–111
☐ How to use exponential and logarithmic equations to model and solve real-life applications	112–114
Section 5.5	
☐ How to recognize the five most common types of models involving exponential and logarithmic functions	115–120
☐ How to use exponential growth and decay functions to model and solve real-life problems	121–125
☐ How to use Gaussian functions to model and solve real-life problems	126
☐ How to use logistic growth functions to model and solve real-life problems	127
☐ How to use logarithmic functions to model and solve real-life problems	128, 129

Review Exercises

5.1 In Exercises 1–6, evaluate the expression. Approximate your result to three decimal places.

1. $(6.1)^{2.4}$
2. $-14(5^{-0.8})$
3. $2^{-0.5\pi}$
4. $\sqrt[5]{1278}$
5. $60^{\sqrt{3}}$
6. $7^{-\sqrt{11}}$

In Exercises 7–10, match the function with its graph. [The graphs are labeled (a) through (d).]

(a)
(b)
(c)
(d)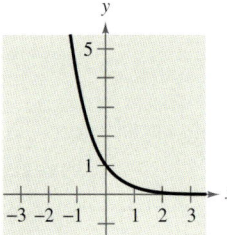

7. $f(x) = 4^x$
8. $f(x) = 4^{-x}$
9. $f(x) = -4^x$
10. $f(x) = 4^x + 1$

In Exercises 11–18, use a graphing utility to construct a table of values. Then sketch the graph of the function.

11. $f(x) = 4^{-x} + 4$
12. $f(x) = -4^x - 3$
13. $f(x) = -2.65^{x+1}$
14. $f(x) = 2.65^{x-1}$
15. $f(x) = 5^{x-2} + 4$
16. $f(x) = 2^{x-6} - 5$
17. $f(x) = \left(\frac{1}{2}\right)^{-x} + 3$
18. $f(x) = \left(\frac{1}{8}\right)^{x+2} - 5$

In Exercises 19–22, evaluate the expression. Approximate your result to three decimal places.

19. e^8
20. $e^{5/8}$
21. $e^{-1.7}$
22. $e^{0.278}$

In Exercises 23–26, use a graphing utility to construct a table of values. Then sketch the graph of the function.

23. $h(x) = e^{-x/2}$
24. $h(x) = 2 - e^{-x/2}$
25. $f(x) = e^{x+2}$
26. $s(t) = 4e^{-2/t}, \quad t > 0$

In Exercises 27 and 28, complete the table to determine the balance A for P dollars invested at rate r for t years and compounded n times per year.

n	1	2	4	12	365	Continuous
A						

27. $P = \$3500$, $r = 6.5\%$, $t = 10$ years
28. $P = \$2000$, $r = 5\%$, $t = 30$ years

In Exercises 29 and 30, complete the table to determine the amount P that should be invested at rate r to produce a balance of $\$200{,}000$ in t years.

t	1	10	20	30	40	50
P						

29. $r = 8\%$, compounded continuously
30. $r = 6\%$, compounded monthly

31. *Waiting Times* The average time between incoming calls at a switchboard is 3 minutes. The probability of waiting less than t minutes until the next incoming call is approximated by the model

$$F(t) = 1 - e^{-t/3}.$$

If a call has just come in, find the probability that the next call will be within

(a) $\frac{1}{2}$ minute. (b) 2 minutes. (c) 5 minutes.

32. *Depreciation* After t years, the value of a car that cost $\$14{,}000$ is

$$V(t) = 14{,}000\left(\tfrac{3}{4}\right)^t.$$

(a) Use a graphing utility to graph the function.
(b) Find the value of the car 2 years after it was purchased.
(c) According to the model, when does the car depreciate most rapidly? Is this realistic? Explain.

33. *Finance* On the day a person was born, a deposit of $\$50{,}000$ was made in a trust fund that pays 8.75% interest, compounded continuously.
(a) Find the balance on the person's 35th birthday.
(b) How much longer would the person have to wait to get twice as much?

34. Fuel Efficiency A certain automobile gets 28 miles per gallon of gasoline for speeds up to 50 miles per hour. Over 50 miles per hour, the number of miles per gallon drops at a rate of 12% for each additional 10 miles per hour. If s is the speed and y is the number of miles per gallon, then

$$y = 28e^{0.6 - 0.012s}, \quad s \geq 50.$$

Use this model to complete the table.

s	50	55	60	65	70
y					

5.2 In Exercises 35 and 36, write the exponential equation in logarithmic form.

35. $4^3 = 64$
36. $25^{3/2} = 125$

In Exercises 37–40, evaluate the expression by hand.

37. $\log_{10} 1000$
38. $\log_9 3$
39. $\log_2 \frac{1}{8}$
40. $\log_a \frac{1}{a}$

In Exercises 41–46, sketch the graph of the function. Identify any asymptotes.

41. $g(x) = \log_7 x$
42. $g(x) = \log_5 x$
43. $f(x) = \log_{10}\left(\frac{x}{3}\right)$
44. $f(x) = 6 + \log_{10} x$
45. $f(x) = 4 - \log_{10}(x + 5)$
46. $f(x) = \log_{10}(x - 3) + 1$

In Exercises 47–52, use your calculator to evaluate each expression. Approximate your result to three decimal places if necessary.

47. $\ln 22.6$
48. $\ln 0.98$
49. $\ln e^{-12}$
50. $\ln e^7$
51. $\ln(\sqrt{7} + 5)$
52. $\ln\left(\frac{\sqrt{3}}{8}\right)$

In Exercises 53–56, sketch the graph of the function. Identify any asymptotes.

53. $f(x) = \ln x + 3$
54. $f(x) = \ln(x - 3)$
55. $h(x) = \ln(x^2)$
56. $f(x) = \frac{1}{4} \ln x$

57. Snow Removal The number of miles s of roads cleared of snow is approximated by the model

$$s = 25 - \frac{13 \ln(h/12)}{\ln 3}, \quad 2 \leq h \leq 15$$

where h is the depth of the snow in inches. Use this model to find s when $h = 10$ inches.

5.3 In Exercises 58–61, evaluate the logarithm using the change-of-base formula. Do each problem twice, once with common logarithms and once with natural logarithms. Approximate the results to three decimal places.

58. $\log_4 9$
59. $\log_{12} 200$
60. $\log_{1/2} 5$
61. $\log_3 0.28$

In Exercises 62–66, verify each statement using the properties of logarithms.

62. $\ln 8 + \ln 5 = \ln 40$
63. $-\ln\left(\frac{1}{12}\right) = \ln 12$
64. $\ln \sqrt[4]{\frac{x}{y}} = \frac{1}{4} \ln x - \frac{1}{4} \ln y$
65. $\log_8\left(\frac{\sqrt{x}}{y^3}\right) = \frac{1}{2} \log_8 x - 3 \log_8 y$
66. $\log_{10}\left(\frac{p^2 q^3}{r}\right) = 2 \log_{10} p + 3 \log_{10} q - \log_{10} r$

In Exercises 67–70, use the properties of logarithms to write the expression as a sum, difference, and/or multiple of logarithms.

67. $\log_5 5x^2$
68. $\log_7 \frac{\sqrt{x}}{4}$
69. $\log_{10} \frac{5\sqrt{y}}{x^2}$
70. $\ln \left|\frac{x - 1}{x + 1}\right|$

In Exercises 71–74, write the expression as the logarithm of a single quantity.

71. $\log_2 5 + \log_2 x$
72. $\log_6 y - 2 \log_6 z$
73. $\frac{1}{2} \ln|2x - 1| - 2 \ln|x + 1|$
74. $5 \ln|x - 2| - \ln|x + 2| - 3 \ln|x|$

75. Climb Rate The time t, in minutes, for a small plane to climb to an altitude of h feet is modeled by

$$t = 50 \log_{10} \frac{18{,}000}{18{,}000 - h}$$

where 18,000 feet is the plane's absolute ceiling.

(a) Determine the domain of the function appropriate for the context of the problem.

(b) Use a graphing utility to graph the time function and identify any asymptotes.

(c) As the plane approaches its absolute ceiling, what can be said about the time required to increase its altitude further?

(d) Find the time for the plane to climb to an altitude of 4000 feet.

5.4 In Exercises 76–81, solve for x.

76. $8^x = 512$
77. $3^x = 729$
78. $6^x = \frac{1}{216}$
79. $6^{x-2} = 1296$
80. $\log_7 x = 4$
81. $\log_x 243 = 5$

In Exercises 82–91, solve the exponential equation. Approximate your result to three decimal places.

82. $e^x = 12$
83. $e^{3x} = 25$
84. $3e^{-5x} = 132$
85. $14e^{3x+2} = 560$
86. $e^x + 13 = 35$
87. $e^x - 28 = -8$
88. $-4(5^x) = -68$
89. $2(12^x) = 190$
90. $e^{2x} - 7e^x + 10 = 0$
91. $e^{2x} - 6e^x + 8 = 0$

In Exercises 92–95, use a graphing utility to solve the equation. Approximate the result to two decimal places.

92. $2^{0.6x} - 3x = 0$
93. $4^{-0.2x} + x = 0$
94. $25e^{-0.3x} = 12$
95. $4e^{1.2x} = 9$

In Exercises 96–107, solve the logarithmic equation. Approximate the result to three decimal places.

96. $\ln 3x = 8.2$
97. $\ln 5x = 7.2$
98. $2 \ln 4x = 15$
99. $4 \ln 3x = 15$
100. $\ln x - \ln 3 = 2$
101. $\ln \sqrt{x + 8} = 3$
102. $\ln \sqrt{x + 1} = 2$
103. $\ln x - \ln 5 = 4$
104. $\log_{10}(x - 1) = \log_{10}(x - 2) - \log_{10}(x + 2)$
105. $\log_{10}(x + 2) - \log_{10} x = \log_{10}(x + 5)$
106. $\log_{10}(1 - x) = -1$
107. $\log_{10}(-x - 4) = 2$

In Exercises 108–111, use a graphing utility to solve the equation. Approximate the result to two decimal places.

108. $2 \ln(x + 3) + 3x = 8$
109. $6 \log_{10}(x^2 + 1) - x = 0$
110. $4 \ln(x + 5) - x = 10$
111. $x - 2 \log_{10}(x + 4) = 0$

112. *Finance* $7550 is deposited in an account that pays 7.25% interest, compounded continuously. How long will it take the money to triple?

113. *Finance* $2440 is deposited in an account that pays 6.5% interest, compounded continuously. How long will it take the money to quadruple?

114. *Economics* The demand equation for a certain product is modeled by

$$p = 500 - 0.5e^{0.004x}.$$

Find the demand x for a price of (a) $p = \$450$ and (b) $p = \$400$.

5.5 In Exercises 115–120, match the function with its graph. [The graphs are labeled (a) through (f).]

(a)

(b)

(c)

(d)

(e)

(f)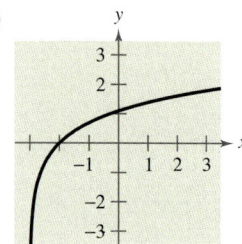

115. $y = 3e^{-2x/3}$
116. $y = 4e^{2x/3}$
117. $y = \ln(x + 3)$
118. $y = 7 - \log_{10}(x + 3)$
119. $y = 2e^{-(x+4)^2/3}$
120. $y = \dfrac{6}{1 + 2e^{-2x}}$

121. *Population* The population of a town is modeled by

$$P = 12{,}620e^{0.0118t}$$

where $t = 0$ represents the year 2000. According to this model, when will the population reach 17,000?

122. *Radioactive Decay* The half-life of radioactive uranium II (^{234}U) is 250,000 years. What percent of a present amount of radioactive uranium II will remain after 5000 years?

123. *Finance* A deposit of $10,000 is made in a savings account for which the interest is compounded continuously. The balance will double in 5 years.

(a) What is the annual interest rate for this account?

(b) Find the balance after 1 year.

In Exercises 124 and 125, find the exponential function $y = ae^{bx}$ that passes through the points.

124. $(0, 2)$, $(4, 3)$

125. $\left(0, \tfrac{1}{2}\right)$, $(5, 5)$

126. *Test Scores* The test scores for a biology test follow a normal distribution modeled by

$$y = 0.0499e^{-(x-71)^2/128}, \quad 40 \le x \le 100$$

where x is the test score.

(a) Use a graphing utility to sketch the graph of the equation.

(b) From the graph, estimate the average test score.

127. *Typing Speed* In a typing class, the average number of words per minute typed after t weeks of lessons was found to be

$$N = \frac{157}{1 + 5.4e^{-0.12t}}.$$

Find the time necessary to type (a) 50 words per minute and (b) 75 words per minute.

128. *Physics* The relationship between the number of decibels β and the intensity of a sound I in watts per square centimeter is

$$\beta = 10 \log_{10}\left(\frac{I}{10^{-16}}\right).$$

Determine the intensity of a sound in watts per square centimeter if the decibel level is 125.

129. *Geology* On the Richter scale, the magnitude R of an earthquake of intensity I is

$$R = \log_{10} \frac{I}{I_0}$$

where $I_0 = 1$ is the minimum intensity used for comparison. Find the intensity per unit of area for the following values of R.

(a) $R = 8.4$

(b) $R = 6.85$

(c) $R = 9.1$

Synthesis

True or False? In Exercises 130–135, determine whether the equation or statement is true or false. Justify your answer.

130. $\log_b b^{2x} = 2x$

131. $e^{x-1} = \dfrac{e^x}{e}$

132. $\ln(x + y) = \ln x + \ln y$

133. $\ln(x + y) = \ln(x \cdot y)$

134. $\log_{10}\left(\dfrac{10}{x}\right) = 1 - \log_{10} x$

135. The domain of the function $f(x) = \ln x$ is the set of all real numbers.

136. The graphs of $y = e^{kt}$ are shown for $k = a, b, c,$ and d. Use the graphs to order $a, b, c,$ and d. Which of the four values are negative? Which are positive?

(a) (b)

(c) 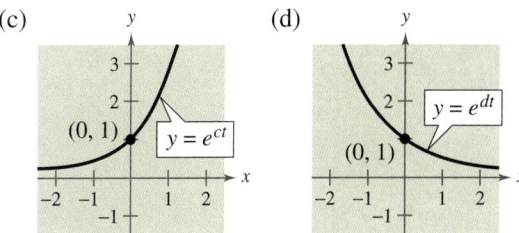 (d)

Chapter Project ▶ Graphical Approach to Compound Interest

A graphing utility can be used to investigate the rates of growth of different types of compound interest.

Example ▶ Comparing Balances

You are depositing $1000 in a savings account. Which of the following will produce the largest balance?

a. 6% annual interest rate, compounded annually
b. 6% annual interest rate, compounded continuously
c. 6.25% annual interest rate, compounded quarterly

Solution

One way to compare all three options is to sketch their graphs in the same viewing window.

Option (a)   Option (b)   Option (c)

$A = 1000(1 + 0.06)^t$ $A = 1000e^{0.06t}$ $A = 1000\left(1 + \dfrac{0.0625}{4}\right)^{4t}$

The graphs are shown at the left. From the graphs, you can conclude that option (c) is better than option (b), and option (b) is better than option (a). Note that for the first 50 years, there is little difference in the graphs. Between 50 and 100 years, however, the balances obtained begin to differ significantly. At the end of 100 years, the balances are (a) $339,302, (b) $403,429, and (c) $493,575.

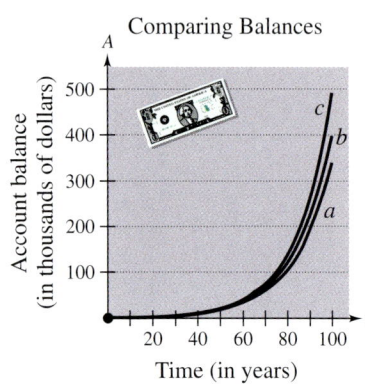

Comparing Balances

Chapter Project Investigations

1. Which would produce a larger balance: an annual interest rate of 8.05% compounded monthly or an annual interest rate of 8% compounded continuously? Explain.

2. You deposit $1000 in each of two savings accounts. The interest for the accounts is paid according to the two options described in Question 1. How long would it take for the balance in one of the accounts to exceed the balance in the other account by $100? By $100,000?

3. No income tax is due on the interest earned in some types of investments. You deposit $25,000 in an account. Which of the following plans is better? Explain.

   (a) *Tax-free* The account pays 5% compounded annually. There is no income tax due on the earned interest.

   (b) *Tax-deferred* The account pays 7% compounded annually. At maturity, the earned interest is taxable at a rate of 40%.

Chapter Test

The *Interactive* CD-ROM and *Internet* versions of this text provide answers to the Chapter Tests and Cumulative Tests. They also offer Chapter Pre-Tests (which test key skills and concepts covered in previous chapters) and Chapter Post-Tests, both of which have randomly generated exercises with diagnostic capabilities.

Take this test as you would take a test in class. After you are done, check your work against the answers given in the back of the book.

In Exercises 1–4, evaluate the expression. Approximate your result to three decimal places.

1. $12.4^{2.79}$ 2. $4^{3\pi/2}$ 3. $e^{-7/10}$ 4. $e^{3.1}$

In Exercises 5–7, construct a table of values. Then sketch the graph of the function.

5. $f(x) = 10^{-x}$ 6. $f(x) = -6^{x-2}$ 7. $f(x) = 1 - e^{2x}$

8. Evaluate (a) $\log_7 7^{-0.89}$ and (b) $4.6 \ln e^2$.

In Exercises 9–11, construct a table of values. Then sketch the graph of the function. Identify any asymptotes.

9. $f(x) = -\log_{10} x - 6$ 10. $f(x) = \ln(x - 4)$ 11. $f(x) = 1 + \ln(x + 6)$

In Exercises 12–14, evaluate the expression. Approximate your result to three decimal places.

12. $\log_7 44$ 13. $\log_{2/5} 0.9$ 14. $\log_{24} 68$

In Exercises 15 and 16, use the properties of logarithms to write the expression as a sum, difference, and/or multiple of logarithms.

15. $\log_2 3a^4$ 16. $\ln \dfrac{5\sqrt{x}}{6}$

In Exercises 17 and 18, rewrite the expression as the logarithm of a single quantity.

17. $\log_3 13 + \log_3 y$ 18. $4 \ln x - 4 \ln y$

In Exercises 19 and 20, solve the equation algebraically. Approximate your result to three decimal places.

19. $\dfrac{1025}{8 + e^{4x}} = 5$ 20. $\log_{10} x - \log_{10}(8 - 5x) = 2$

FIGURE FOR 21

21. Find an exponential growth model for the graph in the figure.

22. The half-life of radioactive actinium (^{227}Ac) is 22 years. What percent of a present amount of radioactive actinium will remain after 19 years?

23. A model that can be used for predicting the height H (in centimeters) of a child based on his or her age is

$$H = 70.228 + 5.104x + 9.222 \ln x, \quad \tfrac{1}{4} \le x \le 6$$

where x is the age of the child in years. (Source: Snapshots of Applications in Mathematics)

(a) Construct a table of values. Then sketch the graph of the model.

(b) Use the graph from part (a) to estimate the height of a 4-year-old child. Then calculate the actual height using the model.

- 6.1 ▶ **Lines**
- 6.2 ▶ **Introduction to Conics: Parabolas**
- 6.3 ▶ **Ellipses**
- 6.4 ▶ **Hyperbolas**
- 6.5 ▶ **Rotation of Conics**
- 6.6 ▶ **Parametric Equations**
- 6.7 ▶ **Polar Coordinates**
- 6.8 ▶ **Graphs of Polar Equations**
- 6.9 ▶ **Polar Equations of Conics**

The space shuttle orbits earth at about 200 miles. By comparison, geosynchronous satellites orbit at about 24,000 miles, and Global Positioning System (GPS) satellites orbit at about 11,000 miles. *(Source: Howstuffworks.com, Inc.)*

VGC/FPG International

6 Topics in Analytic Geometry

▶ How to Study This Chapter

The Big Picture
In this chapter you will learn the following skills and concepts.

▶ How to find the inclination of a line, the angle between two lines, and the distance between a point and a line

▶ How to write the standard forms of the equations of a parabola, an ellipse, and a hyperbola

▶ How to eliminate the xy-term in the equation of a conic and use the discriminant to identify a conic

▶ How to rewrite a set of parametric equations as a rectangular equation and find a set of parametric equations for a graph

▶ How to write equations in polar form and graph polar equations

Important Vocabulary
As you encounter each new vocabulary term in this chapter, add the term and its definition to your notebook glossary.

Inclination (p. 406)
Angle between two lines (p. 407)
Conic section or conic (pp. 413, 470)
Parabola (pp. 414, 470)
Directrix (p. 414)
Focus or foci (pp. 414, 422)
Focal chord (p. 416)
Latus rectum (p. 116)
Tangent (p. 416)
Ellipse (pp. 422, 470)
Vertices (pp. 422, 431)
Major axis (p. 422)
Minor axis (p. 422)
Center (pp. 422, 431)
Eccentricity (pp. 426, 435, 470)

Hyperbola (pp. 431, 470)
Branches (p. 431)
Transverse axis (p. 431)
Asymptotes (p. 433)
Conjugate axis (p. 433)
Invariant under rotation (p. 444)
Discriminant (p. 444)
Parameter (p. 448)
Parametric equations (p. 448)
Plane curve (p. 448)
Orientation (p. 449)
Polar coordinate system (p. 456)
Pole or origin (p. 456)
Polar axis (p. 456)
Polar coordinates (p. 456)

Study Tools
- Learning objectives at the beginning of each section
- Chapter Summary (p. 477)
- Review Exercises (pp. 478–481)
- Chapter Test (p. 483)
- Cumulative Test for Chapters 4–6 (pp. 484–485)

Additional Resources
- Study and Solutions Guide
- Interactive Trigonometry
- Videotapes for Chapter 6
- Trigonometry Website
- Student Success Organizer

STUDY T!P
Once a test has begun, read the directions carefully and work at a reasonable pace. If you finish early, take a few moments to clear your thoughts and then go over your work.

6.1 Lines

▶ **What you should learn**
- How to find the inclination of a line
- How to find the angle between two lines
- How to find the distance between a point and a line

▶ **Why you should learn it**

The inclination of a line can be used to measure heights indirectly. For instance, Exercise 56 on page 412 shows how the inclination of a line can be used to determine the heights of two mountain peaks.

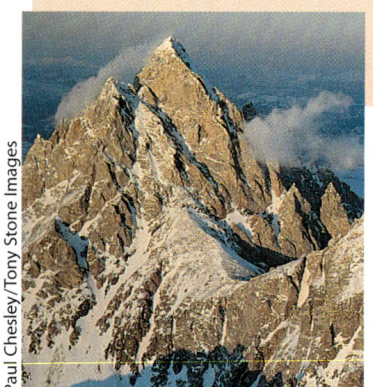

Inclination of a Line

In a previous course, you should have learned that the graph of the linear equation

$$y = mx + b$$

is a nonvertical line with slope m and y-intercept at $(0, b)$. There, the slope of a line was described as the rate of change in y with respect to x. In this section, you will look at the slope of a line in terms of the angle of inclination of the line.

Every nonhorizontal line must intersect the x-axis. The angle formed by such an intersection determines the **inclination** of the line, as specified in the following definition.

Definition of Inclination

The **inclination** of a nonhorizontal line is the positive angle θ (less than π) measured counterclockwise from the x-axis to the line. (See Figure 6.1.)

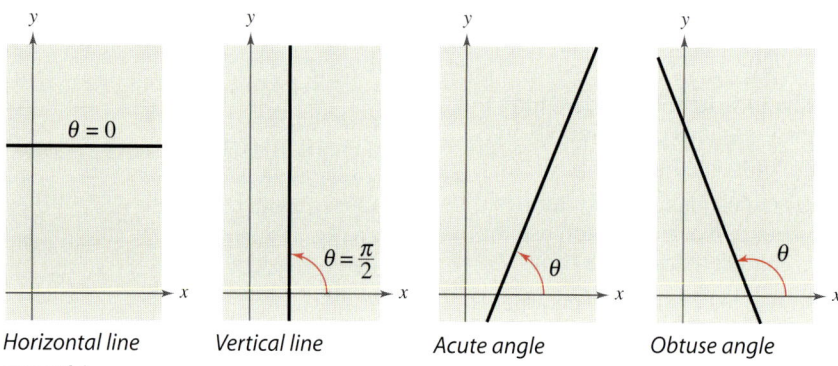

Horizontal line   *Vertical line*   *Acute angle*   *Obtuse angle*
FIGURE 6.1

The inclination of a line is related to its slope in the following manner.

Inclination and Slope

If a nonvertical line has inclination θ and slope m, then

$$m = \tan \theta.$$

If the line has a positive slope, it will intersect the x-axis. Call this point $(x_1, 0)$. If (x_2, y_2) is a second point on the line, the slope is

$$m = \frac{y_2 - 0}{x_2 - x_1} = \frac{y_2}{x_2 - x_1}$$

$$= \tan \theta.$$

Try proving the case in which the line has a negative slope.

FIGURE 6.2

A computer simulation of this concept appears in the *Interactive* CD-ROM and *Internet* versions of this text.

Example 1 ▶ Finding the Inclination of a Line

Find the inclination of the line given by $2x + 3y = 6$.

Solution

The slope of this line is $m = -\frac{2}{3}$. So, its inclination is determined from the equation

$$\tan \theta = -\frac{2}{3}.$$

From Figure 6.2, it follows that $\frac{\pi}{2} < \theta < \pi$. This means that

$$\theta = \pi + \arctan\left(-\frac{2}{3}\right)$$

$$\approx \pi + (-0.588)$$

$$= \pi - 0.588$$

$$\approx 2.554.$$

The angle of inclination is about 2.554 radians or about 146.3°.

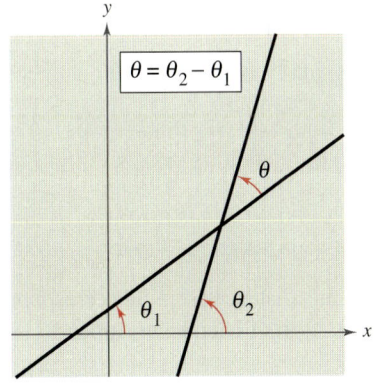

FIGURE 6.3

The Angle Between Two Lines

Two distinct lines in a plane are either parallel or intersecting. If they intersect, their intersection forms two pairs of opposite angles. One pair is acute and the other pair is obtuse. The smaller of these angles is called the **angle between the two lines**. As shown in Figure 6.3, you can use the inclinations of the two lines to find the angle between the two lines. Specifically, if two lines have inclinations θ_1 and θ_2, the angle between the two lines is

$$\theta = \theta_2 - \theta_1$$

where $\theta_1 < \theta_2$. You can use the formula for the tangent of the difference of two angles

$$\tan \theta = \tan(\theta_2 - \theta_1)$$

$$= \frac{\tan \theta_2 - \tan \theta_1}{1 + \tan \theta_1 \tan \theta_2}$$

to obtain the formula for the angle between two lines.

Angle Between Two Lines

If two nonperpendicular lines have slopes m_1 and m_2, the angle between the two lines is

$$\tan \theta = \left| \frac{m_2 - m_1}{1 + m_1 m_2} \right|.$$

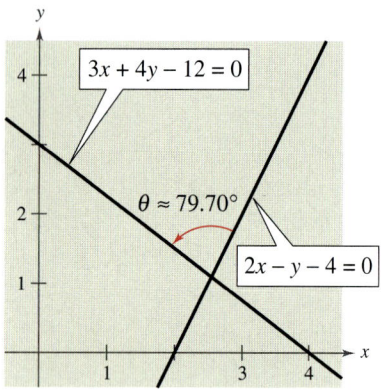

FIGURE 6.4

Example 2 ▶ Finding the Angle Between Two Lines

Find the angle between the two lines.

Line 1: $2x - y - 4 = 0$ Line 2: $3x + 4y - 12 = 0$

Solution

The two lines have slopes of $m_1 = 2$ and $m_2 = -\frac{3}{4}$, respectively. So, the tangent of the angle between the two lines is

$$\tan\theta = \left|\frac{m_2 - m_1}{1 + m_1 m_2}\right| = \left|\frac{(-3/4) - 2}{1 + (2)(-3/4)}\right| = \left|\frac{-11/4}{-2/4}\right| = \frac{11}{2}.$$

Finally, you can conclude that the angle is

$$\theta = \arctan\frac{11}{2} \approx 1.391 \text{ radians} \approx 79.70°$$

as shown in Figure 6.4.

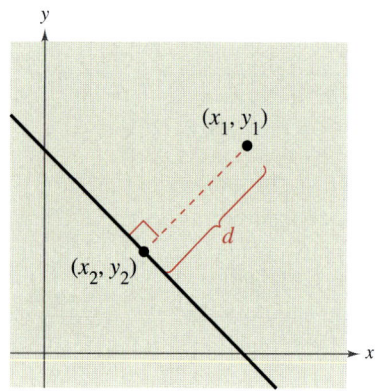

FIGURE 6.5

The Distance Between a Point and a Line

Finding the distance between a line and a point not on the line is an application of perpendicular lines. This distance is defined to be the length of the perpendicular line segment joining the point to the given line, as shown in Figure 6.5.

Distance Between a Point and a Line

The distance between the point (x_1, y_1) and the line $Ax + By + C = 0$ is

$$d = \frac{|Ax_1 + By_1 + C|}{\sqrt{A^2 + B^2}}.$$

Remember that the values of A, B, and C in this distance formula correspond to the general equation of a line, $Ax + By + C = 0$. A proof is given in Appendix A.

Example 3 ▶ Finding the Distance Between a Point and a Line

Find the distance between the point $(4, 1)$ and the line $y = 2x + 1$.

Solution

The general form of the equation is

$$-2x + y - 1 = 0.$$

So, the distance between the point and the line is

$$d = \frac{|-2(4) + 1(1) - 1|}{\sqrt{(-2)^2 + 1^2}} = \frac{8}{\sqrt{5}} \approx 3.58.$$

The line and the point are shown in Figure 6.6.

FIGURE 6.6

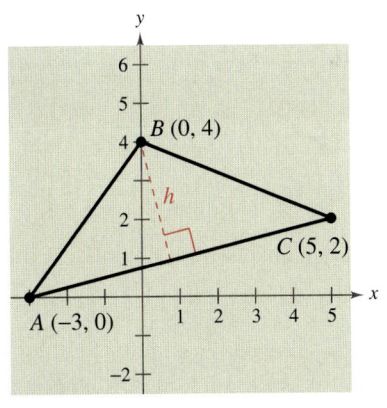

FIGURE 6.7

Example 4 ▶ An Application of Two Distance Formulas

Figure 6.7 shows a triangle with vertices $A(-3, 0)$, $B(0, 4)$, and $C(5, 2)$.

a. Find the altitude from vertex B to side AC.

b. Find the area of the triangle.

Solution

a. To find the altitude, use the formula for the distance between line AC and the point $(0, 4)$. The equation of line AC is obtained as follows.

$$\text{Slope: } m = \frac{2 - 0}{5 + 3} = \frac{1}{4}$$

$$\text{Equation: } \quad y - 0 = \frac{1}{4}(x + 3)$$

$$4y = x + 3$$

$$x - 4y + 3 = 0$$

Therefore, the distance between this line and the point $(0, 4)$ is

$$\text{Altitude} = h = \frac{|1(0) - 4(4) + 3|}{\sqrt{1^2 + (-4)^2}} = \frac{13}{\sqrt{17}}.$$

b. Using the formula for the distance between two points, you can find the length of the base AC to be

$$b = \sqrt{(5 + 3)^2 + (2 - 0)^2}$$

$$= \sqrt{68}$$

$$= 2\sqrt{17}.$$

Finally, the area of the triangle in Figure 6.7 is

$$A = \frac{1}{2}bh$$

$$= \frac{1}{2}(2\sqrt{17})\left(\frac{13}{\sqrt{17}}\right)$$

$$= 13 \text{ square units.}$$

The *Interactive* CD-ROM and *Internet* versions of this text show every example with its solution; clicking on the *Try It!* button brings up similar problems. Guided Examples and Integrated Examples show step-by-step solutions to additional examples. Integrated Examples are related to several concepts in the section.

Writing ABOUT MATHEMATICS

Inclination and the Angle Between Two Lines Discuss why the inclination of a line can be an angle that is larger than $\pi/2$, but the angle between two lines cannot be larger than $\pi/2$. Decide whether the following statement is true or false: "The inclination of a line is the angle between the line and the *x*-axis." Explain.

6.1 Exercises

The Interactive CD-ROM and Internet versions of this text contain step-by-step solutions to all odd-numbered Section and Review Exercises. They also provide Tutorial Exercises that link to Guided Examples for additional help.

In Exercises 1–8, find the slope of the line with inclination θ.

1.
2.
3.
4.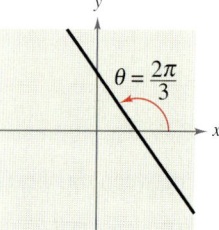

5. $\theta = \dfrac{\pi}{3}$ radians
6. $\theta = \dfrac{5\pi}{6}$ radians
7. $\theta = 1.27$ radians
8. $\theta = 2.88$ radians

In Exercises 9–14, find, in radians and degrees, the inclination θ of the line with a slope of m.

9. $m = -1$
10. $m = -2$
11. $m = 1$
12. $m = 2$
13. $m = \tfrac{3}{4}$
14. $m = -\tfrac{5}{2}$

In Exercises 15–18, find, in radians and degrees, the inclination θ of the line passing through the points.

15. $(6, 1), (10, 8)$
16. $(12, 8), (-4, -3)$
17. $(-2, 20), (10, 0)$
18. $(0, 100), (50, 0)$

In Exercises 19–22, find, in radians and degrees, the inclination θ of the line.

19. $6x - 2y + 8 = 0$
20. $4x + 5y - 9 = 0$
21. $5x + 3y = 0$
22. $x - y - 10 = 0$

In Exercises 23–32, find, in radians and degrees, the angle θ between the lines.

23. $3x + y = 3$
    $x - y = 2$

24. $x + 3y = 2$
    $x - 2y = -3$

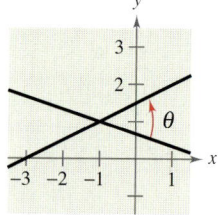

25. $x - y = 0$
    $3x - 2y = -1$

26. $2x - y = 2$
    $4x + 3y = 24$

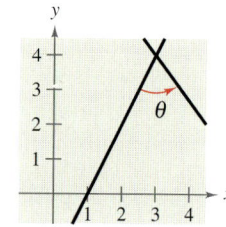

27. $x - 2y = 7$
    $6x + 2y = 5$

28. $5x + 2y = 16$
    $3x - 5y = -1$

29. $x + 2y = 8$
    $x - 2y = 2$

30. $3x - 5y = 3$
    $3x + 5y = 12$

31. $0.05x - 0.03y = 0.21$
    $0.07x + 0.02y = 0.16$

32. $0.02x - 0.05y = -0.19$
    $0.03x + 0.04y = 0.52$

Angle Measurement In Exercises 33–36, find the slope of each side of the triangle and use the slopes to find the measures of the interior angles.

33.
34.

35. 36.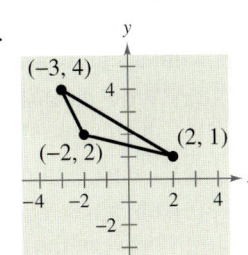

In Exercises 37–44, find the distance between the point and the line.

Point	Line
37. $(0, 0)$	$4x + 3y = 0$
38. $(0, 0)$	$2x - y = 4$
39. $(2, 3)$	$4x + 3y = 10$
40. $(-2, 1)$	$x - y = 2$
41. $(6, 2)$	$x + 1 = 0$
42. $(10, 8)$	$y - 4 = 0$
43. $(0, 8)$	$6x - y = 0$
44. $(4, 2)$	$x - y = 20$

Area In Exercises 45–48, (a) find the altitude from vertex B of the triangle to side AC, and (b) find the area of the triangle.

45. $A = (0, 0)$, $B = (1, 4)$, $C = (4, 0)$
46. $A = (0, 0)$, $B = (4, 5)$, $C = (5, -2)$
47. $A = \left(-\frac{1}{2}, \frac{1}{2}\right)$, $B = (2, 3)$, $C = \left(\frac{5}{2}, 0\right)$
48. $A = (-4, -5)$, $B = (3, 10)$, $C = (6, 12)$

In Exercises 49 and 50, find the distance between the parallel lines.

49. $x + y = 1$
    $x + y = 5$

50. $3x - 4y = 1$
    $3x - 4y = 10$

51. *Grade of a Road* A straight road rises with an inclination of 0.10 radian from the horizontal. Find the slope of the road and the change in elevation over a 2-mile stretch of the road.

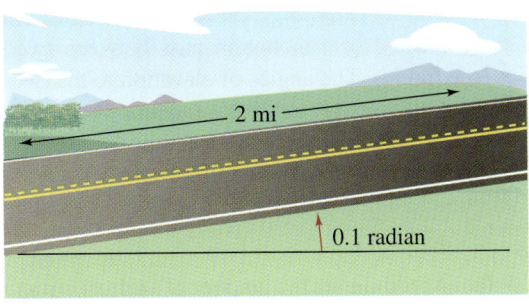

FIGURE FOR 51

52. *Grade of a Road* A straight road rises with an inclination of 0.20 radian from the horizontal. Find the slope of the road and the change in elevation over a 1-mile stretch of the road.

53. *Conveyor Design* A moving conveyor is built so that it rises 1 meter for each 3 meters of horizontal travel.
    (a) Find the inclination of the conveyor.
    (b) The conveyor runs between two floors in a factory. The distance between the floors is 5 meters. Find the length of the conveyor.

54. *Pitch of a Roof* A roof has a rise of 3 feet for every horizontal change of 5 feet (see figure). Find the inclination of the roof.

55. *Truss* Find the angles α and β shown in the drawing of the roof truss.

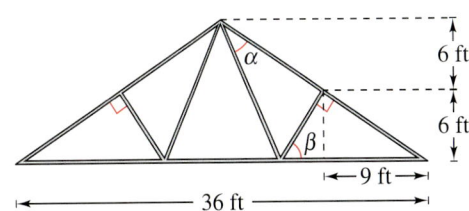

56. *Mountain Climbing* Several mountain climbers are located in a mountain pass between two peaks (see figure). The angles of elevation to the two peaks are 0.84 radian and 1.10 radians. A range finder shows that the distances to the peaks are 3250 feet and 6700 feet, respectively.

(a) Find the angle between the two lines of sight to the peaks.

(b) Approximate the amount of vertical climb that is necessary to reach the summit of each peak.

Synthesis

True or False? In Exercises 57 and 58, determine whether the statement is true or false. Justify your answer.

57. A line that has an inclination greater than $\pi/2$ radians has a negative slope.

58. To find the angle between two lines whose angles of inclination θ_1 and θ_2 are known, substitute θ_1 and θ_2 for m_1 and m_2, respectively, in the formula for the angle between two lines.

59. *Exploration* Consider a line with slope m and y-intercept $(0, 4)$.

(a) Write the distance d between the origin and the line as a function of m.

(b) Graph the function in part (a).

(c) Find the slope that yields the maximum distance between the origin and the line.

(d) Find the asymptote of the graph in part (b) and interpret its meaning in the context of the problem.

60. *Exploration* Consider a line with slope m and y-intercept $(0, 4)$.

(a) Write the distance d between the point $(3, 1)$ and the line as a function of m.

(b) Graph the function in part (a).

(c) Find the slope that yields the maximum distance between the point and the line.

(d) Is it possible for the distance to be 0? If so, what is the slope of the line that yields a distance of 0?

(e) Find the asymptote of the graph in part (b) and interpret its meaning in the context of the problem.

Review

In Exercises 61–66, find all *x*-intercepts and *y*-intercepts of the quadratic function.

61. $f(x) = (x - 7)^2$

62. $f(x) = (x + 9)^2$

63. $f(x) = (x - 5)^2 - 5$

64. $f(x) = (x + 11)^2 + 12$

65. $f(x) = x^2 - 7x - 1$

66. $f(x) = x^2 + 9x - 22$

In Exercises 67–72, write the quadratic function in standard form by completing the square. Identify the vertex of the function.

67. $f(x) = 3x^2 + 2x - 16$

68. $f(x) = 2x^2 - x - 21$

69. $f(x) = 5x^2 + 34x - 7$

70. $f(x) = -x^2 - 8x - 15$

71. $f(x) = 6x^2 - x - 12$

72. $f(x) = -8x^2 - 34x - 21$

6.2 Introduction to Conics: Parabolas

▶ **What you should learn**

- How to recognize a conic as the intersection of a plane and a double-napped cone
- How to write the standard form of the equation of a parabola
- How to use the reflective property of parabolas to solve real-life problems

▶ **Why you should learn it**

Parabolas can be used to model and solve many types of real-life problems. For instance, in Exercise 62 on page 419, a parabola is used to model the cables of a suspension bridge.

Adam Woolfitt/CORBIS

A computer animation of this concept appears in the *Interactive* CD-ROM and *Internet* versions of this text.

Conics

Conic sections were discovered during the classical Greek period, 600 to 300 B.C. The early Greeks were concerned largely with the geometric properties of conics. In the early 17th century the broad applicability of conics became apparent, and they then played a prominent role in the early development of calculus.

Each **conic section** (or simply **conic**) is the intersection of a plane and a double-napped cone. Notice in Figure 6.8(a) that in the formation of the four basic conics, the intersecting plane does not pass through the vertex of the cone. When the plane does pass through the vertex, the resulting figure is a *degenerate conic*, as shown in Figure 6.8(b).

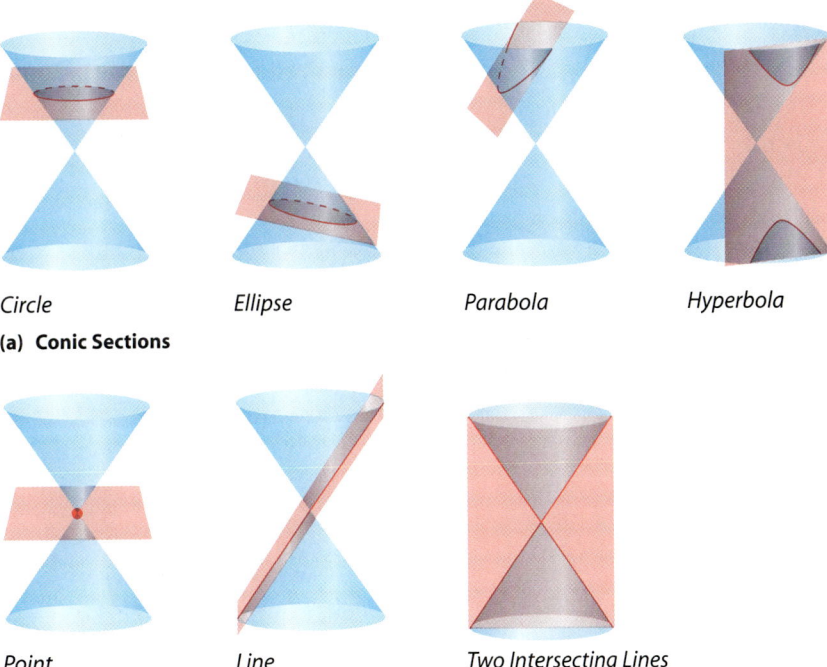

Circle   Ellipse   Parabola   Hyperbola

(a) Conic Sections

Point   Line   Two Intersecting Lines

(b) Degenerate Conics

FIGURE 6.8

There are several ways to approach a study of conics. You could define conics in terms of the intersections of planes and cones, as the Greeks did, or you could define them algebraically in terms of the general second-degree equation

$$Ax^2 + Bxy + Cy^2 + Dx + Ey + F = 0.$$

However, you will study a third approach, in which each of the conics is defined as a **locus** (collection) of points satisfying a geometric property. For example, a circle is defined as the collection of all points (x, y) that are equidistant from a fixed point (h, k). This leads to the standard equation of a circle

$$(x - h)^2 + (y - k)^2 = r^2. \qquad \text{Equation of circle}$$

Parabolas

The first type of conic is called a **parabola,** and is defined as follows.

Definition of Parabola

A **parabola** is the set of all points (x, y) that are equidistant from a fixed line **(directrix)** and a fixed point **(focus)** not on the line.

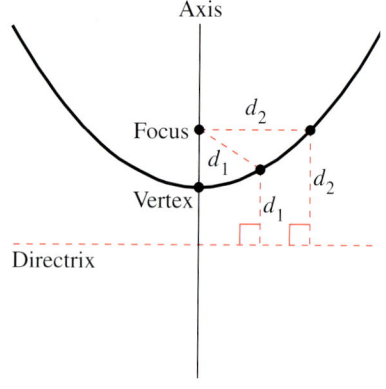

FIGURE 6.9

The midpoint between the focus and the directrix is called the **vertex,** and the line passing through the focus and the vertex is called the **axis** of the parabola. Note in Figure 6.9 that a parabola is symmetric with respect to its axis. Using the definition of a parabola, you can derive the following **standard form** of the equation of a parabola whose directrix is parallel to the x-axis or to the y-axis. A proof is given in Appendix A.

Standard Equation of a Parabola

The **standard form** of the equation of a parabola with vertex at (h, k) is as follows.

$$(x - h)^2 = 4p(y - k), \quad p \neq 0 \quad \text{Vertical axis, directrix: } y = k - p$$

$$(y - k)^2 = 4p(x - h), \quad p \neq 0 \quad \text{Horizontal axis, directrix: } x = h - p$$

The focus lies on the axis p units (*directed distance*) from the vertex. If the vertex is at the origin $(0, 0)$, the equation takes one of the following forms.

$$x^2 = 4py \quad \text{Vertical axis}$$

$$y^2 = 4px \quad \text{Horizontal axis}$$

See Figure 6.10.

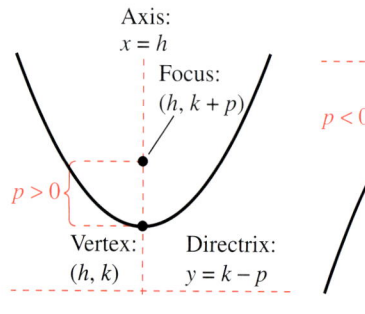

(a) $(x - h)^2 = 4p(y - k)$
Vertical axis: $p > 0$

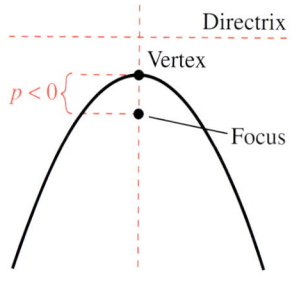

(b) $(x - h)^2 = 4p(y - k)$
Vertical axis: $p < 0$

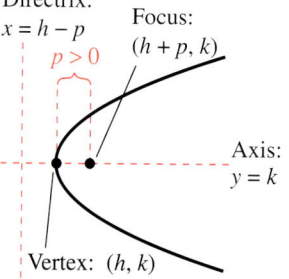

(c) $(y - k)^2 = 4p(x - h)$
Horizontal axis: $p > 0$

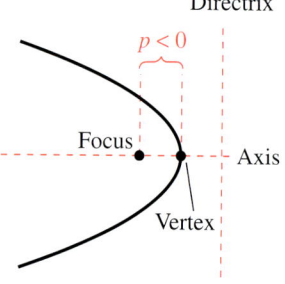

(d) $(y - k)^2 = 4p(x - h)$
Horizontal axis: $p < 0$

FIGURE 6.10

Section 6.2 ▶ Introduction to Conics: Parabolas 415

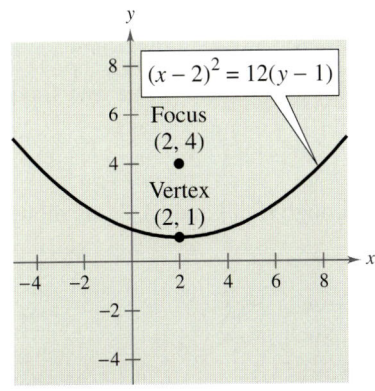

FIGURE 6.11

STUDY TIP

You may want to review the technique of completing the square found in Section P.2, which will be used to rewrite each of the conics in standard form.

Example 1 ▶ **Finding the Standard Equation of a Parabola**

Find the standard form of the equation of the parabola with vertex (2, 1) and focus (2, 4).

Solution

Because the axis of the parabola is vertical, consider the equation

$$(x - h)^2 = 4p(y - k)$$

where $h = 2$, $k = 1$, and $p = 4 - 1 = 3$. So, the standard form is

$$(x - 2)^2 = 12(y - 1).$$

You can obtain the more common quadratic form as follows.

$x^2 - 4x + 4 = 12y - 12$	Multiply.
$x^2 - 4x + 16 = 12y$	Add 12 to each side.
$\frac{1}{12}(x^2 - 4x + 16) = y$	Divide each side by 12.

The graph of this parabola is shown in Figure 6.11.

Example 2 ▶ **Finding the Focus of a Parabola**

Find the focus of the parabola

$$y = -\frac{1}{2}x^2 - x + \frac{1}{2}.$$

Solution

To find the focus, convert to standard form by completing the square.

$y = -\frac{1}{2}x^2 - x + \frac{1}{2}$	Write original equation.
$-2y = x^2 + 2x - 1$	Multiply each side by −2.
$1 - 2y = x^2 + 2x$	Add 1 to each side.
$1 + 1 - 2y = x^2 + 2x + 1$	Complete the square.
$2 - 2y = x^2 + 2x + 1$	Combine like terms.
$-2(y - 1) = (x + 1)^2$	Standard form

Comparing this equation with

$$(x - h)^2 = 4p(y - k),$$

you can conclude that $h = -1$, $k = 1$, and $p = -\frac{1}{2}$. Because p is negative, the parabola opens downward, as shown in Figure 6.12. Therefore, the focus of the parabola is

$$(h, k + p) = \left(-1, \frac{1}{2}\right). \quad \text{Focus}$$

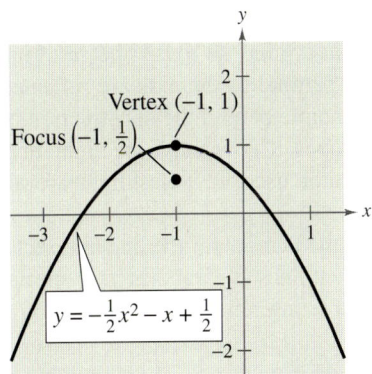

FIGURE 6.12

Technology

Try using a graphing utility to confirm the equation found in Example 3. To do this, it helps to split the equation into two parts:

$y_1 = \sqrt{8x}$ (upper part)

and

$y_2 = -\sqrt{8x}$ (lower part).

Example 3 Vertex at the Origin

Find the standard equation of the parabola with vertex at the origin and focus $(2, 0)$.

Solution

The axis of the parabola is horizontal, passing through $(0, 0)$ and $(2, 0)$, as shown in Figure 6.13.

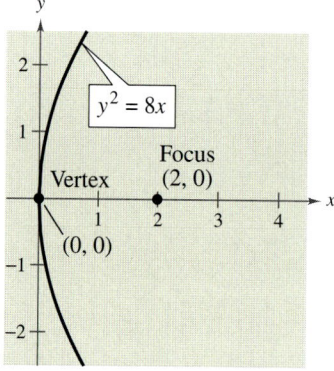

FIGURE 6.13

So, the standard form is $y^2 = 4px$, where $h = k = 0$ and $p = 2$. Therefore, the equation is $y^2 = 8x$.

Application

A line segment that passes through the focus of a parabola and has endpoints on the parabola is called a **focal chord**. The specific focal chord perpendicular to the axis of the parabola is called the **latus rectum**.

Parabolas occur in a wide variety of applications. For instance, a parabolic reflector can be formed by revolving a parabola around its axis. The resulting surface has the property that all incoming rays parallel to the axis are reflected through the focus of the parabola; this is the principle behind the construction of the parabolic mirrors used in reflecting telescopes. Conversely, the light rays emanating from the focus of a parabolic reflector used in a flashlight are all parallel to one another, as shown in Figure 6.14.

A line is **tangent** to a parabola at a point on the parabola if the line intersects, but does not cross, the parabola at the point. Tangent lines to parabolas have special properties related to the use of parabolas in constructing reflective surfaces.

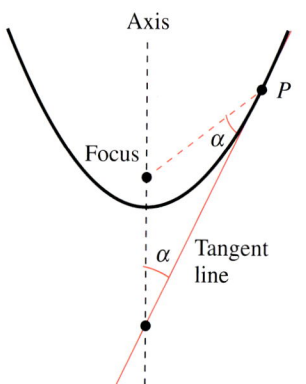

FIGURE 6.14

Reflective Property of a Parabola

The tangent line to a parabola at a point P makes equal angles with the following two lines (see Figure 6.15).

1. The line passing through P and the focus
2. The axis of the parabola

FIGURE 6.15

Section 6.2 ▶ Introduction to Conics: Parabolas 417

Example 4 ▶ Finding the Tangent Line at a Point on a Parabola

Find the equation of the tangent line to the parabola given by $y = x^2$ at the point $(1, 1)$.

Solution

For this parabola, $p = \frac{1}{4}$ and the focus is $\left(0, \frac{1}{4}\right)$, as shown in Figure 6.16. You can find the y-intercept $(0, b)$ of the tangent line by equating the lengths of the two sides of the isosceles triangle shown in Figure 6.16:

$$d_1 = \frac{1}{4} - b$$

and

$$d_2 = \sqrt{(1-0)^2 + [1-(1/4)]^2} = \frac{5}{4}.$$

Setting $d_1 = d_2$ produces

$$\frac{1}{4} - b = \frac{5}{4}$$

$$b = -1.$$

So, the slope of the tangent line is

$$m = \frac{1 - (-1)}{1 - 0} = 2$$

and its slope-intercept equation is

$$y = 2x - 1.$$

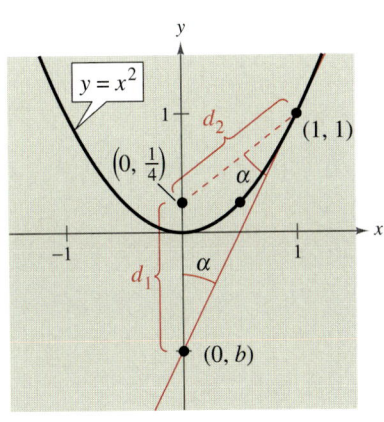

FIGURE 6.16

Technology

Try using a graphing utility to confirm the result of Example 4. By graphing

$y = x^2$ and $y = 2x - 1$

in the same viewing window, you should be able to see that the line touches the parabola at the point $(1, 1)$.

 The *Interactive* CD-ROM and *Internet* versions of this text offer a built-in graphing calculator, which can be used with the Examples, Explorations, Technology notes, and Exercises.

Writing ABOUT MATHEMATICS

Television Antenna Dishes Cross sections of television antenna dishes are parabolic in shape. Write a paragraph explaining why these dishes are parabolic.

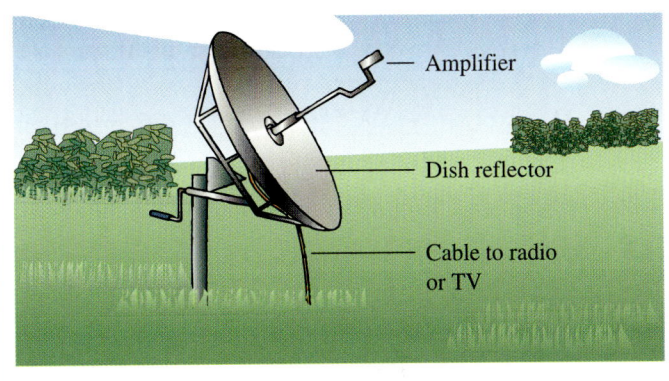

6.2 Exercises

In Exercises 1–4, describe in words how a plane could intersect with the double-napped cone shown to form the conic section.

1. Circle
2. Ellipse
3. Parabola
4. Hyperbola

In Exercises 5–10, match the equation with its graph. [The graphs are labeled (a), (b), (c), (d), (e), and (f).]

(a)
(b)
(c)
(d)
(e)
(f)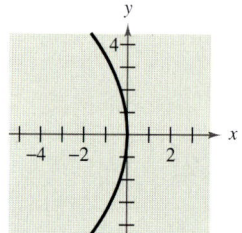

5. $y^2 = -4x$
6. $x^2 = 2y$
7. $x^2 = -8y$
8. $y^2 = -12x$
9. $(y-1)^2 = 4(x-3)$
10. $(x+3)^2 = -2(y-1)$

In Exercises 11–24, find the vertex, focus, and directrix of the parabola and sketch its graph.

11. $y = \frac{1}{2}x^2$
12. $y = -2x^2$
13. $y^2 = -6x$
14. $y^2 = 3x$
15. $x^2 + 6y = 0$
16. $x + y^2 = 0$
17. $(x-1)^2 + 8(y+2) = 0$
18. $(x+5) + (y-1)^2 = 0$
19. $\left(x + \frac{3}{2}\right)^2 = 4(y-2)$
20. $\left(x + \frac{1}{2}\right)^2 = 4(y-1)$
21. $y = \frac{1}{4}(x^2 - 2x + 5)$
22. $x = \frac{1}{4}(y^2 + 2y + 33)$
23. $y^2 + 6y + 8x + 25 = 0$
24. $y^2 - 4y - 4x = 0$

In Exercises 25–28, find the vertex, focus, and directrix of the parabola. Use a graphing utility to graph the parabola.

25. $x^2 + 4x + 6y - 2 = 0$
26. $x^2 - 2x + 8y + 9 = 0$
27. $y^2 + x + y = 0$
28. $y^2 - 4x - 4 = 0$

In Exercises 29 and 30, the equations of a parabola and a tangent line to the parabola are given. Use a graphing utility to graph both equations in the same viewing window. Determine the coordinates of the point of tangency.

Parabola	Tangent Line
29. $y^2 - 8x = 0$	$x - y + 2 = 0$
30. $x^2 + 12y = 0$	$x + y - 3 = 0$

In Exercises 31–42, find the standard form of the equation of the parabola with its vertex at the origin.

31.
32.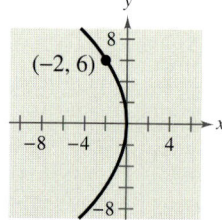

33. Focus: $\left(0, -\frac{3}{2}\right)$
34. Focus: $(2, 0)$
35. Focus: $(-2, 0)$

36. Focus: $(0, -2)$
37. Directrix: $y = -1$
38. Directrix: $y = 3$
39. Directrix: $x = 2$
40. Directrix: $x = -3$
41. Horizontal axis and passes through the point $(4, 6)$
42. Vertical axis and passes through the point $(-3, -3)$

In Exercises 43–52, find the standard form of the equation of the parabola.

43.
44.
45.
46.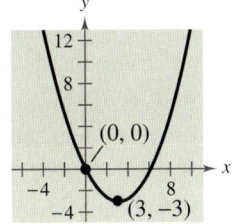

47. Vertex: $(5, 2)$; Focus: $(3, 2)$
48. Vertex: $(-1, 2)$; Focus: $(-1, 0)$
49. Vertex: $(0, 4)$; Directrix: $y = 2$
50. Vertex: $(-2, 1)$; Directrix: $x = 1$
51. Focus: $(2, 2)$; Directrix: $x = -2$
52. Focus: $(0, 0)$; Directrix: $y = 8$

In Exercises 53 and 54, change the equation so that its graph matches the description.

53. $(y - 3)^2 = 6(x + 1)$; upper half of parabola
54. $(y + 1)^2 = 2(x - 4)$; lower half of parabola

In Exercises 55–58, find an equation of the tangent line to the parabola at the given point, and find the x-intercept of the line.

55. $x^2 = 2y$, $(4, 8)$
56. $x^2 = 2y$, $\left(-3, \frac{9}{2}\right)$
57. $y = -2x^2$, $(-1, -2)$
58. $y = -2x^2$, $(2, -8)$

59. *Revenue* The revenue R generated by the sale of x units of a product is
$$R = 265x - \frac{5}{4}x^2.$$
Use a graphing utility to graph the function and approximate the number of sales that will maximize revenue.

60. *Revenue* The revenue R generated by the sale of x units of a product is
$$R = 378x - \frac{7}{5}x^2.$$
Use a graphing utility to graph the function and approximate the number of sales that will maximize revenue.

61. *Satellite Antenna* The receiver in a parabolic television dish antenna is 4.5 feet from the vertex and is located at the focus (see figure). Find an equation of a cross section of the reflector. (Assume that the dish is directed upward and the vertex is at the origin.)

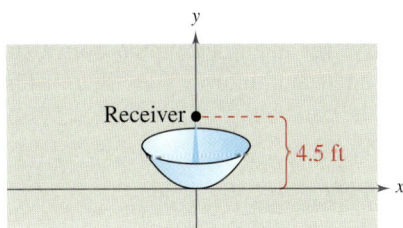

62. *Suspension Bridge* Each cable of a suspension bridge is suspended (in the shape of a parabola) between two towers that are 120 meters apart and whose tops are 20 meters above the roadway. The cables touch the roadway midway between the towers.

    (a) Create a sketch of the bridge. Draw a rectangular coordinate system on the bridge with the center of the bridge at the origin. Identify the coordinates of the known points.
    (b) Find an equation for the parabolic shape of each cable.
    (c) Complete the table by finding the heights y of the suspension cables over the roadway at distances of x meters from the center of the bridge.

x	0	20	40	60
y				

63. *Road Design* Roads are often designed with parabolic surfaces to allow rain to drain off. A particular road that is 32 feet wide is 0.4 foot higher in the center than it is on the sides (see figure).

(a) Find an equation of the parabola that models the road surface. (Assume that the origin is at the center of the road.)

(b) How far from the center of the road is the road surface 0.1 foot lower than in the middle?

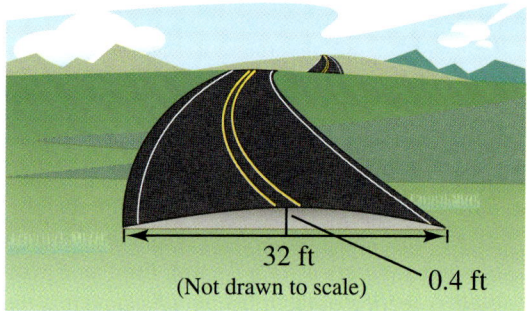

Cross section of road surface

64. *Highway Design* Highway engineers design a parabolic curve for an entrance ramp from a straight street to an interstate highway (see figure). Find an equation of the parabola.

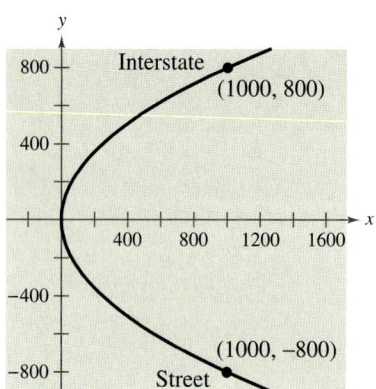

65. *Satellite Orbit* An earth satellite in a 100-mile-high circular orbit around the earth has a velocity of approximately 17,500 miles per hour. If this velocity is multiplied by $\sqrt{2}$, the satellite will have the minimum velocity necessary to escape the earth's gravity and it will follow a parabolic path with the center of the earth as the focus (see figure).

(a) Find the escape velocity of the satellite.

(b) Find an equation of its path (assume that the radius of the earth is 4000 miles).

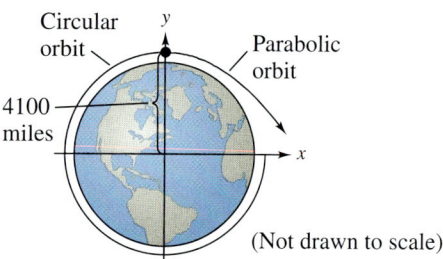

(Not drawn to scale)

FIGURE FOR **65**

66. *Path of a Projectile* The path of a softball is given by the equation

$$y = -0.08x^2 + x + 4.$$

The coordinates x and y are measured in feet, with $x = 0$ corresponding to the position from which the ball was thrown.

(a) Use a graphing utility to graph the trajectory of the softball.

(b) Move the cursor along the path to approximate the highest point. Approximate the range of the trajectory.

67. *Projectile Motion* A bomber is flying at an altitude of 30,000 feet and a speed of 540 miles per hour (792 feet per second). When should a bomb be dropped so that it will hit the target if the path of the bomb is modeled by

$$y = 30,000 - \frac{x^2}{39,204}?$$

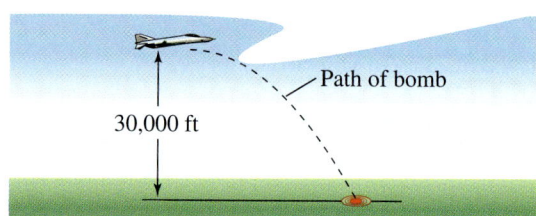

Synthesis

True or False? In Exercises 68 and 69, determine whether the statement is true or false. Justify your answer.

68. It is possible for a parabola to intersect its directrix.

69. If the vertex and focus of a parabola are on a horizontal line, then the directrix of the parabola is vertical.

70. *Exploration* Consider the parabola $x^2 = 4py$.

(a) Use a graphing utility to graph the parabola for $p = 1$, $p = 2$, $p = 3$, and $p = 4$. Describe the effect on the graph when p increases.

(b) Locate the focus for each parabola in part (a).

(c) For each parabola in part (a), find the length of the chord passing through the focus and parallel to the directrix. How can the length of this chord be determined directly from the standard form of the equation of the parabola?

(d) Explain how the result of part (c) can be used as a sketching aid when graphing parabolas.

71. *Area* The area of the shaded region in the figure is

$$A = \frac{4}{3}pb^{3/2}.$$

(a) Find the area if $p = 2$ and $b = 4$.

(b) Give a geometric explanation of why the area approaches 0 as p approaches 0.

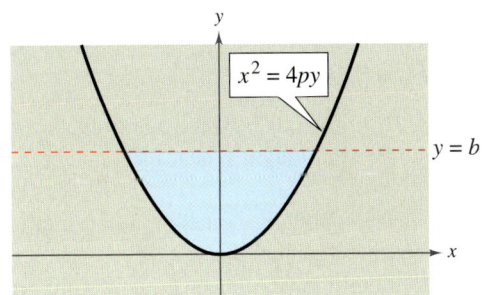

72. *Exploration* Let (x_1, y_1) be the coordinates of a point on the parabola $x^2 = 4py$. The equation of the line tangent to the parabola at the point is

$$y - y_1 = \frac{x_1}{2p}(x - x_1).$$

What is the slope of the tangent line?

Review

In Exercises 73–76, solve as a review of the skills and problem-solving techniques you learned in previous sections.

73. Find a polynomial with integer coefficients that has the zeros 3, $2 + i$, and $2 - i$.

74. Find all the zeros of

$$f(x) = 2x^3 - 3x^2 + 50x - 75$$

if one of the zeros is $x = \frac{3}{2}$.

75. Find all the zeros of the function

$$g(x) = 6x^4 + 7x^3 - 29x^2 - 28x + 20$$

if two of the zeros are $x = \pm 2$.

76. Use a graphing utility to graph the function

$$h(x) = 2x^4 + x^3 - 19x^2 - 9x + 9.$$

Use the graph to approximate the zeros of h.

6.3 Ellipses

What you should learn

- How to write the standard form of the equation of an ellipse
- How to use properties of ellipses to model and solve real-life problems
- How to find the eccentricity of an ellipse

Why you should learn it

Ellipses can be used to model and solve many types of real-life problems. For instance, in Exercise 52 on page 429, an ellipse is used to model the orbit of Halley's comet.

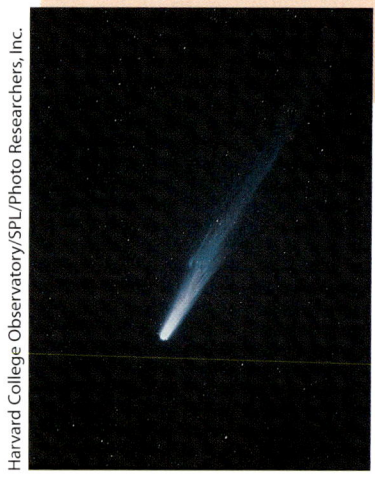

Introduction

The second type of conic is called an **ellipse**, and is defined as follows.

Definition of Ellipse

An **ellipse** is the set of all points (x, y) the sum of whose distances from two distinct fixed points (**foci**) is constant. (See Figure 6.17.)

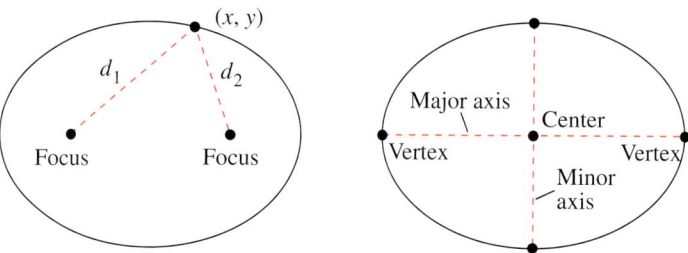

$d_1 + d_2$ is constant.

FIGURE 6.17

The line through the foci intersects the ellipse at two points called **vertices.** The chord joining the vertices is the **major axis,** and its midpoint is the **center** of the ellipse. The chord perpendicular to the major axis at the center is the **minor axis** of the ellipse.

To derive the standard form of the equation of an ellipse, consider the ellipse in Figure 6.18 with the following points: center, (h, k); vertices, $(h \pm a, k)$; foci, $(h \pm c, k)$. Note that the center is the midpoint of the segment joining the foci.

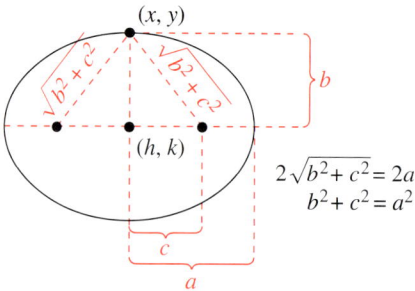

FIGURE 6.18

The sum of the distances from any point on the ellipse to the two foci is constant. Using a vertex point, this constant sum is

$$(a + c) + (a - c) = 2a \qquad \text{Length of major axis}$$

or simply the length of the major axis. Now, if you let (x, y) be *any* point on the ellipse, the sum of the distances between (x, y) and the two foci must also be $2a$.

That is,

$$\sqrt{[x-(h-c)]^2+(y-k)^2}+\sqrt{[x-(h+c)]^2+(y-k)^2}=2a.$$

Finally, in Figure 6.18, you can see that $b^2=a^2-c^2$, which implies that the equation of the ellipse is

$$b^2(x-h)^2+a^2(y-k)^2=a^2b^2$$

$$\frac{(x-h)^2}{a^2}+\frac{(y-k)^2}{b^2}=1.$$

You would obtain a similar equation in the derivation by starting with a vertical major axis. Both results are summarized as follows.

Standard Equation of an Ellipse

The standard form of the equation of an ellipse, with center (h, k) and major and minor axes of lengths $2a$ and $2b$, where $0 < b < a$, is

$$\frac{(x-h)^2}{a^2}+\frac{(y-k)^2}{b^2}=1 \qquad \text{Major axis is horizontal.}$$

$$\frac{(x-h)^2}{b^2}+\frac{(y-k)^2}{a^2}=1. \qquad \text{Major axis is vertical.}$$

The foci lie on the major axis, c units from the center, with $c^2 = a^2 - b^2$. If the center is at the origin $(0, 0)$, the equation takes one of the following forms.

$$\frac{x^2}{a^2}+\frac{y^2}{b^2}=1 \qquad \text{Major axis is horizontal.}$$

$$\frac{x^2}{b^2}+\frac{y^2}{a^2}=1 \qquad \text{Major axis is vertical.}$$

Figure 6.19 shows both the vertical and horizontal orientations for an ellipse.

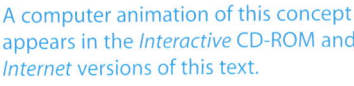
A computer animation of this concept appears in the *Interactive* CD-ROM and *Internet* versions of this text.

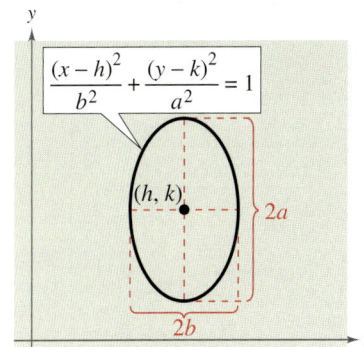

FIGURE 6.19

You can visualize the definition of an ellipse by imagining two thumbtacks placed at the foci, as shown in Figure 6.20. If the ends of a fixed length of string are fastened to the thumbtacks and the string is drawn taut with a pencil, the path traced by the pencil will be an ellipse.

FIGURE 6.20

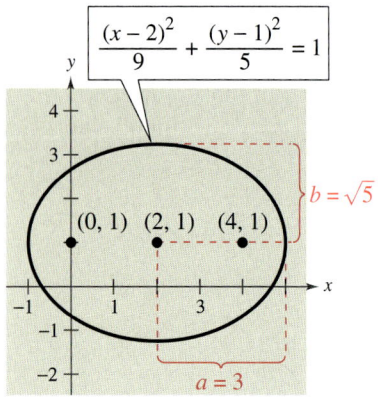

FIGURE 6.21

Example 1 ▶ Finding the Standard Equation of an Ellipse

Find the standard form of the equation of the ellipse having foci at $(0, 1)$ and $(4, 1)$ and a major axis of length 6, as shown in Figure 6.21.

Solution

Because the foci occur at $(0, 1)$ and $(4, 1)$, the center of the ellipse is $(2, 1)$ and the distance from the center to one of the foci is $c = 2$. Because $2a = 6$, you know that $a = 3$. Now, from $c^2 = a^2 - b^2$, you have

$$b^2 = a^2 - c^2$$
$$= 3^2 - 2^2$$
$$= 5.$$

Because the major axis is horizontal, the standard equation is

$$\frac{(x - 2)^2}{9} + \frac{(y - 1)^2}{5} = 1.$$

In Example 1, note the use of the equation $c^2 = a^2 - b^2$. Don't confuse this equation with the Pythagorean Theorem—there is a difference in sign.

Example 2 ▶ Writing an Equation in Standard Form

Sketch the graph of the ellipse whose equation is

$$x^2 + 4y^2 + 6x - 8y + 9 = 0.$$

Solution

Begin by writing the given equation in standard form. In the fourth step, note that 9 and 4 are added to *both* sides of the equation when completing the squares.

$x^2 + 4y^2 + 6x - 8y + 9 = 0$ Write original equation.

$(x^2 + 6x +) + (4y^2 - 8y +) = -9$ Group terms.

$(x^2 + 6x +) + 4(y^2 - 2y +) = -9$ Factor 4 out of y-terms.

$(x^2 + 6x + 9) + 4(y^2 - 2y + 1) = -9 + 9 + 4(1)$

$(x + 3)^2 + 4(y - 1)^2 = 4$ Completed square form

$$\frac{(x + 3)^2}{4} + \frac{(y - 1)^2}{1} = 1$$ Standard form

From the equation, you can see that the center occurs at $(h, k) = (-3, 1)$. Because the denominator of the x-term is $a^2 = 2^2$, you can locate the endpoints of the major axis 2 units to the right and left of the center. Similarly, because the denominator of the y-term is $b^2 = 1^2$, you can locate the endpoints of the minor axis 1 unit up and down from the center. The graph of this ellipse is shown in Figure 6.22.

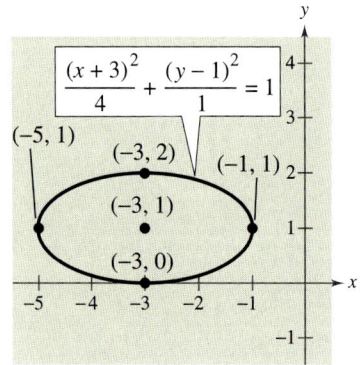

FIGURE 6.22

Example 3 ▶ Analyzing an Ellipse

Find the center, vertices, and foci of the ellipse
$$4x^2 + y^2 - 8x + 4y - 8 = 0.$$

Solution

By completing the square, you can write the given equation in standard form.

$$4x^2 + y^2 - 8x + 4y - 8 = 0$$
$$(4x^2 - 8x +) + (y^2 + 4y +) = 8$$
$$4(x^2 - 2x +) + (y^2 + 4y +) = 8$$
$$4(x^2 - 2x + 1) + (y^2 + 4y + 4) = 8 + 4(1) + 4$$
$$4(x - 1)^2 + (y + 2)^2 = 16$$
$$\frac{(x - 1)^2}{4} + \frac{(y + 2)^2}{16} = 1$$

So, the major axis is vertical, where $h = 1$, $k = -2$, $a = 4$, $b = 2$, and
$$c = \sqrt{16 - 4} = \sqrt{12} = 2\sqrt{3}.$$

Therefore, you have the following.

Center: $(1, -2)$ Vertices: $(1, -6)$ Foci: $(1, -2 - 2\sqrt{3})$
                              $(1, 2)$ $(1, -2 + 2\sqrt{3})$

The graph of the ellipse is shown in Figure 6.23.

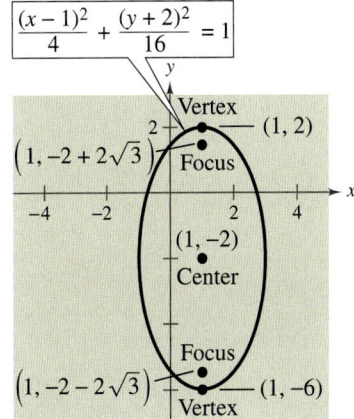

FIGURE 6.23

Technology

You can use a graphing utility to graph an ellipse by graphing the upper and lower portions in the same viewing window. For instance, to graph the ellipse in Example 3, first solve for y to get

$$y_1 = -2 + 4\sqrt{1 - \frac{(x - 1)^2}{4}}$$

and

$$y_2 = -2 - 4\sqrt{1 - \frac{(x - 1)^2}{4}}.$$

Use a viewing window in which $-6 \leq x \leq 9$ and $-7 \leq y \leq 3$. You should obtain the graph shown below.

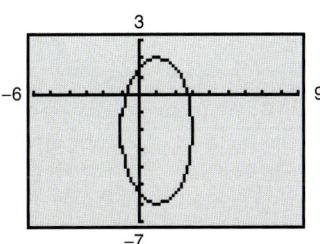

Application

Ellipses have many practical and aesthetic uses. For instance, machine gears, supporting arches, and acoustic designs often involve elliptical shapes. The orbits of satellites and planets are also ellipses. Example 4 investigates the elliptical orbit of the moon about the earth.

Example 4 ▶ An Application Involving an Elliptical Orbit

The moon travels about the earth in an elliptical orbit with the earth at one focus, as shown in Figure 6.24. The major and minor axes of the orbit have lengths of 768,806 kilometers and 767,746 kilometers, respectively. Find the greatest and smallest distances (the *apogee* and *perigee*) from the earth's center to the moon's center.

Solution

Because $2a = 768,806$ and $2b = 767,746$, you have $a = 384,403$ and $b = 383,873$, which implies that

$$c = \sqrt{a^2 - b^2}$$
$$= \sqrt{384,403^2 - 383,873^2}$$
$$\approx 20,179.$$

Therefore, the greatest distance between the center of the earth and the center of the moon is

$$a + c \approx 404,582 \text{ kilometers}$$

and the smallest distance is

$$a - c \approx 364,224 \text{ kilometers.}$$

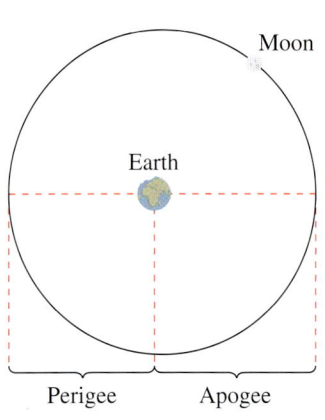

FIGURE 6.24

Eccentricity

One of the reasons it was difficult for early astronomers to detect that the orbits of the planets are ellipses is that the foci of the planetary orbits are relatively close to their centers, and so the orbits are nearly circular. To measure the ovalness of an ellipse, you can use the concept of **eccentricity.**

Definition of Eccentricity

The **eccentricity** e of an ellipse is given by the ratio

$$e = \frac{c}{a}.$$

Note that $0 < e < 1$ for *every* ellipse.

To see how this ratio is used to describe the shape of an ellipse, note that because the foci of an ellipse are located along the major axis between the vertices and the center, it follows that

$$0 < c < a.$$

For an ellipse that is nearly circular, the foci are close to the center and the ratio c/a is small, as shown in Figure 6.25(a). On the other hand, for an elongated ellipse, the foci are close to the vertices, and the ratio c/a is close to 1, as shown in Figure 6.25(b).

(a)

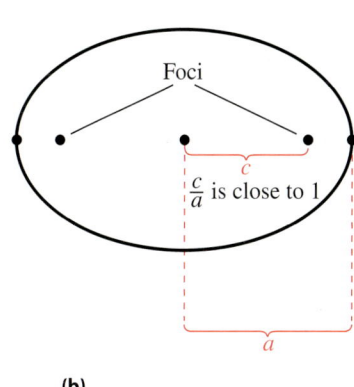
(b)

FIGURE 6.25

The orbit of the moon has an eccentricity of $e \approx 0.0525$, and the eccentricities of the nine planetary orbits are as follows.

Mercury: $e \approx 0.2056$	Saturn: $e \approx 0.0543$
Venus: $e \approx 0.0068$	Uranus: $e \approx 0.0460$
Earth: $e \approx 0.0167$	Neptune: $e \approx 0.0082$
Mars: $e \approx 0.0934$	Pluto:
Jupiter: $e \approx 0.0484$	

The time it takes Saturn to orbit the sun is equal to 29.5 earth years.

Writing ABOUT MATHEMATICS

Ellipses and Circles

a. Show that the equation of an ellipse can be written as

$$\frac{(x-h)^2}{a^2} + \frac{(y-k)^2}{a^2(1-e^2)} = 1.$$

b. For the equation in part (a), let $a = 4$, $h = 1$, and $k = 2$, and use a graphing utility to graph the ellipse for $e = 0.95$, $e = 0.75$, $e = 0.5$, $e = 0.25$, and $e = 0.1$. Discuss the changes in the shape of the ellipse as e approaches 0.

c. Make a conjecture about the shape of the graph in part (b) when $e = 0$. What is the equation of this ellipse? What is another name for an ellipse with an eccentricity of 0?

6.3 Exercises

In Exercises 1–6, match the equation with its graph. [The graphs are labeled (a), (b), (c), (d), (e), and (f).]

(a)

(b)

(c)

(d)

(e)

(f)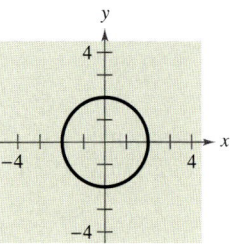

1. $\dfrac{x^2}{4} + \dfrac{y^2}{9} = 1$
2. $\dfrac{x^2}{9} + \dfrac{y^2}{4} = 1$
3. $\dfrac{x^2}{4} + \dfrac{y^2}{25} = 1$
4. $\dfrac{x^2}{4} + \dfrac{y^2}{4} = 1$
5. $\dfrac{(x-2)^2}{16} + (y+1)^2 = 1$
6. $\dfrac{(x+2)^2}{9} + \dfrac{(y+2)^2}{4} = 1$

In Exercises 7–22, find the center, vertices, foci, and eccentricity of the ellipse, then sketch its graph.

7. $\dfrac{x^2}{25} + \dfrac{y^2}{16} = 1$
8. $\dfrac{x^2}{81} + \dfrac{y^2}{144} = 1$
9. $\dfrac{x^2}{5} + \dfrac{y^2}{9} = 1$
10. $\dfrac{x^2}{64} + \dfrac{y^2}{28} = 1$
11. $\dfrac{(x+3)^2}{16} + \dfrac{(y-5)^2}{25} = 1$
12. $\dfrac{(x-4)^2}{12} + \dfrac{(y+3)^2}{16} = 1$
13. $\dfrac{(x+5)^2}{9/4} + (y-1)^2 = 1$
14. $(x+2)^2 + \dfrac{(y+4)^2}{1/4} = 1$
15. $9x^2 + 4y^2 + 36x - 24y + 36 = 0$
16. $9x^2 + 4y^2 - 54x + 40y + 37 = 0$
17. $x^2 + 5y^2 - 8x - 30y - 39 = 0$
18. $3x^2 + y^2 + 18x - 2y - 8 = 0$
19. $6x^2 + 2y^2 + 18x - 10y + 2 = 0$
20. $x^2 + 4y^2 - 6x + 20y - 2 = 0$
21. $16x^2 + 25y^2 - 32x + 50y + 16 = 0$
22. $9x^2 + 25y^2 - 36x - 50y + 60 = 0$

In Exercises 23–26, use a graphing utility to graph the ellipse. Find the center, foci, and vertices. (Recall that it may be necessary to solve the equation for y and obtain two functions.)

23. $5x^2 + 3y^2 = 15$
24. $3x^2 + 4y^2 = 12$
25. $12x^2 + 20y^2 - 12x + 40y - 37 = 0$
26. $36x^2 + 9y^2 + 48x - 36y - 72 = 0$

In Exercises 27–34, find the standard form of the equation of the ellipse with center at the origin.

27.

28.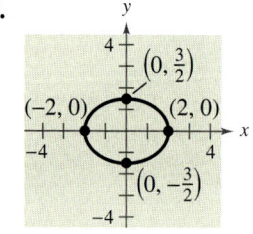

29. Vertices: $(\pm 6, 0)$; Foci: $(\pm 2, 0)$
30. Vertices: $(0, \pm 8)$; Foci: $(0, \pm 4)$
31. Foci: $(\pm 5, 0)$; Major axis of length 12
32. Foci: $(\pm 2, 0)$; Major axis of length 8
33. Vertices: $(0, \pm 5)$; Passes through the point $(4, 2)$
34. Major axis vertical; Passes through the points $(0, 4)$ and $(2, 0)$

In Exercises 35–46, find the standard form of the equation of the specified ellipse.

35.

36.

37.

38.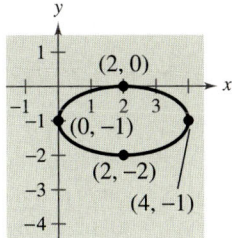

39. Vertices: $(0, 4)$, $(4, 4)$; Minor axis of length 2
40. Foci: $(0, 0)$, $(4, 0)$; Major axis of length 8
41. Foci: $(0, 0)$, $(0, 8)$; Major axis of length 16
42. Center: $(2, -1)$; Vertex: $(2, \frac{1}{2})$; Minor axis of length 2
43. Vertices: $(3, 1)$, $(3, 9)$; Minor axis of length 6
44. Center: $(3, 2)$; $a = 3c$; Foci: $(1, 2)$, $(5, 2)$
45. Center: $(0, 4)$; $a = 2c$; Vertices: $(-4, 4)$, $(4, 4)$
46. Vertices: $(5, 0)$, $(5, 12)$; Endpoints of the minor axis: $(1, 6)$, $(9, 6)$

47. Find an equation of the ellipse with vertices $(\pm 5, 0)$ and eccentricity $e = \frac{3}{5}$.
48. Find an equation of the ellipse with vertices $(0, \pm 8)$ and eccentricity $e = \frac{1}{2}$.
49. *Fireplace Arch* A fireplace arch is to be constructed in the shape of a semiellipse. The opening is to have a height of 2 feet at the center and a width of 6 feet along the base (see figure). The contractor draws the outline of the ellipse using tacks as described at the beginning of this section. Give the required positions of the tacks and the length of the string.

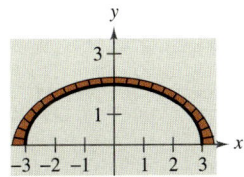

50. *Mountain Tunnel* A semielliptical arch over a tunnel for a road through a mountain has a major axis of 80 feet and a height at the center of 30 feet.
    (a) Draw a rectangular coordinate system on a sketch of the tunnel with the center of the road entering the tunnel at the origin. Identify the coordinates of the known points.
    (b) Find an equation of the elliptical tunnel.
    (c) Determine the height of the arch 5 feet from the edge of the tunnel.

51. *Geometry* The area of the ellipse in the figure is twice the area of the circle. What is the length of the major axis? (*Hint:* $A = \pi ab$ for an ellipse.)

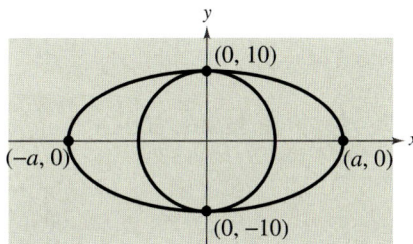

52. *Comet Orbit* Halley's comet has an elliptical orbit, with the sun at one focus. The eccentricity of the orbit is approximately 0.97. The length of the major axis of the orbit is approximately 36.18 astronomical units. (An astronomical unit is about 93 million miles.) Find an equation of the orbit. Place the center of the orbit at the origin, and place the major axis on the x-axis.

53. *Comet Orbit* The comet Encke has an elliptical orbit, with the sun at one focus. Encke ranges from 0.34 to 4.08 astronomical units from the sun. Find an equation of the orbit. Place the center of the orbit at the origin, and place the major axis on the x-axis.

54. *Satellite Orbit* The first artificial satellite to orbit earth was Sputnik I (launched by Russia in 1957). Its highest point above earth's surface was 938 kilometers, and its lowest point was 212 kilometers (see figure). The radius of earth is 6378 kilometers. Find the eccentricity of the orbit.

55. *Geometry* A line segment through a focus of an ellipse with endpoints on the ellipse and perpendicular to the major axis is called a **latus rectum** of the ellipse. Therefore, an ellipse has two latera recta. Knowing the length of the latera recta is helpful in sketching an ellipse because it yields other points on the curve (see figure). Show that the length of each latus rectum is $2b^2/a$.

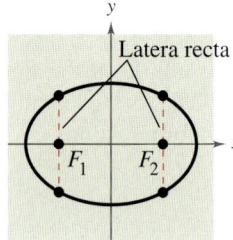

In Exercises 56–59, sketch the graph of the ellipse, making use of the latera recta (see Exercise 55).

56. $\dfrac{x^2}{4} + \dfrac{y^2}{1} = 1$

57. $\dfrac{x^2}{9} + \dfrac{y^2}{16} = 1$

58. $9x^2 + 4y^2 = 36$

59. $5x^2 + 3y^2 = 15$

Synthesis

True or False? In Exercises 60–63, determine whether the statement is true or false. Justify your answer.

60. The graph of $(x^2/4) + y^4 = 1$ is an ellipse.

61. It is easier to distinguish the graph of an ellipse from the graph of a circle if the eccentricity of the ellipse is large (close to 1).

62. The area of a circle with diameter $d = 2r = 8$ is greater than the area of an ellipse with major axis $2a = 8$.

63. It is possible for the foci of an ellipse to occur outside the ellipse.

64. *Think About It* At the beginning of this section it was noted that an ellipse can be drawn using two thumbtacks, a string of fixed length (greater than the distance between the two tacks), and a pencil. If the ends of the string are fastened at the tacks and the string is drawn taut with a pencil, the path traced by the pencil is an ellipse.

(a) What is the length of the string in terms of a?

(b) Explain why the path is an ellipse.

65. *Exploration* The area A of the ellipse
$$\dfrac{x^2}{a^2} + \dfrac{y^2}{b^2} = 1$$
is $A = \pi ab$. Let $a + b = 20$.

(a) Write the area of the ellipse as a function of a.

(b) Find the equation of an ellipse with an area of 264 square centimeters.

(c) Complete the table and make a conjecture about the shape of the ellipse with a maximum area.

a	8	9	10	11	12	13
A						

(d) Use a graphing utility to graph the area function, and use the graph to make a conjecture about the shape of the ellipse that yields a maximum area.

Review

In Exercises 66–69, solve the equation.

66. $x^2 - 10x = 0$

67. $100 - (x - 5)^2 = 0$

68. $x^3 - x = 0$

69. $16x^2 - 40x + 25 = 0$

In Exercises 70–73, evaluate the function at each specified value of the independent variable and simplify.

70. $f(x) = 5x - 8$

(a) $f(9)$ (b) $f(-4)$ (c) $f(x - 7)$

71. $f(x) = x^2 - 10x$

(a) $f(4)$ (b) $f(-8)$ (c) $f(x - 4)$

72. $f(x) = \sqrt{x - 12} - 9$

(a) $f(12)$ (b) $f(40)$ (c) $f(-\sqrt{36})$

73. $f(x) = x^4 - x - 5$

(a) $f(-1)$ (b) $f(\tfrac{1}{2})$ (c) $f(2\sqrt{3})$

In Exercises 74 and 75, find the difference quotient and simplify your answer.

74. $f(x) = x^2 - 2x + 9$, $\dfrac{f(3 + h) - f(3)}{h}$, $h \neq 0$

75. $f(x) = 5 + 6x - x^2$, $\dfrac{f(6 + h) - f(6)}{h}$, $h \neq 0$

6.4 Hyperbolas

What you should learn
- How to write the standard form of the equation of a hyperbola
- How to find the asymptotes of a hyperbola
- How to use properties of hyperbolas to solve real-life problems
- How to classify a conic from its general equation

Why you should learn it
Hyperbolas can be used to model and solve many types of real-life problems. For instance, in Exercise 39 on page 439, hyperbolas are used to locate the position of an explosion that was recorded by three listening stations.

Introduction
The definition of a hyperbola parallels that of an ellipse. The difference is that for an ellipse the *sum* of the distances between the foci and a point on the ellipse is fixed, whereas for a hyperbola the *difference* of these distances is fixed.

Definition of Hyperbola
A **hyperbola** is the set of all points (x, y) the difference of whose distances from two distinct fixed points (foci) is a positive constant. (See Figure 6.26.)

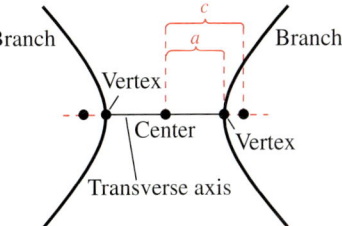

FIGURE 6.26

Every hyperbola has two disconnected **branches**. The line through the two foci intersects a hyperbola at its two **vertices**. The line segment connecting the vertices is called the **transverse axis**, and the midpoint of the transverse axis is called the **center** of the hyperbola. The development of the standard form of the equation of a hyperbola is similar to that of an ellipse.

Standard Equation of a Hyperbola
The standard form of the equation of a hyperbola with center at (h, k) is

$$\frac{(x-h)^2}{a^2} - \frac{(y-k)^2}{b^2} = 1 \quad \text{Transverse axis is horizontal.}$$

$$\frac{(y-k)^2}{a^2} - \frac{(x-h)^2}{b^2} = 1. \quad \text{Transverse axis is vertical.}$$

The vertices are a units from the center, and the foci are c units from the center. Moreover, $c^2 = a^2 + b^2$. If the center of the hyperbola is at the origin $(0, 0)$, the equation takes one of the following forms.

$$\frac{x^2}{a^2} - \frac{y^2}{b^2} = 1 \quad \text{Transverse axis is horizontal.} \qquad \frac{y^2}{a^2} - \frac{x^2}{b^2} = 1 \quad \text{Transverse axis is vertical.}$$

Note that a, b, and c are related differently for hyperbolas than for ellipses.

Figure 6.27 shows both the horizontal and vertical orientations for a hyperbola.

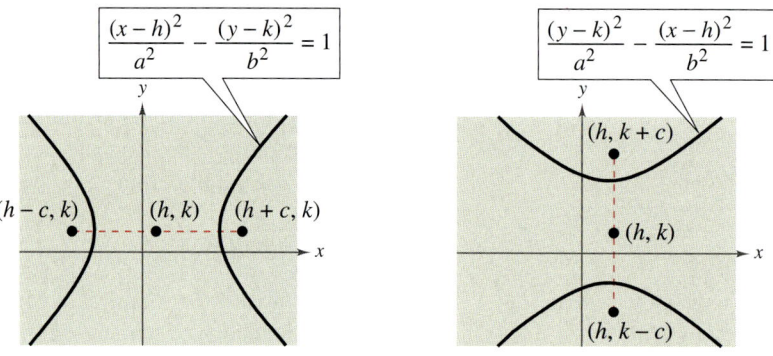

FIGURE 6.27

Example 1 ▶ Finding the Standard Equation of a Hyperbola

Find the standard form of the equation of the hyperbola with foci at $(-1, 2)$ and $(5, 2)$ and vertices at $(0, 2)$ and $(4, 2)$.

Solution

By the Midpoint Formula, the center of the hyperbola occurs at the point $(2, 2)$. Furthermore, $c = 3$ and $a = 2$, and it follows that

$$b^2 = c^2 - a^2$$
$$= 3^2 - 2^2$$
$$= 9 - 4$$
$$= 5.$$

So, the equation of the hyperbola is

$$\frac{(x-2)^2}{4} - \frac{(y-2)^2}{5} = 1.$$

Figure 6.28 shows the hyperbola.

◀ **Exploration** ▶

Most graphing utilities have a parametric mode. Try using parametric mode to graph the hyperbola $x = 2 + 2\sec t$ and $y = 2 + \sqrt{5}\tan t$. How does the result compare with the graph given in Figure 6.28? (Let the parameter vary from Tmin $= 0$ to Tmax $= 6.28$ with Tstep $= 0.13$.)

FIGURE 6.28

Section 6.4 ▶ Hyperbolas   433

Asymptotes of a Hyperbola

Conjugate axis
(h, k + b)
(h, k)
(h, k − b)
(h − a, k)   (h + a, k)
Asymptote
Asymptote

FIGURE 6.29

Each hyperbola has two **asymptotes** that intersect at the center of the hyperbola, as shown in Figure 6.29. The asymptotes pass through the vertices of a rectangle of dimensions $2a$ by $2b$, with its center at (h, k). The line segment of length $2b$ joining $(h, k + b)$ and $(h, k - b)$ [or $(h + b, k)$ and $(h - b, k)$] is the **conjugate axis** of the hyperbola.

Asymptotes of a Hyperbola

The equations for the asymptotes of a hyperbola are

$$y = k \pm \frac{b}{a}(x - h) \qquad \text{Asymptotes for horizontal transverse axis}$$

$$y = k \pm \frac{a}{b}(x - h). \qquad \text{Asymptotes for vertical transverse axis}$$

Example 2 ▶ Using Asymptotes to Sketch a Hyperbola

Sketch the hyperbola whose equation is $4x^2 - y^2 = 16$.

Solution

Divide both sides of the original equation by 16, and rewrite the equation.

$$\frac{x^2}{4} - \frac{y^2}{16} = 1 \qquad \text{Standard form}$$

From this, you can conclude that $a = 2$, $b = 4$, and the transverse axis is horizontal. So, the vertices occur at $(-2, 0)$ and $(2, 0)$, and the ends of the conjugate axis occur at $(0, -4)$ and $(0, 4)$. Using these four points, you are able to sketch the rectangle shown in Figure 6.30(a). Finally, after drawing the asymptotes through the corners of this rectangle, you can complete the sketch, as shown in Figure 6.30(b).

(a)

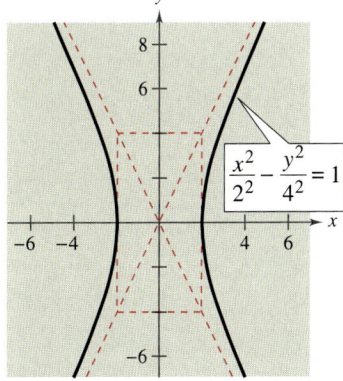

(b)

FIGURE 6.30

Example 3 ▶ Finding the Asymptotes of a Hyperbola

Sketch the hyperbola given by $4x^2 - 3y^2 + 8x + 16 = 0$ and find the equations of its asymptotes.

Solution

$$4x^2 - 3y^2 + 8x + 16 = 0 \qquad \text{Write original equation.}$$
$$4(x^2 + 2x) - 3y^2 = -16 \qquad \text{Subtract 16 from each side and factor.}$$
$$4(x^2 + 2x + 1) - 3y^2 = -16 + 4 \qquad \text{Add 4 to each side.}$$
$$4(x + 1)^2 - 3y^2 = -12 \qquad \text{Complete the square.}$$
$$\frac{y^2}{4} - \frac{(x + 1)^2}{3} = 1 \qquad \text{Standard form}$$

From this equation you can conclude that the hyperbola is centered at $(-1, 0)$, has vertices at $(-1, 2)$ and $(-1, -2)$, and has a conjugate axis with ends at $(-1 - \sqrt{3}, 0)$ and $(-1 + \sqrt{3}, 0)$. To sketch the hyperbola, draw a rectangle through these four points. The asymptotes are the lines passing through the corners of the rectangle, as shown in Figure 6.31. Finally, using $a = 2$ and $b = \sqrt{3}$, you can conclude that the equations of the asymptotes are

$$y = \frac{2}{\sqrt{3}}(x + 1) \quad \text{and} \quad y = -\frac{2}{\sqrt{3}}(x + 1).$$

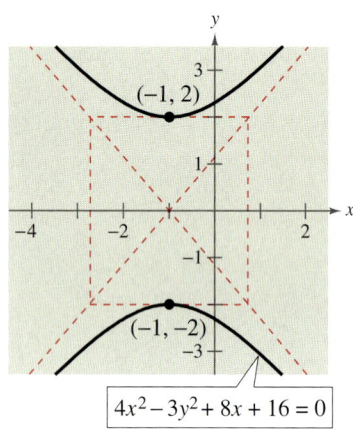

FIGURE 6.31

If the constant term F in the equation in Example 3 had been 4 instead of 16, you would have obtained the following degenerate case.

Two Intersecting Lines: $\dfrac{y^2}{4} - \dfrac{(x + 1)^2}{3} = 0$

Technology

You can use a graphing utility to graph a hyperbola by graphing the upper and lower portions in the same viewing window. For instance, to graph the hyperbola in Example 3, first solve for y to get

$$y_1 = 2\sqrt{1 + \frac{(x + 1)^2}{3}} \quad \text{and} \quad y_2 = -2\sqrt{1 + \frac{(x + 1)^2}{3}}.$$

Use a viewing window in which $-9 \leq x \leq 9$ and $-6 \leq y \leq 6$. You should obtain the graph shown below.

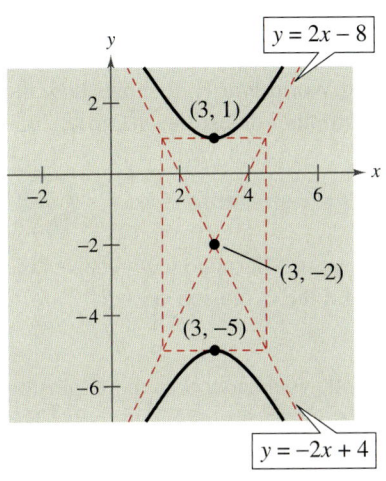

FIGURE 6.32

Example 4 ▶ Using Asymptotes to Find the Standard Equation

Find the standard form of the equation of the hyperbola having vertices at $(3, -5)$ and $(3, 1)$ and having asymptotes

$$y = 2x - 8 \quad \text{and} \quad y = -2x + 4$$

as shown in Figure 6.32.

Solution

According to the Midpoint Formula, the center of the hyperbola is at $(3, -2)$. Furthermore, the hyperbola has a vertical transverse axis with $a = 3$. From the given equations, you can determine the slopes of the asymptotes to be

$$m_1 = 2 = \frac{a}{b} \quad \text{and} \quad m_2 = -2 = -\frac{a}{b}$$

and, because $a = 3$,

$$2 = \frac{a}{b} \quad \Longrightarrow \quad 2 = \frac{3}{b} \quad \Longrightarrow \quad b = \frac{3}{2}.$$

So, the standard equation is

$$\frac{(y + 2)^2}{9} - \frac{(x - 3)^2}{9/4} = 1.$$

As with ellipses, the **eccentricity** of a hyperbola is

$$e = \frac{c}{a} \qquad \textcolor{red}{\text{Eccentricity}}$$

and because $c > a$ it follows that $e > 1$. If the eccentricity is large, the branches of the hyperbola are nearly flat, as shown in Figure 6.33(a). If the eccentricity is close to 1, the branches of the hyperbola are more pointed, as shown in Figure 6.33(b).

(a)

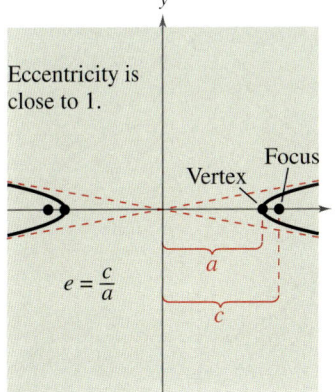

(b)

FIGURE 6.33

Applications

The following application was developed during World War II. It shows how the properties of hyperbolas can be used in radar and other detection systems.

Example 5 ▶ An Application Involving Hyperbolas

Two microphones, 1 mile apart, record an explosion. Microphone A receives the sound 2 seconds before microphone B. Where did the explosion occur?

Solution

Assuming sound travels at 1100 feet per second, you know that the explosion took place 2200 feet farther from B than from A, as shown in Figure 6.34. The locus of all points that are 2200 feet closer to A than to B is one branch of the hyperbola

$$\frac{x^2}{a^2} - \frac{y^2}{b^2} = 1$$

where

$$c = \frac{5280}{2} = 2640$$

and

$$a = \frac{2200}{2} = 1100.$$

So, $b^2 = c^2 - a^2 = 5{,}759{,}600$, and you conclude that the explosion occurred somewhere on the right branch of the hyperbola

$$\frac{x^2}{1{,}210{,}000} - \frac{y^2}{5{,}759{,}600} = 1.$$

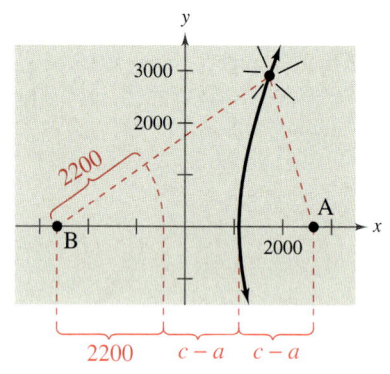

FIGURE 6.34

Another interesting application of conic sections involves the orbits of comets in our solar system. Of the 610 comets identified prior to 1970, 245 have elliptical orbits, 295 have parabolic orbits, and 70 have hyperbolic orbits. The center of the sun is a focus of each of these orbits, and each orbit has a vertex at the point where the comet is closest to the sun, as shown in Figure 6.35. Undoubtedly, there have been many comets with parabolic or hyperbolic orbits that were not identified. We only get to see such comets *once*. Comets with elliptical orbits, such as Halley's comet, are the only ones that remain in our solar system.

If p is the distance between the vertex and the focus in meters, and v is the velocity of the comet at the vertex in meters per second, the type of orbit is determined as follows.

1. *Ellipse:* $v < \sqrt{2GM/p}$
2. *Parabola:* $v = \sqrt{2GM/p}$
3. *Hyperbola:* $v > \sqrt{2GM/p}$

In each of these equations, $M \approx 1.991 \times 10^{30}$ kilograms (the mass of the sun) and $G \approx 6.67 \times 10^{-11}$ cubic meters per kilogram-second squared.

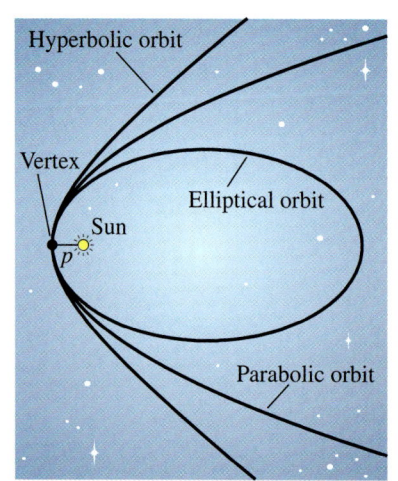

FIGURE 6.35

General Equations of Conics

Classifying a Conic from Its General Equation

The graph of $Ax^2 + Cy^2 + Dx + Ey + F = 0$ is one of the following.

1. **Circle:** $A = C$
2. **Parabola:** $AC = 0$ $A = 0$ or $C = 0$, but not both.
3. **Ellipse:** $AC > 0$ A and C have like signs.
4. **Hyperbola:** $AC < 0$ A and C have unlike signs.

The test above is valid if the graph is a conic. The test does not apply to equations such as $x^2 + y^2 = -1$, which is not a conic.

Example 6 ▶ Classifying Conics from General Equations

Classify each graph.

a. $4x^2 - 9x + y - 5 = 0$
b. $4x^2 - y^2 + 8x - 6y + 4 = 0$
c. $2x^2 + 4y^2 - 4x + 12y = 0$
d. $2x^2 + 2y^2 - 8x + 12y + 26 = 0$

Solution

a. For the equation $4x^2 - 9x + y - 5 = 0$, you have

$$AC = 4(0) = 0. \qquad \text{Parabola}$$

So, the graph is a parabola.

b. For the equation $4x^2 - y^2 + 8x - 6y + 4 = 0$, you have

$$AC = 4(-1) < 0. \qquad \text{Hyperbola}$$

So, the graph is a hyperbola.

c. For the equation $2x^2 + 4y^2 - 4x + 12y = 0$, you have

$$AC = 2(4) > 0. \qquad \text{Ellipse}$$

So, the graph is an ellipse.

d. For the equation $2x^2 + 2y^2 - 8x + 12y + 2 = 0$, you have

$$A = C = 2. \qquad \text{Circle}$$

So, the graph is a circle.

Historical Note
Caroline Herschel (1750–1848) was the first woman to be credited with detecting a new comet. During her long life, this English astronomer discovered a total of eight new comets.

Writing ABOUT MATHEMATICS

Sketching Conics Sketch each of the conics described in Example 6. Write a paragraph describing the procedures that allow you to sketch the conics efficiently.

6.4 Exercises

In Exercises 1–4, match the equation with its graph. [The graphs are labeled (a), (b), (c), and (d).]

(a)

(b)

(c)

(d)
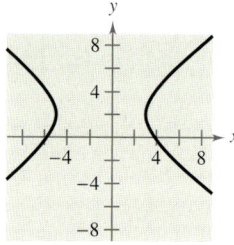

1. $\dfrac{y^2}{9} - \dfrac{x^2}{25} = 1$

2. $\dfrac{y^2}{25} - \dfrac{x^2}{9} = 1$

3. $\dfrac{(x-1)^2}{16} - \dfrac{y^2}{4} = 1$

4. $\dfrac{(x+1)^2}{16} - \dfrac{(y-2)^2}{9} = 1$

In Exercises 5–16, find the center, vertices, foci, and the equations of the asymptotes of the hyperbola, and sketch its graph.

5. $x^2 - y^2 = 1$

6. $\dfrac{x^2}{9} - \dfrac{y^2}{25} = 1$

7. $\dfrac{y^2}{25} - \dfrac{x^2}{81} = 1$

8. $\dfrac{x^2}{36} - \dfrac{y^2}{4} = 1$

9. $\dfrac{(x-1)^2}{4} - \dfrac{(y+2)^2}{1} = 1$

10. $\dfrac{(x+3)^2}{144} - \dfrac{(y-2)^2}{25} = 1$

11. $\dfrac{(y+6)^2}{1/9} - \dfrac{(x-2)^2}{1/4} = 1$

12. $\dfrac{(y-1)^2}{1/4} - \dfrac{(x+3)^2}{1/16} = 1$

13. $9x^2 - y^2 - 36x - 6y + 18 = 0$

14. $x^2 - 9y^2 + 36y - 72 = 0$

15. $x^2 - 9y^2 + 2x - 54y - 80 = 0$

16. $16y^2 - x^2 + 2x + 64y + 63 = 0$

In Exercises 17–20, find the center, vertices, foci, and the equations of the asymptotes of the hyperbola. Use a graphing utility to graph the hyperbola and its asymptotes.

17. $2x^2 - 3y^2 = 6$

18. $6y^2 - 3x^2 = 18$

19. $9y^2 - x^2 + 2x + 54y + 62 = 0$

20. $9x^2 - y^2 + 54x + 10y + 55 = 0$

In Exercises 21–26, find the standard form of the equation of the specified hyperbola with center at the origin.

21. Vertices: $(0, \pm 2)$; Foci: $(0, \pm 4)$
22. Vertices: $(\pm 4, 0)$; Foci: $(\pm 6, 0)$
23. Vertices: $(\pm 1, 0)$; Asymptotes: $y = \pm 5x$
24. Vertices: $(0, \pm 3)$; Asymptotes: $y = \pm 3x$
25. Foci: $(0, \pm 8)$; Asymptotes: $y = \pm 4x$
26. Foci: $(\pm 10, 0)$; Asymptotes: $y = \pm \frac{3}{4}x$

In Exercises 27–38, find the standard form of the equation of the specified hyperbola.

27. Vertices: $(2, 0), (6, 0)$; Foci: $(0, 0), (8, 0)$
28. Vertices: $(2, 3), (2, -3)$; Foci: $(2, 6), (2, -6)$
29. Vertices: $(4, 1), (4, 9)$; Foci: $(4, 0), (4, 10)$
30. Vertices: $(-2, 1), (2, 1)$; Foci: $(-3, 1), (3, 1)$
31. Vertices: $(2, 3), (2, -3)$;
    Passes through the point $(0, 5)$
32. Vertices: $(-2, 1), (2, 1)$;
    Passes through the point $(5, 4)$
33. Vertices: $(0, 4), (0, 0)$;
    Passes through the point $(\sqrt{5}, -1)$
34. Vertices: $(1, 2), (1, -2)$;
    Passes through the point $(0, \sqrt{5})$
35. Vertices: $(1, 2), (3, 2)$
    Asymptotes: $y = x, \; y = 4 - x$
36. Vertices: $(3, 0), (3, 6)$
    Asymptotes: $y = 6 - x, \; y = x$
37. Vertices: $(0, 2), (6, 2)$;
    Asymptotes: $y = \frac{2}{3}x, \; y = 4 - \frac{2}{3}x$
38. Vertices: $(3, 0), (3, 4)$;
    Asymptotes: $y = \frac{2}{3}x, \; y = 4 - \frac{2}{3}x$

39. *Sound Location* Three listening stations located at (3300, 0), (3300, 1100), and (−3300, 0) monitor an explosion. If the last two stations detect the explosion 1 second and 4 seconds after the first, respectively, determine the coordinates of the explosion. (Assume that the coordinate system is measured in feet and that sound travels at 1100 feet per second.)

40. *LORAN* Long distance radio navigation for aircraft and ships uses synchronized pulses transmitted by widely separated transmitting stations. These pulses travel at the speed of light (186,000 miles per second). The difference in the times of arrival of these pulses at an aircraft or ship is constant on a hyperbola having the transmitting stations as foci. Assume that two stations, 300 miles apart, are positioned on the rectangular coordinate system at points with coordinates (−150, 0) and (150, 0), and that a ship is traveling on a path with coordinates $(x, 75)$ (see figure). Find the x-coordinate of the position of the ship if the time difference between the pulses from the transmitting stations is 1000 microseconds (0.001 second).

41. *Hyperbolic Mirror* A hyperbolic mirror (used in some telescopes) has the property that a light ray directed at a focus will be reflected to the other focus (see figure). The focus of a hyperbolic mirror has coordinates (24, 0). Find the vertex of the mirror if its mount has coordinates (24, 24).

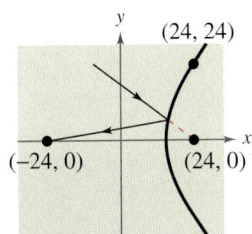

In Exercises 42–49, classify the graph of the equation as a circle, a parabola, an ellipse, or a hyperbola.

42. $x^2 + y^2 - 6x + 4y + 9 = 0$
43. $x^2 + 4y^2 - 6x + 16y + 21 = 0$
44. $4x^2 - y^2 - 4x - 3 = 0$
45. $y^2 - 6y - 4x + 21 = 0$
46. $4x^2 + 3y^2 + 8x - 24y + 51 = 0$
47. $4y^2 - 2x^2 - 4y - 8x - 15 = 0$
48. $25x^2 - 10x - 200y - 119 = 0$
49. $4y^2 + 4x^2 - 24x + 35 = 0$

Synthesis

True or False? In Exercises 50 and 51, determine whether the statement is true or false. Justify your answer.

50. In the standard form of the equation of a hyperbola, the larger the ratio of b to a, the larger the eccentricity of the hyperbola.

51. In the standard form of the equation of a hyperbola, the trivial solution of two intersecting lines occurs when $b = 0$.

52. Consider a hyperbola centered at the origin with a horizontal transverse axis. Use the definition of a hyperbola to derive its standard form.

53. Explain how the central rectangle of a hyperbola can be used to sketch its asymptotes.

Review

In Exercises 54–57, perform the indicated polynomial operation.

54. Subtract: $(x^3 - 3x^2) - (6 - 2x - 4x^2)$
55. Multiply: $(3x - \frac{1}{2})(x + 4)$
56. Divide: $\dfrac{x^3 - 3x + 4}{x + 2}$
57. Expand: $[(x + y) + 3]^2$

In Exercises 58–63, factor the given polynomial.

58. $x^3 - 16x$
59. $x^2 + 14x + 49$
60. $2x^3 - 24x^2 + 72x$
61. $6x^3 - 11x^2 - 10x$
62. $16x^3 + 54$
63. $4 - x + 4x^2 - x^3$

6.5 Rotation of Conics

What you should learn
- How to rotate the coordinate axes to eliminate the *xy*-term in the equation of a conic
- How to use the discriminant to classify a conic

Why you should learn it
As illustrated in Exercises 7–18 on page 446, rotation of the coordinate axes can help you identify the graph of a general second-degree equation.

Rotation

In the preceding section you learned that the equation of a conic with axes parallel to one of the coordinate axes has a standard form that can be written in the general form

$$Ax^2 + Cy^2 + Dx + Ey + F = 0.\qquad \text{Horizontal or vertical axis}$$

In this section you will study the equations of conics whose axes are rotated so that they are not parallel to either the *x*-axis or the *y*-axis. The general equation for such conics contains an *xy*-term.

$$Ax^2 + Bxy + Cy^2 + Dx + Ey + F = 0 \qquad \text{Equation in } xy\text{-plane}$$

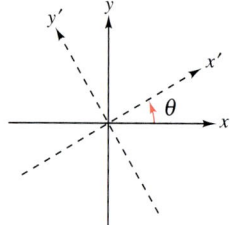

FIGURE 6.36

To eliminate this *xy*-term, you can use a procedure called **rotation of axes.** The objective is to rotate the *x*- and *y*-axes until they are parallel to the axes of the conic. The rotated axes are denoted as the x'-axis and the y'-axis, as shown in Figure 6.36. After the rotation, the equation of the conic in the new $x'y'$-plane will have the form

$$A'(x')^2 + C'(y')^2 + D'x' + E'y' + F' = 0. \qquad \text{Equation in } x'y'\text{-plane}$$

Because this equation has no *xy*-term, you can obtain a standard form by completing the square. The following theorem identifies how much to rotate the axes to eliminate the *xy*-term and also the equations for determining the new coefficients A', C', D', E', and F'.

Rotation of Axes to Eliminate an *xy*-Term
The general second-degree equation $Ax^2 + Bxy + Cy^2 + Dx + Ey + F = 0$ can be rewritten as

$$A'(x')^2 + C'(y')^2 + D'x' + E'y' + F' = 0$$

by rotating the coordinate axes through an angle θ, where

$$\cot 2\theta = \frac{A - C}{B}.$$

The coefficients of the new equation are obtained by making the substitutions

$$x = x' \cos \theta - y' \sin \theta \quad \text{and} \quad y = x' \sin \theta + y' \cos \theta.$$

STUDY TIP

Remember that the substitutions

$$x = x'\cos\theta - y'\sin\theta$$

and

$$y = x'\sin\theta + y'\cos\theta$$

were developed to eliminate the $x'y'$-term in the rotated system. You can use this as a check on your work. In other words, if your final equation contains an $x'y'$-term, you know that you made a mistake.

Example 1 ▶ Rotation of Axes for a Hyperbola

Write the equation $xy - 1 = 0$ in standard form.

Solution

Because $A = 0$, $B = 1$, and $C = 0$, you have

$$\cot 2\theta = \frac{A - C}{B} = 0 \quad\Longrightarrow\quad 2\theta = \frac{\pi}{2} \quad\Longrightarrow\quad \theta = \frac{\pi}{4}$$

which implies that

$$\begin{aligned}
x &= x'\cos\frac{\pi}{4} - y'\sin\frac{\pi}{4} \\
&= x'\left(\frac{\sqrt{2}}{2}\right) - y'\left(\frac{\sqrt{2}}{2}\right) \\
&= \frac{x' - y'}{\sqrt{2}}
\end{aligned}$$

and

$$\begin{aligned}
y &= x'\sin\frac{\pi}{4} + y'\cos\frac{\pi}{4} \\
&= x'\left(\frac{\sqrt{2}}{2}\right) + y'\left(\frac{\sqrt{2}}{2}\right) \\
&= \frac{x' + y'}{\sqrt{2}}.
\end{aligned}$$

The equation in the $x'y'$-system is obtained by substituting these expressions in the equation $xy - 1 = 0$.

$$\left(\frac{x' - y'}{\sqrt{2}}\right)\left(\frac{x' + y'}{\sqrt{2}}\right) - 1 = 0$$

$$\frac{(x')^2 - (y')^2}{2} - 1 = 0$$

$$\frac{(x')^2}{(\sqrt{2})^2} - \frac{(y')^2}{(\sqrt{2})^2} = 1 \qquad \text{Standard form}$$

In the $x'y'$-system, this is a hyperbola centered at the origin with vertices at $(\pm\sqrt{2}, 0)$, as shown in Figure 6.37. To find the coordinates of the vertices in the xy-system, substitute the coordinates $(\pm\sqrt{2}, 0)$ in the equations

$$x = \frac{x' - y'}{\sqrt{2}} \qquad \text{and} \qquad y = \frac{x' + y'}{\sqrt{2}}.$$

This substitution yields the vertices $(1, 1)$ and $(-1, -1)$ in the xy-system. Note also that the asymptotes of the hyperbola have equations $y' = \pm x'$, which correspond to the original x- and y-axes.

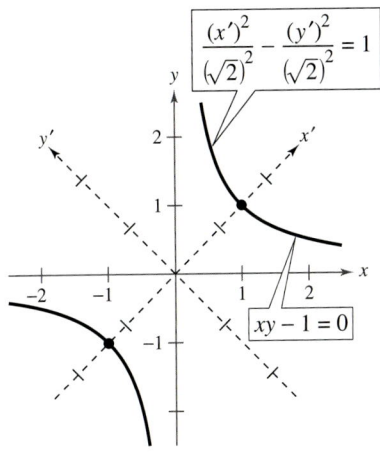

Vertices:
In $x'y'$-system: $(\sqrt{2}, 0), (-\sqrt{2}, 0)$
In xy-system: $(1, 1), (-1, -1)$
FIGURE 6.37

Example 2 Rotation of Axes for an Ellipse

Sketch the graph of $7x^2 - 6\sqrt{3}xy + 13y^2 - 16 = 0$.

Solution

Because $A = 7$, $B = -6\sqrt{3}$, and $C = 13$, you have

$$\cot 2\theta = \frac{A - C}{B}$$

$$= \frac{7 - 13}{-6\sqrt{3}}$$

$$= \frac{1}{\sqrt{3}}$$

which implies that $\theta = \pi/6$. The equation in the $x'y'$-system is obtained by making the substitutions

$$x = x' \cos \frac{\pi}{6} - y' \sin \frac{\pi}{6}$$

$$= x'\left(\frac{\sqrt{3}}{2}\right) - y'\left(\frac{1}{2}\right)$$

$$= \frac{\sqrt{3}x' - y'}{2}$$

and

$$y = x' \sin \frac{\pi}{6} + y' \cos \frac{\pi}{6}$$

$$= x'\left(\frac{1}{2}\right) + y'\left(\frac{\sqrt{3}}{2}\right)$$

$$= \frac{x' + \sqrt{3}y'}{2}$$

in the original equation. So, you have

$$7x^2 - 6\sqrt{3}xy + 13y^2 - 16 = 0$$

$$7\left(\frac{\sqrt{3}x' - y'}{2}\right)^2 - 6\sqrt{3}\left(\frac{\sqrt{3}x' - y'}{2}\right)\left(\frac{x' + \sqrt{3}y'}{2}\right) + 13\left(\frac{x' + \sqrt{3}y'}{2}\right)^2 - 16 = 0$$

which simplifies to

$$4(x')^2 + 16(y')^2 - 16 = 0$$

$$4(x')^2 + 16(y')^2 = 16$$

$$\frac{(x')^2}{4} + \frac{(y')^2}{1} = 1. \qquad \text{Standard form}$$

This is the equation of an ellipse centered at the origin with vertices $(\pm 2, 0)$ in the $x'y'$-system, as shown in Figure 6.38.

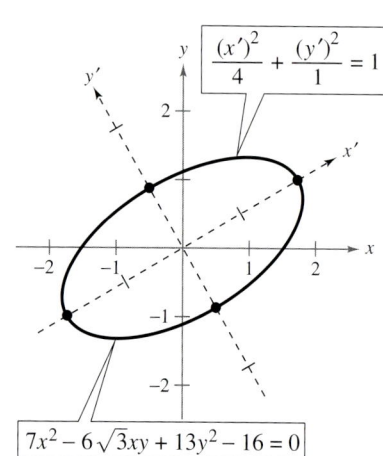

$7x^2 - 6\sqrt{3}xy + 13y^2 - 16 = 0$

Vertices:

In $x'y'$-system: $(\pm 2, 0)$, $(0, \pm 1)$

In xy-system: $(\sqrt{3}, 1)$, $(-\sqrt{3}, -1)$,

$\left(\dfrac{1}{2}, -\dfrac{\sqrt{3}}{2}\right)$, $\left(-\dfrac{1}{2}, \dfrac{\sqrt{3}}{2}\right)$

FIGURE 6.38

Example 3 ▶ Rotation of Axes for a Parabola

Sketch the graph of $x^2 - 4xy + 4y^2 + 5\sqrt{5}y + 1 = 0$.

Solution

Because $A = 1$, $B = -4$, and $C = 4$, you have

$$\cot 2\theta = \frac{A - C}{B} = \frac{1 - 4}{-4} = \frac{3}{4}.$$

Using the identity $\cot 2\theta = (\cot^2 \theta - 1)/(2 \cot \theta)$ produces

$$\cot 2\theta = \frac{3}{4} = \frac{\cot^2 \theta - 1}{2 \cot \theta}$$

from which you obtain the equation

$$4 \cot^2 \theta - 4 = 6 \cot \theta$$

$$4 \cot^2 \theta - 6 \cot \theta - 4 = 0$$

$$(2 \cot \theta - 4)(2 \cot \theta + 1) = 0.$$

Considering $0 < \theta < \pi/2$, you have $2 \cot \theta = 4$. So,

$$\cot \theta = 2 \quad \Longrightarrow \quad \theta \approx 26.6°.$$

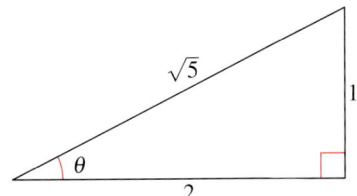

FIGURE 6.39

From the triangle in Figure 6.39, you obtain $\sin \theta = 1/\sqrt{5}$ and $\cos \theta = 2/\sqrt{5}$. Consequently, you use the substitutions

$$x = x' \cos \theta - y' \sin \theta$$

$$= x'\left(\frac{2}{\sqrt{5}}\right) - y'\left(\frac{1}{\sqrt{5}}\right) = \frac{2x' - y'}{\sqrt{5}}$$

$$y = x' \sin \theta + y' \cos \theta$$

$$= x'\left(\frac{1}{\sqrt{5}}\right) + y'\left(\frac{2}{\sqrt{5}}\right) = \frac{x' + 2y'}{\sqrt{5}}.$$

Substituting these expressions in the original equation, you have

$$x^2 - 4xy + 4y^2 + 5\sqrt{5}y + 1 = 0$$

$$\left(\frac{2x' - y'}{\sqrt{5}}\right)^2 - 4\left(\frac{2x' - y'}{\sqrt{5}}\right)\left(\frac{x' + 2y'}{\sqrt{5}}\right) + 4\left(\frac{x' + 2y'}{\sqrt{5}}\right)^2 + 5\sqrt{5}\left(\frac{x' + 2y'}{\sqrt{5}}\right) + 1 = 0$$

which simplifies as follows.

$$5(y')^2 + 5x' + 10y' + 1 = 0$$

$$5(y' + 1)^2 = -5x' + 4 \qquad \text{Complete the square.}$$

$$(y' + 1)^2 = (-1)\left(x' - \frac{4}{5}\right) \qquad \text{Standard form}$$

The graph of this equation is a parabola with vertex at $\left(\frac{4}{5}, -1\right)$. Its axis is parallel to the x'-axis in the $x'y'$-system, as shown in Figure 6.40.

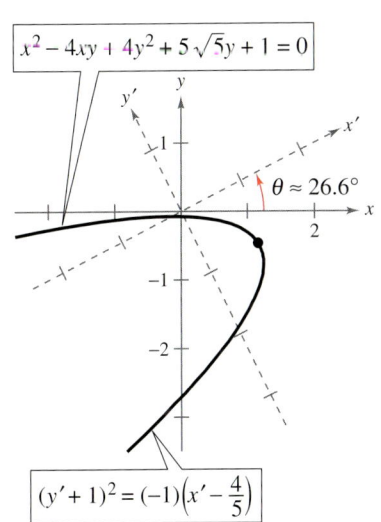

Vertex:
In $x'y'$-system: $\left(\frac{4}{5}, -1\right)$

In xy-system: $\left(\frac{13}{5\sqrt{5}}, -\frac{6}{5\sqrt{5}}\right)$

FIGURE 6.40

Invariants Under Rotation

In the rotation of axes theorem listed at the beginning of this section, note that the constant term is the same in both equations, $F' = F$. Such quantities are **invariant under rotation.** The next theorem lists some other rotation invariants.

Rotation Invariants

The rotation of the coordinate axes through an angle θ that transforms the equation $Ax^2 + Bxy + Cy^2 + Dx + Ey + F = 0$ into the form

$$A'(x')^2 + C'(y')^2 + D'x' + E'y' + F' = 0$$

has the following rotation invariants.

1. $F = F'$
2. $A + C = A' + C'$
3. $B^2 - 4AC = (B')^2 - 4A'C'$

You can use the results of this theorem to classify the graph of a second-degree equation *with* an xy-term in much the same way you do for a second-degree equation *without* an xy-term. Note that because $B' = 0$, the invariant $B^2 - 4AC$ reduces to

$$B^2 - 4AC = -4A'C'. \qquad \text{Discriminant}$$

This quantity is called the **discriminant** of the equation

$$Ax^2 + Bxy + Cy^2 + Dx + Ey + F = 0.$$

Now, from the classification procedure given in Section 6.4, you know that the sign of $A'C'$ determines the type of graph for the equation

$$A'(x')^2 + C'(y')^2 + D'x' + E'y' + F' = 0.$$

Consequently, the sign of $B^2 - 4AC$ will determine the type of graph for the original equation, as given in the following classification.

Classification of Conics by the Discriminant

The graph of the equation $Ax^2 + Bxy + Cy^2 + Dx + Ey + F = 0$ is, except in degenerate cases, determined by its discriminant as follows.

1. *Ellipse or circle:* $B^2 - 4AC < 0$
2. *Parabola:* $\qquad \;\; B^2 - 4AC = 0$
3. *Hyperbola:* $\qquad B^2 - 4AC > 0$

For example, in the general equation

$$3x^2 + 7xy + 5y^2 - 6x - 7y + 15 = 0$$

you have $A = 3$, $B = 7$, and $C = 5$. So the discriminant is

$$B^2 - 4AC = 7^2 - 4(3)(5) = 49 - 60 = -11.$$

Because $-11 < 0$, the graph of the equation is an ellipse or a circle.

Section 6.5 ▶ Rotation of Conics   **445**

FIGURE 6.41

FIGURE 6.42

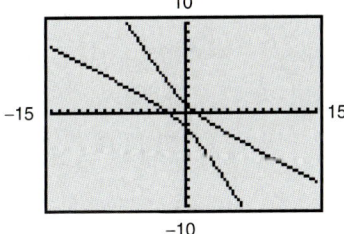

FIGURE 6.43

Example 4 ▶ Rotations and Graphing Utilities

For each of the following, classify the graph, use the Quadratic Formula to solve for y, and then use a graphing utility to graph the equation.

a. $2x^2 - 3xy + 2y^2 - 2x = 0$ **b.** $x^2 - 6xy + 9y^2 - 2y + 1 = 0$
c. $3x^2 + 8xy + 4y^2 - 7 = 0$

Solution

a. Because $B^2 - 4AC = 9 - 16 < 0$, the graph is a circle or an ellipse. Solve for y as follows.

$$2x^2 - 3xy + 2y^2 - 2x = 0 \quad \text{Write original equation.}$$

$$2y^2 - 3xy + (2x^2 - 2x) = 0 \quad \text{Quadratic form } ay^2 + by + c = 0$$

$$y = \frac{-(-3x) \pm \sqrt{(-3x)^2 - 4(2)(2x^2 - 2x)}}{2(2)}$$

Graph both of the equations to obtain the ellipse in Figure 6.41.

$$y = \frac{3x + \sqrt{9x^2 - 16(x^2 - x)}}{4} \quad \text{Top half of ellipse}$$

$$y = \frac{3x - \sqrt{9x^2 - 16(x^2 - x)}}{4} \quad \text{Bottom half of ellipse}$$

b. Because $B^2 - 4AC = 36 - 36 = 0$, the graph is a parabola.

$$x^2 - 6xy + 9y^2 - 2y + 1 = 0 \quad \text{Write original equation.}$$

$$9y^2 - (6x + 2)y + (x^2 + 1) = 0 \quad \text{Quadratic form } ay^2 + by + c = 0$$

$$y = \frac{(6x + 2) \pm \sqrt{(6x + 2)^2 - 4(9)(x^2 + 1)}}{18}$$

Graphing the resulting two equations gives the parabola in Figure 6.42.

c. Because $B^2 - 4AC = 64 - 48 > 0$, the graph is a hyperbola.

$$3x^2 + 8xy + 4y^2 - 7 = 0 \quad \text{Write original equation.}$$

$$4y^2 + 8xy + (3x^2 - 7) = 0 \quad \text{Quadratic form } ay^2 + by + c = 0$$

$$y = \frac{-8x \pm \sqrt{(8x)^2 - 4(4)(3x^2 - 7)}}{8}$$

The graphs of the resulting two equations yield the hyperbola in Figure 6.43.

Writing ABOUT MATHEMATICS

Classifying a Graph as a Hyperbola The graph of $f(x) = 1/x$ is a hyperbola. Discuss how you could use the techniques in this section to verify this, and then do so. Compare your statement with that of another student.

6.5 Exercises

In Exercises 1–6, the $x'y'$-coordinate system has been rotated θ degrees from the xy-coordinate system. The coordinates of a point in the xy-coordinate system are given. Find the coordinates of the point in the rotated coordinate system.

1. $\theta = 90°, (0, 3)$
2. $\theta = 45°, (3, 3)$
3. $\theta = 30°, (1, 3)$
4. $\theta = 60°, (3, 1)$
5. $\theta = 45°, (2, 1)$
6. $\theta = 30°, (2, 4)$

In Exercises 7–18, rotate the axes to eliminate the xy-term. Sketch the graph of the resulting equation, showing both sets of axes.

7. $xy + 1 = 0$
8. $xy - 2 = 0$
9. $x^2 - 2xy + y^2 - 1 = 0$
10. $xy + x - 2y + 3 = 0$
11. $xy - 2y - 4x = 0$
12. $2x^2 - 3xy - 2y^2 + 10 = 0$
13. $5x^2 - 6xy + 5y^2 - 12 = 0$
14. $13x^2 + 6\sqrt{3}xy + 7y^2 - 16 = 0$
15. $3x^2 - 2\sqrt{3}xy + y^2 + 2x + 2\sqrt{3}y = 0$
16. $16x^2 - 24xy + 9y^2 - 60x - 80y + 100 = 0$
17. $9x^2 + 24xy + 16y^2 + 90x - 130y = 0$
18. $9x^2 + 24xy + 16y^2 + 80x - 60y = 0$

In Exercises 19–26, use a graphing utility to graph the conic. Determine the angle θ through which the axes are rotated. Explain how you used the graphing utility to obtain the graph.

19. $x^2 + 2xy + y^2 = 20$
20. $x^2 - 4xy + 2y^2 = 6$
21. $17x^2 + 32xy - 7y^2 = 75$
22. $40x^2 + 36xy + 25y^2 = 52$
23. $32x^2 + 48xy + 8y^2 = 50$
24. $24x^2 + 18xy + 12y^2 = 34$
25. $4x^2 - 12xy + 9y^2 + (4\sqrt{13} - 12)x - (6\sqrt{13} + 8)y = 91$
26. $6x^2 - 4xy + 8y^2 + (5\sqrt{5} - 10)x - (7\sqrt{5} + 5)y = 80$

In Exercises 27–32, match the graph with its equation. [The graphs are labeled (a), (b), (c), (d), (e), and (f).]

(a)
(b)
(c)
(d)
(e)
(f)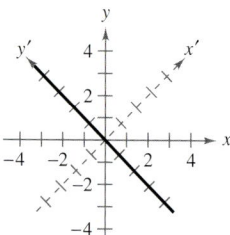

27. $xy + 2 = 0$
28. $x^2 + 2xy + y^2 = 0$
29. $-2x^2 + 3xy + 2y^2 + 3 = 0$
30. $x^2 - xy + 3y^2 - 5 = 0$
31. $3x^2 + 2xy + y^2 - 10 = 0$
32. $x^2 - 4xy + 4y^2 + 10x - 30 = 0$

In Exercises 33–40, use the discriminant to classify the graph, use the Quadratic Formula to solve for y, and then use a graphing utility to graph the equation.

33. $16x^2 - 8xy + y^2 - 10x + 5y = 0$
34. $x^2 - 4xy - 2y^2 - 6 = 0$
35. $12x^2 - 6xy + 7y^2 - 45 = 0$
36. $2x^2 + 4xy + 5y^2 + 3x - 4y - 20 = 0$
37. $x^2 - 6xy - 5y^2 + 4x - 22 = 0$
38. $36x^2 - 60xy + 25y^2 + 9y = 0$
39. $x^2 + 4xy + 4y^2 - 5x - y - 3 = 0$
40. $x^2 + xy + 4y^2 + x + y - 4 = 0$

In Exercises 41–44, sketch (if possible) the graph of the degenerate conic.

41. $y^2 - 9x^2 = 0$
42. $x^2 + y^2 - 2x + 6y + 10 = 0$
43. $x^2 + 2xy + y^2 - 1 = 0$
44. $x^2 - 10xy + y^2 = 0$

In Exercises 45–58, find any points of intersection of the graphs algebraically and then verify using a graphing utility.

45. $-x^2 + y^2 + 4x - 6y + 4 = 0$
    $x^2 + y^2 - 4x - 6y + 12 = 0$
46. $-x^2 - y^2 - 8x + 20y - 7 = 0$
    $x^2 + 9y^2 + 8x + 4y + 7 = 0$
47. $-4x^2 - y^2 - 16x + 24y - 16 = 0$
    $4x^2 + y^2 + 40x - 24y + 208 = 0$
48. $x^2 - 4y^2 - 20x - 64y - 172 = 0$
    $16x^2 + 4y^2 - 320x + 64y + 1600 = 0$
49. $x^2 - y^2 - 12x + 16y - 64 = 0$
    $x^2 + y^2 - 12x - 16y + 64 = 0$
50. $x^2 + 4y^2 - 2x - 8y + 1 = 0$
    $-x^2 + 2x - 4y - 1 = 0$
51. $-16x^2 - y^2 + 24y - 80 = 0$
    $16x^2 + 25y^2 - 400 = 0$
52. $16x^2 - y^2 + 16y - 128 = 0$
    $y^2 - 48x - 16y - 32 = 0$
53. $x^2 + y^2 - 4 = 0$
    $3x - y^2 = 0$
54. $4x^2 + 9y^2 - 36y = 0$
    $x^2 + 9y - 27 = 0$
55. $x^2 + 2y^2 - 4x + 6y - 5 = 0$
    $-x + y - 4 = 0$
56. $x^2 + 2y^2 - 4x + 6y - 5 = 0$
    $x^2 - 4x - y + 4 = 0$
57. $xy + x - 2y + 3 = 0$
    $x^2 + 4y^2 - 9 = 0$
58. $5x^2 - 2xy + 5y^2 - 12 = 0$
    $x + y - 1 = 0$

Synthesis

True or False? **In Exercises 59 and 60, determine whether the statement is true or false. Justify your answer.**

59. The graph of the equation
    $$x^2 + xy + ky^2 + 6x + 10 = 0$$
    where k is any constant less than $\frac{1}{4}$, is a hyperbola.

60. After a rotation of axes is used to eliminate the xy-term from an equation of the form
    $$Ax^2 + Bxy + Cy^2 + Dx + Ey + F = 0$$
    the coefficients of the x^2- and y^2-terms remain A and C, respectively.

61. Show that the equation $x^2 + y^2 = r^2$ is invariant under rotation of axes.

62. Find the lengths of the major and minor axes of the ellipse graphed in Exercise 14.

Review

In Exercises 63–68, find the zeros of the function.

63. $f(x) = \sqrt{6x - 5}$
64. $f(x) = \sqrt{2x - 9} - 3$
65. $g(x) = 3x^2 - 14x + 8$
66. $g(x) = 2x^2 - 11x - 6$
67. $h(x) = x^3 - 7x^2 + 17x - 15$
68. $h(x) = x^4 - 11x^3 + 39x^2 - 29x$

In Exercises 69–76, graph the function.

69. $f(x) = |x - 4|$
70. $f(x) = |x + 3| + 1$
71. $g(x) = \sqrt{4 - x^2}$
72. $g(x) = \sqrt{3x - 2}$
73. $h(x) = x^2 - 10x + 21$
74. $h(x) = 2x^2 - x - 6$
75. $f(t) = -(t - 2)^3 + 3$
76. $f(t) = \frac{1}{2}(t + 4)^3 - 1$

6.6 Parametric Equations

▶ **What you should learn**
- How to evaluate a set of parametric equations for a given value of the parameter
- How to sketch the curve that is represented by a set of parametric equations
- How to rewrite a set of parametric equations as a single rectangular equation
- How to find a set of parametric equations for a graph

▶ **Why you should learn it**
Parametric equations are useful for modeling the path of an object. For instance, in Exercise 60 on page 455, you will use a set of parametric equations to model the path of an arrow.

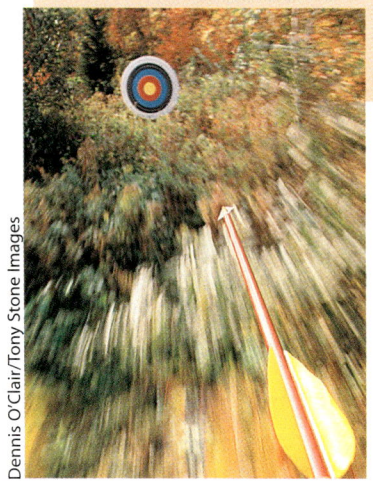

A computer animation of this concept appears in the *Interactive* CD-ROM and *Internet* versions of this text.

Plane Curves

Up to this point you have been representing a graph by a single equation involving the *two* variables x and y. In this section, you will study situations in which it is useful to introduce a *third* variable to represent a curve in the plane.

To see the usefulness of this procedure, consider the path followed by an object that is propelled into the air at an angle of $45°$. If the initial velocity of the object is 48 feet per second, it can be shown that the object follows the parabolic path

$$y = -\frac{x^2}{72} + x \qquad \text{Rectangular equation}$$

as shown in Figure 6.44. However, this equation does not tell the whole story. Although it does tell you *where* the object has been, it doesn't tell you *when* the object was at a given point (x, y) on the path. To determine this time, you can introduce a third variable t, called a **parameter.** It is possible to write both x and y as functions of t to obtain the **parametric equations**

$$x = 24\sqrt{2}\,t \qquad \text{Parametric equation for } x$$
$$y = -16t^2 + 24\sqrt{2}\,t. \qquad \text{Parametric equation for } y$$

From this set of equations you can determine that at time $t = 0$, the object is at the point $(0, 0)$. Similarly, at time $t = 1$, the object is at the point $(24\sqrt{2},\ 24\sqrt{2} - 16)$, and so on.

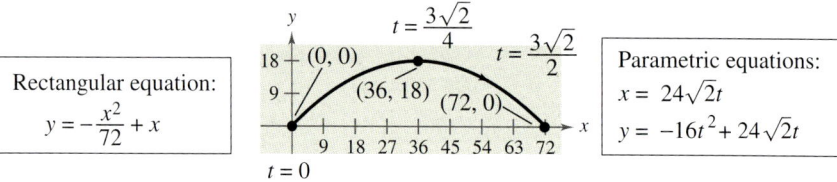

FIGURE 6.44 *Curvilinear Motion: Two Variables for Position, One Variable for Time*

For this particular motion problem, x and y are continuous functions of t, and the resulting path is a **plane curve.** (For this text, it is sufficient to think of a *continuous function* as one whose graph can be traced without lifting the pencil from the paper.)

Definition of Plane Curve

If f and g are continuous functions of t on an interval I, the set of ordered pairs $(f(t), g(t))$ is a **plane curve** C. The equations

$$x = f(t) \quad \text{and} \quad y = g(t)$$

are **parametric equations** for C, and t is the **parameter.**

Sketching a Plane Curve

When sketching a curve represented by a pair of parametric equations, you still plot points in the xy-plane. Each set of coordinates (x, y) is determined from a value chosen for the parameter t. Plotting the resulting points in the order of *increasing* values of t traces the curve in a specific direction. This is called the **orientation** of the curve.

 A computer animation of this example appears in the *Interactive* CD-ROM and *Internet* versions of this text.

Example 1 ▶ Sketching a Curve

Sketch the curve described by the parametric equations

$$x = t^2 - 4 \quad \text{and} \quad y = \frac{t}{2}, \quad -2 \leq t \leq 3.$$

Solution

Using values of t in the given interval, the parametric equations yield the points (x, y) shown in the table.

t	-2	-1	0	1	2	3
x	0	-3	-4	-3	0	5
y	-1	$-1/2$	0	$1/2$	1	$3/2$

By plotting these points in the order of increasing t, you obtain the curve C shown in Figure 6.45. Note that the arrows on the curve indicate its orientation as t increases from -2 to 3.

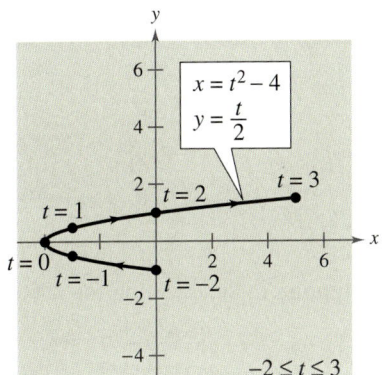

FIGURE 6.45

Note that the graph shown in Figure 6.45 does not define y as a function of x. This points out one benefit of parametric equations—they can be used to represent graphs that are more general than graphs of functions.

It often happens that two different sets of parametric equations have the same graph. For example, the set of parametric equations

$$x = 4t^2 - 4 \quad \text{and} \quad y = t, \quad -1 \leq t \leq \frac{3}{2}$$

has the same graph as the set given in Example 1. However, by comparing the values of t in Figures 6.45 and 6.46, you see that this second graph is traced out more *rapidly* (considering t as time) than the first graph. So, in applications, different parametric representations can be used to represent various *speeds* at which objects travel along a given path.

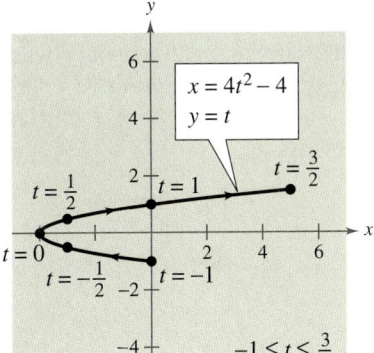

FIGURE 6.46

Eliminating the Parameter

Example 1 uses simple point plotting to sketch the given curve. This tedious process can sometimes be simplified by finding a rectangular equation (in x and y) that has the same graph. This process is called **eliminating the parameter**.

Parametric equations	⇒	Solve for t in one equation.	⇒	Substitute in other equation.	⇒	Rectangular equation
$x = t^2 - 4$ $y = t/2$		$t = 2y$		$x = (2y)^2 - 4$		$x = 4y^2 - 4$

Now you can recognize that the equation $x = 4y^2 - 4$ represents a parabola with a horizontal axis and a vertex at $(-4, 0)$.

When converting equations from parametric to rectangular form, you may need to alter the domain of the rectangular equation so that its graph matches the graph of the parametric equations. Such a situation is demonstrated in Example 2.

Example 2 ▶ Eliminating the Parameter

Sketch the curve represented by the equations $x = 1/\sqrt{t+1}$ and $y = t/(t+1)$ by eliminating the parameter and adjusting the domain of the resulting rectangular equation.

Solution

Solving for t in the equation for x, you have

$$x = \frac{1}{\sqrt{t+1}} \quad \Rightarrow \quad x^2 = \frac{1}{t+1}$$

which implies that $t = (1 - x^2)/x^2$. Now, substituting in the equation for y, you obtain

$$y = \frac{t}{t+1} = \frac{\frac{(1-x^2)}{x^2}}{\left[\frac{(1-x^2)}{x^2}\right] + 1} = \frac{\frac{1-x^2}{x^2}}{\frac{1-x^2}{x^2} + 1} \cdot \frac{x^2}{x^2} = 1 - x^2.$$

The rectangular equation, $y = 1 - x^2$, is defined for all values of x, but from the parametric equation for x you can see that the curve is defined only when $t > -1$. This implies that you should restrict the domain of x to positive values, as shown in Figure 6.47.

◀ **Exploration** ▶

Most graphing utilities have a parametric graphing mode. If yours does, try entering the parametric equations given in Example 2. Over what values should you let t vary to obtain the graph shown in Figure 6.47?

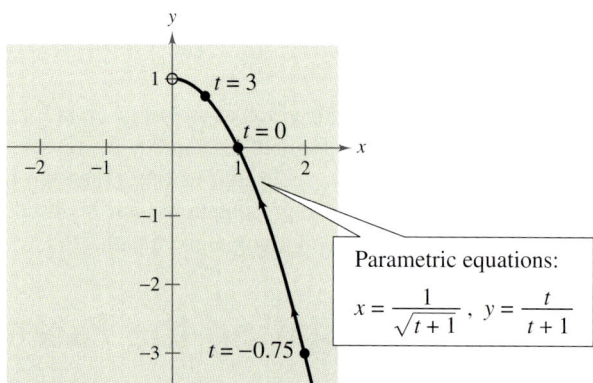

FIGURE 6.47

It is not necessary for the parameter in a set of parametric equations to represent time. The next example uses an *angle* as the parameter.

Example 3 ▶ Eliminating the Parameter

Sketch the curve represented by

$$x = 3 \cos \theta \quad \text{and} \quad y = 4 \sin \theta, \quad 0 \leq \theta \leq 2\pi$$

by eliminating the parameter.

Solution

Begin by solving for $\cos \theta$ and $\sin \theta$ in the equations.

$$\cos \theta = \frac{x}{3} \quad \text{and} \quad \sin \theta = \frac{y}{4} \qquad \text{Solve for } \cos \theta \text{ and } \sin \theta.$$

Make use of the identity $\sin^2 \theta + \cos^2 \theta = 1$ to form an equation involving only x and y.

$$\cos^2 \theta + \sin^2 \theta = 1 \qquad \text{Trigonometric identity}$$

$$\left(\frac{x}{3}\right)^2 + \left(\frac{y}{4}\right)^2 = 1 \qquad \text{Substitute } \frac{x}{3} \text{ for } \cos \theta \text{ and } \frac{y}{4} \text{ for } \sin \theta.$$

$$\frac{x^2}{9} + \frac{y^2}{16} = 1 \qquad \text{Rectangular equation}$$

From this rectangular equation, you can see that the graph is an ellipse centered at $(0, 0)$, with vertices at $(0, 4)$ and $(0, -4)$ and minor axis of length $2b = 6$, as shown in Figure 6.48. Note that the elliptic curve is traced out *counterclockwise* as θ varies from 0 to 2π.

In Examples 2 and 3 it is important to realize that eliminating the parameter is primarily an *aid to curve sketching*. If the parametric equations represent the path of a moving object, the graph alone is not sufficient to describe the object's motion. You still need the parametric equations to tell you the *position, direction,* and *speed* at a given time.

Finding Parametric Equations for a Graph

You have been studying techniques for sketching the graph represented by a set of parametric equations. Now consider the reverse problem—that is, how can you find a set of parametric equations for a given graph or a given physical description? From the discussion following Example 1, you know that such a representation is not unique. That is, the equations

$$x = 4t^2 - 4 \quad \text{and} \quad y = t, \quad -1 \leq t \leq \frac{3}{2}$$

produced the same graph as the equations

$$x = t^2 - 4 \quad \text{and} \quad y = \frac{t}{2}, \quad -2 \leq t \leq 3.$$

This is further demonstrated in Example 4.

STUDY TIP

To eliminate the parameter in equations involving trigonometric functions, try using the identities

$$\sin^2 \theta + \cos^2 \theta = 1$$

$$\sec^2 \theta - \tan^2 \theta = 1$$

as done in Example 3.

FIGURE 6.48

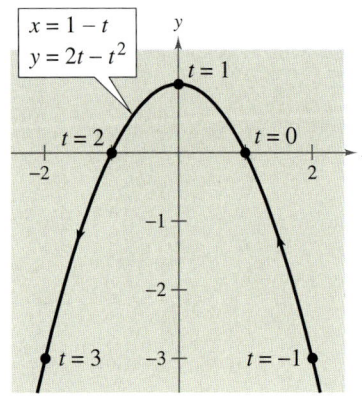

FIGURE 6.49

Example 4 ▶ Finding Parametric Equations for a Given Graph

Find a set of parametric equations to represent the graph of $y = 1 - x^2$, using the following parameters.

a. $t = x$ **b.** $t = 1 - x$

Solution

a. Letting $t = x$, you obtain the parametric equations
$$x = t \quad \text{and} \quad y = 1 - x^2 = 1 - t^2.$$

b. Letting $t = 1 - x$, you obtain
$$x = 1 - t \quad \text{and} \quad y = 1 - (1 - t)^2 = 2t - t^2.$$

In Figure 6.49, note how the resulting curve is oriented by the increasing values of t. For part (a), the curve would have the opposite orientation.

Example 5 ▶ Parametric Equations for a Cycloid

Describe the **cycloid** traced out by a point P on the circumference of a circle of radius a as the circle rolls along a straight line in a plane.

Solution

As the parameter, let θ be the measure of the circle's rotation, and let the point $P = (x, y)$ begin at the origin. When $\theta = 0$, P is at the origin; when $\theta = \pi$, P is at a maximum point $(\pi a, 2a)$; and when $\theta = 2\pi$, P is back on the x-axis at $(2\pi a, 0)$. From Figure 6.50, you can see that $\angle APC = 180° - \theta$. So, you have

$$\sin \theta = \sin(180° - \theta) = \sin(\angle APC) = \frac{AC}{a} = \frac{BD}{a}$$

$$\cos \theta = -\cos(180° - \theta) = -\cos(\angle APC) = \frac{AP}{-a}$$

which implies that $AP = -a \cos \theta$ and $BD = a \sin \theta$. Because the circle rolls along the x-axis, you know that $OD = \overset{\frown}{PD} = a\theta$. Furthermore, because $BA = DC = a$, you have

$$x = OD - BD = a\theta - a \sin \theta \quad \text{and} \quad y = BA + AP = a - a \cos \theta.$$

Therefore, the parametric equations are

$$x = a(\theta - \sin \theta) \quad \text{and} \quad y = a(1 - \cos \theta).$$

STUDY TIP

In Example 5, $\overset{\frown}{PD}$ represents the arc of the circle between points P and D.

Technology

Use a graphing utility in parametric mode to obtain a graph similar to Figure 6.50 by graphing the following equations.

$X_{1T} = T - \sin T$

$Y_{1T} = 1 - \cos T$

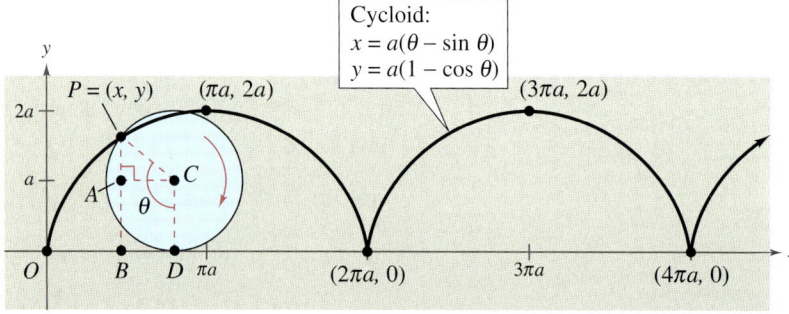

FIGURE 6.50

A computer animation of this example appears in the *Interactive* CD-ROM and *Internet* versions of this text.

6.6 Exercises

1. Consider the parametric equations $x = \sqrt{t}$ and $y = 3 - t$.
   (a) Create a table of x- and y-values using $t = 0, 1, 2, 3,$ and 4.
   (b) Plot the points (x, y) generated in part (a), and sketch a graph of the parametric equations.
   (c) Find the rectangular equation by eliminating the parameter. Sketch its graph. How do the graphs differ?

2. Consider the parametric equations $x = 4\cos^2\theta$ and $y = 2\sin\theta$.
   (a) Create a table of x- and y-values using $\theta = -\pi/2, -\pi/4, 0, \pi/4,$ and $\pi/2$.
   (b) Plot the points (x, y) generated in part (a), and sketch a graph of the parametric equations.
   (c) Find the rectangular equation by eliminating the parameter. Sketch its graph. How do the graphs differ?

In Exercises 3–22, sketch the curve represented by the parametric equations (indicate the direction of the curve) by eliminating the parameter and adjusting the domain of the resulting rectangular equation.

3. $x = 3t - 3$
   $y = 2t + 1$

4. $x = 3 - 2t$
   $y = 2 + 3t$

5. $x = \frac{1}{4}t$
   $y = t^2$

6. $x = t$
   $y = t^3$

7. $x = t + 2$
   $y = t^2$

8. $x = \sqrt{t}$
   $y = 1 - t$

9. $x = t + 1$
   $y = t/(t + 1)$

10. $x = t - 1$
    $y = t/(t - 1)$

11. $x = 2(t + 1)$
    $y = |t - 2|$

12. $x = |t - 1|$
    $y = t + 2$

13. $x = 3\cos\theta$
    $y = 3\sin\theta$

14. $x = 2\cos\theta$
    $y = 3\sin\theta$

15. $x = 4\sin 2\theta$
    $y = 2\cos 2\theta$

16. $x = \cos\theta$
    $y = 2\sin 2\theta$

17. $x = 4 + 2\cos\theta$
    $y = -1 + \sin\theta$

18. $x = 4 + 2\cos\theta$
    $y = 2 + 3\sin\theta$

19. $x = e^{-t}$
    $y = e^{3t}$

20. $x = e^{2t}$
    $y = e^t$

21. $x = t^3$
    $y = 3\ln t$

22. $x = \ln 2t$
    $y = 2t^2$

In Exercises 23 and 24, determine how the plane curves differ from each other.

23. (a) $x = t$
       $y = 2t + 1$
    (b) $x = \cos\theta$
        $y = 2\cos\theta + 1$
    (c) $x = e^{-t}$
        $y = 2e^{-t} + 1$
    (d) $x = e^t$
        $y = 2e^t + 1$

24. (a) $x = t$
       $y = t^2 - 1$
    (b) $x = t^2$
        $y = t^4 - 1$
    (c) $x = \sin t$
        $y = \sin^2 t - 1$
    (d) $x = e^t$
        $y = e^{2t} - 1$

In Exercises 25–28, eliminate the parameter and obtain the standard form of the rectangular equation.

25. Line through (x_1, y_1) and (x_2, y_2):
    $x = x_1 + t(x_2 - x_1), \ y = y_1 + t(y_2 - y_1)$

26. Circle:
    $x = h + r\cos\theta, \ y = k + r\sin\theta$

27. Ellipse:
    $x = h + a\cos\theta, \ y = k + b\sin\theta$

28. Hyperbola:
    $x = h + a\sec\theta, \ y = k + b\tan\theta$

In Exercises 29–36, use the results of Exercises 25–28 to find a set of parametric equations for the line or conic.

29. Line: Passes through $(0, 0)$ and $(6, -3)$
30. Line: Passes through $(2, 3)$ and $(6, -3)$
31. Circle: Center: $(3, 2)$; Radius: 4
32. Circle: Center: $(-3, 2)$; Radius: 5
33. Ellipse: Vertices: $(\pm 4, 0)$; Foci: $(\pm 3, 0)$
34. Ellipse: Vertices: $(4, 7), (4, -3)$;
    Foci: $(4, 5), (4, -1)$
35. Hyperbola: Vertices: $(\pm 4, 0)$; Foci: $(\pm 5, 0)$
36. Hyperbola: Vertices: $(0, \pm 2)$; Foci: $(0, \pm 4)$

In Exercises 37–44, find a set of parametric equations for the given rectangular equation using (a) $t = x$ **and (b)** $t = 2 - x$.

37. $y = 3x - 2$
38. $x = 3y - 2$
39. $y = x^2$
40. $y = x^3$
41. $y = x^2 + 1$
42. $y = 2 - x$
43. $y = \dfrac{1}{x}$
44. $y = \dfrac{1}{2x}$

 In Exercises 45–52, use a graphing utility to obtain a graph of the curve represented by the parametric equations.

45. Cycloid: $x = 4(\theta - \sin \theta)$, $y = 4(1 - \cos \theta)$
46. Cycloid: $x = \theta + \sin \theta$, $y = 1 - \cos \theta$
47. Prolate cycloid: $x = \theta - \frac{3}{2} \sin \theta$, $y = 1 - \frac{3}{2} \cos \theta$
48. Prolate cycloid: $x = 2\theta - 4 \sin \theta$, $y = 2 - 4 \cos \theta$
49. Hypocycloid: $x = 3 \cos^3 \theta$, $y = 3 \sin^3 \theta$
50. Curtate cycloid: $x = 8\theta - 4 \sin \theta$, $y = 8 - 4 \cos \theta$
51. Witch of Agnesi: $x = 2 \cot \theta$, $y = 2 \sin^2 \theta$
52. Folium of Descartes: $x = \dfrac{3t}{1 + t^3}$, $y = \dfrac{3t^2}{1 + t^3}$

In Exercises 53–56, match the parametric equations with the correct graph and describe the domain and range. [The graphs are labeled (a), (b), (c), and (d).]

(a)

(b)

(c)

(d)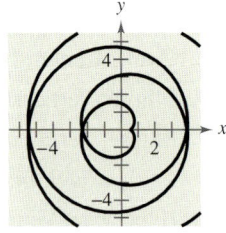

53. Lissajous curve: $x = 2 \cos \theta$
$y = \sin 2\theta$
54. Evolute of ellipse: $x = 4 \cos^3 \theta$
$y = 6 \sin^3 \theta$
55. Involute of circle: $x = \frac{1}{2}(\cos \theta + \theta \sin \theta)$
$y = \frac{1}{2}(\sin \theta - \theta \cos \theta)$
56. Serpentine curve: $x = \frac{1}{2} \cot \theta$
$y = 4 \sin \theta \cos \theta$

Projectile Motion A projectile is launched at a height of h feet above the ground and at an angle θ with the horizontal. If the initial velocity is v_0 feet per second, the path of the projectile is modeled by the parametric equations

$$x = (v_0 \cos \theta)t \quad \text{and} \quad y = h + (v_0 \sin \theta)t - 16t^2.$$

In Exercises 57 and 58, use a graphing utility to graph the paths of a projectile launched from ground level at the specified values of θ and v_0. For each case, use the graph to approximate the maximum height and the range of the projectile.

57. (a) $\theta = 60°$, $v_0 = 88$ ft/sec
    (b) $\theta = 60°$, $v_0 = 132$ ft/sec
    (c) $\theta = 45°$, $v_0 = 88$ ft/sec
    (d) $\theta = 45°$, $v_0 = 132$ ft/sec
58. (a) $\theta = 15°$, $v_0 = 60$ ft/sec
    (b) $\theta = 15°$, $v_0 = 100$ ft/sec
    (c) $\theta = 30°$, $v_0 = 60$ ft/sec
    (d) $\theta = 30°$, $v_0 = 100$ ft/sec

59. **Baseball** The center field fence in a ballpark is 10 feet high and 400 feet from home plate. The baseball is hit 3 feet above the ground. It leaves the bat at an angle of θ degrees with the horizontal at a speed of 100 miles per hour (see figure).

    (a) Write a set of parametric equations for the path of the baseball.
    (b) Use a graphing utility to sketch the path of the baseball if $\theta = 15°$. Is the hit a home run?
    (c) Use a graphing utility to sketch the path of the baseball if $\theta = 23°$. Is the hit a home run?
    (d) Find the minimum angle required for the hit to be a home run.

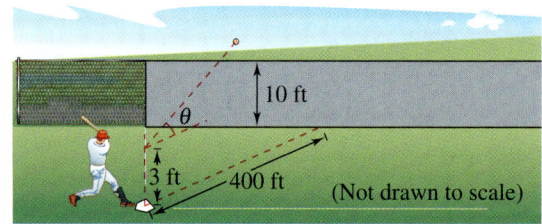

60. Archery An archer releases an arrow from a bow 5 feet above the ground. The arrow leaves the bow at an angle of 10° with the horizontal and at an initial speed of 240 feet per second.

(a) Write a set of parametric equations for the path of the arrow.

(b) Assuming the ground is level, find the distance the arrow travels before it hits the ground. (Ignore air resistance.)

(c) Use a graphing utility to graph the path of the arrow and approximate its maximum height.

(d) Find the time the arrow is in the air.

61. Projectile Motion Eliminate the parameter t from the position function for the motion of a projectile to show that the rectangular equation is

$$y = -\frac{16 \sec^2 \theta}{v_0^2} x^2 + (\tan \theta)x + h.$$

62. Path of a Projectile The path of a projectile is given by the rectangular equation

$$y = 7 + x - 0.02x^2.$$

(a) Use the result of Exercise 61 to find h, v_0, and θ. Find the parametric equations of the path.

(b) Use a graphing utility to graph the rectangular equation for the path of the projectile. Confirm your answer in part (a) by sketching the curve represented by the parametric equations.

(c) Use a graphing utility to approximate the maximum height of the projectile and its range.

63. Curtate Cycloid A wheel of radius a rolls along a straight line without slipping. The curve traced by a point P that is b units from the center ($b < a$) is called a **curtate cycloid** (see figure). Use the angle θ shown in the figure to find a set of parametric equations for the curve.

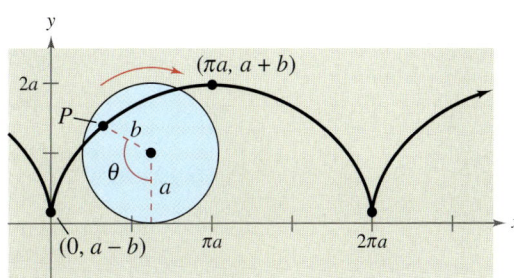

64. Epicycloid A circle of radius 1 rolls around the outside of a circle of radius 2 without slipping. The curve traced by a point on the circumference of the smaller circle is called an **epicycloid** (see figure). Use the angle θ shown in the figure to find a set of parametric equations for the curve.

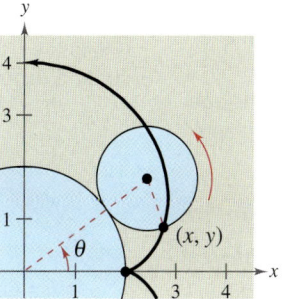

Synthesis

True or False? In Exercises 65 and 66, determine whether the statement is true or false. Justify your answer.

65. The two sets of parametric equations $x = t$, $y = t^2 + 1$ and $x = 3t$, $y = 9t^2 + 1$ have the same rectangular equation.

66. The graph of the parametric equations $x = t^2$ and $y = t^2$ is the line $y = x$.

Review

In Exercises 67–70, find the equation of the parabola

$$y = ax^2 + bx + c$$

that passes through the given points.

67. $(0, 0)$, $(3, -3)$, $(6, 0)$

68. $(0, -6)$, $(2, -2)$, $(4, -6)$

69. $(4, 12)$, $(8, -2)$, $(12, 12)$

70. $(-8, -6)$, $(-7, -8)$, $(-6, -6)$

6.7 Polar Coordinates

What you should learn
- How to plot points in the polar coordinate system
- How to convert points from rectangular to polar form and vice versa
- How to convert equations from rectangular to polar form and vice versa

Why you should learn it
Polar coordinates offer a different mathematical perspective on graphing. For instance, in Exercises 1–8 on page 460, you see that a polar coordinate can be written in more than one way.

A computer animation of this concept appears in the *Interactive* CD-ROM and *Internet* versions of this text.

Introduction

So far, you have been representing graphs of equations as collections of points (x, y) on the rectangular coordinate system, where x and y represent the directed distances from the coordinate axes to the point (x, y). In this section you will study a different system called the **polar coordinate system.**

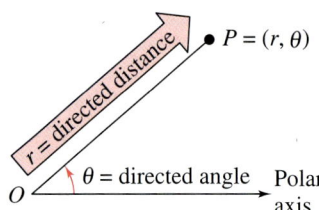

FIGURE 6.51

To form the polar coordinate system in the plane, fix a point O, called the **pole** (or **origin**), and construct from O an initial ray called the **polar axis,** as shown in Figure 6.51. Then each point P in the plane can be assigned **polar coordinates** (r, θ) as follows.

1. $r = $ *directed distance* from O to P
2. $\theta = $ *directed angle*, counterclockwise from polar axis to segment $\overline{OP}$

Example 1 ▶ Plotting Points on the Polar Coordinate System

a. The point $(r, \theta) = (2, \pi/3)$ lies 2 units from the pole on the terminal side of the angle $\theta = \pi/3$, as shown in Figure 6.52(a).
b. The point $(r, \theta) = (3, -\pi/6)$ lies 3 units from the pole on the terminal side of the angle $\theta = -\pi/6$, as shown in Figure 6.52(b).
c. The point $(r, \theta) = (3, 11\pi/6)$ coincides with the point $(3, -\pi/6)$, as shown in Figure 6.52(c).

(a)

(b)

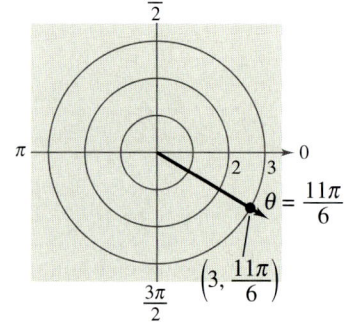
(c)

FIGURE 6.52

Section 6.7 ▶ Polar Coordinates   457

Exploration

Most graphing calculators have a *polar* graphing mode. If yours does, try graphing the equation $r = 3$. (Use a setting of $-6 \le x \le 6$ and $-4 \le y \le 4$.) You should obtain a circle of radius 3.

a. Use the trace feature to cursor around the circle. Can you locate the point $(3, 5\pi/4)$?

b. Can you find other polar representations of the point $(3, 5\pi/4)$? If so, explain how you did it.

In rectangular coordinates, each point (x, y) has a unique representation. This is not true for polar coordinates. For instance, the coordinates (r, θ) and $(r, \theta + 2\pi)$ represent the same point, as illustrated in Example 1. Another way to obtain multiple representations of a point is to use negative values for r. Because r is a *directed distance*, the coordinates (r, θ) and $(-r, \theta + \pi)$ represent the same point. In general, the point (r, θ) can be represented as

$$(r, \theta) = (r, \theta \pm 2n\pi) \qquad \text{or} \qquad (r, \theta) = (-r, \theta \pm (2n+1)\pi)$$

where n is any integer. Moreover, the pole is represented by $(0, \theta)$, where θ is any angle.

Example 2 ▶ Multiple Representation of Points

Plot the point $(3, -3\pi/4)$ and find three additional polar representations of this point, using $-2\pi < \theta < 2\pi$.

Solution

The point is shown in Figure 6.53. Three other representations are as follows.

$$\left(3, -\frac{3\pi}{4} + 2\pi\right) = \left(3, \frac{5\pi}{4}\right) \qquad \text{Add } 2\pi \text{ to } \theta.$$

$$\left(-3, -\frac{3\pi}{4} - \pi\right) = \left(-3, -\frac{7\pi}{4}\right) \qquad \text{Replace } r \text{ by } -r; \text{ subtract } \pi \text{ from } \theta.$$

$$\left(-3, -\frac{3\pi}{4} + \pi\right) = \left(-3, \frac{\pi}{4}\right) \qquad \text{Replace } r \text{ by } -r; \text{ add } \pi \text{ to } \theta.$$

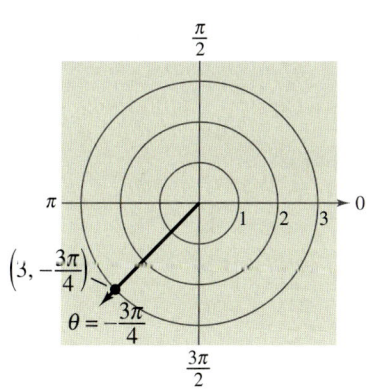

$\left(3, -\frac{3\pi}{4}\right) = \left(3, \frac{5\pi}{4}\right) = \left(-3, -\frac{7\pi}{4}\right) = \left(-3, \frac{\pi}{4}\right) = \ldots$

FIGURE 6.53

Coordinate Conversion

To establish the relationship between polar and rectangular coordinates, let the polar axis coincide with the positive x-axis and the pole with the origin, as shown in Figure 6.54. Because (x, y) lies on a circle of radius r, it follows that $r^2 = x^2 + y^2$. Moreover, for $r > 0$, the definitions of the trigonometric functions imply that

$$\tan \theta = \frac{y}{x}, \qquad \cos \theta = \frac{x}{r}, \qquad \text{and} \qquad \sin \theta = \frac{y}{r}.$$

If $r < 0$, you can show that the same relationships hold.

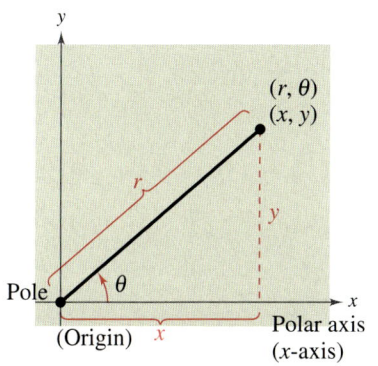

FIGURE 6.54

Coordinate Conversion

The polar coordinates (r, θ) are related to the rectangular coordinates (x, y) as follows.

$$x = r \cos \theta \qquad \text{and} \qquad \tan \theta = \frac{y}{x}$$

$$y = r \sin \theta \qquad \qquad r^2 = x^2 + y^2$$

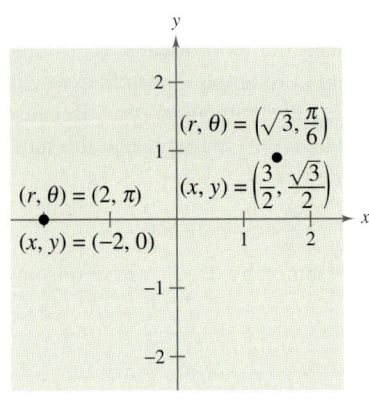

FIGURE 6.55

Example 3 ▶ Polar-to-Rectangular Conversion

Convert the points to rectangular coordinates. (See Figure 6.55.)

a. $(2, \pi)$ **b.** $(\sqrt{3}, \pi/6)$

Solution

a. For the point $(r, \theta) = (2, \pi)$, you have

$$x = r\cos\theta = 2\cos\pi = -2$$

and

$$y = r\sin\theta = 2\sin\pi = 0.$$

The rectangular coordinates are $(x, y) = (-2, 0)$.

b. For the point $(r, \theta) = (\sqrt{3}, \pi/6)$, you have

$$x = \sqrt{3}\cos\frac{\pi}{6} = \sqrt{3}\left(\frac{\sqrt{3}}{2}\right) = \frac{3}{2}$$

and

$$y = \sqrt{3}\sin\frac{\pi}{6} = \sqrt{3}\left(\frac{1}{2}\right) = \frac{\sqrt{3}}{2}.$$

The rectangular coordinates are $(x, y) = (3/2, \sqrt{3}/2)$.

Example 4 ▶ Rectangular-to-Polar Conversion

Convert the points to polar coordinates.

a. $(-1, 1)$ **b.** $(0, 2)$

Solution

a. For the second-quadrant point $(x, y) = (-1, 1)$, you have

$$\tan\theta = \frac{y}{x}$$

$$\tan\theta = -1$$

$$\theta = \frac{3\pi}{4}.$$

Because θ lies in the same quadrant as (x, y), use positive r.

$$r = \sqrt{x^2 + y^2} = \sqrt{(-1)^2 + (1)^2} = \sqrt{2}$$

So, *one* set of polar coordinates is $(r, \theta) = (\sqrt{2}, 3\pi/4)$, as shown in Figure 6.56(a).

b. Because the point $(x, y) = (0, 2)$ lies on the positive y-axis, choose

$$\theta = \frac{\pi}{2} \quad \text{and} \quad r = 2.$$

This implies that *one* set of polar coordinates is $(r, \theta) = (2, \pi/2)$, as shown in Figure 6.56(b).

(a)

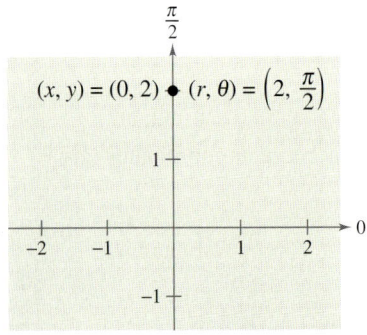

(b)

FIGURE 6.56

Equation Conversion

By comparing Examples 3 and 4, you can see that point conversion from the polar to the rectangular system is straightforward, whereas point conversion from the rectangular to the polar system is more involved. For equations, the opposite is true. To convert a rectangular equation to polar form, you simply replace x by $r\cos\theta$ and y by $r\sin\theta$. For instance, the rectangular equation $y = x^2$ can be written in polar form as follows.

$$y = x^2 \qquad \text{Rectangular equation}$$
$$r\sin\theta = (r\cos\theta)^2 \qquad \text{Polar equation}$$
$$r = \sec\theta\tan\theta \qquad \text{Simplest form}$$

On the other hand, converting a polar equation to rectangular form requires considerable ingenuity.

Example 5 demonstrates several polar-to-rectangular conversions that enable you to sketch the graphs of some polar equations.

Example 5 ▶ Converting Polar Equations to Rectangular Form

Describe the graph of each polar equation and find the corresponding rectangular equation.

a. $r = 2$ **b.** $\theta = \dfrac{\pi}{3}$ **c.** $r = \sec\theta$

Solution

a. The graph of the polar equation $r = 2$ consists of all points that are 2 units from the pole. In other words, this graph is a circle centered at the origin with a radius of 2, as shown in Figure 6.57(a). You can confirm this by converting to rectangular form, using the relationship $r^2 = x^2 + y^2$.

$$\underbrace{r = 2}_{\text{Polar equation}} \quad \Longrightarrow \quad r^2 = 2^2 \quad \Longrightarrow \quad \underbrace{x^2 + y^2 = 2^2}_{\text{Rectangular equation}}$$

b. The graph of the polar equation $\theta = \pi/3$ consists of all points on the line that makes an angle of $\pi/3$ with the positive polar axis, as shown in Figure 6.57(b). To convert to rectangular form, make use of the relationship $\tan\theta = y/x$.

$$\underbrace{\theta = \dfrac{\pi}{3}}_{\text{Polar equation}} \quad \Longrightarrow \quad \tan\theta = \sqrt{3} \quad \Longrightarrow \quad \underbrace{y = \sqrt{3}x}_{\text{Rectangular equation}}$$

c. The graph of the polar equation $r = \sec\theta$ is not evident by simple inspection, so you convert to rectangular form by using the relationship $r\cos\theta = x$.

$$\underbrace{r = \sec\theta}_{\text{Polar equation}} \quad \Longrightarrow \quad r\cos\theta = 1 \quad \Longrightarrow \quad \underbrace{x = 1}_{\text{Rectangular equation}}$$

Now you see that the graph is a vertical line, as shown in Figure 6.57(c).

(a)

(b)

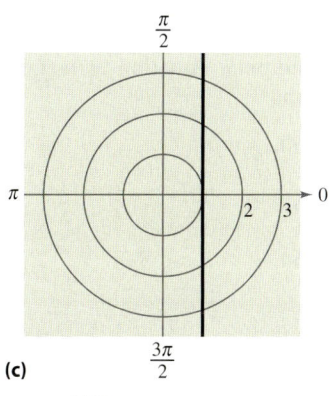
(c)

FIGURE 6.57

6.7 Exercises

In Exercises 1–8, plot the point given in polar coordinates and find two additional polar representations.

1. $(4, -\pi/3)$
2. $(-1, -3\pi/4)$
3. $(0, -7\pi/6)$
4. $(16, 5\pi/2)$
5. $(\sqrt{2}, 2.36)$
6. $(-3, -1.57)$
7. $(2\sqrt{2}, 4.71)$
8. $(-5, -2.36)$

In Exercises 9–16, a point in polar coordinates is given. Convert the point to rectangular coordinates.

9. $\left(3, \dfrac{\pi}{2}\right)$
10. $\left(3, \dfrac{3\pi}{2}\right)$

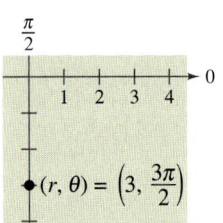

11. $\left(-1, \dfrac{5\pi}{4}\right)$
12. $(0, -\pi)$

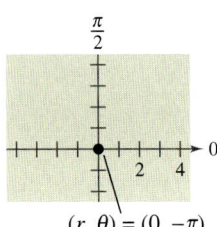

13. $(2, 3\pi/4)$
14. $(-2, 7\pi/6)$
15. $(-2.5, 1.1)$
16. $(8.25, 3.5)$

In Exercises 17–26, a point in rectangular coordinates is given. Convert the point to polar coordinates.

17. $(1, 1)$
18. $(-3, -3)$
19. $(-6, 0)$
20. $(0, -5)$
21. $(-3, 4)$
22. $(3, -1)$
23. $(-\sqrt{3}, -\sqrt{3})$
24. $(\sqrt{3}, -1)$
25. $(6, 9)$
26. $(5, 12)$

In Exercises 27–32, use a graphing utility to find one set of polar coordinates of the point given in rectangular coordinates.

27. $(3, -2)$
28. $(-5, 2)$
29. $(\sqrt{3}, 2)$
30. $(3\sqrt{2}, 3\sqrt{2})$
31. $\left(\dfrac{5}{2}, \dfrac{4}{3}\right)$
32. $\left(\dfrac{7}{4}, \dfrac{3}{2}\right)$

In Exercises 33–48, convert the rectangular equation to polar form.

33. $x^2 + y^2 = 9$
34. $x^2 + y^2 = 16$
35. $y = 4$
36. $x = 10$
37. $3x - y + 2 = 0$
38. $3x + 5y - 2 = 0$
39. $xy = 16$
40. $y = x$
41. $y^2 - 8x - 16 = 0$
42. $(x^2 + y^2)^2 = 9(x^2 - y^2)$
43. $x^2 + y^2 = a^2$
44. $x^2 + y^2 = 9a^2$
45. $y = b$
46. $x = 4a$
47. $x^2 + y^2 - 2ax = 0$
48. $x^2 + y^2 - 2ay = 0$

In Exercises 49–58, convert the polar equation to rectangular form.

49. $r = 4 \sin \theta$
50. $r = 3 \cos \theta$
51. $\theta = \dfrac{2\pi}{3}$
52. $r = 4$
53. $r = 2 \csc \theta$
54. $r^2 = \sin 2\theta$
55. $r = 2 \sin 3\theta$
56. $r = \dfrac{1}{1 - \cos \theta}$
57. $r = \dfrac{6}{2 - 3 \sin \theta}$
58. $r = \dfrac{6}{2 \cos \theta - 3 \sin \theta}$

In Exercises 59–64, convert the polar equation to rectangular form and sketch its graph.

59. $r = 6$
60. $r = 8$
61. $\theta = \dfrac{\pi}{6}$
62. $\theta = \dfrac{3\pi}{4}$
63. $r = 3 \sec \theta$
64. $r = 2 \csc \theta$

Synthesis

True or False? **In Exercises 65 and 66, determine whether the statement is true or false. Justify your answer.**

65. If $\theta_1 = \theta_2 + 2\pi n$ for some integer n, then (r, θ_1) and (r, θ_2) represent the same point on the polar coordinate system.

66. If $|r_1| = |r_2|$, then (r_1, θ) and (r_2, θ) represent the same point on the polar coordinate system.

67. Convert the polar equation

$$r = 2(h\cos\theta + k\sin\theta)$$

to rectangular form and verify that it is the equation of a circle. Find the radius and the rectangular coordinates of the center of the circle.

68. Convert the polar equation $r = \cos\theta + 3\sin\theta$ to rectangular form and identify the graph.

69. ***Think About It***
    (a) Show that the distance between the points (r_1, θ_1) and (r_2, θ_2) is

    $$\sqrt{r_1^2 + r_2^2 - 2r_1 r_2 \cos(\theta_1 - \theta_2)}.$$

    (b) Describe the positions of the points relative to each other if $\theta_1 = \theta_2$. Simplify the Distance Formula for this case. Is the simplification what you expected? Explain.

    (c) Simplify the Distance Formula if $\theta_1 - \theta_2 = 90°$. Is the simplification what you expected? Explain.

    (d) Choose two points on the polar coordinate system and find the distance between them. Then choose different polar representations of the same two points and apply the Distance Formula again. Discuss the result.

70. ***Exploration***
    (a) Set the window format of your graphing utility on rectangular coordinates and locate the cursor at any position off the coordinate axes. Move the cursor horizontally and observe any changes in the displayed coordinates of the points. Explain the changes. Now repeat the process moving the cursor vertically.

    (b) Set the window format of your graphing utility on polar coordinates and locate the cursor at any position off the coordinate axes. Move the cursor horizontally and observe any changes in the displayed coordinates of the points. Explain the changes. Now repeat the process moving the cursor vertically.

    (c) Explain why the results of parts (a) and (b) are not the same.

Review

In Exercises 71–76, use the Law of Cosines to solve the triangle.

71. $A = 22°$, $b = 9$, $c = 16$
72. $A = 101°$, $b = 6$, $c = 3$
73. $a = 5$, $b = 8$, $c = 10$
74. $a = 21$, $b = 17$, $c = 29$
75. $C = 38°$, $a = 4$, $b = 15$
76. $C = 49°$, $a = 12$, $b = 19$

In Exercises 77–80, use Heron's Area Formula to find the area of the triangle.

77. $a = 4$, $b = 8$, $c = 11$
78. $a = 11$, $b = 12$, $c = 21$
79. $a = 31$, $b = 48$, $c = 48$
80. $a = 27$, $b = 16$, $c = 12$

6.8 Graphs of Polar Equations

▶ **What you should learn**
- How to graph a polar equation by point plotting
- How to use symmetry, zeros, and maximum *r*-values as graphing aids
- How to recognize special polar graphs

▶ **Why you should learn it**

Equations of several common figures are simpler in polar form than in rectangular form. For instance, Exercise 6 on page 468 shows the graph of a circle and its polar equation.

Introduction

In previous chapters you spent a lot of time learning how to sketch graphs on rectangular coordinate systems. You began with the basic point-plotting method, which was then enhanced by sketching aids such as symmetry, intercepts, asymptotes, periods, and shifts. This section approaches curve sketching on the polar coordinate system similarly, beginning with a demonstration of point plotting.

Example 1 ▶ **Graphing a Polar Equation by Point Plotting**

Sketch the graph of the polar equation $r = 4 \sin \theta$.

Solution

The sine function is periodic, so you can get a full range of *r*-values by considering values of θ in the interval $0 \leq \theta \leq 2\pi$, as shown in the following table.

θ	0	$\frac{\pi}{6}$	$\frac{\pi}{3}$	$\frac{\pi}{2}$	$\frac{2\pi}{3}$	$\frac{5\pi}{6}$	π	$\frac{7\pi}{6}$	$\frac{3\pi}{2}$	$\frac{11\pi}{6}$	2π
r	0	2	$2\sqrt{3}$	4	$2\sqrt{3}$	2	0	-2	-4	-2	0

If you plot these points as shown in Figure 6.58, it appears that the graph is a circle of radius 2 whose center is at the point $(x, y) = (0, 2)$. Try confirming this by squaring both sides of the polar equation and converting the result to rectangular form.

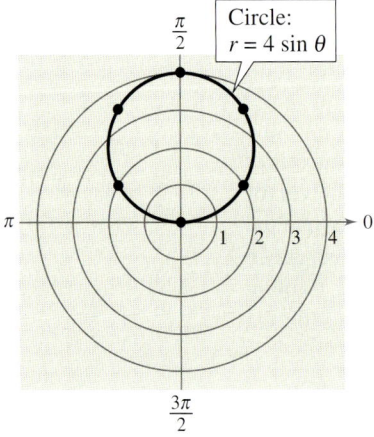

FIGURE 6.58

Symmetry

In Figure 6.58, note that as θ increases from 0 to 2π the graph is traced out twice. Moreover, note that the graph is *symmetric with respect to the line* $\theta = \pi/2$. Had you known about this symmetry and retracing ahead of time, you could have used fewer points.

Symmetry with respect to the line $\theta = \pi/2$ is one of three important types of symmetry to consider in polar curve sketching. (See Figure 6.59.)

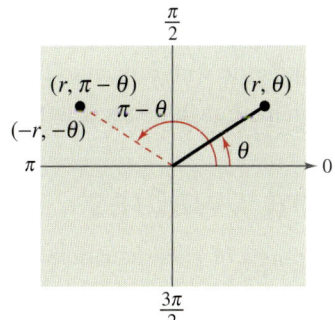
Symmetry with Respect to the Line $\theta = \dfrac{\pi}{2}$

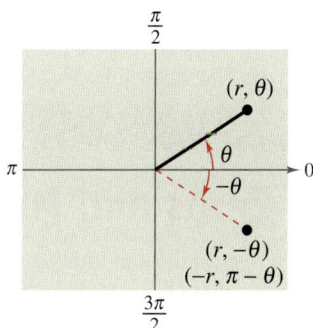
Symmetry with Respect to the Polar Axis

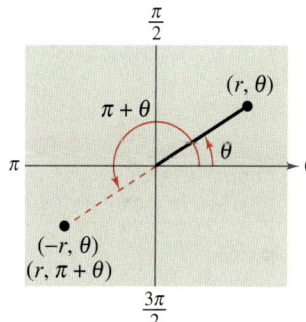
Symmetry with Respect to the Pole

FIGURE 6.59

Tests for Symmetry in Polar Coordinates

The graph of a polar equation is symmetric with respect to the following if the given substitution yields an equivalent equation.

1. *The line* $\theta = \pi/2$: Replace (r, θ) by $(r, \pi - \theta)$ or $(-r, -\theta)$.
2. *The polar axis:* Replace (r, θ) by $(r, -\theta)$ or $(-r, \pi - \theta)$.
3. *The pole:* Replace (r, θ) by $(r, \pi + \theta)$ or $(-r, \theta)$.

Example 2 ▶ Using Symmetry to Sketch a Polar Graph

Use symmetry to sketch the graph of $r = 3 + 2\cos\theta$.

Solution

Replacing (r, θ) by $(r, -\theta)$ produces $r = 3 + 2\cos(-\theta) = 3 + 2\cos\theta$. So, you can conclude that the curve is symmetric with respect to the polar axis. Plotting the points in the table and using polar axis symmetry, you obtain the graph of a **limaçon,** as shown in Figure 6.60.

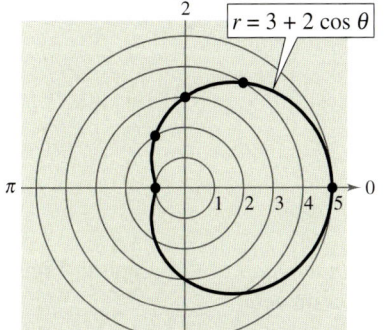

FIGURE 6.60

θ	0	$\dfrac{\pi}{3}$	$\dfrac{\pi}{2}$	$\dfrac{2\pi}{3}$	π
r	5	4	3	2	1

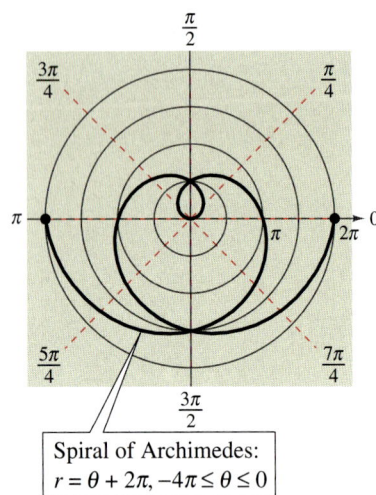

Spiral of Archimedes:
$r = \theta + 2\pi, -4\pi \leq \theta \leq 0$

FIGURE 6.61

The three tests for symmetry in polar coordinates listed on page 463 are sufficient to guarantee symmetry, but they are not necessary. For instance, Figure 6.61 shows the graph of $r = \theta + 2\pi$ to be symmetric with respect to the line $\theta = \pi/2$, and yet the tests on page 463 fail to indicate symmetry.

The equations discussed in Examples 1 and 2 are of the form

$$r = 4 \sin \theta = f(\sin \theta) \quad \text{and} \quad r = 3 + 2 \cos \theta = g(\cos \theta).$$

The graph of the first equation is symmetric with respect to the line $\theta = \pi/2$, and the graph of the second equation is symmetric with respect to the polar axis. This observation can be generalized to yield the following *quick tests for symmetry*.

1. The graph of $r = f(\sin \theta)$ is symmetric with respect to the line $\theta = \pi/2$.
2. The graph of $r = g(\cos \theta)$ is symmetric with respect to the polar axis.

Zeros and Maximum *r*-Values

Two additional aids to sketching graphs of polar equations involve knowing the θ-values for which $|r|$ is maximum and knowing the θ-values for which $r = 0$. For instance, in Example 1, the maximum value of $|r|$ for $r = 4 \sin \theta$ is $|r| = 4$, and this occurs when $\theta = \pi/2$, as shown in Figure 6.58. Moreover, $r = 0$ when $\theta = 0$.

Example 3 ▶ Sketching a Polar Graph

Sketch the graph of

$$r = 1 - 2 \cos \theta.$$

Solution

From the equation $r = 1 - 2 \cos \theta$, you can obtain the following.

Symmetry: With respect to the polar axis
Maximum value of $|r|$: $r = 3$ when $\theta = \pi$
Zero of r: $r = 0$ when $\theta = \pi/3$

The table shows several θ-values in the interval $[0, \pi]$. By plotting the corresponding points, you can sketch the graph shown in Figure 6.62.

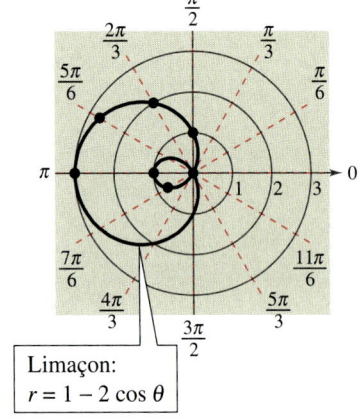

Limaçon:
$r = 1 - 2 \cos \theta$

FIGURE 6.62

θ	0	$\dfrac{\pi}{6}$	$\dfrac{\pi}{3}$	$\dfrac{\pi}{2}$	$\dfrac{2\pi}{3}$	$\dfrac{5\pi}{6}$	π
r	-1	-0.73	0	1	2	2.73	3

Note how the negative *r*-values determine the *inner loop* of the graph in Figure 6.62. This graph, like the one in Figure 6.60, is a limaçon.

Some curves reach their zeros and maximum r-values at more than one point. Example 4 shows how to handle this situation.

Example 4 ▶ Sketching a Polar Graph

Sketch the graph of $r = 2 \cos 3\theta$.

Solution

Symmetry: With respect to the polar axis

Maximum value of $|r|$: $|r| = 2$ when $3\theta = 0, \pi, 2\pi, 3\pi$ or $\theta = 0, \pi/3, 2\pi/3, \pi$

Zeros of r: $r = 0$ when $3\theta = \pi/2, 3\pi/2, 5\pi/2$ or $\theta = \pi/6, \pi/2, 5\pi/6$

θ	0	$\dfrac{\pi}{12}$	$\dfrac{\pi}{6}$	$\dfrac{\pi}{4}$	$\dfrac{\pi}{3}$	$\dfrac{5\pi}{12}$	$\dfrac{\pi}{2}$
r	2	$\sqrt{2}$	0	$-\sqrt{2}$	-2	$-\sqrt{2}$	0

A computer animation of this example appears in the *Interactive* CD-ROM and *Internet* versions of this text.

By plotting these points and using the specified symmetry, zeros, and maximum values, you can obtain the graph shown in Figure 6.63. This graph is called a **rose curve,** and each of the loops on the graph is called a *petal* of the rose curve. Note how the entire curve is generated as θ increases from 0 to π.

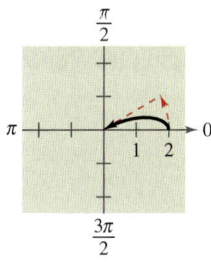

$0 \leq \theta \leq \dfrac{\pi}{6}$

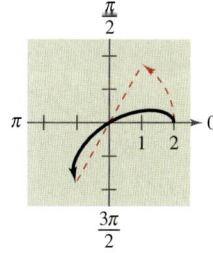

$0 \leq \theta \leq \dfrac{\pi}{3}$

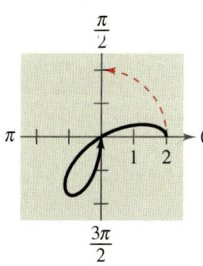

$0 \leq \theta \leq \dfrac{\pi}{2}$

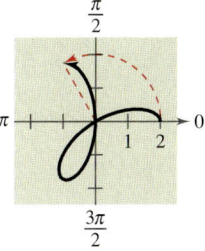

$0 \leq \theta \leq \dfrac{2\pi}{3}$

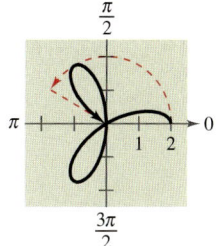

$0 \leq \theta \leq \dfrac{5\pi}{6}$

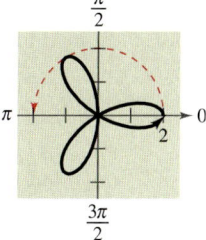

$0 \leq \theta \leq \pi$

FIGURE 6.63

Technology

Use a graphing utility in polar mode to verify the graph of $r = 2 \cos 3\theta$ shown in Figure 6.63.

Special Polar Graphs

Several important types of graphs have equations that are simpler in polar form than in rectangular form. For example, the circle

$$r = 4 \sin \theta$$

in Example 1 has the more complicated rectangular equation

$$x^2 + (y - 2)^2 = 4.$$

Several other types of graphs that have simple polar equations are shown below.

Limaçons
$r = a \pm b \cos \theta$
$r = a \pm b \sin \theta$
$(a > 0, b > 0)$

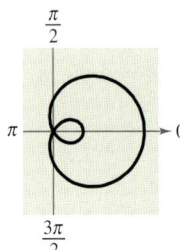

$\dfrac{a}{b} < 1$

Limaçon with inner loop

$\dfrac{a}{b} = 1$

Cardioid (heart-shaped)

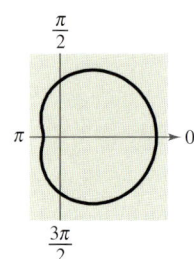

$1 < \dfrac{a}{b} < 2$

Dimpled limaçon

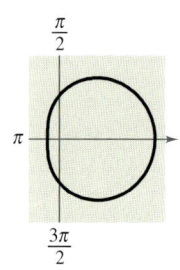

$\dfrac{a}{b} \geq 2$

Convex limaçon

Rose Curves
n petals if n is odd
$2n$ petals if n is even
$(n \geq 2)$

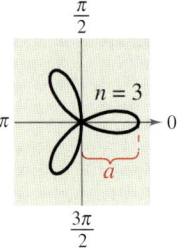

$n = 3$

$r = a \cos n\theta$
Rose curve

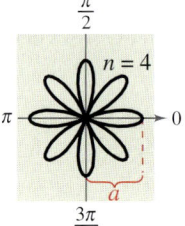

$n = 4$

$r = a \cos n\theta$
Rose curve

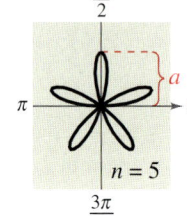

$n = 5$

$r = a \sin n\theta$
Rose curve

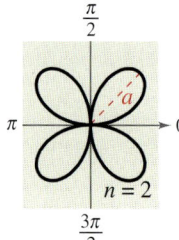

$n = 2$

$r = a \sin n\theta$
Rose curve

Circles and Lemniscates

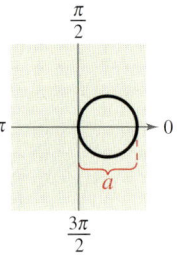

$r = a \cos \theta$
Circle

$r = a \sin \theta$
Circle

$r^2 = a^2 \sin 2\theta$
Lemniscate

$r^2 = a^2 \cos 2\theta$
Lemniscate

Section 6.8 ▶ Graphs of Polar Equations   467

Example 5 ▶ Sketching a Rose Curve

Sketch the graph of

$$r = 3 \cos 2\theta.$$

Solution

Type of curve: Rose curve with $2n = 4$ petals

Symmetry: With respect to polar axis, the line $\theta = \pi/2$, and the pole

Maximum value of $|r|$: $|r| = 3$ when $\theta = 0, \pi/2, \pi, 3\pi/2$

Zeros of r: $r = 0$ when $\theta = \pi/4, 3\pi/4$

Using this information together with the additional points shown in the following table, you obtain the graph shown in Figure 6.64.

θ	0	$\dfrac{\pi}{6}$	$\dfrac{\pi}{4}$	$\dfrac{\pi}{3}$
r	3	$\dfrac{3}{2}$	0	$-\dfrac{3}{2}$

Example 6 ▶ Sketching a Lemniscate

Sketch the graph of

$$r^2 = 9 \sin 2\theta.$$

Solution

Type of curve: Lemniscate

Symmetry: With respect to the pole

Maximum value of $|r|$: $|r| = 3$ when $\theta = \pi/4$

Zeros of r: $r = 0$ when $\theta = 0, \pi/2$

If $\sin 2\theta < 0$, this equation has no solution points. So, you restrict the values of θ to those for which $\sin 2\theta \geq 0$.

$$0 \leq \theta \leq \frac{\pi}{2} \quad \text{or} \quad \pi \leq \theta \leq \frac{3\pi}{2}$$

Moreover, using symmetry, you need to consider only the first of these two intervals. By finding a few additional points, you can obtain the graph shown in Figure 6.65.

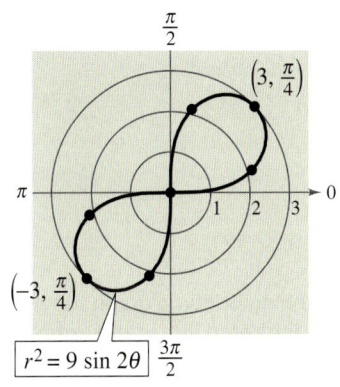

FIGURE 6.65

θ	0	$\dfrac{\pi}{12}$	$\dfrac{\pi}{4}$	$\dfrac{5\pi}{12}$	$\dfrac{\pi}{2}$
$r = \pm 3\sqrt{\sin 2\theta}$	0	$\dfrac{\pm 3}{\sqrt{2}}$	± 3	$\dfrac{\pm 3}{\sqrt{2}}$	0

6.8 Exercises

In Exercises 1–6, identify the type of polar graph.

1.
2.
3.
4.
5.
6.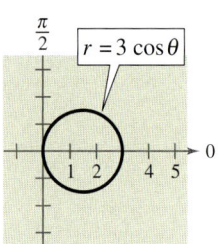

In Exercises 7–12, test for symmetry with respect to $\theta = \pi/2$, the polar axis, and the pole.

7. $r = 5 + 4\cos\theta$
8. $r = 16\cos 3\theta$
9. $r = \dfrac{2}{1 + \sin\theta}$
10. $r = \dfrac{3}{2 + \cos\theta}$
11. $r^2 = 16\cos 2\theta$
12. $r^2 = 36\sin 2\theta$

In Exercises 13–16, find the maximum value of $|r|$ and any zeros of r.

13. $r = 10(1 - \sin\theta)$
14. $r = 6 + 12\cos\theta$
15. $r = 4\cos 3\theta$
16. $r = 3\sin 2\theta$

In Exercises 17–40, sketch the graph of the polar equation.

17. $r = 5$
18. $r = 2$
19. $r = \dfrac{\pi}{6}$
20. $r = -\dfrac{3\pi}{4}$
21. $r = 3\sin\theta$
22. $r = 4\cos\theta$
23. $r = 3(1 - \cos\theta)$
24. $r = 4(1 - \sin\theta)$
25. $r = 4(1 + \sin\theta)$
26. $r = 2(1 + \cos\theta)$
27. $r = 3 + 6\sin\theta$
28. $r = 4 - 3\sin\theta$
29. $r = 1 - 2\sin\theta$
30. $r = 1 - 2\cos\theta$
31. $r = 3 - 4\cos\theta$
32. $r = 4 + 3\cos\theta$
33. $r = 5\sin 2\theta$
34. $r = 3\cos 2\theta$
35. $r = 2\sec\theta$
36. $r = 5\csc\theta$
37. $r = \dfrac{3}{\sin\theta - 2\cos\theta}$
38. $r = \dfrac{6}{2\sin\theta - 3\cos\theta}$
39. $r^2 = 9\cos 2\theta$
40. $r^2 = 4\sin\theta$

In Exercises 41–46, use a graphing utility to graph the polar equation.

41. $r = 8\cos\theta$
42. $r = \cos 2\theta$
43. $r = 3(2 - \sin\theta)$
44. $r = 2\cos(3\theta - 2)$
45. $r = 8\sin\theta\cos^2\theta$
46. $r = 2\csc\theta + 5$

In Exercises 47–52, use a graphing utility to graph the polar equation. Find an interval for θ for which the graph is traced *only once*.

47. $r = 3 - 4\cos\theta$
48. $r = 5 + 4\cos\theta$
49. $r = 2\cos\left(\dfrac{3\theta}{2}\right)$
50. $r = 3\sin\left(\dfrac{5\theta}{2}\right)$
51. $r^2 = 9\sin 2\theta$
52. $r^2 = \dfrac{1}{\theta}$

In Exercises 53–56, use a graphing utility to graph the polar equation and show that the indicated line is an asymptote of the graph.

Name of Graph	Polar Equation	Asymptote
53. Conchoid	$r = 2 - \sec\theta$	$x = -1$
54. Conchoid	$r = 2 + \csc\theta$	$y = 1$
55. Hyperbolic spiral	$r = \dfrac{3}{\theta}$	$y = 3$
56. Strophoid	$r = 2\cos 2\theta \sec\theta$	$x = -2$

Synthesis

True or False? In Exercises 57 and 58, determine whether the statement is true or false. Justify your answer.

57. In the polar coordinate system, if a graph that has symmetry with respect to the polar axis were folded on the line $\theta = 0$, the portion of the graph above the polar axis would coincide with the portion of the graph below the polar axis.

58. In the polar coordinate system, if a graph that has symmetry with respect to the pole were folded on the line $\theta = 3\pi/4$, the portion of the graph on one side of the fold would coincide with the portion of the graph on the other side of the fold.

59. ***Exploration*** Sketch the graph of $r = 6 \cos \theta$ over each interval. Describe the part of the graph obtained in each case.

   (a) $0 \le \theta \le \dfrac{\pi}{2}$ (b) $\dfrac{\pi}{2} \le \theta \le \pi$

   (c) $-\dfrac{\pi}{2} \le \theta \le \dfrac{\pi}{2}$ (d) $\dfrac{\pi}{4} \le \theta \le \dfrac{3\pi}{4}$

60. ***Graphical Reasoning*** Use a graphing utility to graph the polar equation

   $r = 6[1 + \cos(\theta - \phi)]$

   for (a) $\phi = 0$, (b) $\phi = \pi/4$, and (c) $\phi = \pi/2$. Use the graphs to describe the effect of the angle ϕ. Write the equation as a function of $\sin \theta$ for part (c).

61. The graph of $r = f(\theta)$ is rotated about the pole through an angle ϕ. Show that the equation of the rotated graph is $r = f(\theta - \phi)$.

62. Consider the graph of $r = f(\sin \theta)$.

   (a) Show that if the graph is rotated counterclockwise $\pi/2$ radians about the pole, the equation of the rotated graph is $r = f(-\cos \theta)$.

   (b) Show that if the graph is rotated counterclockwise π radians about the pole, the equation of the rotated graph is $r = f(-\sin \theta)$.

   (c) Show that if the graph is rotated counterclockwise $3\pi/2$ radians about the pole, the equation of the rotated graph is $r = f(\cos \theta)$.

In Exercises 63–66, use the results of Exercises 61 and 62.

63. Write an equation for the limaçon $r = 2 - \sin \theta$ after it has been rotated by the given amount.

   (a) $\dfrac{\pi}{4}$ (b) $\dfrac{\pi}{2}$ (c) π (d) $\dfrac{3\pi}{2}$

64. Write an equation for the rose curve $r = 2 \sin 2\theta$ after it has been rotated by the given amount.

   (a) $\dfrac{\pi}{6}$ (b) $\dfrac{\pi}{2}$ (c) $\dfrac{2\pi}{3}$ (d) π

65. Sketch the graph of each equation.

   (a) $r = 1 - \sin \theta$ (b) $r = 1 - \sin\left(\theta - \dfrac{\pi}{4}\right)$

66. Sketch the graph of each equation.

   (a) $r = 3 \sec \theta$ (b) $r = 3 \sec\left(\theta - \dfrac{\pi}{4}\right)$

   (c) $r = 3 \sec\left(\theta + \dfrac{\pi}{3}\right)$ (d) $r = 3 \sec\left(\theta - \dfrac{\pi}{2}\right)$

67. ***Exploration*** Use a graphing utility to graph and identify $r = 2 + k \sin \theta$ for $k = 0, 1, 2,$ and 3.

68. ***Exploration*** Consider the equation $r = 3 \sin k\theta$.

   (a) Use a graphing utility to graph the equation for $k = 1.5$. Find the interval for θ over which the graph is traced only once.

   (b) Use a graphing utility to graph the equation for $k = 2.5$. Find the interval for θ over which the graph is traced only once.

   (c) Is it possible to find an interval for θ over which the graph is traced only once for any rational number k? Explain.

Review

In Exercises 69–74, solve the equation algebraically. Round the result to three decimal places.

69. $e^x = 19$ 70. $6e^x = 47$

71. $10^x = 84$ 72. $5^{2x} = 60$

73. $\ln x = 4$ 74. $4 \ln 4x = 18$

6.9 Polar Equations of Conics

▶ **What you should learn**
- How to define a conic in terms of eccentricity
- How to write equations of conics in polar form
- How to use equations of conics in polar form to model real-life problems

▶ **Why you should learn it**

The elliptical orbits of planets and satellites can be modeled with polar equations. For instance, Exercise 49 on page 475 uses a polar equation to model a planet's orbit.

Alternative Definition of Conics

In Sections 6.3 and 6.4, you learned that the rectangular equations of ellipses and hyperbolas take simple forms when the origin lies at their *centers*. As it happens, there are many important applications of conics in which it is more convenient to use one of the *foci* as the origin of the coordinate system. For example, the sun lies at a focus of the earth's orbit. In this section you will learn that polar equations of conics take simple forms if one of the foci lies at the pole.

To begin, consider the following alternative definition of conic that uses the concept of eccentricity.

Alternative Definition of Conic

The locus of a point in the plane that moves so that its distance from a fixed point (focus) is in a constant ratio to its distance from a fixed line (directrix) is a **conic**. The constant ratio is the **eccentricity** of the conic and is denoted by e. Moreover, the conic is an **ellipse** if $e < 1$, a **parabola** if $e = 1$, and a **hyperbola** if $e > 1$.

In Figure 6.66, note that for each type of conic, the pole corresponds to the fixed point (focus) given in the definition.

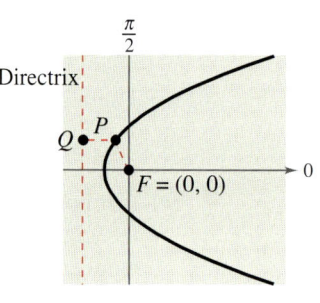

Parabola: $e = 1$

Ellipse: $0 < e < 1$
$\dfrac{PF}{PQ} < 1$

Hyperbola: $e > 1$
$\dfrac{PF}{PQ} = \dfrac{P'F}{P'Q'} > 1$

FIGURE 6.66

 A computer simulation of this concept appears in the *Interactive* CD-ROM and *Internet* versions of this text.

Polar Equations of Conics

The benefit of locating a focus of a conic at the pole is that the equation of the conic takes on a simpler form. A proof of the polar form is given in Appendix A.

Polar Equations of Conics

The graph of a polar equation of the form

1. $r = \dfrac{ep}{1 \pm e \cos \theta}$ 2. $r = \dfrac{ep}{1 \pm e \sin \theta}$

is a conic, where $e > 0$ is the eccentricity and $|p|$ is the distance between the focus (pole) and the directrix.

The equations

$$r = \frac{ep}{1 \pm e \cos \theta}$$ Vertical directrix

correspond to conics with vertical directrices, and the equations

$$r = \frac{ep}{1 \pm e \sin \theta}$$ Horizontal directrix

correspond to conics with horizontal directrices. Moreover, the converse is also true—that is, any conic with a focus at the pole and having a horizontal or vertical directrix can be represented by one of the given equations.

Example 1 ▶ Determining a Conic from Its Equation

Identify the conic and sketch its graph.

$$r = \frac{15}{3 - 2 \cos \theta}$$

Solution

To identify the type of conic, rewrite the equation as

$$r = \frac{15}{3 - 2 \cos \theta} = \frac{5}{1 - (2/3) \cos \theta}.$$ Divide numerator and denominator by 3.

From this form you can conclude that the graph is an ellipse with $e = \frac{2}{3}$. You can sketch the upper half of the ellipse by plotting points from $\theta = 0$ to $\theta = \pi$, as shown in Figure 6.67. Using symmetry with respect to the polar axis, you can sketch the lower half.

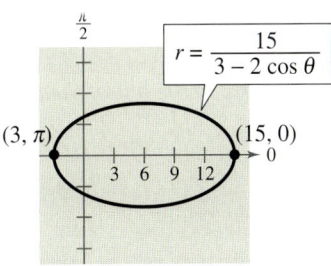

FIGURE 6.67

For the ellipse in Figure 6.67, the major axis is horizontal and the vertices lie at $(15, 0)$ and $(3, \pi)$. So, the length of the *major* axis is $2a = 18$. To find the length of the *minor* axis, you can use the equations $e = c/a$ and $b^2 = a^2 - c^2$ to conclude that

$$b^2 = a^2 - c^2 = a^2 - (ea)^2 = a^2(1 - e^2).$$ Ellipse

Because $e = \frac{2}{3}$, you have $b^2 = 9^2 \left[1 - \left(\frac{2}{3}\right)^2\right] = 45$, which implies that $b = \sqrt{45} = 3\sqrt{5}$. So, the length of the minor axis is $2b = 6\sqrt{5}$. A similar analysis for hyperbolas yields

$$b^2 = c^2 - a^2 = (ea)^2 - a^2 = a^2(e^2 - 1).$$ Hyperbola

Example 2 ▶ Sketching a Conic from Its Polar Equation

Identify the conic $r = 32/(3 + 5 \sin \theta)$ and sketch its graph.

Solution

Dividing each term by 3, you have

$$r = \frac{32/3}{1 + (5/3) \sin \theta}.$$

Because $e = \frac{5}{3} > 1$, the graph is a hyperbola. The transverse axis of the hyperbola lies on the line $\theta = \pi/2$, and the vertices occur at $(4, \pi/2)$ and $(-16, 3\pi/2)$. Because the length of the transverse axis is 12, you can see that $a = 6$. To find b, write

$$b^2 = a^2(e^2 - 1) = 6^2 \left[\left(\frac{5}{3} \right)^2 - 1 \right] = 64.$$

Therefore, $b = 8$. Finally, you can use a and b to determine the asymptotes of the hyperbola and obtain the sketch shown in Figure 6.68.

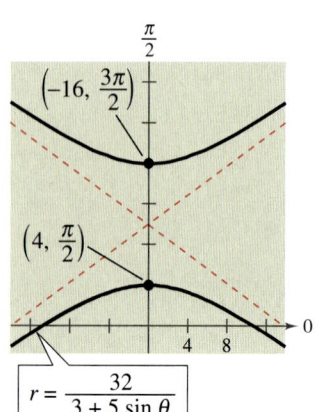

FIGURE 6.68

In the next example you are asked to find a polar equation of a specified conic. To do this, let p be the distance between the pole and the directrix.

1. *Horizontal directrix above the pole:* $\quad r = \dfrac{ep}{1 + e \sin \theta}$

2. *Horizontal directrix below the pole:* $\quad r = \dfrac{ep}{1 - e \sin \theta}$

3. *Vertical directrix to the right of the pole:* $r = \dfrac{ep}{1 + e \cos \theta}$

4. *Vertical directrix to the left of the pole:* $\quad r = \dfrac{ep}{1 - e \cos \theta}$

Technology

Most graphing utilities have a polar mode. Try using a graphing utility set in polar mode to verify the four orientations shown at the right. Remember that e must be positive, but p can be positive or negative.

Example 3 ▶ Finding the Polar Equation of a Conic

Find the polar equation of the parabola whose focus is the pole and whose directrix is the line $y = 3$.

Solution

From Figure 6.69, you can see that the directrix is horizontal and above the pole. So, you can choose an equation of the form

$$r = \frac{ep}{1 + e \sin \theta}.$$

Moreover, because the eccentricity of a parabola is $e = 1$ and the distance between the pole and the directrix is $p = 3$, you have the equation

$$r = \frac{3}{1 + \sin \theta}.$$

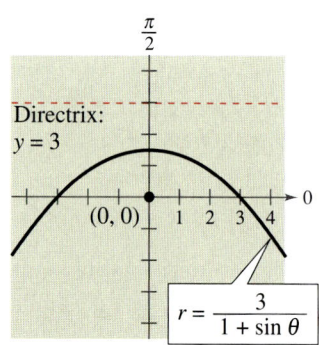

FIGURE 6.69

Applications

Kepler's Laws (listed below), named after the German astronomer Johannes Kepler (1571–1630), can be used to describe the orbits of the planets about the sun.

1. Each planet moves in an elliptical orbit with the sun at one focus.
2. A ray from the sun to the planet sweeps out equal areas of the ellipse in equal times.
3. The square of the period is proportional to the cube of the mean distance between the planet and the sun.

Although Kepler simply stated these laws on the basis of observation, they were later validated by Isaac Newton (1642–1727). In fact, Newton was able to show that each law can be deduced from a set of universal laws of motion and gravitation that govern the movement of all heavenly bodies, including comets and satellites. This is illustrated in the next example, which involves the comet named after the English mathematician and physicist Edmund Halley (1656–1742).

If you use earth as a reference with a period of 1 year and a distance of 1 astronomical unit, the proportionality constant in Kepler's third law is 1. For example, because Mars has a mean distance to the sun of $d = 1.523$ astronomical units, its period P is given by $d^3 = P^2$. So, the period of Mars is $P = 1.88$ years.

Example 4 ▶ Halley's Comet

Halley's comet has an elliptical orbit with an eccentricity of $e \approx 0.97$. The length of the major axis of the orbit is approximately 36.18 astronomical units. (An *astronomical unit* is defined as the mean distance between earth and the sun, or about 93 million miles.) Find a polar equation for the orbit. How close does Halley's comet come to the sun?

Solution

Using a vertical axis, as shown in Figure 6.70, choose an equation of the form $r = ep/(1 + e \sin\theta)$. Because the vertices of the ellipse occur when $\theta = \pi/2$ and $\theta = 3\pi/2$, you can determine the length of the major axis to be the sum of the r-values of the vertices. That is,

$$2a = \frac{0.97p}{1 + 0.97} + \frac{0.97p}{1 - 0.97} \approx 32.83p \approx 36.18.$$

So, $p \approx 1.102$ and $ep \approx (0.97)(1.102) \approx 1.069$. Using this value of ep in the equation, you have

$$r = \frac{1.069}{1 + 0.97 \sin \theta}$$

where r is measured in astronomical units. To find the closest point to the sun (the focus), substitute $\theta = \pi/2$ in this equation to obtain

$$r = \frac{1.069}{1 + 0.97 \sin(\pi/2)} \approx 0.54 \text{ astronomical unit} \approx 50{,}000{,}000 \text{ miles.}$$

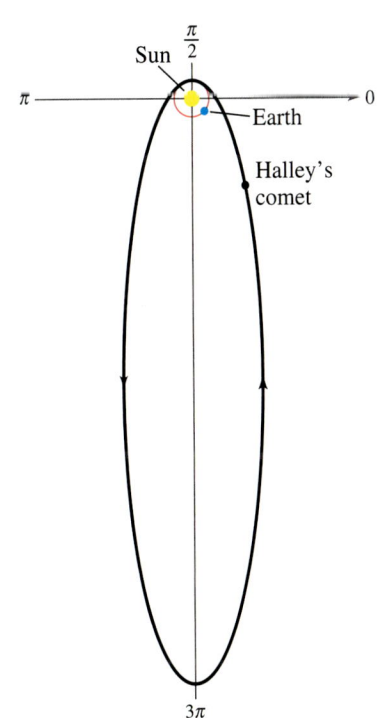

FIGURE 6.70

6.9 Exercises

In Exercises 1–4, use a graphing utility to graph the polar equation for $e = 1$, $e = 0.5$, and $e = 1.5$. What can you conclude?

1. $r = \dfrac{4e}{1 + e \cos \theta}$
2. $r = \dfrac{4e}{1 - e \cos \theta}$
3. $r = \dfrac{4e}{1 - e \sin \theta}$
4. $r = \dfrac{4e}{1 + e \sin \theta}$

In Exercises 5–10, match the polar equation with its graph. [The graphs are labeled (a), (b), (c), (d), (e), and (f).]

(a)

(b)

(c)

(d)

(e)

(f)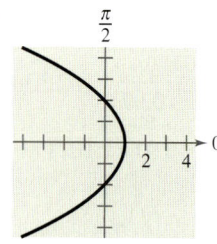

5. $r = \dfrac{2}{1 + \cos \theta}$
6. $r = \dfrac{3}{2 - \cos \theta}$
7. $r = \dfrac{3}{1 + 2 \sin \theta}$
8. $r = \dfrac{2}{1 - \sin \theta}$
9. $r = \dfrac{4}{2 + \cos \theta}$
10. $r = \dfrac{4}{1 - 3 \sin \theta}$

In Exercises 11–24, identify the conic and sketch its graph.

11. $r = \dfrac{2}{1 - \cos \theta}$
12. $r = \dfrac{3}{1 + \sin \theta}$
13. $r = \dfrac{5}{1 + \sin \theta}$
14. $r = \dfrac{6}{1 + \cos \theta}$
15. $r = \dfrac{2}{2 - \cos \theta}$
16. $r = \dfrac{3}{3 + \sin \theta}$
17. $r = \dfrac{6}{2 + \sin \theta}$
18. $r = \dfrac{9}{3 - 2 \cos \theta}$
19. $r = \dfrac{3}{2 + 4 \sin \theta}$
20. $r = \dfrac{5}{-1 + 2 \cos \theta}$
21. $r = \dfrac{3}{2 - 6 \cos \theta}$
22. $r = \dfrac{3}{2 + 6 \sin \theta}$
23. $r = \dfrac{4}{2 - \cos \theta}$
24. $r = \dfrac{2}{2 + 3 \sin \theta}$

In Exercises 25–28, use a graphing utility to graph the polar equation. Identify the graph.

25. $r = \dfrac{-1}{1 - \sin \theta}$
26. $r = \dfrac{-5}{2 + 4 \sin \theta}$
27. $r = \dfrac{3}{-4 + 2 \cos \theta}$
28. $r = \dfrac{4}{1 - 2 \cos \theta}$

In Exercises 29–32, use a graphing utility to graph the rotated conic.

29. $r = \dfrac{2}{1 - \cos(\theta - \pi/4)}$ (See Exercise 11.)
30. $r = \dfrac{3}{3 + \sin(\theta - \pi/3)}$ (See Exercise 16.)
31. $r = \dfrac{6}{2 + \sin(\theta + \pi/6)}$ (See Exercise 17.)
32. $r = \dfrac{5}{-1 + 2 \cos(\theta + 2\pi/3)}$ (See Exercise 20.)

Section 6.9 ▸ Polar Equations of Conics

In Exercises 33–48, find a polar equation of the conic with its focus at the pole.

Conic	Eccentricity	Directrix
33. Parabola	$e = 1$	$x = -1$
34. Parabola	$e = 1$	$y = -2$
35. Ellipse	$e = \frac{1}{2}$	$y = 1$
36. Ellipse	$e = \frac{3}{4}$	$y = -3$
37. Hyperbola	$e = 2$	$x = 1$
38. Hyperbola	$e = \frac{3}{2}$	$x = -1$

Conic	Vertex or Vertices
39. Parabola	$(1, -\pi/2)$
40. Parabola	$(6, 0)$
41. Parabola	$(5, \pi)$
42. Parabola	$(10, \pi/2)$
43. Ellipse	$(2, 0), (10, \pi)$
44. Ellipse	$(2, \pi/2), (4, 3\pi/2)$
45. Ellipse	$(20, 0), (4, \pi)$
46. Hyperbola	$(2, 0), (8, 0)$
47. Hyperbola	$(1, 3\pi/2), (9, 3\pi/2)$
48. Hyperbola	$(4, \pi/2), (-1, 3\pi/2)$

49. *Planetary Motion* The planets travel in elliptical orbits with the sun at one focus. Assume that the focus is at the pole, the major axis lies on the polar axis, and the length of the major axis is $2a$ (see figure). Show that the polar equation of the orbit is

$$r = \frac{(1 - e^2)a}{1 - e \cos \theta}$$

where e is the eccentricity.

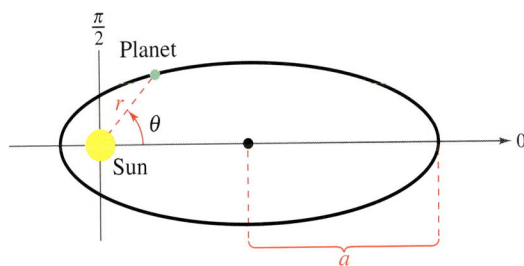

50. *Planetary Motion* Use the result of Exercise 49 to show that the minimum distance (*perihelion distance*) from the sun to the planet is $r = a(1 - e)$ and the maximum distance (*aphelion distance*) is $r = a(1 + e)$.

In Exercises 51–56, use the results of Exercises 49 and 50 to find the polar equation of the planet's orbit and the perihelion and aphelion distances.

51. Earth $a = 92.960 \times 10^6$ miles
           $e = 0.0167$

52. Saturn $a = 1.429 \times 10^9$ kilometers
           $e = 0.0543$

53. Pluto $a = 5.900 \times 10^9$ kilometers
           $e = 0.2481$

54. Mercury $a = 35.98 \times 10^6$ miles
           $e = 0.2056$

55. Mars $a = 141.00 \times 10^6$ miles
           $e = 0.0934$

56. Jupiter $a = 778.40 \times 10^6$ kilometers
           $e = 0.0484$

57. *Satellite Tracking* A satellite in a 100-mile-high circular orbit around earth has a velocity of approximately 17,500 miles per hour. If this velocity is multiplied by $\sqrt{2}$, the satellite will have the minimum velocity necessary to escape the earth's gravity and it will follow a parabolic path with the center of earth as the focus (see figure). Find a polar equation of the parabolic path of the satellite (assume the radius of earth is 4000 miles). Find the distance between the surface of the earth and the satellite when $\theta = 30°$.

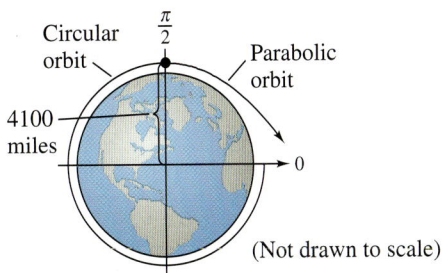

(Not drawn to scale)

58. Use the equation found in Exercise 57. Find the distance between the surface of the earth and the satellite when $\theta = 60°$.

Synthesis

True or False? In Exercises 59 and 60, determine whether the statement is true or false. Justify your answer.

59. If $r = \dfrac{ep}{1 \pm e \sin \theta}$ is the equation of an ellipse with $e < 1$, changing the value of p will affect the length of the major axis, but will not affect the length of the minor axis.

60. For a given value of $e > 1$ over the interval $\theta = 0$ to $\theta = 2\pi$, the graph of $r = \dfrac{ex}{1 - e \cos \theta}$ is the same as the graph of $r = \dfrac{e(-x)}{1 + e \cos \theta}$.

61. Show that the polar equation of the ellipse $\dfrac{x^2}{a^2} + \dfrac{y^2}{b^2} = 1$ is $r^2 = \dfrac{b^2}{1 - e^2 \cos^2 \theta}$.

62. Show that the polar equation of the hyperbola $\dfrac{x^2}{a^2} - \dfrac{y^2}{b^2} = 1$ is $r^2 = \dfrac{-b^2}{1 - e^2 \cos^2 \theta}$.

In Exercises 63–68, use the results of Exercises 61 and 62 to write the polar form of the equation of the conic.

63. $\dfrac{x^2}{169} + \dfrac{y^2}{144} = 1$

64. $\dfrac{x^2}{25} + \dfrac{y^2}{16} = 1$

65. $\dfrac{x^2}{9} - \dfrac{y^2}{16} = 1$

66. $\dfrac{x^2}{36} - \dfrac{y^2}{4} = 1$

67. Hyperbola One focus: $(5, \pi/2)$
Vertices: $(4, \pi/2), (4, -\pi/2)$

68. Ellipse One focus: $(4, 0)$
Vertices: $(5, 0), (5, \pi)$

Review

In Exercises 69–74, solve the trigonometric equation.

69. $4\sqrt{3} \tan \theta - 3 = 1$

70. $6 \cos x - 2 = 1$

71. $12 \sin^2 \theta = 9$

72. $9 \csc^2 x - 10 = 2$

73. $2 \cot x = 5 \cos \dfrac{\pi}{2}$

74. $\sqrt{2} \sec \theta = 2 \csc \dfrac{\pi}{4}$

In Exercises 75–78, find the value of the trigonometric function given that u and v are in Quadrant IV and $\sin u = -\tfrac{3}{5}$ and $\cos v = 1/\sqrt{2}$.

75. $\cos(u + v)$

76. $\sin(u + v)$

77. $\cos(u - v)$

78. $\sin(u - v)$

In Exercises 79 and 80, find the exact values of sin 2u, cos 2u, and tan 2u using the double-angle formulas.

79. $\sin u = \dfrac{4}{5}, \dfrac{\pi}{2} < u < \pi$

80. $\tan u = -\sqrt{3}, \dfrac{3\pi}{2} < u < 2\pi$

Chapter Summary

What did you learn?

Section 6.1
	Review Exercises
☐ How to find the inclination of a line and the angle between two lines	1–8
☐ How to find the distance between a point and a line	9, 10

Section 6.2
☐ How to recognize a conic as the intersection of a plane and a cone	11, 12
☐ How to write the standard form of the equation of a parabola	13–16
☐ How to use the reflective property of parabolas to solve real-life problems	17–20

Section 6.3
☐ How to write the standard form of the equation of an ellipse	21–24
☐ How to use properties of ellipses to model and solve real-life problems	25, 26
☐ How to find the eccentricity of an ellipse	27–30

Section 6.4
☐ How to write the standard form of the equation of a hyperbola	31–34
☐ How to find the asymptotes of a hyperbola	35–38
☐ How to use properties of hyperbolas to solve real-life problems	39, 40
☐ How to classify a conic from its general equation	41, 42

Section 6.5
☐ How to rotate coordinate axes to eliminate the xy-term of a conic	43–46
☐ How to use the discriminant to classify a conic	47–50

Section 6.6
☐ How to evaluate a set of parametric equations for a given value of the parameter	51–54
☐ How to sketch the curve that is represented by a set of parametric equations and rewrite the equations as a single rectangular equation	55–60
☐ How to find a set of parametric equations for a graph	61–64

Section 6.7
☐ How to plot points in the polar coordinate system	65–68
☐ How to convert points from rectangular to polar form and vice versa	69–72
☐ How to convert equations from rectangular to polar form and vice versa	73–78

Section 6.8
☐ How to graph a polar equation by point plotting	79–84
☐ How to use symmetry, zeros, and maximum r-values as graphing aids	85–88
☐ How to recognize special polar graphs	89–92

Section 6.9
☐ How to define a conic in terms of eccentricity	93–96
☐ How to write equations of conics in polar form	97–100
☐ How to use equations of conics in polar form to model real-life problems	101, 102

Review Exercises

6.1 In Exercises 1–4, find the inclination θ of the line with the given characteristics.

1. Slope: $m = \frac{3}{5}$
2. Passes through the points $(3, 4)$ and $(-2, 7)$
3. Equation: $y = 2x + 4$
4. Equation: $6x - 7y - 5 = 0$

In Exercises 5–8, find the angle θ between the lines.

5. $4x + y = 2$
   $-5x + y = -1$
6. $-5x + 3y = 3$
   $-2x + 3y = 1$

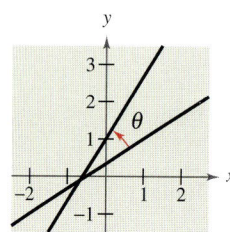

7. $2x - 7y = 8$
   $0.4x + y = 0$
8. $0.02x + 0.07y = 0.18$
   $0.09x - 0.04y = 0.17$

In Exercises 9 and 10, find the distance between the point and the line.

Point	Line
9. $(1, 2)$	$x - y - 3 = 0$
10. $(0, 4)$	$x + 2y - 2 = 0$

6.2 In Exercises 11 and 12, state what type of conic is formed by the intersection of the plane and the double-napped cone.

11.

12.

In Exercises 13–16, find the standard form of the equation of the parabola.

13. Vertex: $(4, 2)$
    Focus: $(4, 0)$
14. Vertex: $(2, 0)$
    Focus: $(0, 0)$
15. Vertex: $(0, 2)$
    Directrix: $x = -3$
16. Vertex: $(2, 2)$
    Directrix: $y = 0$

In Exercises 17 and 18, find an equation of a tangent line to the parabola at the given point, and find the x-intercept of the line.

17. $x^2 = -2y$, $(2, -2)$
18. $x^2 = -2y$, $(-4, -8)$

19. *Parabolic Archway* A parabolic archway is 12 meters high at the vertex. At a height of 10 meters, the width of the archway is 8 meters (see figure). How wide is the archway at ground level?

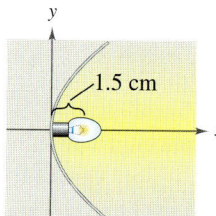

FIGURE FOR 19   FIGURE FOR 20

20. *Flashlight* The light bulb in a flashlight is at the focus of its parabolic reflector, 1.5 centimeters from the vertex of the reflector (see figure). Write an equation for a cross section of the flashlight's reflector with its focus on the positive x-axis and its vertex at the origin.

6.3 In Exercises 21–24, find the standard form of the equation of the ellipse.

21. Vertices: $(-3, 0), (7, 0)$; Foci: $(0, 0), (4, 0)$
22. Vertices: $(2, 0), (2, 4)$; Foci: $(2, 1), (2, 3)$
23. Vertices: $(0, \pm 6)$; Passes through $(2, 2)$
24. Vertices: $(0, 1), (4, 1)$;
    Endpoints of the minor axis: $(2, 0), (2, 2)$

25. *Semielliptical Archway* A semielliptical archway is set on pillars that are 10 feet apart. Its height (atop the pillars) is 4 feet. Where should the foci be placed in order to sketch the semielliptical arch?

26. Wading Pool You are building a wading pool that is in the shape of an ellipse. Your plans give an equation for the elliptical shape of the pool measured in feet as

$$\frac{x^2}{324} + \frac{y^2}{196} = 1.$$

Find the longest distance across the pool, the shortest distance, and the distance between the foci.

In Exercises 27–30, find the center, vertices, foci, and eccentricity of the ellipse.

27. $16x^2 + 9y^2 - 32x + 72y + 16 = 0$
28. $4x^2 + 25y^2 + 16x - 150y + 141 = 0$
29. $\dfrac{(x+2)^2}{81} + \dfrac{(y-1)^2}{100} = 1$
30. $\dfrac{(x-5)^2}{1} + \dfrac{(y+3)^2}{36} = 1$

6.4 In Exercises 31–34, find the standard form of the equation of the hyperbola.

31. Vertices: $(0, \pm 1)$; Foci: $(0, \pm 3)$
32. Vertices: $(2, 2), (-2, 2)$; Foci: $(4, 2), (-4, 2)$
33. Foci: $(0, 0), (8, 0)$; Asymptotes: $y = \pm 2(x - 4)$
34. Foci: $(3, \pm 2)$; Asymptotes: $y = \pm 2(x - 3)$

In Exercises 35–38, find the center, vertices, foci, and equations of the asymptotes of the hyperbola. Then sketch its graph.

35. $9x^2 - 16y^2 - 18x - 32y - 151 = 0$
36. $-4x^2 + 25y^2 - 8x + 150y + 121 = 0$
37. $\dfrac{(x-3)^2}{16} - \dfrac{(y+5)^2}{4} = 1$
38. $\dfrac{(y-1)^2}{4} - x^2 = 1$

39. Loran Radio transmitting station A is located 200 miles east of transmitting station B. A ship is in an area to the north and 40 miles west of station A. Synchronized radio pulses transmitted at 186,000 miles per second by the two stations are received 0.0005 second sooner from station A than from station B. How far north is the ship?

40. Locating an Explosion Two of your friends live 4 miles apart and on the same "east-west" street, and you live halfway between them. You are talking on a three-way phone call when you hear an explosion. Six seconds later your friend to the east hears the explosion, and your friend to the west hears it 8 seconds after you do. Find equations of two hyperbolas that would locate the explosion. (Sound travels at a rate of 1100 feet per second.)

In Exercises 41 and 42, classify the conic from its general equation.

41. $5x^2 - 2y^2 + 10x - 4y + 17 = 0$
42. $-4y^2 + 5x + 3y + 7 = 0$

6.5 In Exercises 43–46, rotate the axes to eliminate the xy-term. Sketch the graph of the resulting equation, showing both sets of axes.

43. $xy - 4 = 0$
44. $x^2 - 10xy + y^2 + 1 = 0$
45. $5x^2 - 2xy + 5y^2 - 12 = 0$
46. $4x^2 + 8xy + 4y^2 + 7\sqrt{2}x + 9\sqrt{2}y = 0$

In Exercises 47–50, use the discriminant to classify the graph, use the Quadratic Formula to solve for y, and then use a graphing utility to graph the equation.

47. $16x^2 - 24xy + 9y^2 - 30x - 40y = 0$
48. $13x^2 - 8xy + 7y^2 - 45 = 0$
49. $x^2 + y^2 + 2xy + 2\sqrt{2}x - 2\sqrt{2}y + 2 = 0$
50. $x^2 - 10xy + y^2 + 1 = 0$

6.6 In Exercises 51–54, evaluate the parametric equations $x = 3\cos\theta$ and $y = 2\sin^2\theta$ for the given value of θ.

51. $\theta = 0$
52. $\theta = \dfrac{\pi}{3}$
53. $\theta = \dfrac{\pi}{6}$
54. $\theta = -\dfrac{\pi}{4}$

In Exercises 55–60, sketch the curve represented by the parametric equations and, where possible, write the corresponding rectangular equation by eliminating the parameter. Verify your result with a graphing utility.

55. $x = 2t$
    $y = 4t$
56. $x = 1 + 4t$
    $y = 2 - 3t$
57. $x = t^2$
    $y = \sqrt{t}$
58. $x = t + 4$
    $y = t^2$

59. $x = 6 \cos \theta$
$y = 6 \sin \theta$

60. $x = 3 + 3 \cos \theta$
$y = 2 + 5 \sin \theta$

61. Find a parametric representation of the ellipse with center at $(-3, 4)$, major axis horizontal and 8 units in length, and minor axis 6 units in length.

62. Find a parametric representation of the hyperbola with vertices $(0, \pm 4)$ and foci $(0, \pm 5)$.

63. *Rotary Engine* The rotary engine was developed by Felix Wankel in the 1950s. It features a rotor that is basically a modified equilateral triangle. The rotor moves in a chamber that, in two dimensions, is an epitrochoid. Use a graphing utility to graph the chamber modeled by the parametric equations $x = \cos 3\theta + 5 \cos \theta$ and $y = \sin 3\theta + 5 \sin \theta$.

64. *Involute of a Circle* The *involute* of a circle is described by the endpoint P of a string that is held taut as it is unwound from a spool (see figure). The spool does not rotate. Show that a parametric representation of the involute of a circle is

$x = r(\cos \theta + \theta \sin \theta)$

$y = r(\sin \theta - \theta \cos \theta)$.

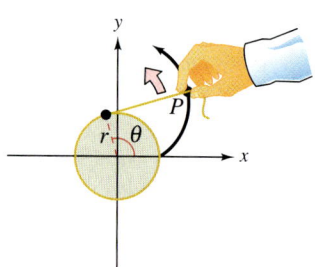

6.7 In Exercises 65–68, plot the point given in polar coordinates and find the corresponding rectangular coordinates of the point.

65. $\left(2, \dfrac{\pi}{4}\right)$

66. $\left(-5, -\dfrac{\pi}{3}\right)$

67. $(-7, 4.19)$

68. $\left(\sqrt{3}, 2.62\right)$

In Exercises 69–72, the rectangular coordinates of a point are given. Find two sets of polar coordinates of the point for $0 \leq \theta \leq 2\pi$.

69. $(0, 2)$

70. $\left(-\sqrt{5}, \sqrt{5}\right)$

71. $(4, 6)$

72. $(3, -4)$

In Exercises 73–76, convert the polar equation to rectangular form.

73. $r = 3 \cos \theta$

74. $r = 10$

75. $r = \dfrac{2}{1 + \sin \theta}$

76. $r^2 = \cos 2\theta$

In Exercises 77 and 78, convert the rectangular equation to polar form.

77. $(x^2 + y^2)^2 = ax^2 y$

78. $x^2 + y^2 - 4x = 0$

6.8 In Exercises 79–84, identify and sketch the graph of the polar equation.

79. $r = 4$

80. $r = 2\theta$

81. $r = 4 \sin 2\theta$

82. $r = \cos 5\theta$

83. $r = -2(1 + \cos \theta)$

84. $r = 3 - 4 \cos \theta$

In Exercises 85–88, determine the symmetry of r, the maximum value of $|r|$, and any zeros of r. Then sketch the graph of the equation.

85. $r = 2 + 6 \sin \theta$

86. $r = 5 - 5 \cos \theta$

87. $r = -3 \cos 2\theta$

88. $r^2 = \cos 2\theta$

In Exercises 89–92, identify the type of polar graph.

89. $r = 3(2 - \cos \theta)$

90. $r = 3(1 - 2 \cos \theta)$

91. $r = 4 \cos 3\theta$

92. $r^2 = 9 \cos 2\theta$

6.9 In Exercises 93–96, state the eccentricity of the conic and identify the conic from its eccentricity. Sketch a graph of the conic.

93. $r = \dfrac{1}{1 + 2 \sin \theta}$

94. $r = \dfrac{2}{1 - \sin \theta}$

95. $r = \dfrac{4}{5 - 3 \cos \theta}$

96. $r = \dfrac{16}{4 + 5 \cos \theta}$

In Exercises 97–100, find a polar equation of the conic.

97. Parabola   Vertex: $(2, \pi)$
            Focus: $(0, 0)$

98. Parabola   Vertex: $(2, \pi/2)$
            Focus: $(0, 0)$

99. Ellipse   Vertices: $(5, 0), (1, \pi)$
            One focus: $(0, 0)$

100. Hyperbola   Vertices: $(1, 0), (7, 0)$
            One focus: $(0, 0)$

101. *Explorer 18* On November 26, 1963, the United States launched Explorer 18. Its low and high points above the surface of earth were 119 miles and 122,000 miles, respectively (see figure). The center of earth is at one focus of the orbit. Find the polar equation of the orbit and find the distance between the surface of the earth (assume a radius of 4000 miles) and the satellite when $\theta = \pi/3$ radians.

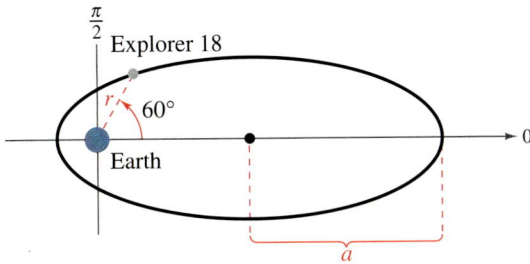

102. *Asteroid* An asteroid takes a parabolic path with earth as its focus. It is about 6,000,000 miles from earth at its closest approach. Write the polar equation of the path of the asteroid with its vertex at $\theta = \pi/2$. Find the distance between the asteroid and earth when $\theta = -\pi/3$.

Synthesis

True or False? In Exercises 103–106, determine whether the statement is true or false. Justify your answer.

103. When $B = 0$ in an equation of the form $Ax^2 + Bxy + Cy^2 + Dx + Ey + F = 0$, the graph of the equation can be a parabola only if $C = 0$ also.

104. The graph of $(x^2/4) - y^4 = 1$ is a hyperbola.

105. Only one set of parametric equations can represent the line $y = 3 - 2x$.

106. There is a unique polar coordinate representation of each point in the plane.

107. Consider an ellipse with the major axis horizontal and 10 units in length. The number b in the standard form of the equation of the ellipse must be less than what real number? Explain the change in the shape of the ellipse as b approaches this number.

108. The graph of the parametric equations $x = 2 \sec t$ and $y = 3 \tan t$ is given in the figure. Would the graph change for the equations $x = 2 \sec(-t)$ and $y = 3 \tan(-t)$? If so, how would it change?

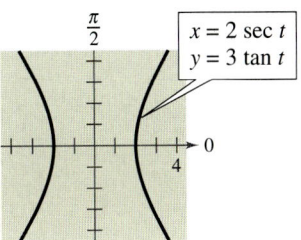

FIGURE FOR 108

109. A moving object is modeled by the parametric equations $x = 4 \cos t$ and $y = 3 \sin t$, where t is time (see figure). How would the orbit change for the following?

(a) $x = 4 \cos 2t, \quad y = 3 \sin 2t$

(b) $x = 5 \cos t, \quad y = 3 \sin t$

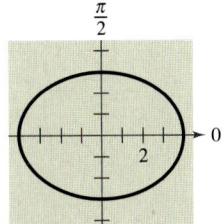

110. Identify the type of symmetry each of the following polar points has with the point in the figure.

(a) $\left(-4, \dfrac{\pi}{6}\right)$

(b) $\left(4, -\dfrac{\pi}{6}\right)$

(c) $\left(-4, -\dfrac{\pi}{6}\right)$

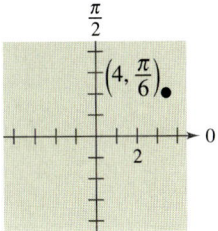

111. What is the relationship between the graphs of the rectangular and polar equations?

(a) $x^2 + y^2 = 25, \quad r = 5$

(b) $x - y = 0, \quad \theta = \dfrac{\pi}{4}$

Chapter Project ▶ Polar, Rectangular, and Parametric Forms

In this project, you will compare the polar, rectangular, and parametric forms of equations for conics.

(a) Consider the polar equation

$$r = \frac{15}{3 - 2\cos\theta}.$$

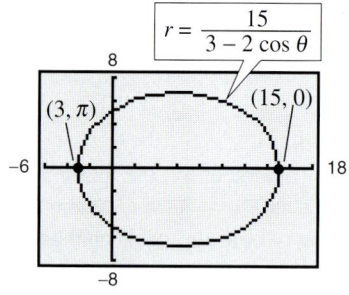

In Section 6.9, you learned that the graph of this polar equation is an ellipse and that one of the ellipse's foci is at the pole, as shown in the graph at the left. To find the rectangular equation of this ellipse, begin by rewriting the equation as

$$3r - 2r\cos\theta = 15.$$

Then use the substitutions $r = \sqrt{x^2 + y^2}$ and $r\cos\theta = x$ to find the rectangular equation. After you find the rectangular equation, write it in the standard form

$$\frac{(x-h)^2}{a^2} + \frac{(y-k)^2}{b^2} = 1.$$

(b) Use the standard form of the ellipse in part (a) to find the center, foci, and eccentricity of the ellipse. Compare your results with those obtained in Example 1 in Section 6.9. Use the function mode of a graphing utility to graph the ellipse.

(c) Use the standard form of the ellipse and the identity $\sin^2\theta + \cos^2\theta = 1$ to write parametric equations for the ellipse. Then use the parametric mode of a graphing utility to graph the ellipse.

Chapter Project Investigations

1. Sketch the graph of the polar equation

$$r = \frac{32}{3 - 5\sin\theta}.$$

Which type of conic is this?

2. Write the standard form of the rectangular equation of the conic in Question 1. Then sketch the graph using the *function* mode of a graphing utility. Does your graph agree with the graph obtained in Question 1? Which of the two graphs is easier to obtain? Explain your reasoning.

3. Write parametric equations for the conic given in Question 1. Then use the *parametric* mode of a graphing utility to graph the conic. Does your graph agree with the graph obtained in Question 1?

4. Consider the parametric equations

$$x = 4 + 5\cos t \quad \text{and} \quad y = 3\sin t.$$

With appropriate scaling of the coordinate system, could these equations represent the motion of a comet about the sun? Explain your reasoning.

Chapter Test

The *Interactive* CD-ROM and *Internet* versions of this text provide answers to the Chapter Tests and Cumulative Tests. They also offer Chapter Pre-Tests (which test key skills and concepts covered in previous chapters) and Chapter Post-Tests, both of which have randomly generated exercises with diagnostic capabilities.

Take this test as you would take a test in class. After you are done, check your work against the answers given in the back of the book.

1. Find the inclination of the line $2x - 7y + 3 = 0$.
2. Find the angle between the lines $3x + 2y - 4 = 0$ and $4x - y + 6 = 0$.
3. Find the distance between the point $(7, 5)$ and the line $y = 5 - x$.

In Exercises 4–7, classify the conic and write the equation in standard form. Identify the center, vertices, foci, and asymptotes (if any). Then sketch the graph of the conic.

4. $y^2 - 4x + 4 = 0$
5. $x^2 - 4y^2 - 4x = 0$
6. $9x^2 + 16y^2 + 54x - 32y - 47 = 0$
7. $2x^2 + 2y^2 - 8x - 4y + 9 = 0$

8. Find the standard form of the equation of the parabola with vertex $(3, -2)$, with vertical axis, and passing through the point $(0, 4)$.
9. Find the standard form of the equation of the hyperbola with foci $(0, 0)$ and $(0, 4)$ and asymptotes $y = \pm\frac{1}{2}x + 2$.
10. (a) Determine the number of degrees the axis must be rotated to eliminate the xy-term of the conic $x^2 + 6xy + y^2 - 6 = 0$.
    (b) Graph the conic and use a graphing utility to confirm your result.
11. Sketch the curve represented by the parametric equations $x = 2 + 3\cos\theta$ and $y = 2\sin\theta$. Eliminate the parameter and write the corresponding rectangular equation.
12. Find a set of parametric equations of the line passing through the points $(2, -3)$ and $(6, 4)$. (The answer is not unique.)
13. Convert the polar coordinate $(-2, 5\pi/6)$ to rectangular form.
14. Convert the rectangular coordinate $(2, -2)$ to polar form and find two additional representations for this point.
15. Convert the rectangular equation $x^2 + y^2 - 4y = 0$ to polar form.

In Exercises 16–19, sketch the graph of the polar equation. Identify the type of graph.

16. $r = \dfrac{4}{1 + \cos\theta}$
17. $r = \dfrac{4}{2 + \cos\theta}$
18. $r = 2 + 3\sin\theta$
19. $r = 3\sin 2\theta$

20. A straight road rises with an inclination of 0.15 radian from the horizontal. Find the slope of the road and the change in elevation over a 1-mile stretch of the road.
21. A baseball is hit 3 feet above the ground toward the left field fence. The fence is 10 feet high and 375 feet from home plate. The path of the baseball can be modeled by the parametric equations $x = (115\cos\theta)t$ and $y = 3 + (115\sin\theta)t - 16t^2$. Does the baseball go over the fence when the baseball is hit at an angle of $\theta = 30°$? Does the baseball go over the fence when $\theta = 35°$?

Cumulative Test for Chapters 4–6

Take this test to review material from earlier chapters. After you are done, check your work against the answers given in the back of the book.

1. Write the complex number $3 - \sqrt{-25}$ in standard form.

In Exercises 2–5, perform the operations and write the result in standard form.

2. $6i - (2 + \sqrt{-81})$
3. $(2i - 3)^2$
4. $(\sqrt{3} + i)(\sqrt{3} - i)$
5. $\dfrac{4i}{1 + 2i}$

In Exercises 6 and 7, find all the zeros of the function.

6. $f(x) = x^3 + 2x^2 + 4x + 8$
7. $f(x) = x^4 + 4x^3 - 21x^2$

8. Find a polynomial with integer coefficients that has -5, -2, and $2 + \sqrt{3}i$ as its zeros.

9. Find the trigonometric form of the complex number $-2 + 2i$.

10. Find the product of $[4(\cos 30° + i \sin 30°)][6(\cos 120° + i \sin 120°)]$. Write the answer in standard form.

11. Use DeMoivre's Theorem to find the indicated power of the complex number.

$$\left[2\left(\cos \frac{2\pi}{3} + i \sin \frac{2\pi}{3}\right)\right]^4$$

12. Find the three cube roots of 1.

13. Write all the solutions of the equation $x^4 - 81i = 0$.

In Exercises 14 and 15, use the graph of f to describe the transformation that yields the graph of g. Use a graphing utility to graph both equations in the same viewing window.

14. $f(x) = \left(\frac{2}{5}\right)^x$, $g(x) = -\left(\frac{2}{5}\right)^{-x+3}$
15. $f(x) = 2.2^x$, $g(x) = -2.2^x + 4$

In Exercises 16–19, use a calculator to evaluate each expression. Approximate your result to three decimal places.

16. $\log_{10} 98$
17. $\log_{10}\left(\frac{6}{7}\right)$
18. $\ln \sqrt{31}$
19. $\ln\left(\sqrt{40} - 5\right)$

In Exercises 20–22, evaluate the logarithm using the change-of-base formula. Approximate your answer to three decimal places.

20. $\log_7 1.8$
21. $\log_3 0.149$
22. $\log_{1/2} 17$

23. Use the logarithmic properties to expand $\ln\left(\dfrac{x^2 - 16}{x^4}\right)$, where $x > 4$.

24. Write $2 \ln x - \frac{1}{2} \ln(x + 5)$ as a logarithm of a single quantity.

▶ Cumulative Test for Chapters 4–6   **485**

In Exercises 25 and 26, solve the equation.

25. $6e^{2x} = 72$

26. $\log_2 x + \log_2 5 = 6$

27. Use a graphing utility to graph $f(x) = \dfrac{1000}{1 + 4e^{-0.2x}}$ and determine the horizontal asymptotes.

28. The number of bacteria N in a culture is given by the model

$$N = 175e^{kt}$$

where t is the time in hours. If $N = 420$ when $t = 8$, estimate the time required for the population to double in size.

29. Find the angle between the lines $2x + y - 3 = 0$ and $x - 3y + 6 = 0$.

30. Find the distance between the point $(6, -3)$ and the line $y = 2x - 4$.

In Exercises 31–34, classify the conic and write the equation in standard form. Identify the center, vertices, foci, and asymptotes (if any). Then sketch the graph.

31. $9x^2 + 4y^2 - 36x + 8y + 4 = 0$

32. $4x^2 - y^2 - 4 = 0$

33. $x^2 + y^2 + 2x - 6y - 12 = 0$

34. $y^2 + 2x + 2 = 0$

35. Find an equation in rectangular coordinates of the circle with center $(2, -4)$ and passing through the point $(0, 4)$.

36. Find an equation in rectangular coordinates of the hyperbola with foci $(0, 0)$ and $(0, 6)$ and asymptotes $y = \pm \dfrac{2\sqrt{5}}{5} x + 3$.

37. (a) Determine the number of degrees the axes must be rotated to eliminate the xy-term of the conic $x^2 + xy + y^2 + 2x - 3y - 30 = 0$.

(b) Graph the conic and use a graphing utility to confirm your result.

38. Sketch the curve represented by the parametric equations $x = 3 + 4 \cos \theta$ and $y = \sin \theta$. Eliminate the parameter and write the corresponding rectangular equation.

39. Find a set of parametric equations of the line passing through the points $(3, -2)$ and $(-3, 4)$. (The answer is not unique.)

40. Convert the rectangular equation $x^2 + y^2 - 6y = 0$ to polar form.

In Exercises 41 and 42, sketch the graph of the polar equation.

41. $r = \dfrac{3}{2 + \cos \theta}$

42. $r = \dfrac{4}{1 + \sin \theta}$

43. Match the polar equation with its graph at the left.

(a) $r = 2 + 3 \sin \theta$

(b) $r = 3 \sin \theta$

(c) $r = 3 \sin 2\theta$

(i)

(ii)

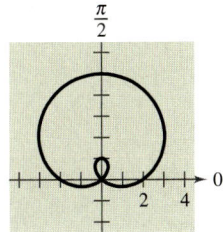
(iii)

FIGURE FOR 43

Appendix A Proofs of Selected Theorems

SECTION P.3, PAGE 28

The Midpoint Formula
The midpoint of the segment joining the points (x_1, y_1) and (x_2, y_2) is given by the Midpoint Formula

$$\text{Midpoint} = \left(\frac{x_1 + x_2}{2}, \frac{y_1 + y_2}{2}\right).$$

Proof
Using the figure, you must show that

$$d_1 = d_2 \quad \text{and} \quad d_1 + d_2 = d_3.$$

By the Distance Formula, you obtain

$$d_1 = \sqrt{\left(\frac{x_1 + x_2}{2} - x_1\right)^2 + \left(\frac{y_1 + y_2}{2} - y_1\right)^2} = \frac{1}{2}\sqrt{(x_2 - x_1)^2 + (y_2 - y_1)^2}$$

$$d_2 = \sqrt{\left(x_2 - \frac{x_1 + x_2}{2}\right)^2 + \left(y_2 - \frac{y_1 + y_2}{2}\right)^2} = \frac{1}{2}\sqrt{(x_2 - x_1)^2 + (y_2 - y_1)^2}$$

$$d_3 = \sqrt{(x_2 - x_1)^2 + (y_2 - y_1)^2}.$$

So, it follows that $d_1 = d_2$ and $d_1 + d_2 = d_3$.

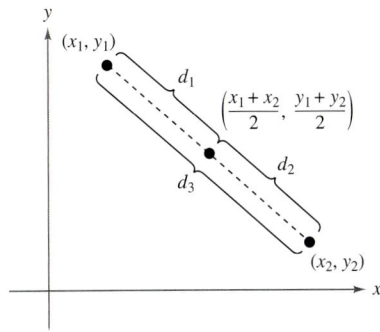

Midpoint Formula

SECTION 2.4, PAGE 236

Sum and Difference Formulas

$\sin(u + v) = \sin u \cos v + \cos u \sin v$

$\sin(u - v) = \sin u \cos v - \cos u \sin v$

$\cos(u + v) = \cos u \cos v - \sin u \sin v$

$\cos(u - v) = \cos u \cos v + \sin u \sin v$

$\tan(u + v) = \dfrac{\tan u + \tan v}{1 - \tan u \tan v}$

$\tan(u - v) = \dfrac{\tan u - \tan v}{1 + \tan u \tan v}$

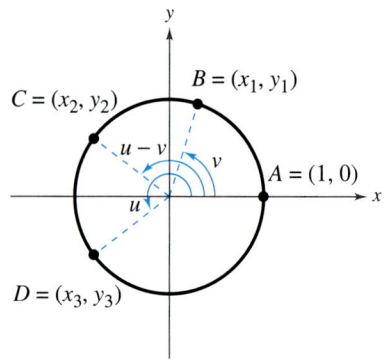

Proof

Here are proofs for the formulas for $\cos(u \pm v)$. In the top figure, let A be the point $(1, 0)$ and then use u and v to locate the points $B = (x_1, y_1)$, $C = (x_2, y_2)$, and $D = (x_3, y_3)$ on the unit circle. So, $x_i^2 + y_i^2 = 1$ for $i = 1, 2,$ and 3. For convenience, assume that $0 < v < u < 2\pi$. In the bottom figure, note that arcs AC and BD have the same length. So, line segments AC and BD are also equal in length, which implies that

$$\sqrt{(x_2 - 1)^2 + (y_2 - 0)^2} = \sqrt{(x_3 - x_1)^2 + (y_3 - y_1)^2}$$

$$x_2^2 - 2x_2 + 1 + y_2^2 = x_3^2 - 2x_1x_3 + x_1^2 + y_3^2 - 2y_1y_3 + y_1^2$$

$$(x_2^2 + y_2^2) + 1 - 2x_2 = (x_3^2 + y_3^2) + (x_1^2 + y_1^2) - 2x_1x_3 - 2y_1y_3$$

$$1 + 1 - 2x_2 = 1 + 1 - 2x_1x_3 - 2y_1y_3$$

$$x_2 = x_3x_1 + y_3y_1.$$

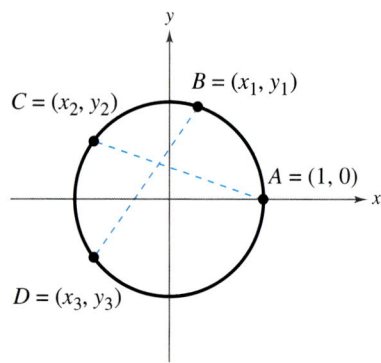

Finally, by substituting the values $x_2 = \cos(u - v)$, $x_3 = \cos u$, $x_1 = \cos v$, $y_3 = \sin u$, and $y_1 = \sin v$, you obtain $\cos(u - v) = \cos u \cos v + \sin u \sin v$.

The formula for $\cos(u + v)$ can be established by considering $u + v = u - (-v)$ and using the formula just derived to obtain

$$\cos(u + v) = \cos[u - (-v)]$$

$$= \cos u \cos(-v) + \sin u \sin(-v)$$

$$= \cos u \cos v - \sin u \sin v.$$

SECTION 2.5, PAGE 243

Double-Angle Formulas

$\sin 2u = 2 \sin u \cos u$ $\tan 2u = \dfrac{2 \tan u}{1 - \tan^2 u}$

$\cos 2u = \cos^2 u - \sin^2 u$

$ = 2 \cos^2 u - 1 = 1 - 2 \sin^2 u$

Proof

To prove the first formula, let $v = u$ in the formula for $\sin(u + v)$.

$$\sin 2u = \sin(u + u)$$

$$= \sin u \cos u + \cos u \sin u$$

$$= 2 \sin u \cos u$$

To prove the second formula, let $v = u$ in the formula for $\cos(u + v)$.

$$\cos 2u = \cos(u + u)$$

$$= \cos u \cos u - \sin u \sin u$$

$$= \cos^2 u - \sin^2 u$$

The tangent double-angle formula can be proven in a similar way.

SECTION 2.5, PAGE 245

Power-Reducing Formulas

$$\sin^2 u = \frac{1 - \cos 2u}{2} \qquad \cos^2 u = \frac{1 + \cos 2u}{2} \qquad \tan^2 u = \frac{1 - \cos 2u}{1 + \cos 2u}$$

Proof

The first two formulas can be verified by solving for $\sin^2 u$ and $\cos^2 u$, respectively, in the double-angle formulas

$$\cos 2u = 1 - 2\sin^2 u \qquad \text{and} \qquad \cos 2u = 2\cos^2 u - 1.$$

The third formula can be verified using the fact that

$$\tan^2 u = \frac{\sin^2 u}{\cos^2 u}.$$

SECTION 2.5, PAGE 248

Sum-to-Product Formulas

$$\sin x + \sin y = 2\sin\left(\frac{x+y}{2}\right)\cos\left(\frac{x-y}{2}\right)$$

$$\sin x - \sin y = 2\cos\left(\frac{x+y}{2}\right)\sin\left(\frac{x-y}{2}\right)$$

$$\cos x + \cos y = 2\cos\left(\frac{x+y}{2}\right)\cos\left(\frac{x-y}{2}\right)$$

$$\cos x - \cos y = -2\sin\left(\frac{x+y}{2}\right)\sin\left(\frac{x-y}{2}\right)$$

Proof

To prove the first formula, let $x = u + v$ and $y = u - v$. Then substitute $u = (x + y)/2$ and $v = (x - y)/2$ in the product-to-sum formula.

$$\sin u \cos v = \frac{1}{2}[\sin(u + v) + \sin(u - v)]$$

$$\sin\left(\frac{x+y}{2}\right)\cos\left(\frac{x-y}{2}\right) = \frac{1}{2}(\sin x + \sin y)$$

$$2\sin\left(\frac{x+y}{2}\right)\cos\left(\frac{x-y}{2}\right) = \sin x + \sin y$$

SECTION 3.1, PAGE 262

Law of Sines

If ABC is a triangle with sides a, b, and c, then

$$\frac{a}{\sin A} = \frac{b}{\sin B} = \frac{c}{\sin C}.$$

A is acute.

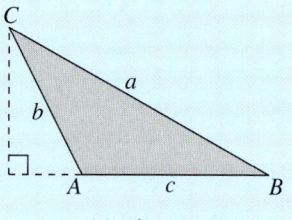

A is obtuse.

Proof

Let h be the altitude of either triangle found in the figure above. Then you have

$$\sin A = \frac{h}{b} \quad \text{or} \quad h = b \sin A$$

$$\sin B = \frac{h}{a} \quad \text{or} \quad h = a \sin B.$$

Equating these two values of h, you have

$$a \sin B = b \sin A \quad \text{or} \quad \frac{a}{\sin A} = \frac{b}{\sin B}.$$

Note that $\sin A \neq 0$ and $\sin B \neq 0$ because no angle of a triangle can have a measure of $0°$ or $180°$. In a similar manner, by constructing an altitude from vertex B to side AC (extended), you can show that

$$\frac{a}{\sin A} = \frac{c}{\sin C}.$$

So, the Law of Sines is established.

SECTION 3.2, PAGE 271

Law of Cosines

Standard Form	Alternative Form
$a^2 = b^2 + c^2 - 2bc \cos A$	$\cos A = \dfrac{b^2 + c^2 - a^2}{2bc}$
$b^2 = a^2 + c^2 - 2ac \cos B$	$\cos B = \dfrac{a^2 + c^2 - b^2}{2ac}$
$c^2 = a^2 + b^2 - 2ab \cos C$	$\cos C = \dfrac{a^2 + b^2 - c^2}{2ab}$

Proof

Consider a triangle that has three acute angles, as shown in the figure. Note that vertex B has coordinates $(c, 0)$. Furthermore, C has coordinates (x, y), where $x = b \cos A$ and $y = b \sin A$. Because a is the distance from vertex C to vertex B, it follows that

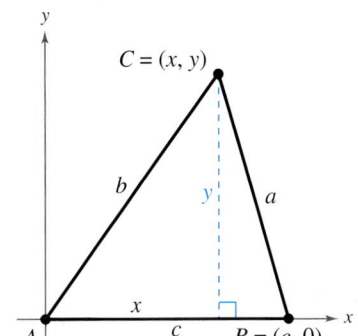

$$a = \sqrt{(x - c)^2 + (y - 0)^2}$$
$$a^2 = (b \cos A - c)^2 + (b \sin A)^2$$
$$a^2 = b^2 \cos^2 A - 2bc \cos A + c^2 + b^2 \sin^2 A$$
$$a^2 = b^2(\sin^2 A + \cos^2 A) + c^2 - 2ab \cos A$$
$$a^2 = b^2 + c^2 - 2bc \cos A. \qquad \sin^2 A + \cos^2 A = 1$$

Similar arguments can be used to establish the other two equations.

SECTION 3.2, PAGE 274

> **Heron's Area Formula**
>
> Given any triangle with sides of lengths a, b, and c, the area of the triangle is
>
> $$\text{Area} = \sqrt{s(s-a)(s-b)(s-c)}$$
>
> where $s = (a+b+c)/2$.

Proof

From Section 3.1, you know that

$$\begin{aligned}
\text{Area} &= \frac{1}{2}bc \sin A \\
&= \sqrt{\frac{1}{4}b^2c^2 \sin^2 A} \\
&= \sqrt{\frac{1}{4}b^2c^2(1-\cos^2 A)} \\
&= \sqrt{\left[\frac{1}{2}bc(1+\cos A)\right]\left[\frac{1}{2}bc(1-\cos A)\right]}.
\end{aligned}$$

Using the Law of Cosines, you can show that

$$\frac{1}{2}bc(1+\cos A) = \frac{a+b+c}{2} \cdot \frac{-a+b+c}{2}$$

and

$$\frac{1}{2}bc(1-\cos A) = \frac{a-b+c}{2} \cdot \frac{a+b-c}{2}.$$

Letting $s = (a+b+c)/2$, these two equations can be rewritten as

$$\frac{1}{2}bc(1+\cos A) = s(s-a)$$

and

$$\frac{1}{2}bc(1-\cos A) = (s-b)(s-c).$$

So, you can conclude that

$$\text{Area} = \sqrt{s(s-a)(s-b)(s-c)}.$$

SECTION 3.4, PAGE 292

Properties of the Dot Product
Let **u**, **v**, and **w** be vectors in the plane or in space and let c be a scalar.

1. $\mathbf{u} \cdot \mathbf{v} = \mathbf{v} \cdot \mathbf{u}$
2. $\mathbf{0} \cdot \mathbf{v} = 0$
3. $\mathbf{u} \cdot (\mathbf{v} + \mathbf{w}) = \mathbf{u} \cdot \mathbf{v} + \mathbf{u} \cdot \mathbf{w}$
4. $\mathbf{v} \cdot \mathbf{v} = \|\mathbf{v}\|^2$
5. $c(\mathbf{u} \cdot \mathbf{v}) = c\mathbf{u} \cdot \mathbf{v} = \mathbf{u} \cdot c\mathbf{v}$

Proof

To prove Property 1, let $\mathbf{u} = \langle u_1, u_2 \rangle$ and $\mathbf{v} = \langle v_1, v_2 \rangle$. Then

$$\begin{aligned} \mathbf{u} \cdot \mathbf{v} &= u_1 v_1 + u_2 v_2 \\ &= v_1 u_1 + v_2 u_2 \\ &= \mathbf{v} \cdot \mathbf{u}. \end{aligned}$$

To prove Property 4, let $\mathbf{v} = \langle v_1, v_2 \rangle$. Then

$$\begin{aligned} \mathbf{v} \cdot \mathbf{v} &= v_1^2 + v_2^2 \\ &= \left(\sqrt{v_1^2 + v_2^2}\right)^2 \\ &= \|\mathbf{v}\|^2. \end{aligned}$$

To prove part of Property 5, let $\mathbf{u} = \langle u_1, u_2 \rangle$ and $\mathbf{v} = \langle v_1, v_2 \rangle$ and let c be a scalar. Then,

$$\begin{aligned} c(\mathbf{u} \cdot \mathbf{v}) &= c(\langle u_1, u_2 \rangle \cdot \langle v_1, v_2 \rangle) \\ &= c(u_1 v_1 + u_2 v_2) \\ &= (cu_1)v_1 + (cu_2)v_2 \\ &= \langle cu_1, cu_2 \rangle \cdot \langle v_1, v_2 \rangle \\ &= c\mathbf{u} \cdot \mathbf{v}. \end{aligned}$$

SECTION 3.4, PAGE 293

Angle Between Two Vectors

If θ is the angle between two nonzero vectors $\mathbf{u}$ and $\mathbf{v}$, then
$$\cos\theta = \frac{\mathbf{u}\cdot\mathbf{v}}{\|\mathbf{u}\|\,\|\mathbf{v}\|}.$$

Proof

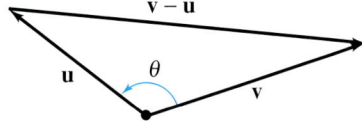

Consider the triangle determined by vectors $\mathbf{u}$, $\mathbf{v}$, and $\mathbf{v}-\mathbf{u}$, as shown in the figure. By the Law of Cosines, you can write

$$\|\mathbf{v}-\mathbf{u}\|^2 = \|\mathbf{u}\|^2 + \|\mathbf{v}\|^2 - 2\|\mathbf{u}\|\,\|\mathbf{v}\|\cos\theta$$
$$(\mathbf{v}-\mathbf{u})\cdot(\mathbf{v}-\mathbf{u}) = \|\mathbf{u}\|^2 + \|\mathbf{v}\|^2 - 2\|\mathbf{u}\|\,\|\mathbf{v}\|\cos\theta$$
$$(\mathbf{v}-\mathbf{u})\cdot\mathbf{v} - (\mathbf{v}-\mathbf{u})\cdot\mathbf{u} = \|\mathbf{u}\|^2 + \|\mathbf{v}\|^2 - 2\|\mathbf{u}\|\,\|\mathbf{v}\|\cos\theta$$
$$\mathbf{v}\cdot\mathbf{v} - \mathbf{u}\cdot\mathbf{v} - \mathbf{v}\cdot\mathbf{u} + \mathbf{u}\cdot\mathbf{u} = \|\mathbf{u}\|^2 + \|\mathbf{v}\|^2 - 2\|\mathbf{u}\|\,\|\mathbf{v}\|\cos\theta$$
$$\|\mathbf{v}\|^2 - 2\mathbf{u}\cdot\mathbf{v} + \|\mathbf{u}\|^2 = \|\mathbf{u}\|^2 + \|\mathbf{v}\|^2 - 2\|\mathbf{u}\|\,\|\mathbf{v}\|\cos\theta$$
$$-2\mathbf{u}\cdot\mathbf{v} = -2\|\mathbf{u}\|\,\|\mathbf{v}\|\cos\theta$$
$$\cos\theta = \frac{\mathbf{u}\cdot\mathbf{v}}{\|\mathbf{u}\|\,\|\mathbf{v}\|}.$$

SECTION 5.3, PAGE 368

Properties of Logarithms

Let a be a positive number such that $a \neq 1$, and let n be a real number. If u and v are positive real numbers, the following properties are true.

1. $\log_a(uv) = \log_a u + \log_a v$
2. $\log_a \dfrac{u}{v} = \log_a u - \log_a v$
3. $\log_a u^n = n\log_a u$

1. $\ln(uv) = \ln u + \ln v$
2. $\ln \dfrac{u}{v} = \ln u - \ln v$
3. $\ln u^n = n\ln u$

Proof

To prove Property 1, let $x = \log_a u$ and $y = \log_a v$. The corresponding exponential forms of these two equations are

$$a^x = u \quad \text{and} \quad a^y = v.$$

Multiplying u and v produces $uv = a^x a^y = a^{x+y}$. The corresponding logarithmic form of $uv = a^{x+y}$ is $\log_a(uv) = x + y$. So, $\log_a(uv) = \log_a u + \log_a v$.

SECTION 6.1, PAGE 408

> **Distance Between a Point and a Line**
>
> The distance between the point (x_1, y_1) and the line $Ax + By + C = 0$ is
>
> $$d = \frac{|Ax_1 + By_1 + C|}{\sqrt{A^2 + B^2}}.$$

Proof

For simplicity's sake, assume that the given line is neither horizontal nor vertical. By writing the equation $Ax + By + C = 0$ in slope-intercept form

$$y = -\frac{A}{B}x - \frac{C}{B}$$

you can see that the line has a slope of $m = -A/B$. So, the slope of the line passing through (x_1, y_1) and perpendicular to the given line is B/A, and its equation is $y - y_1 = (B/A)(x - x_1)$. These two lines intersect at the point (x_2, y_2), where

$$x_2 = \frac{B(Bx_1 - Ay_1) - AC}{A^2 + B^2} \quad \text{and} \quad y_2 = \frac{A(-Bx_1 + Ay_1) - BC}{A^2 + B^2}.$$

Finally, the distance between (x_1, y_1) and (x_2, y_2) is

$$\begin{aligned}
d &= \sqrt{(x_2 - x_1)^2 + (y_2 - y_1)^2} \\
&= \sqrt{\left(\frac{B^2 x_1 - ABy_1 - AC}{A^2 + B^2} - x_1\right)^2 + \left(\frac{-ABx_1 + A^2 y_1 - BC}{A^2 + B^2} - y_1\right)^2} \\
&= \sqrt{\frac{A^2(Ax_1 + By_1 + C)^2 + B^2(Ax_1 + By_1 + C)^2}{(A^2 + B^2)^2}} \\
&= \frac{|Ax_1 + By_1 + C|}{\sqrt{A^2 + B^2}}.
\end{aligned}$$

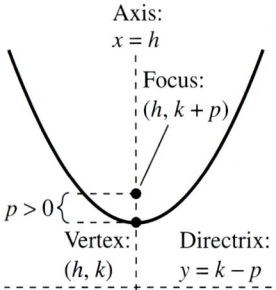

SECTION 6.2, PAGE 414

Standard Equation of a Parabola

The standard form of the equation of a parabola with vertex at (h, k) is as follows.

$$(x - h)^2 = 4p(y - k), \quad p \neq 0 \qquad \text{Vertical axis, directrix: } y = k - p$$

$$(y - k)^2 = 4p(x - h), \quad p \neq 0 \qquad \text{Horizontal axis, directrix: } x = h - p$$

The focus lies on the axis p units (*directed distance*) from the vertex. If the vertex is at the origin $(0, 0)$, the equation takes one of the following forms.

$$x^2 = 4py \qquad \text{Vertical axis}$$

$$y^2 = 4px \qquad \text{Horizontal axis}$$

Proof

The case for which the directrix is parallel to the x-axis and the focus lies above the vertex, as shown in the figure, is proven here. If (x, y) is any point on the parabola, then, by definition, it is equidistant from the focus $(h, k + p)$ and the directrix $y = k - p$, and you have

$$\sqrt{(x - h)^2 + [y - (k + p)]^2} = y - (k - p)$$

$$(x - h)^2 + [y - (k + p)]^2 = [y - (k - p)]^2$$

$$(x - h)^2 + y^2 - 2y(k + p) + (k + p)^2 = y^2 - 2y(k - p) + (k - p)^2$$

$$(x - h)^2 - 2py + 2pk = 2py - 2pk$$

$$(x - h)^2 = 4p(y - k).$$

The case in which the directrix is parallel to the y-axis and the focus lies to the right of the vertex is left for you to prove.

SECTION 6.9, PAGE 470

Polar Equations of Conics

The graph of a polar equation of the form

1. $r = \dfrac{ep}{1 \pm e \cos \theta}$

2. $r = \dfrac{ep}{1 \pm e \sin \theta}$

is a conic, where $e > 0$ is the eccentricity and $|p|$ is the distance between the focus (pole) and the directrix.

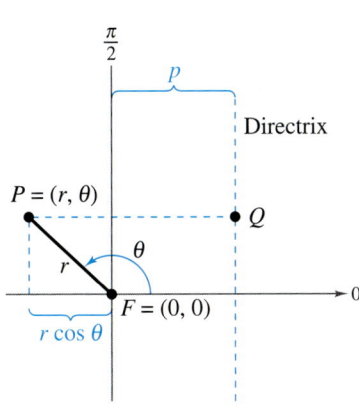

Proof

A proof for $r = ep/(1 + e \cos \theta)$ with $p > 0$ is listed here. The proofs of the other cases are similar. In the figure, consider a vertical directrix, p units to the right of the focus $F = (0, 0)$. If $P = (r, \theta)$ is a point on the graph of

$$r = \frac{ep}{1 + e \cos \theta}$$

the distance between P and the directrix is

$$\begin{aligned} PQ &= |p - x| \\ &= |p - r \cos \theta| \\ &= \left| p - \left(\frac{ep}{1 + e \cos \theta}\right) \cos \theta \right| \\ &= \left| p\left(1 - \frac{e \cos \theta}{1 + e \cos \theta}\right) \right| \\ &= \left| \frac{p}{1 + e \cos \theta} \right| \\ &= \left| \frac{r}{e} \right|. \end{aligned}$$

Moreover, because the distance between P and the pole is simply $PF = |r|$, the ratio of PF to PQ is

$$\begin{aligned} \frac{PF}{PQ} &= \frac{|r|}{|r/e|} \\ &= |e| \\ &= e \end{aligned}$$

and, by definition, the graph of the equation must be a conic.

Appendix B Concepts in Statistics

B.1 Representing Data

▶ **What you should learn**
- How to use line plots to order and analyze data
- How to use stem-and-leaf plots to organize and compare data
- How to use histograms to represent frequency distributions

▶ **Why you should learn it**

Line plots, stem-and-leaf plots, and histograms are quick methods for determining which elements in a set of data occur with the greatest frequency. For instance, in Exercise 9 on page A15, you are asked to construct a frequency distribution and a histogram of employees' monthly salary contributions to a retirement plan.

Line Plots

Statistics is the branch of mathematics that studies techniques for collecting, organizing, and interpreting data. In this section, you will study several ways to organize data. The first is a **line plot,** which uses a portion of a real number line to order numbers. Line plots are especially useful for ordering small sets of numbers (about 50 or less) by hand.

Example 1 ▶ Constructing a Line Plot

Use a line plot to organize the following test scores. Which number occurs with the greatest frequency?

93, 70, 76, 67, 86, 93, 82, 78, 83, 86, 64, 78, 76, 66, 83,
83, 96, 74, 69, 76, 64, 74, 79, 76, 88, 76, 81, 82, 74, 70

Solution

Begin by scanning the data to find the smallest and largest numbers. For this data, the smallest number is 64 and the largest is 96. Next, draw a portion of a real number line that includes the interval [64, 96]. To create the line plot, start with the first number, 93, and enter an × above 93 on the number line. Continue recording ×'s for each number in the list until you obtain the line plot shown in Figure B.1. From the line plot, you can see that 76 had the greatest frequency.

FIGURE B.1

Test Scores
93, 70, 76, 58, 86, 93, 82, 78,
83, 86, 64, 78, 76, 66, 83, 83,
96, 74, 69, 76, 64, 74, 79, 76,
88, 76, 81, 82, 74, 70

Stems	Leaves
5	8
6	4 4 6 9
7	0 0 4 4 4 6 6 6 6 8 9
8	1 2 2 3 3 3 6 6 8
9	3 3 6

Stem-and-Leaf Plots

Another type of plot that can be used to organize sets of numbers by hand is a **stem-and-leaf plot.** A set of test scores and the corresponding stem-and-leaf plot are shown at the left.

Note that the *leaves* represent the units digits of the numbers and the *stems* represent the tens digits. Stem-and-leaf plots can also be used to compare two sets of data, as shown in the following example.

Example 2 ▶ Comparing Two Sets of Data

Use a stem-and-leaf plot to compare the test scores given on page A13 with the following test scores. Which set of test scores is better?

90, 81, 70, 62, 64, 73, 81, 92, 73, 81, 92, 93, 83, 75, 76,
83, 94, 96, 86, 77, 77, 86, 96, 86, 77, 86, 87, 87, 79, 88

Solution

Begin by ordering the second set of scores.

62, 64, 70, 73, 73, 75, 76, 77, 77, 77, 79, 81, 81, 81, 83,
83, 86, 86, 86, 86, 87, 87, 88, 90, 92, 92, 93, 94, 96, 96

Now that the data has been ordered, you can construct a *double* stem-and-leaf plot by letting the leaves to the right of the stems represent the units digits for the first group of test scores and letting the leaves to the left of the stems represent the units digits for the second group of test scores.

Leaves (2nd Group)	Stems	Leaves (1st Group)
	5	8
4 2	6	4 4 6 9
9 7 7 7 6 5 3 3 0	7	0 0 4 4 4 6 6 6 6 6 8 8 9
8 7 7 6 6 6 6 3 3 1 1 1	8	1 2 2 3 3 3 6 6 8
6 6 4 3 2 2 0	9	3 3 6

By comparing the two sets of leaves, you can see that the second group of test scores is better than the first group.

Example 3 ▶ Using a Stem-and-Leaf Plot

The table at the left above shows the percent of the population of each state and the District of Columbia that was at least 65 years old in 1997. Use a stem-and-leaf plot to organize the data. *(Source: U.S. Bureau of the Census)*

Solution

Begin by ordering the numbers, as shown below.

5.3, 8.7, 9.9, 10.1, 10.1, 11.1, 11.2, 11.2, 11.3, 11.3, 11.4, 11.5, 11.5, 11.5,
12.1, 12.1, 12.2, 12.3, 12.3, 12.4, 12.5, 12.5, 12.5, 12.5, 12.5, 12.9, 13.0,
13.2, 13.2, 13.2, 13.2, 13.3, 13.4, 13.4, 13.4, 13.5, 13.7, 13.7, 13.7, 13.9,
13.9, 14.1, 14.3, 14.3, 14.4, 14.4, 15.0, 15.1, 15.8, 15.8, 18.5

Next construct the stem-and-leaf plot using the leaves to represent the digits to the right of the decimal points, as shown at the left. From the stem-and-leaf plot, you can see that Alaska has the lowest percent and Florida has the highest percent.

AK	5.3	MT	13.2
AL	13.0	NC	12.5
AR	14.3	ND	14.4
AZ	13.2	NE	13.7
CA	11.1	NH	12.1
CO	10.1	NJ	13.7
CT	14.4	NM	11.2
DC	13.9	NV	11.5
DE	12.9	NY	13.4
FL	18.5	OH	13.4
GA	9.9	OK	13.4
HI	13.2	OR	13.3
IA	15.0	PA	15.8
ID	11.3	RI	15.8
IL	12.5	SC	12.1
IN	12.5	SD	14.3
KS	13.5	TN	12.5
KY	12.5	TX	10.1
LA	11.4	UT	8.7
MA	14.1	VA	11.2
MD	11.5	VT	12.3
ME	13.9	WA	11.5
MI	12.4	WI	13.2
MN	12.3	WV	15.1
MO	13.7	WY	11.3
MS	12.2		

Stems	Leaves
5.	3
6.	
7.	
8.	7
9.	9
10.	1 1
11.	1 2 2 3 3 4 5 5 5
12.	1 1 2 3 3 4 5 5 5 5 5 9
13.	0 2 2 2 2 3 4 4 4 5 7 7 7 9 9
14.	1 3 3 4 4
15.	0 1 8 8
16.	
17.	
18.	5

Histograms and Frequency Distributions

With data such as that given in Example 3, it is useful to group the numbers into intervals and plot the frequency of the data in each interval. For instance, the **frequency distribution** and **histogram** shown in Figure B.2 represent the data given in Example 3.

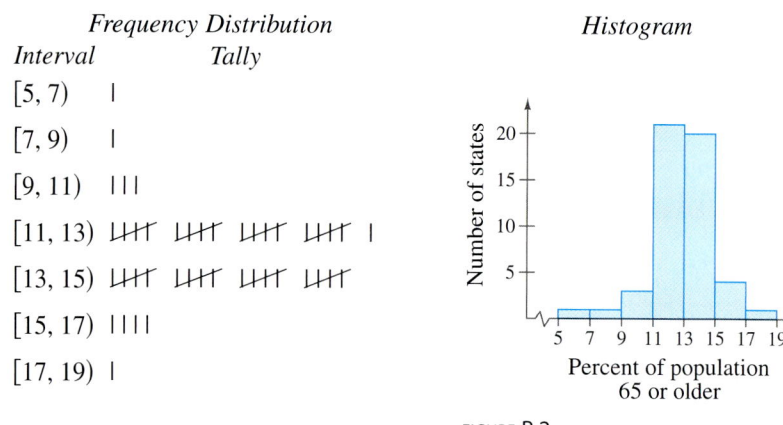

Frequency Distribution

Interval	Tally
[5, 7)	I
[7, 9)	I
[9, 11)	III
[11, 13)	LHT LHT LHT LHT I
[13, 15)	LHT LHT LHT LHT
[15, 17)	IIII
[17, 19)	I

FIGURE B.2

A histogram has a portion of a real number line as its horizontal axis. A histogram is similar to a bar graph, except that the rectangles (bars) in a bar graph can be either horizontal or vertical and the labels of the bars are not necessarily numbers.

Another difference between a bar graph and a histogram is that the bars in a bar graph are usually separated by spaces, whereas the bars in a histogram are not separated by spaces.

Example 4 ▶ Constructing a Histogram

A company has 48 sales representatives who sold the following numbers of units during the first quarter of 2000. Construct a grouped frequency distribution for this data.

107	162	184	170	177	102	145	141
105	193	167	149	195	127	193	191
150	153	164	167	171	163	141	129
109	171	150	138	100	164	147	153
171	163	118	142	107	144	100	132
153	107	124	162	192	134	187	177

Solution

To begin constructing a grouped frequency distribution, you must first decide on the number of groups. There are several ways to group this data. However, because the smallest number is 100 and the largest is 195, it seems that 10 groups of 10 each would be appropriate. The first group would be 100–109, the second group would be 110–119, and so on. By tallying the data into the 10 groups, you obtain the distribution shown at the left above. A histogram for the distribution is shown in Figure B.3.

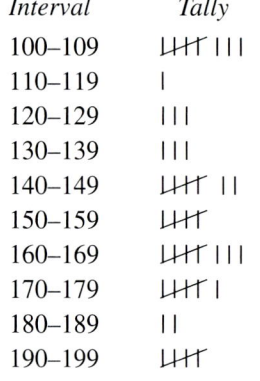

Interval	Tally
100–109	LHT III
110–119	I
120–129	III
130–139	III
140–149	LHT II
150–159	LHT
160–169	LHT III
170–179	LHT I
180–189	II
190–199	LHT

FIGURE B.3

B.1 Exercises

1. *Gasoline Prices* The line plot shows a sample of prices of unleaded regular gasoline from 25 different cities.

   (a) What price occurred with the greatest frequency?
   (b) What is the range of prices?

2. *Livestock Weights* The line plot shows the weights (to the nearest hundred pounds) of 30 head of cattle sold by a rancher.

   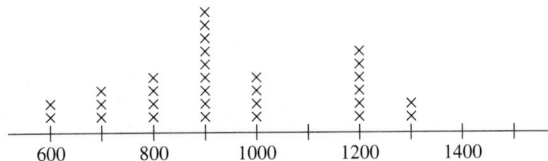

   (a) What weight occurred with the greatest frequency?
   (b) What is the range of weights?

   Quiz and Exam Scores In Exercises 3–6, use the following scores from a math class of 30 students. The scores are for two 25-point quizzes and two 100-point exams.

   Quiz #1 20, 15, 14, 20, 16, 19, 10, 21, 24, 15, 15, 14, 15, 21, 19, 15, 20, 18, 18, 22, 18, 16, 18, 19, 21, 19, 16, 20, 14, 12

   Quiz #2 22, 22, 23, 22, 21, 24, 22, 19, 21, 23, 23, 25, 24, 22, 22, 23, 23, 23, 23, 22, 24, 23, 22, 24, 21, 24, 16, 21, 16, 14

   Exam #1 77, 100, 77, 70, 83, 89, 87, 85, 81, 84, 81, 78, 89, 78, 88, 85, 90, 92, 75, 81, 85, 100, 98, 81, 78, 75, 85, 89, 82, 75

   Exam #2 76, 78, 73, 59, 70, 81, 71, 66, 66, 73, 68, 67, 63, 67, 77, 84, 87, 71, 78, 78, 90, 80, 77, 70, 80, 64, 74, 68, 68, 68

3. Construct a line plot for each quiz. For each quiz, which score occurred with the greatest frequency?

4. Construct a line plot for each exam. For each exam, which score occurred with the greatest frequency?

5. Construct a stem-and-leaf plot for Exam #1.

6. Construct a double stem-and-leaf plot to compare the scores for Exam #1 and Exam #2. Which set of scores is higher?

7. *Insurance Coverage* The table shows the total numbers of persons (in thousands) without health insurance coverage in the 50 states and the District of Columbia in 1996. Use a stem-and-leaf plot to organize the data. (Source: U.S. Bureau of the Census)

AK 89	AL 550	AR 566	AZ 1159	CA 6514
CO 644	CT 368	DC 80	DE 98	FL 2722
GA 1319	HI 101	IA 335	ID 196	IL 1337
IN 600	KS 292	KY 601	LA 890	MA 766
MD 581	ME 146	MI 857	MN 480	MO 700
MS 518	MT 124	NC 1160	ND 62	NE 190
NH 109	NJ 1317	NM 412	NV 255	NY 3132
OH 1292	OK 570	OR 496	PA 1133	RI 93
SC 634	SD 67	TN 841	TX 4680	UT 240
VA 811	VT 65	WA 761	WI 438	WV 261
WY 66				

8. *Snowfall* The data below shows the seasonal snowfall (in inches) at Lincoln, Nebraska for the years 1968 through 1997 (the amounts are listed in order by year). How would you organize this data? Explain your reasoning. (Source: University of Nebraska–Lincoln)

   39.8, 26.2, 49.0, 21.6, 29.2, 33.6, 42.1, 21.1, 21.8, 31.0, 34.4, 23.3, 13.0, 32.3, 38.0, 47.5, 21.5, 18.9, 15.7, 13.0, 19.1, 18.7, 25.8, 23.8, 32.1, 21.3, 21.8, 30.7, 29.0, 44.6

9. *Retirement Contributions* The employees of a company must contribute 7% of their monthly salaries to a company-sponsored retirement plan. The contributed amounts (in dollars) for the company's 35 employees are as follows.

   100, 200, 130, 136, 161, 156, 209, 126, 135, 98, 114, 117, 168, 133, 140, 124, 172, 127, 143, 157, 124, 152, 104, 126, 155, 92, 194, 115, 120, 136, 148, 112, 116, 146, 96

   (a) Construct a frequency distribution using groups of 20. The first group should be 90–109.
   (b) Construct a histogram for this frequency distribution.

B.2 Measures of Central Tendency and Dispersion

▶ What you should learn
- How to find and interpret the mean, median, and mode of a set of data
- How to determine the measure of central tendency that best represents a set of data
- How to find the standard deviation of a set of data
- How to use box-and-whisker plots

▶ Why you should learn it
Measures of central tendency and dispersion provide a convenient way to describe and compare sets of data. For instance, in Exercise 36 on page A24, the mean and standard deviation are used to analyze the price of gold for the years 1978 through 1997.

Mean, Median, and Mode

In many real-life situations, it is helpful to describe data by a single number that is most representative of the entire collection of numbers. Such a number is called a **measure of central tendency**. The most commonly used measures are as follows.

1. The **mean**, or **average**, of n numbers is the sum of the numbers divided by n.
2. The **median** of n numbers is the middle number when the numbers are written in order. If n is even, the median is the average of the two middle numbers.
3. The **mode** of n numbers is the number that occurs most frequently. If two numbers tie for most frequent occurrence, the collection has two modes and is called **bimodal**.

Example 1 ▶ Comparing Measures of Central Tendency

On an interview for a job, the interviewer tells you that the average annual income of the company's 25 employees is $60,849. The actual annual incomes of the 25 employees are shown below. What are the mean, median, and mode of the incomes? Was the person telling you the truth?

$17,305,	$478,320,	$45,678,	$18,980,	$17,408,
$25,676,	$28,906,	$12,500,	$24,540,	$33,450,
$12,500,	$33,855,	$37,450,	$20,432,	$28,956,
$34,983,	$36,540,	$250,921,	$36,853,	$16,430,
$32,654,	$98,213,	$48,980,	$94,024,	$35,671

Solution

The mean of the incomes is

$$\text{Mean} = \frac{17{,}305 + 478{,}320 + 45{,}678 + 18{,}980 + \cdots + 35{,}671}{25}$$

$$= \frac{1{,}521{,}225}{25} = \$60{,}849.$$

To find the median, order the incomes as follows.

$12,500,	$12,500,	$16,430,	$17,305,	$17,408,
$18,980,	$20,432,	$24,540,	$25,676,	$28,906,
$28,956,	$32,654,	$33,450,	$33,855,	$34,983,
$35,671,	$36,540,	$36,853,	$37,450,	$45,678,
$48,980,	$94,024,	$98,213,	$250,921,	$478,320

From this list, you can see that the median (the middle number) is $33,450. From the same list, you can see that $12,500 is the only income that occurs more than once. So, the mode is $12,500. Technically, the person was telling the truth because the average is (generally) defined to be the mean. However, of the three measures of central tendency *Mean:* $60,849 *Median:* $33,450 *Mode:* $12,500 it seems clear that the median is most representative. The mean is inflated by the two highest salaries.

Choosing a Measure of Central Tendency

Which of the three measures of central tendency is the most representative? The answer is that it depends on the distribution of the data *and* the way in which you plan to use the data.

For instance, in Example 1, the mean salary of $60,849 does not seem very representative to a potential employee. To a city income tax collector who wants to estimate 1% of the total income of the 25 employees, however, the mean is precisely the right measure.

Example 2 ▶ Choosing a Measure of Central Tendency

Which measure of central tendency is the most representative of the data given in each of the following frequency distributions?

a. Number	Tally	b. Number	Tally	c. Number	Tally
1	7	1	9	1	6
2	20	2	8	2	1
3	15	3	7	3	2
4	11	4	6	4	3
5	8	5	5	5	5
6	3	6	6	6	5
7	2	7	7	7	4
8	0	8	8	8	3
9	15	9	9	9	0

Solution

a. For this data, the mean is 4.23, the median is 3, and the mode is 2. Of these, the mode is probably the most representative.

b. For this data, the mean and median are each 5 and the modes are 1 and 9 (the distribution is bimodal). Of these, the mean or median is the most representative.

c. For this data, the mean is 4.59, the median is 5, and the mode is 1. Of these, the mean or median is the most representative.

Variance and Standard Deviation

Very different sets of numbers can have the same mean. You will now study two **measures of dispersion,** which give you an idea of how much the numbers in a set differ from the mean of the set. These two measures are called the *variance* of the set and the *standard deviation* of the set.

Definitions of Variance and Standard Deviation

Consider a set of numbers $\{x_1, x_2, \ldots, x_n\}$ with a mean of $\bar{x}$. The **variance** of the set is

$$v = \frac{(x_1 - \bar{x})^2 + (x_2 - \bar{x})^2 + \cdots + (x_n - \bar{x})^2}{n}$$

and the **standard deviation** of the set is $\sigma = \sqrt{v}$ (σ is the lowercase Greek letter *sigma*).

The standard deviation of a set is a measure of how much a typical number in the set differs from the mean. The greater the standard deviation, the more the numbers in the set *vary* from the mean. For instance, each of the following sets has a mean of 5.

$$\{5, 5, 5, 5\}, \quad \{4, 4, 6, 6\}, \quad \text{and} \quad \{3, 3, 7, 7\}$$

The standard deviations of the sets are 0, 1, and 2.

$$\sigma_1 = \sqrt{\frac{(5-5)^2 + (5-5)^2 + (5-5)^2 + (5-5)^2}{4}}$$

$$= 0$$

$$\sigma_2 = \sqrt{\frac{(4-5)^2 + (4-5)^2 + (6-5)^2 + (6-5)^2}{4}}$$

$$= 1$$

$$\sigma_3 = \sqrt{\frac{(3-5)^2 + (3-5)^2 + (7-5)^2 + (7-5)^2}{4}}$$

$$= 2$$

Example 3 ▶ Estimations of Standard Deviation

Consider the three sets of data represented by the bar graphs in Figure B.4. Which set has the smallest standard deviation? Which has the largest?

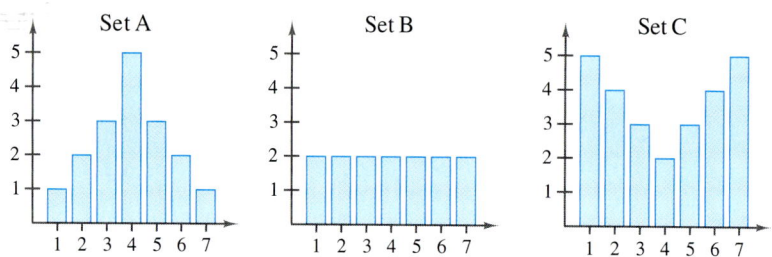

FIGURE B.4

Solution

Of the three sets, the numbers in set A are grouped most closely to the center and the numbers in set C are the most dispersed. So, set A has the smallest standard deviation and set C has the largest standard deviation.

Example 4 ▶ Finding Standard Deviation

Find the standard deviation of each set shown in Example 3.

Solution

Because of the symmetry of each bar graph, you can conclude that each has a mean of $\bar{x} = 4$. The standard deviation of set A is

$$\sigma = \sqrt{\frac{(-3)^2 + 2(-2)^2 + 3(-1)^2 + 5(0)^2 + 3(1)^2 + 2(2)^2 + (3)^2}{17}}$$

$$\approx 1.53.$$

The standard deviation of set B is

$$\sigma = \sqrt{\frac{2(-3)^2 + 2(-2)^2 + 2(-1)^2 + 2(0)^2 + 2(1)^2 + 2(2)^2 + 2(3)^2}{14}}$$

$$= 2.$$

The standard deviation of set C is

$$\sigma = \sqrt{\frac{5(-3)^2 + 4(-2)^2 + 3(-1)^2 + 2(0)^2 + 3(1)^2 + 4(2)^2 + 5(3)^2}{26}}$$

$$\approx 2.22.$$

These values confirm the results of Example 3. That is, set A has the smallest standard deviation and set C has the largest.

The following alternative formula provides a more efficient way to compute the standard deviation.

Alternative Formula for Standard Deviation

The standard deviation of $\{x_1, x_2, \ldots, x_n\}$ is

$$\sigma = \sqrt{\frac{x_1^2 + x_2^2 + \cdots + x_n^2}{n} - \bar{x}^2}.$$

Because of messy computations, this formula is difficult to verify. Conceptually, however, the process is straightforward. It consists of showing that the expressions

$$\sqrt{\frac{(x_1 - \bar{x})^2 + (x_2 - \bar{x})^2 + \cdots + (x_n - \bar{x})^2}{n}}$$

and

$$\sqrt{\frac{x_1^2 + x_2^2 + \cdots + x_n^2}{n} - \bar{x}^2}$$

are equivalent. Try verifying this equivalence for the set $\{x_1, x_2, x_3\}$ with $\bar{x} = (x_1 + x_2 + x_3)/3$.

AK	66	MT	62
AL	40	NC	42
AR	39	ND	47
AZ	51	NE	63
CA	62	NH	59
CO	69	NJ	77
CT	80	NM	45
DC	94	NV	49
DE	44	NY	73
FL	50	OH	55
GA	46	OK	47
HI	80	OR	70
IA	55	PA	61
ID	53	RI	56
IL	61	SC	41
IN	47	SD	49
KS	51	TN	53
KY	53	TX	47
LA	45	UT	66
MA	74	VA	54
MD	68	VT	57
ME	47	WA	68
MI	62	WI	65
MN	67	WV	43
MO	53	WY	52
MS	37		

Example 5 ▶ Using the Alternative Formula

Use the alternative formula for standard deviation to find the standard deviation of the following set of numbers.

5, 6, 6, 7, 7, 8, 8, 8, 9, 10

Solution

Begin by finding the mean of the set, which is 7.4. So, the standard deviation is

$$\sigma = \sqrt{\frac{5^2 + 2(6^2) + 2(7^2) + 3(8^2) + 9^2 + 10^2}{10} - (7.4)^2}$$

$$= \sqrt{\frac{568}{10} - 54.76}$$

$$= \sqrt{2.04}$$

$$\approx 1.43.$$

You can use the statistical features of a graphing utility to check this result.

A well-known theorem in statistics, called *Chebychev's Theorem*, states that at least

$$1 - \frac{1}{k^2}$$

of the numbers in a distribution must lie within k standard deviations of the mean. So, 75% of the numbers in a collection must lie within two standard deviations of the mean, and at least 88.9% of the numbers must lie within three standard deviations of the mean. For most distributions, these percentages are low. For instance, in all three distributions shown in Example 3, 100% of the numbers lie within two standard deviations of the mean.

Example 6 ▶ Describing a Distribution

The table at the left above shows the number of dentists (per 100,000 people) in each state and the District of Columbia. Find the mean and standard deviation of the numbers. What percent of the numbers lie within two standard deviations of the mean? (Source: American Dental Association)

Solution

Begin by entering the numbers into a graphing utility that has a standard deviation program. After running the program, you should obtain

$$\bar{x} \approx 56.76 \quad \text{and} \quad \sigma = 12.14.$$

The interval that contains all numbers that lie within two standard deviations of the mean is

$$[56.76 - 2(12.14), 56.76 + 2(12.14)] \quad \text{or} \quad [32.48, 81.04].$$

From the histogram in Figure B.5, you can see that all but one of the numbers (98%) lie in this interval—all but the number that corresponds to the number of dentists (per 100,000 people) in Washington, DC.

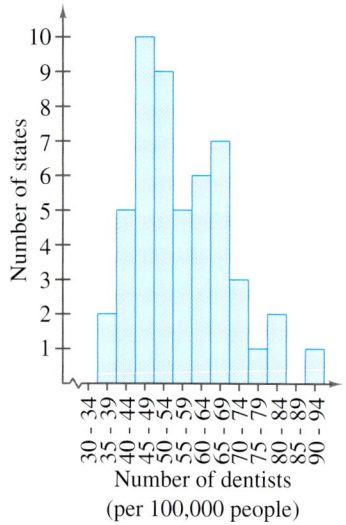

FIGURE B.5

Section B.2 ▶ Measures of Central Tendency and Dispersion **A21**

Box-and-Whisker Plots

Standard deviation is the measure of dispersion that is associated with the mean. **Quartiles** measure dispersion associated with the median.

> **Definition of Quartiles**
>
> Consider an ordered set of numbers whose median is m. The **lower quartile** is the median of the numbers that occur before m. The **upper quartile** is the median of the numbers that occur after m.

Example 7 ▶ Finding Quartiles of a Set

Find the lower and upper quartiles for the following set.

$$34, 14, 24, 16, 12, 18, 20, 24, 16, 26, 13, 27$$

Solution

Begin by ordering the set.

The median of the entire set is 19. The median of the six numbers that are less than 19 is 15. So, the lower quartile is 15. The median of the six numbers that are greater than 19 is 25. So, the upper quartile is 25.

Quartiles are represented graphically by a **box-and-whisker plot,** as shown in Figure B.6. In the plot, notice that five numbers are listed: the smallest number, the lower quartile, the median, the upper quartile, and the largest number. Also notice that the numbers are spaced proportionally, as though they were on a real number line.

FIGURE B.6

The next example shows how to find quartiles when the number of elements in a set is not divisible by 4.

Example 8 ▶ Sketching Box-and-Whisker Plots

Sketch a box-and-whisker plot for each of the given sets.

a. 27, 28, 30, 42, 45, 50, 50, 61, 62, 64, 66
b. 82, 82, 83, 85, 87, 89, 90, 94, 95, 95, 96, 98, 99
c. 11, 13, 13, 15, 17, 18, 20, 24, 24, 27

Solution

a. This set has 11 numbers. The median is 50 (the sixth number). The lower quartile is 30 (the median of the first five numbers). The upper quartile is 62 (the median of the last five numbers).

b. This set has 13 numbers. The median is 90 (the seventh number). The lower quartile is 84 (the median of the first six numbers). The upper quartile is 95.5 (the median of the last six numbers).

c. This set has 10 numbers. The median is 17.5 (the average of the fifth and sixth numbers). The lower quartile is 13 (the median of the first five numbers). The upper quartile is 24 (the median of the last five numbers).

B.2 Exercises

In Exercises 1–6, find the mean, median, and mode of the set of measurements.

1. 5, 12, 7, 14, 8, 9, 7
2. 30, 37, 32, 39, 33, 34, 32
3. 5, 12, 7, 24, 8, 9, 7
4. 20, 37, 32, 39, 33, 34, 32
5. 5, 12, 7, 14, 9, 7
6. 30, 37, 32, 39, 34, 32

7. *Reasoning* Compare your answers for Exercises 1 and 3 with those for Exercises 2 and 4. Which of the measures of central tendency is sensitive to extreme measurements? Explain your reasoning.

8. *Reasoning*
   (a) Add 6 to each measurement in Exercise 1 and calculate the mean, median, and mode of the revised measurements. How are the measures of central tendency changed?
   (b) If a constant k is added to each measurement in a set of data, how will the measures of central tendency change?

9. *Electric Bills* A person had the following monthly bills for electricity. What are the mean and median of the collection of bills?

January	$67.92	February	$59.84
March	$52.00	April	$52.50
May	$57.99	June	$65.35
July	$81.76	August	$74.98
September	$87.82	October	$83.18
November	$65.35	December	$57.00

10. *Car Rental* A car rental company kept the following record of the numbers of miles a rental car was driven. What are the mean, median, and mode of this data?

Monday	410	Tuesday	260
Wednesday	320	Thursday	320
Friday	460	Saturday	150

11. *Six-Child Families* A study was done on families having six children. The table gives the numbers of families in the study with the indicated numbers of girls. Determine the mean, median, and mode of this set of data.

Number of girls	0	1	2	3	4	5	6
Frequency	1	24	45	54	50	19	7

12. *Baseball* A baseball fan examined the records of a favorite baseball player's performance during his last 50 games. The numbers of games in which the player had 0, 1, 2, 3, and 4 hits are recorded in the table.

Number of hits	0	1	2	3	4
Frequency	14	26	7	2	1

(a) Determine the average number of hits per game.
(b) Determine the player's batting average if he had 200 at-bats during the 50-game series.

13. *Think About It* Construct a collection of numbers that has the following properties. If this is not possible, explain why it is not.

Mean = 6, median = 4, mode = 4

14. *Think About It* Construct a collection of numbers that has the following properties. If this is not possible, explain why it is not.

Mean = 6, median = 6, mode = 4

15. *Test Scores* A professor records the following scores for a 100-point exam.

99, 64, 80, 77, 59, 72, 87, 79, 92, 88,
90, 42, 20, 89, 42, 100, 98, 84, 78, 91

Which measure of central tendency best describes these test scores?

16. *Shoe Sales* A salesman sold eight pairs of a certain style of men's shoes. The sizes of the eight pairs were as follows: $10\frac{1}{2}$, 8, 12, $10\frac{1}{2}$, 10, $9\frac{1}{2}$, 11, and $10\frac{1}{2}$. Which measure (or measures) of central tendency best describes the typical shoe size for this data?

In Exercises 17–24, find the mean ($\bar{x}$), variance (v), and standard deviation (σ) of the numbers.

17. 4, 10, 8, 2
18. 3, 15, 6, 9, 2
19. 0, 1, 1, 2, 2, 2, 3, 3, 4
20. 2, 2, 2, 2, 2, 2
21. 1, 2, 3, 4, 5, 6, 7
22. 1, 1, 1, 5, 5, 5
23. 49, 62, 40, 29, 32, 70
24. 1.5, 0.4, 2.1, 0.7, 0.8

In Exercises 25–30, use the alternative formula to find the standard deviation of the numbers.

25. 2, 4, 6, 6, 13, 5
26. 10, 25, 50, 26, 15, 33, 29, 4
27. 246, 336, 473, 167, 219, 359
28. 6.0, 9.1, 4.4, 8.7, 10.4
29. 8.1, 6.9, 3.7, 4.2, 6.1
30. 9.0, 7.5, 3.3, 7.4, 6.0

In Exercises 31 and 32, line plots of sets of data are given. Determine the mean and standard deviation of each set.

31. (a)

(b)

(c)

(d)

32. (a)

(b)

(c)

(d)

33. *Reasoning* Without calculating the standard deviation, explain why the set {4, 4, 20, 20} has a standard deviation of 8.

34. *Reasoning* If the standard deviation of a set of numbers is 0, what does this imply about the set?

35. *Test Scores* An instructor adds five points to each student's exam score. Will this change the mean or standard deviation of the exam scores? Explain.

36. *Price of Gold* The following data represents the average prices of gold (in dollars per fine ounce) for the years 1978 to 1997. Use a computer or calculator to find the mean, variance, and standard deviation of the data. What percent of the data lies within two standard deviations of the mean? (Source: U.S. Bureau of Mines)

194, 308, 613, 460, 376,
424, 361, 318, 368, 448,
438, 383, 385, 363, 345,
361, 385, 386, 389, 333

37. *Think About It* The histograms represent the test scores of two classes of a college course in mathematics. Which histogram has the smaller standard deviation?

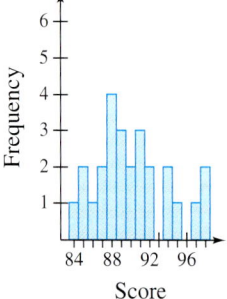

In Exercises 38–41, sketch a box-and-whisker plot for the data without the aid of a graphing utility.

38. 23, 15, 14, 23, 13, 14, 13, 20, 12
39. 11, 10, 11, 14, 17, 16, 14, 11, 8, 14, 20
40. 46, 48, 48, 50, 52, 47, 51, 47, 49, 53
41. 25, 20, 22, 28, 24, 28, 25, 19, 27, 29, 28, 21

In Exercises 42–45, use a graphing utility to create a box-and-whisker plot for the data.

42. 19, 12, 14, 9, 14, 15, 17, 13, 19, 11, 10, 19
43. 9, 5, 5, 5, 6, 5, 4, 12, 7, 10, 7, 11, 8, 9, 9
44. 20.1, 43.4, 34.9, 23.9, 33.5, 24.1, 22.5, 42.4, 25.7, 17.4, 23.8, 33.3, 17.3, 36.4, 21.8
45. 78.4, 76.3, 107.5, 78.5, 93.2, 90.3, 77.8, 37.1, 97.1, 75.5, 58.8, 65.6

46. *Product Lifetime* A company has redesigned a product in an attempt to increase the lifetime of the product. The two sets of data list the lifetimes (in months) of 20 units with the original design and 20 units with the new design. Create a box-and-whisker plot for each set of data, and then comment on the differences between the plots.

Original Design

15.1	78.3	56.3	68.9	30.6
27.2	12.5	42.7	72.7	20.2
53.0	13.5	11.0	18.4	85.2
10.8	38.3	85.1	10.0	12.6

New Design

55.8	71.5	25.6	19.0	23.1
37.2	60.0	35.3	18.9	80.5
46.7	31.1	67.9	23.5	99.5
54.0	23.2	45.5	24.8	87.8

B.3 Least Squares Regression

▶ **What you should learn**
- How to use the sum of squared differences to determine a least squares regression line
- How to find a least squares regression line for a set of data
- How to find a least squares regression parabola for a set of data

▶ **Why you should learn it**
The method of least squares provides a way of creating mathematical models for a set of data, which can then be analyzed.

In many of the examples and exercises in this text, you have been asked to use the regression capabilities of a graphing utility to find mathematical models for sets of data. Another way to find a mathematical model for a set of data is to use the **method of least squares**. As a measure of how well a model fits a set of data points

$$\{(x_1, y_1), (x_2, y_2), (x_3, y_3), \ldots, (x_n, y_n)\}$$

you can add the squares of the differences between the actual y-values and the values given by the model to obtain the **sum of the squared differences.** For instance, the table shows the heights x (in feet) and the diameters y (in inches) of eight trees. The table also shows the values of a linear model $y^* = 0.69x - 40$ for each x-value. The sum of squared differences for the model is 53.3.

x	70	72	75	76	85	78	77	80
y	8.3	10.5	11.0	11.4	12.9	14.0	16.3	18.0
y^*	8.3	9.68	11.75	12.44	18.65	13.82	13.13	15.2

The model that has the least sum of squared differences is the **least squares regression line** for the data. The least squares regression line for the data in the table is $y \approx 0.43x - 20.3$. The sum of squared differences is 43.32.

To find the least squares regression line $y = ax + b$ for the points $\{(x_1, y_1), (x_2, y_2), (x_3, y_3), \ldots, (x_n, y_n)\}$ algebraically, you need to solve the following system for a and b.

$$\begin{cases} nb + \left(\sum_{i=1}^{n} x_i\right) a = \sum_{i=1}^{n} y_i \\ \left(\sum_{i=1}^{n} x_i\right) b + \left(\sum_{i=1}^{n} x_i^2\right) a = \sum_{i=1}^{n} x_i y_i \end{cases}$$

In the system,

$$\sum_{i=1}^{n} x_i = x_1 + x_2 + \cdots + x_n$$

$$\sum_{i=1}^{n} y_i = y_1 + y_2 + \cdots + y_n$$

$$\sum_{i=1}^{n} x_i^2 = x_1^2 + x_2^2 + \cdots + x_n^2$$

$$\sum_{i=1}^{n} x_i y_i = x_1 y_1 + x_2 y_2 + \cdots + x_n y_n.$$

Example 1 ▶ Finding a Least Squares Regression Line

Find the least squares regression line for the points $(-3, 0)$, $(-1, 1)$, $(0, 2)$, and $(2, 3)$.

Solution

Begin by constructing a table like that shown below.

x	y	xy	x^2
-3	0	0	9
-1	1	-1	1
0	2	0	0
2	3	6	4
$\sum_{i=1}^{n} x_i = -2$	$\sum_{i=1}^{n} y_i = 6$	$\sum_{i=1}^{n} x_i y_i = 5$	$\sum_{i=1}^{n} x_i^2 = 14$

FIGURE B.7

Applying the system for the least squares regression line with $n = 4$ produces

$$\begin{cases} nb + \left(\sum_{i=1}^{n} x_i\right) a = \sum_{i=1}^{n} y_i \\ \left(\sum_{i=1}^{n} x_i\right) b + \left(\sum_{i=1}^{n} x_i^2\right) a = \sum_{i=1}^{n} x_i y_i \end{cases} \implies \begin{cases} 4b - 2a = 6 \\ -2b + 14a = 5 \end{cases}.$$

Solving this system of equations produces $a = \frac{8}{13}$ and $b = \frac{47}{26}$. So, the least squares regression line is $y = \frac{8}{13}x + \frac{47}{26}$, as shown in Figure B.7.

The least squares regression parabola $y = ax^2 + bx + c$ for the points

$$\{(x_1, y_1), (x_2, y_2), (x_3, y_3), \ldots, (x_n, y_n)\}$$

is obtained in a similar manner by solving the following system of three equations in three unknowns for a, b, and c.

$$\begin{cases} nc + \left(\sum_{i=1}^{n} x_i\right) b + \left(\sum_{i=1}^{n} x_i^2\right) a = \sum_{i=1}^{n} y_i \\ \left(\sum_{i=1}^{n} x_i\right) c + \left(\sum_{i=1}^{n} x_i^2\right) b + \left(\sum_{i=1}^{n} x_i^3\right) a = \sum_{i=1}^{n} x_i y_i \\ \left(\sum_{i=1}^{n} x_i^2\right) c + \left(\sum_{i=1}^{n} x_i^3\right) b + \left(\sum_{i=1}^{n} x_i^4\right) a = \sum_{i=1}^{n} x_i^2 y_i \end{cases}$$

Fortunately, graphing utilities have built-in least squares regression capabilities.

Looking for solutions to your math problems? These study aids offer more than just the answers!

Two technology options offer complete solutions to every odd-numbered problem, help you practice and assess your skills, and provide problem-solving tools.

Interactive Trigonometry 2.0 CD-ROM

- Enhance your understanding with step-by-step solutions to all odd-numbered exercises in the text.
- Assess your skill levels with diagnostic pre- and post-tests for each chapter.
- Strengthen your skills with tutorial exercises accompanied by examples and diagnostics.
- Visualize and graph with a built-in Meridian Graphing Calculator Emulator.
- Enjoy learning with additional interactive features.

Internet Trigonometry 1.0

A subscription to this web site offers all of the CD-ROM features listed above, plus:
- Chat rooms for peer support
- Bulletin boards for sharing ideas and insights on each chapter

To purchase Interactive Trigonometry 2.0 CD-ROM:

- Visit Houghton Mifflin's College Store at **college.hmco.com** or contact your campus bookstore.

To subscribe online to Internet Trigonometry 1.0:

- Visit Houghton Mifflin's College Division web site at **college.hmco.com** and select mathematics.

Use these print supplements for added practice and convenient support.

Study and Solutions Guide to accompany *Trigonometry*, 5th Edition

- Work through step-by-step solutions for all odd-numbered exercises in the text.
- Test your skills by taking practice tests with accompanying solutions.
- Find useful study strategies designed to help you succeed.

Student Success Organizer
Ask your instructor about this new study aid.

- Use its practical format to guide you step-by-step through difficult concepts.
- Enhance your organizational skills for approaching problems and assignments.

To purchase these print supplements:

- Visit Houghton Mifflin's College Store at **college.hmco.com** or contact your campus bookstore.

Larson • Hostetler

Answers to Odd-Numbered Exercises and Tests

Chapter P

Section P.1 *(page 9)*

1. (a) 5, 1, 2 (b) $-9, 5, 0, 1, -4, 2, -11$
   (c) $-9, -\frac{7}{2}, 5, \frac{2}{3}, 0, 1, -4, 2, -11$ (d) $\sqrt{2}$
3. (a) 1 (b) $-13, 1, -6$
   (c) $2.01, 0.666\ldots, -13, 1, -6$ (d) $0.010110111\ldots$
5. (a) $\frac{6}{3}, 8$ (b) $\frac{6}{3}, -1, 8, -22$
   (c) $-\frac{1}{3}, \frac{6}{3}, -7.5, -1, 8, -22$ (d) $-\pi, \frac{1}{2}\sqrt{2}$
7. 0.625   9. $0.\overline{123}$ 11. $\frac{41}{10}$
13. $\frac{92}{9}$ 15. $-\frac{1811}{900}$ 17. $-1 < 2.5$
19. $-4 > -8$ 21. $\frac{3}{2} < 7$

23. $\frac{5}{6} > \frac{2}{3}$

25. $x \le 5$ denotes the set of all real numbers less than or equal to 5. Unbounded

27. $x < 0$ denotes the set of all negative real numbers. Unbounded

29. $x \ge 4$ denotes the set of all real numbers greater than or equal to 4. Unbounded

31. $-2 < x < 2$ denotes the set of all real numbers greater than -2 and less than 2. Bounded

33. $-1 \le x < 0$ denotes the set of all negative real numbers greater than or equal to -1. Bounded

35. $\frac{127}{90}, \frac{584}{413}, \frac{7071}{5000}, \sqrt{2}, \frac{47}{33}$ 37. $-2 < x \le 4$ 39. $y \ge 0$
41. $10 \le t \le 22$ 43. $W > 65$
45. This interval consists of all real numbers greater than or equal to 0 and less than 8.
47. This interval consists of all real numbers greater than -6.

49. 10   51. If $x \le 3, 3 - x$. If $x > 3, x - 3$.   53. -1
55. -1 57. -1 59. $|-3| > -|-3|$
61. $-5 = -|5|$ 63. $-|-2| = -|2|$ 65. 4
67. 51   69. $\frac{5}{2}$ 71. $\frac{128}{75}$ 73. (a) Negative (b) Negative
75. $|7 - 18| = 11$ miles   77. $|60 - 23| = 37°$
79. $|x - 5| \le 3$ 81. $|y| \ge 6$
83. $|\$113{,}356 - \$112{,}700| = \$656 > \$500$
    $0.05(\$112{,}700) = \$5635$
    Because the actual expenses differ from the budget by more than $500, there is failure to meet the "budget variance test."
85. $|\$37{,}335 - \$37{,}640| = \$305 < \$500$
    $0.05(\$37{,}640) = \$1882$
    Because the difference between the actual expenses and the budget is less than $500 and less than 5% of the budgeted amount, there is compliance with the "budget variance test."
87. $|92.5 - 92.2| = 0.3$
    There was a surplus of $0.3 billion.
89. $|1032.0 - 1253.2| = 221.2$
    There was a deficit of $221.2 billion.
91. $7x$ and 4 are the terms; 7 is the coefficient.
93. $\sqrt{3}x^2, -8x,$ and -11 are the terms; $\sqrt{3}$ and -8 are the coefficients.
95. $4x^3, x/2,$ and -5 are the terms; 4 and $\frac{1}{2}$ are the coefficients.
97. (a) -10 (b) -6 99. (a) 14 (b) 2
101. (a) Division by 0 is undefined. (b) 0
103. Commutative Property of Addition
105. Multiplicative Inverse Property
107. Distributive Property
109. Multiplicative Identity Property
111. Associative and Commutative Properties of Multiplication
113. $\frac{1}{2}$ 115. $\frac{3}{8}$ 117. 48   119. $\frac{5x}{12}$
121. -2.57 123. 1.56
125. (a)

n	1	0.5	0.01	0.0001	0.000001
$5/n$	5	10	500	50,000	5,000,000

(b) The value of $5/n$ approaches infinity as n approaches 0.

127. (a) No. If one variable is negative while the other is positive, the expressions are unequal.

(b) $|u + v| \le |u| + |v|$

The expressions are equal when u and v have the same sign. If u and v differ in sign, $|u + v|$ is less than $|u| + |v|$.

129. The only even prime number is 2, because its only factors are itself and 1.

131. False. The denominators cannot be added when adding fractions.

133. Yes. $|a| = -a$ if $a < 0$.

Section P.2 (page 21)

1. Identity **3.** Conditional equation **5.** Identity
7. Conditional equation **9.** Conditional equation
11. 4 **13.** -9 **15.** 5 **17.** 9 **19.** No solution
21. -4 **23.** $-\frac{6}{5}$ **25.** 9
27. No solution. The x-terms sum to zero. **29.** 10
31. 4 **33.** 3
35. No solution. The variable is divided out. **37.** $\frac{5}{3}$
39. No solution. The solution is extraneous. **41.** 5
43. No solution. The solution is extraneous. **45.** 0
47. All real numbers **49.** $2x^2 + 8x - 3 = 0$
51. $x^2 - 6x + 6 = 0$ **53.** $3x^2 - 90x - 10 = 0$
55. $0, -\frac{1}{2}$ **57.** $4, -2$ **59.** -5 **61.** $3, -\frac{1}{2}$
63. $2, -6$ **65.** $-\frac{20}{3}, -4$ **67.** $-a$ **69.** $\pm 7; \pm 7.00$
71. $\pm\sqrt{11}; \pm 3.32$ **73.** $\pm 3\sqrt{3}; \pm 5.20$
75. $8, 16; 8.00, 16.00$ **77.** $-2 \pm \sqrt{14}; 1.74, -5.74$
79. $\frac{1 \pm 3\sqrt{2}}{2}; 2.62, -1.62$ **81.** $2; 2.00$ **83.** $0, 2$
85. $4, -8$ **87.** $-3 \pm \sqrt{7}$ **89.** $1 \pm \frac{\sqrt{6}}{3}$
91. $2 \pm 2\sqrt{3}$ **93.** $\frac{1}{2}, -1$ **95.** $\frac{1}{4}, -\frac{3}{4}$ **97.** $1 \pm \sqrt{3}$
99. $-7 \pm \sqrt{5}$ **101.** $-4 \pm 2\sqrt{5}$ **103.** $\frac{2}{3} \pm \frac{\sqrt{7}}{3}$
105. $-\frac{4}{3}$ **107.** $-\frac{1}{2} \pm \sqrt{2}$ **109.** $\frac{2}{7}$ **111.** $2 \pm \frac{\sqrt{6}}{2}$
113. $6 \pm \sqrt{11}$ **115.** $-\frac{3}{8} \pm \frac{\sqrt{265}}{8}$
117. $0.976, -0.643$ **119.** $1.355, -14.071$
121. $1.687, -0.488$ **123.** $-0.290, -2.200$
125. $1 \pm \sqrt{2}$ **127.** $6, -12$ **129.** $\frac{1}{2} \pm \sqrt{3}$
131. $-\frac{1}{2}$ **133.** $\frac{3}{4} \pm \frac{\sqrt{97}}{4}$ **135.** $0, \pm\frac{3\sqrt{2}}{2}$
137. ± 3 **139.** -6 **141.** $-3, 0$ **143.** $3, 1, -1$
145. ± 1 **147.** $\pm\sqrt{3}, \pm 1$ **149.** $\pm\frac{1}{2}, \pm 4$
151. $1, -2$ **153.** 50 **155.** 26 **157.** -16
159. $2, -5$ **161.** 0 **163.** 9
165. $-3 \pm 16\sqrt{2}$ **167.** $\pm\sqrt{14}$
169. 1 **171.** $4, -5$ **173.** $\frac{-3 \pm \sqrt{21}}{6}$
175. $2, -\frac{3}{2}$ **177.** $1, -3$ **179.** $3, -2$
181. $\sqrt{3}, -3$ **183.** $3, \frac{-1 - \sqrt{17}}{2}$
185. 61.2 inches **187.** 23,437.5 miles
189. (a)

(b) $w(w + 14) = 1632$

(c) $w = 34$ feet, $l = 48$ feet

191. $\frac{5\sqrt{2}}{2} \approx 3.54$ centimeters

193. ≈ 550 miles per hour and 600 miles per hour

195. 50,000 units **197.** 24.725 pounds per square inch

199. False. The product must equal zero for the Zero-Factor Property to be used.

201. The student should have subtracted $15x$ from both sides so that the equation was equal to zero. By factoring out an x, there are two solutions, $x = 0$ and $x = 6$.

203. Remove symbols of grouping, combine like terms, reduce fractions.

Add (or subtract) the same quantity to (from) both sides of the equation.

Multiply (or divide) both sides of the equation by the same nonzero quantity.

Interchange the two sides of the equation.

205. (a) $0, -\frac{b}{a}$ (b) $0, 1$

207. Isolate the absolute value by subtracting x from both sides of the equation. The expression inside the absolute value signs can be positive or negative, so two separate equations must be solved.

Section P.3 (page 34)

1.   **3.**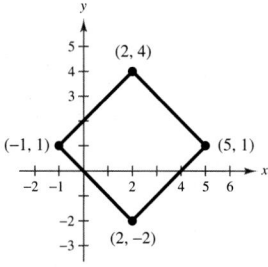

5. $A: (2, 6)$, $B: (-6, -2)$, $C: (4, -4)$, $D: (-3, 2)$
7. $(-3, 4)$ **9.** $(-5, -5)$
11. Quadrant IV   **13.** Quadrant II
15. Quadrants III and IV   **17.** Quadrant III
19. Quadrants I and III
21.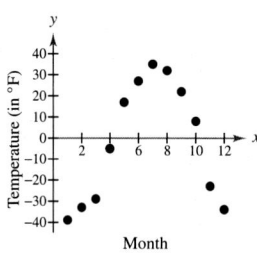

23. (a) 4, 3, 5   **25.** (a) 10, 3, $\sqrt{109}$
   (b) $4^2 + 3^2 = 5^2$ (b) $10^2 + 3^2 = (\sqrt{109})^2$

27. (a) 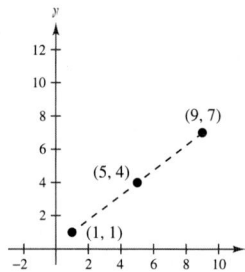   (b) 10
   (c) $(5, 4)$

29. (a)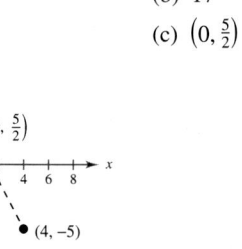
   (b) 17
   (c) $(0, \frac{5}{2})$

31. (a) 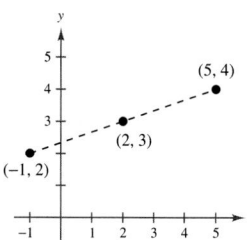   (b) $2\sqrt{10}$
   (c) $(2, 3)$

33. (a) 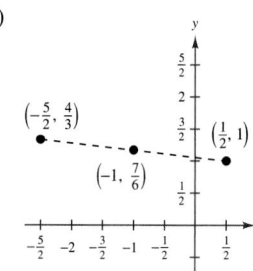   (b) $\dfrac{\sqrt{82}}{3}$
   (c) $\left(-1, \frac{7}{6}\right)$

35. (a) 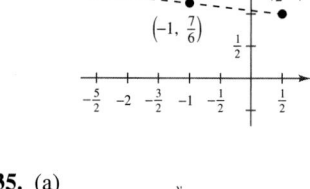   (b) $\sqrt{110.97}$
   (c) $(1.25, 3.6)$

37. (a) 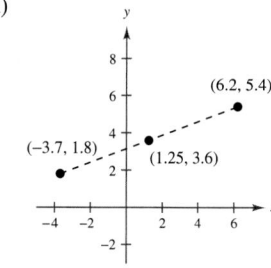   (b) $6\sqrt{277}$
   (c) $(6, -45)$

39. (a) The number of artists elected each year seems to be nearly steady except for the first few years. Between 6 and 8 artists will be elected in 2001.
   (b) The Rock and Roll Hall of Fame began in 1986.
41. $5\sqrt{74} \approx 43$ yards
43. $1002.6 million
45. (a) Yes   (b) Yes   **47.** (a) No   (b) Yes

49.

x	-2	0	1	$\frac{4}{3}$	2
y	$-\frac{5}{2}$	-1	$-\frac{1}{4}$	0	$\frac{1}{2}$

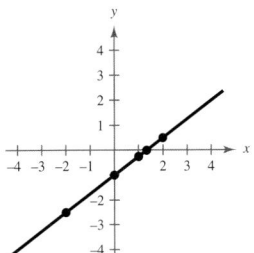

51. x-intercepts: $(\pm 2, 0)$
    y-intercept: $(0, 16)$
53. x-intercepts: $(0, 0), \left(\frac{5}{2}, 0\right)$
    y-intercept: $(0, 0)$

55.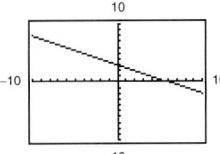

Intercepts: $(6, 0), (0, 3)$

57.

Intercept: $(0, 0)$

59. y-axis symmetry   **61.** Origin symmetry
63. Origin symmetry   **65.** x-axis symmetry

67.   **69.**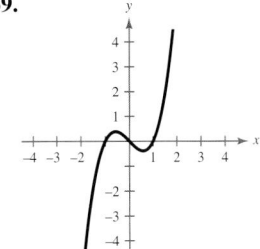

71. c   **73.** b

75.   **77.**

79.   **81.**

83.   **85.**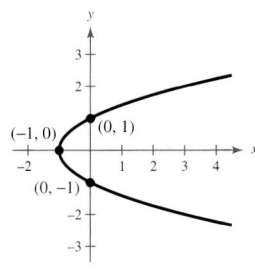

87. $x^2 + y^2 = 9$
89. $(x - 2)^2 + (y + 1)^2 = 16$
91. $(x + 1)^2 + (y - 2)^2 = 5$
93. $(x - 3)^2 + (y - 4)^2 = 25$
95. Center: $(0, 0)$; Radius: 5

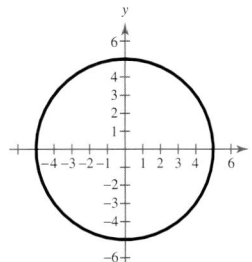

97. Center: $(1, -3)$; Radius: 3

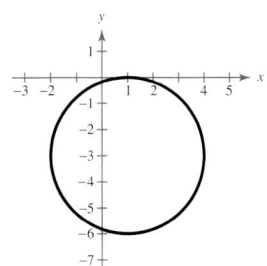

99. Center: $\left(\frac{1}{2}, \frac{1}{2}\right)$; Radius: $\frac{3}{2}$

101.

103. (a)

(b) Answers will vary.

(c)

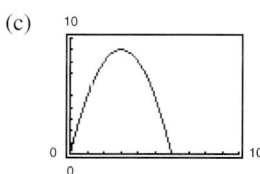

(d) $x = 3$, $w = 3$

105. (a)

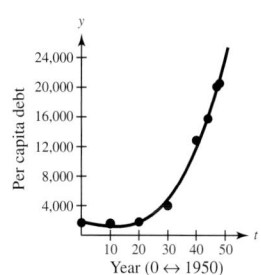

(b) The curve seems to be a good fit for the data.

(c) 2002: $27,142; 2004: $30,589

107. True. Two sides of the triangle have lengths of $\sqrt{149}$, and the third side has a length of $\sqrt{18}$.

109. True. The graph of the linear equation can have only one x-intercept because equations of lines do not curve and extend to infinity.

111. Use the Midpoint Formula to prove that the diagonals of the parallelogram bisect each other.

$$\left(\frac{b+a}{2}, \frac{c+0}{2}\right) = \left(\frac{a+b}{2}, \frac{c}{2}\right)$$

$$\left(\frac{a+b+0}{2}, \frac{c+0}{2}\right) = \left(\frac{a+b}{2}, \frac{c}{2}\right)$$

Section P.4 (page 47)

1. (a) L_2 (b) L_3 (c) L_1

3.

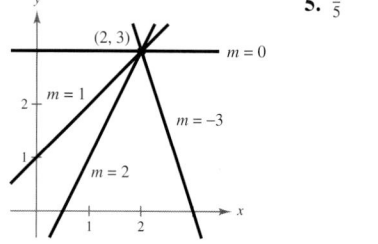

5. $\frac{8}{5}$ **7.** 0   **9.** -4

11. $m = 2$ **13.** m is undefined.

   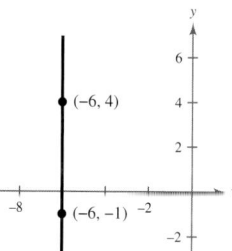

15. $m = -\frac{1}{7}$ **17.** $m = 0.15$

   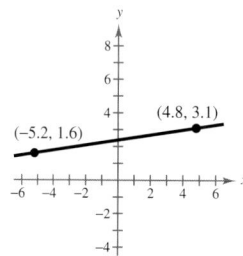

19. $(0, 1)$, $(3, 1)$, $(-1, 1)$ **21.** $(6, -5)$, $(7, -4)$, $(8, -3)$

23. $(-8, 0)$, $(-8, 2)$, $(-8, 3)$

25. $(-4, 6)$, $(-3, 8)$, $(-2, 10)$

27. $(9, -1)$, $(11, 0)$, $(13, 1)$ **29.** Perpendicular

31. Parallel

33. (a) Sales increasing 135 units per year

(b) No change in sales

(c) Sales decreasing 40 units per year

35. (a) Greatest increase: 1990 to 1991 and 1996 to 1997
Greatest decrease: 1997 to 1998
(b) $m = 0.037$
(c) Each year, the earnings per share increase by \$0.037.

37. (a) and (b)

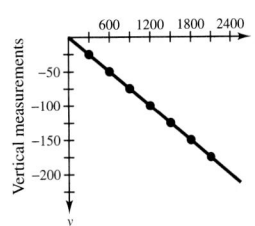

(c) $y = -\frac{1}{12}x$
(d) For every 12 horizontal measurements, the vertical measurement decreases by 1.
(e) "8.3% grade"

39. 12 feet

41. $m = 1$; Intercept: $(0, -10)$

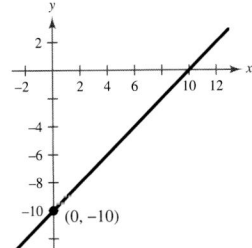

43. $m = 0$; Intercept: $\left(0, -\frac{5}{3}\right)$ **45.** $m = -\frac{2}{3}$; Intercept: $(0, 3)$

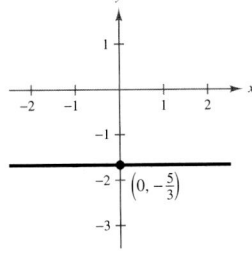

47. $y = -x + 10$ **49.** $y = 4x$

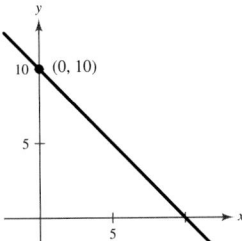

51. $y = \frac{3}{4}x - \frac{7}{2}$ **53.** $y = 4$

 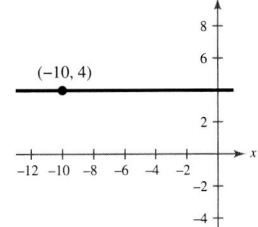

55. $y = -3x$ **57.** $y = -2.5x - 2.75$

 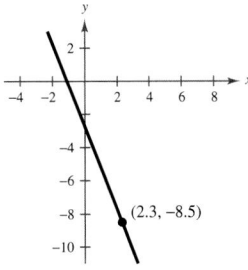

59. $y = \frac{7}{8}x - \frac{1}{2}$ **61.** $y = 4$

 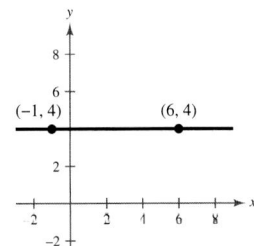

63. $y = -\frac{1}{3}x + \frac{4}{3}$ **65.** $y = -\frac{3}{25}x + \frac{159}{100}$

 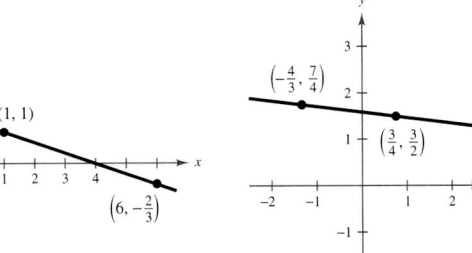

67. $y = 0.3x - 1.8$

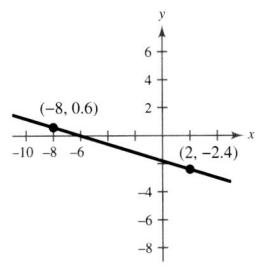

69. $4x - 3y + 12 = 0$ **71.** $3x - y - 2 = 0$
73. $x + y - 1 = 0$
75. (a) $y = -x - 1$ (b) $y = x + 5$
77. (a) $y = -\frac{5}{3}x + \frac{53}{24}$ (b) $y = \frac{3}{5}x + \frac{9}{40}$
79. (a) $x = 2$ (b) $y = 5$
81. (a) $y = -3x - 13.1$ (b) $y = \frac{1}{3}x - 0.1$
83. (a) is parallel to (c). (b) is perpendicular to (a) and (c).

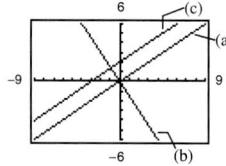

85. (a) is parallel to (b). (c) is perpendicular to (a) and (b).

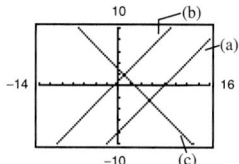

87. $V = 4.5t + 151.5$
89. c; The slope is 2, which represents the hourly wage per unit produced.
91. d; The slope is -100, which represents the decrease in the value of the word processor each year.
93. $y = -\frac{5}{13}x - \frac{2}{13}$ **95.** $y = -\frac{16}{21}x - \frac{13}{56}$
97. $y = 54t + 3927$; 1999: 4089 stores, 2000: 4143 stores
99.

C	$-17.8°$	$-10°$	$10°$	$20°$	$32.2°$	$177°$
F	$0°$	$14°$	$50°$	$68°$	$90°$	$350.6°$

101. 3014 students   **103.** $V = 25,000 - 2300t$
105. $W = 0.75x + 11.50$
107. (a) $x = -\frac{1}{15}p + \frac{266}{3}$ (b) 45   (c) 49
109. $W = 0.07S + 2500$
111. (a) $12,000 - x$ (b) $y = -0.015x + 480$

(c)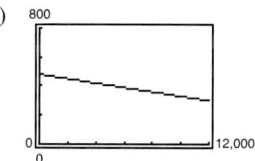

(d) As the amount invested at the lower interest rate increases, the annual interest decreases.

113. (a) and (b)

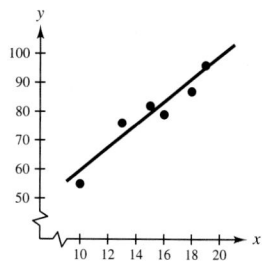

(c) $y = 4x + 19$ (d) 87
(e) Vertical shift 4 units upward

115. False. The slope of the first line is $\frac{2}{7}$ and the slope of the second line is $-\frac{11}{7}$.

117. The slope of a vertical line is undefined because division by zero is undefined.

119. The line with a slope of -4 is steeper. The slope with the greatest magnitude corresponds to the steepest line.

121. Yes. The rate of change remains the same on a line.

Section P.5   *(page 61)*

1. Yes   **3.** No
5. Yes, each input value has exactly one output value.
7. No, the input value of 7 has two output values, 6 and 12.
9. (a) Function
(b) Not a function, because the element 1 in A corresponds to two elements, -2 and 1, in B.
(c) Function
(d) Not a function, because not every element in A is matched with an element in B.

11. Each is a function. For each year there corresponds one and only one circulation.

13. Not a function   **15.** Function   **17.** Function
19. Not a function   **21.** Function
23. (a) 4   (b) 0   (c) $4x$ (d) $(x + c)$
25. (a) -1 (b) -9 (c) $2x - 5$
27. (a) 36π (b) $\frac{9}{2}\pi$ (c) $\frac{32}{3}\pi r^3$
29. (a) 1   (b) 2.5   (c) $3 - 2|x|$
31. (a) $-\frac{1}{9}$ (b) Undefined   (c) $\dfrac{1}{y^2 + 6y}$
33. (a) 1   (b) -1 (c) $\dfrac{|x - 1|}{x - 1}$
35. (a) -1 (b) 2   (c) 6

37.

x	-2	-1	0	1	2
$f(x)$	1	-2	-3	-2	1

39.

t	-5	-4	-3	-2	-1
$h(t)$	1	$\frac{1}{2}$	0	$\frac{1}{2}$	1

41.

x	-2	-1	0	1	2
$f(x)$	5	$\frac{9}{2}$	4	1	0

43. 5 **45.** $\frac{4}{3}$ **47.** ± 3 **49.** $0, \pm 1$
51. $2, -1$ **53.** $3, 0$ **55.** All real numbers x
57. All real numbers $t \ne 0$ **59.** $y \ge 10$
61. $-1 \le x \le 1$ **63.** All real numbers $x \ne 0, -2$
65. $s \ge 1, s \ne 4$ **67.** All real numbers $x \ne 0$
69. $\{(-2, 4), (-1, 1), (0, 0), (1, 1), (2, 4)\}$
71. $\{(-2, 0), (-1, 1), (0, \sqrt{2}), (1, \sqrt{3}), (2, 2)\}$
73. $g(x) = -2x^2$ **75.** $r(x) = \dfrac{32}{x}$ **77.** $3 + h, h \ne 0$
79. $3x^2 + 3xc + c^2, c \ne 0$ **81.** $3, x \ne 3$
83. $\dfrac{\sqrt{5x} - 5}{x - 5}$ **85.** $A = \dfrac{p^2}{16}$ **87.** $A = 8\sqrt{s^2 - 64}$

89. (a)

Height x	Width	Volume V
1	$24-2(1)$	$1[24-2(1)]^2 = 484$
2	$24-2(2)$	$2[24-2(2)]^2 = 800$
3	$24-2(3)$	$3[24-2(3)]^2 = 972$
4	$24-2(4)$	$4[24-2(4)]^2 = 1024$
5	$24-2(5)$	$5[24-2(5)]^2 = 980$
6	$24-2(6)$	$6[24-2(6)]^2 = 864$

Maximum when $x = 4$

(b)
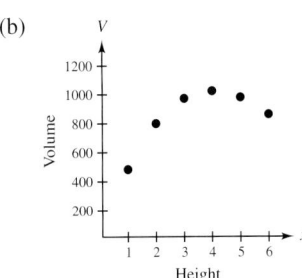
V is a function of x.

(c) $V = x(24 - 2x)^2, 0 < x < 12$

91. $A = \dfrac{x^2}{2(x-2)}, x > 2$

93. 1978: \$15,198; 1988: \$25,558;
1993: \$30,800; 1997: \$41,008
95. (a) $C = 12.30x + 98{,}000$
(b) $R = 17.98x$
(c) $P = 5.68x - 98{,}000$
97. (a) $R = \dfrac{240n - n^2}{20}, n \ge 80$

(b)

n	90	100	110	120	130	140	150
$R(n)$	\$675	\$700	\$715	\$720	\$715	\$700	\$675

The revenue is maximum when 120 people take the trip.

99. (a)
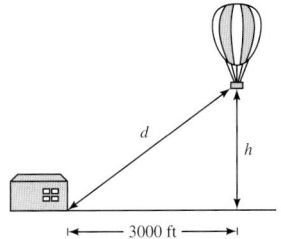

(b) $h = \sqrt{d^2 - 3000^2}, d \ge 3000$
101. Yes, the ball will be at a height of 6 feet.
103. True. Each x-value corresponds to one y-value.
105. The domain is the set of inputs of the function, and the range is the set of outputs.

Section P.6 (page 75)

1. Domain: all real numbers
Range: all real numbers
3. Domain: all real numbers
Range: $(-\infty, 1]$
5. Domain: $(-\infty, -1], [1, \infty)$ **7.** Domain: $[-4, 4]$
Range: $[0, \infty)$ Range: $[0, 4]$
9. Function   **11.** Not a function   **13.** Function
15. $-\frac{5}{2}, 6$ **17.** 0   **19.** $0, \pm\sqrt{2}$ **21.** $\pm\frac{1}{2}, 6$
23. $-\frac{5}{3}$ **25.** $-\frac{11}{2}$

27. $\frac{1}{3}$

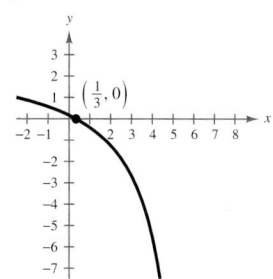

29. (a) Increasing on $(-\infty, \infty)$ (b) Odd
31. (a) Increasing on $(-\infty, 0)$ and $(2, \infty)$
   Decreasing on $(0, 2)$
   (b) Neither even nor odd
33. (a)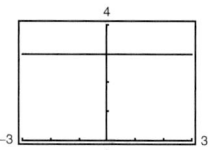
   Constant on $(-\infty, \infty)$

(b)
x	-2	-1	0	1	2
$f(x)$	3	3	3	3	3

35. (a)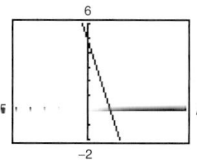
   Decreasing on $(-\infty, \infty)$

(b)
x	-2	-1	0	1	2
$f(x)$	11	8	5	2	-1

37. (a)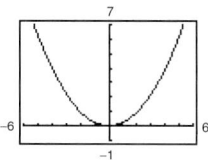
   Decreasing on $(-\infty, 0)$; Increasing on $(0, \infty)$

(b)
s	-4	-2	0	2	4
$g(s)$	4	1	0	1	4

39. (a)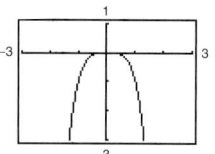
   Increasing on $(-\infty, 0)$; Decreasing on $(0, \infty)$

(b)
t	-2	-1	0	1	2
$f(t)$	-16	-1	0	-1	-16

41. (a)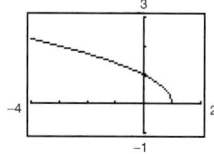
   Decreasing on $(-\infty, 1)$

(b)
x	-3	-2	-1	0	1
$f(x)$	2	$\sqrt{3}$	$\sqrt{2}$	1	0

43. (a)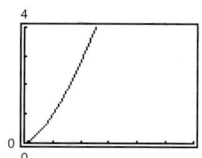
   Increasing on $(0, \infty)$

(b)
x	0	1	2	3	4
$f(x)$	0	1	2.82	5.2	8

45. (a)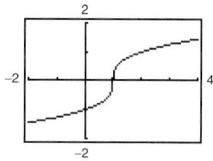
   Increasing on $(-\infty, \infty)$

(b)
t	-2	-1	0	1	2
$g(t)$	-1.44	-1.26	-1	0	1

47. (a)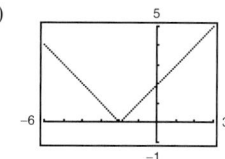
   Decreasing on $(-\infty, -2)$; Increasing on $(-2, \infty)$

(b)
x	-6	-4	-2	0	2
$f(x)$	4	2	0	2	4

49. (a)

Increasing on $(-\infty, 0)$ and $(2, \infty)$; Constant on $(0, 2)$

(b)
x	-2	-1	0	1	2	3	4
$f(x)$	1	2	3	3	3	5	7

51.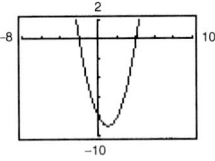

Relative minimum: $(1, -9)$

53.

Relative maximum: $(-1.79, 8.21)$

Relative minimum: $(1.12, -4.06)$

55.

Relative maximum: $(-0.33, -0.30)$

Relative minimum: $(2, -13)$

57.   **59.**

61.   **63.**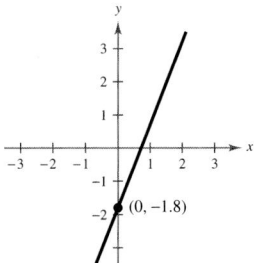

65. $f(x) = -2x + 6$

67. $f(x) = -3x + 11$

69. $f(x) = \frac{2}{5}x - 3$

71. $f(x) = \frac{6}{7}x - \frac{45}{7}$

73.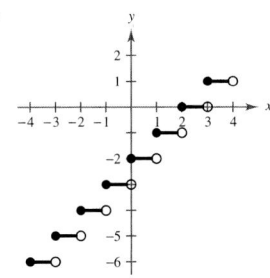

Vertical shift 2 units downward

75.   **77.**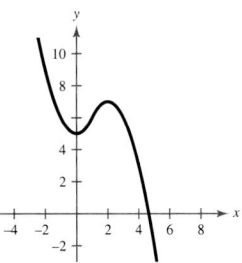

79. $(-\infty, 4]$ **81.** $(-\infty, -3], [3, \infty)$

   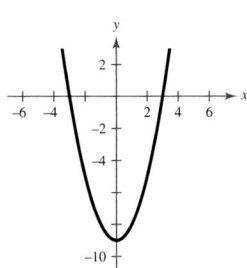

83. $[-1, 1]$ **85.** $(-\infty, \infty)$ **105.**

  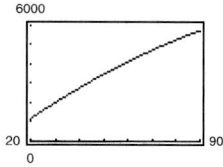

30 watts

107. $h = 3 - 4x + x^2$ **109.** $h = 2 - \sqrt[3]{x}$

111. $L = 2 - \sqrt[3]{2y}$ **113.** $L = \dfrac{2}{y}$

87. $f(x) < 0$ for all x.   **89.** $(-2, 8]$

115. (a) $A = 64 - 2x^2,\ 0 \le x \le 4$

(b)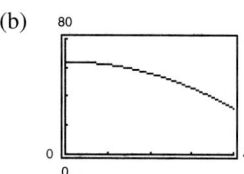

$32 \le A \le 64$

(c) Square with sides $4\sqrt{2}$ m

117.

Interval	Inside Pipe	Drainpipe 1	Drainpipe 2
$[0, 5]$	Open	Closed	Closed
$[5, 10]$	Open	Open	Closed
$[10, 20]$	Closed	Closed	Closed
$[20, 30]$	Closed	Closed	Open
$[30, 40]$	Open	Open	Open
$[40, 45]$	Open	Closed	Open
$[45, 50]$	Open	Open	Open
$[50, 60]$	Open	Open	Closed

91.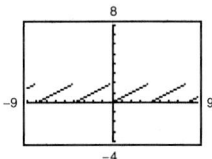

Domain: $(-\infty, \infty)$

Range: $[0, 2)$

Sawtooth pattern

93. Even   **95.** Odd   **97.** Neither even nor odd

99. (a) $\left(\tfrac{3}{2}, 4\right)$ (b) $\left(\tfrac{3}{2}, -4\right)$

101. (a) $(-4, 9)$ (b) $(-4, -9)$

103. (a) C_2 is the appropriate model, because the cost does not increase until after the next minute of conversation has started.

(b)

$7.89

119. False. A piecewise-defined function is a function that is defined by two or more equations over a specified domain. That domain may or may not include x- and y-intercepts.

121. Answers will vary.

123. Yes. For each value of y there corresponds one and only one value of x.

125. (a)   (b)

(c)   (d)

(e)   (f)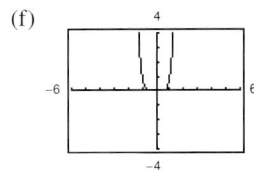

All the graphs pass through the origin. The graphs of the odd powers of x are symmetric with respect to the origin, and the graphs of the even powers are symmetric with respect to the y-axis. As the powers increase, the graphs become flatter in the interval $-1 < x < 1$.

Section P.7 (page 86)

1. (a)   (b)

3. (a)

(c)

5. (a)   (b)

(c)   (d)

(e)   (f)

7. (a)   (b)

(c)   (d)

(e)   (f)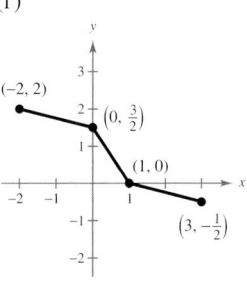

9. (a) $y = x^2 - 1$ (b) $y = 1 - (x + 1)^2$
(c) $y = -(x - 2)^2 + 6$ (d) $y = (x - 5)^2 - 3$

11. (a) $y = |x| + 5$ (b) $y = -|x + 3|$
(c) $y = |x - 2| - 4$ (d) $y = -|x - 6| - 1$

13. Horizontal shift of $y = x^3$; $y = (x - 2)^3$

15. Reflection in the x-axis of $y = x^2$; $y = -x^2$

17. Reflection in the x-axis and vertical shift of $y = \sqrt{x}$; $y = 1 - \sqrt{x}$

19. Reflection of $f(x) = x^2$ in x-axis and vertical shift of 12 units upward

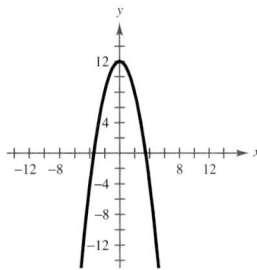

21. Vertical shift of $f(x) = x^3$ seven units upward

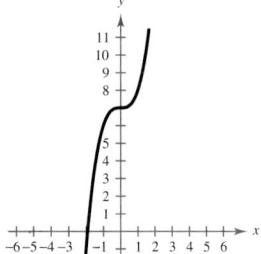

23. Reflection of $f(x) = x^2$ in x-axis, vertical shift of 2 units upward, and horizontal shift of 5 units to the left

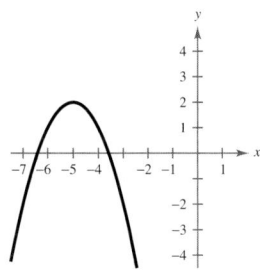

25. Vertical shift of $f(x) = x^3$ 2 units upward and horizontal shift of 1 unit to the right

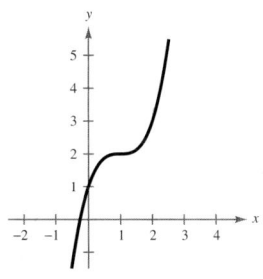

27. Reflection of $f(x) = |x|$ in x-axis and vertical shift of 2 units downward

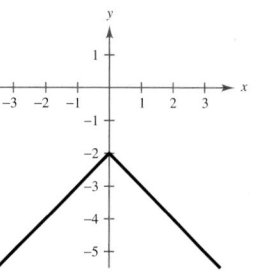

29. Reflection of $f(x) = |x|$ in x-axis, vertical shift of 8 units upward, and horizontal shift of 4 units to the left

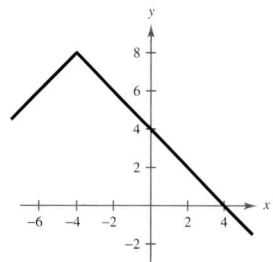

31. Horizontal shift of $f(x) = \sqrt{x}$ 9 units to the right

33. Reflection of $f(x) = \sqrt{x}$ in y-axis, vertical shift of 2 units downward, and horizontal shift of 7 units to the right

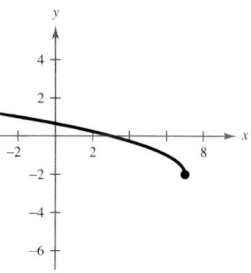

35. $f(x) = (x - 2)^2 - 8$ **37.** $f(x) = (x - 13)^3$
39. $f(x) = -|x| - 10$ **41.** $f(x) = -\sqrt{-x + 6}$
43. (a) $y = -3x^2$ (b) $y = 4x^2 + 3$
45. (a) $y = -\frac{1}{2}|x|$ (b) $y = 3|x| - 3$
47. Vertical stretch of $y = x^3$; $y = 2x^3$

49. Reflection in x-axis and vertical shrink of $y = x^2$; $y = -\frac{1}{2}x^2$

51. Reflection in y-axis and vertical shrink of $y = \sqrt{x}$; $y = \frac{1}{2}\sqrt{-x}$

53. $y = -(x-2)^3 + 2$ **55.** $y = -\sqrt{x} - 3$

57. (a)   (b)

(c)   (d)

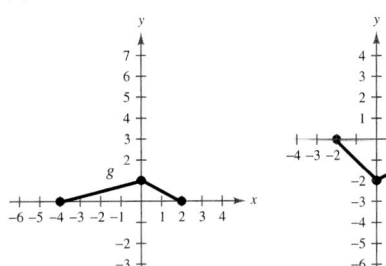

59. (a) Vertical shrink of 0.04 and vertical shift of 20.46 units upward

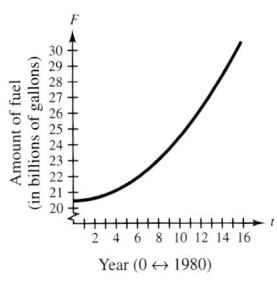

(b) $f(t) = 20.46 + 0.04(t + 10)^2$. By shifting the graph 10 units to the left, you obtain $t = 0$ represents 1990.

61. True. $|-x| = |x|$

63. (a) $g(t) = \frac{3}{4}f(t)$ (b) $g(t) = f(t) + 10,000$
(c) $g(t) = f(t - 2)$

65. $(-2, 0), (-1, 1), (0, 2)$

Section P.8 (page 96)

1.   **3.**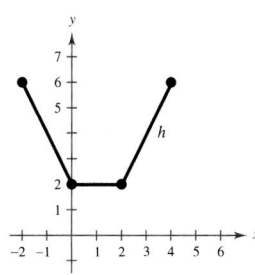

5. (a) $2x$ (b) 4 (c) $x^2 - 4$ (d) $\dfrac{x+2}{x-2}$; $x \neq 2$

7. (a) $x^2 - x + 2$ (b) $x^2 + x - 2$
(c) $2x^2 - x^3$ (d) $\dfrac{x^2}{2-x}$; $x \neq 2$

9. (a) $x^2 + 6 + \sqrt{1-x}$ (b) $x^2 + 6 - \sqrt{1-x}$
(c) $(x^2 + 6)\sqrt{1-x}$ (d) $\dfrac{(x^2+6)\sqrt{1-x}}{1-x}$; $x < 1$

11. (a) $\dfrac{x+1}{x^2}$ (b) $\dfrac{x-1}{x^2}$ (c) $\dfrac{1}{x^3}$ (d) x; $x \neq 0$

13. 3   **15.** 5   **17.** $9t^2 - 3t + 5$ **19.** 74

21. 26   **23.** $\frac{3}{5}$

25.   **27.**

29.

$f(x), g(x)$

31. $T = \frac{3}{4}x + \frac{1}{15}x^2$

33. $y_1 = -0.59t^2 + 7.66t - 144.90$

$y_2 = 16.58t + 245.06$

$y_3 = 1.85t + 21.88$

35. (a) For each time t there corresponds one and only one temperature T.

(b) 60°, 72°

(c) All the temperature changes would occur 1 hour later.

(d) The temperature would be decreased by 1 degree.

37. (a) $(x - 1)^2$ (b) $x^2 - 1$ (c) x^4

39. (a) $20 - 3x$ (b) $-3x$ (c) $9x + 20$

41. (a) $\sqrt{x^2 + 4}$ (b) $x + 4$

Domain of f and $g \circ f$: $x \geq -4$;

Domain of g and $f \circ g$: all real numbers

43. (a) $x - \frac{8}{3}$ (b) $x - 8$

Domain of f, g, $f \circ g$, and $g \circ f$: all real numbers

45. (a) x^{16} (b) x^{16}

Domain of f, g, $f \circ g$, and $g \circ f$: all real numbers

47. (a) $|x + 6|$ (b) $|x| + 6$

Domain of f, g, $f \circ g$, and $g \circ f$: all real numbers

49. (a) $\dfrac{1}{x + 3}$ (b) $\dfrac{1}{x} + 3$

Domain of f and $g \circ f$: all real numbers $x \neq 0$

Domain of g: all real numbers

Domain of $f \circ g$: all real numbers $x \neq -3$

51. (a) 3 (b) 0 **53.** (a) 0 (b) 4

55. $f(x) = x^2$, $g(x) = 2x + 1$

57. $f(x) = \sqrt[3]{x}$, $g(x) = x^2 - 4$

59. $f(x) = \dfrac{1}{x}$, $g(x) = x + 2$

61. $f(x) = \dfrac{x + 3}{4 + x}$, $g(x) = -x^2$

63. (a) $r(x) = \dfrac{x}{2}$ (b) $A(r) = \pi r^2$

(c) $(A \circ r)(x) = \pi \left(\dfrac{x}{2}\right)^2$; $(A \circ r)(x)$ represents the area of the circular base of the tank on the square foundation with side length x.

65. $(C \circ x)(t) = 3000t + 750$; $(C \circ x)(t)$ represents the cost after t production hours.

67. True. The range of g must be a subset of the domain of f for $(f \circ g)(x)$ to be defined.

69. Answers will vary.

Section P.9 (page 105)

1. c **3.** a **5.** $f^{-1}(x) = \frac{1}{6}x$ **7.** $f^{-1}(x) = x - 9$

9. $f^{-1}(x) = \dfrac{x - 1}{3}$ **11.** $f^{-1}(x) = x^3$

13. (a) $f(g(x)) = f\left(\dfrac{x}{2}\right) = 2\left(\dfrac{x}{2}\right) = x$

$g(f(x)) = g(2x) = \dfrac{(2x)}{2} = x$

(b)

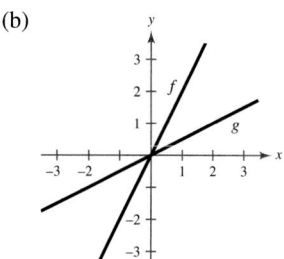

15. (a) $f(g(x)) = f\left(\dfrac{x - 1}{5}\right) = 5\left(\dfrac{x - 1}{5}\right) + 1 = x$

$g(f(x)) = g(5x + 1) = \dfrac{(5x + 1) - 1}{5} = x$

(b)

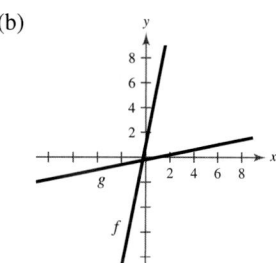

17. (a) $f(g(x)) = f(\sqrt[3]{x}) = (\sqrt[3]{x})^3 = x$

$g(f(x)) = g(x^3) = \sqrt[3]{x^3} = x$

(b)

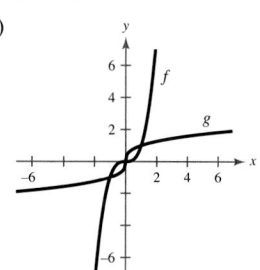

19. (a) $f(g(x)) = f(x^2 + 4)$, $x \geq 0$

$= \sqrt{(x^2 + 4) - 4} = x$

$g(f(x)) = g(\sqrt{x - 4})$

$= (\sqrt{x - 4})^2 + 4 = x$

(b)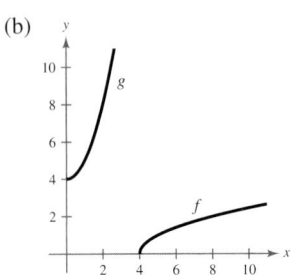

21. (a) $f(g(x)) = f(\sqrt{9-x}), \ x \leq 9$
    $= 9 - (\sqrt{9-x})^2 = x$
    $g(f(x)) = g(9-x^2), \ x \geq 0$
    $= \sqrt{9-(9-x^2)} = x$

(b)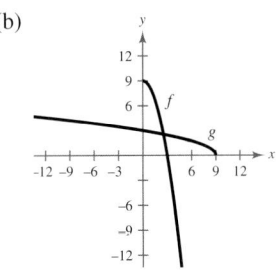

23. (a) $f(g(x)) = f\left(-\dfrac{5x+1}{x-1}\right) = \dfrac{-\left(\dfrac{5x+1}{x-1}\right)-1}{-\left(\dfrac{5x+1}{x-1}\right)+5}$

    $= \dfrac{-5x-1-x+1}{-5x-1+5x-5} = x$

    $g(f(x)) = g\left(\dfrac{x-1}{x+5}\right) = \dfrac{-5\left(\dfrac{x-1}{x+5}\right)-1}{\dfrac{x-1}{x+5}-1}$

    $= \dfrac{-5x+5-x-5}{x-1-x-5} = x$

(b)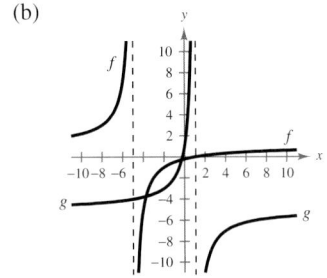

25. No

27.
x	-2	0	2	4	6	8
$f^{-1}(x)$	-2	-1	0	1	2	3

29. Yes   31. No

33.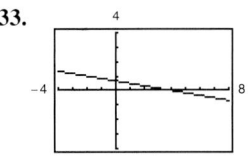

The function has an inverse.

35.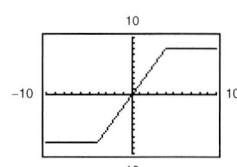

The function does not have an inverse.

37.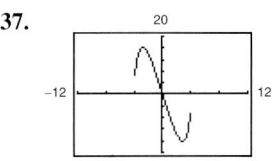

The function does not have an inverse.

39. $f^{-1}(x) = \dfrac{x+3}{2}$ 41. $f^{-1}(x) = \sqrt[5]{x+2}$

   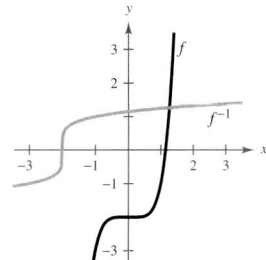

43. $f^{-1}(x) = x^2, \ x \geq 0$

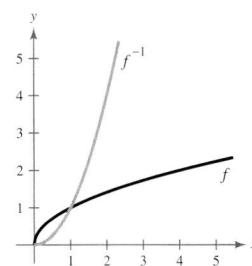

45. $f^{-1}(x) = \sqrt{4-x^2}, \ 0 \leq x \leq 2$

47. $f^{-1}(x) = \dfrac{4}{x}$

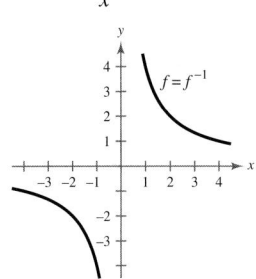

49. $f^{-1}(x) = \dfrac{2x+1}{x-1}$ **51.** $f^{-1}(x) = x^3 + 1$

53. $f^{-1}(x) = \dfrac{5x-4}{6-4x}$

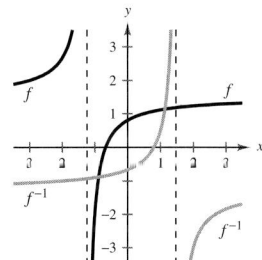

55. No inverse   **57.** $g^{-1}(x) = 8x$ **59.** No inverse

61. $f^{-1}(x) = \sqrt{x} - 3$ **63.** No inverse

65. $h^{-1}(x) = \dfrac{1}{x}$ **67.** $f^{-1}(x) = \dfrac{x^2 - 3}{2},\ x \geq 0$ **69.** 32

71. 600   **73.** $2\sqrt[3]{x+3}$ **75.** $\dfrac{x+1}{2}$ **77.** $\dfrac{x+1}{2}$

79. (a) $y = \dfrac{x-8}{0.75}$

(b) y = number of units produced; x = hourly wage

(c) 19 units

81. (a) $y = \sqrt{\dfrac{x - 245.50}{0.03}},\ 245.5 < x < 545.5$

x = degrees Fahrenheit; y = % load

(b)   (c) $0 < x < 92.11$

83. No. The function would not pass the Horizontal Line Test.

85. (a) Yes. f^{-1} yields the year for a given per capita regular soft drink consumption.

(b) 5

87. True. If $f(x) = x - 6$ and $f^{-1}(x) = x + 6$, then the y-intercept of f is $(0, -6)$ and the x-intercept of f^{-1} is $(-6, 0)$.

89. False. $f(x) = \dfrac{1}{x} = f^{-1}(x)$.

91.

x	-2	-1	1	3
y	-5	-2	2	3

x	-5	-2	2	3
$f^{-1}(x)$	-2	-1	1	3

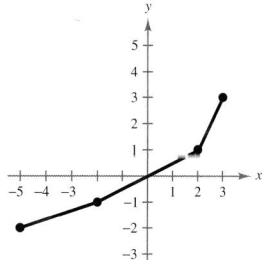

93.

x	-4	-2	0	3
y	3	4	0	-1

The graph of f does not pass the Horizontal Line Test, so $f^{-1}(x)$ does not exist.

Review Exercises (page 111)

1. (a) 11   (b) 11, -14

(c) 11, -14, $-\tfrac{8}{9}$, $\tfrac{5}{2}$, 0.4   (d) $\sqrt{6}$

3. (a) $0.8\overline{3}$ (b) 0.875

5. The set consists of all real numbers less than or equal to 7.

7. 155 **9.** $|x - 25| \leq 10$ **11.** (a) -1 (b) -3
13. Associative Property of Addition
15. Multiplicative Inverse Property
17. -11 **19.** $\frac{1}{12}$ **21.** -144 **23.** Identity
25. 20 **27.** $-\frac{1}{2}$ **29.** $2\frac{6}{7}$ liters **31.** $-\frac{7}{2}, 4$
33. $8 \pm \sqrt{15}$ **35.** $-\frac{5}{4} \pm \frac{\sqrt{241}}{4}$ **37.** $0, \frac{3}{2}$ **39.** 2
41. 79 **43.** 2, 3
45.

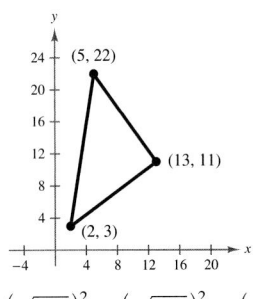

$(\sqrt{185})^2 + (\sqrt{185})^2 = (\sqrt{370})^2$

47. Quadrant IV **49.** Quadrant IV
51. (a) (b) 5

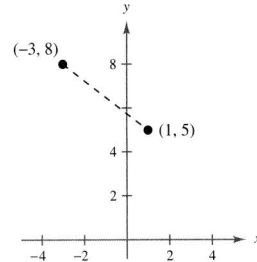

53. (a) (b) $\left(1, \frac{3}{2}\right)$

55. **57.**

 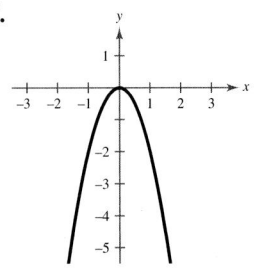

59. x-intercept: $\left(\frac{9}{2}, 0\right)$
 y-intercept: $(0, -9)$
61. x-intercepts: $(0, 0), (\pm 3, 0)$
 y-intercept: $(0, 0)$
63. **65.**

 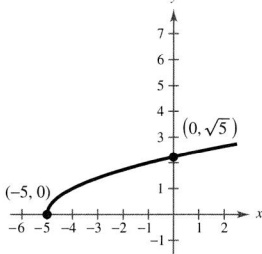

67. Center: $(-2, 0)$; Radius: 4

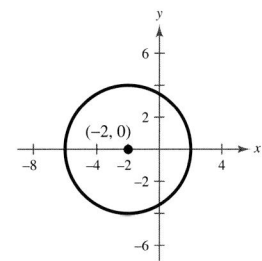

69. Center: $\left(\frac{1}{2}, -1\right)$; Radius: 6

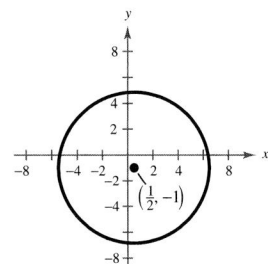

71. $(x - 2)^2 + (y + 3)^2 = 13$
73. **75.**

 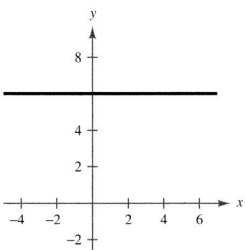

Answers to Odd-Numbered Exercises and Tests **A47**

77.
79.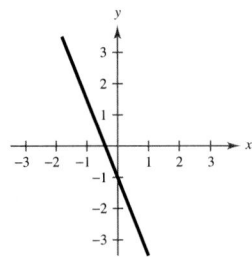
111. $-5 \leq x \leq 5$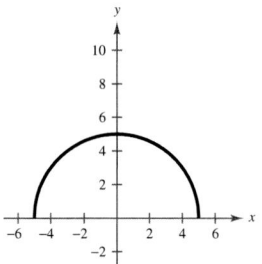
113. All real numbers $s \neq 3$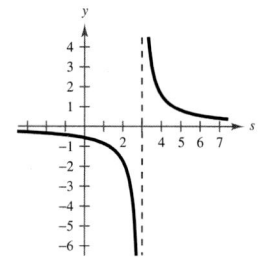

81. $(6, 0), (10, 1), (-2, -2)$

83.
85.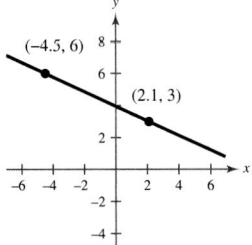

115. All real numbers $x \neq 3, -2$
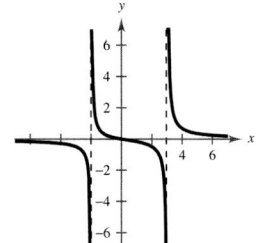

$m = -\frac{1}{2}$ $m = -\frac{5}{11}$

117. (a) 16 feet per second
(b) 1.5 seconds
(c) -16 feet per second

119. (a) $A = x(12 - x)$
(b) $0 < x < 12$

87. $t = \frac{7}{3}$ **89.** $x = 0$ **91.** $4x + 3y - 8 = 0$
93. $3x - 2y - 10 = 0$ **95.** $x + 2y - 4 = 0$

121. Function   **123.** Not a function
125. $\frac{7}{3}, 3$ **127.** $-\frac{3}{8}$
129.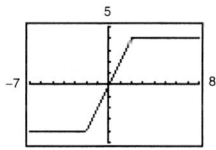

Increasing on $(-2, 2)$;
Constant on $(-\infty, -2)$ and $(2, \infty)$

131.

Increasing on $(-\infty, 3)$;
Decreasing on $(3, \infty)$

97. (a) $5x - 4y - 23 = 0$ (b) $4x + 5y - 2 = 0$
99. $V = 850t + 12{,}500, 0 \leq t \leq 5$
101. \$210,000
103. (a) Not a function because 20 in the domain corresponds to two values in the range.
(b) Function
(c) Function
(d) Not a function because 30 in A is not matched with any element in B.
105. No   **107.** Yes
109. (a) 5   (b) 17   (c) $t^4 + 1$ (d) $-x^2 - 1$

133. $f(x) = -3x$
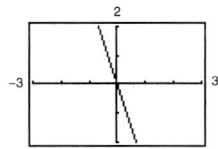

135. $f(x) = \frac{5}{3}x + \frac{10}{3}$

137.

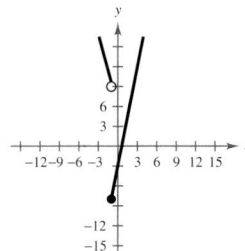

139. Neither even nor odd   **141.** Odd

143. Common function: $y = x^3$; Transformation: vertical shift of 4 units upward and horizontal shift of 4 units to the left

145. Common function: $y = |x|$; Transformation: horizontal shift of 6 units to the right

147. Vertical shift of 9 units downward

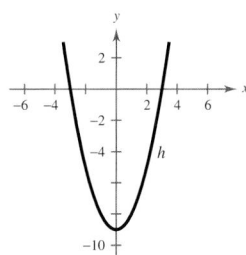

149. Horizontal shift of 7 units to the right

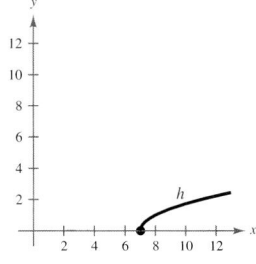

151. Reflection in x-axis, horizontal shift of 3 units to the left, and vertical shift of 1 unit upward

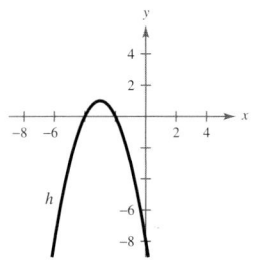

153. Reflection in x-axis, horizontal shift of 1 unit to the left, and vertical shift of 9 units upward

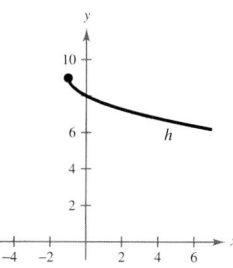

155. Reflection in x-axis, horizontal shift of 1 unit to the left, vertical shift of 3 units downward, and vertical stretch

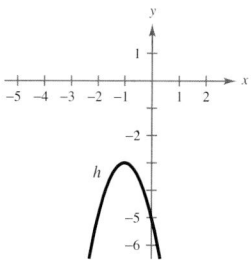

157. Reflection in x-axis, vertical stretch, and horizontal shift of 4 units to the left

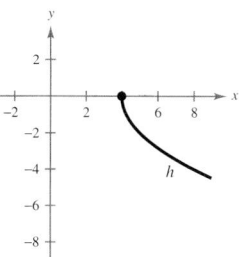

159. -7 **161.** 5   **163.** 23   **165.** $\sqrt{7}$

167. $y_1 = 0.80t^2 + 3.34t + 24.23$
$y_2 = -0.43t^2 + 18.14t + 62.89$

169. $f^{-1}(x) = \dfrac{x}{6}$

$f(f^{-1}(x)) = 6\left(\dfrac{x}{6}\right) = x$

$f^{-1}(f(x)) = \dfrac{6x}{6} = x$

171. $f^{-1}(x) = x + 7$

$f(f^{-1}(x)) = x + 7 - 7 = x$

$f^{-1}(f(x)) = x - 7 + 7 = x$

173. **175.**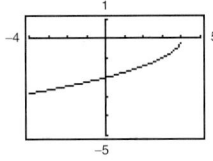

The function has an inverse.   The function has an inverse.

177.

The function does not have an inverse.

179. (a) $f^{-1}(x) = 2x + 6$

(b)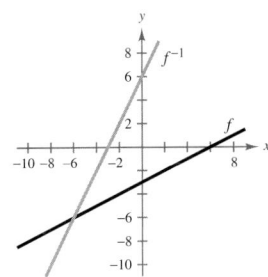

(c) $f^{-1}(f(x)) = 2(\frac{1}{2}x - 3) + 6 = x - 6 + 6 = x$
$f(f^{-1}(x)) = \frac{1}{2}(2x + 6) - 3 = x + 3 - 3 = x$

181. (a) $f^{-1}(x) = x^2 - 1, \ x \geq 0$

(b)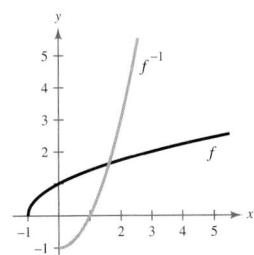

(c) $f^{-1}(f(x)) = f^{-1}(\sqrt{x + 1})$
$= (x + 1) - 1$
$= x$
$f(f^{-1}(x)) = f(x^2 - 1), \ x \geq 0$
$= \sqrt{x^2 - 1 + 1}$
$= x$

183. $x \geq 4; \ f^{-1}(x) = \sqrt{\dfrac{x}{2} + 4}$

185. False. $\dfrac{x^3 - 1}{x - 1} = x^2 + x + 1, \ x \neq 1$

187. False. The graph is reflected in the x-axis, shifted 9 units to the left, and then shifted 13 units downward.

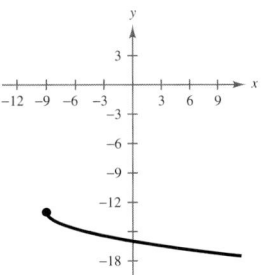

189. There is no solution. The absolute value of an expression is always nonnegative.

191. The Vertical Line Test is used to determine if the graph of y is a function of x. The Horizontal Line Test is used to determine if a function has an inverse function.

Chapter Test (page 117)

1. $-\dfrac{10}{3} > -|-4|$ **2.** 9.15

3.   **4.**

5.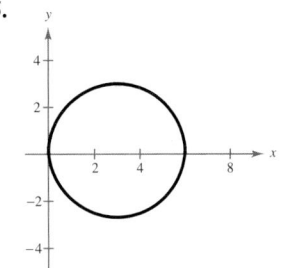

6. $\dfrac{128}{11}$ **7.** $-4, 5$ **8.** 4   **9.** $-2, \dfrac{8}{3}$

10. (a) $4x - 7y + 44 = 0$ (b) $7x + 4y - 53 = 0$

11. (a) -9 (b) 1   (c) $|x - 4| - 15$

12. $-10 \leq x \leq 10$ **13.** All real numbers

14. (a)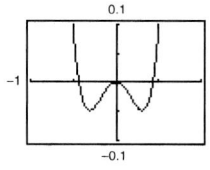

(b) Increasing on $(-0.31, 0), (0.31, \infty)$

Decreasing on $(-\infty, -0.31), (0, 0.31)$

(c) Even

15. (a)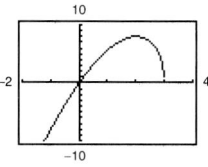

(b) Increasing on $(-\infty, 2)$

Decreasing on $(2, 3)$

(c) Neither even nor odd

16. (a)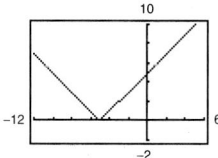

(b) Increasing on $(-5, \infty)$

Decreasing on $(-\infty, -5)$

(c) Neither even nor odd

17.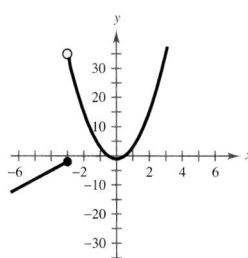

18. Common function: $y = x^3$

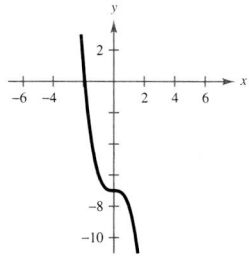

19. Common function: $y = \sqrt{x}$

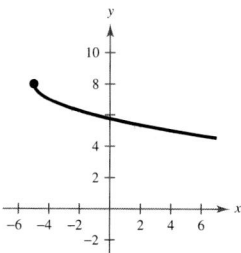

20. Common function: $y = |x|$

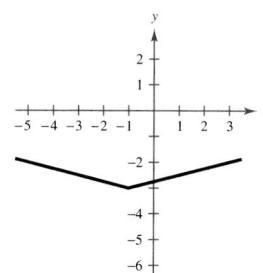

21. -2 **22.** -35 **23.** 5 **24.** $f^{-1}(x) = \sqrt[3]{x - 8}$

25. No inverse   **26.** $f^{-1}(x) = \left(\frac{8}{3}x\right)^{2/3}, x \geq 0$ **27.** $153

Chapter 1
Section 1.1 *(page 127)*

1. 2   **3.** 3   **5.** 1

7. (a) Quadrant I   (b) Quadrant III

9. (a) Quadrant IV   (b) Quadrant II

11. (a) Quadrant III   (b) Quadrant II

13. (a)   (b)

15. (a)   (b)

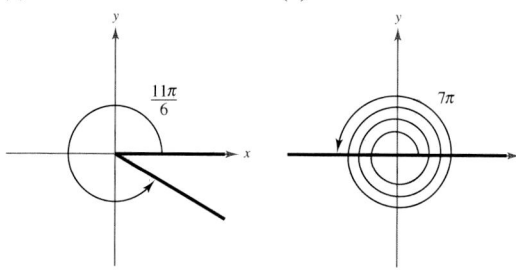

Answers to Odd-Numbered Exercises and Tests **A51**

17. (a) $\frac{13\pi}{6}, -\frac{11\pi}{6}$ (b) $\frac{17\pi}{6}, -\frac{7\pi}{6}$

19. (a) $\frac{8\pi}{3}, -\frac{4\pi}{3}$ (b) $\frac{25\pi}{12}, -\frac{23\pi}{12}$

21. (a) Complement: $\frac{\pi}{6}$; Supplement: $\frac{2\pi}{3}$

    (b) Complement: none; Supplement: $\frac{\pi}{4}$

23. (a) Complement: $\frac{\pi}{2} - 1 \approx 0.57$;

    Supplement: $\pi - 1 \approx 2.14$

    (b) Complement: none; Supplement: $\pi - 2 \approx 1.14$

25. (a) $\frac{\pi}{6}$ (b) $\frac{5\pi}{6}$ 27. (a) $-\frac{\pi}{9}$ (b) $-\frac{4\pi}{3}$

29. 2.007   31. -3.776 33. 9.285   35. -0.014
37. (a) 270°   (b) 210°   39. (a) 420°   (b) $-66°$
41. 25.714°   43. 337.5°   45. $-756°$
47. $-114.592°$ 49. 210°   51. $-60°$ 53. 165°
55. (a) Quadrant II   (b) Quadrant IV
57. (a) Quadrant III   (b) Quadrant I
59. (a)   (b)

61. (a)   (b)

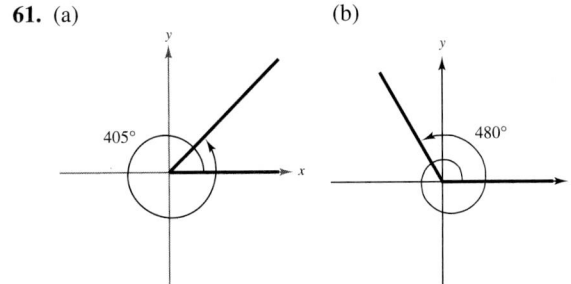

63. (a) 405°, $-315°$ (b) 324°, $-396°$
65. (a) 600°, $-120°$ (b) 180°, $-540°$
67. (a) Complement: 72°; Supplement: 162°
    (b) Complement: none; Supplement: 65°
69. (a) Complement: 11°; Supplement: 101°
    (b) Complement: none; Supplement: 30°

71. (a) 54.75°   (b) $-128.5°$
73. (a) 85.308°   (b) 330.007°
75. (a) 240° 36′   (b) $-145°$ 48′
77. (a) 2° 30′   (b) $-3°$ 34′ 48″   79. $\frac{6}{5}$ radians
81. $\frac{32}{7}$ radians   83. $\frac{2}{9}$ radian   85. $\frac{50}{29}$ radians
87. 15π inches ≈ 47.12 inches   89. 3 meters
91. 591.72 miles   93. 1141.02 miles
95. 0.063 radian $\approx 3.59°$ 97. $\frac{5}{12}$ radian
99. (a) 728.3 revolutions per minute
    (b) 4576 radians per minute
101. 20.16π inches per second
103. False. A measurement of 4π radians corresponds to two complete revolutions from the initial to the terminal side of an angle.
105. (a) The vertex is at the origin and the initial side is on the positive x-axis.
    (b) Clockwise rotation of the terminal side
    (c) Two angles in standard position where the terminal sides coincide
    (d) The magnitude of the angle is between 90° and 180°.
107. Radian. 1 radian $\approx 57.3°$ 109. Answers will vary.
111.   113.

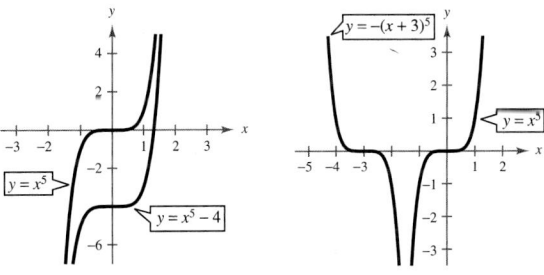

115. $\frac{2\sqrt{3}}{3}$ 117. $\frac{5\sqrt{2}}{4}$ 119. $6\sqrt{13}$ 121. $4\sqrt{13}$

Section 1.2 (page 136)

1. $\sin\theta = \frac{15}{17}$ 3. $\sin\theta = -\frac{5}{13}$
   $\cos\theta = -\frac{8}{17}$ $\cos\theta = \frac{12}{13}$
   $\tan\theta = -\frac{15}{8}$ $\tan\theta = -\frac{5}{12}$
   $\csc\theta = \frac{17}{15}$ $\csc\theta = -\frac{13}{5}$
   $\sec\theta = -\frac{17}{8}$ $\sec\theta = \frac{13}{12}$
   $\cot\theta = -\frac{8}{15}$ $\cot\theta = -\frac{12}{5}$

5. $\left(\frac{\sqrt{2}}{2}, \frac{\sqrt{2}}{2}\right)$ 7. $\left(-\frac{\sqrt{3}}{2}, -\frac{1}{2}\right)$ 9. $\left(-\frac{1}{2}, -\frac{\sqrt{3}}{2}\right)$

11. $(0, -1)$

13. $\sin\dfrac{\pi}{4} = \dfrac{\sqrt{2}}{2}$

$\cos\dfrac{\pi}{4} = \dfrac{\sqrt{2}}{2}$

$\tan\dfrac{\pi}{4} = 1$

15. $\sin\left(-\dfrac{\pi}{6}\right) = -\dfrac{1}{2}$

$\cos\left(-\dfrac{\pi}{6}\right) = \dfrac{\sqrt{3}}{2}$

$\tan\left(-\dfrac{\pi}{6}\right) = -\dfrac{\sqrt{3}}{3}$

17. $\sin\left(-\dfrac{7\pi}{4}\right) = \dfrac{\sqrt{2}}{2}$

$\cos\left(-\dfrac{7\pi}{4}\right) = \dfrac{\sqrt{2}}{2}$

$\tan\left(-\dfrac{7\pi}{4}\right) = 1$

19. $\sin\dfrac{11\pi}{6} = -\dfrac{1}{2}$

$\cos\dfrac{11\pi}{6} = \dfrac{\sqrt{3}}{2}$

$\tan\dfrac{11\pi}{6} = -\dfrac{\sqrt{3}}{3}$

21. $\sin\left(-\dfrac{3\pi}{2}\right) = 1$

$\cos\left(-\dfrac{3\pi}{2}\right) = 0$

$\tan\left(-\dfrac{3\pi}{2}\right)$ is undefined.

23. $\sin\dfrac{3\pi}{4} = \dfrac{\sqrt{2}}{2}$

$\cos\dfrac{3\pi}{4} = -\dfrac{\sqrt{2}}{2}$

$\tan\dfrac{3\pi}{4} = -1$

$\csc\dfrac{3\pi}{4} = \sqrt{2}$

$\sec\dfrac{3\pi}{4} = -\sqrt{2}$

$\cot\dfrac{3\pi}{4} = -1$

25. $\sin\dfrac{\pi}{2} = 1$

$\cos\dfrac{\pi}{2} = 0$

$\tan\dfrac{\pi}{2}$ is undefined.

$\csc\dfrac{\pi}{2} = 1$

$\sec\dfrac{\pi}{2}$ is undefined.

$\cot\dfrac{\pi}{2} = 0$

27. $\sin\left(-\dfrac{\pi}{3}\right) = -\dfrac{\sqrt{3}}{2}$

$\cos\left(-\dfrac{\pi}{3}\right) = \dfrac{1}{2}$

$\tan\left(-\dfrac{\pi}{3}\right) = -\sqrt{3}$

$\csc\left(-\dfrac{\pi}{3}\right) = -\dfrac{2\sqrt{3}}{3}$

$\sec\left(-\dfrac{\pi}{3}\right) = 2$

$\cot\left(-\dfrac{\pi}{3}\right) = -\dfrac{\sqrt{3}}{3}$

29. $\sin 5\pi = \sin \pi = 0$ **31.** $\cos\dfrac{8\pi}{3} = \cos\dfrac{2\pi}{3} = -\dfrac{1}{2}$

33. $\cos(-3\pi) = \cos(-\pi) = -1$

35. $\sin\left(-\dfrac{9\pi}{4}\right) = \sin\dfrac{7\pi}{4} = -\dfrac{\sqrt{2}}{2}$

37. (a) $-\dfrac{1}{3}$ (b) -3 **39.** (a) $-\dfrac{1}{5}$ (b) -5

41. (a) $\dfrac{4}{5}$ (b) $-\dfrac{4}{5}$ **43.** 0.7071   **45.** 1.0378

47. -0.1288 **49.** 1.3940   **51.** -1.4486

53. (a) -1 (b) -0.4

55. (a) 0.25, 2.89   (b) 1.82, 4.46

57. $0.0707 = \cos 1.5 \neq 2\cos 0.75 = 1.4634$

59. (a) 0.2500 foot   (b) 0.0177 foot   (c) -0.2475 foot

61. False. $\sin(-t) = -\sin t$ means that the function is odd, not that the sine of a negative angle is a negative number.

63. (a) y-axis   (b) $\sin t_1 = \sin(\pi - t_1)$

(c) $\cos(\pi - t_1) = -\cos t_1$

65. Answers will vary.   **67.** Even

69. $f^{-1}(x) = \sqrt[3]{4(x-1)}$ **71.** $f^{-1}(x) = \dfrac{x}{2-x}$, $x < 2$

Section 1.3 *(page 145)*

1. $\sin\theta = \dfrac{3}{5}$

$\cos\theta = \dfrac{4}{5}$

$\tan\theta = \dfrac{3}{4}$

$\csc\theta = \dfrac{5}{3}$

$\sec\theta = \dfrac{5}{4}$

$\cot\theta = \dfrac{4}{3}$

3. $\sin\theta = \dfrac{9}{41}$

$\cos\theta = \dfrac{40}{41}$

$\tan\theta = \dfrac{9}{40}$

$\csc\theta = \dfrac{41}{9}$

$\sec\theta = \dfrac{41}{40}$

$\cot\theta = \dfrac{40}{9}$

5. $\sin\theta = \dfrac{1}{3}$ $\csc\theta = 3$

$\cos\theta = \dfrac{2\sqrt{2}}{3}$ $\sec\theta = \dfrac{3\sqrt{2}}{4}$

$\tan\theta = \dfrac{\sqrt{2}}{4}$ $\cot\theta = 2\sqrt{2}$

The triangles are similar, and corresponding sides are proportional.

7. $\sin\theta = \dfrac{3}{5}$ $\csc\theta = \dfrac{5}{3}$

$\cos\theta = \dfrac{4}{5}$ $\sec\theta = \dfrac{5}{4}$

$\tan\theta = \dfrac{3}{4}$ $\cot\theta = \dfrac{4}{3}$

The triangles are similar, and corresponding sides are proportional.

9.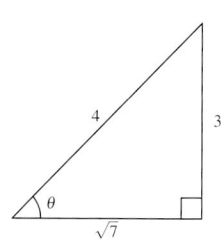

$\cos\theta = \dfrac{\sqrt{7}}{4}$

$\tan\theta = \dfrac{3\sqrt{7}}{7}$

$\csc\theta = \dfrac{4}{3}$

$\sec\theta = \dfrac{4\sqrt{7}}{7}$

$\cot\theta = \dfrac{\sqrt{7}}{3}$

11. 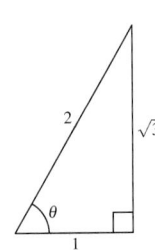 $\sin\theta = \dfrac{\sqrt{3}}{2}$

$\cos\theta = \dfrac{1}{2}$

$\tan\theta = \sqrt{3}$

$\csc\theta = \dfrac{2\sqrt{3}}{3}$

$\cot\theta = \dfrac{\sqrt{3}}{3}$

13. 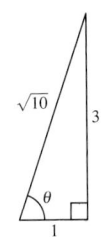 $\sin\theta = \dfrac{3\sqrt{10}}{10}$

$\cos\theta = \dfrac{\sqrt{10}}{10}$

$\csc\theta = \dfrac{\sqrt{10}}{3}$

$\sec\theta = \sqrt{10}$

$\cot\theta = \dfrac{1}{3}$

15. 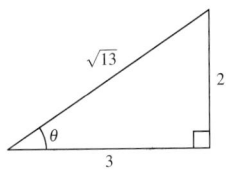 $\sin\theta = \dfrac{2\sqrt{13}}{13}$

$\cos\theta = \dfrac{3\sqrt{13}}{13}$

$\tan\theta = \dfrac{2}{3}$

$\csc\theta = \dfrac{\sqrt{13}}{2}$

$\sec\theta = \dfrac{\sqrt{13}}{3}$

17. (a) $\sqrt{3}$ (b) $\dfrac{1}{2}$ (c) $\dfrac{\sqrt{3}}{2}$ (d) $\dfrac{\sqrt{3}}{3}$

19. (a) $\dfrac{2\sqrt{13}}{13}$ (b) $\dfrac{3\sqrt{13}}{13}$ (c) $\dfrac{2}{3}$ (d) $\dfrac{\sqrt{13}}{2}$

21. (a) 3 (b) $\dfrac{2\sqrt{2}}{3}$ (c) $\dfrac{\sqrt{2}}{4}$ (d) $\dfrac{1}{3}$

23. (a) $\dfrac{1}{2}$ (b) 2 (c) $\sqrt{3}$

25. (a) $\dfrac{\sqrt{2}}{2}$ (b) $\dfrac{\sqrt{3}}{2}$ (c) $\dfrac{\sqrt{3}}{3}$

27. (a) 0.1736 (b) 0.1736   **29.** (a) 0.2815 (b) 3.5523

31. (a) 1.3499 (b) 1.3432   **33.** (a) 5.0273 (b) 0.1989

35. (a) 1.8527 (b) 0.9817

37. (a) $30° = \dfrac{\pi}{6}$ (b) $30° = \dfrac{\pi}{6}$

39. (a) $60° = \dfrac{\pi}{3}$ (b) $45° = \dfrac{\pi}{4}$

41. (a) $60° = \dfrac{\pi}{3}$ (b) $45° = \dfrac{\pi}{4}$

43. (a) $0.83° \approx 0.015$ (b) $27° \approx 0.474$

45. (a) $0.72° \approx 0.012$ (b) $67° \approx 1.169$

47. $30\sqrt{3}$ **49.** $\dfrac{32\sqrt{3}}{3}$ **51.–59.** Answers will vary.

61. (a) 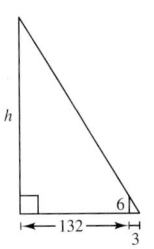   (b) $\dfrac{6}{3} = \dfrac{h}{135}$

(c) 270 feet

63. (a) 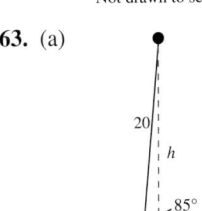   (b) $\sin 85° = \dfrac{h}{20}$

(c) 19.9 meters

65. 572 feet   **67.** $(x_1, y_1) = (28\sqrt{3},\ 28)$

$(x_2, y_2) = (28,\ 28\sqrt{3})$

69. $\sin 20° \approx 0.34$ $\csc 20° \approx 2.92$

$\cos 20° \approx 0.94$ $\sec 20° \approx 1.06$

$\tan 20° \approx 0.36$ $\cot 20° \approx 2.75$

71. True, $\csc x = \dfrac{1}{\sin x}$ **73.** False, $\dfrac{\sqrt{2}}{2} + \dfrac{\sqrt{2}}{2} \neq 1$

75. False, $1.7321 \neq 0.0349$

77. Corresponding sides of similar triangles are proportional.

79. (a)

θ	0.1	0.2	0.3	0.4	0.5
$\sin\theta$	0.0998	0.1987	0.2955	0.3894	0.4794

(b) As θ approaches 0, $\sin\theta$ approaches θ.

81. $\dfrac{x}{x-2},\ x \neq \pm 6$ **83.** $\dfrac{2(x^2 - 5x - 10)}{(x-2)(x+2)^2}$

85. $x = \dfrac{11 \pm \sqrt{409}}{6}$ **87.** $x = \dfrac{2}{3}$

Section 1.4 (page 155)

1. (a) $\sin\theta = \frac{3}{5}$ (b) $\sin\theta = -\frac{15}{17}$
$\cos\theta = \frac{4}{5}$ $\cos\theta = \frac{8}{17}$
$\tan\theta = \frac{3}{4}$ $\tan\theta = -\frac{15}{8}$
$\csc\theta = \frac{5}{3}$ $\csc\theta = -\frac{17}{15}$
$\sec\theta = \frac{5}{4}$ $\sec\theta = \frac{17}{8}$
$\cot\theta = \frac{4}{3}$ $\cot\theta = -\frac{8}{15}$

3. (a) $\sin\theta = -\frac{1}{2}$ (b) $\sin\theta = \frac{\sqrt{17}}{17}$
$\cos\theta = -\frac{\sqrt{3}}{2}$ $\cos\theta = -\frac{4\sqrt{17}}{17}$
$\tan\theta = \frac{\sqrt{3}}{3}$ $\tan\theta = -\frac{1}{4}$
$\csc\theta = -2$ $\csc\theta = \sqrt{17}$
$\sec\theta = -\frac{2\sqrt{3}}{3}$ $\sec\theta = -\frac{\sqrt{17}}{4}$
$\cot\theta = \sqrt{3}$ $\cot\theta = -4$

5. $\sin\theta = \frac{24}{25}$ $\csc\theta = \frac{25}{24}$
$\cos\theta = \frac{7}{25}$ $\sec\theta = \frac{25}{7}$
$\tan\theta = \frac{24}{7}$ $\cot\theta = \frac{7}{24}$

7. $\sin\theta = \frac{5\sqrt{29}}{29}$ $\csc\theta = \frac{\sqrt{29}}{5}$
$\cos\theta = -\frac{2\sqrt{29}}{29}$ $\sec\theta = -\frac{\sqrt{29}}{2}$
$\tan\theta = -\frac{5}{2}$ $\cot\theta = -\frac{2}{5}$

9. $\sin\theta = \frac{68\sqrt{5849}}{5849} \approx 0.9$ $\csc\theta = \frac{\sqrt{5849}}{68} \approx 1.1$
$\cos\theta = -\frac{35\sqrt{5849}}{5849} \approx -0.5$ $\sec\theta = -\frac{\sqrt{5849}}{35} \approx -2.2$
$\tan\theta = -\frac{68}{35} \approx -1.9$ $\cot\theta = -\frac{35}{68} \approx -0.5$

11. Quadrant III **13.** Quadrant II

15. $\sin\theta = \frac{3}{5}$ $\csc\theta = \frac{5}{3}$
$\cos\theta = -\frac{4}{5}$ $\sec\theta = -\frac{5}{4}$
$\tan\theta = -\frac{3}{4}$ $\cot\theta = -\frac{4}{3}$

17. $\sin\theta = -\frac{15}{17}$ $\csc\theta = -\frac{17}{15}$
$\cos\theta = \frac{8}{17}$ $\sec\theta = \frac{17}{8}$
$\tan\theta = -\frac{15}{8}$ $\cot\theta = -\frac{8}{15}$

19. $\sin\theta = -\frac{\sqrt{10}}{10}$ $\csc\theta = -\sqrt{10}$
$\cos\theta = \frac{3\sqrt{10}}{10}$ $\sec\theta = \frac{\sqrt{10}}{3}$
$\tan\theta = -\frac{1}{3}$ $\cot\theta = -3$

21. $\sin\theta = \frac{\sqrt{3}}{2}$ $\csc\theta = \frac{2\sqrt{3}}{3}$
$\cos\theta = -\frac{1}{2}$ $\sec\theta = -2$
$\tan\theta = -\sqrt{3}$ $\cot\theta = -\frac{\sqrt{3}}{3}$

23. $\sin\theta = 0$ $\csc\theta$ is undefined.
$\cos\theta = -1$ $\sec\theta = -1$
$\tan\theta = 0$ $\cot\theta$ is undefined.

25. $\sin\theta = \frac{\sqrt{2}}{2}$ $\csc\theta = \sqrt{2}$
$\cos\theta = -\frac{\sqrt{2}}{2}$ $\sec\theta = -\sqrt{2}$
$\tan\theta = -1$ $\cot\theta = -1$

27. $\sin\theta = -\frac{2\sqrt{5}}{5}$ $\csc\theta = -\frac{\sqrt{5}}{2}$
$\cos\theta = -\frac{\sqrt{5}}{5}$ $\sec\theta = -\sqrt{5}$
$\tan\theta = 2$ $\cot\theta = \frac{1}{2}$

29. -1 **31.** -1 **33.** Undefined **35.** 0

37. $\theta' = 23°$ **39.** $\theta' = 65°$

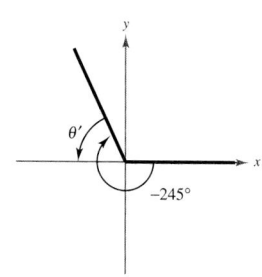

41. $\theta' = \frac{\pi}{3}$ **43.** $\theta' = 3.5 - \pi$

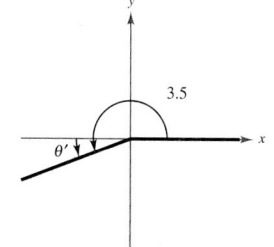

45. $\sin 225° = -\dfrac{\sqrt{2}}{2}$ **47.** $\sin 750° = \dfrac{1}{2}$

$\cos 225° = -\dfrac{\sqrt{2}}{2}$ $\cos 750° = \dfrac{\sqrt{3}}{2}$

$\tan 225° = 1$ $\tan 750° = \dfrac{\sqrt{3}}{3}$

49. $\sin(-150°) = -\dfrac{1}{2}$ **51.** $\sin \dfrac{4\pi}{3} = -\dfrac{\sqrt{3}}{2}$

$\cos(-150°) = -\dfrac{\sqrt{3}}{2}$ $\cos \dfrac{4\pi}{3} = -\dfrac{1}{2}$

$\tan(-150°) = \dfrac{\sqrt{3}}{3}$ $\tan \dfrac{4\pi}{3} = \sqrt{3}$

53. $\sin\left(-\dfrac{\pi}{6}\right) = -\dfrac{1}{2}$ **55.** $\sin \dfrac{11\pi}{4} = \dfrac{\sqrt{2}}{2}$

$\cos\left(-\dfrac{\pi}{6}\right) = \dfrac{\sqrt{3}}{2}$ $\cos \dfrac{11\pi}{4} = -\dfrac{\sqrt{2}}{2}$

$\tan\left(-\dfrac{\pi}{6}\right) = -\dfrac{\sqrt{3}}{3}$ $\tan \dfrac{11\pi}{4} = -1$

57. $\sin\left(-\dfrac{3\pi}{2}\right) = 1$ **59.** 0.1736 **61.** -0.3420

$\cos\left(-\dfrac{3\pi}{2}\right) = 0$

$\tan\left(-\dfrac{3\pi}{2}\right)$ is undefined.

63. 4.6373 **65.** 0.3640 **67.** -0.6052

69. (a) $30° = \dfrac{\pi}{6}$, $150° = \dfrac{5\pi}{6}$ (b) $210° = \dfrac{7\pi}{6}$, $330° = \dfrac{11\pi}{6}$

71. (a) $60° = \dfrac{\pi}{3}$, $120° = \dfrac{2\pi}{3}$ (b) $135° = \dfrac{3\pi}{4}$, $315° = \dfrac{7\pi}{4}$

73. (a) $45° = \dfrac{\pi}{4}$, $225° = \dfrac{5\pi}{4}$ (b) $150° = \dfrac{5\pi}{6}$, $330° = \dfrac{11\pi}{6}$

75. 54.99°, 125.01° **77.** 115.89°, 244.11°

79. 0.175, 6.109 **81.** 0.873, 4.014 **83.** 1.955, 4.328

85. $\dfrac{4}{5}$ **87.** $-\dfrac{\sqrt{13}}{2}$ **89.** $\dfrac{8}{5}$

91. (a) 25.2°F
(b) 65.1°F
(c) 50.8°F

93. (a) 2 centimeters
(b) 0.14 centimeter
(c) -1.98 centimeters

95. (a) 156.25 feet (b) 239.39 feet (c) 270.63 feet

97. False. In each of the four quadrants, the signs of the secant function and cosine function will be the same, because these functions are reciprocals of each other.

99. As θ increases from 0° to 90°, x decreases from 12 cm to 0 cm and y increases from 0 cm to 12 cm. Therefore, $\sin\theta = y/12$ increases from 0 to 1 and $\cos\theta = x/12$ decreases from 1 to 0. Thus, $\tan\theta = y/x$ and increases without bound. When $\theta = 90°$, the tangent is undefined.

101. First, determine a positive coterminal angle. Then determine the trigonometric function of the reference angle and prefix the appropriate sign.

103. 8 **105.** $\dfrac{2}{3}$ **107.** $x + y - 10 = 0$

109. $10x + 18y - 49 = 0$

Section 1.5 (page 165)

1. Period: π **3.** Period: 4π **5.** Period: 6

Amplitude: 3 Amplitude: $\dfrac{5}{2}$ Amplitude: $\dfrac{1}{2}$

7. Period: 2π **9.** Period: $\dfrac{\pi}{5}$

Amplitude: 2 Amplitude: 3

11. Period: 3π **13.** Period: 1

Amplitude: $\dfrac{1}{2}$ Amplitude: $\dfrac{1}{4}$

15. g is a shift of f π units to the right.

17. g is a reflection of f in the x-axis.

19. The period of f is twice the period of g.

21. g is a shift of f 3 units upward.

23. The graph of g has twice the amplitude of the graph of f.

25. The graph of g is a horizontal shift of the graph of f π units to the right.

27.   **29.**

31. **33.**

35. **37.**

39.   **41.**

43.   **45.**

47.   **49.**

51.   **53.**

55.   **57.**

59.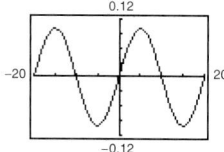

61. $a = 2, d = 1$ **63.** $a = -4, d = 4$

65. $a = -3, b = 2, c = 0$ **67.** $a = 2, b = 1, c = -\dfrac{\pi}{4}$

69.

$x = -\dfrac{\pi}{6}, -\dfrac{5\pi}{6}, \dfrac{7\pi}{6}, \dfrac{11\pi}{6}$

71.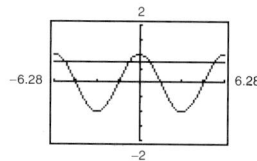

$x = \pm\dfrac{\pi}{4}, \pm\dfrac{7\pi}{4}$

73. (a) 6 seconds   (b) 10 cycles per minute

(c)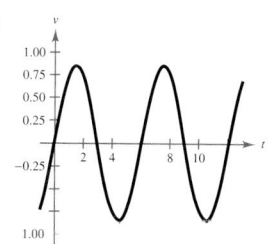

75. (a) $\dfrac{1}{440}$ second   (b) 440 cycles per second

77. (a) $C(t) = 56.35 + 27.35 \sin\left(\dfrac{\pi t}{6} + 4.19\right)$

(b)

The model is a good fit.

(c)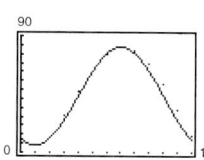

The model is a good fit.

(d) Honolulu: 84.40°; Chicago: 56.35°.
The constant term gives average annual temperature.

(e) 12. Yes. One full period is 1 year.

(f) Chicago; amplitude

79.

81. (a) and (c)

The model is a good fit.

(b) $y = \dfrac{1}{2} + \dfrac{1}{2}\sin\left[\dfrac{\pi}{15}(x - 76)\right]$ (d) 0

83. False. The function $y = \frac{1}{2}\cos 2x$ has an amplitude that is one-half that of $y = \cos x$. For $y = a \cos bx$, the amplitude is $|a|$.

85.

Vertical translations

87.

Horizontal shifts

89.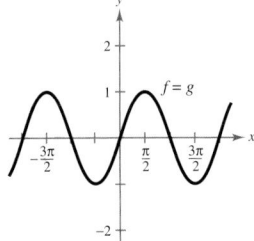

Conjecture: $\sin x = -\cos\left(x + \dfrac{\pi}{2}\right)$

91.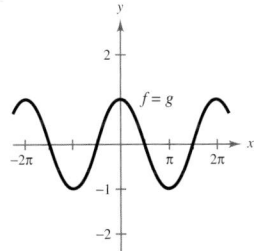

Conjecture: $\cos x = -\cos(x - \pi)$

93. (a) 0.4794, 0.4794   (b) 0.8417, 0.8415   (c) 0.5, 0.5
(d) 0.8776, 0.8776   (e) 0.5417, 0.5403
(f) 0.7074, 0.7071

The error increases as x moves farther away from 0.

95. (a) Even   (b) Even   **97.** $-\dfrac{20}{(x+5)(x-5)}$

99. $\dfrac{3x-5}{2(x-5)}$ **101.** $x \geq 3, x \neq 8$

103. All real numbers

Section 1.6 (page 176)

1. e, π **3.** a, 1   **5.** d, 2π

7.   **9.**

11.   **13.**

15.   **17.**

19.   **21.**

23.   **25.**

27.

29.   **31.**

33.   **35.**

37.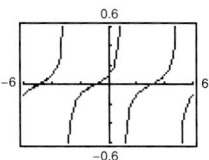

39. $-\dfrac{7\pi}{4}, -\dfrac{3\pi}{4}, \dfrac{\pi}{4}, \dfrac{5\pi}{4}$ **41.** $-\dfrac{4\pi}{3}, -\dfrac{\pi}{3}, \dfrac{2\pi}{3}, \dfrac{5\pi}{3}$

43. $-\dfrac{4\pi}{3}, -\dfrac{2\pi}{3}, \dfrac{2\pi}{3}, \dfrac{4\pi}{3}$ **45.** $-\dfrac{7\pi}{4}, -\dfrac{5\pi}{4}, \dfrac{\pi}{4}, \dfrac{3\pi}{4}$

47. Even

49. (a) 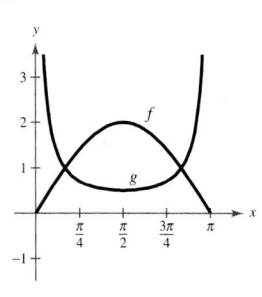   (b) $\dfrac{\pi}{6} < x < \dfrac{5\pi}{6}$

(c) Sine approaches 0 and cosecant approaches $+\infty$ because the cosecant is the reciprocal of the sine.

51.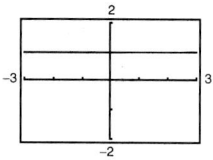

The expressions are equivalent except that when $x = 0$, y_1 is undefined.

53.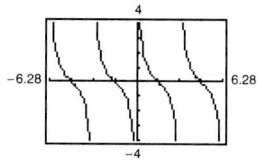

The expressions are equivalent.

55. d, $f \to 0$ as $x \to 0$.   **57.** b, $g \to 0$ as $x \to 0$.

59.   **61.**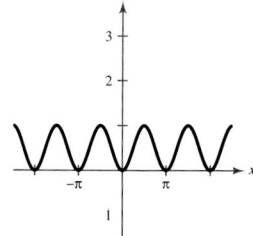

The functions are equal.   The functions are equal.

63.   **65.**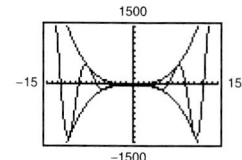

As $x \to \infty$, $f(x)$ oscillates and approaches $-\infty$ and ∞.

As $x \to \infty$, $g(x)$ oscillates and approaches $-\infty$ and ∞.

67.   **69.**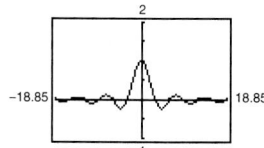

As $x \to 0$, $y \to \infty$.   As $x \to 0$, $g(x) \to 1$.

71.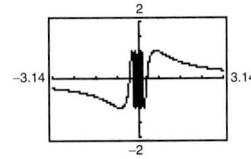

As $x \to 0$, $f(x)$ oscillates between 1 and -1.

73. $d = 7 \cot x$ **75.**

 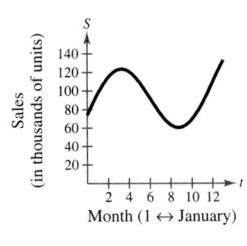

77. (a) 12
(b) Summer; winter
(c) 1 month

79. True. For a given value of x, the y-coordinate of $\csc x$ is the reciprocal of the y-coordinate of $\sin x$.

81. As x approaches $\pi/2$ from the left, f approaches ∞. As x approaches $\pi/2$ from the right, f approaches $-\infty$.

83. (a)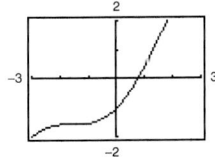

0.739

(b) 1, 0.5403, 0.8576, 0.6543, 0.7935, 0.7014, 0.7640, 0.7221, 0.7504, 0.7314, . . .

85.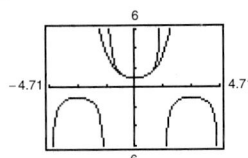

The graphs appear to coincide on the interval $-1.1 \le x \le 1.1$.

87. ± 8 **89.** $\frac{3}{2}$ **91.** $3 \pm \sqrt{5}$ **93.** $-\frac{10}{3}, 5$

Section 1.7 (page 186)

1. $\dfrac{\pi}{6}$ **3.** $\dfrac{\pi}{3}$ **5.** $\dfrac{\pi}{3}$ **7.** $\dfrac{5\pi}{6}$ **9.** $-\dfrac{\pi}{3}$ **11.** $\dfrac{2\pi}{3}$

13. $\dfrac{\pi}{3}$ **15.** 0   **17.** 1.29   **19.** -0.85 **21.** -1.25

23. 0.32   **25.** 1.99   **27.** 0.74   **29.** 0.85

31. 1.29   **33.** $-\dfrac{\pi}{3}, -\dfrac{\sqrt{3}}{3}, 1$

35.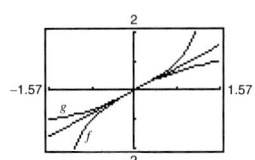

37. $\theta = \arctan \dfrac{x}{4}$ **39.** $\theta = \arcsin \dfrac{x+2}{5}$

41. $\theta = \arccos \dfrac{x+3}{2x}$ **43.** 0.3   **45.** -0.1 **47.** 0

49. $\dfrac{3}{5}$ **51.** $\dfrac{\sqrt{5}}{5}$ **53.** $\dfrac{12}{13}$ **55.** $\dfrac{\sqrt{34}}{5}$

57. $\dfrac{\sqrt{5}}{3}$ **59.** $\dfrac{1}{x}$ **61.** $\sqrt{1-4x^2}$

63. $\sqrt{1-x^2}$ **65.** $\dfrac{\sqrt{9-x^2}}{x}$ **67.** $\dfrac{\sqrt{x^2+2}}{x}$

69.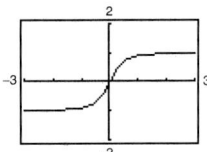

Asymptotes: $y = \pm 1$

71. $\dfrac{9}{\sqrt{x^2+81}}, x > 0; \dfrac{9}{\sqrt{x^2+81}}, x < 0$

73. $\dfrac{|x-1|}{\sqrt{x^2-2x+10}}$

75.   **77.**

79.   **81.**

83. **85.**

87.

89. $3\sqrt{2}\sin\left(2t + \dfrac{\pi}{4}\right)$

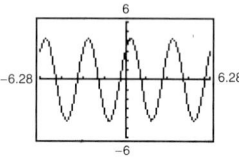

The graph implies that the identity is true.

91. (a) $\theta = \arcsin\dfrac{5}{s}$ (b) 0.13, 0.25

93. (a)

(b) 2 feet

(c) $\beta = 0$; As x increases, β approaches 0.

95. (a) $\theta = \arctan\dfrac{x}{20}$ (b) 0.24, 0.54

97. False. $\dfrac{5\pi}{4}$ is not in the range of the arctangent.

99. Domain: $(-\infty, \infty)$
Range: $(0, \pi)$

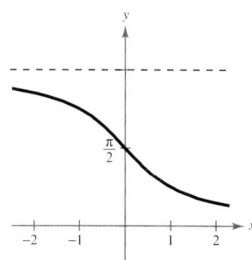

101. Domain: $(-\infty, -1] \cup [1, \infty)$
Range: $[-\pi/2, 0) \cup (0, \pi/2]$

103.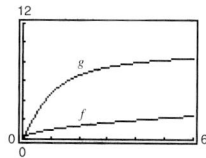

As x increases to infinity, g approaches 3π, but f has no maximum.

$a \approx 87.54$

105.–109. Answers will vary.

111. $\sin\theta = \dfrac{\sqrt{3}}{2}$ **113.** $\sin\theta = \dfrac{\sqrt{2}}{2}$

$\cos\theta = \dfrac{1}{2}$ $\cos\theta = \dfrac{\sqrt{2}}{2}$

$\tan\theta = \sqrt{3}$ $\tan\theta = 1$

$\csc\theta = \dfrac{2\sqrt{3}}{3}$ $\csc\theta = \sqrt{2}$

$\sec\theta = 2$ $\sec\theta = \sqrt{2}$

$\cot\theta = \dfrac{\sqrt{3}}{3}$ $\cot\theta = 1$

115.   **117.**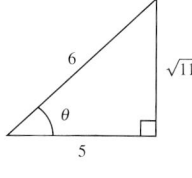

119. 8 people

Section 1.8 (page 196)

1. $a \approx 3.64$ **3.** $a \approx 8.26$ **5.** $c \approx 11.66$
$c \approx 10.64$ $c \approx 25.38$ $A \approx 30.96°$
$B = 70°$ $A = 19°$ $B \approx 59.04°$

7. $a \approx 49.48$
$A \approx 72.08°$
$B = 17.92°$

9. $a \approx 91.34$
$b \approx 420.70$
$B = 77°45'$

11. 2.56 inches

13. 19.99 inches **15.** 107.2 feet **17.** 19.7 feet

19. (a)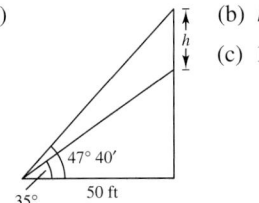
(b) $h = 50(\tan 47°\,40' - \tan 35°)$
(c) 19.9 feet

21. 2236.8 feet **23.** 56.3° **25.** 1.84° **27.** 5099 feet
29. 0.73 mile **31.** 554 miles north; 709 miles east
33. (a) N 58° E (b) 68.82 meters **35.** N 56.3° W
37. 1933.3 feet **39.** ≈ 3.23 miles or $\approx 17{,}054$ feet
41. 78.7° **43.** 35.3° **45.** 29.4 inches
47. $y = \sqrt{3}\,r$ **49.** $a \approx 12.2$, $b \approx 7$
51. (a) 4 (b) 4 (c) $\frac{1}{16}$
53. (a) $\frac{1}{16}$ (b) 60 (c) $\frac{1}{120}$ **55.** $d = 4\sin(\pi t)$
57. $d = 3\cos\left(\dfrac{4\pi t}{3}\right)$ **59.** $\omega = 528\pi$

61. (a)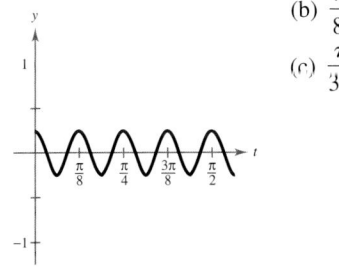
(b) $\dfrac{\pi}{8}$
(c) $\dfrac{\pi}{32}$

63. False. One period is the time for one complete cycle of the motion.

65. (a) and (b)

Base 1	Base 2	Altitude	Area
8	$8 + 16\cos 10°$	$8\sin 10°$	22.1
8	$8 + 16\cos 20°$	$8\sin 20°$	42.5
8	$8 + 16\cos 30°$	$8\sin 30°$	59.7
8	$8 + 16\cos 40°$	$8\sin 40°$	72.7
8	$8 + 16\cos 50°$	$8\sin 50°$	80.5
8	$8 + 16\cos 60°$	$8\sin 60°$	83.1
8	$8 + 16\cos 70°$	$8\sin 70°$	80.7

≈ 83.1 square feet when $\theta = 60°$

(c) $A = 64(1 + \cos\theta)(\sin\theta)$
(d)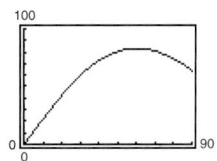

≈ 83.1 square feet when $\theta = 60°$

67. (a)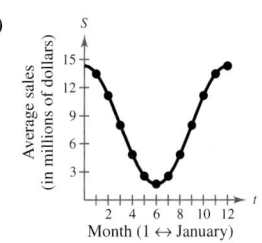

(b) $S = 8 + 6.3\cos\left(\dfrac{\pi}{6}t\right)$ or $S = 8 + 6.3\sin\left(\dfrac{\pi}{6}t + \dfrac{\pi}{2}\right)$

The model is a good fit.

(c) 12. Yes, sales of outerwear are seasonal.

(d) Maximum displacement from average sales of $8 million

69.   **71.**

73.   **75.**

77.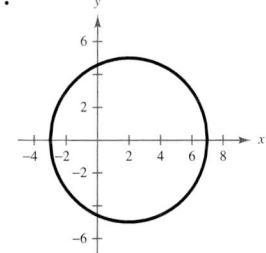

Review Exercises (page 202)

1. 0.5 **3.** 4.5

5.

$\frac{3\pi}{4}, -\frac{5\pi}{4}$

7.

$\frac{2\pi}{3}, -\frac{10\pi}{3}$

9.

430°, −290°

11.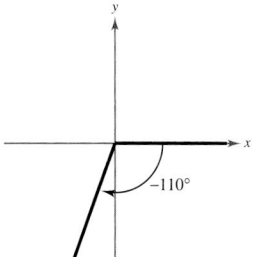

250°, −470°

13. 128.57° **15.** −200.54°
17. 8.3776 **19.** −0.5890
21. (a) $66\frac{2}{3}\pi$ radians per minute
(b) 400π inches per minute
23. $\left(-\frac{1}{2}, \frac{\sqrt{3}}{2}\right)$ **25.** $\left(-\frac{\sqrt{3}}{2}, \frac{1}{2}\right)$

27. $\sin\frac{7\pi}{6} = -\frac{1}{2}$ **29.** $\sin\left(-\frac{2\pi}{3}\right) = -\frac{\sqrt{3}}{2}$

$\cos\frac{7\pi}{6} = -\frac{\sqrt{3}}{2}$ $\cos\left(-\frac{2\pi}{3}\right) = -\frac{1}{2}$

$\tan\frac{7\pi}{6} = \frac{\sqrt{3}}{3}$ $\tan\left(-\frac{2\pi}{3}\right) = \sqrt{3}$

$\csc\frac{7\pi}{6} = -2$ $\csc\left(-\frac{2\pi}{3}\right) = -\frac{2\sqrt{3}}{3}$

$\sec\frac{7\pi}{6} = -\frac{2\sqrt{3}}{3}$ $\sec\left(-\frac{2\pi}{3}\right) = -2$

$\cot\frac{7\pi}{6} = \sqrt{3}$ $\cot\left(-\frac{2\pi}{3}\right) = \frac{\sqrt{3}}{3}$

31. $\sin\frac{11\pi}{4} = \sin\frac{3\pi}{4} = \frac{\sqrt{2}}{2}$

33. $\sin\left(-\frac{17\pi}{6}\right) = \sin\left(-\frac{5\pi}{6}\right) = -\frac{1}{2}$

35. −75.31 **37.** 3.24

39. $\sin\theta = \frac{4\sqrt{41}}{41}$ **41.** $\sin\theta = \frac{1}{2}$

$\cos\theta = \frac{5\sqrt{41}}{41}$ $\cos\theta = \frac{\sqrt{3}}{2}$

$\tan\theta = \frac{4}{5}$ $\tan\theta = \frac{\sqrt{3}}{3}$

$\csc\theta = \frac{\sqrt{41}}{4}$ $\csc\theta = 2$

$\sec\theta = \frac{\sqrt{41}}{5}$ $\sec\theta = \frac{2\sqrt{3}}{3}$

$\cot\theta = \frac{5}{4}$ $\cot\theta = \sqrt{3}$

43. (a) 3 (b) $\frac{2\sqrt{2}}{3}$ (c) $\frac{3\sqrt{2}}{4}$ (d) $\frac{\sqrt{2}}{4}$

45. (a) $\frac{1}{4}$ (b) $\frac{\sqrt{15}}{4}$ (c) $\frac{4\sqrt{15}}{15}$ (d) $\frac{\sqrt{15}}{15}$

47. 0.65 **49.** 0.56 **51.** 3.67 **53.** 0.07 kilometer

55. $\sin\theta = \frac{4}{5}$ $\csc\theta = \frac{5}{4}$

$\cos\theta = \frac{3}{5}$ $\sec\theta = \frac{5}{3}$

$\tan\theta = \frac{4}{3}$ $\cot\theta = \frac{3}{4}$

57. $\sin\theta = \frac{15\sqrt{241}}{241}$ $\csc\theta = \frac{\sqrt{241}}{15}$

$\cos\theta = \frac{4\sqrt{241}}{241}$ $\sec\theta = \frac{\sqrt{241}}{4}$

$\tan\theta = \frac{15}{4}$ $\cot\theta = \frac{4}{15}$

59. $\sin\theta = 1$ $\csc\theta = 1$

$\cos\theta = -0.1$ $\sec\theta = -9$

$\tan\theta = -9$ $\cot\theta = -0.1$

61. $\sin\theta = \frac{4\sqrt{17}}{17}$ $\csc\theta = \frac{\sqrt{17}}{4}$

$\cos\theta = \frac{\sqrt{17}}{17}$ $\sec\theta = \sqrt{17}$

$\tan\theta = 4$ $\cot\theta = \frac{1}{4}$

63. $\sin\theta = -\frac{\sqrt{11}}{6}$ **65.** $\cos\theta = -\frac{\sqrt{55}}{8}$

$\cos\theta = \frac{5}{6}$ $\tan\theta = -\frac{3\sqrt{55}}{55}$

$\tan\theta = -\frac{\sqrt{11}}{5}$ $\csc\theta = \frac{8}{3}$

$\csc\theta = -\frac{6\sqrt{11}}{11}$ $\sec\theta = -\frac{8\sqrt{55}}{55}$

$\cot\theta = -\frac{5\sqrt{11}}{11}$ $\cot\theta = -\frac{\sqrt{55}}{3}$

67. $\sin\theta = \dfrac{\sqrt{21}}{5}$

$\tan\theta = -\dfrac{\sqrt{21}}{2}$

$\csc\theta = \dfrac{5\sqrt{21}}{21}$

$\sec\theta = -\dfrac{5}{2}$

$\cot\theta = -\dfrac{2\sqrt{21}}{21}$

69. $\sqrt{3}$ **71.** $\dfrac{1}{2}$ **73.** $-\dfrac{\sqrt{2}}{2}$ **75.** -0.76

77. 0.06 **79.** 0 **81.** 3.24

83.

85.

87.

89.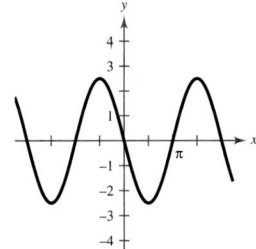

91. (a) $y = 2\sin 528\pi x$

(b) 264 cycles per second

93.

95.

97.

99.

101.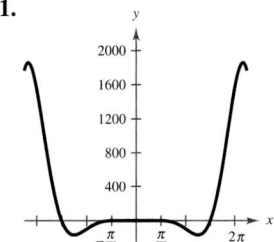

103. $-\dfrac{\pi}{6}$ **105.** 0.41

107. -0.46 **109.** $\dfrac{\pi}{6}$ **111.** π **113.** 1.24

115. 0.12 **117.** 1.40 **119.** -0.98 **121.** 0.72

123. 0 **125.** $\dfrac{4}{5}$ **127.** $\dfrac{13}{5}$ **129.** 66.8°

131. 1221 miles, N 85.6° E

133. False. The sine or cosine function is often useful for modeling simple harmonic motion.

135. False. For each θ there corresponds exactly one value of y.

137. d; the period is 2π and the amplitude is 3.

139. b; the period is 2 and the amplitude is 2.

141. Undefined because $\sec\theta = 1/\cos\theta$.

143. Their ranges are $(-\infty, \infty)$ or $(-\infty, -1] \cup [1, \infty)$.

145. (a)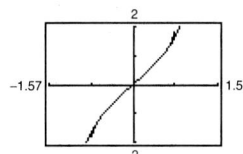

The approximation is accurate over the interval $-1 \le x \le 1$.

(b)

The approximation is accurate over the interval $-1 \le x \le 1$.

(c) $\dfrac{x^9}{9}$

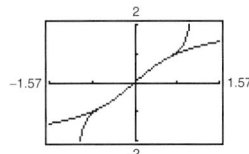

The accuracy improved.

Chapter Test (page 207)

1. (a)   (b) $\dfrac{13\pi}{4},\ -\dfrac{3\pi}{4}$

(c) $225°$

2. 3000 radians per minute

3. $\sin\theta = \dfrac{3\sqrt{10}}{10}$ $\csc\theta = \dfrac{\sqrt{10}}{3}$

$\cos\theta = -\dfrac{\sqrt{10}}{10}$ $\sec\theta = -\sqrt{10}$

$\tan\theta = -3$ $\cot\theta = -\dfrac{1}{3}$

4. For $0 \le \theta < \dfrac{\pi}{2}$:   For $\pi \le \theta < \dfrac{3\pi}{2}$:

$\sin\theta = \dfrac{3\sqrt{13}}{13}$ $\sin\theta = -\dfrac{3\sqrt{13}}{13}$

$\cos\theta = \dfrac{2\sqrt{13}}{13}$ $\cos\theta = -\dfrac{2\sqrt{13}}{13}$

$\csc\theta = \dfrac{\sqrt{13}}{3}$ $\csc\theta = -\dfrac{\sqrt{13}}{3}$

$\sec\theta = \dfrac{\sqrt{13}}{2}$ $\sec\theta = -\dfrac{\sqrt{13}}{2}$

$\cot\theta = \dfrac{2}{3}$ $\cot\theta = \dfrac{2}{3}$

5. $\theta' = 70°$

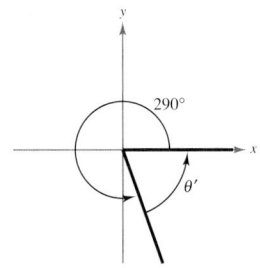

6. Quadrant III   **7.** $150°,\ 210°$ **8.** 1.33, 1.81

9. $\sin\theta = -\dfrac{4}{5}$ **10.** $\sin\theta = \dfrac{15}{17}$

$\tan\theta = -\dfrac{4}{3}$ $\cos\theta = -\dfrac{8}{17}$

$\csc\theta = -\dfrac{5}{4}$ $\tan\theta = -\dfrac{15}{8}$

$\sec\theta = \dfrac{5}{3}$ $\csc\theta = \dfrac{17}{15}$

$\cot\theta = -\dfrac{3}{4}$ $\cot\theta = -\dfrac{8}{15}$

11.   **12.**

13.   **14.**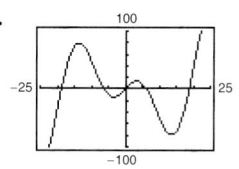

Period: 2

15. $y = -2\sin\left(\dfrac{x}{2} - \dfrac{\pi}{4}\right)$ **16.** $\dfrac{\sqrt{5}}{2}$

17.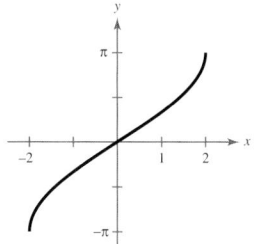

18. N 49.9° W   **19.** $d = -6\cos\pi t$

Chapter 2
Section 2.1 (page 215)

1. $\tan x = -\sqrt{3}$ **3.** $\cos\theta = \dfrac{\sqrt{2}}{2}$

$\csc x = \dfrac{2\sqrt{3}}{3}$ $\tan\theta = -1$

$\sec x = -2$ $\csc\theta = -\sqrt{2}$

$\cot x = -\dfrac{\sqrt{3}}{3}$ $\cot\theta = -1$

5. $\sin x = -\dfrac{5}{13}$
   $\cos x = -\dfrac{12}{13}$
   $\csc x = -\dfrac{13}{5}$
   $\cot x = \dfrac{12}{5}$

7. $\sin \phi = -\dfrac{\sqrt{5}}{3}$
   $\cos \phi = \dfrac{2}{3}$
   $\tan \phi = -\dfrac{\sqrt{5}}{2}$
   $\cot \phi = -\dfrac{2\sqrt{5}}{5}$

9. $\sin x = \dfrac{1}{3}$
   $\cos x = -\dfrac{2\sqrt{2}}{3}$
   $\csc x = 3$
   $\sec x = -\dfrac{3\sqrt{2}}{4}$
   $\cot x = -2\sqrt{2}$

11. $\sin \theta = -\dfrac{2\sqrt{5}}{5}$
    $\cos \theta = -\dfrac{\sqrt{5}}{5}$
    $\csc \theta = -\dfrac{\sqrt{5}}{2}$
    $\sec \theta = -\sqrt{5}$
    $\cot \theta = \dfrac{1}{2}$

13. $\cos \theta = 0$
    $\tan \theta$ is undefined.
    $\csc \theta = -1$
    $\sec \theta$ is undefined.

15. d   17. b   19. e   21. b   23. f   25. e
27. $\csc \theta$ 29. $\cos^2 \phi$ 31. $\cos x$ 33. $\sin^2 x$
35. 1   37. $\tan x$ 39. $1 + \sin y$ 41. $\sec \beta$
43. $\cos u + \sin u$ 45. $\sin^2 x$ 47. $\sin^2 x \tan^2 x$
49. $\sec x + 1$ 51. $\sec^4 x$ 53. $\sin^2 x - \cos^2 x$
55. $\cot^2 x(\csc x - 1)$ 57. $1 + 2\sin x \cos x$
59. $4\cot^2 x$ 61. $2\csc^2 x$ 63. $2\sec x$
65. $1 + \cos y$ 67. $3(\sec x + \tan x)$

69.

x	0.2	0.4	0.6	0.8	1.0
y_1	0.1987	0.3894	0.5646	0.7174	0.8415
y_2	0.1987	0.3894	0.5646	0.7174	0.8415

x	1.2	1.4
y_1	0.9320	0.9854
y_2	0.9320	0.9854

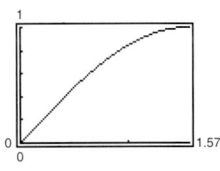

$y_1 = y_2$

71.

x	0.2	0.4	0.6	0.8	1.0
y_1	1.2230	1.5085	1.8958	2.4650	3.4082
y_2	1.2230	1.5085	1.8958	2.4650	3.4082

x	1.2	1.4
y_1	5.3319	11.6814
y_2	5.3319	11.6814

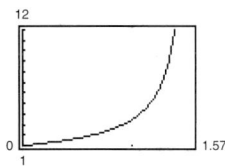

$y_1 = y_2$

73. $\csc x$ 75. $\tan x$ 77. $3 \sin \theta$ 79. $3 \tan \theta$
81. $5 \sec \theta$ 83. $3 \cos \theta = 3; \sin \theta = 0; \cos \theta = 1$
85. $4 \sin \theta = 2\sqrt{2}; \sin \theta = \dfrac{\sqrt{2}}{2}; \cos \theta = \dfrac{\sqrt{2}}{2}$
87. $0 \le \theta \le \pi$ 89. $0 \le \theta < \dfrac{\pi}{2}, \dfrac{3\pi}{2} < \theta < 2\pi$

91. (a) $\csc^2 132° - \cot^2 132° \approx 1.8107 - 0.8107 = 1$
    (b) $\csc^2 \dfrac{2\pi}{7} - \cot^2 \dfrac{2\pi}{7} \approx 1.6360 - 0.6360 = 1$

93. (a) $\cos(90° - 80°) = \sin 80° \approx 0.9848$
    (b) $\cos\left(\dfrac{\pi}{2} - 0.8\right) = \sin 0.8 \approx 0.7174$

95. (a) $\cot 25° = \dfrac{\cos 25°}{\sin 25°} \approx 2.1445$
    (b) $\cot \dfrac{\pi}{8} = \dfrac{\cos \pi/8}{\sin \pi/8} \approx 2.4142$

97. (a) $\tan(90° - 5°) = \cot 5° \approx 11.4301$
    (b) $\tan\left(\dfrac{\pi}{2} - \dfrac{11\pi}{12}\right) = \cot \dfrac{11\pi}{12} \approx -3.7321$

99. $\tan \theta$
101. False. A cofunction identity can be used to transform a tangent function so that it can be represented by a cotangent function.
103. 1, 1   105. 0, $-\infty$
107. Not an identity because $\cot \theta = \pm\sqrt{\csc^2 \theta - 1}$
109. Not an identity because $\dfrac{5}{\cos \theta} \ne \dfrac{1}{5 \cos \theta}$
111. Not an identity because θ and ϕ may not be equal
113. $\sin \theta = \pm\sqrt{1 - \cos^2 \theta}$

$\tan\theta = \pm\dfrac{\sqrt{1-\cos^2\theta}}{\cos\theta}$

$\csc\theta = \pm\dfrac{1}{\sqrt{1-\cos^2\theta}}$

$\sec\theta = \dfrac{1}{\cos\theta}$

$\cot\theta = \pm\dfrac{\cos\theta}{\sqrt{1-\cos^2\theta}}$

115. $\sqrt{5v}$ **117.** $10x(\sqrt{30}+5)$

119.   **121.**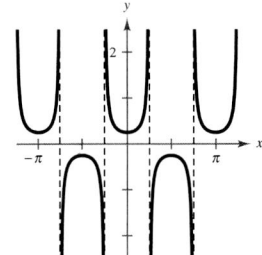

Section 2.2 (page 223)

1.–55. Answers will vary.   **57.** 1   **59.** 2

61. Answers will vary.

63. True. An identity is an equation that is true for all real values in the domain of the variable.

65. Not an identity because $\tan\theta = \pm\sqrt{\sec^2\theta - 1}$;

Possible answer: $\dfrac{3\pi}{4}$

67. Not an identity because $\sqrt{\sin^2 x + \cos^2 x} \neq \sin x + \cos x$ (the left side is 1 for any x, but the right side is not necessarily 1);

Possible answer: $\dfrac{\pi}{4}$

69. Answers will vary.   **71.** $4, -9$ **73.** $5, -\tfrac{3}{2}$

75. $\pm\tfrac{1}{3}$ **77.** $\dfrac{\pi}{3}, 60°$ **79.** $\dfrac{\pi}{6}, 30°$ **81.** $\dfrac{2\pi}{3}, 120°$

Section 2.3 (page 232)

1.–5. Answers will vary.   **7.** $\dfrac{2\pi}{3} + 2n\pi, \dfrac{4\pi}{3} + 2n\pi$

9. $\dfrac{\pi}{3} + 2n\pi, \dfrac{2\pi}{3} + 2n\pi$ **11.** $\dfrac{\pi}{6} + n\pi, \dfrac{5\pi}{6} + n\pi$

13. $n\pi, \dfrac{3\pi}{2} + 2n\pi$ **15.** $\dfrac{\pi}{3} + n\pi, \dfrac{2\pi}{3} + n\pi$

17. $\dfrac{\pi}{8} + n\pi, \dfrac{3\pi}{8} + n\pi, \dfrac{5\pi}{8} + n\pi, \dfrac{7\pi}{8} + n\pi$

19. $\dfrac{n\pi}{3}, \dfrac{\pi}{4} + n\pi$ **21.** $0, \dfrac{\pi}{2}, \pi, \dfrac{3\pi}{2}$

23. $0, \pi, \dfrac{\pi}{6}, \dfrac{5\pi}{6}, \dfrac{7\pi}{6}, \dfrac{11\pi}{6}$ **25.** $\dfrac{\pi}{3}, \dfrac{5\pi}{3}, \pi$

27. No solution   **29.** $\pi, \dfrac{\pi}{3}, \dfrac{5\pi}{3}$ **31.** $\dfrac{\pi}{6}, \dfrac{5\pi}{6}, \dfrac{7\pi}{6}, \dfrac{11\pi}{6}$

33. $\dfrac{\pi}{6} + n\pi, \dfrac{5\pi}{6} + n\pi$ **35.** $\dfrac{\pi}{12} + \dfrac{n\pi}{3}$

37. $\dfrac{\pi}{2} + 4n\pi, \dfrac{7\pi}{2} + 4n\pi$ **39.** $-1, 3$ **41.** ± 2

43. $\tfrac{2}{3}, \tfrac{3}{2}$; $0.8411 + 2n\pi, 5.4421 + 2n\pi$

45. $2.6779, 5.8195$ **47.** $1.0472, 5.2360$

49. $0.8603, 3.4256$ **51.** $0, 2.6779, 3.1416, 5.8195$

53. $0.9828, 1.7682, 4.1244, 4.9098$

55. $0.3398, 0.8481, 2.2935, 2.8018$

57. $1.9357, 2.7767, 5.0773, 5.9183$

59. $\dfrac{\pi}{4}, \dfrac{5\pi}{4}$, arctan 5, arctan $5 + \pi$ **61.** $\dfrac{\pi}{3}, \dfrac{5\pi}{3}$

63. (a)   (b) $\dfrac{\pi}{4}, \dfrac{5\pi}{4}$

Maximum: $\left(\dfrac{\pi}{4}, \sqrt{2}\right)$

Minimum: $\left(\dfrac{5\pi}{4}, -\sqrt{2}\right)$

65. 1

67. (a) All real numbers except $x = 0$

(b) y-axis symmetry; horizontal asymptote: $y = 1$

(c) Oscillates

(d) Infinitely many solutions

(e) Yes, 0.6366

69. 0.04 second, 0.43 second, 0.83 second

71. February, March, and April   **73.** $1.9°$

75. (a) $\tfrac{10}{3}$

(b)

For $3.5 \leq x \leq 6$, the approximation appears to be good.

(c) $3.46, 8.81$

3.46 is close to the zero of f on the interval $[0, 6]$.

77. True. The first equation has a smaller period than the second equation, so it will have more solutions in the interval $[0, 2\pi)$.

79. 1 **81.** $B = 45°$ **83.** $C = 24°$
$b \approx 11.31$ $a \approx 54.8$
$c \approx 11.31$ $b \approx 50.1$

85. $\sin \theta = \dfrac{1}{2}$ **87.** $\sin \theta = \dfrac{\sqrt{2}}{2}$

$\cos \theta = \dfrac{\sqrt{3}}{2}$ $\cos \theta = -\dfrac{\sqrt{2}}{2}$

$\tan \theta = \dfrac{\sqrt{3}}{3}$ $\tan \theta = -1$

89. $\sin \theta = -\dfrac{\sqrt{2}}{2}$ **91.** 30 feet

$\cos \theta = \dfrac{\sqrt{2}}{2}$

$\tan \theta = -1$

Section 2.4 (page 240)

1. (a) $\dfrac{\sqrt{2} - \sqrt{6}}{4}$ (b) $\dfrac{\sqrt{2} + 1}{2}$

3. (a) $\dfrac{1}{2}$ (b) $\dfrac{-\sqrt{3} - 1}{2}$

5. (a) $\dfrac{-\sqrt{2} - \sqrt{6}}{4}$ (b) $\dfrac{-1 + \sqrt{2}}{2}$

7. $\sin 105° = \dfrac{\sqrt{2}}{4}(\sqrt{3} + 1)$

$\cos 105° = \dfrac{\sqrt{2}}{4}(1 - \sqrt{3})$

$\tan 105° = -2 - \sqrt{3}$

9. $\sin 195° = \dfrac{\sqrt{2}}{4}(1 - \sqrt{3})$

$\cos 195° = -\dfrac{\sqrt{2}}{4}(\sqrt{3} + 1)$

$\tan 195° = 2 - \sqrt{3}$

11. $\sin \dfrac{11\pi}{12} = \dfrac{\sqrt{2}}{4}(\sqrt{3} - 1)$

$\cos \dfrac{11\pi}{12} = -\dfrac{\sqrt{2}}{4}(\sqrt{3} + 1)$

$\tan \dfrac{11\pi}{12} = -2 + \sqrt{3}$

13. $\sin \dfrac{17\pi}{12} = -\dfrac{\sqrt{2}}{4}(\sqrt{3} + 1)$

$\cos \dfrac{17\pi}{12} = \dfrac{\sqrt{2}}{4}(1 - \sqrt{3})$

$\tan \dfrac{17\pi}{12} = 2 + \sqrt{3}$

15. $\sin 285° = -\dfrac{\sqrt{2}}{4}(\sqrt{3} + 1)$

$\cos 285° = \dfrac{\sqrt{2}}{4}(\sqrt{3} - 1)$

$\tan 285° = -(2 + \sqrt{3})$

17. $\sin(-165°) = -\dfrac{\sqrt{2}}{4}(\sqrt{3} - 1)$

$\cos(-165°) = -\dfrac{\sqrt{2}}{4}(1 + \sqrt{3})$

$\tan(-165°) = 2 - \sqrt{3}$

19. $\sin \dfrac{13\pi}{12} = \dfrac{\sqrt{2}}{4}(1 - \sqrt{3})$

$\cos \dfrac{13\pi}{12} = -\dfrac{\sqrt{2}}{4}(1 + \sqrt{3})$

$\tan \dfrac{13\pi}{12} = 2 - \sqrt{3}$

21. $\sin\left(-\dfrac{13\pi}{12}\right) = \dfrac{\sqrt{2}}{4}(\sqrt{3} - 1)$

$\cos\left(-\dfrac{13\pi}{12}\right) = -\dfrac{\sqrt{2}}{4}(\sqrt{3} + 1)$

$\tan\left(-\dfrac{13\pi}{12}\right) = -2 + \sqrt{3}$

23. $\cos 40°$ **25.** $\tan 239°$ **27.** $\sin 1.8$ **29.** $\tan 3x$

31. $-\dfrac{\sqrt{3}}{2}$ **33.** $\dfrac{\sqrt{3}}{2}$ **35.** -1 **37.** $-\dfrac{63}{65}$

39. $\dfrac{16}{65}$ **41.** $-\dfrac{63}{16}$ **43.** $\dfrac{65}{56}$ **45.** $\dfrac{3}{5}$ **47.** $-\dfrac{44}{117}$

49. $\dfrac{5}{3}$ **51.** 1 **53.** 0 **55.–63.** Answers will vary.

65. $-\sin x$ **67.** $-\cos \theta$ **69.** $\dfrac{\pi}{2}$ **71.** $\dfrac{5\pi}{4}, \dfrac{7\pi}{4}$

73. $\dfrac{\pi}{4}, \dfrac{7\pi}{4}$ **75.** (a) $y = \dfrac{5}{12}\sin(2t + 0.6435)$

(b) $\dfrac{5}{12}$ feet (c) $\dfrac{1}{\pi}$ cycle per second

77. False. $\sin(u \pm v) = \sin u \cos v \pm \cos u \sin v$

79. False.

$\cos\left(x - \dfrac{\pi}{2}\right) = \cos x \cos \dfrac{\pi}{2} + \sin x \sin \dfrac{\pi}{2} = \sin x$

81. and 83. Answers will vary.

85. (a) $\sqrt{2}\sin\left(\theta + \dfrac{\pi}{4}\right)$ (b) $\sqrt{2}\cos\left(\theta - \dfrac{\pi}{4}\right)$

87. (a) $13\sin(3\theta + 0.3948)$ (b) $13\cos(3\theta - 1.1760)$

89. $2\cos\theta$ **91.** $15°$

93.

[graph from -6.28 to 6.28, -3 to 3]

$\sin^2\left(\theta + \dfrac{\pi}{4}\right) + \sin^2\left(\theta - \dfrac{\pi}{4}\right) = 1$

95. Answers will vary.   **97.** $f^{-1}(x) = \dfrac{x+15}{5}$

99. Because f is not one-to-one, f^{-1} does not exist.

Section 2.5 (page 250)

1. $\dfrac{\sqrt{17}}{17}$ **3.** $\dfrac{15}{17}$ **5.** $\dfrac{8}{15}$ **7.** $\dfrac{17}{8}$

9. $0, \dfrac{\pi}{3}, \pi, \dfrac{5\pi}{3}$ **11.** $\dfrac{\pi}{12}, \dfrac{5\pi}{12}, \dfrac{13\pi}{12}, \dfrac{17\pi}{12}$

13. $0, \dfrac{2\pi}{3}, \dfrac{4\pi}{3}$ **15.** $\dfrac{\pi}{2}, \dfrac{\pi}{6}, \dfrac{5\pi}{6}, \dfrac{7\pi}{6}, \dfrac{3\pi}{2}, \dfrac{11\pi}{6}$

17. $0, \dfrac{\pi}{2}, \pi, \dfrac{3\pi}{2}$ **19.** $3\sin 2x$ **21.** $4\cos 2x$

23. $\sin 2u = \dfrac{24}{25}$ **25.** $\sin 2u = \dfrac{24}{25}$
$\cos 2u = -\dfrac{7}{25}$ $\quad\quad\quad\;\; \cos 2u = \dfrac{7}{25}$
$\tan 2u = -\dfrac{24}{7}$ $\quad\quad\quad\;\; \tan 2u = \dfrac{24}{7}$

27. $\sin 2u = -\dfrac{4\sqrt{21}}{25}$
$\cos 2u = -\dfrac{17}{25}$
$\tan 2u = \dfrac{4\sqrt{21}}{17}$

29. $\dfrac{1}{8}(3 + 4\cos 2x + \cos 4x)$ **31.** $\dfrac{1}{8}(1 - \cos 4x)$

33. $\dfrac{1}{16}(1 + \cos 2x - \cos 4x - \cos 2x \cos 4x)$

35. $\dfrac{4\sqrt{17}}{17}$ **37.** $\dfrac{1}{4}$ **39.** $\sqrt{17}$

41. $\sin 75° = \dfrac{1}{2}\sqrt{2 + \sqrt{3}}$
$\cos 75° = \dfrac{1}{2}\sqrt{2 - \sqrt{3}}$
$\tan 75° = 2 + \sqrt{3}$

43. $\sin 112°\,30' = \dfrac{1}{2}\sqrt{2 + \sqrt{2}}$
$\cos 112°\,30' = -\dfrac{1}{2}\sqrt{2 - \sqrt{2}}$
$\tan 112°\,30' = -1 - \sqrt{2}$

45. $\sin\dfrac{\pi}{8} = \dfrac{1}{2}\sqrt{2 - \sqrt{2}}$ **47.** $\sin\dfrac{3\pi}{8} = \dfrac{1}{2}\sqrt{2 + \sqrt{2}}$
$\cos\dfrac{\pi}{8} = \dfrac{1}{2}\sqrt{2 + \sqrt{2}}$ $\quad\quad \cos\dfrac{3\pi}{8} = \dfrac{1}{2}\sqrt{2 - \sqrt{2}}$
$\tan\dfrac{\pi}{8} = \sqrt{2} - 1$ $\quad\quad\quad\;\; \tan\dfrac{3\pi}{8} = \sqrt{2} + 1$

49. $\sin\dfrac{u}{2} = \dfrac{5\sqrt{26}}{26}$ **51.** $\sin\dfrac{u}{2} = \sqrt{\dfrac{89 - 8\sqrt{89}}{178}}$
$\cos\dfrac{u}{2} = \dfrac{\sqrt{26}}{26}$ $\quad\quad\;\; \cos\dfrac{u}{2} = -\sqrt{\dfrac{89 + 8\sqrt{89}}{178}}$
$\tan\dfrac{u}{2} = 5$ $\quad\quad\quad\quad\;\; \tan\dfrac{u}{2} = \dfrac{8 - \sqrt{89}}{5}$

53. $\sin\dfrac{u}{2} = \dfrac{3\sqrt{10}}{10}$ **55.** $|\sin 3x|$ **57.** $-|\tan 4x|$
$\cos\dfrac{u}{2} = -\dfrac{\sqrt{10}}{10}$
$\tan\dfrac{u}{2} = -3$

59. π **61.** $\dfrac{\pi}{3}, \pi, \dfrac{5\pi}{3}$ **63.** $3\left(\sin\dfrac{\pi}{2} + \sin 0\right)$

65. $\dfrac{1}{2}[\sin 10\theta - \sin(-2\theta)]$ **67.** $\dfrac{5}{2}(\cos 8\beta + \cos 2\beta)$

69. $\dfrac{1}{2}(\cos 2y - \cos 2x)$ **71.** $\dfrac{1}{2}[\sin 2\theta - \sin(-2\pi)]$

73. $5(\cos 60° + \cos 90°)$ **75.** $2\sin 45° \cos 15°$

77. $-2\sin\dfrac{\pi}{2}\sin\dfrac{\pi}{4}$ **79.** $2\cos 4\theta \sin\theta$

81. $2\cos 4x \cos 2x$ **83.** $2\cos\alpha \sin\beta$

85. $-2\sin\theta \sin\dfrac{\pi}{2}$ **87.** $0, \dfrac{\pi}{4}, \dfrac{\pi}{2}, \dfrac{3\pi}{4}, \pi, \dfrac{5\pi}{4}, \dfrac{3\pi}{2}, \dfrac{7\pi}{4}$

89. $\dfrac{\pi}{6}, \dfrac{5\pi}{6}$ **91.** $\dfrac{25}{169}$ **93.** $\dfrac{4}{13}$

95.–109. Answers will vary.

111.

113.

115.

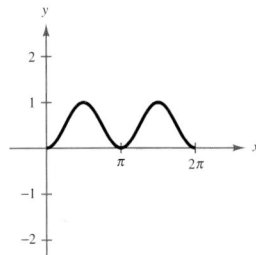

117. $2x\sqrt{1-x^2}$

119. (a) $A = 100 \sin \dfrac{\theta}{2} \cos \dfrac{\theta}{2}$

(b) $A = 50 \sin \theta$

The area is maximum when $\theta = \pi/2$.

121. 0.4482

123. False. For $u < 0$,

$\sin 2u = -\sin(-2u)$
$= -2 \sin(-u) \cos(-u)$
$= -2(-\sin u) \cos u$
$= 2 \sin u \cos u$.

125. (a)

(b) π

Maximum: $(\pi, 3)$

127. (a) $\frac{1}{4}(3 + \cos 4x)$

(b) $2\cos^4 x - 2\cos^2 x + 1$

(c) $1 - 2\sin^2 x \cos^2 x$

(d) $1 - \frac{1}{2}\sin^2 2x$

(e) No. There is often more than one way to rewrite a trigonometric expression.

129. $(x-3)^2 = (y-4)$ **131.** $(x-1)^2 = \frac{1}{2}(y-1)$

Vertex: $(3, 4)$ Vertex: $(1, 1)$

133. Vertex: $(5, -8)$

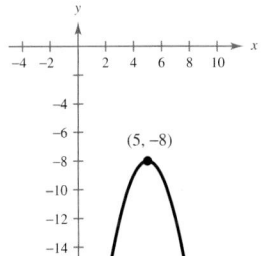

135. September: $235,000
October: $272,600

137. 15.7 gallons

Review Exercises (page 255)

1. $\sec x$ **3.** $\cos x$ **5.** $\cot x$

7. $\tan x = \dfrac{3}{4}$ **9.** $\cos x = \dfrac{\sqrt{2}}{2}$

$\csc x = \dfrac{5}{3}$ $\tan x = -1$

$\sec x = \dfrac{5}{4}$ $\csc x = -\sqrt{2}$

$\cot x = \dfrac{4}{3}$ $\sec x = \sqrt{2}$

$\cot x = -1$

11. $\sin^2 x$ **13.** 1   **15.** $\cot \theta$ **17.** $\cot^2 x$

19. $\sec x + 2 \sin x$ **21.** $-2 \tan^2 \theta$

23.–31. Answers will vary.   **33.** $\dfrac{\pi}{3} + 2n\pi, \dfrac{5\pi}{3} + 2n\pi$

35. $\dfrac{\pi}{3} + 2n\pi, \dfrac{5\pi}{3} + 2n\pi$ **37.** $\dfrac{\pi}{6} + n\pi, \dfrac{5\pi}{6} + n\pi$

39. $\dfrac{\pi}{6}, \dfrac{\pi}{2}, \dfrac{5\pi}{6}$ **41.** 0   **43.** $0, \dfrac{\pi}{3}, \dfrac{2\pi}{3}, \pi, \dfrac{4\pi}{3}, \dfrac{5\pi}{3}$

45. No solution   **47.** $\dfrac{\pi}{2}, \dfrac{3\pi}{2}$

49. $\dfrac{3\pi}{4}, \dfrac{7\pi}{4}$, $\arctan(-5) + \pi$, $\arctan(-5) + 2\pi$

51. $\sin 345° = \dfrac{\sqrt{2}}{4}(1 - \sqrt{3})$

$\cos 345° = \dfrac{\sqrt{2}}{4}(1 + \sqrt{3})$

$\tan 345° = -2 + \sqrt{3}$

53. $\sin \dfrac{19\pi}{12} = -\dfrac{\sqrt{2}}{4}(\sqrt{3} + 1)$

$\cos \dfrac{19\pi}{12} = \dfrac{\sqrt{2}}{4}(\sqrt{3} - 1)$

$\tan \dfrac{19\pi}{12} = -2 - \sqrt{3}$

55. $\cos 165°$ **57.** $\tan(-47°)$ **59.** $\dfrac{960 + 507\sqrt{7}}{1121}$

61. $\dfrac{12\sqrt{7} - 15}{52}$ **63.** $\dfrac{-960 + 507\sqrt{7}}{1121}$ **65.** $\dfrac{3\pi}{2}$

67. $0, \pi$

69.

71. $\sin 2u = -\dfrac{4}{5}$

$\cos 2u = \dfrac{3}{5}$

$\tan 2u = -\dfrac{4}{3}$

73. $\dfrac{1-\cos 4x}{1+\cos 4x}$ **75.** $\dfrac{3-4\cos 2x+\cos 4x}{4(1+\cos 2x)}$

77. $\sin(-75°) = -\tfrac{1}{2}\sqrt{2+\sqrt{3}}$
$\cos(-75°) = \tfrac{1}{2}\sqrt{2-\sqrt{3}}$
$\tan(-75°) = -2-\sqrt{3}$

79. $\sin\dfrac{19\pi}{12} = -\tfrac{1}{2}\sqrt{2+\sqrt{3}}$
$\cos\dfrac{19\pi}{12} = \tfrac{1}{2}\sqrt{2-\sqrt{3}}$
$\tan\dfrac{19\pi}{12} = -2-\sqrt{3}$

81. $-|\cos 5x|$

83. (a) $V = \sin\dfrac{\theta}{2}\cos\dfrac{\theta}{2}$ cubic meters

(b) $V = \sin\theta$ cubic meters
The volume is maximum when $\theta = \pi/2$.

85. $3[\cos(-30°) - \cos 60°]$ **87.** $2(\sin 5\alpha + \sin \alpha)$

89. $2\cos\dfrac{5\theta}{2}\cos\dfrac{\theta}{2}$ **91.** $2\cos x \sin\dfrac{\pi}{4}$

93. False. If $\dfrac{\pi}{2} < \theta < \pi$, then $\cos\dfrac{\theta}{2} > 0$. The sign of $\cos\dfrac{\theta}{2}$ depends on the quadrant in which $\dfrac{\theta}{2}$ lies.

95. True. $4\sin(-x)\cos(-x) = 4(-\sin x)\cos x$
$= -4\sin x \cos x$
$= -2(2\sin x \cos x)$
$= -2\sin 2x$

97. Reciprocal identities: $\sin\theta = \dfrac{1}{\csc\theta}$, $\cos\theta = \dfrac{1}{\sec\theta}$,
$\tan\theta = \dfrac{1}{\cot\theta}$, $\csc\theta = \dfrac{1}{\sin\theta}$, $\sec\theta = \dfrac{1}{\cos\theta}$,
$\cot\theta = \dfrac{1}{\tan\theta}$

Quotient identities: $\tan\theta = \dfrac{\sin\theta}{\cos\theta}$, $\cot\theta = \dfrac{\cos\theta}{\sin\theta}$

Pythagorean identities: $\sin^2\theta + \cos^2\theta = 1$,
$1 + \tan^2\theta = \sec^2\theta$, $1 + \cot^2\theta = \csc^2\theta$

99. No. For an equation to be an identity, the equation must be true for all real numbers x. $\sin\theta = \tfrac{1}{2}$ has an infinite number of solutions but is not an identity.

101. $y_1 = y_2 + 1$

103. $-1.8431, 2.1758, 3.9903, 8.8935, 9.8820$

Chapter Test (page 259)

1. $\sin\theta = -\dfrac{3\sqrt{13}}{13}$ $\csc\theta = -\dfrac{\sqrt{13}}{3}$
$\cos\theta = -\dfrac{2\sqrt{13}}{13}$ $\sec\theta = -\dfrac{\sqrt{13}}{2}$
$\cot\theta = \dfrac{2}{3}$

2. 1   **3.** 1   **4.** $\csc\theta \sec\theta$

5. $\theta = 0, \dfrac{\pi}{2} < \theta \le \pi, \dfrac{3\pi}{2} < \theta < 2\pi$

6.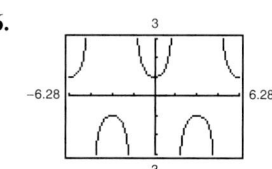

$y_1 = y_2$

7.–12. Answers will vary.

13. $\dfrac{1}{16}\left(\dfrac{10-15\cos 2x+6\cos 4x-\cos 6x}{1+\cos 2x}\right)$ **14.** $\tan 2\theta$

15. $2(\sin 6\theta + \sin 2\theta)$ **16.** $-2\cos\dfrac{7\theta}{2}\sin\dfrac{\theta}{2}$

17. $0, \dfrac{3\pi}{4}, \pi, \dfrac{7\pi}{4}$ **18.** $\dfrac{\pi}{6}, \dfrac{\pi}{2}, \dfrac{5\pi}{6}, \dfrac{3\pi}{2}$

19. $\dfrac{\pi}{6}, \dfrac{5\pi}{6}, \dfrac{7\pi}{6}, \dfrac{11\pi}{6}$ **20.** $\dfrac{\pi}{6}, \dfrac{5\pi}{6}, \dfrac{3\pi}{2}$

21. $-2.938, -2.663, 1.170$

22. $|\cos^2 x + \cos x| \le 2$ for all x

23. $\dfrac{\sqrt{2}-\sqrt{6}}{4}$ **24.** $\sin 2u = \dfrac{4}{5}, \tan 2u = -\dfrac{4}{3}$

25. $\theta \approx 76.5°$

Chapter 3
Section 3.1 (page 268)

1. $C = 105°, b \approx 28.28, c \approx 38.64$

3. $C = 120°, b \approx 4.8, c \approx 7.2$

5. $B \approx 21.55°, C \approx 122.45°, c \approx 11.49$

7. $B = 60.9°, b \approx 19.3, c \approx 6.4$

9. $B = 42°4', a \approx 22.05, b \approx 14.88$

11. $A \approx 10°11', C \approx 154°19', c \approx 11.03$

13. $A \approx 25.57°, B \approx 9.43°, a \approx 10.53$

15. $B \approx 18°13', C \approx 51°32', c \approx 40.05$

17. $C = 83°, a \approx 0.62, b \approx 0.51$ **19.** No solution

21. No solution   **23.** No solution   **25.** No solution

27. (a) $b \leq 5$, $b = \dfrac{5}{\sin 36°}$
(b) $5 < b < \dfrac{5}{\sin 36°}$
(c) $b > \dfrac{5}{\sin 36°}$

29. (a) $b \leq 10.8$, $b = \dfrac{10.8}{\sin 10°}$
(b) $10.8 < b < \dfrac{10.8}{\sin 10°}$
(c) $b > \dfrac{10.8}{\sin 10°}$

31. 10.4 **33.** 1675.2 **35.** 3204.5 **37.** 15.3 meters
39. 16.1° **41.** 77 meters

43. (a)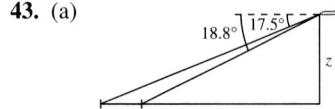
(b) 22.6 miles
(c) 21.4 miles
(d) 38,443 feet

45. 3.2 miles **47.** $d = \dfrac{2 \sin \theta}{\sin(\phi - \theta)}$

49. False. Two sides and one opposite angle do not necessarily determine a unique triangle.

51. (a) $A = 20\left[15 \sin \dfrac{3\theta}{2} - 4 \sin \dfrac{\theta}{2} - 6 \sin \theta\right]$

(b)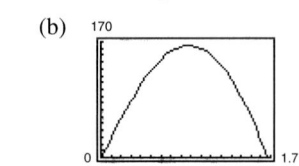

(c) Domain: $0 \leq \theta \leq 1.6690$

The area would increase and the domain would increase in length.

53. $\sin x = -\dfrac{2\sqrt{6}}{5}$
$\tan x = -2\sqrt{6}$
$\csc x = -\dfrac{5\sqrt{6}}{12}$
$\sec x = 5$
$\cot x = -\dfrac{\sqrt{6}}{12}$

55. $\sin x = \dfrac{5\sqrt{26}}{26}$
$\cos x = -\dfrac{\sqrt{26}}{26}$
$\csc x = \dfrac{\sqrt{26}}{5}$
$\sec x = -\sqrt{26}$
$\cot x = -\dfrac{1}{5}$

57. $\tan x$ **59.** $\sec^2 x$

Section 3.2 (page 275)

1. $A \approx 23.07°$, $B \approx 34.05°$, $C \approx 122.88°$
3. $B \approx 23.8°$, $C \approx 126.2°$, $a \approx 18.6$
5. $A \approx 31.98°$, $B \approx 42.38°$, $C \approx 105.63°$
7. $A \approx 92.94°$, $B \approx 43.53°$, $C \approx 43.53°$
9. $B \approx 13.45°$, $C \approx 31.55°$, $a \approx 12.16$
11. $A \approx 141°45'$, $C \approx 27°40'$, $b \approx 11.9$
13. $A = 27°10'$, $C = 27°10'$, $b \approx 56.9$
15. $A \approx 33.8°$, $B \approx 103.2°$, $c \approx 0.5448$

	a	b	c	d	θ	ϕ
17.	5	8	12.07	5.69	45°	135°
19.	10	14	20	13.86	68.2°	111.8°
21.	15	16.96	25	20	77.2°	102.8°

23. 16.25 **25.** 10.44 **27.** 52.11

29.

N 37.1° E, S 63.1° E

31. 373.3 meters **33.** 72.3° **35.** 43.3 miles
37. (a) N 58.4° W (b) S 81.5° W **39.** 63.7 feet
41. 24.2 miles **43.** $\overline{PQ} \approx 9.4$, $\overline{QS} = 5$, $\overline{RS} \approx 12.8$

45.

d (inches)	9	10	12	13	14
θ (degrees)	60.9°	69.5°	88.0°	98.2°	109.6°
s (inches)	20.88	20.28	18.99	18.28	17.48

d (inches)	15	16
θ (degrees)	122.9°	139.8°
s (inches)	16.55	15.37

47. 46,837.5 square feet

49. False. For s to be the average of the lengths of the three sides of the triangle, s would be equal to $(a + b + c)/3$.

51. Answers will vary. **53.** 405.2 feet
55. Answers will vary. **57.** $\dfrac{\pi}{2}$ **59.** $-\dfrac{\pi}{3}$ **61.** $\dfrac{5\pi}{6}$

63. $\dfrac{\sqrt{1 - 9x^2}}{3x}$ **65.** $\dfrac{\sqrt{4 - (x - 1)^2}}{2}$

67. $\sin \theta = -\dfrac{\sqrt{2}}{2}$
$\sec \theta = \sqrt{2}$
$\csc \theta = -\sqrt{2}$

69. $12 = 6 \sec \theta$
$\sec \theta = 2$
$\csc \theta = \pm\dfrac{2\sqrt{3}}{3}$

Section 3.3 (page 288)

1. $\mathbf{v} = \langle 3, 2 \rangle$; $\|\mathbf{v}\| = \sqrt{13}$ **3.** $\mathbf{v} = \langle -3, 2 \rangle$; $\|\mathbf{v}\| = \sqrt{13}$
5. $\mathbf{v} = \langle 0, 5 \rangle$; $\|\mathbf{v}\| = 5$ **7.** $\mathbf{v} = \langle 16, 7 \rangle$; $\|\mathbf{v}\| = \sqrt{305}$
9. $\mathbf{v} = \langle 8, 6 \rangle$; $\|\mathbf{v}\| = 10$ **11.** $\mathbf{v} = \langle -9, -12 \rangle$; $\|\mathbf{v}\| = 15$

A72 Answers to Odd-Numbered Exercises and Tests

13.

15.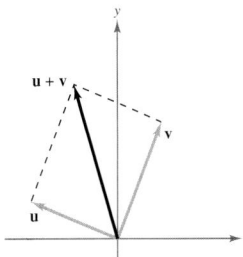

(c) $\langle -10, 6 \rangle$

17.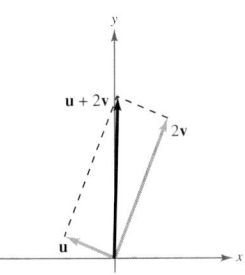

23. (a) $3\mathbf{i} - 2\mathbf{j}$ (b) $-\mathbf{i} + 4\mathbf{j}$
 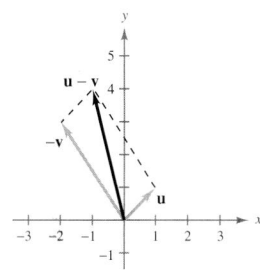

19. (a) $\langle 3, 4 \rangle$ (b) $\langle 1, -2 \rangle$

(c) $-4\mathbf{i} + 11\mathbf{j}$
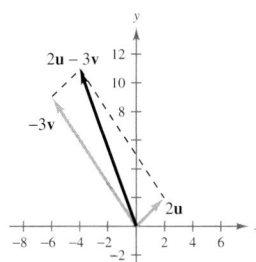

(c) $\langle 1, -7 \rangle$
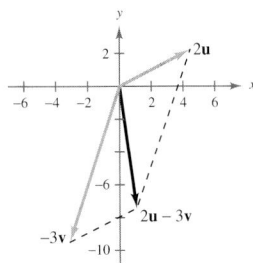

25. (a) $2\mathbf{i} + \mathbf{j}$ (b) $2\mathbf{i} - \mathbf{j}$

(c) $4\mathbf{i} - 3\mathbf{j}$
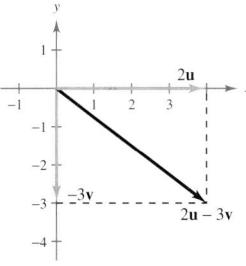

21. (a) $\langle -5, 3 \rangle$ (b) $\langle -5, 3 \rangle$
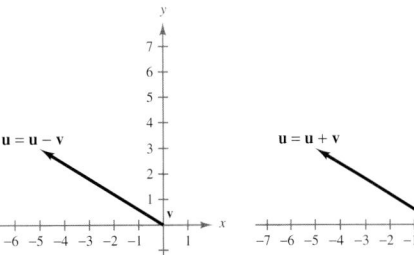

Answers to Odd-Numbered Exercises and Tests **A73**

27. $\langle 1, 0 \rangle$ **29.** $\langle -\frac{1}{\sqrt{2}}, \frac{1}{\sqrt{2}} \rangle$ **31.** $\frac{3}{\sqrt{10}}\mathbf{i} - \frac{1}{\sqrt{10}}\mathbf{j}$

33. $\mathbf{j}$ **35.** $\frac{1}{\sqrt{5}}\mathbf{i} - \frac{2}{\sqrt{5}}\mathbf{j}$ **37.** $\langle \frac{5}{\sqrt{2}}, \frac{5}{\sqrt{2}} \rangle$

39. $\langle \frac{18}{\sqrt{29}}, \frac{45}{\sqrt{29}} \rangle$

41. $\mathbf{v} = \langle 3, -\frac{3}{2} \rangle$ **43.** $\mathbf{v} = \langle 4, 3 \rangle$

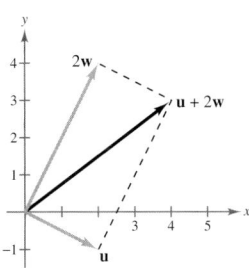

45. $\mathbf{v} = \langle \frac{7}{2}, -\frac{1}{2} \rangle$

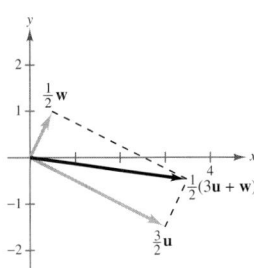

47. $\|\mathbf{v}\| = 3; \theta = 60°$ **49.** $\|\mathbf{v}\| = 6\sqrt{2}; \theta = 315°$

51. $\mathbf{v} = \langle 3, 0 \rangle$ **53.** $\mathbf{v} = \langle -\frac{7\sqrt{3}}{4}, \frac{7}{4} \rangle$

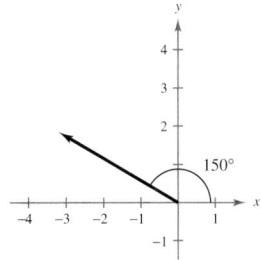

55. $\mathbf{v} = \langle -\frac{3\sqrt{6}}{2}, \frac{3\sqrt{2}}{2} \rangle$ **57.** $\mathbf{v} = \langle \frac{\sqrt{10}}{5}, \frac{3\sqrt{10}}{5} \rangle$

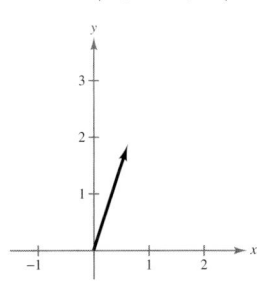

59. $\langle 5, 5 \rangle$ **61.** $\langle 10\sqrt{2} - 50, 10\sqrt{2} \rangle$ **63.** 90°
65. 63.4° **67.** 62.7° **69.** 12.8°; 398.32 newtons
71. 71.3°; 228.5 pounds
73. Vertical component: 70 sin 35° ≈ 40.15 feet per second
Horizontal component: 70 cos 35° ≈ 57.34 feet per second
75. $T_{AC} \approx 1758.8$ pounds **77.** 3154.4 pounds
$T_{BC} \approx 1305.4$ pounds
79. N 21.4° E; 138.7 kilometers per hour
81. 1928.4 foot-pounds **83.** True. See Example 1.
85. False. $a = b = 0$
87. (a) 0° (b) 180°
(c) No. The magnitude is at most equal to the sum when the angle between the vectors is 0°.
89. Answers will vary. **91.** $\langle 1, 3 \rangle$ or $\langle -1, -3 \rangle$
93. $8 \tan \theta$ **95.** $6 \sec \theta$
97. $\frac{\pi}{2} + n\pi, \pi + 2n\pi$
99. $n\pi, \frac{\pi}{6} + 2n\pi, \frac{11\pi}{6} + 2n\pi$ **101.** $\frac{n\pi}{2}$

Section 3.4 *(page 298)*

1. −9 **3.** 6 **5.** 8; scalar **7.** $\langle -6, 8 \rangle$; vector
9. 13 **11.** $5\sqrt{41}$ **13.** 6 **15.** 90° **17.** 143.13°
19. 60.26° **21.** 90° **23.** $\frac{5\pi}{12}$ **25.** 26.6°, 63.4°, 90°
27. 41.63°, 53.13°, 85.24° **29.** −20 **31.** $2592\sqrt{2}$
33. Parallel **35.** Neither **37.** Orthogonal
39. $\frac{1}{37}\langle 84, 14 \rangle, \frac{1}{37}\langle -10, 60 \rangle$ **41.** $\frac{45}{229}\langle 2, 15 \rangle, \frac{6}{229}\langle -15, 2 \rangle$
43. $\langle -5, 3 \rangle, \langle 5, -3 \rangle$ **45.** $\frac{2}{3}\mathbf{i} + \frac{1}{2}\mathbf{j}, -\frac{2}{3}\mathbf{i} - \frac{1}{2}\mathbf{j}$ **47.** 32
49. $58,762.50
This value gives the total revenue that can be earned by selling all of the units.
51. (a) 2614.7 pounds (b) 29,885.9 pounds
53. 735 newton-meters **55.** 779.4 foot-pounds
57. False. Work is represented by a scalar.
59. (a) $\theta = \frac{\pi}{2}$ (b) $0 \leq \theta < \frac{\pi}{2}$ (c) $\frac{\pi}{2} < \theta \leq \pi$
61. and 63. Answers will vary. **65.** $0, \frac{\pi}{6}, \pi, \frac{11\pi}{6}$
67. $0, \pi$ **69.** $-\frac{253}{325}$ **71.** $\frac{204}{325}$

Review Exercises (page 301)

1. $C = 74°, b \approx 13.19, c \approx 13.41$
3. $A = 26°, a \approx 24.89, c \approx 56.23$
5. $C = 66°, a \approx 2.53, b \approx 9.11$
7. $B = 108°, a \approx 11.76, c \approx 21.49$
9. $A \approx 20.41°, C \approx 9.59°, a \approx 20.92$
11. $B \approx 39.48°, C \approx 65.52°, c \approx 48.24$
13. 7.945   15. 33.547   17. 31.1 meters
19. 31.01 feet   21. $A \approx 53.13°, B \approx 36.87°, C \approx 90°$
23. $A \approx 101.47°, B \approx 31.73°, C \approx 46.8°$
25. $A \approx 9.90°, C \approx 20.10°, b \approx 29.09$
27. $B \approx 35.20°, C \approx 82.8°, a \approx 17.37$
29.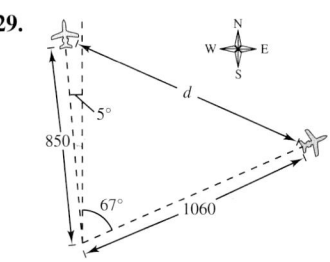

1135 miles

31. 36.979   33. 242.630

35.   37.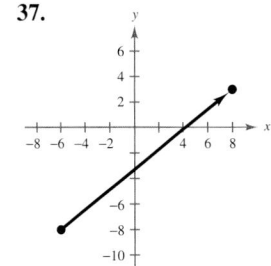

39. $\langle 6, \frac{5}{2} \rangle$ 41. $\langle 14, 4 \rangle$ 43. $\langle -\frac{\sqrt{2}}{4}, -\frac{\sqrt{2}}{4} \rangle$
45. $\langle -26, -35 \rangle$ 47. $\langle 5, \frac{3}{2} \rangle$

   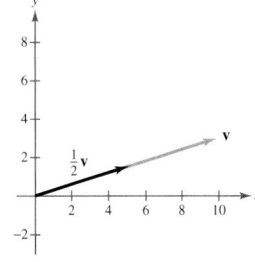

49. $-6\mathbf{i} - 8\mathbf{j}$ 51. $7\mathbf{i} - 16\mathbf{j}$
53. $\sqrt{17}(\mathbf{i} \cos 346° + \mathbf{j} \sin 346°)$ 55. $\|\mathbf{v}\| = 3; \theta = 150°$
57. $\|\mathbf{v}\| = \sqrt{65}; \theta = 119.7°$ 59. $\|\mathbf{v}\| = \sqrt{65}; \theta = 352.9°$

61. 180 pounds   63. 45   65. -2 67. 50, scalar
69. $\langle 6, -8 \rangle$, vector   71. $\frac{11\pi}{12}$ 73. 160.5°
75. Orthogonal   77. Neither   79. $-\frac{13}{17}\langle 4, 1 \rangle, \frac{16}{17}\langle -1, 4 \rangle$
81. $\frac{5}{2}\langle -1, 1 \rangle, \frac{9}{2}\langle 1, 1 \rangle$ 83. 48
85. True. $\sin 90°$ is defined in the Law of Sines.
87. $\dfrac{a}{\sin A} = \dfrac{b}{\sin B} = \dfrac{c}{\sin C}$ 89. Pythagorean Theorem
91. A, C
93. If $k > 0$, the direction is the same and the magnitude is k times as great.

If $k < 0$, the result is a vector in the opposite direction and the magnitude is k times as great.

Chapter Test (page 305)

1. $C = 88°, b \approx 27.81, c \approx 29.98$
2. $A = 43°, b \approx 25.75, c \approx 14.45$
3. Two solutions:
   $B \approx 29.12°, C \approx 126.88°, c \approx 22.03$
   $B \approx 150.88°, C \approx 5.12°, c \approx 2.46$
4. No solution   5. $A \approx 39.96°, C \approx 40.04°, c \approx 15.02$
6. $A \approx 23.44°, B \approx 33.56°, c \approx 86.46$
7. 2052.5 square meters   8. 606.3 miles, N 29° E
9. $\langle 14, -23 \rangle$ 10. $\left\langle \dfrac{18\sqrt{34}}{17}, -\dfrac{30\sqrt{34}}{17} \right\rangle$
11. $\langle -4, 6 \rangle$ 12. $\langle 10, 4 \rangle$

   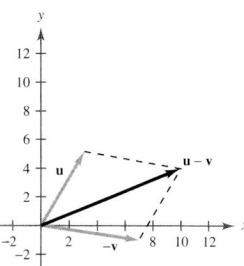

13. $\langle 36, 22 \rangle$ 14. $\langle \frac{4}{5}, -\frac{3}{5} \rangle$

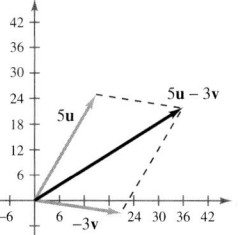

15. 14.9°; 250.15 pounds   16. 135°   17. No

18. $\frac{37}{26}\langle 5, 1\rangle$; $\frac{29}{26}\langle -1, 5\rangle$

Cumulative Test for Chapters 1–3
(page 306)

1. (a) 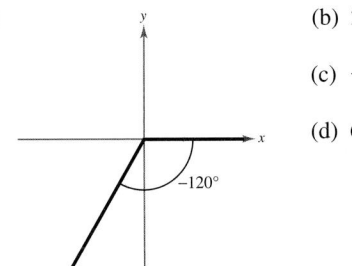 (b) $240°$ (c) $-\frac{2\pi}{3}$ (d) $60°$

(e) $\sin(-120°) = -\frac{\sqrt{3}}{2}$ $\csc(-120°) = -\frac{2\sqrt{3}}{3}$

$\cos(-120°) = -\frac{1}{2}$ $\sec(-120°) = -2$

$\tan(-120°) = \sqrt{3}$ $\cot(-120°) = \frac{\sqrt{3}}{3}$

2. $134.6°$ **3.** $\frac{3}{5}$

4. Period: 2; Amplitude: 2 **5.** Period: π; Amplitude: $\frac{1}{2}$

 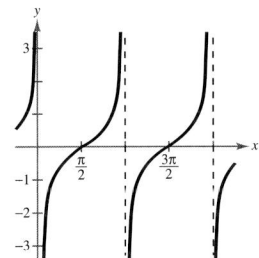

6. $a = -3, b = \pi, c = 0$

7.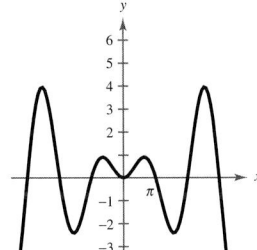

8. 6.7 **9.** $\frac{3}{4}$ **10.** $\sqrt{1 - 4x^2}$ **11.** 1

12. $2 \tan \theta$ **13.–15.** Answers will vary.

16. $\frac{\pi}{3}, \frac{\pi}{2}, \frac{3\pi}{2}, \frac{5\pi}{3}$ **17.** $\frac{\pi}{6}, \frac{5\pi}{6}, \frac{7\pi}{6}, \frac{11\pi}{6}$ **18.** $\frac{3\pi}{2}$

19. $\frac{16}{63}$ **20.** $\frac{4}{3}$ **21.** $\frac{\sqrt{5}}{5}, \frac{2\sqrt{5}}{5}$

22. $\frac{5}{2}\left(\sin\frac{5\pi}{2} - \sin \pi\right)$

23. $B \approx 26.4°, C \approx 123.6°, c \approx 15.0$
24. $B \approx 52.5°, C \approx 97.5°, a \approx 5.0$
25. $B = 60°, a \approx 5.8, c \approx 11.5$
26. $A = 26.4°, B \approx 62.7°, C \approx 90.9°$
27. 36.4 square inches **28.** 85.2 square inches
29. $3\mathbf{i} + 5\mathbf{j}$ **30.** -5 **31.** $-\frac{1}{13}\langle 1, 5\rangle; \frac{21}{13}\langle 5, -1\rangle$
32. 5 feet
33. 90π radians per minute; 848.23 inches per minute
34. $22.6°$ **35.** $d = 4\cos\frac{\pi}{4}t$
36. N $32.6°$ E; 543.9 kilometers per hour

Chapter 4
Section 4.1 *(page 315)*

1. $a = -10, b = 6$ **3.** $a = 6, b = 5$ **5.** $4 + 3i$
7. $2 - 3\sqrt{3}i$ **9.** $5\sqrt{3}i$ **11.** 8 **13.** $-1 - 6i$
15. $0.3i$ **17.** $11 - i$ **19.** 4 **21.** $3 - 3\sqrt{2}i$
23. $-14 + 20i$ **25.** $\frac{1}{6} + \frac{7}{6}i$ **27.** $-2\sqrt{3}$ **29.** -10
31. $5 + i$ **33.** $12 + 30i$ **35.** 24 **37.** $-9 + 40i$
39. -10 **41.** $6 - 3i, 45$ **43.** $-1 + \sqrt{5}i, 6$
45. $-2\sqrt{5}i, 20$ **47.** $\sqrt{8}, 8$ **49.** $-5i$
51. $\frac{8}{41} + \frac{10}{41}i$ **53.** $\frac{4}{5} + \frac{3}{5}i$ **55.** $-5 - 6i$
57. $-\frac{120}{1681} - \frac{27}{1681}i$ **59.** $-\frac{1}{2} - \frac{5}{2}i$ **61.** $\frac{62}{949} + \frac{297}{949}i$
63. $1 \pm i$ **65.** $-2 \pm \frac{1}{2}i$ **67.** $-\frac{3}{2}, -\frac{5}{2}$
69. $2 \pm \sqrt{2}i$ **71.** $\frac{5}{7} \pm \frac{5\sqrt{15}}{7}$
73. (a) 1 (b) i (c) -1 (d) $-i$
75. $-4 + 2i$ **77.** i **79.** -8 **81.** $\frac{1}{8}i$
83. (a) 16 (b) 16 (c) 16 (d) 16
85. False. If the complex number is real, the number equals its conjugate.
87. False.
$i^{44} + i^{150} - i^{74} - i^{109} + i^{61} = 1 - 1 + 1 - i + i = 1$
89. Answers will vary. **91.** $-x^2 - 3x + 12$
93. $3x^2 + \frac{23}{2}x - 2$ **95.** -31 **97.** $\frac{27}{2}$
99. $a = \frac{\sqrt{3V\pi b}}{2\pi b}$ **101.** 1 liter

Section 4.2 *(page 322)*

1. Three solutions **3.** Four solutions
5. No real solutions **7.** Two real solutions
9. Two real solutions **11.** No real solutions

13. $\pm\sqrt{5}$ 15. $-5 \pm \sqrt{6}$ 17. 4   19. $-1 \pm 2i$
21. $\frac{1}{2} \pm i$ 23. $20 \pm 2\sqrt{215}$ 25. $-3 \pm 2\sqrt{2}i$
27. $4, \pm i$; same   29. $\pm\sqrt{2}i$; no x-intercepts
31. $\pm 5i$; $(x + 5i)(x - 5i)$
33. $2 \pm \sqrt{3}$; $(x - 2 - \sqrt{3})(x - 2 + \sqrt{3})$
35. $\pm 3, \pm 3i$; $(x + 3)(x - 3)(x + 3i)(x - 3i)$
37. $2, 2 \pm i$; $(x - 2)(x - 2 + i)(x - 2 - i)$
39. $-2, 1 \pm \sqrt{2}i$; $(x + 2)(x - 1 + \sqrt{2}i)(x - 1 - \sqrt{2}i)$
41. $-\frac{1}{5}, 1 \pm \sqrt{5}i$; $(5x + 1)(x - 1 + \sqrt{5}i)(x - 1 - \sqrt{5}i)$
43. $2, \pm 2i$; $(x - 2)^2(x + 2i)(x - 2i)$
45. $\pm i, \pm 3i$; $(x - i)(x + i)(x - 3i)(x + 3i)$
47. $-\frac{3}{2}, \pm 5i$ 49. $\pm 2i, 1, -\frac{1}{2}$ 51. $-3 \pm i, \frac{1}{4}$
53. $2, -3 \pm \sqrt{2}i, 1$ 55. $x^3 - x^2 + 25x - 25$
57. $x^3 + 4x^2 - 31x - 174$
59. $3x^4 - 17x^3 + 25x^2 + 23x - 22$
61. No. Setting $h = 64$ and solving the resulting equation yields imaginary roots.
63. False. The most complex zeros it can have is two and the Linear Factorization Theorem guarantees that there are three linear factors, so one zero must be real.
65. r_1, r_2, r_3 67. $5 + r_1, 5 + r_2, 5 + r_3$
69. The zeros cannot be determined.
71. (a) $0 < k < 4$ (b) $k = 4$ (c) $k < 0$ (d) $k > 4$
73. (a) $x^2 + b$ (b) $x^2 - 2ax + a^2 + b^2$
75. $-11 + 9i$ 77. $20 + 40i$ 79. i

81.
83.

85.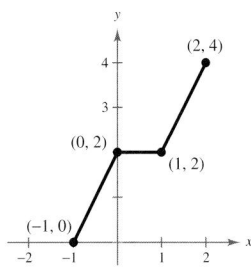

Section 4.3 (page 329)

1.

3.

   7    $4\sqrt{2}$

5.

   $\sqrt{85}$

7. $3\left(\cos \dfrac{\pi}{2} + i \sin \dfrac{\pi}{2}\right)$ 9. $\sqrt{10}(\cos 5.96 + i \sin 5.96)$

11.

13.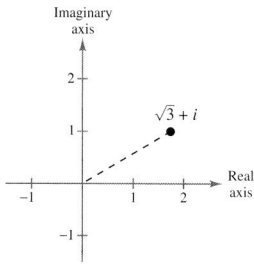

$3\sqrt{2}\left(\cos \dfrac{7\pi}{4} + i \sin \dfrac{7\pi}{4}\right)$ $2\left(\cos \dfrac{\pi}{6} + i \sin \dfrac{\pi}{6}\right)$

15.

17.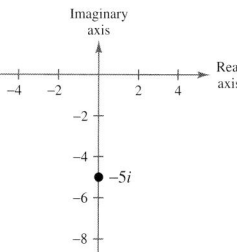

$4\left(\cos \dfrac{4\pi}{3} + i \sin \dfrac{4\pi}{3}\right)$ $5\left(\cos \dfrac{3\pi}{2} + i \sin \dfrac{3\pi}{2}\right)$

19.
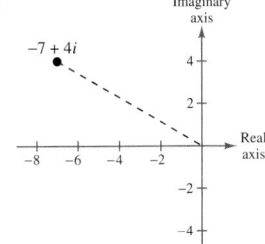

$\sqrt{65}(\cos 2.62 + i \sin 2.62)$

21.
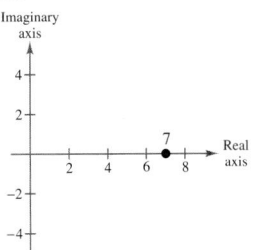

$7(\cos 0 + i \sin 0)$

23.
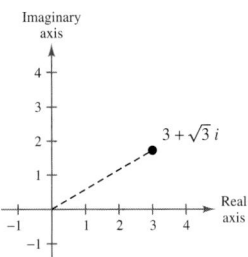

$2\sqrt{3}\left(\cos \dfrac{\pi}{6} + i \sin \dfrac{\pi}{6}\right)$

25.
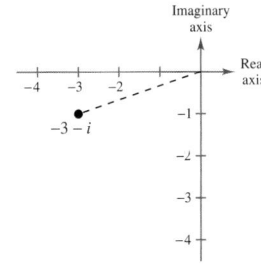

$\sqrt{10}(\cos 3.46 + i \sin 3.46)$

27. $5.39(\cos 0.38 + i \sin 0.38)$
29. $3.16(\cos 2.82 + i \sin 2.82)$
31. $8.19(\cos 5.26 + i \sin 5.26)$
33. $11.79(\cos 3.97 + i \sin 3.97)$

35.

$-\dfrac{3}{2} + \dfrac{3\sqrt{3}}{2}i$

37.

$\dfrac{3}{4} - \dfrac{3\sqrt{3}}{4}i$

39.

$-\dfrac{15\sqrt{2}}{8} + \dfrac{15\sqrt{2}}{8}i$

41.

$8i$

43.
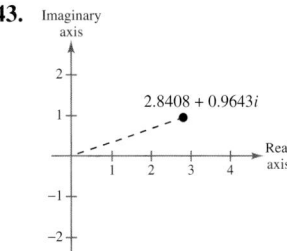

$2.8408 + 0.9643i$

45. $4.70 + 1.71i$ **47.** $-2.90 + 0.75i$
49. $12\left(\cos \dfrac{\pi}{3} + i \sin \dfrac{\pi}{3}\right)$ **51.** $\dfrac{10}{9}(\cos 200° + i \sin 200°)$
53. $0.27(\cos 150° + i \sin 150°)$ **55.** $\cos 30° + i \sin 30°$
57. $\cos \dfrac{2\pi}{3} + i \sin \dfrac{2\pi}{3}$ **59.** $4[\cos(302°) + i \sin(302°)]$
61. (a) $\left[2\sqrt{2}\left(\cos \dfrac{\pi}{4} + i \sin \dfrac{\pi}{4}\right)\right]\left\{\sqrt{2}\left[\cos\left(\dfrac{7\pi}{4}\right) + i \sin\left(\dfrac{7\pi}{4}\right)\right]\right\}$
(b) $4(\cos 0 + i \sin 0) = 4$
(c) 4
63. (a) $2\left[\cos\left(\dfrac{3\pi}{2}\right) + i \sin\left(\dfrac{3\pi}{2}\right)\right]\left[\sqrt{2}\left(\cos \dfrac{\pi}{4} + i \sin \dfrac{\pi}{4}\right)\right]$
(b) $2\sqrt{2}\left[\cos\left(\dfrac{7\pi}{4}\right) + i \sin\left(\dfrac{7\pi}{4}\right)\right] = 2 - 2i$
(c) $-2i - 2i^2 = -2i + 2 = 2 - 2i$
65. (a) $[5(\cos 0.93 + i \sin 0.93)] \div \left[2\left(\cos \dfrac{5\pi}{3} + i \sin \dfrac{5\pi}{3}\right)\right]$
(b) $\dfrac{5}{2}[\cos(1.97) + i \sin(1.97)] = -0.982 + 2.299i$
(c) $\approx -0.982 + 2.299i$
67. (a) $[5(\cos 0 + i \sin 0)] \div \left[\sqrt{13}(\cos 0.98 + i \sin 0.98)\right]$
(b) $\dfrac{5}{\sqrt{13}}[\cos(5.30) + i \sin(5.30)] \approx 0.769 - 1.154i$
(c) $\dfrac{10}{13} - \dfrac{15}{13}i \approx 0.769 - 1.154i$

69.   **71.**   (b)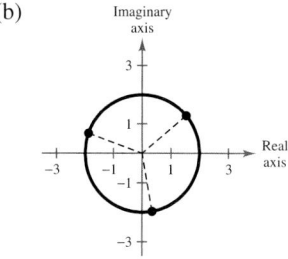

73. True, by the definition of the absolute value of a complex number.

75. Answers will vary.

77. (a) r^2 (b) $\cos 2\theta + i\sin 2\theta$

79. $B = 68°, b \approx 19.80, c \approx 21.36$

81. $B = 60°, a \approx 65.01, c \approx 130.02$

83. $B = 47°45', a \approx 7.53, b \approx 8.29$

85. 16; 2   **87.** $\frac{1}{16}$; 0

Section 4.4 *(page 335)*

1. $-4 - 4i$ **3.** $-32i$ **5.** $-128\sqrt{3} - 128i$

7. $\frac{125}{2} + \frac{125\sqrt{3}}{2}i$ **9.** -1 **11.** $608.02 + 144.69i$

13. $-597 - 122i$ **15.** $-43\sqrt{5} + 4i$

17. $\frac{81}{2} + \frac{81\sqrt{3}}{2}i$ **19.** $32.3524 - 120.7407i$

21. $32i$ **23.** 27

25. (a) $\sqrt{5}(\cos 60° + i\sin 60°)$
$\sqrt{5}(\cos 240° + i\sin 240°)$

(b)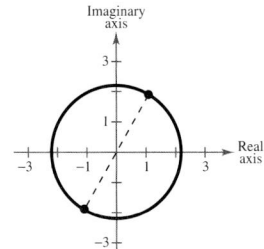

(c) $\frac{\sqrt{5}}{2} + \frac{\sqrt{15}}{2}i, -\frac{\sqrt{5}}{2} - \frac{\sqrt{15}}{2}i$

27. (a) $2\left(\cos\frac{2\pi}{9} + i\sin\frac{2\pi}{9}\right)$

$2\left(\cos\frac{8\pi}{9} + i\sin\frac{8\pi}{9}\right)$

$2\left(\cos\frac{14\pi}{9} + i\sin\frac{14\pi}{9}\right)$

(c) $1.5321 + 1.2856i, -1.8794 + 0.6840i,$
$0.3473 - 1.9696i$

29. (a) $3\left(\cos\frac{\pi}{30} + i\sin\frac{\pi}{30}\right)$

$3\left(\cos\frac{13\pi}{30} + i\sin\frac{13\pi}{30}\right)$

$3\left(\cos\frac{5\pi}{6} + i\sin\frac{5\pi}{6}\right)$

$3\left(\cos\frac{37\pi}{30} + i\sin\frac{37\pi}{30}\right)$

$3\left(\cos\frac{49\pi}{30} + i\sin\frac{49\pi}{30}\right)$

(b)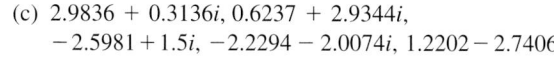

(c) $2.9836 + 0.3136i, 0.6237 + 2.9344i,$
$-2.5981 + 1.5i, -2.2294 - 2.0074i, 1.2202 - 2.7406i$

31. (a) $5\left(\cos\frac{3\pi}{4} + i\sin\frac{3\pi}{4}\right)$

$5\left(\cos\frac{7\pi}{4} + i\sin\frac{7\pi}{4}\right)$

(b)

(c) $-\frac{5\sqrt{2}}{2} + \frac{5\sqrt{2}}{2}i, \frac{5\sqrt{2}}{2} - \frac{5\sqrt{2}}{2}i$

33. (a) $3\left(\cos\dfrac{\pi}{8} + i\sin\dfrac{\pi}{8}\right)$

$3\left(\cos\dfrac{5\pi}{8} + i\sin\dfrac{5\pi}{8}\right)$

$3\left(\cos\dfrac{9\pi}{8} + i\sin\dfrac{9\pi}{8}\right)$

$3\left(\cos\dfrac{13\pi}{8} + i\sin\dfrac{13\pi}{8}\right)$

(b)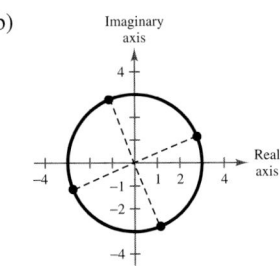

(c) $2.7716 + 1.1481i,\ -1.1481 + 2.7716i,$
$-2.7716 - 1.1481i,\ 1.1481 - 2.7716i$

35. (a) $5\left(\cos\dfrac{4\pi}{9} + i\sin\dfrac{4\pi}{9}\right)$

$5\left(\cos\dfrac{10\pi}{9} + i\sin\dfrac{10\pi}{9}\right)$

$5\left(\cos\dfrac{16\pi}{9} + i\sin\dfrac{16\pi}{9}\right)$

(b)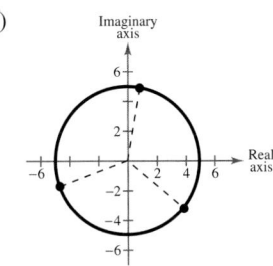

(c) $0.8682 + 4.924i,\ -4.6985 - 1.7101i,$
$3.8302 - 3.214i$

37. (a) $2(\cos 0 + i\sin 0)$

$2\left(\cos\dfrac{\pi}{2} + i\sin\dfrac{\pi}{2}\right)$

$2(\cos\pi + i\sin\pi)$

$2\left(\cos\dfrac{3\pi}{2} + i\sin\dfrac{3\pi}{2}\right)$

(b)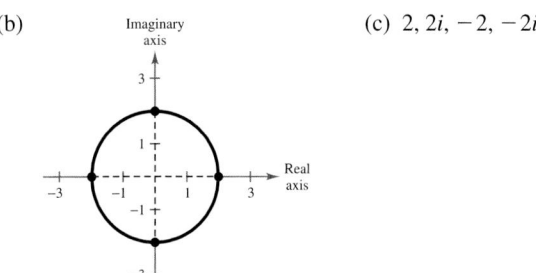

(c) $2, 2i, -2, -2i$

39. (a) $\cos 0 + i\sin 0$

$\cos\dfrac{2\pi}{5} + i\sin\dfrac{2\pi}{5}$

$\cos\dfrac{4\pi}{5} + i\sin\dfrac{4\pi}{5}$

$\cos\dfrac{6\pi}{5} + i\sin\dfrac{6\pi}{5}$

$\cos\dfrac{8\pi}{5} + i\sin\dfrac{8\pi}{5}$

(b)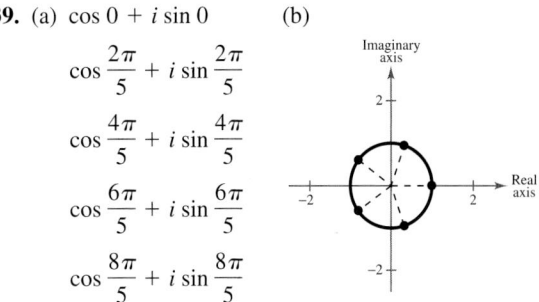

(c) $1, 0.3090 + 0.9511i,\ -0.8090 + 0.5878i,$
$-0.8090 - 0.5878i,\ 0.3090 - 0.9511i$

41. (a) $5\left(\cos\dfrac{\pi}{3} + i\sin\dfrac{\pi}{3}\right)$

$5(\cos\pi + i\sin\pi)$

$5\left(\cos\dfrac{5\pi}{3} + i\sin\dfrac{5\pi}{3}\right)$

(b)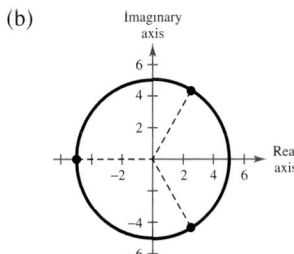

(c) $\dfrac{5}{2} + \dfrac{5\sqrt{3}}{2}i,\ -5,\ \dfrac{5}{2} - \dfrac{5\sqrt{3}}{2}i$

43. (a) $2\sqrt[5]{4\sqrt{2}}\left(\cos\dfrac{3\pi}{20} + i\sin\dfrac{3\pi}{20}\right)$

$2\sqrt[5]{4\sqrt{2}}\left(\cos\dfrac{11\pi}{20} + i\sin\dfrac{11\pi}{20}\right)$

$2\sqrt[5]{4\sqrt{2}}\left(\cos\dfrac{19\pi}{20} + i\sin\dfrac{19\pi}{20}\right)$

$2\sqrt[5]{4\sqrt{2}}\left(\cos\dfrac{27\pi}{20} + i\sin\dfrac{27\pi}{20}\right)$

$2\sqrt[5]{4\sqrt{2}}\left(\cos\dfrac{7\pi}{4} + i\sin\dfrac{7\pi}{4}\right)$

(b)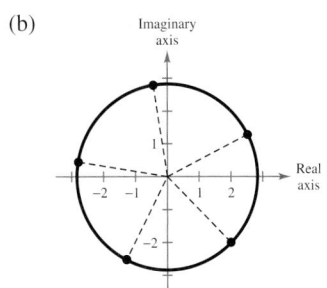

(c) $2.5201 + 1.2841i$, $-0.4425 + 2.7936i$,
$-2.7936 + 0.4425i$, $-1.2841 - 2.5201i$, $2 - 2i$

45. $\cos\dfrac{\pi}{8} + i\sin\dfrac{\pi}{8}$

$\cos\dfrac{5\pi}{8} + i\sin\dfrac{5\pi}{8}$

$\cos\dfrac{9\pi}{8} + i\sin\dfrac{9\pi}{8}$

$\cos\dfrac{13\pi}{8} + i\sin\dfrac{13\pi}{8}$

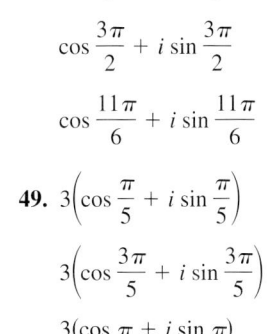

47. $\cos\dfrac{\pi}{6} + i\sin\dfrac{\pi}{6}$

$\cos\dfrac{\pi}{2} + i\sin\dfrac{\pi}{2}$

$\cos\dfrac{5\pi}{6} + i\sin\dfrac{5\pi}{6}$

$\cos\dfrac{7\pi}{6} + i\sin\dfrac{7\pi}{6}$

$\cos\dfrac{3\pi}{2} + i\sin\dfrac{3\pi}{2}$

$\cos\dfrac{11\pi}{6} + i\sin\dfrac{11\pi}{6}$

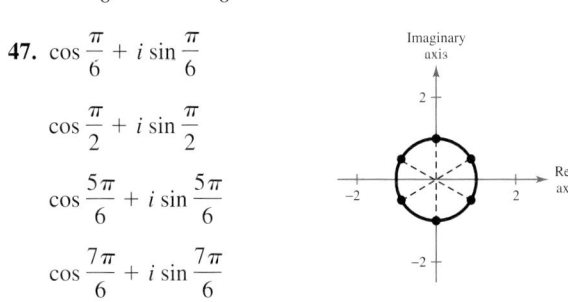

49. $3\left(\cos\dfrac{\pi}{5} + i\sin\dfrac{\pi}{5}\right)$

$3\left(\cos\dfrac{3\pi}{5} + i\sin\dfrac{3\pi}{5}\right)$

$3(\cos\pi + i\sin\pi)$

$3\left(\cos\dfrac{7\pi}{5} + i\sin\dfrac{7\pi}{5}\right)$

$3\left(\cos\dfrac{9\pi}{5} + i\sin\dfrac{9\pi}{5}\right)$

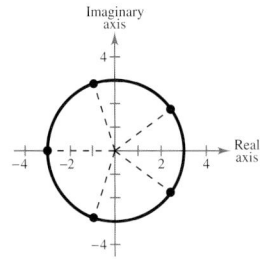

51. $2(\cos 0 + i\sin 0)$

$2\left(\cos\dfrac{2\pi}{5} + i\sin\dfrac{2\pi}{5}\right)$

$2\left(\cos\dfrac{4\pi}{5} + i\sin\dfrac{4\pi}{5}\right)$

$2\left(\cos\dfrac{6\pi}{5} + i\sin\dfrac{6\pi}{5}\right)$

$2\left(\cos\dfrac{8\pi}{5} + i\sin\dfrac{8\pi}{5}\right)$

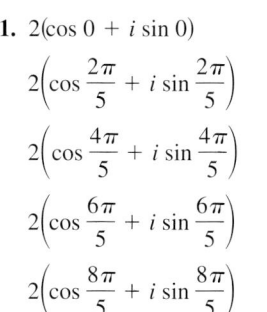

53. $2\left(\cos\dfrac{3\pi}{8} + i\sin\dfrac{3\pi}{8}\right)$

$2\left(\cos\dfrac{7\pi}{8} + i\sin\dfrac{7\pi}{8}\right)$

$2\left(\cos\dfrac{11\pi}{8} + i\sin\dfrac{11\pi}{8}\right)$

$2\left(\cos\dfrac{15\pi}{8} + i\sin\dfrac{15\pi}{8}\right)$

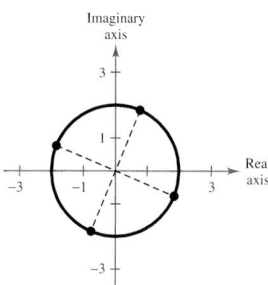

55. $2\left(\cos\dfrac{\pi}{8} + i\sin\dfrac{\pi}{8}\right)$

$2\left(\cos\dfrac{5\pi}{8} + i\sin\dfrac{5\pi}{8}\right)$

$2\left(\cos\dfrac{9\pi}{8} + i\sin\dfrac{9\pi}{8}\right)$

$2\left(\cos\dfrac{13\pi}{8} + i\sin\dfrac{13\pi}{8}\right)$

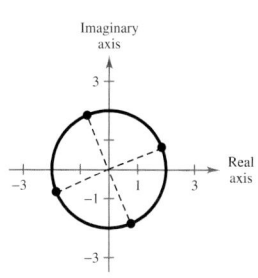

57. $\sqrt[6]{2}\left(\cos\dfrac{7\pi}{12} + i\sin\dfrac{7\pi}{12}\right)$

$\sqrt[6]{2}\left(\cos\dfrac{5\pi}{4} + i\sin\dfrac{5\pi}{4}\right)$

$\sqrt[6]{2}\left(\cos\dfrac{23\pi}{12} + i\sin\dfrac{23\pi}{12}\right)$

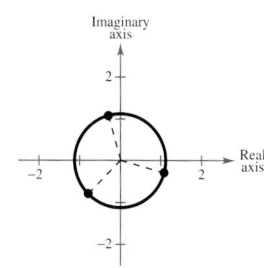

59. $\sqrt[12]{2}\left(\cos\dfrac{5\pi}{24} + i\sin\dfrac{5\pi}{24}\right)$

$\sqrt[12]{2}\left(\cos\dfrac{13\pi}{24} + i\sin\dfrac{13\pi}{24}\right)$

$\sqrt[12]{2}\left(\cos\dfrac{7\pi}{8} + i\sin\dfrac{7\pi}{8}\right)$

$\sqrt[12]{2}\left(\cos\dfrac{29\pi}{24} + i\sin\dfrac{29\pi}{24}\right)$

$\sqrt[12]{2}\left(\cos\dfrac{37\pi}{24} + i\sin\dfrac{37\pi}{24}\right)$

$\sqrt[12]{2}\left(\cos\dfrac{15\pi}{8} + i\sin\dfrac{15\pi}{8}\right)$

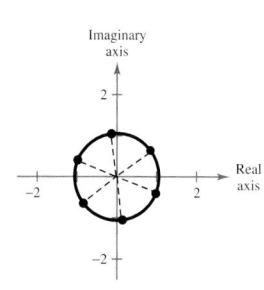

61. False. They are equally spaced along the circle centered at the origin with radius $\sqrt[n]{r}$.

63. Answers will vary.

65. (a) $2(\cos 30° + i \sin 30°)$
$2(\cos 150° + i \sin 150°)$
$2(\cos 270° + i \sin 270°)$
(b) $8i$

67. $\dfrac{\sqrt{34}}{136}(5\sqrt{7} - 9)$ **69.** $\dfrac{3\sqrt{34}}{136}(\sqrt{7} - 5)$

71. $\dfrac{120 + 51\sqrt{7}}{47}$ **73.** $\dfrac{15}{8}$ **75.** $\dfrac{1}{2}\sqrt{\dfrac{4 - \sqrt{7}}{2}}$

77. 5   **79.** 41   **81.** $\sqrt{493}$ **83.** $\sqrt{205}$

Review Exercises (page 338)

1. $6 + 2i$ **3.** $-1 + 3i$ **5.** $3 + 7i$ **7.** $40 + 65i$

9. $-4 - 46i$ **11.** $\dfrac{23 + 10i}{17}$ **13.** $\dfrac{21 - i}{13}$

15. $\pm\sqrt{\dfrac{1}{3}}i$ **17.** $1 \pm 3i$ **19.** Five solutions

21. Four solutions   **23.** Two real solutions

25. No real solutions   **27.** 0, 2   **29.** $-4 \pm \sqrt{6}$

31. $-\dfrac{1}{2} \pm \dfrac{\sqrt{5}}{2}i$ **33.** $1, \dfrac{3}{4}$ **35.** $-1, \dfrac{3}{2}, 3, \dfrac{2}{3}$

37. $-7, 2; (x + 7)(x - 2)^2$

39. $-5, 1 \pm 2i; (x + 5)(x - 1 - 2i)(x - 1 + 2i)$

41. $-\dfrac{1}{2}, 5 \pm 3i; (2x + 1)(x - 5 - 3i)(x - 5 + 3i)$

43. $-2, 3, -3 \pm \sqrt{5}i;$
$(x - 3)(x + 2)(x + 3 - \sqrt{5}i)(x + 3 + \sqrt{5}i)$

45. $12x^4 - 19x^3 + 9x - 2$ **47.** $x^3 - 7x^2 + 13x - 3$

49. $3x^4 - 14x^3 + 17x^2 - 42x + 24$

51. $x^4 + 27x^2 + 50$

53. Yes. A price of $95.41 or $119.59 per unit would yield a profit of 9 million dollars.

55.

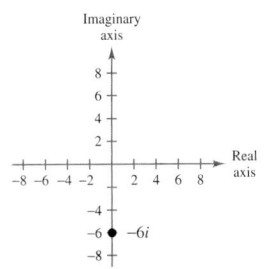

6

57.

Imaginary axis

$-10 - 4i$

$2\sqrt{29}$

59. $13(\cos 1.176 + i \sin 1.176)$ **61.** $7(\cos \pi + i \sin \pi)$

63. (a) $z_1 = 3\sqrt{2}\left(\cos \dfrac{5\pi}{4} + i \sin \dfrac{5\pi}{4}\right)$

$z_2 = 4\left(\cos \dfrac{\pi}{6} + i \sin \dfrac{\pi}{6}\right)$

(b) $z_1 z_2 = 12\sqrt{2}\left(\cos \dfrac{17\pi}{12} + i \sin \dfrac{17\pi}{12}\right)$

$\dfrac{z_1}{z_2} = \dfrac{3\sqrt{2}}{4}\left(\cos \dfrac{13\pi}{12} + i \sin \dfrac{13\pi}{12}\right)$

65. $-16 - 16\sqrt{3}i$ **67.** 16

69. (a) $4(\cos 60° + i \sin 60°)$
$4(\cos 150° + i \sin 150°)$
$4(\cos 240° + i \sin 240°)$
$4(\cos 330° + i \sin 330°)$

(b) $-128 - 128\sqrt{3}i$

71. $4(\cos 0 + i \sin 0) = 4$

$4\left(\cos \dfrac{\pi}{2} + i \sin \dfrac{\pi}{2}\right) = 4i$

$4(\cos \pi + i \sin \pi) = -4$

$4\left(\cos \dfrac{3\pi}{2} + i \sin \dfrac{3\pi}{2}\right) = -4i$

73. $2(\cos 0 + i \sin 0) = 2$

$2\left(\cos \dfrac{2\pi}{5} + i \sin \dfrac{2\pi}{5}\right) = 0.6180 + 1.9021i$

$2\left(\cos \dfrac{4\pi}{5} + i \sin \dfrac{4\pi}{5}\right) = -1.6180 + 1.1756i$

$2\left(\cos \dfrac{6\pi}{5} + i \sin \dfrac{6\pi}{5}\right) = -1.6180 - 1.1756i$

$2\left(\cos \dfrac{8\pi}{5} + i \sin \dfrac{8\pi}{5}\right) = 0.6180 - 1.9021i$

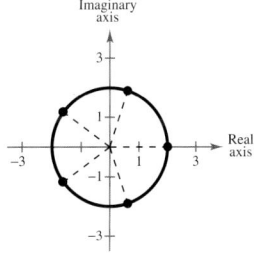

75. $\cos 0 + i \sin 0 = 1$

$\cos \frac{\pi}{2} + i \sin \frac{\pi}{2} = i$

$\cos \frac{2\pi}{3} + i \sin \frac{2\pi}{3} = -\frac{1}{2} + \frac{\sqrt{3}}{2}i$

$\cos \frac{4\pi}{3} + i \sin \frac{4\pi}{3} = -\frac{1}{2} - \frac{\sqrt{3}}{2}i$

$\cos \frac{3\pi}{2} + i \sin \frac{3\pi}{2} = -i$

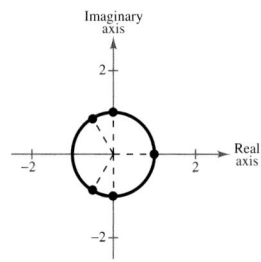

77. False. The equation has no real solution. The solutions are: $\frac{717}{650} \pm \frac{\sqrt{3311}\,i}{650}$.

79. Answers will vary.

81. (a) 3

(b) On the circle 120°, 210°, and 300° from the positive x-axis

Chapter Test (page 341)

1. $-3 + 9i$ **2.** $-3 + 5i$ **3.** $-32 + 24i$

4. 7 **5.** $2 - i$ **6.** $\frac{1}{2} \pm \frac{\sqrt{5}}{2}i$

7. Five solutions **8.** Four solutions

9. $6, \pm\sqrt{5}\,i$ **10.** $\pm\sqrt{6}, \pm 2i$

11. $\pm 2, \pm \sqrt{2}\,i;\ (x + 2)(x - 2)(x + \sqrt{2}\,i)(x - \sqrt{2}\,i)$

12. $\frac{3}{2}, 2 \pm i;\ (2v - 3)(v - 2 - i)(v - 2 + i)$

13. $x^4 - 9x^3 + 28x^2 - 30x$

14. $x^4 - 8x^3 + 28x^2 - 60x + 63$

15. No. If $a + bi$ is a zero, its conjugate $a - bi$ is also a zero.

16. $5\sqrt{2}\left(\cos \frac{7\pi}{4} + i \sin \frac{7\pi}{4}\right)$

17. $-3 + 3\sqrt{3}\,i$

18. $-\frac{6561}{2} - \frac{6561\sqrt{3}}{2}i$

19. $5832i$

20. $4\sqrt[4]{2}\left(\cos \frac{\pi}{12} + i \sin \frac{\pi}{12}\right)$

$4\sqrt[4]{2}\left(\cos \frac{7\pi}{12} + i \sin \frac{7\pi}{12}\right)$

$4\sqrt[4]{2}\left(\cos \frac{13\pi}{12} + i \sin \frac{13\pi}{12}\right)$

$4\sqrt[4]{2}\left(\cos \frac{19\pi}{12} + i \sin \frac{19\pi}{12}\right)$

21. $3\left(\cos \frac{\pi}{6} + i \sin \frac{\pi}{6}\right)$

$3\left(\cos \frac{5\pi}{6} + i \sin \frac{5\pi}{6}\right)$

$3\left(\cos \frac{3\pi}{2} + i \sin \frac{3\pi}{2}\right)$

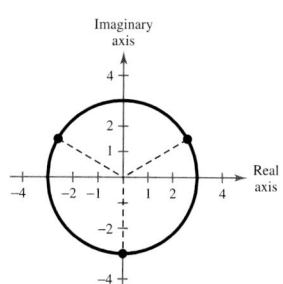

Chapter 5

Section 5.1 (page 352)

1. 946.852 **3.** 7.352 **5.** 0.006 **7.** 673.639

9. 0.472 **11.** d **13.** a

15. Shift the graph of f 4 units to the right.

17. Shift the graph of f 5 units upward.

19. Reflect the graph of f in the x-axis and shift f 4 units to the left.

21. Reflect the graph of f in the x-axis and shift f 5 units upward.

23.

x	-2	-1	0	1	2
$f(x)$	4	2	1	0.5	0.25

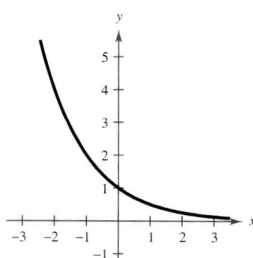

Answers to Odd-Numbered Exercises and Tests **A83**

25.

x	-2	-1	0	1	2
$f(x)$	0.25	0.5	1	2	4

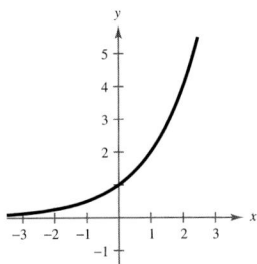

27.

x	-2	-1	0	1	2
$f(x)$	0.125	0.25	0.5	1	2

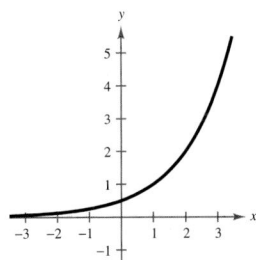

29.

x	-2	-1	0	1	2
$f(x)$	0.135	0.368	1	2.718	7.389

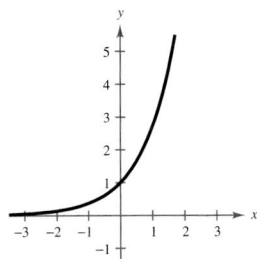

31.

x	-8	-7	-6	-5	-4
$f(x)$	0.055	0.149	0.406	1.104	3

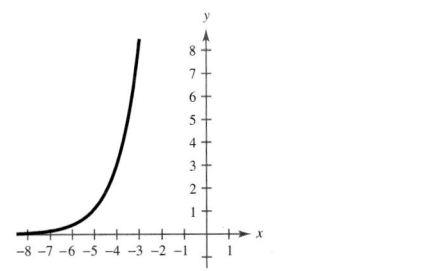

33.

x	-2	-1	0	1	2
$f(x)$	4.037	4.100	4.271	4.736	6

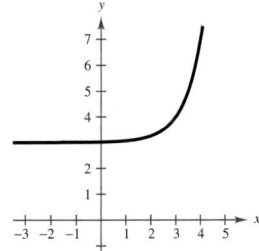

35.

x	-1	0	1	2	3
$f(x)$	3.003	3.016	3.063	3.25	4

37. **39.**

41. **43.**

45. **47.**

49. **51.**

53.
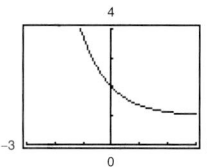

55.

n	1	2	4
A	$5397.31	$5477.81	$5520.10

n	12	365	Continuous
A	$5549.10	$5563.36	$5563.85

57.

n	1	2	4
A	$11,652.39	$12,002.55	$12,188.60

n	12	365	Continuous
A	$12,317.01	$12,380.41	$12,382.58

59.

t	1	10	20
A	$12,999.44	$26,706.49	$59,436.39

t	30	40	50
A	$132,278.12	$294,390.36	$655,177.80

61.

t	1	10	20
A	$12,805.91	$22,986.49	$44,031.56

t	30	40	50
A	$84,344.25	$161,564.86	$309,484.08

63. $222,822.57

65. (a) The steeper curve represents the investment earning compound interest, because compound interest earns more than simple interest.

(b) $A = 500(1.07)^t$

$A = 500(0.07)t + 500$

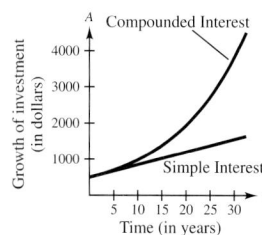

67. $35.45   **69.** (a) 100   (b) 300   (c) 900

71. (a) 25 units   (b) 16.30 units

(c)

73. (a)

(b)

h	0	5	10	15	20
P	102,303	51,570	25,996	13,104	6606

(c) 34,190 pascals   (d) ≈ 11.6 kilometers

75. True. As $x \to -\infty$, $f(x) \to 0$ but never reaches 0.

77. $f(x) = h(x)$ **79.** $f(x) = g(x) = h(x)$

81. (a) $x < 0$ (b) $x > 0$

83. (a) 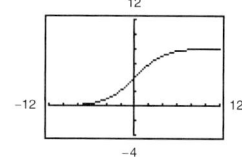   Horizontal asymptotes: $y = 0$, $y = 8$

(b)   Horizontal asymptote: $y = 4$

Vertical asymptote: $x = 0$

85.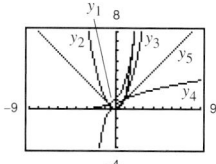

$y = e^x$

87. It usually implies rapid growth.

89. The value of $f(x)$ approaches e^r.

x	1	10	100	200	500
$[1 + (1/x)]^x$	2	2.5937	2.7048	2.7115	2.7156

x	1100	10,000
$[1 + (1/x)]^x$	2.7170	2.718

91. $1 < \sqrt{2} < 2$
$2^1 < 2^{\sqrt{2}} < 2^2$

93. $y_4 = 1 + \dfrac{x}{1!} + \dfrac{x^2}{2!} + \dfrac{x^3}{3!} + \dfrac{x^4}{4!}$

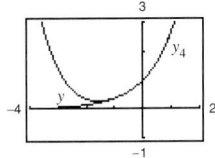

As more terms are added, the polynomial approaches e^x.

95. $y = \frac{1}{7}(2x + 14)$ **97.** $y = \pm\sqrt{25 - x^2}$
99. $f(x) = x^3 - x^2 + 64x - 64$
101. $f(x) = x^3 - 16x^2 + 85x - 148$

Section 5.2 (page 363)

1. $4^3 = 64$ **3.** $7^{-2} = \frac{1}{49}$ **5.** $32^{2/5} = 4$
7. $e^0 = 1$ **9.** $\log_5 125 = 3$ **11.** $\log_{81} 3 = \frac{1}{4}$
13. $\log_6 \frac{1}{36} = -2$ **15.** $\ln 20.0855\ldots = 3$
17. $\ln 1 = 0$ **19.** 4 **21.** $\frac{1}{2}$ **23.** 0
25. -2 **27.** $\frac{5}{3}$ **29.** 3 **31.** 2
33. 2.538 **35.** -0.097 **37.** 2.913 **39.** -3.418
41. 1.005 **43.** -0.405 **45.** c **47.** d **49.** b
51. Domain: $(0, \infty)$
Intercept: $(1, 0)$
Vertical asymptote: $x = 0$

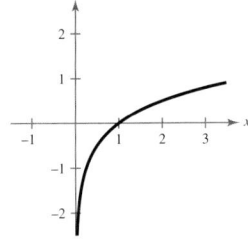

53. Domain: $(0, \infty)$
Intercept: $(9, 0)$
Vertical asymptote: $x = 0$

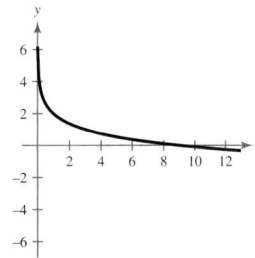

55. Domain: $(-2, \infty)$
Intercept: $(-1, 0)$
Vertical asymptote: $x = -2$

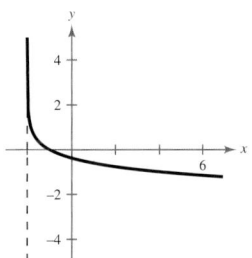

57. Domain: $(0, \infty)$
Intercept: $(5, 0)$
Vertical asymptote: $x = 0$

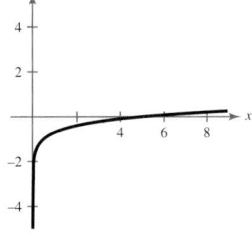

59. Domain: $(2, \infty)$
Intercept: $(3, 0)$
Vertical asymptote: $x = 2$

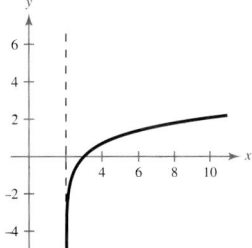

61. Domain: $(-\infty, 0)$

Intercept: $(-1, 0)$

Vertical asymptote: $x = 0$

63.   **65.**

67.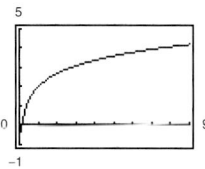

69. (a) 80   (b) 68.1   (c) 62.3

71. (a)

r	0.005	0.01	0.015	0.02	0.025	0.03
t	138.6	69.3	46.2	34.7	27.7	23.1

(b) Answers will vary.

73.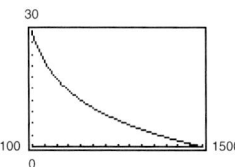

17.66 cubic feet per minute

75. 21,357 foot-pounds   **77.** 30 years

79. Total amount: $396,234

Interest: $246,234

81. (a)

x	1	5	10	10^2
f(x)	0	0.322	0.230	0.046

x	10^4	10^6
f(x)	0.00092	0.0000138

(b) 0

(c)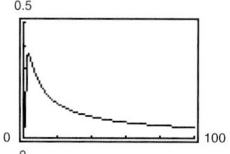

83. True. $\log_3 27 = 3 \Rightarrow 3^3 = 27$

85.   **87.**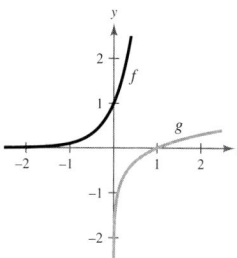

$g = f^{-1}$ $g = f^{-1}$

89. (a) False   (b) True   (c) True   (d) False

91. $y_4 = (x-1) - \frac{1}{2}(x-1)^2 + \frac{1}{3}(x-1)^3 - \frac{1}{4}(x-1)^4$

93. (a)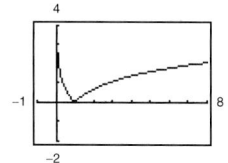

(b) Increasing: $(1, \infty)$

Decreasing: $(0, 1)$

(c) Relative minimum: $(1, 0)$

95. (a)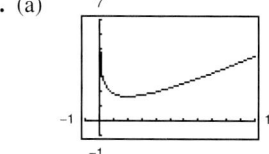

(b) Increasing: $(2, \infty)$

Decreasing: $(0, 2)$

(c) Relative minimum: $\left(2, 1 - \ln\frac{1}{2}\right)$

97. $8n - 3$ **99.** $83.95 + 37.50t$

101. 403.429   **103.** 0.018

105.   **107.**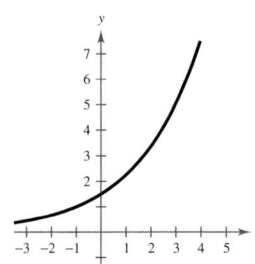

59. $\ln \dfrac{\sqrt[3]{y(y+4)^2}}{y-1}$ **61.** $\ln \dfrac{9}{\sqrt{x^2+1}}$

63. $\log_2 \tfrac{32}{4} = \log_2 32 - \log_2 4$; Property 2

65. 2 **67.** 2.4 **69.** -9 is not in the domain of $\log_3 x$.

71. 2 **73.** -3 **75.** 0 is not in the domain of $\log_{10} x$.

77. 4.5 **79.** $\tfrac{3}{2}$ **81.** $-3 - \log_5 2$ **83.** $6 + \ln 5$

85. (a) 90 (b) 77 (c) 73 (d) 9 months

(e) $90 - \log_{10}(t+1)^{15}$

(f)

Section 5.3 *(page 371)*

1. 1.771 **3.** -2.000 **5.** -0.417 **7.** 2.633

9. (a) $\dfrac{\log_{10} x}{\log_{10} 5}$ (b) $\dfrac{\ln x}{\ln 5}$ **11.** (a) $\dfrac{\log_{10} x}{\log_{10} \tfrac{1}{5}}$ (b) $\dfrac{\ln x}{\ln \tfrac{1}{5}}$

13. (a) $\dfrac{\log_{10} \tfrac{3}{10}}{\log_{10} x}$ (b) $\dfrac{\ln \tfrac{3}{10}}{\ln x}$

15. (a) $\dfrac{\log_{10} x}{\log_{10} 2.6}$ (b) $\dfrac{\ln x}{\ln 2.6}$

17. $f(x) = \dfrac{\log_{10} x}{\log_{10} 2} = \dfrac{\ln x}{\ln 2}$ **19.** $f(x) = \dfrac{\log_{10} x}{\log_{10} \tfrac{1}{2}} = \dfrac{\ln x}{\ln \tfrac{1}{2}}$

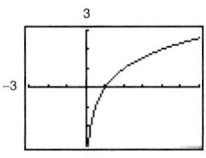

87. False. $\ln 1 = 0$ **89.** False. $\ln(x-2) \neq \ln x - \ln 2$

91. False. $u = v^2$ **93.** Answers will vary.

95.

21. $f(x) = \dfrac{\log_{10} x}{\log_{10} 11.8} = \dfrac{\ln x}{\ln 11.8}$ **23.** $\log_{10} 5 + \log_{10} x$

97.

$f(x) = h(x)$; Property 2

25. $\log_{10} 5 - \log_{10} x$ **27.** $4 \log_8 x$ **29.** $\tfrac{1}{2} \ln z$

99. $\dfrac{3x^4}{2y^3}$, $x \neq 0$ **101.** $1, x \neq 0, y \neq 0$

31. $\ln x + \ln y + \ln z$ **33.** $\tfrac{1}{2}\ln(a-1)$

103. 2502.655 **105.** 0.002 **107.** 4.325

35. $\ln z + 2\ln(z-1)$ **37.** $\tfrac{1}{3}\ln x - \tfrac{1}{3}\ln y$

109. 1.415 **111.** 2.361

39. $4\ln x + \tfrac{1}{2}\ln y - 5\ln z$

41. $2\log_b x - 2\log_b y - 3\log_b z$

Section 5.4 *(page 380)*

1. (a) Yes (b) No

43. $\ln 3x$ **45.** $\log_4 \dfrac{z}{y}$ **47.** $\log_2(x+4)^2$

3. (a) No (b) Yes (c) Yes, approximate

5. (a) Yes, approximate (b) No (c) Yes **7.** 2

49. $\log_3 \sqrt[4]{5x}$ **51.** $\ln \dfrac{x}{(x+1)^3}$ **53.** $\ln \dfrac{x-2}{x+2}$

9. 4 **11.** -2 **13.** -5 **15.** 3 **17.** 4

19. 2 **21.** $\ln 2 \approx 0.693$ **23.** $e^{-1} \approx 0.368$

55. $\ln \dfrac{x}{(x^2-4)^4}$ **57.** $\ln \sqrt[3]{\dfrac{x(x+3)^2}{x^2-1}}$

25. 64 **27.** 100 **29.** $\tfrac{1}{10}$ **31.** (3, 8) **33.** (9, 2)

35. x^2 **37.** $x-2, x > 2$ **39.** $7x+2$

41. $5x + 2, x > -\frac{2}{5}$ **43.** $2x - 1$ **45.** $\ln 10 \approx 2.303$
47. 0 **49.** $\frac{\ln 12}{3} \approx 0.828$ **51.** $\ln \frac{5}{3} \approx 0.511$
53. $\ln 5 \approx 1.609$ **55.** $2 \ln 75 \approx 8.635$
57. $\log_{10} 42 \approx 1.623$ **59.** $\frac{\ln 80}{2 \ln 3} \approx 1.994$
61. 2 **63.** $\frac{\ln 8 - \ln 565}{\ln 2} \approx -6.142$

65.   **67.**

-0.427 3.847

69.   **71.**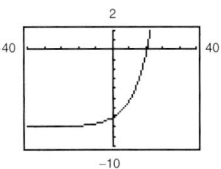

12.207 16.636

73. $\frac{1}{3} \log_{10}\left(\frac{3}{2}\right) \approx 0.059$ **75.** $1 + \frac{\ln 7}{\ln 5} \approx 2.209$
77. $\frac{\ln 4}{365 \ln\left(1 + \frac{0.065}{365}\right)} \approx 21.330$
79. $\frac{\ln 2}{12 \ln\left(1 + \frac{0.10}{12}\right)} \approx 6.960$ **81.** $\frac{\ln 1498}{2} \approx 3.656$
83. $e^{-3} \approx 0.050$ **85.** $\frac{e^{2.4}}{2} \approx 5.512$ **87.** $\frac{e^{10/3}}{5} \approx 5.606$
89. $e^2 - 2 \approx 5.389$
91. $e - 1 \approx 1.718$ and $-e - 1 \approx -3.718$
93. $1 + \sqrt{1 + e} \approx 2.928$ **95.** No solution
97. $10^2 + 3 = 103$ **99.** $2(3^{11/6}) \approx 14.988$
101. $\frac{-1 + \sqrt{17}}{2} \approx 1.562$ **103.** 2
105. $\frac{725 + 125\sqrt{33}}{8} \approx 180.384$

107.   **109.**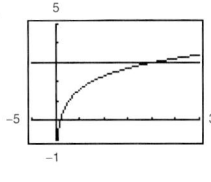

$(2.807, 7)$ $(20.086, 3)$

111. 8.2 years   **113.** 12.9 years
115. (a) 1426 units   (b) 1498 units

117. (a)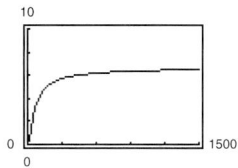

(b) $V = 6.7$. Yield will approach 6.7 million cubic feet per acre.
(c) 29.3 years
119. (a) $y = 100$ and $y = 0$; the range falls between 0% and 100%.
(b) Males: 69.71 inches   Females: 64.51 inches
121. (a) $T = 20$; Room temperature   (b) ≈ 0.81 hour
123. $\log_b uv = \log_b u + \log_b v$
True by Property 1 in Section 5.3.
125. $\log_b(u - v) = \log_b u - \log_b v$
False. $1.95 \approx \log_{10}(100 - 10) \neq \log_{10} 100 - \log_{10} 10 = 1$
127. For $rt < \ln 2$ years, double the amount you invest. For $rt > \ln 2$ years, double the interest rate or double the number of years, because either of these will double the exponent in the exponential function.
129. No. It is dependent on the interest rate.
131. $4\sqrt{2} - 10$ **133.** $\frac{1}{2}\sqrt{10} + 1$ **135.** $t = \frac{k}{s^3}$
137. $x = \frac{k}{b - 3}$ **139.** 1.226 **141.** -5.595

Section 5.5   (page 391)

1. c   **3.** b   **5.** d

Initial Investment	Annual % Rate	Time to Double	Amount after 10 Years
7. $1000	12%	5.78 yr	$3320.12
9. $750	8.9438%	7.75 yr	$1834.37
11. $500	11.0%	6.3 yr	$1505.00
13. $6376.28	4.5%	15.4 yr	$10,000.00

15. $112,087.09
17. (a) 6.642 years   (b) 6.330 years
(c) 6.302 years   (d) 6.301 years

19.

r	2%	4%	6%	8%	10%	12%
t	54.93	27.47	18.31	13.73	10.99	9.16

21.

r	2%	4%	6%	8%	10%	12%
t	55.48	28.01	18.85	14.27	11.53	9.69

23.

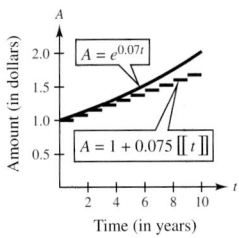

Continuous compounding

Isotope	Half-life (Years)	Initial Quantity	Amount after 1000 Years
25. ^{226}Ra	1620	10 g	6.52 g
27. ^{14}C	5730	2.26 g	2 g
29. ^{239}Pu	24,360	2.16 g	2.1 g

31. $y = e^{0.7675x}$ **33.** $y = 5e^{-0.4024x}$

35. 2023   **37.** $k = 0.0112$; 2796

39. (a) Croatia: $y = 5e^{-0.0018t}$; 4.7 million

Mali: $y = 9.9e^{0.0314t}$; 27.9 million

Singapore: $y = 3.5e^{0.0090t}$; 4.7 million

Sweden: $y = 8.9e^{0.0028t}$; 9.8 million

(b) The greater the rate of growth, the greater the value of b.

(c) b determines whether the population is increasing ($b > 0$) or decreasing ($b < 0$).

41. 61.16 hours   **43.** 15,683 years

45. (a) $V = -750t + 2000$ (b) $V = 2000e^{-0.6931t}$

(c)

Exponential

(d) 1 year:
Straight-line, $1250; Exponential, $1000
3 years:
Straight-line, $-$$250; Exponential, $250

(e) Decreases $750 per year

47. (a) $S = \dfrac{500,000}{1 + 0.6e^{0.053t}}$ (b) 280,771 units

49. $583,275

51. (a)

Asymptotes: $y = 0, y = 1000$. The population size will approach 1000 as time increases.

(b) 203 animals

(c) 13 months

53. (a) 398,107,171   (b) 5,011,872   (c) 50,118,723

55. (a) 30 decibels   (b) 85 decibels

(c) 90 decibels   (d) 115 decibels

57. 97%   **59.** 4.95

61. $10^{-3.2} \approx 6.3 \times 10^{-4}$ moles per liter   **63.** 10

65. (a)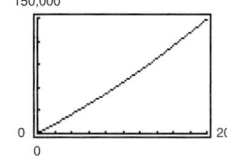

(b) $\approx$ 21 years; Yes

67. 7:30 A.M.

69. False. A logistics growth function never has an x-intercept.

71. (a) Logarithmic   (b) Logistic   (c) Exponential
(d) Linear   (e) None of the above   (f) Exponential

73. Answers will vary.

75.   **77.**

79.   **81.**

83.   **85.**

87.

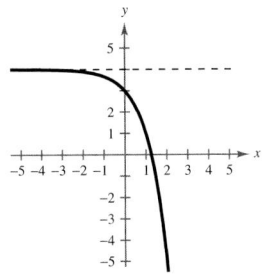

Review Exercises (page 398)

1. 76.699 **3.** 0.337 **5.** 1201.845 **7.** c **9.** a

11.

x	−1	0	1	2	3
f(x)	8	5	4.25	4.063	4.016

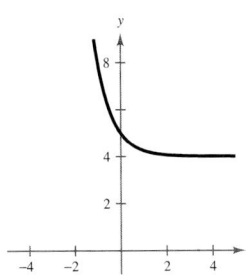

13.

x	−2	−1	0	1	2
f(x)	−0.377	−1	−2.65	−7.023	−18.61

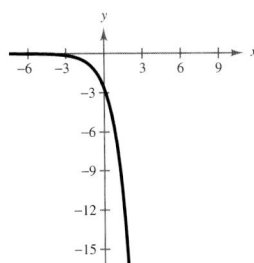

15.

x	−1	0	1	2	3
f(x)	4.008	4.04	4.2	5	9

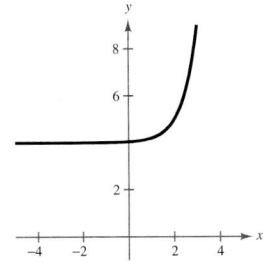

17.

x	−2	−1	0	1	2
f(x)	3.25	3.5	4	5	7

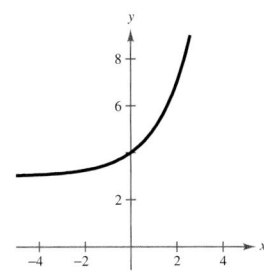

19. 2980.958 **21.** 0.183

23.

x	−2	−1	0	1	2
h(x)	2.72	1.65	1	0.61	0.37

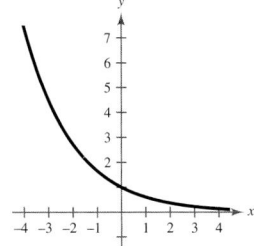

25.

x	−3	−2	−1	0	1
f(x)	0.37	1	2.72	7.39	20.09

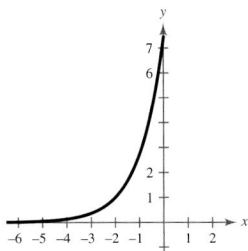

27.

n	1	2	4	12
A	$6569.98	$6635.43	$6669.46	$6692.64

n	365	Continuous
A	$6704.00	$6704.39

29.

t	1	10	20
P	$184,623.27	$89,865.79	$40,379.30

t	30	40	50
P	$18,143.59	$8152.44	$3663.13

31. (a) 0.154 (b) 0.487 (c) 0.811
33. (a) $1,069,047.14 (b) 7.9 years
35. $\log_4 64 = 3$ **37.** 3 **39.** -3

41.
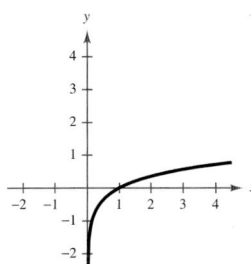
Vertical asymptote: $x = 0$

43.
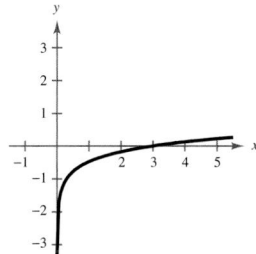
Vertical asymptote: $x = 0$

45.
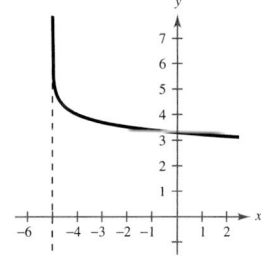
Vertical asymptote: $x = -5$

47. 3.118 **49.** -12 **51.** 2.034

53.
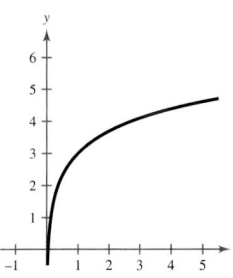
Vertical asymptote: $x = 0$

55.
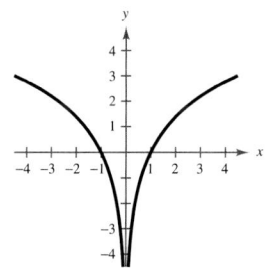
Vertical asymptote: $x = 0$

57. 27.16 miles **59.** 2.132 **61.** -1.159
63. and 65. Answers will vary.
67. $1 + 2\log_5|x|$ **69.** $\log_{10} 5 + \frac{1}{2}\log_{10} y - 2\log_{10}|x|$
71. $\log_2 5x$ **73.** $\ln \dfrac{\sqrt{|2x-1|}}{(x+1)^2}$

75. (a) $0 \le h < 18{,}000$

(b)

Vertical asymptote: $h = 18{,}000$

(c) Climbing at a slower rate, so the time required increases.

(d) 5.46 minutes

77. 6 **79.** 6 **81.** 3 **83.** $\dfrac{\ln 25}{3} \approx 1.073$

85. $\frac{1}{3}(\ln 40 - 2) \approx 0.563$ **87.** $\ln 20 \approx 2.996$

89. $\dfrac{\ln 95}{\ln 12} \approx 1.833$ **91.** $\ln 2 \approx 0.693$, $\ln 4 \approx 1.386$

93. $-7.04, -1.53$ **95.** 0.68 **97.** $\frac{1}{5}e^{7.2} \approx 267.886$
99. $\frac{1}{3}e^{15/4} \approx 14.174$ **101.** $e^6 - 8 \approx 395.429$
103. $5e^4 \approx 272.991$ **105.** $-2 + \sqrt{6} \approx 0.449$
107. -104 **109.** 0, 0.42, 13.63 **111.** $-3.99, 1.48$
113. 21.3 years **115.** e **117.** f **119.** a
121. 2025 **123.** (a) 13.8629% (b) $11,486.98
125. $y = \frac{1}{2}e^{0.4605x}$
127. (a) 7.7 weeks (b) 13.3 weeks
129. (a) 251,188,643 (b) 7,079,458 (c) 1,258,925,412
131. True by properties of exponents.
$$e^{x-1} = e^x \cdot e^{-1} = \dfrac{e^x}{e}$$
133. False. $\ln(x \cdot y) = \ln x + \ln y \ne \ln(x+y)$
135. False. The domain of $f(x) = \ln x$ is $(0, \infty)$.

Chapter Test (page 403)

1. 1123.690 **2.** 687.291 **3.** 0.497 **4.** 22.198

5.

x	-1	$-\frac{1}{2}$	0	$\frac{1}{2}$	1
$f(x)$	10	3.162	1	0.316	0.1

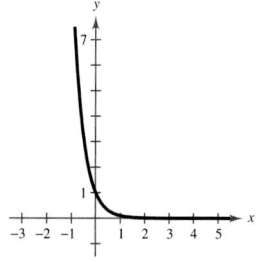

6.

x	-1	0	1	2	3
$f(x)$	-0.005	-0.028	-0.167	-1	-6

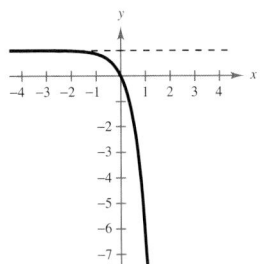

7.

x	-1	$-\frac{1}{2}$	0	$\frac{1}{2}$	1
$f(x)$	0.865	0.632	0	-1.718	-6.389

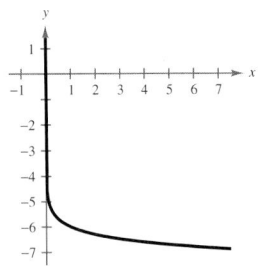

8. (a) -0.89 (b) 9.2

9.

x	$\frac{1}{2}$	1	$\frac{3}{2}$	2	4
$f(x)$	-5.699	-6	-6.176	-6.301	-6.602

Vertical asymptote: $x = 0$

10.

x	5	7	9	11	13
$f(x)$	0	1.099	1.609	1.946	2.197

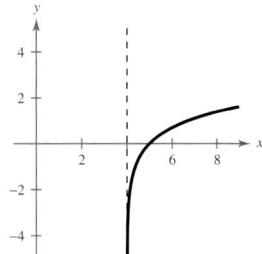

Vertical asymptote: $x = 4$

11.

x	-5	-3	-1	0	1
$f(x)$	1	2.099	2.609	2.792	2.946

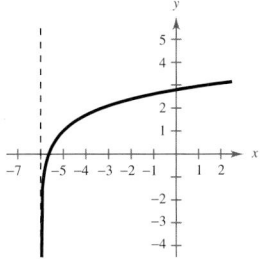

Vertical asymptote: $x = -6$

12. 1.945 **13.** 0.115 **14.** 1.328
15. $\log_2 3 + 4\log_2 a$ **16.** $\ln 5 + \frac{1}{2}\ln x - \ln 6$
17. $\log_3 13y$ **18.** $\ln \dfrac{x^4}{y^4}$ **19.** $\dfrac{\ln 197}{4} \approx 1.321$
20. $\dfrac{800}{501} \approx 1.597$ **21.** $y = 2745e^{0.1570x}$ **22.** 55%
23. (a)

x	$\frac{1}{4}$	1	2	4	5	6
H	58.720	75.332	86.828	103.43	110.59	117.38

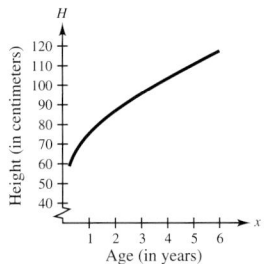

(b) 103 centimeters; 103.43 centimeters

Chapter 6
Section 6.1 (page 410)

1. $\dfrac{\sqrt{3}}{3}$ 3. -1 5. $\sqrt{3}$ 7. 3.2236

9. $\dfrac{3\pi}{4}$ radians, $135°$ 11. $\dfrac{\pi}{4}$ radian, $45°$

13. 0.6435 radian, $36.9°$ 15. 1.0517 radians, $60.3°$
17. 2.1112 radians, $121.0°$ 19. 1.2490 radians, $71.6°$
21. 2.1112 radians, $121.0°$ 23. 1.1071 radians, $63.4°$
25. 0.1974 radian, $11.3°$ 27. 1.4289 radians, $81.9°$
29. 0.9273 radian, $53.1°$ 31. 0.8187 radian, $46.9°$
33. $(2, 1), 42.3°; (4, 4), 78.7°; (6, 2), 59.0°$
35. $(-4, -1), 11.9°; (3, 2), 21.8°; (1, 0), 146.3°$ 37. 0

39. $\dfrac{7}{5}$ 41. 7 43. $\dfrac{8\sqrt{37}}{37} \approx 1.3152$

45. (a) 4 (b) 8 47. (a) $\dfrac{35\sqrt{37}}{74}$ (b) $\dfrac{35}{8}$

49. $2\sqrt{2}$ 51. $0.1003, 1054$ feet
53. (a) $18.4°$ (b) 15.8 meters
55. $\alpha = 33.69°; \beta = 56.31°$

57. True. The inclination of a line is related to its slope by $m = \tan \theta$. If the angle is greater than $\pi/2$ but less than π, then the angle is in the second quadrant, where the tangent function is negative.

59. (a) $d = \dfrac{4}{\sqrt{m^2 + 1}}$

(b)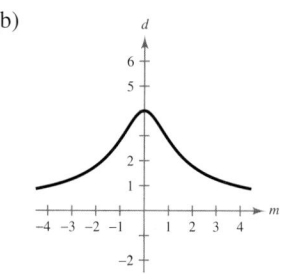

(c) $m = 0$

(d) $d = 0$. As the line approaches the vertical, the distance approaches 0.

61. x-intercept: $(7, 0)$
    y-intercept: $(0, 49)$
63. x-intercepts: $(5 \pm \sqrt{5}, 0)$
    y-intercept: $(0, 20)$
65. x-intercepts: $\left(\dfrac{7 \pm \sqrt{53}}{2}, 0\right)$
    y-intercept: $(0, -1)$
67. $f(x) = 3\left(x + \tfrac{1}{3}\right)^2 - \tfrac{49}{3}$
    Vertex: $\left(-\tfrac{1}{3}, -\tfrac{49}{3}\right)$
69. $f(x) = 5\left(x + \tfrac{17}{5}\right)^2 - \tfrac{324}{5}$
    Vertex: $\left(-\tfrac{17}{5}, -\tfrac{324}{5}\right)$
71. $f(x) = 6\left(x - \tfrac{1}{12}\right)^2 - \tfrac{289}{24}$
    Vertex: $\left(\tfrac{1}{12}, -\tfrac{289}{24}\right)$

Section 6.2 (page 418)

1. A circle is formed when a plane intersects the top or bottom half of a double-napped cone and is perpendicular to the axis of the cone.
3. A parabola is formed when a plane intersects the top or bottom half of a double-napped cone, is parallel to the side of the cone, and does not intersect the vertex.
5. e   7. d   9. a

11. Vertex: $(0, 0)$
    Focus: $\left(0, \tfrac{1}{2}\right)$
    Directrix: $y = -\tfrac{1}{2}$

13. Vertex: $(0, 0)$
    Focus: $\left(-\tfrac{3}{2}, 0\right)$
    Directrix: $x = \tfrac{3}{2}$

   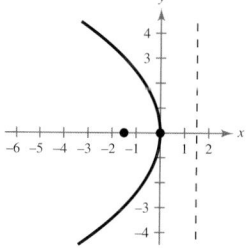

15. Vertex: $(0, 0)$
    Focus: $\left(0, -\tfrac{3}{2}\right)$
    Directrix: $y = \tfrac{3}{2}$

17. Vertex: $(1, -2)$
    Focus: $(1, -4)$
    Directrix: $y = 0$

   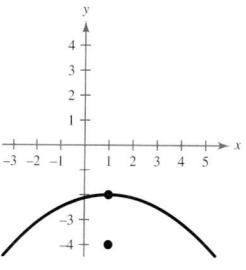

19. Vertex: $\left(-\frac{3}{2}, 2\right)$
Focus: $\left(-\frac{3}{2}, 3\right)$
Directrix: $y = 1$

21. Vertex: $(1, 1)$
Focus: $(1, 2)$
Directrix: $y = 0$

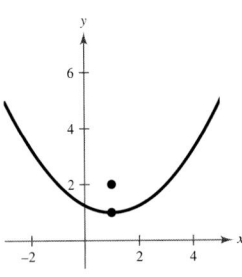

23. Vertex: $(-2, -3)$
Focus: $(-4, -3)$
Directrix: $x = 0$

25. Vertex: $(-2, 1)$
Focus: $\left(-2, -\frac{1}{2}\right)$
Directrix: $y = \frac{5}{2}$

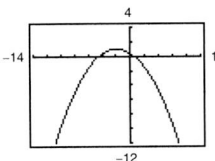

27. Vertex: $\left(\frac{1}{4}, -\frac{1}{2}\right)$
Focus: $\left(0, -\frac{1}{2}\right)$
Directrix: $x = \frac{1}{2}$

29.

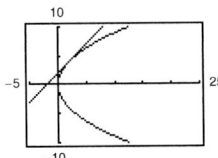

$(2, 4)$

31. $x^2 = \frac{3}{2}y$ **33.** $x^2 = -6y$ **35.** $y^2 = -8x$
37. $x^2 = 4y$ **39.** $y^2 = -8x$ **41.** $y^2 = 9x$
43. $(x - 3)^2 = -(y - 1)$ **45.** $y^2 = 4(x + 4)$
47. $(y - 2)^2 = -8(x - 5)$ **49.** $x^2 = 8(y - 4)$
51. $(y - 2)^2 = 8x$ **53.** $y = \sqrt{6(x + 1)} + 3$
55. $4x - y - 8 = 0$; $(2, 0)$ **57.** $4x - y + 2 = 0$; $\left(-\frac{1}{2}, 0\right)$

59.

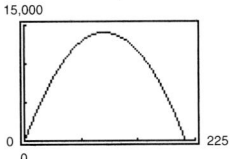

$x = 106$ units

61. $y = \frac{1}{18}x^2$ **63.** (a) $y = -\frac{1}{640}x^2$ (b) 8 feet
65. (a) $17{,}500\sqrt{2}$ miles per hour
  (b) $x^2 = -16{,}400(y - 4100)$
67. 43.3 seconds prior to being over the target
69. True. A parabola is the set of all points that are equidistant from the directrix and the focus.
71. (a) $\frac{64}{3}$
  (b) The parabola becomes narrower for $0 \leq y \leq b$.
73. $y = x^3 - 7x^2 + 17x - 15$ **75.** $\pm 2, \frac{1}{2}, -\frac{5}{3}$

Section 6.3 (page 428)

1. b **3.** d **5.** a

7. Center: $(0, 0)$
Vertices: $(\pm 5, 0)$
Foci: $(\pm 3, 0)$
Eccentricity: $\frac{3}{5}$

9. Center: $(0, 0)$
Vertices: $(0, \pm 3)$
Foci: $(0, \pm 2)$
Eccentricity: $\frac{2}{3}$

11. Center: $(-3, 5)$
Vertices: $(-3, 10), (-3, 0)$
Foci: $(-3, 8), (-3, 2)$
Eccentricity: $\frac{3}{5}$

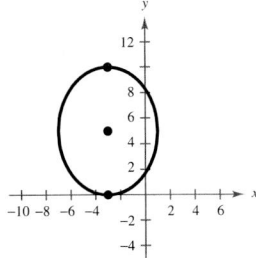

13. Center: $(-5, 1)$

Vertices: $\left(-\frac{7}{2}, 1\right), \left(-\frac{13}{2}, 1\right)$

Foci: $\left(-5 + \frac{\sqrt{5}}{2}, 1\right), \left(-5 - \frac{\sqrt{5}}{2}, 1\right)$

Eccentricity: $\frac{\sqrt{5}}{3}$

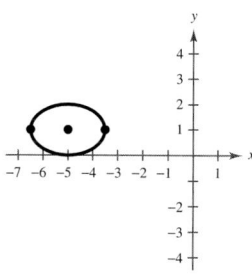

15. Center: $(-2, 3)$

Vertices: $(-2, 6), (-2, 0)$

Foci: $\left(-2, 3 \pm \sqrt{5}\right)$

Eccentricity: $\frac{\sqrt{5}}{3}$

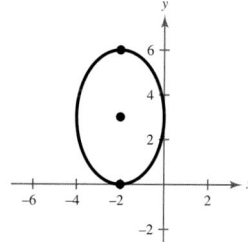

17. Center: $(4, 3)$

Vertices: $(14, 3), (-6, 3)$

Foci: $\left(4 \pm 4\sqrt{5}, 3\right)$

Eccentricity: $\frac{2\sqrt{5}}{5}$

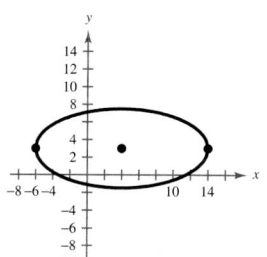

19. Center: $\left(-\frac{3}{2}, \frac{5}{2}\right)$

Vertices: $\left(-\frac{3}{2}, \frac{5}{2} \pm 2\sqrt{3}\right)$

Foci: $\left(-\frac{3}{2}, \frac{5}{2} \pm 2\sqrt{2}\right)$

Eccentricity: $\frac{\sqrt{6}}{3}$

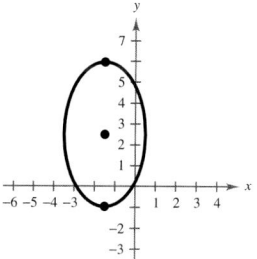

21. Center: $(1, -1)$

Vertices: $\left(\frac{9}{4}, -1\right), \left(-\frac{1}{4}, -1\right)$

Foci: $\left(\frac{7}{4}, -1\right), \left(\frac{1}{4}, -1\right)$

Eccentricity: $\frac{3}{5}$

23.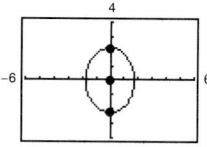

Center: $(0, 0)$

Vertices: $\left(0, \pm\sqrt{5}\right)$

Foci: $\left(0, \pm\sqrt{2}\right)$

25.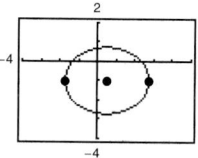

Center: $\left(\frac{1}{2}, -1\right)$

Vertices: $\left(\frac{1}{2} \pm \sqrt{5}, -1\right)$

Foci: $\left(\frac{1}{2} \pm \sqrt{2}, -1\right)$

27. $\dfrac{x^2}{4} + \dfrac{y^2}{16} = 1$ **29.** $\dfrac{x^2}{36} + \dfrac{y^2}{32} = 1$

31. $\dfrac{x^2}{36} + \dfrac{y^2}{11} = 1$ **33.** $\dfrac{21x^2}{400} + \dfrac{y^2}{25} = 1$

35. $\dfrac{(x-2)^2}{1} + \dfrac{(y-3)^2}{9} = 1$

37. $\dfrac{(x+2)^2}{16} + \dfrac{(y-3)^2}{9} = 1$

39. $\dfrac{(x-2)^2}{4} + \dfrac{(y-4)^2}{1} = 1$ **41.** $\dfrac{x^2}{48} + \dfrac{(y-4)^2}{64} = 1$

43. $\dfrac{(x-3)^2}{9} + \dfrac{(y-5)^2}{16} = 1$ **45.** $\dfrac{x^2}{16} + \dfrac{(y-4)^2}{12} = 1$

47. $\dfrac{x^2}{25} + \dfrac{y^2}{16} = 1$

49. Positions of tacks: $\left(\pm\sqrt{5}, 0\right)$; Length of string: 6 feet

51. 40   **53.** $\dfrac{x^2}{4.88} + \dfrac{y^2}{1.39} = 1$ **55.** Answers will vary.

57.

59.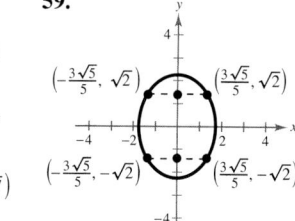

61. True. The ellipse is elongated and the foci are close to the vertices.

63. False. The foci of an ellipse cannot occur outside the ellipse because $0 < c < a$.

65. (a) $A = \pi a(20 - a)$ (b) $\dfrac{x^2}{196} + \dfrac{y^2}{36} = 1$

(c)

a	8	9	10	11	12	13
A	301.6	311.0	314.2	311.0	301.6	285.9

$a = 10$, circle

(d)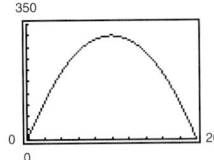

The shape of an ellipse with a maximum area is a circle.

67. $-5, 15$ **69.** $\dfrac{5}{4}$

71. (a) -24 (b) 144 (c) $x^2 - 18x + 56$

73. (a) -3 (b) $-\dfrac{87}{16}$ (c) $139 - 2\sqrt{3}$

75. $-6 - h$

Section 6.4 (page 438)

1. b **3.** a

5. Center: $(0, 0)$
Vertices: $(\pm 1, 0)$
Foci: $(\pm\sqrt{2}, 0)$
Asymptotes: $y = \pm x$

7. Center: $(0, 0)$
Vertices: $(0, \pm 5)$
Foci: $(0, \pm\sqrt{106})$
Asymptotes: $y = \pm\dfrac{5}{9}x$

9. Center: $(1, -2)$
Vertices: $(3, -2), (-1, -2)$
Foci: $(1 \pm \sqrt{5}, -2)$
Asymptotes: $y = -2 \pm \dfrac{1}{2}(x - 1)$

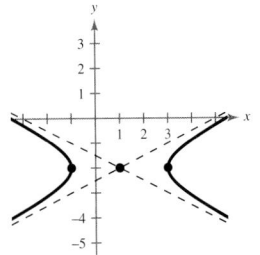

11. Center: $(2, -6)$
Vertices: $\left(2, -\dfrac{17}{3}\right), \left(2, -\dfrac{19}{3}\right)$
Foci: $\left(2, -6 \pm \dfrac{\sqrt{13}}{6}\right)$
Asymptotes: $y = -6 \pm \dfrac{2}{3}(x - 2)$

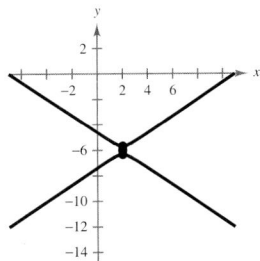

13. Center: $(2, -3)$
Vertices: $(3, -3), (1, -3)$
Foci: $(2 \pm \sqrt{10}, -3)$
Asymptotes: $y = -3 \pm 3(x - 2)$

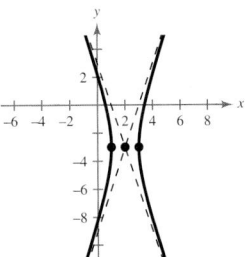

15. The graph of this equation is two lines intersecting at $(-1, -3)$.

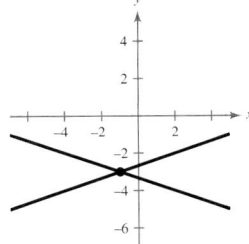

17. Center: $(0, 0)$
Vertices: $(\pm\sqrt{3}, 0)$
Foci: $(\pm\sqrt{5}, 0)$
Asymptotes: $y = \pm\dfrac{\sqrt{6}}{3}x$

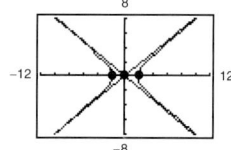

19. Center: $(1, -3)$
Vertices: $(1, -3 \pm \sqrt{2})$
Foci: $(1, -3 \pm 2\sqrt{5})$
Asymptotes: $y = -3 \pm \frac{1}{3}(x - 1)$

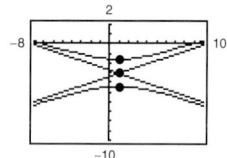

21. $\dfrac{y^2}{4} - \dfrac{x^2}{12} = 1$ **23.** $\dfrac{x^2}{1} - \dfrac{y^2}{25} = 1$

25. $\dfrac{17y^2}{1024} - \dfrac{17x^2}{64} = 1$ **27.** $\dfrac{(x-4)^2}{4} - \dfrac{y^2}{12} = 1$

29. $\dfrac{(y-5)^2}{16} - \dfrac{(x-4)^2}{9} = 1$ **31.** $\dfrac{y^2}{9} - \dfrac{4(x-2)^2}{9} = 1$

33. $\dfrac{(y-2)^2}{4} - \dfrac{x^2}{4} = 1$ **35.** $\dfrac{(x-2)^2}{1} - \dfrac{(y-2)^2}{1} = 1$

37. $\dfrac{(x-3)^2}{9} - \dfrac{(y-2)^2}{4} = 1$ **39.** $(3300, -2750)$

41. $\left(12(\sqrt{5} - 1), 0\right) \approx (14.83, 0)$ **43.** Ellipse

45. Parabola   **47.** Hyperbola   **49.** Circle

51. False. For the trivial solution of two intersecting lines to occur, the standard form of the equation of the hyperbola would be equal to zero,
$\dfrac{(x-h)^2}{a^2} - \dfrac{(y-k)^2}{b^2} = 0$ or $\dfrac{(y-k)^2}{a^2} - \dfrac{(x-h)^2}{b^2} = 0.$

53. The extended diagonals of the central rectangle are asymptotes of the hyperbola.

55. $3x^2 + \frac{23}{2}x - 2$ **57.** $x^2 + 2xy + y^2 + 6x + 6y + 9$

59. $(x + 7)^2$ **61.** $x(3x + 2)(2x - 5)$

63. $(x^2 + 1)(4 - x)$

Section 6.5 (page 446)

1. $(3, 0)$ **3.** $\left(\dfrac{3 + \sqrt{3}}{2}, \dfrac{3\sqrt{3} - 1}{2}\right)$

5. $\left(\dfrac{3\sqrt{2}}{2}, -\dfrac{\sqrt{2}}{2}\right)$

7. $\dfrac{(y')^2}{2} - \dfrac{(x')^2}{2} = 1$ **9.** $y' = \pm\dfrac{\sqrt{2}}{2}$

 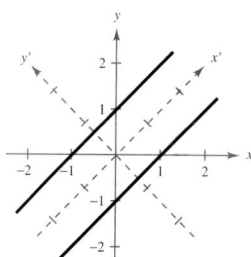

11. $\dfrac{(x' - 3\sqrt{2})^2}{16} - \dfrac{(y' - \sqrt{2})^2}{16} = 1$

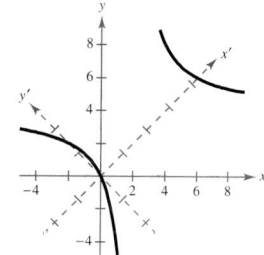

13. $\dfrac{(x')^2}{6} + \dfrac{(y')^2}{\frac{3}{2}} = 1$ **15.** $x' = -(y')^2$

 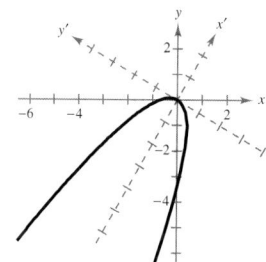

17. $y' = \frac{1}{6}(x')^2 - \frac{1}{3}x'$ **19.**

$\theta = 45°$

21.
$\theta = 26.57°$

23.
$\theta = 31.72°$

25.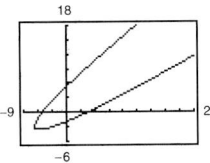
$\theta = 33.69°$

27. e **29.** b **31.** d

33. Parabola
$$y = \frac{(8x-5) \pm \sqrt{(8x-5)^2 - 4(16x^2 - 10x)}}{2}$$
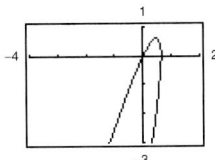

35. Ellipse
$$y = \frac{6x \pm \sqrt{36x^2 - 28(12x^2 - 45)}}{14}$$
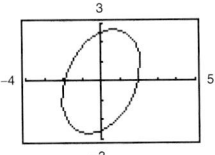

37. Hyperbola
$$y = \frac{6x \pm \sqrt{36x^2 + 20(x^2 + 4x - 22)}}{-10}$$
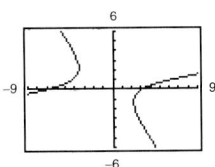

39. Parabola
$$y = \frac{-(4x-1) \pm \sqrt{(4x-1)^2 - 16(x^2 - 5x - 3)}}{8}$$

41.

43.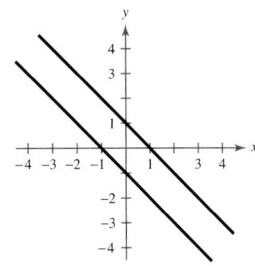

45. $(2, 2), (2, 4)$ **47.** $(-8, 12)$ **49.** $(0, 8), (12, 8)$
51. $(0, 4)$ **53.** $(1, \sqrt{3}), (1, -\sqrt{3})$
55. No solution **57.** $(0, \frac{3}{2}), (-3, 0)$
59. True. The graph of the equation can be classified by finding the discriminant. For a graph to be a hyperbola, the discriminant must be greater than zero. If $k \geq \frac{1}{4}$, then the discriminant would be less than or equal to zero.
61. Answers will vary. **63.** $\frac{5}{6}$ **65.** $\frac{2}{3}, 4$
67. $3, 2 - i, 2 + i$

69.

71.

73.

75.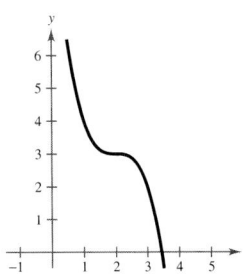

Section 6.6 (page 453)

1. (a)

t	0	1	2	3	4
x	0	1	$\sqrt{2}$	$\sqrt{3}$	2
y	3	2	1	0	-1

(b)

(c) $y = 3 - x^2$
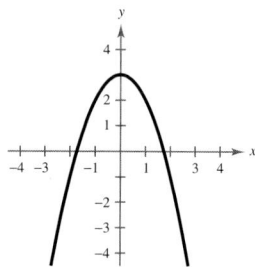

The graph of the rectangular equation shows the entire parabola rather than just the right half.

3.

5.

7.

9.

11.

13.

15.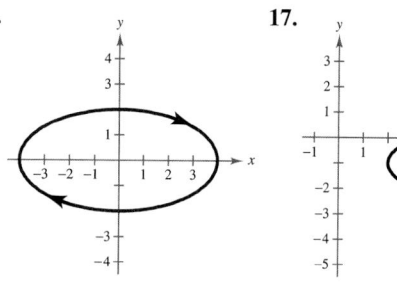

17. (image with 15)

19.

21.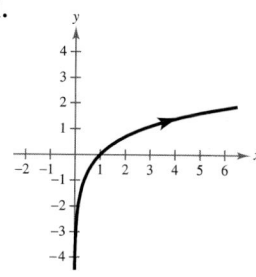

23. Each curve represents a portion of the line $y = 2x + 1$.

	Domain	Orientation
(a)	$(-\infty, \infty)$	Left to right
(b)	$[-1, 1]$	Depends on θ
(c)	$(0, \infty)$	Right to left
(d)	$(0, \infty)$	Left to right

25. $y - y_1 = m(x - x_1)$ **27.** $\dfrac{(x-h)^2}{a^2} + \dfrac{(y-k)^2}{b^2} = 1$

29. $x = 6t$ **31.** $x = 3 + 4\cos\theta$ **33.** $x = 4\cos\theta$
  $y = -3t$ $y = 2 + 4\sin\theta$ $y = \sqrt{7}\sin\theta$

35. $x = 4\sec\theta$ **37.** (a) $x = t,\ y = 3t - 2$
  $y = 3\tan\theta$ (b) $x = -t + 2,\ y = -3t + 4$

39. (a) $x = t,\ y = t^2$
  (b) $x = -t + 2,\ y = t^2 - 4t + 4$

41. (a) $x = t,\ y = t^2 + 1$
  (b) $x = -t + 2,\ y = t^2 - 4t + 5$

43. (a) $x = t,\ y = \dfrac{1}{t}$

  (b) $x = -t + 2,\ y = -\dfrac{1}{t-2}$

45.

47.

49.
51.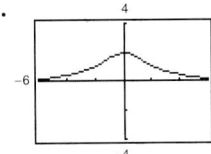

53. b
Domain: $[-2, 2]$
Range: $[-1, 1]$

55. d
Domain: $(-\infty, \infty)$
Range: $(-\infty, \infty)$

57. (a)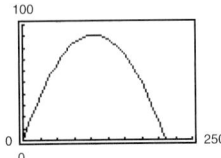
Maximum height: 90.7 feet
Range: 209.6 feet

(b)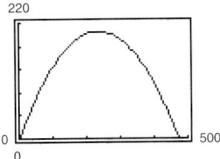
Maximum height: 204.2 feet
Range: 471.6 feet

(c)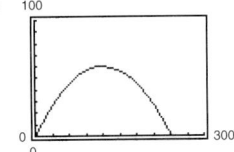
Maximum height: 60.5 feet
Range: 242.0 feet

(d)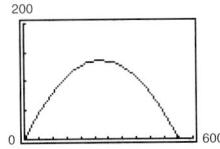
Maximum height: 136.1 feet
Range: 544.5 feet

59. (a) $x = (146.67 \cos \theta)t$
$y = 3 + (146.67 \sin \theta)t - 16t^2$

(b)
No

(c)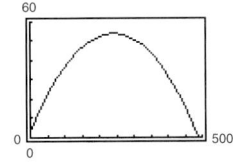
Yes

(d) 19.4°

61. Answers will vary.

63. $x = a\theta - b \sin \theta$
$y = a - b \cos \theta$

65. True
$x = t$
$y = t^2 + 1 \Longrightarrow y = x^2 + 1$
$x = 3t$
$y = 9t^2 + 1 \Longrightarrow y = x^2 + 1$

67. $y = \frac{1}{3}x^2 - 2x$
69. $y = \frac{7}{8}x^2 - 14x + 54$

Section 6.7 (page 460)

1.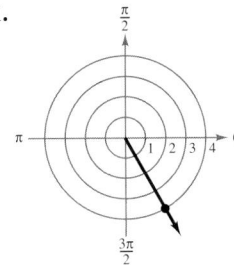
$\left(4, \frac{5\pi}{3}\right), \left(-4, -\frac{4\pi}{3}\right)$

3.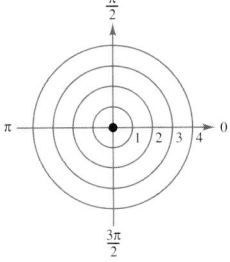
$\left(0, \frac{5\pi}{6}\right), \left(0, -\frac{13\pi}{6}\right)$

5.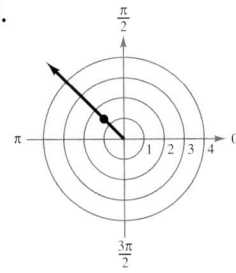
$(\sqrt{2}, 8.64), (-\sqrt{2}, -0.78)$

7.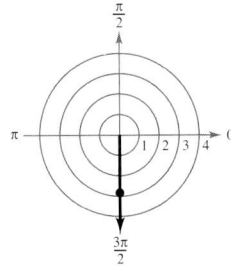
$(2\sqrt{2}, 10.99), (-2\sqrt{2}, 7.85)$

9. $(0, 3)$
11. $\left(\frac{\sqrt{2}}{2}, \frac{\sqrt{2}}{2}\right)$
13. $\left(-\sqrt{2}, \sqrt{2}\right)$
15. $(-1.1, -2.2)$
17. $\left(\sqrt{2}, \frac{\pi}{4}\right)$
19. $(6, \pi)$
21. $(5, 2.2143)$
23. $\left(\sqrt{6}, \frac{5\pi}{4}\right)$
25. $\left(3\sqrt{13}, 0.9828\right)$
27. $\left(\sqrt{13}, -0.5880\right)$
29. $\left(\sqrt{7}, 0.8571\right)$
31. $\left(\frac{17}{6}, 0.4900\right)$
33. $r = 3$
35. $r = 4 \csc \theta$

37. $r = \dfrac{-2}{3\cos\theta - \sin\theta}$

39. $r^2 = 16\sec\theta\csc\theta = 32\csc 2\theta$

41. $r = \dfrac{4}{1-\cos\theta}$ or $-\dfrac{4}{1+\cos\theta}$ **43.** $r = a$

45. $r = b\csc\theta$ **47.** $r = 2a\cos\theta$

49. $x^2 + y^2 - 4y = 0$ **51.** $\sqrt{3}x + y = 0$

53. $y = 2$ **55.** $(x^2 + y^2)^2 = 6x^2y - 2y^3$

57. $4x^2 - 5y^2 - 36y - 36 = 0$

59. $x^2 + y^2 = 36$ **61.** $-\sqrt{3}x + 3y = 0$

63. $x - 3 = 0$

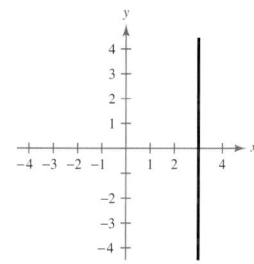

65. True. Because r is a directed distance, the point (r, θ) can be represented as $(r, \theta + 2n\pi)$.

67. $(x-h)^2 + (y-k)^2 = h^2 + k^2$
  Radius: $\sqrt{h^2 + k^2}$
  Center: (h, k)

69. (a) Answers will vary.
  (b) (r_1, θ_1), (r_2, θ_2) and the pole are collinear. The distance between two points on the line $\theta = \theta_1 = \theta_2$.
  $d = \sqrt{r_1^2 + r_2^2 - 2r_1r_2} = |r_1 - r_2|$
  (c) $d = \sqrt{r_1^2 + r_2^2}$
  Pythagorean Theorem
  (d) Example: Points: $(3, \pi/6)$, $(4, \pi/3)$
        Distance: 2.053
        Points: $(-3, 7\pi/6)$, $(-4, 4\pi/3)$
        Distance: 2.053

71. $B \approx 23.7°$, $C \approx 134.3°$, $a \approx 8.36$

73. $A \approx 29.7°$, $B \approx 52.4°$, $C \approx 97.9°$

75. $A \approx 11.7°$, $B \approx 130.3°$, $c \approx 12.10$

77. 12.29   **79.** 704.14

Section 6.8 (page 468)

1. Rose curve   **3.** Limaçon   **5.** Rose curve

7. Polar axis   **9.** $\theta = \dfrac{\pi}{2}$ **11.** $\theta = \dfrac{\pi}{2}$, polar axis, pole

13. Maximum: $|r| = 20$ when $\theta = \dfrac{3\pi}{2}$
  Zero: $r = 0$ when $\theta = \dfrac{\pi}{2}$

15. Maximum: $|r| = 4$ when $\theta = 0, \dfrac{\pi}{3}, \dfrac{2\pi}{3}$
  Zero: $r = 0$ when $\theta = \dfrac{\pi}{6}, \dfrac{\pi}{2}, \dfrac{5\pi}{6}$

17.   **19.**

21.   **23.**

25.   **27.**

29.
31.
33.
35.
37.
39.
41.
43.
45.
47.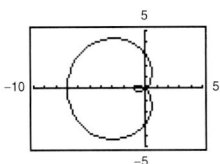
$0 \leq \theta < 2\pi$
49.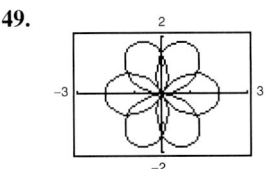
$0 \leq \theta < 4\pi$
51.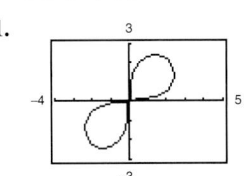
$0 \leq \theta < \pi$

53.
55.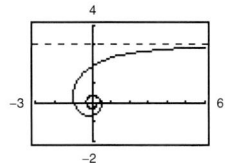

57. True. For a graph to have polar axis symmetry, replace (r, θ) by $(r, -\theta)$ or $(-r, \pi - \theta)$.

59. (a)   (b)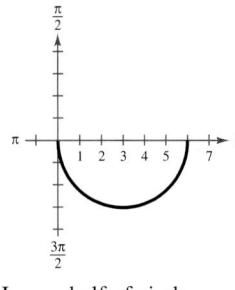

Upper half of circle   Lower half of circle

(c)   (d)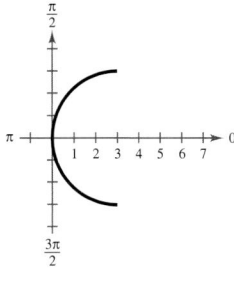

Full circle   Left half of circle

61. Answers will vary.

63. (a) $r = 2 - \sin\left(\theta - \dfrac{\pi}{4}\right)$
(b) $r = 2 + \cos\theta$
(c) $r = 2 + \sin\theta$
(d) $r = 2 - \cos\theta$

65. (a)   (b)

67. 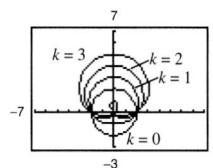 $k = 0$, circle
$k = 1$, limaçon
$k = 2$, cardioid
$k = 3$, limaçon

69. $\ln 19 \approx 2.944$ **71.** $\log_{10} 84 \approx 1.924$
73. $e^4 \approx 54.598$

Section 6.9 (page 474)

1.   **3.**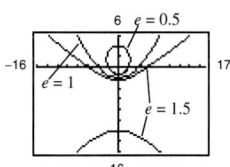

5. f   **7.** d   **9.** a
11. Parabola   **13.** Parabola

15. Ellipse   **17.** Ellipse

19. Hyperbola   **21.** Hyperbola

23. Ellipse

25.   **27.**

Parabola   Ellipse

29.   **31.**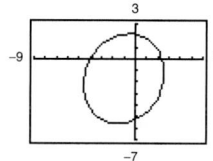

33. $r = \dfrac{1}{1 - \cos\theta}$ **35.** $r = \dfrac{1}{2 + \sin\theta}$

37. $r = \dfrac{2}{1 + 2\cos\theta}$ **39.** $r = \dfrac{2}{1 - \sin\theta}$

41. $r = \dfrac{10}{1 - \cos\theta}$ **43.** $r = \dfrac{10}{3 + 2\cos\theta}$

45. $r = \dfrac{20}{3 - 2\cos\theta}$ **47.** $r = \dfrac{9}{4 - 5\sin\theta}$

49. Answers will vary.

51. $r = \dfrac{9.2934 \times 10^7}{1 - 0.0167\cos\theta}$

Perihelion: 9.1408×10^7 miles
Aphelion: 9.4512×10^7 miles

53. $r = \dfrac{5.5368 \times 10^9}{1 - 0.2481\cos\theta}$

Perihelion: 4.4362×10^9 km
Aphelion: 7.3638×10^9 km

55. $r = \dfrac{1.3977 \times 10^8}{1 - 0.0934\cos\theta}$ **57.** $r = \dfrac{8200}{1 + \sin\theta}$

Perihelion: 1.2783×10^8 miles   1467 miles
Aphelion: 1.5417×10^8 miles

59. False. If e remains fixed and p changes, then the lengths of both the major axis and the minor axis change. For example, graph

$r = \dfrac{5}{1 - \frac{2}{3}\cos\theta}$, with $e = \frac{2}{3}$ and $p = \frac{15}{2}$, and graph

$r = \dfrac{6}{1 - \frac{2}{3}\cos\theta}$, with $e = \frac{2}{3}$ and $p = 9$, on the same set of coordinate axes.

61. Answers will vary.   **63.** $r^2 = \dfrac{24{,}336}{169 - 25\cos^2\theta}$

65. $r^2 = \dfrac{144}{25\cos^2\theta - 9}$ **67.** $r^2 = \dfrac{144}{25\cos^2\theta - 16}$

69. $\dfrac{\pi}{6} + n\pi$ **71.** $\dfrac{\pi}{3} + n\pi, \dfrac{2\pi}{3} + n\pi$

73. $\dfrac{\pi}{2} + n\pi$ **75.** $\dfrac{\sqrt{2}}{10}$ **77.** $\dfrac{7\sqrt{2}}{10}$

79. $\sin 2u = -\dfrac{24}{25}$
$\cos 2u = -\dfrac{7}{25}$
$\tan 2u = \dfrac{24}{7}$

Review Exercises (page 478)

1. 30.96°   **3.** 63.43°   **5.** 25.35°   **7.** 37.75°
9. $2\sqrt{2}$ **11.** Hyperbola   **13.** $(x - 4)^2 = -8(y - 2)$
15. $(y - 2)^2 = 12x$ **17.** $y = -2x + 2$; $(1, 0)$
19. $8\sqrt{6}$ meters   **21.** $\dfrac{(x - 2)^2}{25} + \dfrac{y^2}{21} = 1$

23. $\dfrac{2x^2}{9} + \dfrac{y^2}{36} = 1$

25. The foci occur 3 feet from the center of the arch on a line connecting the tops of the pillars.

27. Center: $(1, -4)$
Vertices: $(1, 0), (1, -8)$
Foci: $\left(1, -4 \pm \sqrt{7}\right)$
Eccentricity: $\dfrac{\sqrt{7}}{4}$

29. Center: $(-2, 1)$
Vertices: $(-2, 11), (-2, -9)$
Foci: $\left(-2, 1 \pm \sqrt{19}\right)$
Eccentricity: $\dfrac{\sqrt{19}}{10}$

31. $y^2 - \dfrac{x^2}{8} = 1$ **33.** $\dfrac{5(x - 4)^2}{16} - \dfrac{5y^2}{64} = 1$

35. Center: $(1, -1)$
Vertices: $(5, -1), (-3, -1)$
Foci: $(6, -1), (-4, -1)$
Asymptotes: $y = -1 \pm \frac{3}{4}(x - 1)$

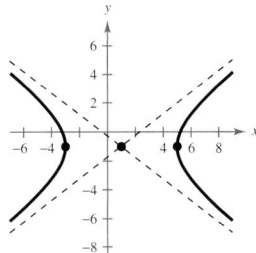

37. Center: $(3, -5)$
Vertices: $(7, -5), (-1, -5)$
Foci: $\left(3 \pm 2\sqrt{5}, -5\right)$
Asymptotes: $y = -5 \pm \frac{1}{2}(x - 3)$

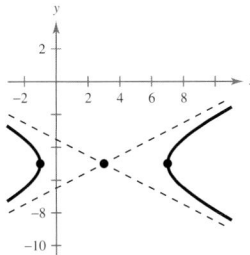

39. 72 miles   **41.** Hyperbola

43. $\dfrac{(x')^2}{8} - \dfrac{(y')^2}{8} = 1$ **45.** $\dfrac{(x')^2}{3} + \dfrac{(y')^2}{2} = 1$

  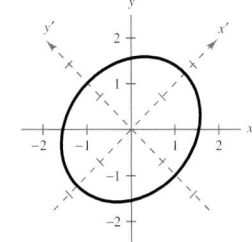

47. Parabola
$y = \dfrac{24x + 40 \pm \sqrt{(24x + 40)^2 - 36(16x^2 - 30x)}}{18}$

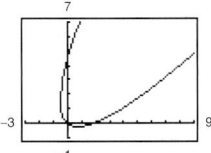

49. Parabola

$$y = \frac{-(2x - 2\sqrt{2}) \pm \sqrt{(2x - 2\sqrt{2})^2 - 4(x^2 + 2\sqrt{2}x + 2)}}{2}$$

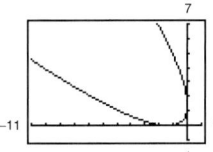

51. $x = 3, y = 0$ **53.** $x = \dfrac{3\sqrt{3}}{2}, y = \dfrac{1}{2}$

55.

$y = 2x$

57.

$y = \sqrt[4]{x}$

59.

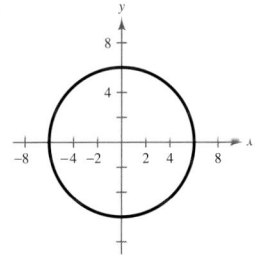

$x^2 + y^2 = 36$

61. $x = -3 + 4\cos\theta$
$y = 4 + 3\sin\theta$

63.

65.

$(\sqrt{2}, \sqrt{2})$

67.

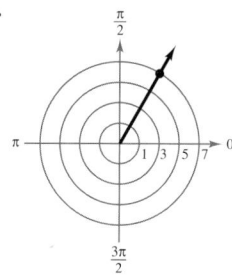

$(3.4927, 6.0664)$

69. $\left(2, \dfrac{\pi}{2}\right), \left(-2, \dfrac{3\pi}{2}\right)$

71. $(7.2111, 0.9828), (-7.211, 4.124)$ **73.** $x^2 + y^2 = 3x$

75. $x^2 + 4y - 4 = 0$ **77.** $r = a\cos^2\theta\sin\theta$

79. Circle   **81.** Rose curve

83. Cardioid

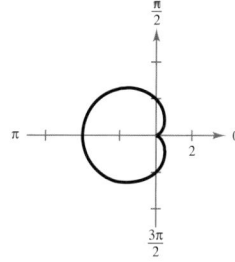

85. Symmetry: $\theta = \dfrac{\pi}{2}$

Maximum value of $|r|$: $|r| = 8$ when $\theta = \dfrac{\pi}{2}$

Zeros of r: $r = 0$ when $\theta = 3.4814, 5.9433$

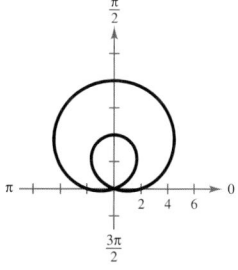

87. Symmetry: $\theta = \dfrac{\pi}{2}$, polar axis

Maximum value of $|r|$: $|r| = 3$ when $\theta = 0, \dfrac{\pi}{2}, \pi, \dfrac{3\pi}{2}$

Zeros of r: $r = 0$ when $\theta = \dfrac{\pi}{4}, \dfrac{3\pi}{4}$

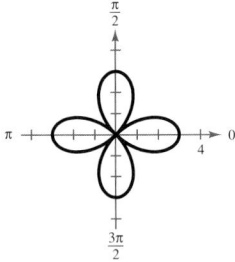

89. Limaçon   **91.** Rose curve

93. Eccentricity: 2; Hyperbola   **95.** Eccentricity: $\tfrac{3}{5}$; Ellipse

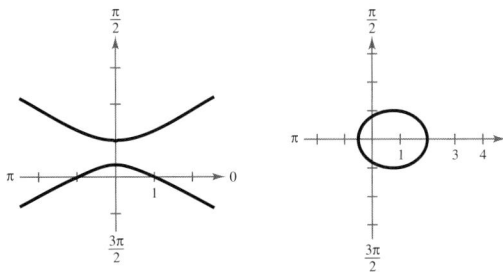

97. $r = \dfrac{4}{1 - \cos\theta}$ **99.** $r = \dfrac{5}{3 - 2\cos\theta}$

101. $r = \dfrac{7977.2}{1 - 0.937\cos\theta}$; 11,008 miles

103. False. When classifying an equation of the form $Ax^2 + Bxy + Cy^2 + Dx + Ey + F = 0$, its graph can be determined by its discriminant. For a graph to be a parabola, its discriminant, $B^2 - 4AC$, must equal zero. So, if $B = 0$, then A or C equals 0.

105. False. The following are two sets of parametric equations for the line.

$x = t,\ y = 3 - 2t$

$x = 3t,\ y = 3 - 6t$

107. 5. The ellipse becomes more circular and approaches a circle of radius 5.

109. (a) The speed would double.

(b) The elliptical orbit would be flatter; the length of the major axis would be greater.

111. (a) The graphs are the same.

(b) The graphs are the same.

Chapter Test   *(page 483)*

1. 15.9°   **2.** 47.7°   **3.** $\dfrac{7\sqrt{2}}{2}$

4. Parabola: $y^2 = 4(x - 1)$

Vertex: $(1, 0)$

Focus: $(2, 0)$

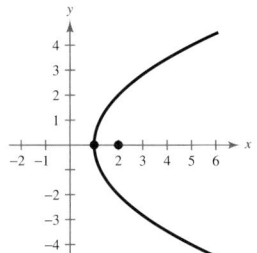

5. Hyperbola: $\dfrac{(x - 2)^2}{4} - y^2 = 1$

Center: $(2, 0)$

Vertices: $(0, 0), (4, 0)$

Foci: $(2 \pm \sqrt{5}, 0)$

Asymptotes: $y = \pm\dfrac{1}{2}(x - 2)$

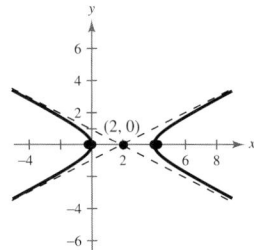

6. Ellipse: $\dfrac{(x + 3)^2}{16} + \dfrac{(y - 1)^2}{9} = 1$

Center: $(-3, 1)$

Vertices: $(1, 1), (-7, 1)$

Foci: $(-3 \pm \sqrt{7}, 1)$

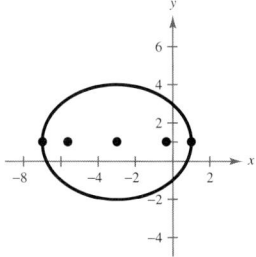

7. Circle: $(x-2)^2 + (y-1)^2 = \frac{1}{2}$
Center: $(2, 1)$

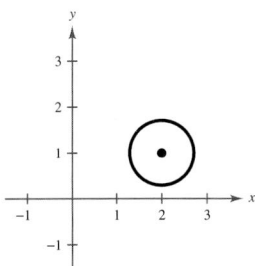

8. $(x-3)^2 = \frac{3}{2}(y+2)$ **9.** $\frac{5(y-2)^2}{4} - \frac{5x^2}{16} = 1$

10. (a) $45°$
(b)

11.
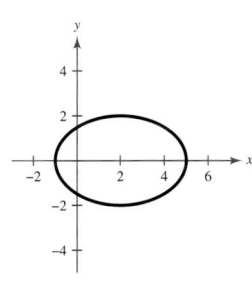

12. $x = 6 + 4t$
$y = 4 + 7t$

$\frac{(x-2)^2}{9} + \frac{y^2}{4} = 1$

13. $(\sqrt{3}, -1)$

14. $\left(2\sqrt{2}, \frac{7\pi}{4}\right), \left(-2\sqrt{2}, \frac{3\pi}{4}\right), \left(2\sqrt{2}, -\frac{\pi}{4}\right)$

15. $r = 4 \sin \theta$

16.

Parabola

17.

Ellipse

18.

Limaçon

19.
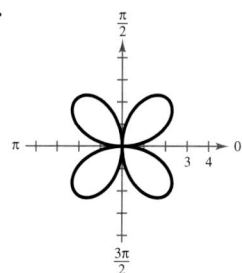
Rose curve

20. Slope: 0.1511; Change in elevation: 789.03 feet

21. No; Yes

Cumulative Test for Chapters 4–6
(page 484)

1. $3 - 5i$ **2.** $-2 - 3i$ **3.** $5 - 12i$ **4.** 4
5. $\frac{8}{5} + \frac{4}{5}i$ **6.** $-2, \pm 2i$ **7.** $-7, 0, 3$
8. $x^4 + 3x^3 - 11x^2 + 9x + 70$
9. $2\sqrt{2}\left(\cos \frac{3\pi}{4} + i \sin \frac{3\pi}{4}\right)$ **10.** $-12\sqrt{3} + 12i$
11. $-8 + 8\sqrt{3}i$

12. $\cos 0 + i \sin 0$ **13.** $3\left(\cos \frac{\pi}{8} + i \sin \frac{\pi}{8}\right)$
$\cos \frac{2\pi}{3} + i \sin \frac{2\pi}{3}$ $3\left(\cos \frac{5\pi}{8} + i \sin \frac{5\pi}{8}\right)$
$\cos \frac{4\pi}{3} + i \sin \frac{4\pi}{3}$ $3\left(\cos \frac{9\pi}{8} + i \sin \frac{9\pi}{8}\right)$
$3\left(\cos \frac{13\pi}{8} + i \sin \frac{13\pi}{8}\right)$

14. Reflect f in the x-axis and y-axis, and shift f 3 units to the right.

15. Reflect f in the x-axis and shift f 4 units upward.

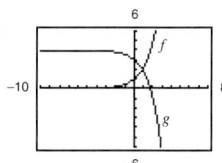

16. 1.991   **17.** -0.067 **18.** 1.717   **19.** 0.281
20. 0.302   **21.** -1.733 **22.** -4.087

23. $\ln(x+4) + \ln(x-4) - 4\ln x, \ x > 4$

24. $\ln \dfrac{x^2}{\sqrt{x+5}}, \ x > 0$ **25.** $\dfrac{\ln 12}{2} \approx 1.242$ **26.** $\dfrac{64}{5}$

27.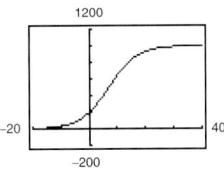

Asymptotes: $y = 0, \ y = 1000$

28. 6.3 hours **29.** $81.87°$ **30.** $\dfrac{11\sqrt{5}}{5}$

31. Ellipse; $\dfrac{(x-2)^2}{4} + \dfrac{(y+1)^2}{9} = 1$

Center: $(2, -1)$

Vertices: $(2, 2), (2, -4)$

Foci: $\left(2, -1 \pm \sqrt{5}\right)$

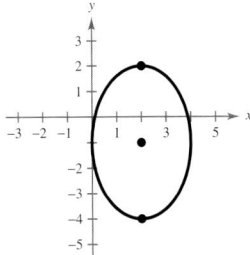

32. Hyperbola; $x^2 - \dfrac{y^2}{4} = 1$

Center: $(0, 0)$

Vertices: $(1, 0), (-1, 0)$

Foci: $\left(\sqrt{5}, 0\right), \left(-\sqrt{5}, 0\right)$

Asymptotes: $y = \pm 2x$

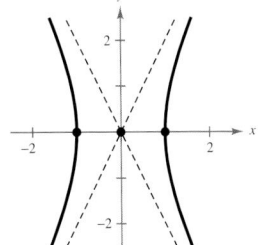

33. Circle; $(x+1)^2 + (y-3)^2 = 22$

Center: $(-1, 3)$

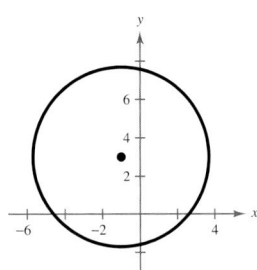

34. Parabola; $y^2 = -2(x+1)$

Vertex: $(-1, 0)$

Focus: $\left(-\tfrac{3}{2}, 0\right)$

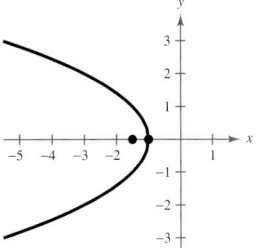

35. $x^2 - 4x + y^2 + 8y - 48 = 0$

36. $5y^2 - 4x^2 - 30y - 55 = 0$

37. (a) $45°$

(b)

38.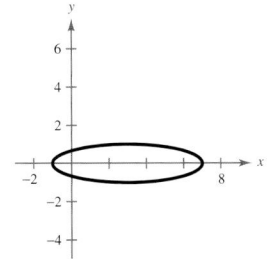

$\dfrac{(x-3)^2}{16} + y^2 = 1$

39. $x = 3 - 6t$
$y = -2 + 6t$

40. $r = 6 \sin \theta$

41.

42.
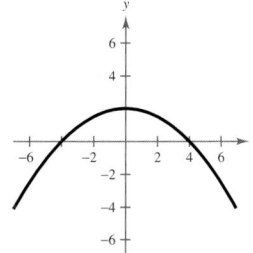

43. (a) iii   (b) i   (c) ii

Appendix B
Section B.1   *(page A16)*

1. (a) $1.109   (b) [$0.999, $1.189]

3. Quiz 1:

```
              ×              15
              ×    × × ×
          × × ×    × × × ×
          × × ×    × × × ×
  ×   ×   × × ×    × × × × ×    ×
 ─┼───┼───┼───┼───┼───┼───┼───┼──
  10  12  14  16  18  20  22  24
```

Quiz 2:

```
                    × ×              22 and 23
                    × ×
                    × ×
                    × × ×
                  × × × ×
                  × × × ×
              ×   × × × ×
      ×   ×   ×   × × × × ×
   ─┼───┼───┼───┼───┼───┼───┼──
     14  16  18  20  22  24  26
```

5.

Stems	Leaves
7	0 5 5 5 7 7 8 8 8
8	1 1 1 1 2 3 4 5 5 5 5 7 8 9 9 9
9	0 2 8
10	0 0

7.

Stems	Leaves
0	89 66 67 65 80 98 62 93
1	09 01 46 24 96 90
2	92 55 40 61
3	68 35
4	12 96 80 38
5	81 18 50 70 66
6	44 00 34 01
7	61 66 00
8	11 57 41 90
9	
10	
11	60 59 33
12	92
13	19 17 37
27	22
31	32
46	80
65	14

9. (a)

Interval	Tally
90–109	|||||
110–129	||||| |||||	
130–149	||||| ||||
150–169	|||||	
170–189		
190–209	|||

(b)
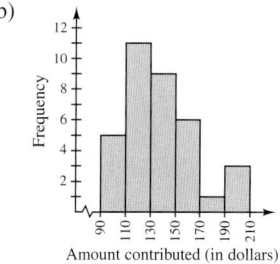

Section B.2   *(page A23)*

1. Mean: 8.86;   median: 8;   mode: 7

3. Mean: 10.29;   median: 8;   mode: 7

5. Mean: 9;   median: 8;   mode: 7

7. The mean is sensitive to extreme values.
9. Mean: $67.14; median: $65.35
11. Mean: 3.07; median: 3; mode: 3
13. One possibility: {4, 4, 10}
15. The median gives the most representative description.
17. $\bar{x} = 6$, $v = 10$, $\sigma = 3.16$
19. $\bar{x} = 2$, $v = \frac{4}{3}$, $\sigma = 1.15$ 21. $\bar{x} = 4$, $v = 4$, $\sigma = 2$
23. $\bar{x} = 47$, $v = 226$, $\sigma = 15.03$ 25. 3.42
27. 101.55   29. 1.65
31. (a) $\bar{x} = 12$; $\sigma = 2.83$ (b) $\bar{x} = 20$; $\sigma = 2.83$
    (c) $\bar{x} = 12$; $\sigma = 1.41$ (d) $\bar{x} = 9$; $\sigma = 1.41$
33. $\bar{x} = 12$ and $|x_i - 12| = 8$ for all x_i
35. It will increase the mean by 5, but the standard deviation will not change.
37. First histogram

39.   41.

43.   45.

Index of Applications

Biology and Life Sciences

Animal population growth, 394
Anthropology, 23
Atmospheric pressure, 354
Bacteria growth, 392, 393
Blood pressure of a person at rest, 167
Flood control, 23
Forensics, 395
Forest yield, 382
Gypsy moth defoliation, 354
Height of a tree, 196, 268, 301
Height of a water wave, 137
Human learning model, 382
Human memory model, 364, 372
Learning curve of factory employees, 393
Lynx population, 66
Number of trees per acre, 382
Percent of the moon's face that is illuminated, 167
Predator-prey model, 178
Respiratory cycle of a person after exercising, 167
Respiratory cycle of a person at rest, 166

Business

America Online, Inc., 50
Break-even point, 51
Budget variance, 10
Change in annual sales, 113
Colgate-Palmolive Company, 48
Cost of airline flights, 24
Demand for a product, 24, 111, 353, 381, 400
Depreciation, 37, 353, 393, 398
Equipment value, 50
Fixed and variable costs, 65, 113
J.C. Penney Company, 50
Kellogg Company, 48
Lands' End, Inc., 35
Maximum profit, 64
Maximum revenue, 419
Monthly sales, 167, 178, 200, 234
Operating costs, 23, 51
Product sales, 156, 167, 178, 234
Production costs, 98
Profit, 338, 393
Restaurant sales, 96
Revenue, 24, 48, 65, 298, 299
Seasonal product sales forecast, 156
Starbucks Corp., 35
Wal-Mart Stores, Inc., 34

Chemistry and Physics

Asteroid orbit, 481
Boiling and freezing temperatures, 50
Coefficient of friction on an inclined plane, 217
Damped harmonic motion, 234
Exhaust temperature of a diesel engine, 107
Fahrenheit and Celsius temperatures, 50
Fluid flow, 78
Force of water against a dam, 65
Frequency of a sound wave, 203
Half-life of radioactive isotopes, 392
Harmonic motion, 241, 257
Hyperbolic telescope mirror, 439
Illumination, 167
Impedance of an electrical circuit, 316
Lumens of a fluorescent lamp, 77
Magnitude of earthquakes, 394, 401
Mixture problems, 111, 114, 316
Motion of a tuning fork, 199
Newton's Law of Cooling, 395
pH of a substance, 394
Planetary motion, 475
Radioactive decay, 353, 393, 401
Satellite antenna, 419
Satellite orbit, 420, 429, 475
Simple harmonic motion, 156, 178, 199, 205
Sound intensity, 364, 372, 394, 401
Standing waves, 241
Temperature change, 382
Temperature of saturated steam, 24
Work, 290, 298, 299, 364

Construction

Angles of a bridge truss, 411
Awning design, 277
Bridge design, 269
Conveyor design, 411
Cross sections of an irrigation canal, 200
Design of an archway, 478
Design of a fireplace arch, 429
Design of a suspension bridge, 419
Design of a tunnel arch, 429
Height of an attic, 49
Highway design, 420
Lengths of a bridge truss, 198, 277
Railroad track design, 269
Road grade, 48, 411, 483
Streetlight design, 276

Consumer

Chartering a bus, 65
Comparison of prices, 50
Cost of commodities, 107
Cost of a telephone calling card, 77
Delivery charges, 77
Hourly wage, 51, 107
Inflation, 113, 353
Renting an apartment, 51
Salary, 50, 51
Sale price, 50

Geometry

Area of an equilateral triangle, 64
Area of an isosceles triangle, 64, 252
Area of a parking lot, 277
Area of a rectangle, 234
Area of a triangular region, 277
Dimensions of a box, 23
Dimensions of a garden, 51
Dimensions of a rectangle, 37, 113
Floor space of a building, 23
Maximum volume of an open box, 64
Pitch of a roof, 411
Postal regulations, 65
Radii of ripples in a pond, 98
Volume of a water trough, 256

Interest Rate

Investment, 51, 353, 364, 381, 391, 400, 401
Mortgage payment, 365, 395
Trust fund, 398

Miscellaneous

Aircraft tracking station, 276, 277
Altitude of an airplane, 197, 198
Angle from a camera to a painting, 188
Angle of depression from a cell phone tower to a mountain top, 197
Angle of depression from a satellite to the horizon, 197
Angle of elevation of a building, 204
Angle of elevation of a plane, 188
Angle of elevation of a ramp, 147
Angle of elevation of a television camera, 187

Angle of elevation of a tower, 196
Angle of elevation of the sun, 196, 269
Angle of a pointer on a scale, 129
Angle of rotation of a electric hoist, 129
Angle of a spotlight on a security car, 188
Angles of a triangular parcel of land, 275
Area of a parking lot, 277
Area of a triangular parcel of lane, 277
Average daily temperature of a city, 156
Bearing between cities, 275, 276
Braking load, 299
Cable tension, 290
College enrollment, 50
Comet orbit, 429
Coordinates of the holes in a steel plate, 147
Course of a boat race, 275
Crumple zone of automobiles, 383
Daily high temperatures for Chicago and Honolulu, 167
Depth of a submarine, 196
Diameter of a shaft, 147
Difference in latitudes, 129
Docking a boat, 187
Engine design, 277
Explorer 18 orbit, 480
Flagpole height, 268
Flight path of an airplane, 269, 275
Flashlight reflector, 478
Force on a hook, 289
Force on a machine part, 289
Force on an object, 289, 302
Fuel consumption on a farm, 167
Fuel efficiency, 399
Glide path of an airplane, 269
Height of a building, 301
Height of bungee jumpers, 196
Height of a church steeple, 196
Height of a football, 204
Height of a kite, 196
Height of a ladder, 196
Height of a mountain, 198
Height of a streetlight, 146
Height of a tethered balloon, 147
Height of a tower, 146
Length of an elevator to a grain bin, 199
Length of a marsh, 275, 301
Length of a shadow, 196
Location of an explosion, 439, 479
Location of a fire, 197, 269
Long-distance radio navigation, 439, 479
Mach number, 252
Monthly high temperatures for Erie, PA, 178

Motion of a bobber, 204
Mountain climbing, 412
Mountain descent, 197
Newspaper circulation, 61
Paper manufacturing, 277
Population growth, 353, 364, 392, 401
Population growth of foreign countries, 392
Programming a thermostat, 97
Public school ventilation, 364
Quiz scores, 52
Record temperatures in Duluth, MN, 34
Rental demand, 51
Rock and Roll Hall of Fame, 35
Rope tension, 290, 291, 302
Rotary engine design, 480
Salaries of major league baseball players, 51
Shadow length, 196
Shape of a wading pool, 479
Snow removal, 399
Sunset times, 200
Test scores, 52, 401
Towing a barge, 290
Tuning a piano, 167
Typing speed, 401
Waiting times at a switchboard, 398
Wave motion, 199
Width of a river, 147

Time and Distance

Airplane speed and direction, 290, 302
Angle of a baseball, 234
Angle of a sharpshooter's gun, 234
Angular speed of a phonograph record, 202
Angular speed of a pulley, 130
Average speed, 316
Climb rate of an airplane, 399
Distance of an arrow, 455
Distance on a baseball diamond, 276
Distance a baseball player runs to make a catch, 276
Distance from a boat to shore, 147, 269
Distance between centers of bolt holes, 198
Distance between cities, 129
Distance between a control tower and an observation post, 270
Distance from a gazebo to a dock, 269
Distance across flat sides of a hexagonal nut, 198
Distance of a guy wire from a pole, 203
Distance from a landmark, 270

Distance a plane flies, 197, 204
Distance of a plane to an observer, 157
Distance from a plane to a radar antenna, 177
Distance a ship travels, 197
Distance across a swamp, 197
Distance from a television camera to a parade, 178
Distance between towns, 198
Distance between two planes, 302
Distance between two ships, 197, 276
Distance traveled by a golf ball, 157
Distance traveled by a model rocket, 157
Flying distance, 35
Flying speed, 24
Height of a balloon, 147
Height of a baseball, 323
Height of a weather balloon, 65
Length of a football pass, 35
Length of guy wires, 276
Path of a ball, 454, 483
Path of a projectile, 157, 420, 454, 455
Range of a projectile, 252
Speed of an automobile, 395
Speed of a bicycle, 130, 202
Speed of a floppy disk, 130
Stopping distance, 96
Velocity of a ball, 113
Velocity of a baseball, 290
Velocity of a bullet, 290
Vertical rise of a railroad grade, 203
Wind speed and direction, 290

U.S. Demographics

Average cellular phone bill, 107
Average heights of males and females, 382
Federal debt, 37
Federal deficit, 10
Health care costs, 97
Life expectancy of a child, 37
Merchandise trade balance of the U.S., 78
New car sales, 107
Outstanding mortgage debt, 89
Per capita soft drink consumption, 108
Price of a mobile home, 64
Truck fuel use, 89
Unemployment rate, 235
Values of U.S. imports from Mexico and Canada, 115

Index

A

Absolute value
  of a complex number, 324
  equations involving, 20
  function, 80
  properties of, 5
  of a real number, 5
Acute angle, 121
Additive identity
  for a complex number, 311
  for a real number, 7
Additive inverse
  for a complex number, 311
  for a real number, 6
Adjacent side
  of a right triangle, 138
Algebraic expression, 6
Algebraic functions, 344
Ambiguous case (SSA), 264
Amplitude
  of simple harmonic motion, 194
  of sine and cosine curves, 160
Amplitude modulation, 195
Angle, 120
  acute, 121
  central, 121
  complementary, 123
  conversions between degrees and radians, 124
  coterminal, 120
  degree, 123
  of depression, 190
  of elevation, 190
  initial side, 120
  measure of, 121
  negative, 120
  obtuse, 121
  of one degree, 123
  positive, 120
  reference, 151
  standard position, 120
  supplementary, 123
  terminal side, 120
  between two lines, 407
  between two nonzero vectors, 293
  vertex, 120
Angular speed of a particle, 125
Arccosine function, 182
Arc length, 125
Arcsine function, 180
Arctangent function, 182

Area of an oblique triangle, 266
Argument
  of a complex number, 325
Arithmetic combination, 91
Associative property of addition
  for complex numbers, 312
  for real numbers, 7
Associative property of multiplication
  for complex numbers, 312
  for real numbers, 7
Asymptote of a hyperbola, 433
Average of n numbers, A16
Axis (axes)
  coordinate, 25
  of an ellipse, 422
  of a hyperbola, 431, 433
  imaginary, 324
  of a parabola, 414
  polar, 456
  real, 324

B

Bar graph, 26
Base
  of an exponential function, 344
  of a logarithmic function, 356
Basic rules of algebra, 7
Bearing, 192
Bell-shaped curve, 388
Bimodal, A16
Bounded interval on the real number line, 3

C

Calculator
  use with exponential expressions, 344
  use with logarithmic functions, 357, 361
  use with trigonometric functions, 135, 142
Cartesian plane, 25
Cardioid, 466
Center
  of a circle, 33
  of an ellipse, 422
  of a hyperbola, 431
Central angle of a circle, 121
Change-of-base formula, 367
Circle
  center of, 33
  classifying by discriminant, 444

classifying by general equation, 437
  radius of, 33
  standard form of equation of, 33, 131
Closed interval, 3
Coefficient of a variable term, 6
Cofunction identities, 210
Common logarithmic function, 357
Commutative property of multiplication
  for complex numbers, 312
  for real numbers, 7
Complementary angles, 123
Complete the square, 15
Complex conjugate, 313
Complex number(s), 310
  absolute value of, 324
  addition of, 311
  conjugate of, 313
  dividing, 313
  equality of, 310
  multiplication of, 312
  nth root of, 332, 333
  nth roots of unity, 334
  polar form of, 325
  powers of, 331
  product of, 312, 327
  quotient of, 313, 327
  standard form of, 310
  subtraction of, 311
  trigonometric form of, 325
Complex plane, 324
  imaginary axis, 324
  real axis, 324
Component of a vector, 295
Component form of a vector **v**, 280
Composite number, 8
Compound interest, 349, 350
Conditional equation, 12
Conic or conic section, 413
  alternative definition, 470
  classifying by discriminant, 444
  classifying by general equation, 437
  degenerate, 413
  locus of, 413
  polar equations of, 470
  rotation of axes of, 440
Conjugate of a complex number, 313
Constant, 6
Constant function, 80
Constant term, 6
Continuous compounding, 349
Conversions between degrees and radians, 124

Coordinate axes, 25, 83
   reflections in, 83
Coordinate system, 25
Correspondence
   one-to-one, 2
Cosecant function, 132
   of any angle, 149
   graph of, 172, 175
   odd, 135
Cosine function, 132
   amplitude of, 160
   of any angle, 149
   common angles, 152
   curve, 158
   domain of, 134
   even, 135
   graph of, 158, 162, 175
   inverse of, 182
   period of, 134, 161
   special angles, 140
Cotangent function, 132
   of any angle, 149
   graph of, 171, 175
   odd, 135
Coterminal angle, 120

D

Damping factor, 174
   graph of, 174
Decimal degree, 124
Decreasing function, 70
Degenerate conic, 413
Degree
   of an angle, 123
   in decimal form, 124
DeMoivre, Abraham (1667–1754), 331
DeMoivre's Theorem, 331
Denominator of a fraction, 6
Dependent variable, 54, 60
Descartes, René (1596–1650), 25, 41
Difference
   of functions, 91
   of a vector, 281
Difference quotient, 59
Difference and sum, formulas, 236
Directed line segment, 279
   initial point, 279
   magnitude (length), 279
   terminal point, 279
Direction angle of a vector, 285
Directrix of a parabola, 414
Discriminant, 318, 444
Distance Formula, 27
Distance between a point and a line, 408
Distance between two points
   in the plane, 27
   on the real number line, 5
Distributive property
   for complex numbers, 312
   for real numbers, 7
Divisor of an integer, 8
Domain, 53, 132
   of a composite function, 94
   of cosine, 134
   defined, 132
   of an exponential function, 346
   implied, 57, 60
   of a logarithmic function, 359
   of sine, 134
   undefined, 60
Dot product
   properties of, 292
   of vectors, 292
Double-angle formulas, 243

E

e, the number, 348
Eccentricity
   of an ellipse, 426
   of a hyperbola, 435
Eliminating the parameter, 449
Ellipse, 422, 470
   center of, 422
   classifying by the discriminant, 444
   classifying by general equation, 437
   eccentricity of, 426
   focus of, 422
   latus rectum of, 430
   major axis of, 422
   minor axis of, 422
   standard form of the equation of, 423
   vertex of, 422
Endpoint of an interval, 3
Equality
   of complex numbers, 310
   of vectors, 280
Equation(s), 12
   conditional, 12
   of an ellipse, 423
   equivalent, 13
   exponential, 373
   graph of, 29
   identity, 12
   involving absolute value, 20
   involving fractional expressions, 14
   involving a radical, 19
   involving a rational exponent, 19
   of a line, 38, 44
      in general form, 44
      in point-slope form, 43
      in slope-intercept form, 38
      in two-point form, 43
   linear, 38
      in one variable, 12
   logarithmic, 373
   of a parabola, 414
   polar conversion of, 459
   quadratic, 15
   of quadratic type, 227
   solution of, 12, 29
Equivalent
   directed line segments, 279
Exponential decay model, 384
Exponential equation, 373
Exponential function, 344
   base of, 344
   graph of, 345
   natural, 348
Exponential growth model, 384

F

Factor(s) of an integer, 8
Factoring
   to solve a polynomial equation, 18
   to solve a quadratic equation, 15
   to solve trigonometric equations, 227
Focal chord of a parabola, 416
Focus
   of an ellipse, 422
   of a hyperbola, 431
   of a parabola, 414
Formula(s)
   change-of-base, 367
   for compound interest, 350
   double-angle, 243
   half-angle, 246
   Heron's area, 274
   multiple-angle, 243
   power-reducing, 245
   product-to-sum, 247
   of projection, 295
   reduction, 238
   sum and difference, 236
   sum-to-product, 248
Frequency distribution, A14
Frequency modulation, 195
Frequency of simple harmonic motion, 194
Function(s), 53, 80
   absolute value, 80
   addition of, 91
   algebraic, 344
   common logarithmic, 357
   composition of, 93
   constant, 70, 80

cubic, 80
decreasing, 70
difference of, 91
division of, 91
domain of, 53, 57, 60
evaluating, 56
even, 74
exponential, 344
family of, 82
graph of, 67
greatest integer, 73
identity, 80
increasing, 70
inverse of, 99, 100, 103
linear, 72
logarithmic, 356
of multiple angles, 230
multiplication of, 91
name of, 55, 60
natural exponential, 348
natural logarithmic, 360
notation, 55, 60
odd, 74
period of, 134
periodic, 134
piecewise-defined, 56, 73
product of, 91
quadratic, 80
quotient of, 91
range of, 53, 60
square root, 80
step, 73
subtraction of, 91
sum of, 91
summary of terminology, 60
testing for, 54
transcendental, 344
trigonometric, 132, 149
   of any angle, 149
value of, 55, 60
vertical line test for, 68
zero of, 69
Fundamental Theorem
  of Algebra, 317
  of Arithmetic, 8
Fundamental trigonometric identities,
  141, 210

G

Gaussian model, 384
General second-degree equation, 413
Graph
  of arccosine function, 182
  of arcsine function, 181
  of arctangent function, 182

of common functions, 80
of cosecant function, 172, 175
of cosine function, 158, 162, 175
of cotangent function, 171, 175
damped trigonometric function, 174
of equation, 29
of exponential function, 345
of function, 67
  reflection of, 83
  shifts in, 81
  shrinks of, 85
  stretches of, 85
  transformations of, 219
of the inverse of a function, 101
of line, 38
of logarithmic function, 358
of the reciprocal functions, 172
of secant function, 172, 175
of sine function, 158, 162, 175
of tangent function, 169, 175
Greatest integer function, 73
Growth and decay, 385

H

Half-angle formulas, 246
Half-life, 351
Harmonic motion, simple, 193, 194
Heron's area formula, 274
Hipparchus, 237
Histogram, A14
Horizontal components of a vector, 284
Horizontal line test, 102
Horizontal shift, 81
Hyperbola, 431, 470
  asymptote of, 433
  branch of, 431
  center of, 431
  classifying by discriminant, 444
  classifying by general equation, 437
  conjugate axis of, 433
  eccentricity of, 435
  focus of, 431
  standard form of the equation of, 431
  transverse axis of, 431
  vertex of, 431
Hypotenuse of a right triangle, 27, 138

I

Identity, 12
Identity function, 80
Identity of trigonometry, 141, 210, 218
Imaginary axis of a complex plane, 324
Imaginary number, 310
Imaginary unit i, 310

Implied domain, 57, 60
Inclination, 406
Inclusive or, 8
Increasing function, 70
Independent variable, 54, 60
Inequality (inequalities), 3
Inequality symbol, 3
Infinity, 4
  negative, 4
  positive, 4
Initial point of a directed line segment,
  279
Initial side of an angle, 120
Integer, 2
  divisors of, 8
  factor of, 8
Intercept of a graph, 30
Interest, 349
Intersect feature of a graphing utility, xix
Interval on the real line, 3
Inverse function, 99
  cosine, 182
  definition of, 100
  exponential, 358
  finding the, 103
  graph of, 101
  horizontal line test for, 102
  sine, 180
    graph of, 181
  tangent, 182
  trigonometric, 182
    properties of, 184
Inverse properties, 373
Irrational number, 2

K

Kepler, Johannes (1571–1630), 473
Kepler's Laws, 473
Key points
  intercepts, 159
  maximum points, 159
  minimum points, 159
  of a trigonometric function, 159

L

Latus rectum
  of an ellipse, 430
  of a parabola, 416
Law of Cosines, 271
Law of Sines, 262
Law of Trichotomy, 4
Length
  of a directed line segment, 279
  of a vector, 280

Limaçon, 466
Line graph, 26
Line(s) in the plane, 38, 406
    equation of, 44
    graph of, 38
    horizontal, 39
    parallel, 45
    perpendicular, 45
    slope of, 38, 41
    vertical, 44
Locus of a conic, 413
Logarithm(s)
    change-of-base formula, 367
    properties of, 357, 368
Logarithmic equation, 373
Logarithmic function, 356
    base of, 356
    common, 357
    graph of, 358
    natural, 360
Logarithmic model, 384
Logistic curve, 389
Logistic growth model, 384

M

Major axis of an ellipse, 422
Mathematical model, 164
Mean of n numbers, A16
Measure of an angle, 121
Measure of central tendency, A16
Measure of dispersion, A17
Median of n numbers, A16
Midpoint, 28
Midpoint Formula, 28
Minor axis of an ellipse, 422
Mode of n numbers, A16
Model, 164
Modulus of a complex number in trigonometric form, 325
Multiple-angle formula, 243
Multiplicative identity
    of a real number, 7
Multiplicative inverse, 6
    of a nonzero real number, 6
    property, 7

N

Natural base e, 348
Natural exponential function, 348
Natural logarithmic function, 360
Negation, properties of, 7
Negative
    angle, 120
    infinity, 4

real number, 2
    of a vector, 281
Nonnegative real number, 2
Normally distributed populations, 388
nth root
    of a complex number, 332, 333
    of unity, 334
Number
    complex, 310
    composite, 8
    imaginary, 310
    irrational, 2
    negative, 2
    nonnegative, 2
    positive, 2
    prime, 8
    rational, 2
    real, 2
Numerator of a fraction, 6

O

Oblique triangle, 262
    area of, 266
Obtuse angle, 121
One cycle of a sine curve, 158
One degree of an angle, 123
One-to-one properties, 373
Open interval, 3
Opposite side of a right triangle, 138
Order on the real number line, 3
Ordered pair, 25
Origin
    in Cartesian plane, 25
    on real number line, 2
Orthogonal vectors, 294

P

Parabola, 414, 470
    axis of, 414
    classifying by discriminant, 444
    classifying by general equation, 437
    directrix of, 414
    focal chord of, 416
    focus of, 414
    latus rectum of, 416
    reflective property of, 416
    standard form of the equation of, 414
    tangent line, 416
    vertex of, 414
Parallel lines, 45
Parallelogram Law, 281
Parameter, 448
Parametric equation, 448
Parametric mode, 449

Period
    of a function, 134
    of simple harmonic motion, 194
    of sine and cosine curves, 161
Periodic function, 134
Perpendicular lines, 45
Phase shift for a trigonometric function, 162
Plane curve, 448
    orientation of, 449
Point-plotting method of graphing, 29
Polar axis, 456
Polar coordinate, 456
    conversion of, 457
    test of symmetry in, 463
Polar coordinate system, 456
Polar equation, 459
    of conics, 470
    conversion, 459
    zeros and maximum r-values, 464
Polar form of a complex number, 325
Pole, 456
Polynomial function
    constant, 80
    zero of, 317
Positive
    angle, 120
    infinity, 4
    real number, 2
Power-reducing formulas, 245
Prime number, 8
Principal square root of a negative number, 314
Product of functions, 91
Product-to-sum formulas, 247
Projection
    formula of, 295
    of a vector, 295
Properties
    of absolute values, 5
    of equality, 7
    of fractions, 8
    of inverse of trigonometric functions, 184
    of logarithms, 357, 368
    of natural logarithms, 360
    of negation, 7
    of vector(s)
        addition, 283
        scalar multiplication, 283
    of zero, 8
Pythagorean identities of trigonometry, 141, 210
Pythagorean Theorem, 27

Q

Quadrant, 25
Quadratic equation, 15
   complex solution of, 314
   solution of, 15
Quadratic Formula, 15
Quotient
   of functions, 91
   of two complex numbers, 313, 327

R

r-values of polar equations, 464
Radian, 121
Radical(s),
   equations involving, 19
Radius of a circle, 33
Range of a function, 53, 60
Rate or rate of change, 40
Ratio, 40
Rational exponent(s), 19
   equations involving, 19
Rational number, 2
Real axis of a complex plane, 324
Real number(s), 2
   absolute value of, 5
   addition of, 6
   division of, 6
   multiplication of, 6
   negative, 2
   nonnegative, 2
   positive, 2
   subtraction of, 5
Real number line, 2
   bounded intervals on, 3
   distance between two points on, 5
   interval on, 3
   unbounded intervals on, 4
Reciprocal function, graphs of, 172
Rectangular coordinate system, 25
Reduction formula, 238
Reference angle, 151
Reflection
   in the coordinate axis, 83
   in the x-axis, 161
Reflective property of a parabola, 416
Regression capabilities of a graphing
     utility, xx
Relative maximum, 71
Relative minimum, 71
Resultant of a vector, 281
Right triangle, 138
   adjacent side, 138

definition of trigonometric functions,
    138
   hypotenuse, 138
   opposite side, 138
   solving, 143
Root(s)
   of a complex number, 332, 333
Root feature of a graphing utility, xxi
Rose curve, 466
Rotation
   of axes, 440
   invariant under, 444

S

Scalar, 281
Scalar multiplication
   of a vector, 281
     properties of, 283
Scatter plot, 26
Secant function, 132
   of any angle, 149
   even, 135
   graph of, 172, 175
Second-degree polynomial equation
    in x, 15
Shrinking of a graph, 85
Sigmoidal curve, 389
Simple harmonic motion, 193, 194
   amplitude of, 194
   frequency of, 194
   period of, 194
Sine function, 132
   amplitude of, 160
   of any angle, 149
   common angles, 152
   curve, 158
   domain of, 134
   odd, 135
   graph of, 158, 162, 175
   inverse of, 180
   period of, 134, 161
   special angles, 140
Slope of a line, 38, 41
Solution
   checking, 13
   of an equation, 12
   extraneous, 14
   of a quadratic equation, 15, 314
   of a right triangle, 143
   of a trigonometric equation, 225
Solution point, 29
Solving
   an equation, 12
   an equation involving absolute

     value, 20
   an equation involving radicals, 19
   an equation involving rational
     exponents, 19
   exponential and logarithmic
     equations, 374, 376
   a polynomial equation, 18
   a quadratic equation, 15
     by completing the square, 15
     by extracting square roots, 15
     by factoring, 15
     by the Quadratic Formula, 15
   a right triangle, 143
   a trigonometric equation, 225
Speed, angular, 125
Speed of a particle, 125
Square root, 15
   function, 80
   of a negative number, 310
   principle, 15
Square setting of a graphing utility, xvii
Standard deviation, A17
   alternate formula, A19
Standard form(s)
   of a complex number, 310
   of the equation of a circle, 33, 131
   of the equation of an ellipse, 423
   of the equation of a hyperbola, 431
   of the equation of a parabola, 414
Standard position
   of an angle, 120
   of a vector, 280
Standard unit vector, 284
Stem-and-leaf plot, A12
Step functions, 73
Stretching of a graph, 85
Substitution Principle, 6
Sum(s)
   of functions, 91
   of vectors, 281
Sum and difference formulas, 236
Sum-to-product formulas, 248
Supplementary angles, 123
Symmetry
   with respect to the origin, 31
   with respect to the x-axis, 31
   with respect to the y-axis, 31
   tests for, 32

T

Table feature of a graphing utility, xvi
Tangent function, 132
   of any angle, 149
   common angles, 152

graph of, 169, 175
inverse of, 182
odd, 135
special angles, 140
Tangent line of a parabola, 416
Term
 of an algebraic expression, 6
 constant, 6
 variable, 6
Terminal point of a directed line segment, 279
Terminal side of an angle, 120
Trace feature of a graphing utility, xvii
Transcendental function, 344
Triangle
 oblique, 262
 right, 138
Trigonometric equation, solution of, 225
Trigonometric form
 of a complex number, 325
  argument of, 325
  modulus of, 325
Trigonometric function
 of any angle, 149
 cosecant, 132
 cosine, 132
 cotangent, 132
 damping factor, 174
 evaluating, 149, 152
 even and odd, 135
 inverse properties of, 184
 key points, 159
  intercepts, 159
  maximum points, 159
  minimum points, 159
 of quadrant angles, 150
 right triangle definitions, 138
 secant, 132
 sine, 132
 tangent, 132
Trigonometric identities, 141, 210
 guidelines for verifying, 218
Trigonometric values of common angles, 152
Trigonometry, 120

U

Unbounded interval on the real number line, 4
Unit circle, 131
Unit vector, 280

V

Value of a function, 55, 60
Variable, 6
 dependent, 54, 60
 independent, 54, 60
 term, 6
Variance, A17
Vector
 addition of, 281
  properties of, 283
 angle between two, 293
 component form of, 280
 difference of, 281
 directed line segment of, 279
 direction angle of, 285
 dot product of, 292
  properties of, 292
 equal, 280
 horizontal component of, 284
 length of, 280
 linear combination of, 284
 magnitude of, 280
 negative of, 281
 orthogonal, 294
 projection, 295
 resultant, 281
 scalar multiplication of, 281
  properties of, 283
 standard position of, 280
 standard unit, 284
 unit, 280
 vertical component of, 284
 zero, 280
Vector components, 295
Vector **v** in the plane, 279
Vertex
 of an angle, 120
 of an ellipse, 422
 of a hyperbola, 431
 of a parabola, 414
Vertical components of a vector, 284
Vertical line test for a function, 68
Vertical shift, 81
Vertical shrink, 85
Vertical stretch, 85
Viewing window of a graphing utility, xvi

W

Work, 297

X

x-axis, 25
x-coordinate, 25
x-intercept of a graph, 30

Y

y-axis, 25
y-coordinate, 25
y-intercept of a graph, 30

Z

Zero, 2
 division by, 8
 of a function, 69
 of a polynomial function, 317
 properties of, 8
Zero-Factor Property, 8
Zero feature of a graphing utility, xviii
Zeros and maximum r-values of polar equations, 464
Zero vector, 280
Zoom feature of a graphing utility, xvii

FORMULAS FROM GEOMETRY

Triangle:
$h = a \sin \theta$
Area $= \dfrac{1}{2} bh$
(Laws of Cosines)
$c^2 = a^2 + b^2 - 2ab \cos \theta$

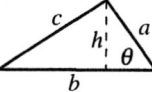

Right Triangle:
(Pythagorean Theorem)
$c^2 = a^2 + b^2$

Equilateral Triangle:
$h = \dfrac{\sqrt{3}s}{2}$
Area $= \dfrac{\sqrt{3}s^2}{4}$

Parallelogram:
Area $= bh$

Trapezoid:
Area $= \dfrac{h}{2}(a + b)$

Circle:
Area $= \pi r^2$
Circumference $= 2\pi r$

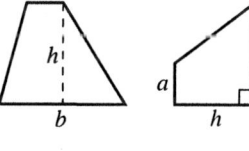

Sector of Circle:
(θ in radians)
Area $= \dfrac{\theta r^2}{2}$
$s = r\theta$

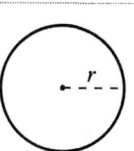

Circular Ring:
(p = average radius,
w = width of ring)
Area $= \pi(R^2 - r^2)$
$= 2\pi pw$

 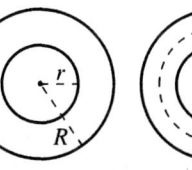

Sector of Circular Ring:
(p = average radius,
w = width of ring,
θ in radians)
Area $= \theta p w$

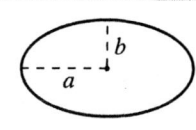

Ellipse:
Area $= \pi ab$
Circumference $\approx 2\pi \sqrt{\dfrac{a^2 + b^2}{2}}$

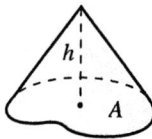

Cone:
(A = area of base)
Volume $= \dfrac{Ah}{3}$

Right Circular Cone:
Volume $= \dfrac{\pi r^2 h}{3}$
Lateral Surface Area $= \pi r \sqrt{r^2 + h^2}$

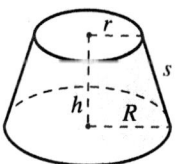

Frustum of Right Circular Cone:
Volume $= \dfrac{\pi(r^2 + rR + R^2)h}{3}$
Lateral Surface Area $= \pi s(R + r)$

Right Circular Cylinder:
Volume $= \pi r^2 h$
Lateral Surface Area $= 2\pi rh$

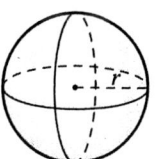

Sphere:
Volume $= \dfrac{4}{3}\pi r^3$
Surface Area $= 4\pi r^2$

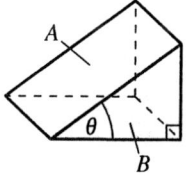

Wedge:
(A = area of upper face,
B = area of base)
$A = B \sec \theta$

GRAPHS OF COMMON FUNCTIONS

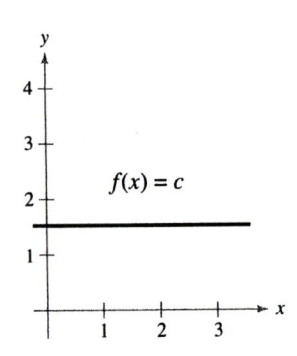
Constant Function
$f(x) = c$

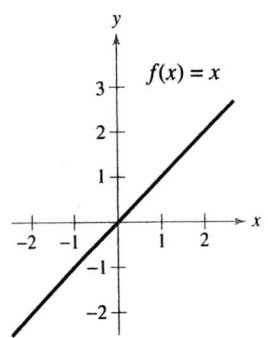
Identity Function
$f(x) = x$

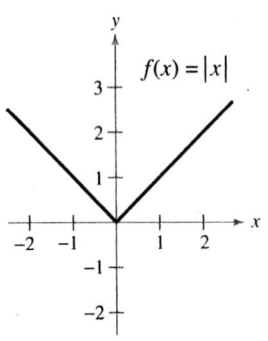
Absolute Value Function
$f(x) = |x|$

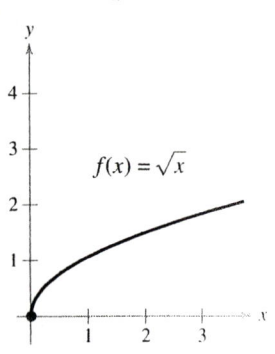
Square Root Function
$f(x) = \sqrt{x}$

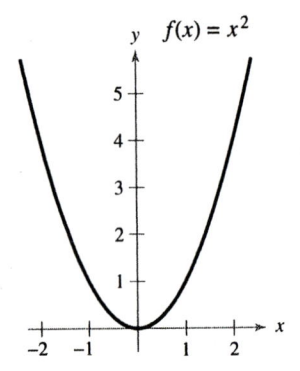
Squaring Function
$f(x) = x^2$

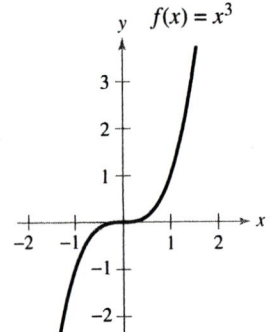
Cubing Function
$f(x) = x^3$

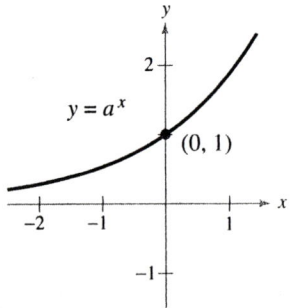
Exponential Function
$y = a^x$, $(0, 1)$

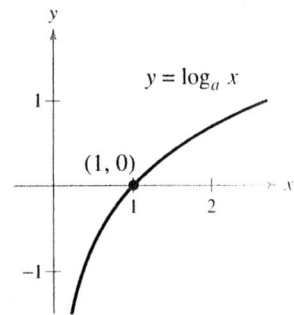
Logarithmic Function
$y = \log_a x$, $(1, 0)$

SYMMETRY

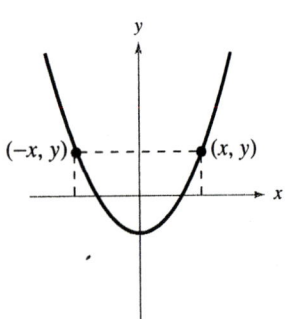
y-Axis Symmetry
$(-x, y)$, (x, y)

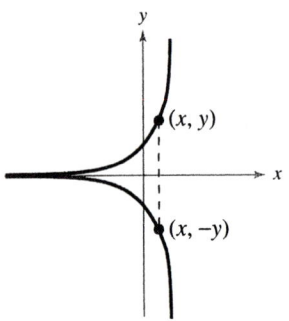
x-Axis Symmetry
(x, y), $(x, -y)$

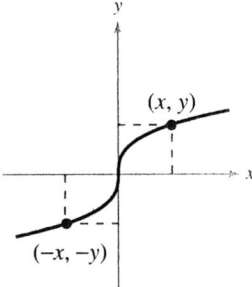
Origin Symmetry
(x, y), $(-x, -y)$